NATURAL PRODUCTS CHEMISTRY

Vol. 1

NATURAL PRODUCTS CHEMISTRY

Vol. 1

Edited by

Koji Nakanishi
DEPT. OF CHEMISTRY
COLUMBIA UNIVERSITY, NEW YORK

Toshio Goto
DEPT. OF AGRICULTURAL CHEMISTRY
NAGOYA UNIVERSITY, NAGOYA

Shô Itô
DEPT. OF CHEMISTRY
TOHOKU UNIVERSITY, SENDAI

Shinsaku Natori
NATIONAL INSTITUTE OF HYGIENIC SCIENCES, TOKYO

Shigeo Nozoe
CHEMISTRY DIVISION
THE INSTITUTE OF APPLIED MICROBIOLOGY,
UNIVERSITY OF TOKYO, TOKYO

KODANSHA LTD.
Tokyo

ACADEMIC PRESS, INC. New York and London
A Subsidiary of Harcourt Brace Jovanovich, Publishers

K KODANSHA SCIENTIFIC BOOKS

Co-published by

KODANSHA LTD.
12-21 Otowa 2-chome, Bunkyo-ku, Tokyo 112
and

ACADEMIC PRESS, INC.
111 Fifth Avenue, New York, New York 10003
United Kingdom Edition published by
ACADEMIC PRESS, INC. (LONDON) LTD.
24/28 Oval Road, London NW 1

INTERNATIONAL STANDARD BOOKS NUMBER: 0-12-513901-2
LIBRARY OF CONGRESS CATALOG CARD NUMBER: 74-6431
KODANSHA EDP NUMBER: 3043-298021-2253

PRINTED IN JAPAN

LIST OF CONTRIBUTORS

Chapter 1

Shinsaku NATORI,   National Institute of Hygienic Sciences, Tokyo

Chapter 2

Koji NAKANISHI,   Department of Chemistry, Columbia University, New York

Chapter 3

Shigeo IWASAKI,   Institute of Applied Microbiology, University of Tokyo, Tokyo
Shigeo NOZOE,   Institute of Applied Microbiology, University of Tokyo, Tokyo

Chapter 4 and 6

Numbers in parentheses indicate the chapter(s) to which the authors' contributions submit.
Steven BLOBSTEIN(4),   College of Physicians and Surgeons, Columbia University, New York
Wan Kit CHAN(4, 6),   Department of Chemistry, Yale University, New Haven, Connecti-
  cut
Sow-mei Lai CHEN(4),   Department of Chemistry, Columbia University, New York
Rosalie CROUCH*(4, 6),   Department of Chemistry, Columbia University, New York
James DILLON(4, 6),   Department of Chemistry, Columbia University, New York
Mamoru ENDO(4, 6),   Suntory Co., Osaka
Aaron FEINBERG(4, 6),   Department of Chemistry, New York University, New York
Hiroshi KAKISAWA(4),   Department of Chemistry, Tokyo Kyoiku University, Tokyo
Masato KOREEDA*(4, 6),   Department of Chemistry, Johns Hopkins University, Baltimore,
  Maryland
Yi-Tsung LIU(4, 6),   Schering Corporation, Bloomfield, New Jersey
Koji NAKANISHI(4, 6),   Department of Chemistry, Columbia University, New York
Philippa H. SOLOMON*(4, 6), Department of Chemistry, Columbia University, New York
Jacob M. TABAK(4, 6),   Department of Chemistry, Columbia University, New York
Masaru TADA(6), Faculty of Sciences and Arts, Waseda University, Tokyo

Akira TERAHARA(6), Sankyo Pharmaceutical Co., Tokyo
George WEISS(4, 6), Department of Chemistry, Columbia University, New York
Yasuji YAMADA, Tokyo College of Pharmacy, Tokyo
Paul ZANNO(6), Department of Chemistry, Columbia University, New York

Chapter 5

Hiroyuki AGETA, Showa College of Pharmaceutical Sciences, Tokyo
Katsuya ENDO,* Department of Chemistry, Tohoku University, Sendai
Yutaka FUJISE,* Department of Chemistry, Tohoku University, Sendai
Shô ITÔ,* Department of Chemistry, Tohoku University, Sendai
Isao KITAGAWA, Faculty of Pharmaceutical Sciences, Osaka University, Toyonaka
Mitsuaki KODAMA,* Department of Chemistry, Tohoku University, Sendai
Shinsaku NATORI, National Institute of Hygienic Sciences, Tokyo
Shigeo NOZOE, Institute of Applied Microbiology, University of Tokyo, Tokyo
Takeo ÔBA, Department of Chemistry, Tohoku University, Sendai
Ushio SANKAWA, Faculty of Pharmaceutical Sciences, University of Tokyo, Tokyo
Yoshisuke TSUDA, Showa College of Pharmaceutical Sciences, Tokyo

PREFACE

Natural products chemistry has lately undergone explosive growth owing to advances in isolation techniques, synthetic methods, physico-chemical measurements, and new concepts. On the other hand, it is precisely the chemistry of natural products which has fostered many of the new developments in these areas, because of the variety of compound types available. We are now keenly aware that two of the most intriguing problems, structure determination and total synthesis, have in many cases become rather routine, and this enables the organic chemist to direct his efforts towards new unexplored areas. Natural products chemistry is thus becoming increasingly diversified and complicated, and the literature is scattered widely in numerous monographs, reviews and papers.

Due to these circumstances, the following principles have been pursued in preparing the present two-volume book.

1) The prime aim is to fill the wide gap between the organic textbooks on the one hand, and the comprehensive treatises on a particular aspect or group of natural products on the other.

2) The natural products are described and discussed according to type classifications. An attempt has been made to cover the various aspects within each particular group— introductory survey, history, structure, synthesis, reactions, and biosynthesis.

   For pedagogical reasons a "historical" approach has been taken to describe some of the early structural studies; this is particularly so in the STEROID chapter.

3) Rapid visual retrieval of information has been emphasized, almost as in a "picture book", by free inclusion of structural formulae and simple abbreviations.

4) The basic attitude in writing each chapter is that of a lecturer presenting a survey on the particular subject matter, with the aid of slides, to an audience consisting not only of natural products chemists, but also of research chemists and graduate students who are not specialists in that subject and who require brief orientation in the subject.

5) Each chapter is made up of topics selected and arranged so that they cover various aspects of the chemistry as well as typical compound types. For ease of reference the format is such that a new topic generally begins at the top of a page. The figures thus correspond to the lecturer's slides, and the brief statements to his comments which focus attention on the salient features of the topics. Physical data are given wherever necessary but these are kept to a minimum.

Although we have attempted to include the more representative and significant topics, rather than to be comprehensive, there is always the possibility that important subjects have inadvertently not been described. The items have been selected from papers published through late 1972/early 1973. Some structural and synthetic aspects have also been included for pedagogical reasons. We sincerely welcome any suggestions or criticism from readers.

We are most grateful to the numerous contributors (see List of Contributors) who undertook the painstaking job of literature survey, writing, and condensing of material. We are also indebted to Mr. T. Yatsunami (University of Tokyo) for his assistance in researching the literature, and to Ms. J. Black (Columbia Univ.) and M. Yuda (Tohoku Univ.) for their work in preparing the typescript.

In the planning stage, none of us realized the vast amount of work involved in a book of this type. It turned out to be an extremely time-consuming project; various unforeseen factors, mostly the responsibility of the editors, contributed to a further delay. Five years have elapsed since the initial planning, and we are indebted to the publishers, Messrs. Y. Haga, W. R. S. Steele, and Ms. Y. Yamada for their patience.

KAMPAI !

March, 1974

Editors
Koji NAKANISHI
Toshio GOTO
Shô ITÔ
Shinsaku NATORI
Shigeo NOZOE

CONTENTS

CONTENTS OF VOLUME 2

DATA CONVENTIONS USED IN THIS BOOK

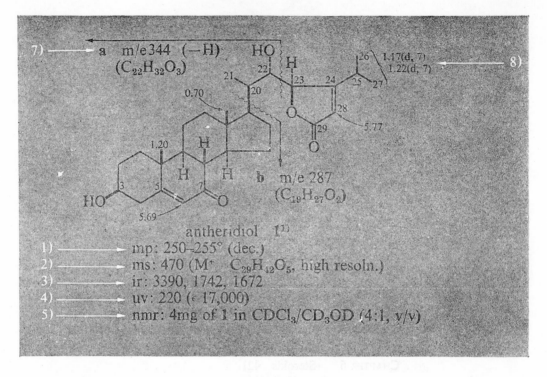

7) ——→ **a** m/e 344 (−H)
($C_{22}H_{32}O_3$)

0.70

1.20

HO

5.69

HO

H

H

H

26 1.17(d, 7)
1.22(d, 7) ——→ 8)

21 22 23 24 25 27
20
O
28
29 5.77
O

b m/e 287
($C_{19}H_{27}O_2$)

antheridiol **1**[1]

1) ——→ mp: 250–255° (dec.)
2) ——→ ms: 470 (M^+ $C_{29}H_{42}O_5$, high resoln.)
3) ——→ ir: 3390, 1742, 1672
4) ——→ uv: 220 (ϵ 17,000)
5) ——→ nmr: 4mg of **1** in $CDCl_3/CD_3OD$ (4:1, v/v)

1) Melting point.

2) Mass spectroscopic data. See also number 7 below.

3) Infrared data: state of measurement, if given, is shown in brackets, e.g. ir ($CHCl_3$): Values (cm^{-1}) are given only for pertinent bands.

4) Ultraviolet/visible spectral data: the solvent, if given, is shown in brackets, e.g. uv (EtOH): The wavelength is given in nanometers with the intensity in brackets (as log ϵ, unless otherwise stated).

5) Nuclear magnetic resonance data: the solvent is usually given in brackets, e.g. nmr ($CDCl_3$): Values are given as ppm from TMS. 1H, 2H, etc. indicate the intensity. s: singlet, d: doublet, t: triplet, m: multiplet. J values are given in Hz, and 14-H indicates a proton attached to C-14. See also number 8 below.

6) Rotation data are given as follows: "α_D (EtOH): +65" indicates the specific rotation at the D line in EtOH. "cd (MeOH): 215 ($\Delta\epsilon$ +13.17)" indicates the circular dichroism extrema in MeOH, 215 nanometers maximum (or minimum) with a $\Delta\epsilon$ value of +13.17 (or −13.17). Rotation data are not given in the above example.

7) Mass spectroscopic data: fragmentation (a) with loss of · H gives m/e 344 ($C_{22}H_{32}O_3$) fragment arising from the C-1 to C-22 portion of the molecule.

8) Nuclear magnetic resonance data: the isopropyl methyls appear at ppm values of 1.17 and 1.22 as doublets with $J=7$ Hz.

CHAPTER **1**

Classification
of Natural Products

The classification of natural products, which cover almost all types of organic molecules, generally follows one of the four schemes outlined below.

1.1   CLASSIFICATION BASED ON CHEMICAL STRUCTURE

This is a formal classification based on the molecular skeleton.[1] Thus:
  i) Open-chain aliphatic or fatty compounds: e.g., fatty acids, sugars, most amino acids.
  ii) Alicyclic or cycloaliphatic compounds: e.g., terpenoids, steroids, some alkaloids.
  iii) Aromatic or benzenoid compounds: e.g., phenolics, quinones.
  iv) Heterocyclic compounds: e.g., alkaloids, flavonoids, nucleic acid bases.

Since this is merely a superficial classification, it is obvious that many closely related natural products will belong to more than one class. For example, geraniol **1**, farnesol **2**, and squalene **3** belong to class i, and thymol **4** to class iii, but because of biogenetic considerations, they are usually treated with other terpenoids and steroids under class ii.[2]

1   **2**   **3**   **4**

1.2   CLASSIFICATION BASED ON PHYSIOLOGICAL ACTIVITY

As exemplified by the discoveries of and ensuing active research on morphine **5** (1806), penicillins **6** (1939), and prostaglandins **7** (1963), our interest in natural products is frequently initiated by attempts to isolate and clarify a physiologically active factor of plant or animal origin. Actually, nearly half the medicines currently in use are natural products, e.g., alkaloids and antibiotics, or synthetic analogs. Therefore, a classification representing physiological activity is frequently employed, as exemplified by hormones, vitamins, antibiotics, and mycotoxins. Although compounds belonging to each group have diverse structures and biosynthetic origins, occasionally a close correlation is found between such aspects and activity.

5   penicillin G **6**   prostaglandin $F_{1\alpha}$ **7**

For example, in spite of the structural variations encountered in steroids, those v h ch exhibit cardiotonic activity (cardenolids and bufadienolides) are characterized by (i) an A/B *cis* ring juncture, (ii) a sugar residue at C-3, and (iii) a 5- or 6-membered conjugated lactone at C-17 (**8** and **9**).

---

1) D. F. Styles, *Rodd's Chemistry of Carbon Compounds* (ed. S. Coffey), vol. 1 (2nd ed.), part A, p. 21, Elsevier, 1964.
2) W. Karrer, *Konstitution und Vorkommen der organischen Pflanzenstoffe*, Birkhäuser, 1958. This classification has mainly been adopted in textbooks and reference books dealing with natural products, with special reference to particular groups of natural products such as flavones, terpenoids and alkaloids.

R=suger residues

1.3 CLASSIFICATION BASED ON TAXONOMY

This is based on comparative morphological studies of plants, i.e., plant taxonomy. In animals and partly in microorganisms, the final metabolites are generally excreted outside the body, while in plants the metabolites are deposited within the plant body. Although some metabolites, once believed to be specific to some plants, are now known to be rather widely distributed among the plant kingdom, many plant constituents such as alkaloids and isoprenoids have been isolated from specific plant species, genera, tribes or families. Even a single species contains numerous constituents which have closely related structures. For example, the "opium" from *Papaver somniferum* contains twenty-odd alkaloids such as morphine **5**, thebaine **10**, codeine **11**, and narcotine **12**, all of which are biosynthesized from the 1-benzylisoquinoline precursor **13** by oxidative coupling; thus alkaloids having these similar structures are characteristic constituents of this plant genus and are designated as opium alkaloids. Similarly, names representing genera and families such as ergot alkaloids, iboga alkaloids, and menispermaceae alkaloids frequently appear in the literature.

Our knowledge of plant constituents has been expanding at a tremendous rate in recent years due to the advance in isolation and microcharacterization methods. This has led to a new field called "chemotaxonomy" or "chemosystematics", which attempts to review plant constituents according to plant taxa.[3-7] Namely, constituents are regarded as markers for evolution and for the classification of plants.

However the number of known compounds for each plant is still quite limited, and no plant has yet been completely analysed. Undoubtedly, a thorough and laborious study starting from a large amount of plant material will lead to clearer picture, although most constituents thus clarified will be known compounds with no biological activity.[8] Nevertheless, organizing knowledge of plant constituents according to taxonomy is an important and actively pursued area.[9-11]

3) C. Mentzer, *Comparative Phytochemistry* (ed. T. Swain), p. 21, Academic Press, 1966.
4) *Chemical Plant Taxonomy* (ed. T. Swain), Academic Press, 1966.
5) R. E. Alston, B. L. Turner, *Biochemical Systematics*, Prentice-Hall, 1963.
6) J. B. Harbourne, *Progress in Phytochemistry* (ed. L. Reinhold, Y. Liwschitz), vol. 1, p. 545, Interscience, 1968.
7) *Recent Advances in Phytochemistry* (ed. T. J. Mabry, R. E. Alstone, V. C. Runeckles), vol. 1, North-Holland, 1968.
8) L. Fowden, *Phytochemistry*, **11**, 2271 (1972).

1.4 CLASSIFICATION BASED ON BIOGENESIS

The constituents of all plants and animals are biosynthesized in organisms by enzymatic reactions. (Although "biogenesis" and "biosynthesis" are sometimes used without distinction, it is customary to use the former term for a hypothesis, and the latter for an experimentally proven route.) The major source of carbon is usually glucose, which is photosynthesized in green plants (autotropic organisms) or obtained from the environment in heterotropic organisms. Recent advances in biochemistry have greatly clarified the interrelated, enzyme-catalysed reactions of primary metabolites (such as sugars, amino acids, fatty acids) and the biopolymers (such as lipids, proteins and nucleic acids). These primary metabolites give rise to the "secondary metabolites", so called because their roles in the metabolism of host organisms are not obvious.

When our knowledge of natural product chemistry had accumulated to some extent in the 1930's, some organic chemists started to form theories of the biogenetic pathways of natural products in living organisms on the basis of their structural regularity. The most notable example was the "isoprene rule" proposed by Ruzicka.[12] He pointed out that all terpenoids are built up from C_5 "isoprene units".

nerol **14**   santonin **15**   oleanolic acid **16**

Robinson[13] also pointed out several structural correlations encountered in groups of natural products; one of his proposals, the "polyketomethylene theory" for phenolic compounds, was the first suggestion of acetogenin (polyketide) biosynthesis.

17   **18**

endocrocin **19**

These early hypotheses naturally lacked experimental verification, but recent developments in organic chemistry and biochemistry have now clarified the biosyntheses of a number of primary and secondary metabolites. The first remarkable success was achieved by Davis using mutants of *Escherichia coli*.[14] These studies established the 'shikimic acid pathway' and showed that biosyntheses of aromatic amino acids and related aromatic compounds proceeded as follows:

9) R. Hegnauer, *Chemotaxonomie der Pflanzen*, vol. 1, Birkhäuser, 1962; see also later volumes.

10) *Annual Index of Reports on Plant Chemistry* (ed. T. Kariyone), Hirokawa (Tokyo), 1957; see also later volumes.

11) *The Lynn Index* (ed. J. W. Schermerhorn, M. W. Quimby), Massachusetts College of Pharmacy, 1957; see also later volumes.

12) L. Ruzicka, *Angew. Chem.*, **51**, 5 (1938); *Proc. Chem. Soc.*, 341 (1959).

13) R. Robinson, *The Structural Relations of Natural Products,* The Clarendon Press, 1955.

14) B. D. Davis, *Amino Acid Metabolism*, p. 799, John Hopkins, 1955; *Advan. Enzymol.*, **16**, 247 (1955); D. B. Sprinson, *Advan. Carbohydrate Chem.*, **15**, 235 (1960).

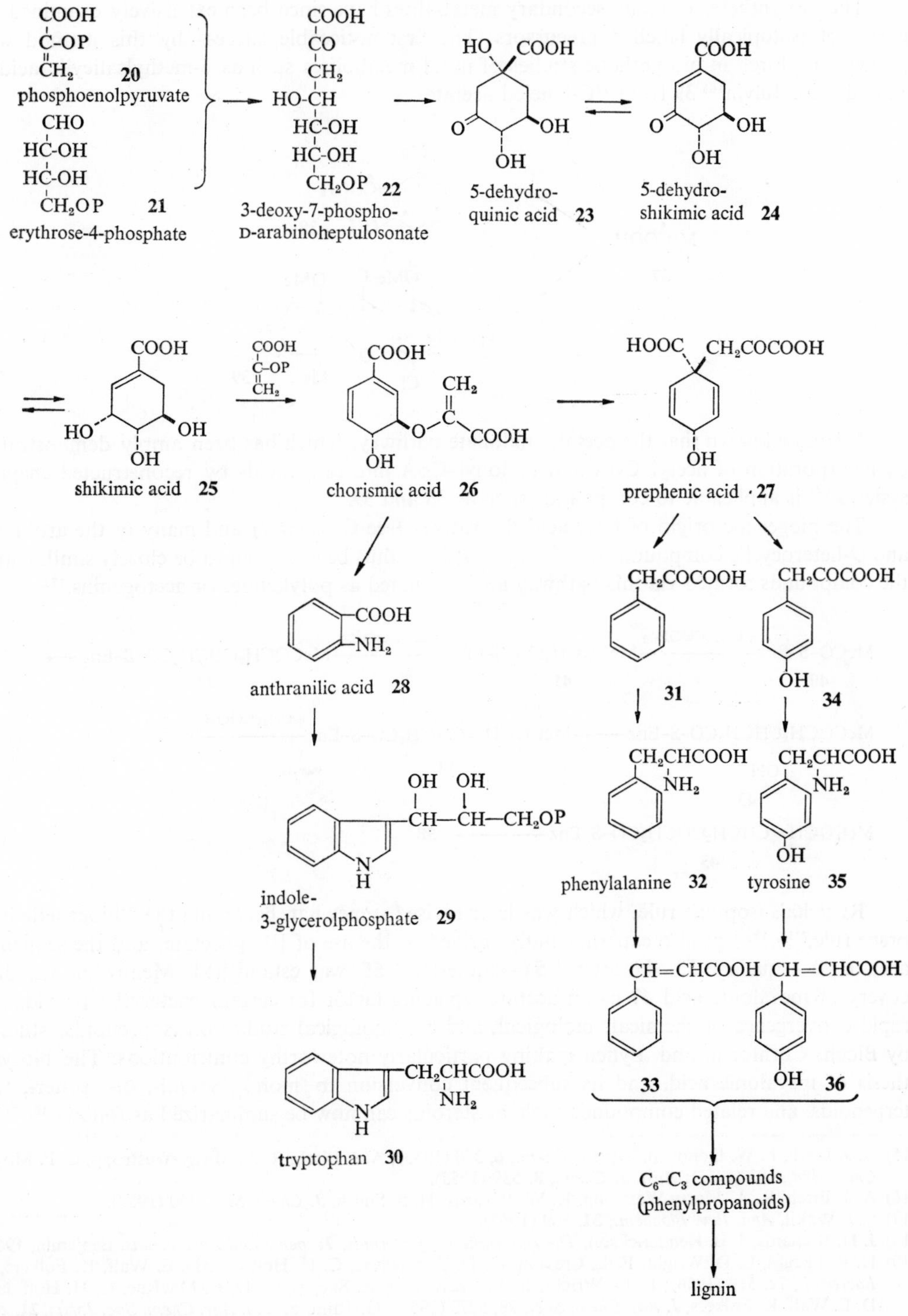

COOH
|
C-OP
‖
CH₂ **20**
phosphoenolpyruvate

CHO
|
HC-OH
|
HC-OH
|
CH₂OP **21**
erythrose-4-phosphate

→

COOH
|
CO
|
CH₂
|
HO-CH
|
HC-OH
|
HC-OH
|
CH₂OP **22**
3-deoxy-7-phospho-
D-arabinoheptulosonate

→

5-dehydro-
quinic acid **23**

⇌

5-dehydro-
shikimic acid **24**

⇌

shikimic acid **25**

→

chorismic acid **26**

→

prephenic acid · **27**

anthranilic acid **28**

↓

indole-
3-glycerolphosphate **29**

↓

tryptophan **30**

CH₂COCOOH

31

↓

phenylalanine **32**

↓

33

CH₂COCOOH

34

↓

tyrosine **35**

↓

36

C₆–C₃ compounds
(phenylpropanoids)

↓

lignin

The biosyntheses of many secondary metabolites have since been extensively examined by means of isotopically labelled precursors. The first noticeable success by this method was achieved by Birch in biosynthetic studies of mold metabolites such as 6-methylsalicylic acid[15] **38** and griseofulvin[16] **39** from ^{14}C-labeled acetate.

It is now known that the acetate-malonate pathway, which has been amply demonstrated by incorporation of acetyl–CoA and malonyl–CoA into fatty acids by reconstructed enzyme systems,[17] is applicable to compounds such as **38** and **39**.

The biogenetic origin of fatty acid derivatives (see Chapter 7) and many of the aromatic and O-heterocyclic compounds (see Chapter 9) have thus been proven to be closely similar and the compounds formed via this pathway are designated as polyketides or acetogenins.[18]

$$\underset{\textbf{40}}{MeCO\text{-}S\text{-}Enz}\xrightarrow{+CH_2(COOH)_2}\underset{\textbf{41}}{MeCOCH_2CO\text{-}S\text{-}Enz}\xrightarrow{+CH_2(COOH)_2}\underset{\textbf{42}}{MeCOCH_2COCH_2CO\text{-}S\text{-}Enz}\longrightarrow$$

$$\underset{\underset{\textbf{43}}{\overset{|}{OH}}}{MeCOCH_2CHCH_2CO\text{-}S\text{-}Enz}\longrightarrow\underset{\textbf{44}}{MeCOCH=CHCH_2CO\text{-}S\text{-}Enz}\xrightarrow{+CH_2(COOH)_2}$$

$$\underset{\textbf{45}}{MeCOCH=CHCH_2COCH_2CO\text{-}S\text{-}Enz}\longrightarrow \textbf{38}$$

Ruzicka's isoprene rule, which was later revised somewhat, becoming the "biogenetic isoprene rule,"[12, 18] was also experimentally verified by the use of [^{14}C]-acetate, and the sequence acetate **40**→squalene **53**→lanosterol **54**→cholesterol **55**, was established. Meanwhile the discovery of mevalonic acid **47** as an acetate-replacing factor for certain bacteria[19] provided a rapid convergence of chemical, biological, and enzymological studies on isoprenoids, studies by Bloch, Cornforth, and Lynen making particularly noteworthy contributions. The biosynthesis of mevalonic acid, and its subsequent conversion to mono-, sesqui-, di-, sester-, triterpenoids, and related compounds such as steroids can now be summarized as follows.[18,20]

15) A. J. Birch, F. W. Donovan, *Aust. J. Chem.*, **6**, 360 (1953); A. J. Birch, R. A. Massy-Westropp, C. P. Moye, *Chem. Ind.*, 683 (1955); *Aust. J. Chem.*, **8**, 539 (1955).
16) A. J. Birch, R. A. Massy-Westropp, R. W. Rikards, H. S. Smith, *J. Chem. Soc.*, 360 (1958).
17) S. J. Wakil, *Ann. Rev. Biochem.*, **31**, 369 (1962).
18) J. H. Richards, J. B. Henderickson, *The Biosynthesis of Steroids, Terpenes and Acetogenins*, Benjamin, 1964.
19) H. R. Skeggs, L. D. Wright, E. L. Cresson, G. D. E. MacRae, C. H. Hoffman, D. E. Walf, K. Folkers, *J. Bacteriol.*, **72**, 519 (1956); L. D. Wright, E. L. Cressen, H. R. Skeggs, G. D. E. MacRae, C. H. Hoffman, D. E. Walf, K. Folkers, *J. Am. Chem. Soc.*, **78**, 5273 (1956); G. Tamura, *Bull. Agr. Chem. Soc. Japan*, **21**, 202 (1957).
20) *Aspects of Terpenoid Chemistry and Biochemistry* (ed. T. W. Goodwin), Academic Press, 1971.

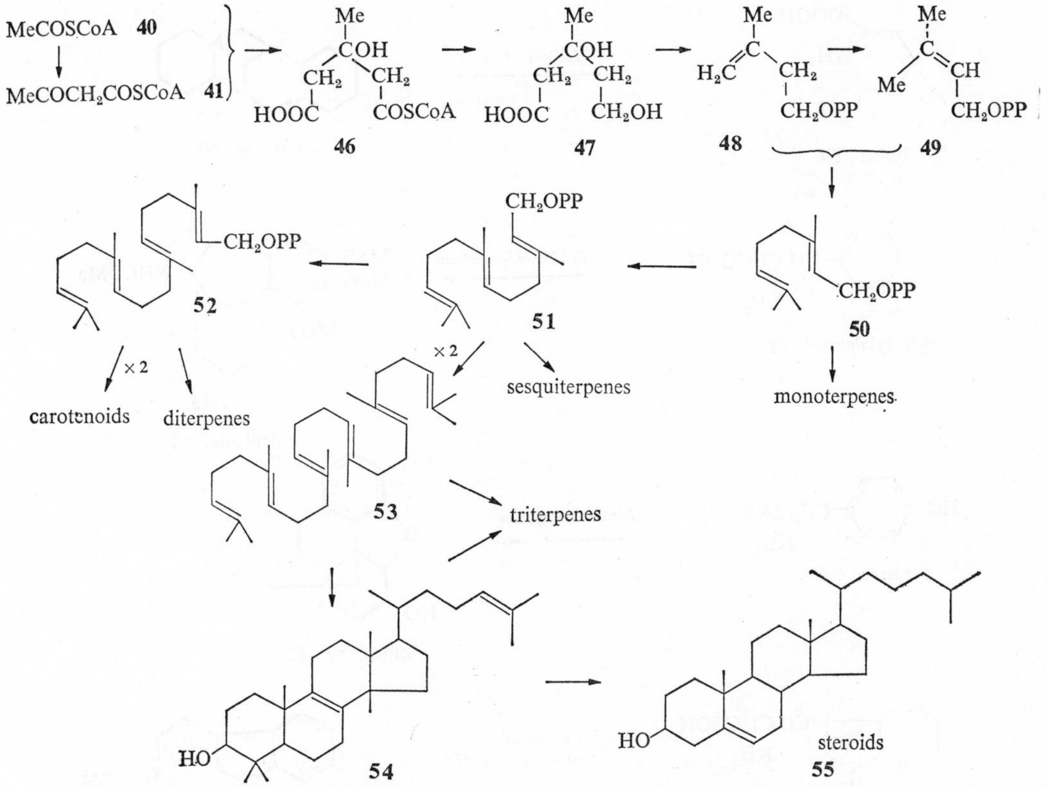
MeCOSCoA 40
MeCOCH2COSCoA 41
Me
COH
CH2
CH2
HOOC
COSCoA
46
Me
COH
CH2
CH2
HOOC
CH2OH
47
Me
C
H2C
CH2
CH2OPP
48
Me
C
Me
CH
CH2OPP
49
CH2OPP
50
monoterpenes
CH2OPP
51
sesquiterpenes
×2
CH2OPP
52
×2
carotenoids
diterpenes
53
triterpenes
HO
54
HO
steroids
55

The structural relationship observed between alkaloids and amino acids, their precursors, was pointed out by Robinson.[13] Precursor-product experiments using isotopes achieved notable success in biosynthetic studies of alkaloids,[21] some examples of which are shown below:

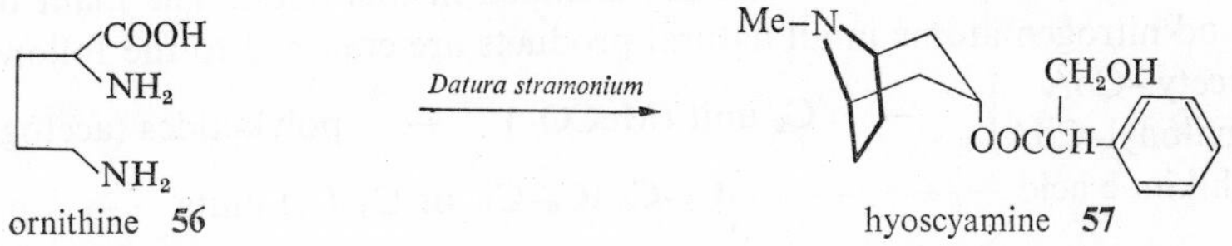
COOH
NH2
NH2
ornithine 56
Datura stramonium
Me–N
CH2OH
OOCCH
hyoscyamine 57

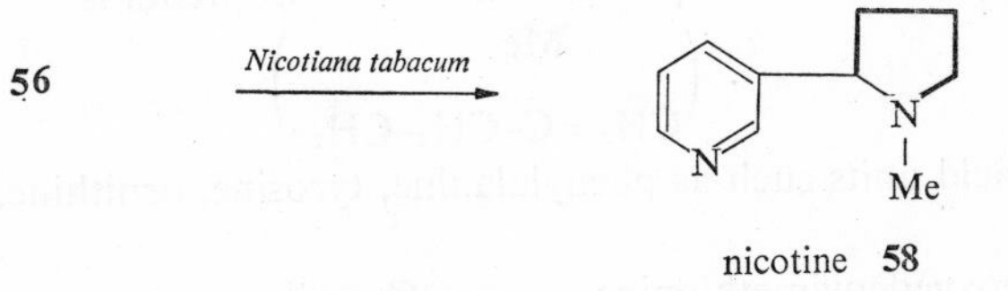
56
Nicotiana tabacum
N
N
Me
nicotine 58

21) A. Battersby, *Proc. Chem. Soc.*, 189 (1963); E. Leete, *The Biogenesis of Natural Compounds* (ed. P. Bernfeld), p. 739, Pergamon, 1963; D. H. R. Barton, *Proc. Chem. Soc.*, 293 (1963); K. Mothes, H. R. Schutte, *Angew. Chem. Intern. Ed.*, 2, 341, 441 (1963).

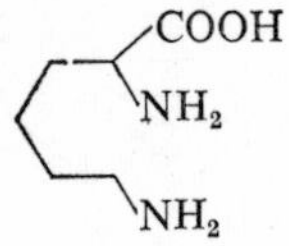

lysine **59**   → *Lupinus luteus* →   sparteine **60**

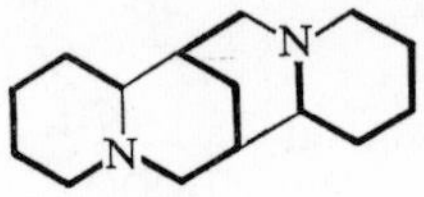

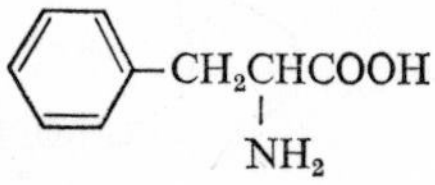

phenylalanine **32**   → *Colchicum byzantinum* →   colchicine **61**

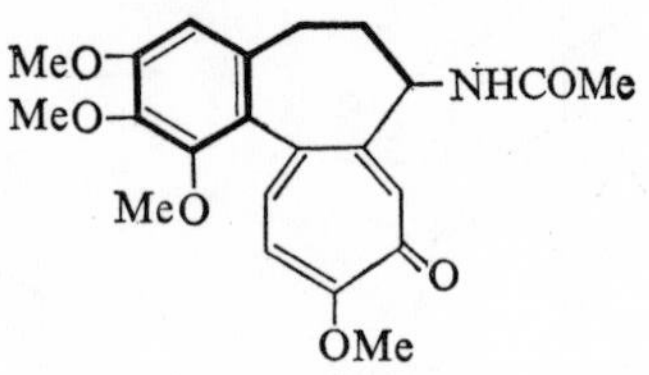

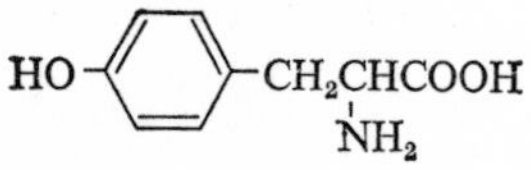

tyrosine **35**   → *Papaver somniferum* →   morphine **62**

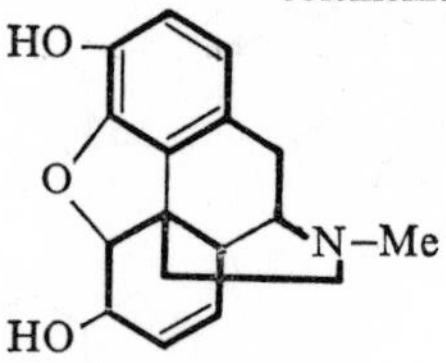

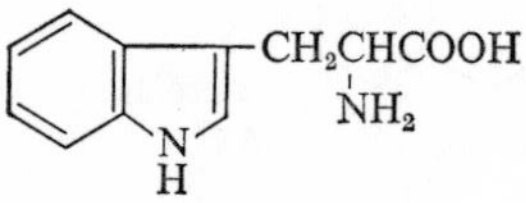

tryptophan **30**   → *Rauwolfia serpentina* →   ajmaline **63**

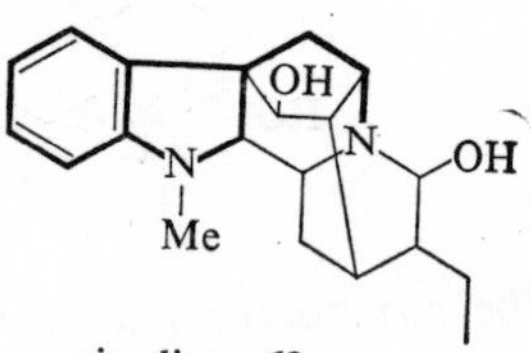

The accumulated data on natural product biosyntheses are reviewed in reference books,[18,22)] and several examples of such studies are included in this book. The main building blocks for carbon and nitrogen atoms in all natural products are confined to the following:

1) acetyl–CoA }
   malonyl–CoA }   → C_2 unit (MeCO–)   →   polyketides (acetogenins)

2) shikimic acid ———————→C_6–C_3 (C_6–C_1, or C_6–C_2) units   →   phenolics
   ↘   ↗   (phenylpropanoids)
       amino acids

3) mevalonic acid   →   prenyl unit   →   isoprenoids

$$\left(\begin{array}{c} \text{Me} \\ | \\ CH_2=C-CH_2-CH_2- \end{array} \right)$$

4) amino acid units such as phenylalanine, tyrosine, ornithine, lysine, and tryptophan
                                                                              →   alkaloids

5) S–5′–deoxyadenylmethionine   → C_1 unit

Although many natural products are formed solely from units 1, 2, or 3, others are formed from combinations of these units. Some examples of mixed biosynthesis of the structural framework are shown in the following:

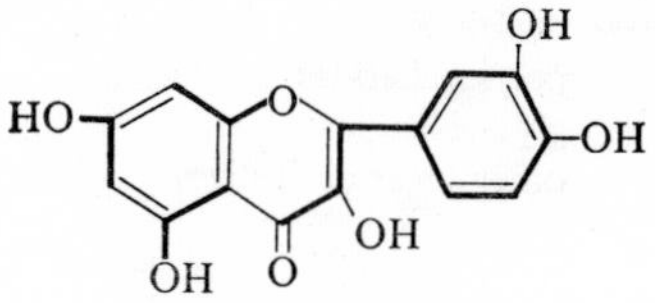

quercetin **64**
(C_6–C_3 + three MeCO)

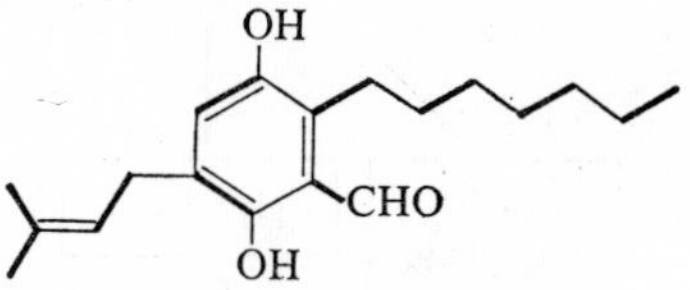

flavoglaucin **65**
(seven MeCO + prenyl)

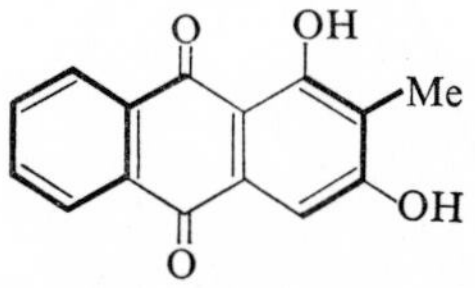

rubiadin **66**
(C_6–C_1 + C_3 + prenyl)

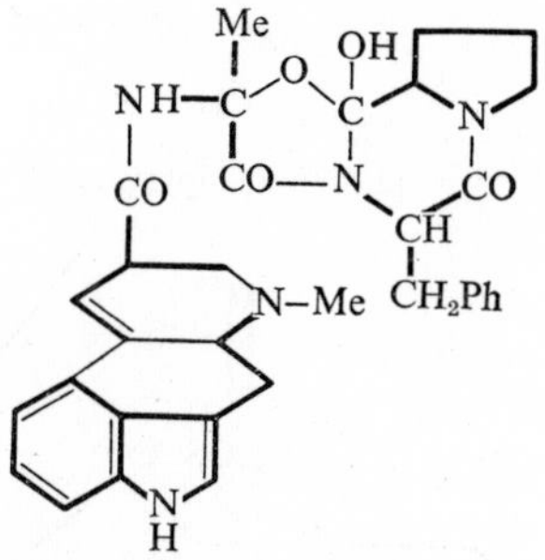

ergotamine **67**
(tryptophan + prenyl + C_1 +
alanine + proline + phenylalanine)

Certain natural products which are generally regarded as secondary metabolites, are wholly or partly derived from primary metabolites not included in the above. Some examples are as follows:

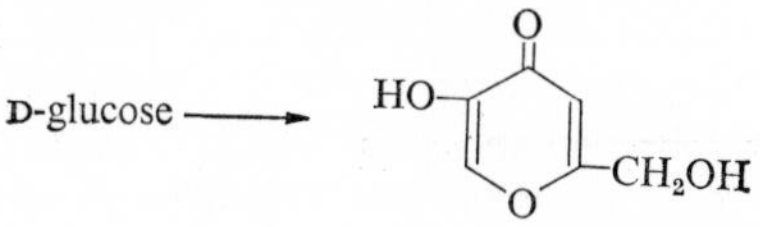

D-glucose ⟶

kojic acid **68**

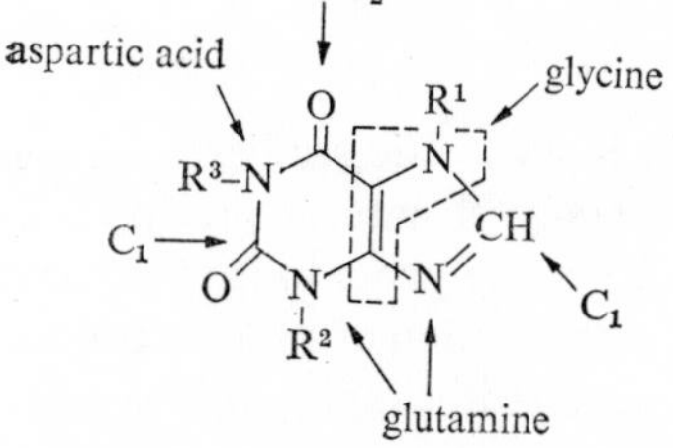

	R^1	R^2	R^3
xanthine **69**	H	H	H
theophiline **70**	H	Me	Me
theobromine **71**	Me	Me	H
caffeine **72**	Me	Me	Me

The general relationship between primary metabolites, secondary metabolites, and interlinking precursors is summarized in the following scheme.[22]

22) *The Biogenesis of Natural Compounds* (ed. P. Bernfeld), Pergamon, 1967; J. D. Bu'Lock, *The Biosynthesis of Natural Products*, McGraw-Hill, 1965; *Biosynthetic Pathways in Higher Plants* (ed. J. B. Pridham, T. Swain), Academic Press, 1965; M. Luckner, *Secondary Metabolism in Plants and Animals*, Chapman and Hall, 1972; *Specialist Periodical Reports, Biosynthesis*, vol. 1, The Chemical Society, 1972; see also later volumes.

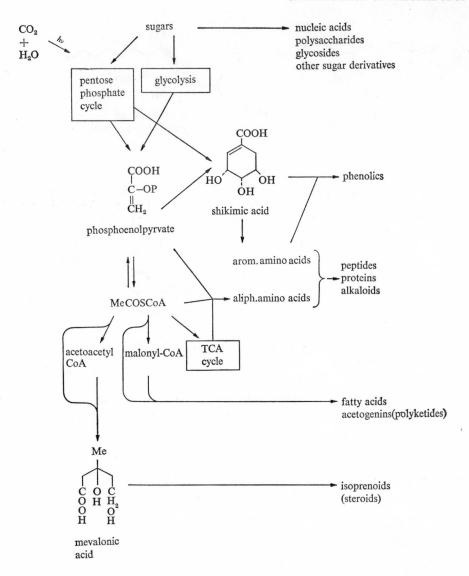

Although the total number of natural products for which biosynthetic studies have been carried out is quite limited compared with the number and diversity of natural products, our accumulation of knowledge in this field makes it possible to predict the gross biogenetic origin of practically all natural products. Consequently the biogenetic classification of natural products is a practical and convenient scheme, and has had considerable success in simplifying the understanding of complex structures.[18,22-24]

23) T. A. Geissman, D. H. G. Crout, *Organic Chemistry of Secondary Plant Metabolism*, Freeman, Cooper & Co., 1969.
24) W. B. Turner, *Fungal Metabolites*, Academic Press, 1971.

CHAPTER 2

Physico-Chemical
Data

11

2.1 INFRARED SPECTROSCOPY[1]

Measurement

1) Range: 4000–500 cm⁻¹ (2.5–20μ)
2) Sampling: The sample can be in any state, i.e., solid, liquid, gas, solution, film, fiber, or even surface coating (by using the attenuated total reflectance (ATR) technique).
3) KBr spectra: The entire ir range of solid samples is most conveniently measured by the KBr method, which also enables one to handle micro amounts of 5–10 μg. Adsorption of moisture during pellet preparation is unavoidable; this results in the appearance of water bands at ca. 3400 and 1640 cm⁻¹, and loss of resolution through the entire range. It can be solved by heating the pellet to 40–50° in 1 mm Hg for several hours, provided the sample is not affected. Nujol mulls do not absorb moisture, but the spectra show paraffin bands at 2919, 2861, 1458, 1378 and 720 cm⁻¹ (w). Usage of beam condensers with micro KBr pellets should be cautioned because the pellet area becomes heated to 60–70°.
4) Solution spectra: Any detailed discussion on band positions should be restricted to spectra taken in dilute solutions of nonpolar solvents, e.g., CCl₄, unless one is specifically looking for effects due to solvation, association, etc. All solvents have several strong absorption bands, the intensities of which depend on cell thickness. Accurate recording cannot be expected where solvent absorption is stronger than 80%.

Vibrations of groups

Group vibrations are divided into stretching (ν) and bending (δ) modes, which are further subdivided, as exemplified by the methylene group.

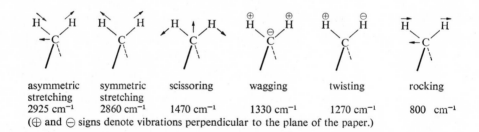

asymmetric stretching	symmetric stretching	scissoring	wagging	twisting	rocking
2925 cm⁻¹	2860 cm⁻¹	1470 cm⁻¹	1330 cm⁻¹	1270 cm⁻¹	800 cm⁻¹

($\oplus$ and $\ominus$ signs denote vibrations perpendicular to the plane of the paper.)

CHARACTERISTIC GROUP FREQUENCIES

Alkanes

1) –Me: 2960,   asym. stretch
          2870,   sym. stretch; weak, but strong in OMe and NMe, 2820 cm⁻¹
          1460,   asym. bend
          1380,   sym. bend; position is for C–Me; position shifts systematically to higher cm⁻¹ with increase in the electronegativity of X in X–Me, e.g., N–Me 1425 cm⁻¹, O–Me 1450 cm⁻¹; shifts to lower cm⁻¹ with increase in the mass of

1) Abstracted from the following with addition of material: K. Nakanishi, *Infrared Spectroscopy, Practical*, Holden-Day/Nankodo, 1962.

X in X–Me, e.g., Si–Me 1260 cm^{-1}; in COMe, 1360 cm^{-1}, sharp, intense; in OCOMe, 1380 band is very strong so that in most acetates this is stronger than the 1470–1460 band envelope; split into doublet when more than one Me is on the same carbon

2) –CH$_2$–: 2925, asym. stretch; sharp in cyclic CH$_2$; 3040 cm^{-1} in cyclopropanes

2850, sym. stretch; weak, but strong in OCH$_2$ and NCH$_2$; in methylenedioxy group, ca. 2780 cm^{-1}

1470, bend; shifted to 1440–1400 cm^{-1} in CH$_2$X, where X is an electron-attracting group, e.g., CO, Ar, C≡C, N$^+$

720, rock; characteristic for (CH$_2$)$_n$ where n > 4; splits into doublet when crystalline; shifted to 740 in Pr and 780 cm^{-1} in Et groups

Alkenes

1) 3080–3020 cm^{-1}: Asym. stretch of =CH; this is higher than 3000 cm^{-1} and is in contrast to the CH stretch of alkanes (lower than 3000 cm^{-1})

Type	Overtone of δCH	νC=C	δCH (in plane)	δCH (out of plane)
mono-	1830 w	1645 m	1420 w	990 s, 910 s
vinylidene	1880 w	1655 m	1415 w	890 s
di-, *cis*		1660 m	1415 w	730–675 s
di-, *trans*		w		965 s
tri-		w		840–800 m

2) νC=C: shifted to 1610 cm^{-1} when conjugated to groups with π electrons. Greatly intensified when linked to O or N, and often split (Fermi resonance with overtone of δCH)

3) δCH (out of plane) is most characteristic of substitution pattern

mono- : 990   CH$_2$ twist; lower when the substituent is –I group

910   CH$_2$ wag; higher when the substituent is an *m*-directing group, and lower when an *o*- or *p*-directing group

vinylidene : 890   CH$_2$ wag; higher when linked to an *m*-directing group, and lower when linked to an *o*, *p*-directing group; at 710 cm^{-1} in dialkoxy vinylidene

cis- : 730–675   Intense in cyclic compounds; 820 cm^{-1} in conj. enones

trans- : 965   CH$_2$ twist; lower when the substituent is –I group (ca. −40 cm^{-1} per –I group); unchanged by conjugation

tri- : 840–800   Weak compared to other δCH bands

Aromatics

1) 3100–3000: C–H stretch and overtones of skeletal vibrations; weak but diagnostic be-

        cause bands differ from saturated CH (lower than 3000 cm^{-1})

2) 2000–1660: Group of weak bands with patterns characteristic of the substitution pattern; due to overtones and combinations of CH (out-of-plane)

3) 1600, 1580, 1500, 1450: In-plane skeletal vibrations
1600 and 1580: Strong with conjugation; forbidden in 1,4- or 1,2,4,5-substitution (center of symmetry), and only one band in 1,3,5- or hexa-substitution by identical or similar groups (C$_3$ axis); 1500 band strong when substituted with *o, p*-directing groups; 1450 band overlaps with CH$_2$ and Me bend

4) 1300–1000: In-plane CH bend; one to several weak but sharp bands; intensified by halogens; non-diagnostic for substitution pattern

5) below 900: Due to in-phase movements (out-of-plane bend) of adjacent H's, bands are intensified with *o, p*-directing groups, but are weakened and shifted to higher frequencies with *m*-directing groups and fail to be characteristic. The following analysis based on adjacent hydrogens is applicable to condensed aromatics and heterocycles (hetero atom is treated as ring substituent):

      5 adjacent H : 750 $\pm$ 30
      4   "   " : 750   "
      3   "   " : 780   "
      2   "   " : 830   "
      isolated H   : 870   " ; weak and hence of limited diagnostic value   **1**
      ring puckering: 700 $\pm$ 20; due to puckering shown in **1**; therefore, this band is present in mono-, 1,3-, 1,3,5-, and 1,2,3-substituted phenyls

X≡Y and X=Y=Z groups

1) –C≡C–H : 3300, strong and sharp, CH stretch
                       2120 w
    –C≡C–   : 2260–2190, very weak or absent when symmetry is high

2) –C≡N   : 2250 m–s, shifted to 2220 and intensified upon conjugation

3) –N$_2^+$: 2260

4) –S–C≡N : 2175–2140

Alcohols and phenols

1) O–H stretch:
      free OH: 3640–3600, sharp
      solvated or intramol. H–bonded OH: 3600–3500 m
      intermol. H–bonded OH: 3400–3200, strong and broad
      chelated OH: 3200–2500, occasionally so wide that detection is difficult

2) C–O stretch:
      prim-OH 1050 ⎫ intense and broad; shifted lower by branching on adjacent
      sec-OH 1100 ⎬ carbon (-15), and by allylic double bond (-30) (effects are
      tert-OH 1150 ⎭ additive)
      Phenolic OH: 1200

3) O–H in-plane bend : 1500–1250, weak and no practical value

Ethers

      C–O–C : 1150–1070 s, asym. stretch
      C–O–C= : 1275–1200 s, asym. stretch, common to all groups containing –O–C=
                      1075–1020 s, sym. stretch

Amines

The shift in νNH upon H-bonding is less than 100 cm^{-1}. The band is sharp and strong when the amino group is "acidic," e.g., in pyrrole, indole, *p*-nitroaniline.

1) N–H stretch:
   RNH$_2$ and ArNH$_2$:   3500 (asym.) and 3400 (sym.) w
   RNHR and ArNHR :   3400–3300 w

2) N–H bend:
   –NH$_2$:   1640–1560, s when H-bonded, in-plane bend
             900–650, m (broad), out-of-plane bend
   –NH– :   1580–1490 w, not diagnostic

3) C–N stretch
   C$_{aliph}$–N :   1280–1030 m
   C$_{arom}$–N :   1360–1250 s

4) –NH$_3^+$:   3000, s (broad), N$^+$H stretch, overlaps with CH band, "ammonium band"
            2500, several, overtones and combinations
            2000,   ″   ″   ″
            1600–1575 and 1500, asym. and sym. NH$_3^+$ bend
   –NH$_2^+$:   2700–2250, s (broad), N$^+$H stretch, may or may not overlap with νCH
            1600–1575, m, N$^+$H$_2$ bend
   –NH$^+$:   2700–2250 s, one to several sharp bands, clearly separated from CH

5) C=N$^+$–H: 2500–2300 s, one to several sharp bands, clearly separated from CH
            2200–1800 m, one to several, "immonium band"
            1680 m C=N$^+$ stretch

6) CH–N–CH: Group of small bands at 2800–2700 cm^{-1} if more than two adjacent C–H bonds are antiperiplanar to the N lone pair electrons, e.g., *trans*-quinolizidine, yohimbine; also called "Bohlmann band" (*Chem. Ber.* **91**, 2157) and used in conformational analysis

Carbonyl groups

All C=O stretching bands (1870–1600 cm^{-1}) are strong; subject to predictable shifts by conjugative, electronic, steric, ring size and H-bonding effects, but spectra should be taken in dilute nonpolar solvents for serious discussions.

1) Ketone : 1715, Pronounced tendency for Fermi resonance to give a doublet when an overtone of the low freq. fundamental vibration equals C=O stretch, e.g., acid chlorides; shifts lower when C–CO–C angle is widened, e.g., di-*t*-Bu ketone, 1690
            1240–1100, strong band in aliph. ketones
            1280, strong band in arom. and α,β-unsat. ketones
   α,β-unsat.: 1675, strong band around 1280
   conj. and cross-conj. dienone: 1665, strong band around 1280
   arom.: 1690, strong band around 1280
   7-memb. and larger: 1705
   6-memb.: 1715
   5-memb.: 1745
   4-memb.: 1780
   α-halo: shift of 0 to +25, longer shift with smaller CO/CX angle; nil when angle exceeds 90°
   CO–CO (s-*trans*): 1720

CO–CO (s-*cis*):   asym./sym. stretch at 1760/1730 (6-memb.) and 1775/1760 (5-memb.)

=C(OH)–CO–   : 1675

CO–CH–CO–   : 1720

CO–C=C(OH)–   : 1650 if OH is free, 1615 if OH is H-bonded (same values for NH); strong C=C stretch at 1605

CO–C=C (OR)–:   1640

quinones (1,2 and 1,4):   1675

2) Aldehyde: 2850–2720, CH stretch, m (sharp); often doublet (Fermi resonance with C=O stretch)

                  1725

  α,β-unsat.:   1685, strong band at ca. 1280

  dienal   : 1675,   "   "

  arom.   : 1700,   "   "

3) Acid (dimer):

    3000–2500, broad, group of small bands, νOH and combinations

    1710, appears at 1760 in monomer

    1420, coupling between in-plane OH bend and C–O stretch of dimer

    1300–1200,   "   "

    920, m (always somewhat broad), out-of-plane OH bend of dimer

  α,β-unsat. and arom. (dimer):   1690

4) Carboxylate :   1610–1550, s, asym. C–O stretch

                 1410, m, sym. stretch

5) Ester   :   1735

                two at 1300–1050, asym. and sym. C–O–C stretch

  α,β-unsat.   : 1720

  arom.   : 1720

  vinylic   : 1760

  Ar–COO–Ar : 1735

  α-keto   : 1745

  β-keto   : 1735 (ester)/1720 for keto form; 1650/1630 for enol form

  δ-lactone   : 1735; α,β-unsat., 1720; enol lactone, 1760

  γ-lactone   : 1770; α,β-unsat., 1750 (doublet at 1785/1755 when α-H is present); enol lactone, 1800

6) Anhydride   :   1820/1760, higher band stronger in acyclic, lower band stronger in cyclic

                 1300–1050, 1 or 2 strong bands from C–O–C stretch

                 1000–900, 1 to several strong broad bands

  succinic, maleic, and phthalic type: 1860/1780

  α,β-unsat. six-memb.:   1780/1735

7) Imide   :   1775 w/1745–1700 very strong, broad

8) Acid chloride   :   1800, lower in –COBr and –COI, higher in –COF

                 950–900, usually strong, broad; overtone couples with C=O stretch

9) Amide: Extensive coupling is present among vibrations but the main contribution is as follows: "Amide I band," C=O stretch; "Amide II band," mainly NH bend mixed with C–N stretch; "Amide III band," mainly C–N stretch, mixed with N–H bend. Values for unassociated state are given in parentheses.

  CO–NH$_2$:   3350–3200, several bands (two at 3500 and 3400), NH s

                1650 (1690), Amide I

                1640 (1600), Amide II

CO–NHR : 3300 (3440), NH s; also 3070 m, overtone of Amide II band

1655 (1680), Amide I

1550 (1530), Amide II, for *trans* amides; in *cis* amides (lactams) appears at 1440, weaker

1300 (1260), Amide III

CO–NR$_2$: 1650 (1650), no H-bonding, hence insensitive to sample state.

C=C–CO–N and CO–N–C=C: shift of $+15$ cm^{-1}

C=C–CO–N–C=C: shift of -15 cm^{-1}

δ-lactam : 1670 (free) when unsubst., 1640 (free) when subst.

γ-lactam : 1700 (free)

β-lactam : 1745 (free)

10) Other carbonyls:

R–O–CO–O–R, 1740; Ar–O–CO–O–R, 1770; Ar–O–CO–O–Ar, 1800

–NH–CO–O–R, 1705; –NH–CO–O–Ar, 1730

R–CO–SH, 1720, R–CO–SR, 1690; Ar–CO–SR, 1665

R–CO–S–Ar, 1710; Ar–CO–S–Ar, 1685

Other groups

nitro, –NO$_2$	:	1560 and 1350, strong
nitrate, –ONO$_2$	:	1640 and 1280, strong
nitroso, –NO	:	1600–1500, strong
nitrite, –ONO	:	two at 1680–1610, strong
oxime, C=N–OH	:	3650–3500, 1685–1650, 960–930 (νN–O)
N-oxide, N→O	:	970–950 (aliph.), 1300–1200 (arom.), very strong
S–H stretch	:	2600–2550
S–CH$_2$	:	2700–2630, C–H stretch; 1420, CH bend
S–Me	:	2570, C–H stretch; 1320, CH bend
C=S stretch	:	1200–1050, strong
S–O stretch	:	900–700, strong
S=O stretch	:	1200–1400, strong
SO$_2$ stretch	:	1400–1310 (asym.) and 1230–1120 (sym.), strong
R–SO$_2$–N	:	1370–1330, 1180–1160
S→O stretch	:	1080–1030, very strong, lowered to 1000 when H-bonded
P–H stretch	:	2440–2350
P–O–C alkyl stretch	:	1050–1030, strong
P–O–C arom. stretch	:	1240–1190, strong
P=O stretch	:	1300–1250, strong
P→O stretch	:	970–930, very strong
Si–Me	:	2150, CH stretch; 1260, CH sym. bend; 860–800, Me rock
Si–O–Si	:	1100–1000, one to several bands, strong
C–F	:	1400–1000, very strong
C–Cl	:	800–600, strong
C–Br	:	600–500, strong
C–I	:	500, strong

2.2   ULTRAVIOLET SPECTROSCOPY[2-5]

Introduction

1) Range: 185–400 nm, ultraviolet; 400–800 nm, visible.

    The 185–800 nm range is usually handled under "ultraviolet" spectroscopy.

2) Beer's law:

$$\text{d or } A = \log \frac{I_0}{I} = \varepsilon cl$$

   d,   optical density,   or   A,   absorbance

   I_0 and I,   intensities of incident and transmitted light

   ε,   mollar extinction coefficient, units of 1000 cm²/mole

   c,   concentration in moles/liter

   l,   cell length in cm

When the molecular weight is unknown, then $E_{1\%}^{1cm}$ is used, the relation between these being:

$$E_{1\%}^{1cm} \times \text{mol. w.} = \varepsilon \times 10$$

3) Relation between energy E and wavelength λ of light:

$$E \text{ (kcal/mole)} = 28.6 \times 1000/\lambda \text{ (nm)}$$

4) Intensity ε: This can be approximated by

$$\varepsilon = 0.87 \times 10^{20} Pa$$

   P,   transition probability (from 0 to 1)

   a,   cross-sectional target area of the chromophore, which is ca. 10 Å² in most cases,
   If P is 1, the strongest ε can be estimated to be ca. 10⁵. Experimentally, the strongest ε
   is of the order of 10⁶.

5) Solvents: All molecules have absorption bands (CH_4 125 nm, far uv). The cut-off wavelength of a solvent is determined by the wavelength and intensity of its longest wavelength
absorption maximum. Cut-off wavelength of some solvents, 10 mm cell (5 nm longer in 1
mm cell):

H_2O, MeCN	190	Et_2O, dioxane	215
cyclohexane	195	CH_2Cl_2	220
n-C_6H_{14}, MeOH	200	$CHCl_3$	240
EtOH, $(MeO)_3PO$	200	CCl_4	260

6) Shifts of bands with solvents: The transitions of polar bonds, such as carbonyls, but not
ethylenes, are affected by solvent polarity in the following directions. Namely, as solvent
polarity is increased, i) $\pi \rightarrow \pi^*$ bands undergo red shifts (excited state is more polar than
the ground state, and hence stabilization is greater relative to the ground state in polar solvents), ii) $n \rightarrow \pi^*$ bands undergo blue shifts (ground state with two n electrons receives
greater stabilization than the excited state with only one n electron).

Example: mesityl oxide **2**; note opposite trend in shifts of two bands.

---

2) K. Hirayama, *Handbook of Ultraviolet and Visible Absorption Spectra of Organic Compounds*, Plenum
   Press Data Division, 1967; K. Hirayama, *Treatise in Experimental Chemistry* (ed. M. Kotake, in Japanese),
   vol. 1., Maruzen, 1957.

3) H. H. Jaffé, M. Orchin, *Theory and Applications of Ultraviolet Spectroscopy*, Wiley 1962.

4) A. I. Scott, *Interpretation of the Ultraviolet Spectra of Natural Products*, Oxford University Press, 1964.

5) Sections on uv in *New Treatise on Experimental Chemistry* (ed. T. Kubota, in Japanese) vol. 5-I (1965),
   vol. 5-II (1966), vol. 5-III (1966), Maruzen.

	Solvent	$\pi \to \pi^*(\varepsilon)$	$n \to \pi^*(\varepsilon)$	Δ
	n-hexane	230 (12,600)	327 (98)	97
	ether	230 (12,600)	326 (96)	96
2	ethanol	237 (12,600)	315 (78)	78
	water	245 (10,000)	305 (60)	60

G. Scheibe, *Chem. Ber.,* **58**, 586 (1925).

7) n→π* band of ketones: This band, ε 10–100, 270–330 nm, for ketones (sat. and α,β-unsat.), is one of the most important of the weak absorption bands. Measurements of such bands require concentrated solutions and also are difficult due to the presence of nearby bands with large ε values; nevertheless, they are frequently important for diagnostic purposes. Since the ketone group in natural products is usually associated with intense n→π* Cotton effects, it is usually simpler to measure cd or ord instead of uv curves in the n→π* region for detection of ketones.

TABLE 1   Saturated Chromophores (mostly in alcohol)

		nm	ε	Origin	Remarks
ROH	alcohol	180	200	n→σ*	
R–N	amine	200–215	3,000	n→σ*	
RSH	thiol	195	1,200	n→σ*	also 225–230 (sh)
–S–	sulfide	195	5,000		
		210–215	1,000–1,500	n→σ*	also 225–240 (sh)
–S–S–	disulfide	255	500		
–S–S–S–	trisulfide[a]	250	1,800		
–S–S–S–S–	tetrasulfide [b]	290	2,500		
–S–S–	cyclic (5-memb.)[c,d]	335	100–160		
–S–S–	cyclic (6-memb.)[c,d]	290	300		
–S–S–	cyclic (8-memb.)[c]	250	475		
$\diagup^S\diagdown$	episulfide	260		n→σ*	
$\big[\!\begin{smallmatrix}S\\S\end{smallmatrix}\!\big\rangle$	dithiolane [e]	245	350	n→σ*	2 cd extrema of opposite signs[e]
$\big[\!\begin{smallmatrix}S\\O\end{smallmatrix}\!\big\rangle$	oxathiolane[f]	245	30	n→σ*	
C–Cl	MeCl	173	200	n→σ*	in hexane
C–Br	MeBr	260	200	n→σ*	vapor
C–I	alkyl iodide	255–270	500	n→σ*	

a) R. E. Moore, *Chem. Commun.*, 1168 (1971).
b) H. P. Koch, *J. Chem. Soc.*, 394 (1949).
c) J. A. Barltrop, P. M. Hayes, M. Calvin, *J. Am. Chem. Soc.*, **76**, 4348 (1954)
d) C. Djerassi, A. Fredga, B. Sjöberg, *Acta. Chem. Scand.*, **15**, 417 (1961); C. Djerassi, H. Wolff, E. Bunnenberg, *J. Am. Chem. Soc* , **84**, 4552 (1962).
e) K. M. Wellman, P. H. A. Laur, W. S. Briggs, A. Moscowitz, C. Djerassi, *J. Am. Chem. Soc.*, **87**, 66 (1965).
f) D. A. Lightner, C. Djerassi, K. Takeda, K. Kuriyama, T. Komeno, *Tetr.*, **21**, 1581 (1965); K. Kuriyama, T. Komeno, K. Takeda, *ibid.*, **22**, 1039 (1966).

TABLE 2 Unsaturated Chromophores (mostly in alcohol)

		nm	ε	Origin	Remarks
C=C	olefin[a]	<190	8,000–10,000	$\pi \rightarrow \pi^*$	
		200–210	5,000–10,000		
C≡C	acetylene	220	100–200		
C≡N	nitrile	190			
C=O	ketone	<190	2,000	$\pi \rightarrow \pi^*$	
		270–300	10–40	$n \rightarrow \pi^*$	See Table 3 for α-ketol, etc.
CH=O	aldehyde	270–300	10–40	$n \rightarrow \pi^*$	
COOH	acid	205–215	50–100	$n \rightarrow \pi^*$	
COOR	ester	200–210	50–60	$n \rightarrow \pi^*$	
COO–	lactone	210–220	100	$n \rightarrow \pi^*$	
CO–O–CO	anhydride	220	60	$n \rightarrow \pi^*$	
CON<	amide	<180		$n \rightarrow \pi^*$	
C=C=C	allene	225	50		
C=C=O	ketene	230	400		
		375	20		
N=C=O	isocyanate	265	15		
N=C=N	carbodiimide	230	4,000		
		270	25		
–N=N–	azo[b]	350	15–400	$n \rightarrow \pi^*$	Cause of yellow color
$C = \overset{+}{N} = \bar{N}$	diazo	250	strong	$\pi \rightarrow \pi^*$	
		380–400	5	$n \rightarrow \pi^*$	Cause of yellow color
–N=N– ↓ O	azoxy[c]	215–225	5,500–7,500		
		275–285 (infl.)	40–60		
CON₃	azide[d]	216	500–540		Bands unaffected by solvent
		285	20–30		Bands unaffected by solvent
C=N–OH	oxime	195	2,000		
C–NO	nitroso	300	100		
		650	20	$n \rightarrow \pi^*$	Cause of blue color
O–NO	nitrite	220–230	1,000–2,000		
		315–385	20–85	$n \rightarrow \pi^*$	6 to 7 bands at 10 nm intervals
N–NO	nitrosamine	235	6,000–7,000		
		350	90–100	$n \rightarrow \pi^*$	Red shift and fine structure decrease with solvent polarity
N–NO₂	nitramine	240	6,000–8,000		

TABLE 2—*Continued*

		nm	ε	Origin	Remarks
N–C̈–N N	guanidinium	265	15		
C=S	thio carbonyl	490	15	$n{\to}\pi^*$	
S→O	sulfoxide	210–215	1,000	$n{\to}\pi^*$	Shoulder
SO₂	sulfone	<180			

a) A. Yogev, J. Sagio, Y. Mazur, *J. Am. Chem. Soc.*, **94** 5123 (1972).
b) H. Rau, *Angew. Chem., Intern. Ed.*, **12**, 224 (1973).
c) B. W. Langley, B. Lythgoe, L. S. Rayner, *J. Chem. Soc.*, 4191 (1952).
d) W. D. Closson, H. B. Gray, *J. Am. Chem. Soc.*, **85**, 290 (1963).

$$\begin{array}{ccc} & X & & X \\ & | & & | \end{array}$$

TABLE 3   Shifts due to X in C–C=O and C–C=C–C=O

X	ir,[a] νC=O. cm⁻¹		uv,[b] $n{\to}\pi^*$, nm		uv,[c] $n{\to}\pi^*$, nm	
	X | C–C=O				X | C–C=C–C=O	
	e	a	e	a	e	a
F	27	18				
Cl	18–25	2–9	−7	+22	+3	+14
Br	15–22	±3	−5	+28	+5	+20
OH			−12	+17	+2	+7
OAc			−5	+10	+2	+10
NH₂	see Table 4					

a) N. L. Allinger, H. M. Blatter, *J. Org. Chem.*, **27**, 1524 (1962).
b) R. C. Cookson, S. H. Dandegaonker, *J. Chem. Soc.*, 282 (1954); *ibid.*, 352 (1955).
c) C. W. Bird, R. C. Cookson, S. H. Dandegaonker, *J. Chem. Soc.*, 3675 (1956).

TABLE 4   Conjugated and Homoconjugated Systems Containing Carbonyls

1, 2-dione, diketo form:		280–300 nm, ε 10–80, $n{\to}\pi^*$; affected only slightly by angle between c=o
		340–470 nm, ε 10–40, $n{\to}\pi^*$, at short ∡ when angle between two carbonyls is close to 90°[a]
	enol form:	cf. Table 6 for calculation
	enolate:	base addition leads to ca. 30 nm red shift from enol
1, 3-dione, diketo form:		270–300 nm, ε 50, $n{\to}\pi^*$; two carbonyls absorb separately
	enol form:	240–270nm, ε 13,000; cf. Table 6 for calculation
	enolate:	280–300 nm, ε 23,000
α-keto acid, keto form:		210 nm, ε 400, $n{\to}\pi^*$ of acid
		330 nm, ε 60, $n{\to}\pi^*$ of ketone
	enol form:	cf. Table 8
α-keto ester, keto form:		CO and COOR absorb separately
β-keto ester, enol form:		245–260 nm, ε 10,000–15,000[b]
	enolate:	base addition leads to ca. 30 nm red shift
α-keto lactone, enol form:		230–240 nm, ε 10,000; red shift (ca. 30 nm) with alkali
α-keto lactone, keto form[c]:		220 nm, ε 200 (lactone) and 380 nm, ε 30 (ketone)

TABLE 4—*Continued*

β-keto lactone, enol form:	230 nm, ε 13,000 (γ-lactone)[d]; 240 nm, ε 12,000 (δ-lactone)[e]
epoxy ketone:	300 nm, ε 20–50

α,β-unsat. ketone and aldehyde: cf. Table 6

α,β-unsat. acid and ester: cf. Table 8

α,β-unsat. lactone[f]:	210–230 nm, ε 8,500–20,000, $\pi\rightarrow\pi^*$
(also exocyclic unsat.)	260–290 nm, ε 30–50, n$\rightarrow\pi^*$
β,γ-unsat. lactone:	230 nm, ε 1,000–5,000, charge transfer band[g,h]
enol lactone:	N$\rightarrow$n absorption above 200 nm
β,γ-unsat. enone[i]:	185–190 nm, ε 2,000–7,000, $\pi\rightarrow\pi^*$
	200–225 nm, ε 1,000–3,000; charge transfer band due to overlap of two orbitals
	270–300 nm, ε 30–500, n$\rightarrow\pi^*$; intense when overlap of oxygen p and olefin π orbitals is efficient
α- and β-amino ketones[j]:	215–245 nm, ε 500–1,500
	280–330 nm, ε 50–80, n$\rightarrow\pi^*$; red shift and enhancement in ε when N lone pair and one π lobe of carbonyl is zig-zag
β-aminocyclobutanone[k]:	two bands at 290–340 nm, ε 45–235

a) N. J. Leonard, *J. Am. Chem. Soc.*, **72**, 484, 5388 (1960).
b) S. J. Rhoads, J. C. Gilbert, A. W. Decora, T. R. Garland, R. J. Spangler, M. J. Urbigkeit, *Tetr.*, **19**, 1625 (1963).
c) Y. Nakadaira, Y. Hirota, K. Nakanishi, *Chem. Commun.*, 1467 (1969).
d) L. J. Haynes, J. R. Plimmer, *Quart. Rev.*, **14**, 292 (1960).
e) E. R. H. Jones, M. C. Whiting, *J. Chem. Soc.*, 1419, 1423 (1949).
f) cf. L. Dorfman, *Chem. Rev.*, **53**, 47 (1953).
g) J. F. Bagli, P. F. Morand, R. Gaudry, *J. Org. Chem.*, **28**, 1207 (1963).
h) T. Kubota, T. Matsuura, T. Tokoroyama T. Kamikawa, T. Matsumoto, *Tetr., Lett.* 325 (1961).
i) R. C. Cookson, N. S. Warylar, *J. Chem. Soc.*, 2302 (1956); H. Labhart, G. Wagniere, *Helv. Chim. Acta*, **42**, 2219 (1959); S. F. Mason, *Quart. Rev.*, **15**, 287 (1961).
j) J. Hudec, *Chem. Commun.*, 829 (1970); C. C. Levin, R. Hoffmann, W. J. Hehre, J. Hudec, *J. Chem. Soc P. II*, 210 (1973).
k) J. M. Conia, J. L. Ripoll, *Bull. Soc. Chim. Fr.*, 755 (1963).

TABLE 5   Diene and Triene $\pi\rightarrow\pi^*$ Bands (calculations)

Parent heteroannular diene	214
Parent homoannular 6-*membered* diene	253
Add for:	
alkyl group or ring residue	5
exocyclic C=C	5
extension of conjugation (not cross-conjugation)	30
substituents: OAc or OBz	0
OR	6
SR	30
Cl, Br	5
NR_2	60
Solvent correction	0
calcd. λ	

From reference 4. This is based on Fieser's modification (L. F. Fieser, M. Fieser, *Steroids*, Reinhold, 1959) of Woodward's diene rule (R. B. Woodward, *J. Am. Chem. Soc.*, **64**, 72 (1942)).

Comments on Table 5

1) The ε values are dependent on the square of the chromophore length, and therefore *trans*-diene maxima are more intense than those of *cis*-dienes.

   ε for planar s-*trans* dienes: 25,000.

   ε for s-*cis* or homoannular dienes: 10,000–15,000.

2) As planarity is destroyed by steric hindrance and/or ring distortion, the following changes occur: i) decrease in ε; ii) decrease in ε and blue shift in λ; iii) disappearance of conjugative effect. Examples (ε values in parentheses):

3 [a)]	4 [a)]	5 [a)]	6 [b)]	7 [c)]
232 (21,500)	236 (12,800)	242 (10,000)	207 (9,500)	R=H
239 (23,500)	243 (14,500)		226 (5,000)	220 (10,050)
248 (16,000)	251 (10,100)		233 (4,350)	R=Me, <220
calcd. 234	calcd. 244	ϵ calcd. 244	calcd. 254	calcd. 234

a) L. Dorfman, *Chem. Rev.*, **53**, 47 (1953).
b) D. H. R. Barton, G. S. Gupta, *J. Chem. Soc.*, 1961 (1962).
c) W. J. Bailey, W. B. Lawson, *J. Am. Chem. Soc.*, **79**, 1444 (1957).

3) The maxima of dienes or trienes containing an s-*trans* diene moiety such as **3**, **4** frequently show fine structure; when band positions are expressed in cm^{-1} (which is proportional to energy) the Δcm^{-1} between each band is ca. 1400–1600 cm^{-1}, i.e., approximately the value of $\nu C=C$ in the ir spectra.

4) Calculations for homoannular dienes are only valid for six-membered rings.

	n	nm		Reference
	5	239	3,400	G. Scheike, *Chem. Ber.*, **59**, 1333 (1926).
	6	256	7,950	V. Henri, L. W. Pickett, *J. Chem. Phys.*, **7**, 439 (1939).
C_{n-4}	7	248	7,400	E. Pesch, S. L. Friess, *J. Am. Chem. Soc.*, **72**, 5756 (1950).
	8	230	6,000	A. C. Cope, L. C. Ester, Jr., *J. Am. Chem. Soc.*, **72**, 1128 (1950).
8	10	223	5,000	R. W. Fawcett, J. O. Harris, *J. Chem. Soc.*, 2673 (1954).
	13	232		M. F. Bartlett, S. K. Figdor, K. Wiesner, *Can. J. Chem.*, **30**, 291 (1952).

5) In cases of cross-conjugated systems, calculations are carried out for each linear-conjugated moiety, and the longest wavelength is taken as the estimated value.

TABLE 6   Enone and Dienone $\pi \rightarrow \pi^*$ Bands, in EtOH (calculations)

Parent enone	215				

		Increments for Substituents			
Add for: exocyclic C=C	5				
substituents (see Table)		Substituent	α	β	γ	δ

Substituent	α	β	γ	δ
alkyl	10	12	18	18
OAc, OBz	6	6	6	6
OR	35	30	17	31
OH	35	30		50
SR		85		
Br	25	30		
Cl	15	12		
NR$_2$		95		

homodiene component 39
extension of conjugation 30
Solvent correction, (Note 1) ——
Calcd. λ^{EtOH}: add values. . . .

(a) Aldehydes: −5 nm from value of ketone
(b) Cyclopentenones with endocyclic double bond: −13 nm
   Cyclopentenones with exocyclic double bond: same as ketone

From reference 4. This is based on Fieser's modification (L. F. Fieser, M. Fieser, *Steroids*, Reinhold, 1959) of Woodward's enone rule (R. B. Woodward, *J. Am. Chem. Soc.*, **63**, 1123 (1941); *ibid.*, **64**, 76 (1942)).

Comments on Table 6

1) Unlike dienes, the enone $\pi \rightarrow \pi^*$ maxima are dependent on solvent polarity (see mesityl oxide **2**). The following solvent corrections (nm) convert into values for ethanol:

    methanol　0　　　　　　chloroform　+1　　　　　ether　　　+7
    hexane　+11　　　　　　water　　　−8　　　　　dioxane　+5
    (L. F. Fieser, M. Fieser, *Steroids*, Reinhold, 1959)

2) ε for s-*trans* enone : usually $>10,000$
    ε for s-*cis* enone　 : usually $<10,000$

3) $n \rightarrow \pi^*$ appear at 310–330 nm, ε 20–100. See Table 3 for the effect of γ-substituents on the $n \rightarrow \pi^*$ band of enones.

4) Take longest calculated λ for cross-conjugated systems.

5) Cd studies of enones have disclosed the presence of another transition at 200–220 nm, i.e., shorter λ than the strong $\pi \rightarrow \pi^*$ band treated in Table 6; however, this band is hidden in the uv.

TABLE 7　2,4-Dinitrophenylhydrazones (2,4-DNPH) (calculations)

		in EtOH		in CHCl$_3$	
		λ	ε	λ	ε
pentan-2-one	(i)	228	21,000		
MeCH$_2$CH$_2$C=N–NH–C$_6$H$_4$(NO$_2$)$_2$	(ii)	255	15,500	260	11,500
|	(iii)	280	8,000	280	9,500
Me	(iv)	362	25,000	368	22,500
propionaldehyde	(i)	228	16,000		
	(ii)	256	13,000	252	14,000
MeCH$_2$CH=N–NH–C$_6$H$_4$(NO$_2$)$_2$	(iii)	278	9,000	280	10,500
	(iv)	357	21,500	361	21,000

Comments on Table 7

Bands (i)–(iii) are fairly constant, independent of whether the hydrazone is derived from a sat. or unsat. carbonyl compound. In contrast, band (iv) undergoes a systematic red shift with substitution, etc., as follows:

Parent		λ in CHCl$_3$ (nm)
2,4-DNPH of acrolein, CH$_2$=CH–CHO		368
2,4-DNPH of vinyl ketone, CH$_2$=CH–CO–R		374
Add for:		
Alkyl substituent		7
Exocyclic double bond		2
Extension of conjugation		16

Total.....

(See K. Hirayama, *Treatise on Experimental Chemistry* (ed. M. Kotake, in Japanese), vol. 1, p. 89, Maruzen, 1957.)

TABLE 8　Unsaturated Acids, Esters and Lactones (in EtOH)

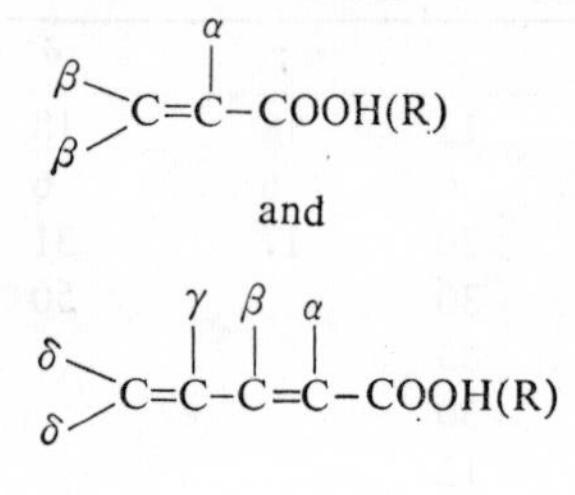

β ⟍　　 α 　　⟍ | 　　 C=C–COOH(R) β ⟋	Parent:　alkyl substituent α or β	208
	$\alpha\beta$ or $\beta\beta$	217
	$\alpha\beta\beta$	225
and	Add for:　exocyclic C=C	5
	endocyclic C=C (in 5- or 7-membered ring)	5
δ ⟍　 γ　β　α 　　⟍ | 　| 　| 　　 C=C–C=C–COOH(R) δ ⟋	γ- or δ-alkyl	18
	α-OH, -OMe, -Br, -Cl	15–20
	β-OMe, O-alkyl	30
	β-NMe$_2$	60
	extension of conjugation	30

calcd. λ^{EtOH} .....

A. T. Nielsen, *J. Org. Chem.*, **22**, 1539 (1957).

TABLE 9   Phenyl Derivatives C_6H_5R' and RC_6H_4R (nm (ε))

Group theory[k]	E_{1u}		B_{1u}		B_{2u}					
Platt classification[l]	^1B_b		^1L_a		^1L_b		Others			
Entry R	λ	ε	λ	ε	λ	ε	λ	ε	Solvent[m]	Ref.
1. H	184 (46,000)		204 (7,400)		254 (204)				W	a
2. Me			207 (7,000)		261 (225)				W	a
3. OH			211 (6,200)		270 (1,450)				W	a
4. OMe			217 (6,400)		269 (1,480)				W	a
5. O⁻			235 (9,400)		287 (2,600)				W	a
6. SH			235 (8,500)		269 (720)				H	b
7. NH_2			230 (8,600)		280 (1,430)				W	a
8. NMe_2			251 (14,000)		298 (2,100)				A	c
9. NH_3^+			203 (7,400)		254 (160)				W	a
10. NHCOMe			238 (10,500)						W	a
11. F			204 (6,300)		254 (1,000)				A	c
12. Cl			210 (7,400)		264 (190)				W	a
13. Br			210 (7,900)		261 (192)				W	a
14. I			207 (7,000)		257 (700)		226 (13,000)		W	a
15. SO_2NH_2			218 (9,700)		265 (740)				W	a
16. COOH			230 (11,600)		273 (970)				W	a
17. COMe			246 (9,800)						W	a
	199 (20,000)		243 (12,600)		278 (1,000)		320 (45)		A	d
18. CHO			250 (11,400)						W	a
			242 (14,000)		280 (1,400)		340 (50)		H	e
19. o-CHO/OMe			253 (10,500)		321 (4,550)				A	f
20. m-CHO/OMe			253 (9,500)		309 (3,000)				A	f
21. p-CHO/OMe			277 (15,500)						A	f
22. NO_2			269 (7,800)						W	a
23. o-NO_2/OH	230 (3,900)		279 (6,000)		351 (3,200)				W	a
24. p-NO_2/OH	223 (6,900)		318 (10,000)						W	a
25. o-NO_2/NH_2	245 (7,000)		283 (5,400)		412 (4,500)				W	a
26. p-NO_2/NH_2			373 (16,800)						W	a
27. C=C	211 (16,000)		244 (12,000)		282 (450)				A	g
28. C=C			235 (16,600)		278 (2,900)				A	h
29. C=C–COOH										
cis			264 (9,500)						A	i
$trans$			273 (21,000)						A	i
30. COPh			254 (18,000)		270 (1,700)		330 (160)		A	j

a) L. Doub, J. M. Vandenbelt, *J. Am. Chem. Soc.*, **69**, 2714 (1947); *ibid.*, **71**, 2414 (1949).
b) K. Bowden, E. A. Braude, E. R. H. Jones, *J. Chem. Soc.*, 948 (1946).
c) K. Bowden, E. A. Braude, *J. Chem. Soc.*, 1068 (1952).
d) R. A. Morton, A. L. Stubbs, *J. Chem. Soc.*, 1347 (1940).
e) E. A. Braude, F. Sondheimer, *J. Chem. Soc.*, 3754 (1955).
f) A. Burawoy, T. Chamberlain, *J. Chem. Soc.*, 2312 (1952).
g) R. A. Morton, A. J. A. deGouveia, *J. Chem. Soc.*, 911 (1934).
h) R. A. Friedel, M. Orchin, *Ultraviolet Spectra of Aromatic Compounds*, Wiley, 1951.
i) E. A. Braude, *Ann. Reports Chem. Soc.*, **42**, 105 (1945).
j) N. Jones, *J. Am. Chem. Soc.*, **67**, 2127 (1945).
k) M. G. Mayer, A. L. Sklar, *J. Chem. Phys.*, **6**, 645 (1938).
l) H. B. Klevens, J. R. Platt, *J. Chem. Phys.*, **17**, 470 (1949); J. B. Platt, *ibid.*, 484 (1949).
m) W: water (containing a trace of MeOH for increasing solubility) was used in reference a.
   A: EtOH. H: hexane.

Comments on Table 9

1) The directions of electric transition moments in monosubstituted benzenes are shown in **9**.

2) Solvent shift: Large red-shift is seen with electron-withdrawing substituents in solvating solvents, presumably due to greater stabilization of the polar excited state. The ^1L_b band of nitrobenzene shifts as follows: vapor (239.1), heptane (251.8), ethanol (259.5), water (265.5). Shift is small with electron-donating groups, e.g., OH, OMe. (See H. H. Jaffé, M. Orchin, *Theory and Applications of Ultraviolet Spectroscopy,* p. 251, Wiley, 1962).

3) Monosubstitution affects the ^1L_a bands to a greater extent than the ^1L_b bands. The stronger electron-donating or attracting groups have a larger interaction with the phenyl ring, and a linear correlation can be seen between the shift of the ^1L_a band and Hammett's σ parameter.

4) Effects of OH and OMe are comparable (nos. 3,4); formation of the phenolate (no. 5) causes a red shift.

5) NR_2 has a large effect (nos. 7,8) but the effect can be nullified by "fixing" of the lone-pair electrons, i.e., protonation (no. 9), N-oxide formation.

6) Disubstitution: Because of the longitudinal direction of the ^1L_a bands they are shifted more by, *p*-substitution; in contrast the transverse ^1L_b bands are shifted more by *o*- and *m*-substituents (methoxybenzaldehydes nos. 19, 20). In *p*-methoxybenzaldehyde (no. 21) the weak ^1L_b transition band is submerged by the intense and strongly displaced 277 nm band.

This red shift is pronounced when the two *p*-substituents have opposing electronic properties, e.g., *p*-nitroaniline (no. 26).
shift for nitro group: 269–204=65 nm (from nos. 22 and 1)
shift for amino group: 230–204=26 nm (from nos. 7 and 1)
shift for *p*-nitroaniline: calcd. 65 + 26=91 nm
obsvd. 373 − 204=169 nm

7) The ^1L_a bands of some carbonyl derivatives, nos. 16, 17, 18, can be calculated (Table 10).

8) "Other absorptions" in Table 9: 226 nm in no. 14 iodobenzene is iodine absorption; ca. 330 nm in C=O compounds nos. 17,18,30, is n→π*.

TABLE 10 ^1L_a **Bands of ArCOR (in EtOH, calculations)**

Parent ArCOR				
R: alkyl or ring residue	246			
R: H	250			
R: OH, O–alkyl	230			
Add for:		*o*	*m*	*p*
alkyl or ring residue		3	3	10
–OH, –OMe, –O–alkyl		7	7	25
–O		11	20	78
–Cl		0	0	10
–Br		2	2	15
–NH₂		13	13	58
–NHAc		20	20	45
–NHMe				73
–NMe₂		20	20	85

From A. I. Scott, *Ultraviolet Spectra of Natural Products,* p. 109, Pergamon Press, 1964; cf. A. I. Scott, *Experientia,* **17**, 68 (1961).

TABLE 11 Condensed Aromatics (nm (ε))[a]

Clar classification[b]			β	*para*	α
Platt classification	^1B_a (trans.)	^1C_b	^1B_b (long.)	^1L_a (trans.)	^1L_b (long.)
benzene			183	208	263
			(46,000)	(6,900)	(220)
naphthalene	167	190	220	289	312
	(30,000)	(10,000)	(133,000)	(9,300)	(280)
anthracene	186	221	256	379	
	(32,000)	(14,500)	(180,000)	(9,000)	
naphthacene	187 211	230	272	474	
	(16,000) (44,000)	(44,000)	(180,000)	(12,500)	
phenanthrene		212	251	293	350
		(33,000)	(65,000)	(16,000)	(350)

a) From H. B. Klevns, J. R. Plaett, *J. Chem. Phys.*, **17**, 470 (1949); H. H. Jaffé, M. Orchin, *Theory and Applications of Ultraviolet Spectroscopy*, pp. 287–337, Wiley, 1962.

b) E. Clar, *Spectrochim. Acta*, **4**, 116 (1950); E. Clar, *Polycyclic Hydrocarbons*, Academic Press, 1964.

Comments on Table 11

1) The directions of polarizations are as shown in **10**.

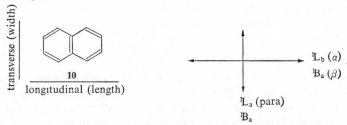

2) Most condensed aromatics show three band groups (above 200 nm), which are frequently associated with vibrational fine structures; the peak at longest wavelength for each band system is given in Table 11.

3) Clar has referred to these band groups, which successively increase in intensity towards shorter wavelengths, as follows:[b]

α band, log ε 2.3–3.2 or ε 200–1,600
para band, log ε 3.6–4.1 or ε 4,000–13,000
β band, log ε 4.5–5.2 or ε 20,000–160,000

4) ^1L_a peak in linear molecules: This peak, with its electric transition moment in the transverse direction, undergoes the most rapid red shift, being already marked in anthracene. In naphthacene, the 474 nm band gives it an orange color. In agreement with the assignment, the ^1L_a peak intensity is increased by substituents which increase the width (see **10**) of the molecule.

5) Linear *vs.* bent molecules: In bent molecules, e.g., phenanthrene, the red shift accompanying an increase in ring number is approximately the same for α, *para*, and β bands, so that the longest α (^1L_a) band is not masked.

6) Position and intensity of the β (^1B_b) band increases linearly with the number of rings in linear molecules.

7) Effects of substituents: The position of a substituent will primarily red shift the transition polarized in that direction. Thus:

	β or ^1B	*para* or ^1L_a	α or ^1L_b
1-naphthylamine	239 (25,000)	323 (5,000)	
2-naphthylamine	236 (63,000)	281 (6,300)	340 (2,000)

The 1-substituent displaces the ^1L_a peak so that the ^1L_b peak is submerged, whereas the longitudinal 2-substituent displaces the ^1L_b band but does not affect the ^1L_a band (R. N. Jones, *J. Am. Chem. Soc.*, **67**, 2127 (1945).)

2.3 OPTICAL ROTATORY DISPERSION AND
CIRCULAR DICHROISM[6-9]

Optical rotatory dispersion (ord)

D-line reading is simply a summation of multiple ord curves.

specific rotation: $[\alpha] = \alpha/1c$ α, angle of rotation

1, cell length in dm; c, concentration in g/ml

molecular rotation: $[\Phi] = ([\alpha] \times \text{mol. wt.})/100$

molecular amplitude: $a = ([\Phi]_1 - [\Phi]_2)/100$ $[\Phi]_1$ and $[\Phi]_2$, values of extrema

Circular dichroism (cd)

specific ellipticity: $[\Psi] = \Psi/1c$ Ψ, angle of ellipticity

1, cell length in dm; c, concentration in g/ml

molecular ellipticity: $[\theta] = ([\Psi] \times \text{mol. wt.})/100 = 3300 \, \Delta\varepsilon$

differential dichroic absorption: $\Delta\varepsilon = \varepsilon_L - \varepsilon_R$ ε_L, ε for left polarized light

The following relation holds for the n→π* band of saturated ketones:

$$a = 40.28 \, \Delta\varepsilon \quad \text{or} \quad a = 0.0122 \, [\theta]$$

This relation can be used for the *approximate* interconversion of cd and ord data other than those of ketones.

Deduction of Absolute Configuration

As a compound has only two possible absolute configurations, a measurement aimed at deducing its absolute configuration will give a definite result, independent of whether the reasoning is correct or not (even if it were erroneous, there is still a 50% chance of arriving at the correct answer!). It is therefore important to be extremely cautious in analysing experimental results.

The D-line rotation *should not be used* for configurational assignments. Even the ketone octant rule,[10] which was one of first rules to be introduced in ord and cd, and which had a great impact on development in this field, should now be used with great caution.

Octant rule (quadrant rule) for ketones[10]

This was originally developed for assessing the contribution of substituents on the n→π* Cotton effect sign of cyclohexanones. The ring is divided into a quadrant by the carbonyl n→π* nodal plane A and the symmetry plane B, which is perpendicular to A and cuts through C-1 and C-4 (**11, 12**). Originally the quadrant was further bisected by a third plane perpendicular to A and B, and running through the carbon of C=O (hence the "octant rule"). However, the third plane is not used any more (see following for discussion on "front octant": D. N. Kirk, W. Klyne, W. P. Mose, *Tetr. Lett.* 1315 (1972)).

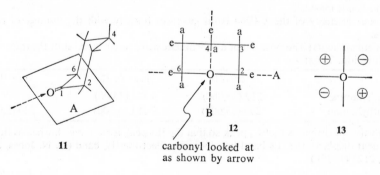

11

12

carbonyl looked at as shown by arrow

13

Quadrant rule for ketones: Atoms or groups in the lower-right or upper-left quadrants make a positive contribution to the Cotton effect; conversely, those in the two other quadrants make negative contributions (13). The contribution of axial substituents decreases in the sequence of:

$$I > Br > Cl > NH_2 > OH > Me > OAc > H$$

The signs of contributions are reversed for F (see below for N^+). Contributions due to equatorial substituents can either be positive or negative depending on their spatial position relative to plane A; those lying exactly on plane A (2e, 6e) or B (4e, 4a) make no contribution. This approach can be extended to other cyclic and acyclic saturated ketones.

Quadrant behavior and antiquadrant behavior

Numerous examples of antiquadrant behavior are being found. It appears that the following treatment of amino ketones by Hudec[11] can be generalized to include alkyl-, hydroxy-, halo-, and other types of ketones.[12] Quadrant behavior holds for amino ketones ($\alpha,\beta,\gamma, \dots$) when the lone pair of N adopts an antiperiplanar zig-zag relation with one of the π lobes of the carbonyl C, either due to its fixed conformation or due to free rotation around the C–N bond, i.e., these are conditions favoring W-type coupling in the nmr if the N lone pair and the carbonyl C lobe were replaced by protons. If the N lone pair cannot adopt the zig-zag relation, then antiquadrant behavior occurs. N^+ always results in antiquadrant behavior; if the amino group is already antiquadrant, N quaternization increases the antiquadrant contribution.

Examples of quadrant (quad.) and antiquadrant (anti.) contributions of ring N or substituents are given.[11,12] $\Delta\Delta\varepsilon$ is difference in $\Delta\varepsilon$ between parent carbocyclic or unsubstituted ketone.

| **14**[11]
quadrant | **15**[11]
antiquadrant | **16**[12]' | **17**[12] |

X	$\Delta\Delta\varepsilon$		X	$\Delta\Delta\varepsilon$	
I	+8.50	quad.	eq-3-Br	−0.97	quad.
Br	+3.97	//	ax-3-Br	+0.37	anti.
Cl	+1.59	//			
OH	+0.18	//			
F	−0.72	anti.			
$NH_3^+Cl^-$	−0.87	//			

6) C. Djerassi, *Optical Rotatory Dispersion*, McGraw-Hill, 1960.
7) L. Velluz, M. Legrand, M. Grosjean, *Optical Circular Dichroism, Principles, Measurements, and Applications*, Verlag Chemie, 1965.
8) *Optical Rotatory Dispersion and Circular Dichroism in Organic Chemistry* (ed. G. Snatzke), Heyden and Sons, 1967.
9) P. Crabbé, *Application de la Dispersion Rotatoire Optique et du Dichroisme Circulaire Optique en Chimie Organique*, Gauthier Villars, 1968; *Ord and Cd in Chemistry and Biochemistry*, Academic Press, 1972.
10) W. Moffitt, R. B. Woodward, A. Moscowitz, W. Klyne, C. Djerassi, *J. Am. Chem. Soc.*, **83**, 4013 (1961).
11) J. Hudec, *Chem. Commun.*, 829 (1970); we are most grateful to John Hudec for a thorough discussion on his approach.
12) M. T. Hughes, J. Hudec, *Chem. Commun.*, 805 (1971); G. P. Powell, J. Hudec, *ibid.*, 806 (1971).

The absolute configuration of the hydrocarbon twistane was reversed[13] after two separate research groups had deduced the opposite configuration on the basis of antiquadrant-behaving synthetic intermediates.[14] Quadrant rules for cyclic, halo, oxido, and other ketones are summarized in ref. 15; they should be treated with the same reservations described above.

Isolated double bond

Monoenes exhibit two Cotton effects of *opposite* sign around 180–200 nm (λ_1) and 200–210 nm (λ_2); λ_2 is used for analysis.

1) Quadrant rule:[16] Mills rule,[17] which was refined by Brewster,[18] has long been used for configurational assignments of 3-substituted cycloalkenes (based on $[\Phi]_D$).

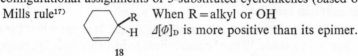

Mills rule[17]   When R = alkyl or OH
                    $\Delta[\Phi]_D$ is more positive than its epimer.

18

Mills rule was analysed as follows by Scott and Wrixon.[16]

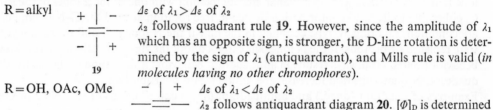

R = alkyl $\Delta\varepsilon$ of $\lambda_1 > \Delta\varepsilon$ of λ_2

λ_2 follows quadrant rule **19**. However, since the amplitude of λ_1 which has an opposite sign, is stronger, the D-line rotation is determined by the sign of λ_1 (antiquardrant), and Mills rule is valid (*in molecules having no other chromophores*).

19

R = OH, OAc, OMe $\Delta\varepsilon$ of $\lambda_1 < \Delta\varepsilon$ of λ_2

λ_2 follows antiquadrant diagram **20**. $[\Phi]_D$ is determined by sign of λ_2 (antiquadrant **20**), and Mills rule is valid.

20

2) Allylic axial chirality:[19-21]

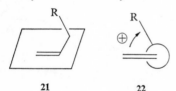

R is an allylic (and homoallylic[22]) axial group, and is H, Me, OR, X or vinyl. The unit **21** or **22** makes a positive contribution to the $\Delta\varepsilon$ of λ_2. It appears that the *chirality of an oxygen substituent* (OH, OMe, OAc) *has a dominating effect*. Projection **22** also incorporates the quadrant rule.

21   **22**

Dienes

diene		if chirality is:	sign of $\pi \rightarrow \pi^*$ cd
homo-annular			positive[23]
hetero-annular			positive[23]
hetero-annular s-*cis*			negative[20]

13) M. Tichý, *Tetr. Lett.*, 2001 (1972).
14) K. Adachi, K. Naemura, M. Nakazaki, *Tetr. Lett.*, 5467 (1968); M. Tichý, J. Sicher, *ibid.*, 4609 (1969). See also, D. Varech, J. Jacques, *Tetr.*, **28**, 5671 (1972). The antiquadrant behavior of bicyclo [2.2.2]octanone systems can now be explained by non-fullfilment of the W-type criterion.
15) G. Snatzke, *Tetr.*, **21**, 413, 421, 439 (1965).
16) A. I. Scott, A. D. Wrixon, *Tetr.*, **27**, 4787 (1971).
17) J. A. Mills, *J. Chem. Soc.*, 4976 (1952).

The chirality of an allylic axial oxygen function overrides diene helicity.[24] The cd has been interpreted by treating the diene as a chromophore with C_2 symmetry,[25] or according to the allylic chirality method.[20] The homoconjugated diene **23** is strongly positive at 220 nm.[26]

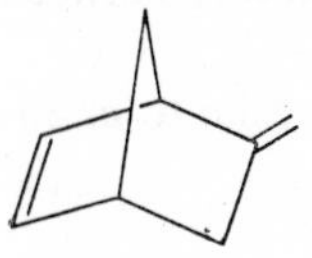

Ord: Φ is $+80,000$ at 220 nm (shortest wavelength recorded; this value roughly corresponds to a cd $\Delta\varepsilon$ of 20)

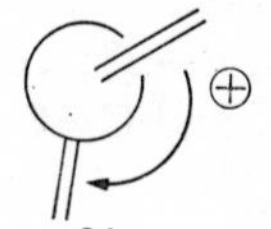

23 $\equiv$ **24**

Enones

The following has been observed for *transoid enones*.

1) 205–220 nm, $\Delta\varepsilon$ 5–15, unassigned: sign agrees with chirality of α'-axial bond[20] (see **25**).
2) 220–250 nm, $\Delta\varepsilon$ 1–15, $\pi\rightarrow\pi^*$: sign agrees with chirality of C–X (X is O, Me or halogen) in $C=C-C-X$.[20,27-29]
3) 320–330 nm, $\Delta\varepsilon$ 1–2, $n\rightarrow\pi^*$: sign generally opposite to chirality of enone[15,30] if no polar allylic group is present; when such group is present, then variable sign.

The following examples from ref. 28 exemplify the variations; nmr showed that conformations of **25–28** are identical (ring A chirality is positive).

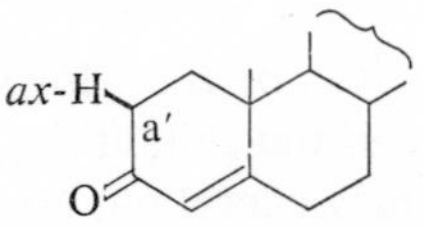

	25	**26**	**27**	**28**
205–220	+ca. 10	+ca. 10	+4.7	+2.7
$\pi\rightarrow\pi^*$	+8.8	−18	−5.2	−10.5
$n\rightarrow\pi^*$	−1.2	+2.2	−1.5	+0.8

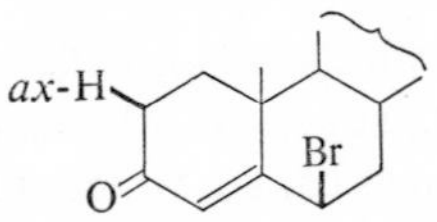

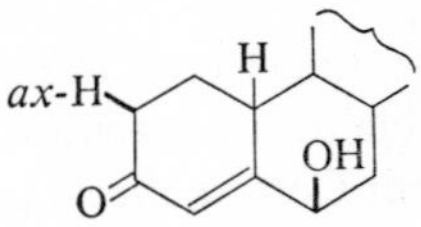

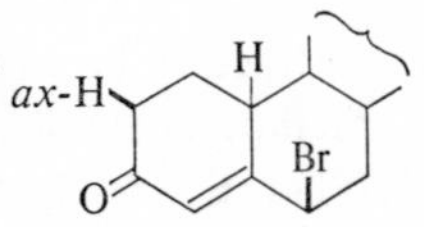

Frequently it is useless to consider the chirality of enones, due to ring flexibility. This is especially so with exomethylene cisoid enones.

Hetero-annular *cisoid enones* have the following characteristics:

1) 200–220nm, $\Delta\varepsilon$ 5–15, unassigned: sign agrees with chirality of α'-axial bond.[20]
2) $\pi\rightarrow\pi^*$:   sign variable with chirality.
3) $n\rightarrow\pi^*$:   sign generally agrees with chirality of enone.[30]

18) J. H. Brewster, *J. Am. Chem. Soc.*, **81**, 5493 (1959).
19) A. Yogev, D. Amar, Y. Mazur, *Chem. Commun.*, 339 (1967)
20) A. W. Burgstahler, R. C. Barkhurst, *J. Am. Chem. Soc.*, **92**, 7601 (1970).
21) N. H. Andersen, C. R. Costin, D. D. Syrdal, D. P. Svedberg, *J. Am. Chem. Soc.*, **95**, 2049 (1973).
22) J. K. Gawronski, M. A. Kielczewski, *Tetr. Lett.*, 2493 (1971).
23) A. Moscowitz, E. Charney, V. Weiss, H. Ziffer, *J. Am. Chem. Soc.*, **83**, 4661 (1961); V. Weiss, H. Ziffer, E. Charney, *Tetr.*, **21**, 3105 (1965); E. Charney, H. Ziffer, V. Weiss, *ibid.*, 3121.
24) A. F. Beecham, A. McL. Mathieson, S. R. Johns, J. A. Lamberton, A. A. Sioumis, T. J. Batterham, I. G. Young, *Tetr.*, **27**, 3725 (1971).
25) W. Hug, G. Wagniére, *Tetr.*, **28**, 1241 (1972).
26) L. S. Forster, A. Moscowitz, J. G. Berger, K. Mislow, *J. Am. Chem. Soc.*, **84** 4353 (1962).
27) C. Djerassi, R. Records, E. Bunnenberg, K. Mislow, A. Moscowitz, *J. Am. Chem. Soc.*, **84**, 870 (1962).
28) K. Kuriyama, M. Moriyama, T. Iwata, K. Tori, *Tetr. Lett.*, 1661 (1968).
29) A. F. Beecham, *Tetr.*, **27**, 5207 (1971).
30) W. B. Whalley, *Chem. Ind.*, 1024 (1962); R. E. Ballard, S. F. Mason, G. W. Vane, *Disc. Faraday Soc.*, **35**, 43 (1963).

α,β-**Unsaturated** δ- **and** γ-**lactones**[31,32]

29

1) 210–230 nm, $\Delta\varepsilon$ 5–30, $\pi\to\pi^*$: if allylic axial OH, OAc, OR is present, cd sign is same as chirality between O–C and C=C.
2) 260 nm, $\Delta\varepsilon$ 2–10, $n\to\pi^*$: cd sign is opposite to chirality of C=C and C=O.

The dibenzoate chirality method[33,34]

When two benzoate groups, not necessarily 1,2-, but more distant (e.g., 1,4-), are close enough in space, the 227 nm bands of the two benzoates interact to give two Cotton effects of the same amplitude ($\Delta\varepsilon$ 9–18) but of opposite sign at 233 nm and 219 nm. If the two benzoates are twisted in a clockwise sense **30** (positive chirality) the 233 nm cd will be positive; conversely, in **31** the 233 nm cd is negative. This type of splitting of Cotton effects due to dipole-dipole

30   **31**

interaction between the electric transition moments is called exciton or Davydov splitting. The method is applicable to tribenzoates, e.g., sugars, in which case, if the chiralities augment each other, the $\Delta\varepsilon$ of split Cotton effects becomes ca. 80, whereas if they cancel out no split cd is observed. Only a small amount of sample is necessary because of the high $\Delta\varepsilon$ values. The Cotton effects can be shifted to longer wavelengths and made even larger by employment of appropriate p-substituted benzoates.

Exciton chirality method[34]

The dibenzoate chirality method can be generalized to include other chromophores.[34] Namely, if one knows the direction of electric transition moments of any chromophoric transition, the chirality between the two moments **32** can be determined from the signs of the split Cotton effects. For example, interactions between the transitions shown in **33–36** (double-headed arrows) give rise to such Davydov split curves with large amplitudes, provided they are close in space[35] and the absorption maxima are not far apart. The interacting chromophores can either be present in the natural product itself, or one or both can be introduced by derivatization or modification. A simple way to determine directions of moments is by linear dichroism measurements using stretched polyethylene films.[36]

The principle of exciton chirality, which is based on well-founded quantum mechanical theories, was first applied to natural products (alkaloids) by Mason and co-workers.[37] The split Cotton effects caused by interaction between two or more isolated chromophores are the strongest of the cd (or ord) extrema; also, determination of absolute configurations based on this method is one of the most straightforward and unambiguous methods.

31) G. Snatzke, *Angew. Chem. Intern. Ed.*, **7**, 14 (1968).
32) A. F. Beecham, *Tetr.*, **28**, 5543 (1972).
33) N. Harada, K. Nakanishi, *J. Am. Chem. Soc.*, **91**, 3989 (1969).
34) N. Harada, K. Nakanishi, *Accounts Chem. Res.*, **5**, 257 (1972); see also: M. Koreeda, N. Harada, K. Nakanishi, *J. Am. Chem. Soc.*, **96**, 266 (1974); N. Harada, S. L. Chan, K. Nakanishi, *ibid.*, **97**, 5345 (1975).
35) The limitation in distance for the appearance of a split cd curve is under study.
36) J. Michl, E. W. Thulstrup, J. H. Eggers, *J. Phys. Chem.*, **74**, 3878 (1970); A. Yogev, J. Sagiv, Y. Mazur, *J. Am. Chem. Soc.*, **94**, 5122 (1972).
37) S. F. Mason, G. W. Vane, *J. Chem. Soc.*, *B*, 370 (1966); S. F. Mason, K. Schofield, R. J. Wells, J. S. Whitehurst, G. W. Vane, *Tetr. Lett.*, 137 (1967).

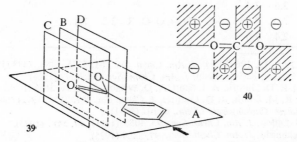

32   **33**   **34**   **35** π, π^* **36**

Benzoate rule (Brewster)[38]

When benzoates are prepared from secondary alcohols with absolute configurations **37** or **38**, a positive shift as compared to the carbinol is observed in the D-line rotation. This simple rule has been applied successfully to numerous natural products. However, here again it should be cautioned that the results rely on $[\alpha]_D$. The method cannot be used if the smaller group happens to be the more polar group.

S: smaller group
L: larger group

MP: more polar group
LP: less polar group

37   **38**

Benzoate sector rule[34]

Benzoates of optically active carbinols have cd maxima at 227 nm, $\Delta\varepsilon$ 2–10. This is due to asymmetric perturbation of the inherently symmetric benzoate chromophore. The benzoate conformation is fixed so that it is flanked by H and the smaller of the remaining two groups, and the rotatory contributions of α,β and β,γ bands are considered according to the sector **39**; bonds falling in the shaded and unshaded sectors make, respectively, negative and positive contributions (**40**). If it is an allylic carbinol, the sector in which the double bond is located will determine the cd sign. The sectors and alternating signs at nodal planes are corroborated by theoretical considerations.

39   **40**

The benzoate sectors. The benzoate is viewed from the direction of the arrow.

Chirality of α-glycols and α-amino alcohols

Several methods have been developed for determining the chirality of α-glycols and α-amino alcohols with dihedral angles around 60° without derivatization.

1) Cuprammonium method for cyclic compounds:[40]

A positive chirality is associated with a positive Cotton effect at 580–600 nm when the compound is measured in aqueous cuprammonium solution. It is an extension of the classical Reeves method[41] used extensively in carbohydrates.

2) Pr(DPM)₃ and Eu(DPM)₃ method for cyclic compounds:[42]

A positive chirality is associated with a positive first Cotton effect (longer wavelength) and a negative second Cotton effect (shorter wavelength) centered around 300 nm when the compound is measured in *dry* chloroform, carbon tetrachloride, etc., containing the lanthanide nmr shift reagents. The split Cotton effect ($\Delta\varepsilon$ 5–10) is presumably due to exciton splitting between the DPM moieties as a result of induced chirality.

3) Ni(AcAc)₂ method for acyclic compounds:[43]

The two functional groups of the α-glycol or α-amino alcohol are set in a skewed conformation in which the more bulky groups are on the remote side of the Newman projection. A positive chirality is associated with a split Cotton effect ($\Delta\varepsilon$ 4–70) centered at 310 nm, in which the longer and shorter wavelength extrema are, respectively, negative and positive; in addition, a weaker and positive cd band appears at the much longer wavelength of 640 nm. The advantage of Ni(AcAc)₂ is that a variety of solvents can be used ranging from CCl_4 to CH_3CN; 0.2M t-BuOH in CCl_4 is usually satisfactory.

2.4 NUCLEAR MAGNETIC RESONANCE SPECTROSCOPY

Peaks usually appear within ± 0.2 ppm of the values quoted in Tables 12 and 13, unless inductive or anisotropic effects of nearby groups are operating.

TABLE 12 Approximate Chemical Shifts[44]

Methyl Protons			Methylene Protons			Methine Protons		
Proton	δ	τ	Proton	δ	τ	Proton	δ	τ
CH₃C	0.9	9.1	–C–CH₂–C–	1.3	8.7	C–CH–C	1.5	8.5
			"(cyclic)	1.5	8.5	"(bridgehead)	2.2	7.8
CH₃–C–C=C	1.1	8.9	–C–CH₂–C–C=C	1.7	8.3			
CH₃–C–O	1.4	8.6	–C–CH₂–C–O	1.9	8.1	–C–CH–C–O	2.0	8.0
CH₃–C=C	1.6	8.4	–C–CH₂–C=C	2.3	7.7			
CH₃–Ar	2.3	7.7	–C–CH₂–Ar	2.7	7.3	–CH–Ar	3.0	7.0
CH₃–CO–R	2.2	7.8	–C–CH₂–CO–R	2.4	7.6	–C–CH–CO–R	2.7	7.3
CH₃–CO–Ar	2.6	7.4						
CH₃–CO–O–R	2.0	8.0	–C–CH₂–CO–O–R	2.2	7.8			
CH₃–CO–O–Ar	2.4	7.6						

38) J. H. Brewster, *Tetr.*, **13**, 106 (1961); *J. Am. Chem. Soc.*, **81**, 5475, 5483, 5493 (1959).
39) N. Harada, M. Ohashi, K. Nakanishi, *J. Am. Chem. Soc.*, **90**, 7349 (1968).
40) S. T. K. Bukhari, R. D. Guthrie, A. I. Scott, A. D. Wrixon, *Tetr.*, **26**, 3653 (1970). Example of application: A. W. Johnson, R. M. Smith, R. D. Guthrie, *J. Chem. Soc. P I*, 2153 (1972) (new amino-sugar).
41) R. E. Reeves, *Advan. Carbohydrate Chem.*, **6**, 107 (1951).
42) K. Nakanishi, J. Dillon, *J. Am. Chem. Soc.*, **93**, 4058 (1971); **97**, 5417 (1975).
43) J. Dillon, K. Nakanishi, *J. Am. Chem. Soc.*, **97**, 5409 (1975).
44) Taken from ref. 1, p. 223, 224.

TABLE 12—*Continued*

Methyl Protons			Methylene Protons			Methine Protons		
Proton	δ	τ	Proton	δ	τ	Proton	δ	τ
CH₃–CO–N–R	2.0	8.0						
CH₃–O–R	3.3	6.7	–C–CH₂–O–R	3.4	6.6	–C–CH–O–R	3.7	6.3
			–C–CH₂–O–H	3.6	6.4	–C–CH–O–H	3.9	6.1
CH₃–OAr	3.8	6.2	–C–CH₂–OAr	4.3	5.7			
CH₃–O–CO–R	3.7	6.3	–C–CH₂–O–CO–R	4.1	5.9	–C–CH–O–CO–R	4.8	5.2
CH₃–N	2.3	7.7	–C–CH₂–N	2.5	7.5	–C–CH–N	2.8	7.2
CH₃–N⁺	3.3	6.7						
CH₃–S	2.1	7.9	–C–CH₂–S	2.4	7.6			
CH₃–C–NO₂	1.6	8.4	–C–CH₂–NO₂	4.4	5.6	–C–CH–NO₂	4.7	5.3
CH₃–C=C–CO	2.0	8.0	–C–CH₂–C–NO₂	2.1	7.9			
–C=C(CH₃)–CO	1.8	8.2	–C–CH₂–C=C–CO	2.4	7.6			
			–C–C(CH₂)–CO	2.4	7.6			

Methylene structure	δ	τ	Methine structure	δ	τ
benzodioxole CH₂	5.9	4.1			
cyclopropyl CH₂	0.3	9.7	cyclopropyl CH–	0.7	9.3
epoxide CH₂	2.6	7.4	epoxide CH–	3.1	6.9

TABLE 13   Chemical Shifts of Miscellaneous Protons

	δ	τ		δ	τ
(cyclic) C=CH₂	4.6ª⁾	5.4	benzene –H	7.246	2.754
–C=CH₂	5.3ª⁾	4.7			
–C=CH–	5.1ª⁾	4.9	toluene CH₃ ←	2.337	7.663
–C=CH–(cyclic)	5.3ª⁾	4.7	toluene H ←	7.095	2.905
Ar–H	9.0–7.0ᵇ⁾	1.0–3.0	H ←	4.65	5.35
–C=CH–CO	5.9	4.1			
–CH=C–CO	6.8	3.2	H, H–O–H ←	6.37	3.63
R–CHO	9.9	0.1		3.97	6.03
Ar–CHO	9.9	0.1			
H–CO–O–	8.0	2.0	cyclohexane H	1.44	8.56
H–CO–N	8.0	2.0	CHCl₃	7.25	2.75
			H₂O	5.0	5.0

a) Olefinic protons are subject to shifts larger than ±0.2.
b) Similar range for heterocyclic aromatics.

2.5 MASS SPECTROMETRY[45-48]

Fragmentation patterns

The following fragmentations occur upon electron impact to give the fragment peaks indicated in the low mass region, or (M–fragment) peaks in the high mass region.

⌒ : movement of one electron

⌒◂ : movement of electron pair

∿ : cleavage with charge on fragment shown by arrow (→)

alkane: $\left[\text{C–C}\substack{\text{C}\\ |\\ \text{C}}\text{–C}\right]^{+\cdot}$ 29, 43, 57, . . .

alkene: $\left[\text{C}\substack{\\ }\text{C–C=C}\right]^{+\cdot}$ β (allylic cleavage): 41 ($\overset{+}{\text{C}}\text{H}_2\text{–CH=CH}_2$), 55, 69, . . .

$\left[\bigcirc\hspace{-0.1em}\right]^{+\cdot}$ retro-Diels-Alder: $\left[\diagup\hspace{-0.3em}\diagdown\right]^{+\cdot}$ or $\left[\,||\,\right]^{+\cdot}$

aliph. alcohol: M–H$_2$O

$\underset{\beta}{\text{C}}\text{–}\underset{\alpha}{\text{C}}\text{–}\overset{+\cdot}{\text{C}}\text{–OH}$ α: 31 ($\text{CH}_2\text{=}\overset{+}{\text{O}}\text{H}$), 45, 59

alkyl halide: M–HX

$\underset{\beta}{\text{C}}\text{–}\underset{\alpha}{\text{C}}\text{–}\overset{+\cdot}{\text{C}}\text{–X}$ α: "alkane" peaks, 29, 43, 57, . . .

aliph. amine: $\underset{\beta}{\text{C}}\text{–}\underset{\alpha}{\text{C}}\text{–}\overset{+\cdot}{\text{C}}\text{–NH}_2$ α: 30 ($\text{CH}_2\text{=}\overset{+\cdot}{\text{N}}\text{H}_2$), 44, 58, . . .

aliph. ether: $\left[\text{O}\substack{\\ }\text{C}\right]^{+\cdot}$ "alkane" peaks, 29, 43, 57, . . .

$\underset{\beta}{\text{C}}\text{–}\underset{\alpha}{\text{C}}\text{–}\overset{+\cdot}{\text{C}}\text{–O–Me}$ α: 45 ($\text{CH}_2\text{=}\overset{+}{\text{O}}\text{–Me}$), 59, 74, . . .

aliph. thioether: M+2 peak is intensified due to 4% natural abundance of ^{34}S.

$\text{C–C}\overset{\alpha}{\text{–}}\overset{+\cdot}{\text{C}}\text{–S–Me}$ α: 61 ($\text{CH}_2\text{=}\overset{+\cdot}{\text{S}}\text{–Me}$), 75, 89, . . .

aliph. *sec-, tert*-amine:

$\left[\text{C–C}\substack{\\ }\text{N–C}\substack{\\ }\text{C}\right]^{+\cdot}$

aliph. aldehyde: $\underset{\beta}{\overset{\gamma}{\text{C}}}\text{–C}\substack{\text{H}\\ }\cdots\overset{+\cdot}{\text{O}}$ β–cleavage with (H·) transfer from γ-C (McLafferty rearrangement): 44, 58, 72, . . .

45) K. Biemann, *Mass Spectrometry*, McGraw-Hill, 1962.
46) F. W. McLafferty, *Mass Spectrometry of Organic Ions*, Academic Press, 1963.
47) H. Budzikiewicz, C. Djerassi, D. H. Williams, *Interpretation of Mass Spectra of Organic Compounds*, and *Structure Elucidation of Natural Products by Mass Spectrometry*, vol. I and vol. II, Holden-Day, 1964.
48) *Biochemical Applications of Mass Spectrometry* (ed. G. R. Waller), Wiley-Interscience, 1972.

aliph. ketone:

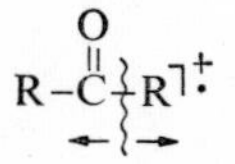

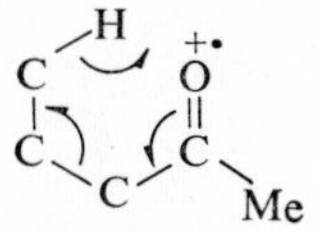

$$R-\overset{\overset{O}{\|}}{C}\!\!+\!\!R\rceil^{\ddot{+}}$$

"$\beta+\gamma(H\cdot)$" series: 58, 72, 86, ...

aliph. ester:

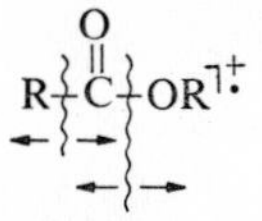

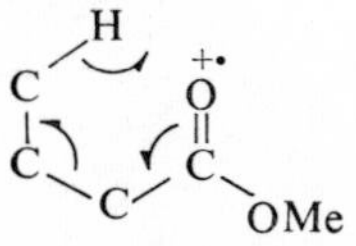

$$R\!+\!\overset{\overset{O}{\|}}{C}\!\!+\!\!OR\rceil^{\ddot{+}}$$

"$\beta+\gamma(H\cdot)$" series: 74, 88, 102, ...

acetate:

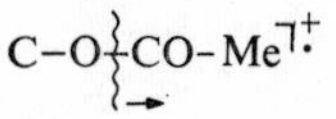

$$C-O\!+\!CO\text{-}Me\rceil^{\ddot{+}}$$ 43

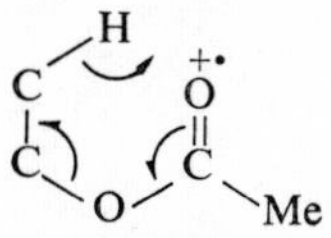

β-cleavage with γ-hydride transfer: M–60

aliph. amides:

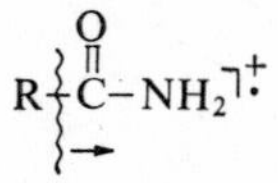

$$R\!+\!\overset{\overset{O}{\|}}{C}\!-\!NH_2\rceil^{\ddot{+}}$$ 44

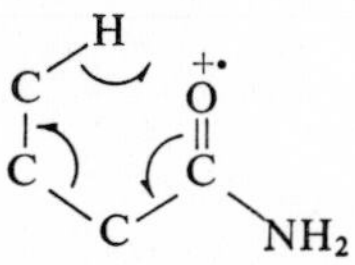

"$\beta+\gamma(H\cdot)$" series: 59, 73, 87, ...

aliph. acids:

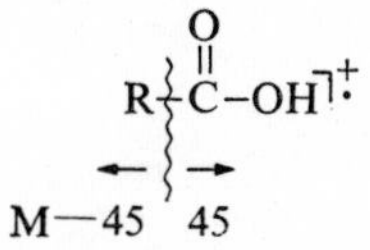

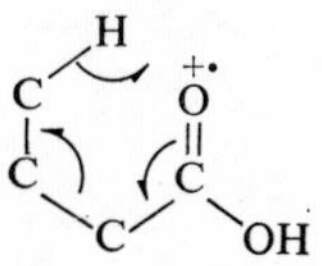

$$R\!+\!\overset{\overset{O}{\|}}{C}\!-\!OH\rceil^{\ddot{+}}$$

M–45   45

"$\beta+\gamma(H\cdot)$" series: 60, 74, 86, ...

aromatics:

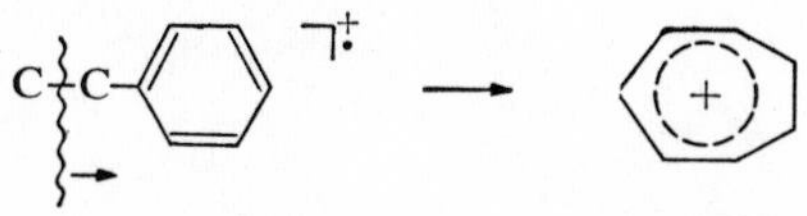

$$C\!+\!C\text{-}\!\!\bigcirc\rceil^{\ddot{+}} \longrightarrow$$ tropylium cation 91, 105, 119, ...

H–A

Mono- and Sesquiterpenes

3.1   INTRODUCTION

The mono- and sesquiterpenoids, substances which are biogenetically derived from two and three isoprene units, respectively, are distributed widely in a variety of living systems such as plants, microorganisms and also insects. Some of these compounds have essential functions within living organisms, and others exhibit important physiological activities. The structural diversity of these compounds has attracted much attention for many years, and provides challenging problems for organic chemists. Research concerned with these compounds, including the isolation of new compounds, structural elucidation, chemical synthesis and biosynthetic studies, is still increasing in volume. This chapter will be mainly concerned with recent problems which are of continuing interest, and therefore some important topics in classical work may be omitted. For details of such work, the reader should refer to comprehensive articles mentioned in the references given throughout this chapter.

The monoterpenes, which are presumed to be derived from geranyl pyrophosphate, have been known for hundreds of years as components of essential oils of higher plants. Typical monoterpene skeletons including acyclic, monocyclic, and bicyclic systems, as shown below.

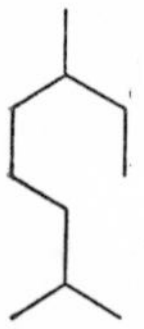

acyclic **1** 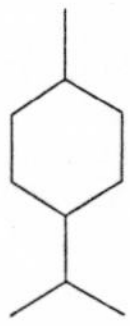   menthane **2** 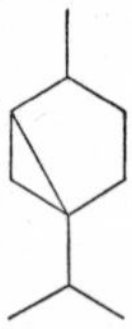   thujane **3** 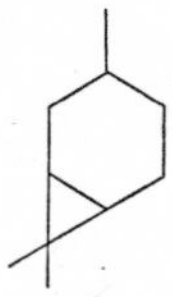   carane **4** 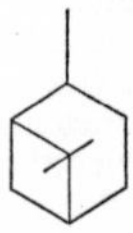   pinane **5** 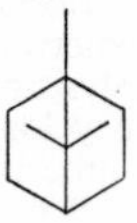   camphane **6**
(bornane)

Monoterpenes containing four- or five-membered rings, and geminal dimethyl cyclohexane rings are also known.

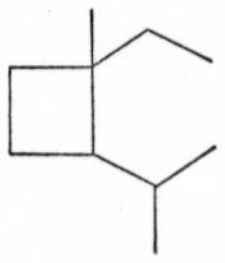

7 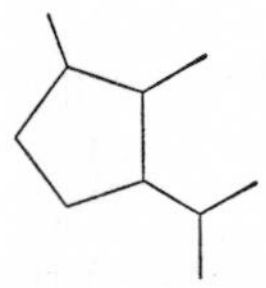   **8** 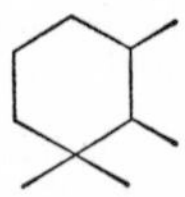   **9**

Some irregular monoterpenes whose carbon skeletons do not obey the isoprene rule are found in Compositae plants.

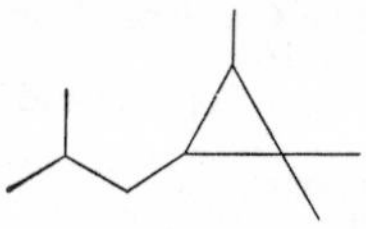

10 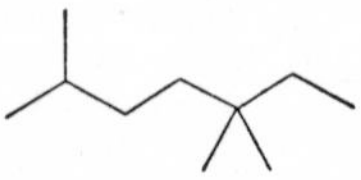   **11** 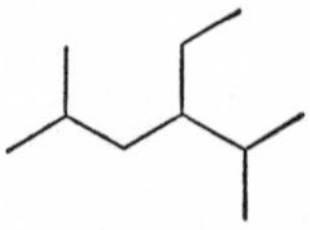   **12** 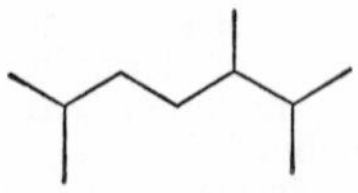   **13**

The sesquiterpenes are C_{15} compounds biogenetically derived from farnesyl pyrophosphate, and they are found mainly in plants and fungi. The diversity of their carbon skeletons is remarkable, compared with other classes of terpenoids. Typical sesquiterpene skeletons and their biogenetic relationships are shown below.

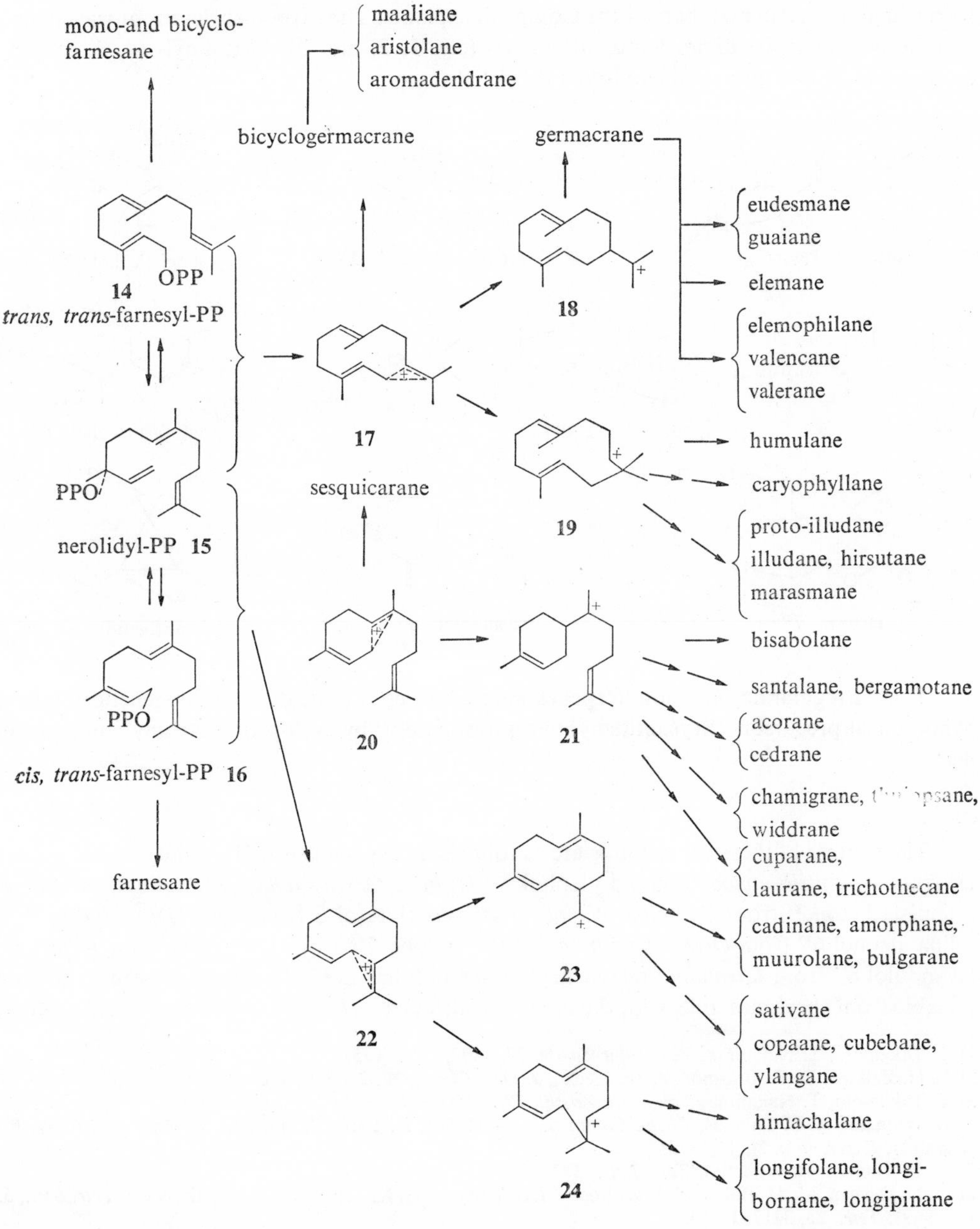

mono-and bicyclo-
farnesane

maaliane
aristolane
aromadendrane

bicyclogermacrane

germacrane

14 OPP
trans, trans-farnesyl-PP

eudesmane
guaiane

elemane

18

elemophilane
valencane
valerane

PPO

nerolidyl-PP **15**

17

sesquicarane

humulane

caryophyllane

19

proto-illudane
illudane, hirsutane
marasmane

bisabolane

20

21

santalane, bergamotane

acorane
cedrane

cis, trans-farnesyl-PP **16**

chamigrane, thujopsane,
widdrane

cuparane,
laurane, trichothecane

farnesane

cadinane, amorphane,
muurolane, bulgarane

23

sativane
copaane, cubebane,
ylangane

22

himachalane

24

longifolane, longi-
bornane, longipinane

Several examples of compounds containing C_{11}–C_{15} skeletons, which are assumed to be derived from carotenoids by oxidative degradation, are also collected at the end of this chapter.

3.2 MONOTERPENES HAVING ARTEMISYL, SANTOLINYL, LAVANDULYL AND CHRYSANTHEMYL SKELETONS

A small group of monoterpenes whose carbon skeletons do not obey the isoprene rule have been found in certain members of the Compositae plants. The carbon skeletons of such irregular monoterpenes can be divided into artemisyl- (e.g. **1**, **2**, and **3**), santolinyl- (**4**, **5**, **6**, and **7**), lavandulyl- (**8**) and chrysanthemyl-type (**9**) skeletons.

artemisia ketone **1**[1,2]   artemisia alcohol **2**[3]   yomogi alcohol **3**[4]

santolinatriene **4**[5]   lyratol **5**[6]   santolina alcohol **6**[7]

lactone **7**[8]   lavandulol **8**[9]   chrysanthemol **9**

These irregular acyclic monoterpenes were considered to be derived, biogenetically, from a hypothetical precursor, chrysanthemyl pyrophosphate,[10] by cyclopropylcarbinyl ion rearrangement.

Remarks

The sources of these compounds are as follows: artemisia ketone **1**[1,2] and artemisia alcohol **2**[3] from *Artemisia annua*; yomogi alcohol **3**[4] from *Artemisia feddei*; santolinatriene **4**[5] and artemisia ketone[2] from *Santolina chamaecyparissus*; lyratol **5** from *Cyathocline lyrata*;[6] santolina alcohol **6**[7] from *Ormenis multicaulis*; the lactone **7**[8] from *Chrysanthemum flosculosum*; lavandulol **8**[9] from *Lavandula vera*. Chrysanthemol **9** has not been found in nature, but is obtainable from chrysanthemic acid, the ester of which occurs in *Chrysanthemum cinerariaefolium*.

1) T. Takemoto, T. Nakajima, *Yakugaku Zasshi*, **77**, 1307, 1339 (1957).
2) L. H. Zalkow, D. R. Brannon, J. W. Uecke, *J. Org. Chem.*, **29**, 2786 (1964).
3) T. Takemoto, T. Nakajima, *Yakugaku Zasshi*, **77**, 1310 (1957).
4) B. Willhalm, A. F. Thomas, *Chem. Commun.*, 1380 (1969); K. Yano, S. Hayashi, T. Matsuura, A. W. Burgstahler, *Experientia*. **26**, 8 (1970).
5) A. F. Thomas, B. Willhalm, *Tetr. Lett.*, 3775 (1964).
6) O. N. Devgan, M. M. Bokadia, A. K. Bose, M. S. Tibbetts, G. K. Trivedi, K. K. Chakravardi *Tetr. Lett.*, 5337 (1969); *Tetr.*, **25**, 3217 (1969).
7) Y. Chretien-Bassiere, L. Peyron, L. Benezet, J. Garnero, *Bull. Soc. Chim. Fr.*, 2018 (1963).
8) F. Bohlmann, M. Grenz, *Tetr. Lett.*, 2413 (1969).
9) M. Soucek, L. Dolejs, *Coll. Czech. Chem. Commun.*, **24**, 3802 (1959).
10) See G. Pattenden, R. Storer, *Tetr. Lett.*, 3473 (1973) (biosynthesis of chrysanthemic acid).

3.3 SOLVOLYSIS OF CHRYSANTHEMYL AND ARTEMISYL DERIVATIVES

The biogenesis of the irregular acyclic monoterpenes has been explained[1,2] in terms of cyclopropylcarbinyl ion rearrangement of chrysanthemyl pyrophosphate. Formal cleavage of the cyclopropane ring in the three ways leads to the artemisyl, santolinyl and lavandulyl skeletons. In fact, several model reactions for interconversion between these skeletons are known.

artemisyl   santolinyl   lavandulyl

Solvolysis of chrysanthemyl derivatives

1) Solvolysis of the chrysanthemyl derivative **1** yields both artemisyl and santolinyl derivatives **2, 3** and **4**. This conversion has been reported as a laboratory model for non-head-to-tail monoterpene biosynthesis.[3,4] Formation of the products was interpreted in terms of the characteristic behavior of the cyclopropylcarbinyl⇌cyclobutyl⇌homoallyl carbonium ion system.

2) Dihydrochrysanthemol rearranges into santolinadiene on treatment with $SOCl_2$ (39% yield).[2]

3) Chrysanthemol **5** or its methyl ether affords artemisiatriene on heating with *p*-TsOH in benzene (55–70%)[2] or treatment with *p*-TsCl in pyridine.[5]

4) From the kinetic data and the solvolysis product composition from chrysanthemyl 2,5-dinitrobenzoate, a homopentadienyl cation was postulated as the intermediate in the reaction.[6]

1) R. B. Bates, S. K. Paknikar, *Tetr. Lett.*, 1453 (1965).
2) L. Crombie, P. A. Firth, R. P. Houghton, D. A. Whiting, D. K. Woods, *J. Chem. Soc.*, Perkin I, 642 (1972); prelim. commun., L. Crombie, R. P. Houghton, D. K. Woods, *Tetr. Lett.*, 4553 (1967).
3) C. D. Poulter, S. G. Moesinger, W. W. Epstein, *Tetr. Lett.*, 67 (1972); C. D. Poulter, *J. Am. Chem. Soc.*, **94**, 5515 (1972).
4) C. D. Poulter, R. J. Goodfellow, W. W. Epstein, *Tetr. Lett.*, 71 (1972).
5) R. B. Bates, D. Feld, *Tetr. Lett.*, 4875 (1967).
6) T. Sasaki, S. Eguchi, M. Ohno, T. Uemura, *Chem. Lett.*, 503 (1972).

Conversion of artemisyl to santolinyl derivatives[7]

Acid treatment of **9** opens the epoxide ring, with participation of the double bond, giving rise to a santolinyl skeleton. The rearrangement may proceed through homoallylic carbonium ions, as indicated.

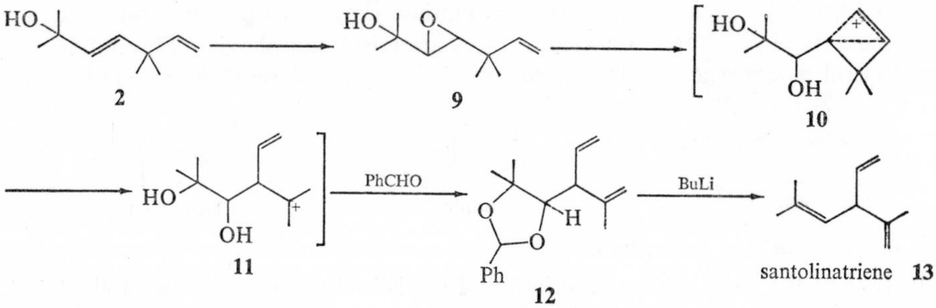

Alcohol solvolysis of 17[8,9]

Alcohol solvolysis of the sulfonium salt **17** affords a mixture of products with **18** and **19** as major products and **20**, **21** and **22** as minor products. This conversion of the sulfonium salt has also been considered as a model for non-head-to-tail monoterpene biosynthesis.

The compound **16** is formed by rearrangement of the ylide generated from the sulfonium salt. This type of rearrangement has attracted much attention in relation to squalene biosynthesis.

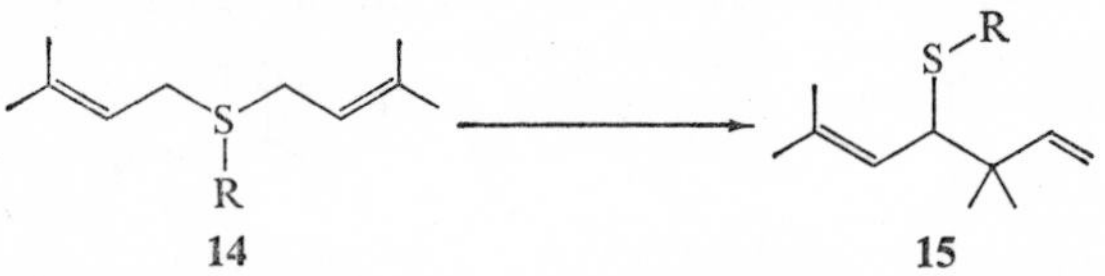

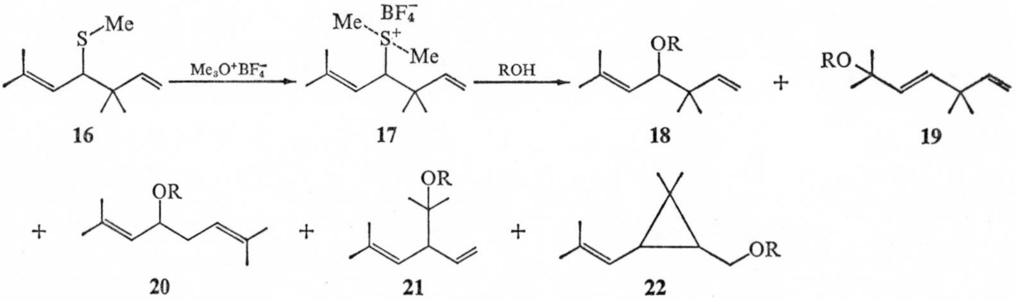

Hydrolysis of 23 and 24[10]

Hydrolysis of the cyclobutyl tosylate **23** in aqueous acetone yielded a mixture of the acyclic dienols **25** and **26**. The same dienol mixture was obtained on hydrolysis of the *trans*-cyclopropylcarbinyl ester **24**.

7) A. F. Thomas, W. Pawlak, *Helv. Chim. Acta*, **54**, 1822 (1971); prelim. commun., A. F. Thomas, *Chem. Commun.*, 1054 (1970).
8) B. M. Trost, R. LaRochelle, *Tetr. Lett.*, 3327 (1968).
9) B. M. Trost, P. Conway, J. S. Stanton, *Chem. Commun.*, 1639 (1971).
10) R. M. Coates, W. H. Robinson, *J. Am. Chem. Soc.*, **94**, 5920 (1972): C. D. Poulter, O. J. Muscio, C. J. Spillner, R. G. Goodfellow, *ibid.*, **94**, 5921 (1972).

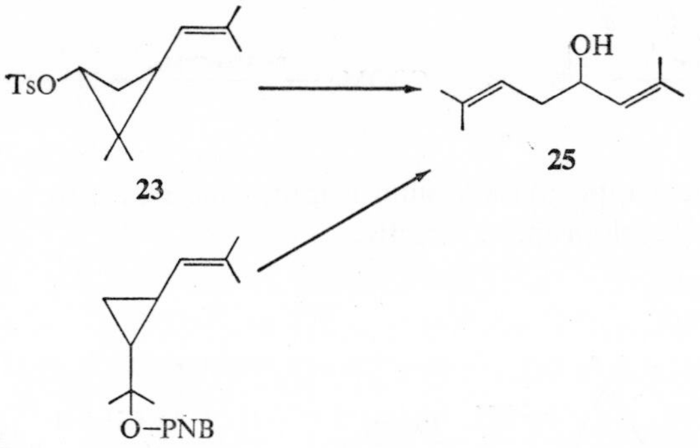

23 25 24

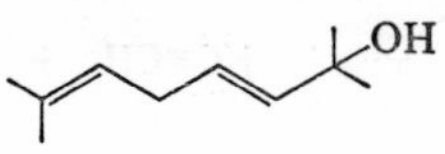

26

3.4 SYNTHESIS OF TRANS-CHRYSANTHEMIC ACID

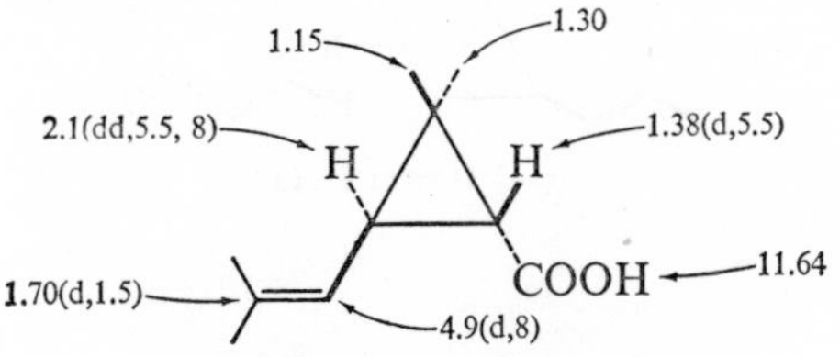

trans-chrysanthemic acid **1**[1)]
uv: 205 (ε 10,000)
nmr in $CDCl_3$

The construction of the cyclopropane ring of chrysanthemic acid **1** has been accomplished in each of the three possible ways involving the addition of a divalent carbon to an olefinic linkage, as shown in Routes I, II (or IV) and III.

Route I[2)]

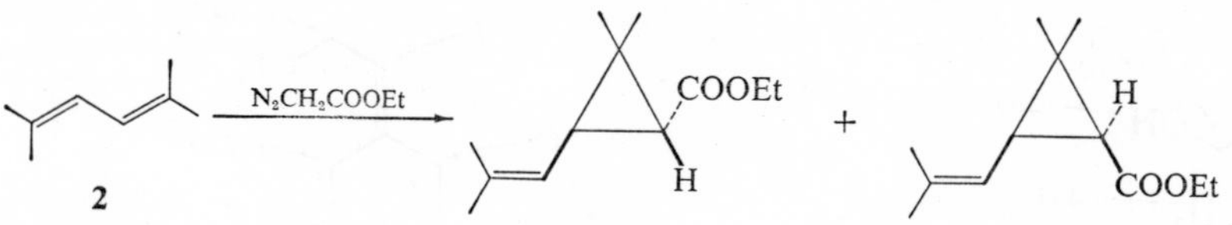

2 *trans*-**1** *cis*-**1**

Route II[3)]

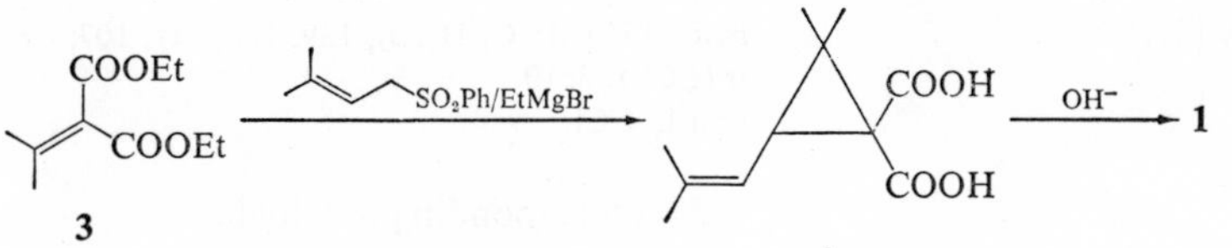

3 4

1) L. Crombie, M. Elliott, *Progress in Chemistry of Natural Products*, **19**, 120 (1961).
2) I. G. M. Campbell, S. H. Harper, *J. Chem. Soc.*, 283 (1945).
3) M. Julia, A. Guy-Rouault, *Bull. Soc. Chim. Fr.*, 1411 (1967).
4) E. J. Corey, M. Jautelat, *J. Am. Chem. Soc.*, **89**, 3912 (1967).
5) R. W. Mills, R. D. H. Murray, R. A. Raphael, *Chem. Commun.*, 555 (1971); *J. Chem. Soc.*, Perkin I, 133 (1973).

Route III[4]

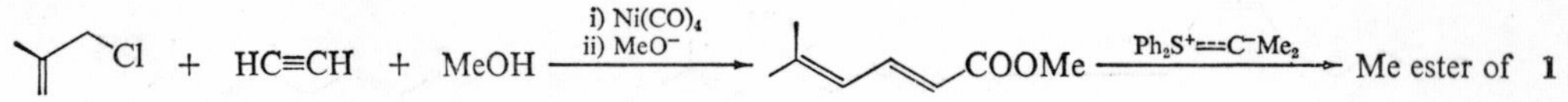

The isopropylidene transfer reagent diphenylsulfonium isopropylide reacts with the α,β-unsaturated ester to yield *gem*-dimethylcyclopropane directly.

Route IV[5]

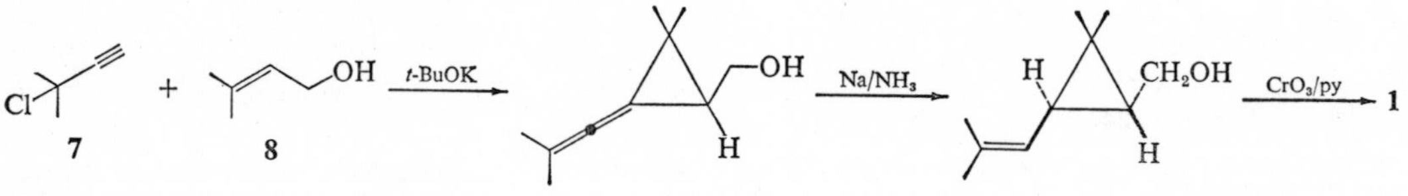

Remarks

Chrysanthemic acid **1** is found in the flower heads of pyrethrum *(Chrysanthemum cinerariae-folium)* as cyclopentanol esters, such as cinerolone, jasmolone and pyrethrolone, which are important natural insecticides (e.g., **11**).

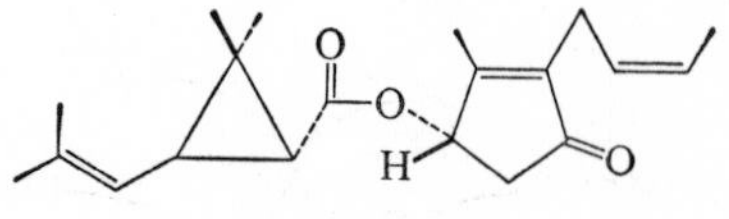

cinerin I   **11**

3.5   STRUCTURE AND SYNTHESIS OF WEEVIL SEX HORMONES

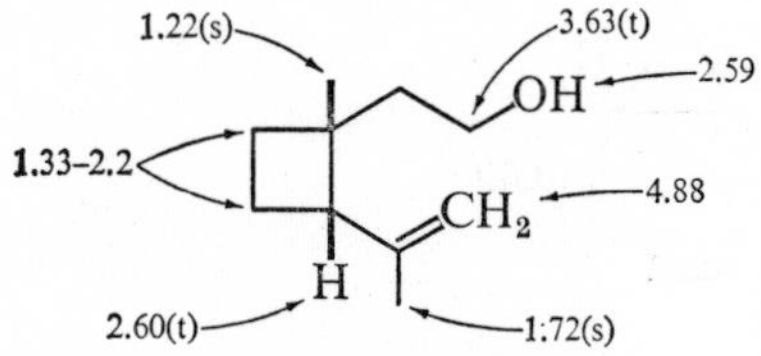

grandisol **1**[1]

ms: 154 (M^+ $C_{10}H_{18}O$), 139, 136, 121, 109, 68
ir (CCl_4): 3630, 3250–3550, 885
nmr in CCl_4

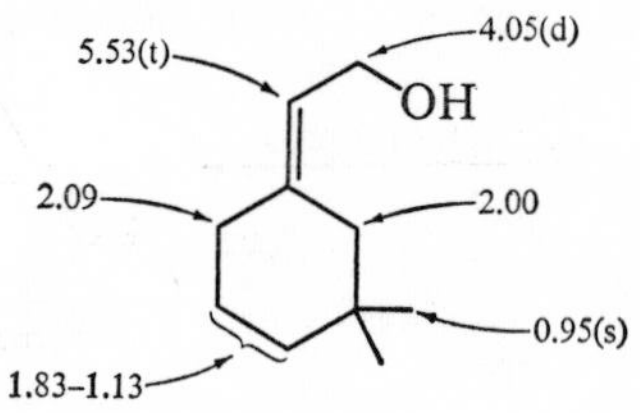

2[1]

ms: 154 (M^+ $C_{10}H_{18}O$), 139, 136, 121, 107, 69
ir (CCl_4): 3610
nmr in CCl_4

3=corresponding aldehyde
4=geometric isomer of **3**

Synthesis of 1[1-4]

Four syntheses are shown, three of which utilize photochemical cycloaddition for construction of the four-membered ring in **1**.

Route I[2)]

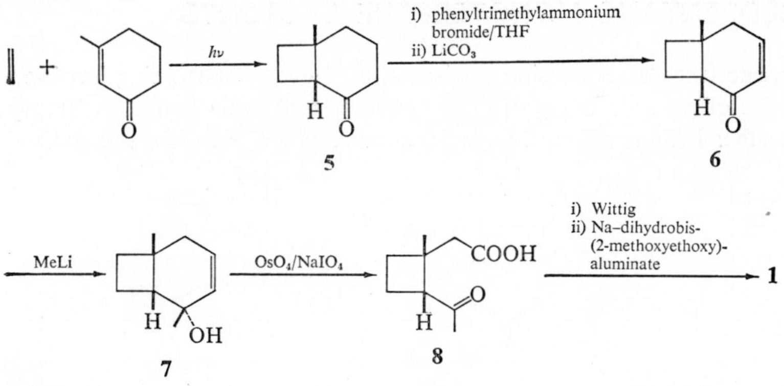

5 → **6**

i) phenyltrimethylammonium bromide/THF
ii) LiCO$_3$

MeLi → **7**

OsO$_4$/NaIO$_4$ → **8**

i) Wittig
ii) Na–dihydrobis-(2-methoxyethoxy)-aluminate → **1**

Route II[1)]

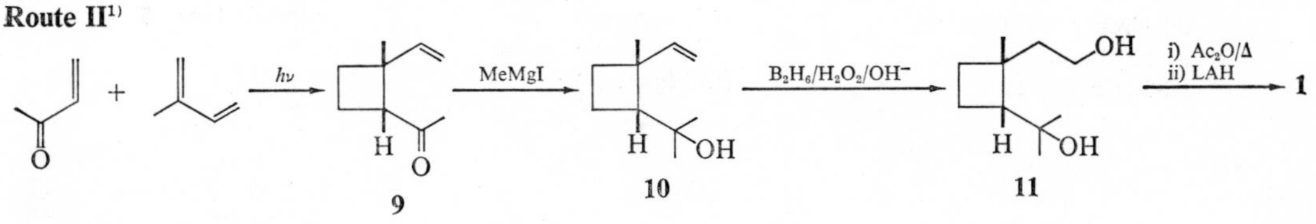

hν → **9**

MeMgI → **10**

B$_2$H$_6$/H$_2$O$_2$/OH$^-$ → **11**

i) Ac$_2$O/Δ
ii) LAH → **1**

Route III[3)]

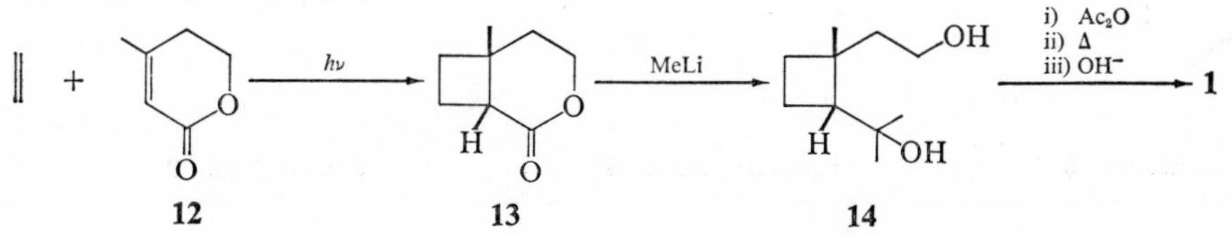

12 + hν → **13**

MeLi → **14**

i) Ac$_2$O
ii) Δ
iii) OH$^-$ → **1**

Route IV[4)]

An alternative synthesis of **1** by a route involving the metal-catalized dimerization of isoprene has been reported.

Synthesis of 2, 3 and 4[1)]

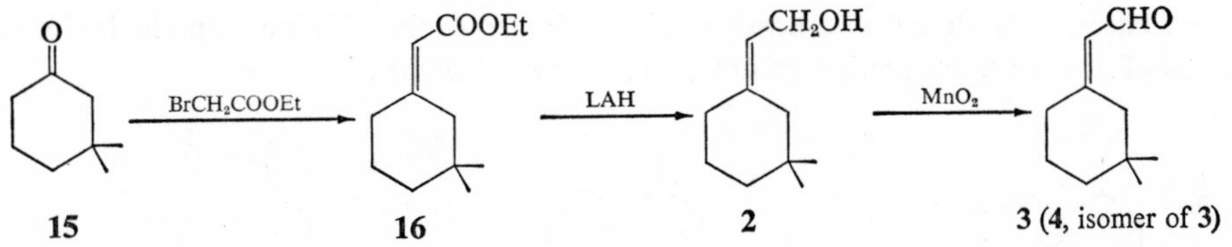

15

BrCH$_2$COOEt → **16**

LAH → **2**

MnO$_2$ → **3 (4, isomer of 3)**

Remarks

1, 2, 3 and **4** were isolated from male boll weevils, *Anthonomus grandis*, and from their fecal material. The compounds were identified as sex attractants of the male weevil. In laboratory bioassays, mixtures of synthetic **1, 2, 3** and **4** elicited a response from female boll weevils to the same extent as the natural compounds.

1) J. H. Tumlinson, D. D. Hardee, R. C. Gueldner, A. C. Thompson, P. A. Hedin, J. P. Minyard, *Science*, **166** 1010 (1969).
2) R. Zurfluh, L. L. Dunham, V. L. Spain, J. B. Siddall, *J. Am. Chem. Soc.*, **92**, 425 (1970).
3) T. D. J. D'Silva, D. W. Peck, *J. Org. Chem.*, **37**, 1828 (1972).
4) W. E. Billups, J. H. Cross, C. V. Smith, *J. Am. Chem. Soc.*, **95**, 3438 (1973).

3.6 CYCLOPENTANO-MONOTERPENE LACTONES

A series of monoterpene lactones possessing a trisubstituted cyclopentane ring has been isolated from certain species of plants.[1] Some of these lactones (**1–5**) show potent excitatory activity towards cats and other Felidae animals. C_9-lactones such as **6–9** were also isolated.

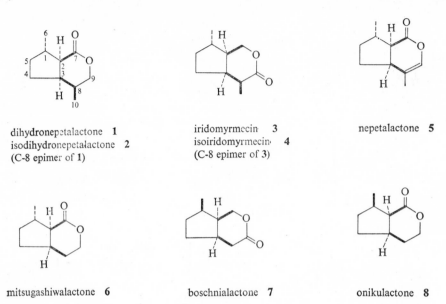

dihydronepetalactone **1**
isodihydronepetalactone **2**
(C-8 epimer of **1**)

iridomyrmecin **3**
isoiridomyrmecin **4**
(C-8 epimer of **3**)

nepetalactone **5**

mitsugashiwalactone **6**   boschnialactone **7**   onikulactone **8**

Remarks

The compounds **1–4** and neonepetalactone **11** were isolated from the essential oil of *Actinidia polygama*.[2] Nepetalactone **5** was obtained from catnip oil, from *Nepeta cataria*.[3] Compounds **7**, **8** and the C-1 epimers of **1-4** were isolated from *Boschnia rossica*[4] and **6** from *Menyanthes trifoliata*.

Two bases, actinidine **9** and boschniakine **10**[4] also excite Felidae animals. Iridomyrmecin **3** was also found in the secretion of ants, *Iridomyrmex humilis*.[5]

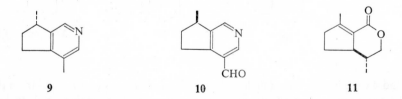

9   **10**   **11**

1) For review, see T. Sakan, F. Murai, S. Isoe, S. B. Hyeon, Y. Hayashi, *Nippon Kagaku Zasshi*, **90**, 507 (1969).
2) T. Sakan, S. Isoe, S. B. Hyeon, R. Katsumura, T. Maeda, J. Wolinsky, D. Dickerson, M. Slabagh, D. Nelson, *Tetr. Lett.*, 4097 (1965).
3) R. B. Bates, E. J. Eisenbraun, S. M. McElvain, *J. Am. Chem. Soc.*, **80**, 3420 (1958).
4) T. Sakan, F. Murai, Y. Hayashi, R. Honda, T. Shono, M. Nakajima, M. Kato, *Tetr.* **23**, 4635 (1967).
5) M. Pavan, *Chim. Ind.* (Milan) **37**, 625, 714 (1955); R. Fusco, R. Trave, A. Vercellone, *ibid.*, **37**, 251, 958 (1955); G. W. K. Cavill, D. L. Ford, H. D. Locksley, *Aust. J. Chem.*, **9**, 128 (1956); G. W. K. Cavill, H. D. Locksley, *ibid.*, **10**, 352 (1957).

3.7 MATATABIOLS AND MATATABIETHERS

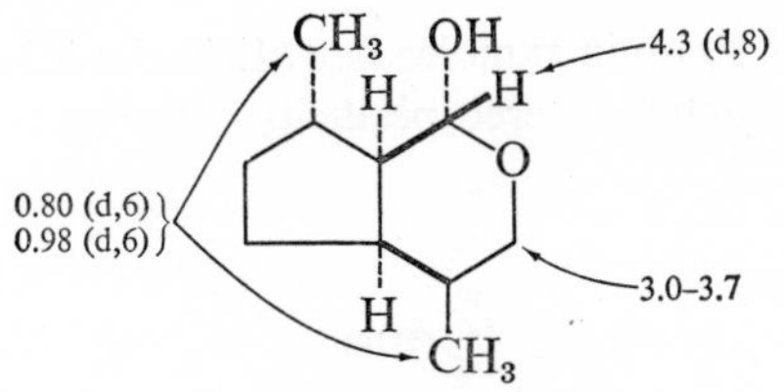

neomatatabiol **1**[1)]
ms: 168 (M[+] $C_{10}H_{16}O_2$)
ir: 3040, 1070

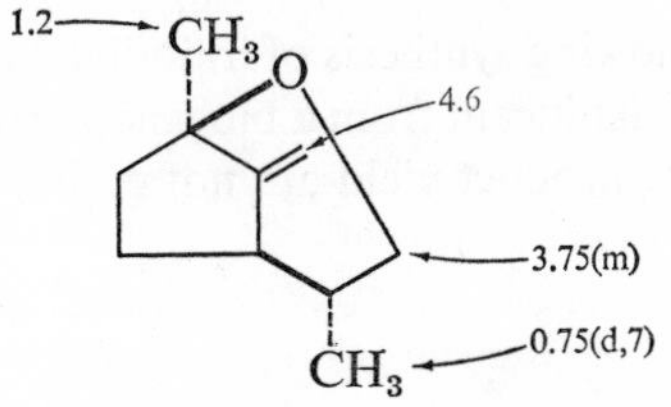

matatabiether **2**[2)]
ms: 152 (M[+] $C_{10}H_{16}O$)
ir: 3100, 1675, 890 (exocyclic methylene),
1045, 1085 (ether)

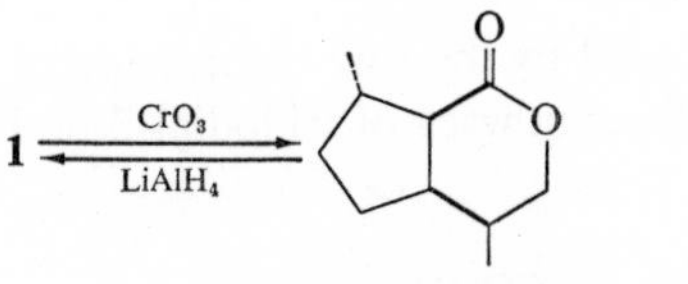

1 $\xrightleftharpoons[\text{LiAlH}_4]{\text{CrO}_3}$

dihydronepetalactone **3**

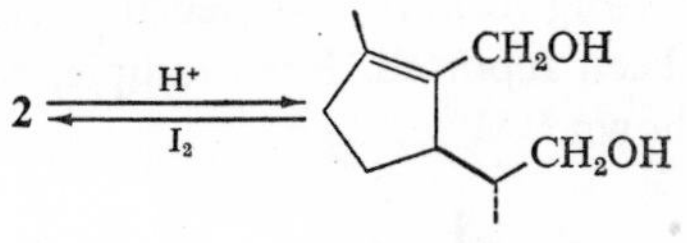

2 $\xrightleftharpoons[\text{I}_2]{\text{H}^+}$ CH$_2$OH CH$_2$OH

dehydroiridodiol **4**

Remarks[3)]

Neomatatabiol **1**, isoneomatatabiol **7**, matatabiol **5**, allomatatabiol **6**, matatabiether **2**, 5-hydroxymatatabiether **8**, 7-hydroxymatatabiether **9** and matatabidiether **10** were isolated from the gall and leaves of *Actinidia polygama*. Some of these compounds show strong attracting ability towards male lacewings, *Chrysopa septemunctata* and *Chrysopa japana* (Chrysopidae). Neomatatabiol **1** is effective in attracting these insects in amounts as low as 10^{-6} μg.

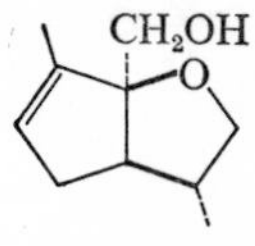

5

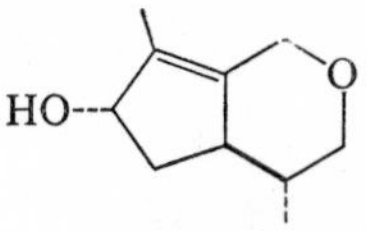

6

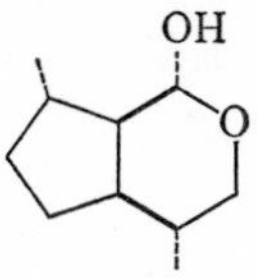

7

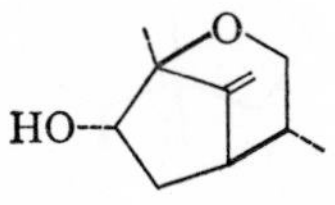

8

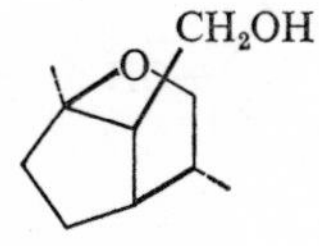

9

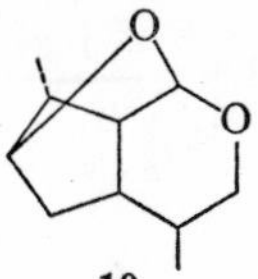

10

1) S. B. Hyeon, S. Isoe, T. Sakan, *Tetr. Lett.*, 5325 (1968).
2) S. Isoe, T. Ono, S. B. Hyeon, T. Sakan, *Tetr. Lett.*, 5319 (1968).
3) T. Sakan, F. Murai, S. Isoe, S. B. Hyeon, Y. Hayashi, *Nippon Kagaku Zasshi*, **90**, 507 (1969).

3.8 SYNTHESIS OF CYCLOPENTANO-MONOTERPENES

Route I[1]

The following synthesis of iridodial **3** from citronellal **1** by an intramolecular Michael addition may be significant from a biogenetic point of view, although the stereospecificity of the reactions and the product yield are not entirely satisfactory.

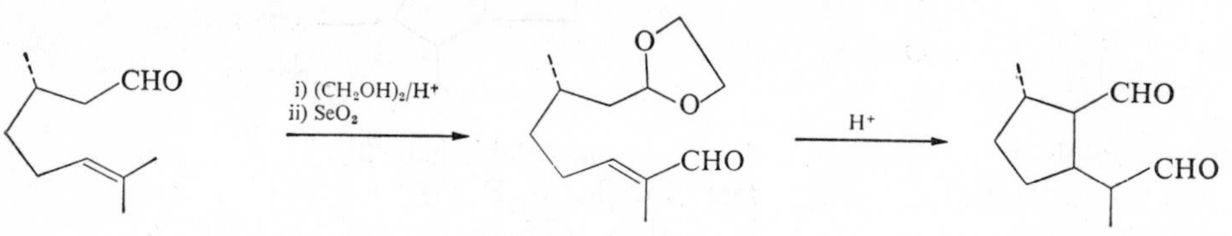

1 **2** **3**

Route II

Many syntheses starting from *trans*-pulegenic acid **6**,[2] prepared by the ring contraction of (+)-pulegone **4**, have been reported. For example, stereospecific syntheses of iridodial **3** and nepetalactone **15** are shown.[3]

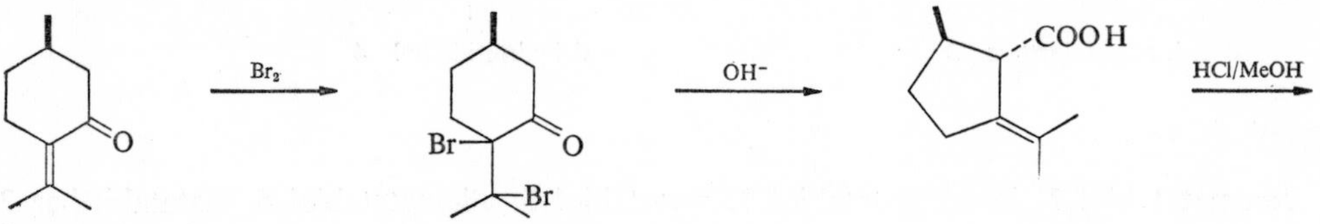

4 **5** **6**

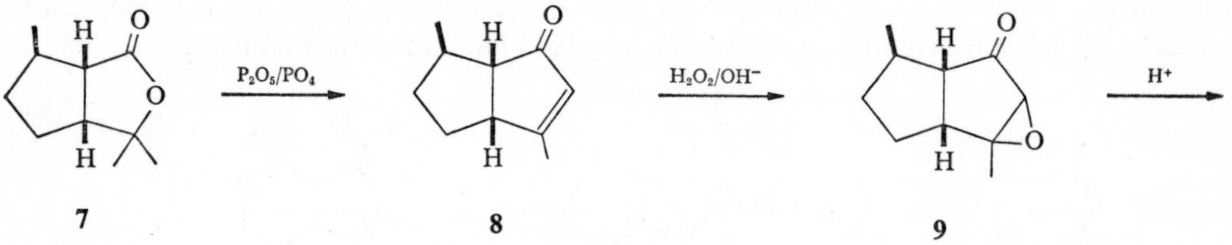

7 **8** **9**

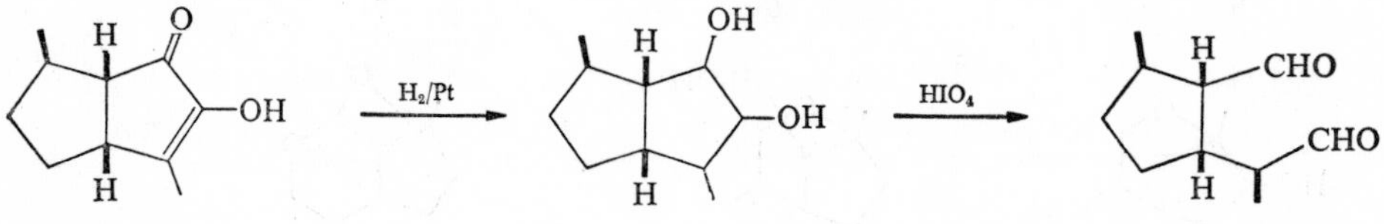

10 **11** iridodial **3**

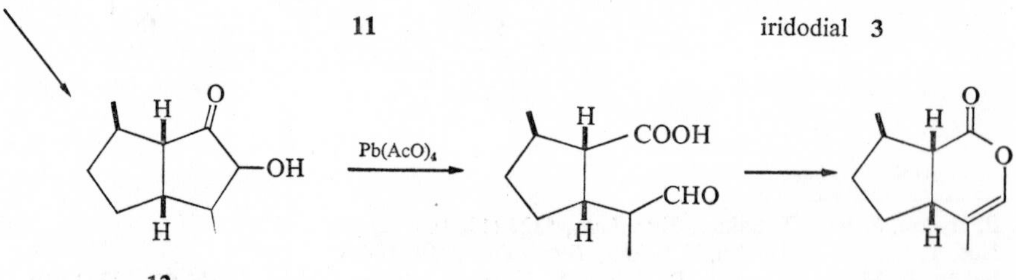

12 **13** nepetalactone **15**

(−)-Iridomyrmecin **20** and (+)-isoiridomyrmecin **21** were also synthesized from *trans*-pulegenic acid **6** by an alternative route.[4]

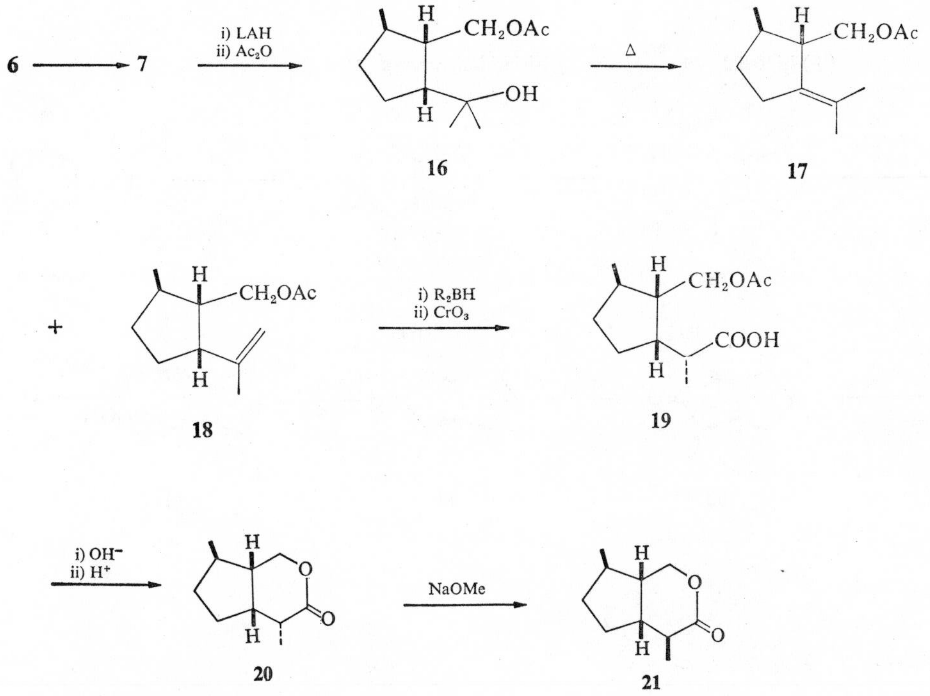

Route III[5]

A series of compounds have been synthesized using 2,6-dimethylbicyclo-[3.3.0]-octan-3-one **25** as a key intermediate. **25** has ideal stereochemistry for the synthesis of these compounds, and was prepared stereospecifically from **22**. The ring junction in **25** is *cis* due to the thermodynamic stability of two fused, five-membered rings. The substituents at the C-1 and C-3 positions must be *trans* oriented during the deethoxycarbonylation stage. The methyl group adjacent to the carbonyl in **25** possesses a stable α configuration (quasi-equatorial).

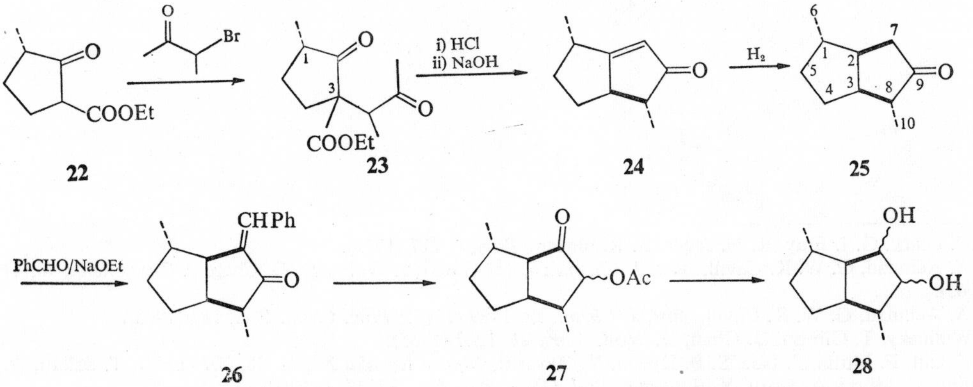

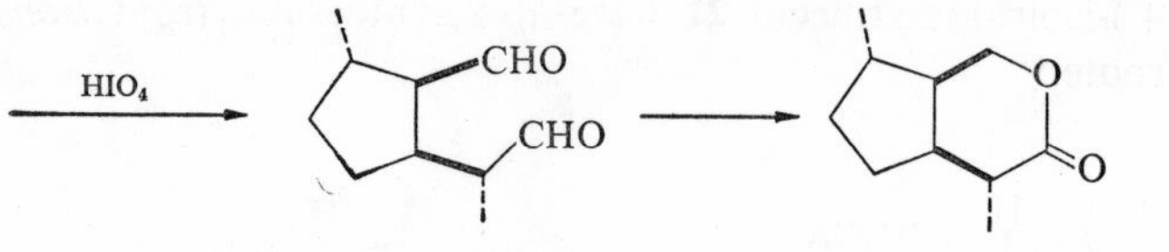

(±)-iridodial **29**   (±)-isoiridomyrmecin **21**

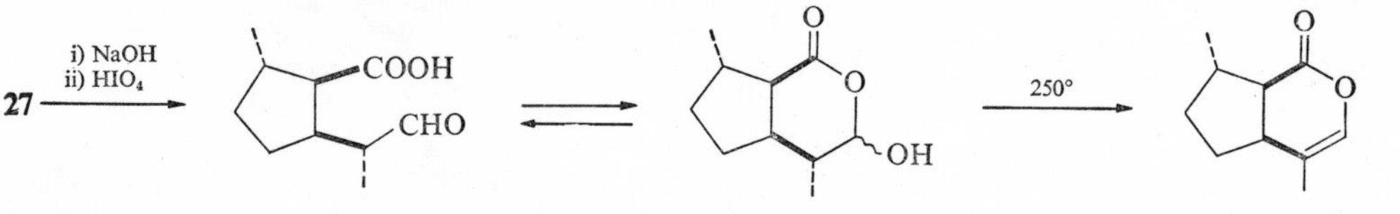

30   **31**   (±)-nepetalactone **15**

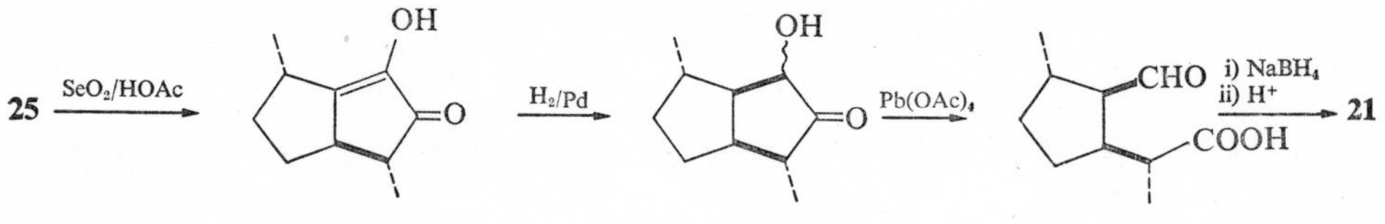

32   **33**   **34**

1) K. J. Clark, G. I. Fray, R. H. Jager, R. Robinson, *Tetr.*, **6**, 217 (1959).
2) S. A. Achmad, G. W. K. Cavill, *Aust. J. Chem.*, **16**, 858 (1963); J. Wolinsky, D. Chan, *J. Org. Chem.*, **30**, 41 (1965).
3) S. A. Achmad, G. W. K. Cavill, *Aust. J. Chem.*, **18**, 1989 (1965); *Proc. Chem. Soc.*, 166 (1963).
4) J. Wolinsky, T. Gibson, D. Chan, H. Wolf, *Tetr.*, **21**, 1247 (1965).
5) T. Sakan, F. Murai, S. Isoe, S. B. Hyeon, Y. Hayashi, *Nippon Kagaku Zasshi*, **90**, 507 (1969); T. Sakan, A. Fujino, F. Murai, A. Suzui, Y. Butsugan, *Bull. Chem. Soc. Jap.*, **33**, 1737 (1960).

3.9   STRUCTURE OF LOGANIN

loganin $1^{1,2)}$

mp: 222–223°
uv: 237 (4.03)
ir: 1710 (unsat. ester)
       1650 (enol ether)
nmr for loganin aglycone **2**
   4.94 (d,5; 1-H); 1.3–2.4 (m,2-H)
   1.12 (d,6.5; -Me); 4.10 (m,5-H)
   1.3–2.4 (6-H); 3.14 (m, 7-H)
   7.39 (9-H); 3.68 (COOCH$_3$)

1) Enzymatic hydrolysis (emulsin) of **1** gives the C_{11} aglycone **2** and glucose.
2) Alkaline hydrolysis of **1** affords norloganin **3**,[3] (free carboxylic acid derivative)
   uv: 234 nm (ε 9360), ir: 2500–2700 cm^{-1}.
3) The penta-*O*-acetyl derivative **4** is formed by acetylation.[3]
4) Jones oxidation of **1** yields a product showing ir absorption at 1751 cm^{-1} (5-membered ring ketone).[3]
5) Chemical interrelation of **1** with verbenalin **8**[4] establishes the absolute stereochemistry at C-2, C-3 and C-7.[1]

6) The nuclear Overhauser effect was studied on *O*-methyl loganin aglycone **11** and its C-5 epimer acetate **12**. The results supported the stereochemical assignment already deduced.[1]
7) X-ray analysis of the bromo derivative of **1** confirmed the structure.[5]

1) A. R. Battersby, E. S. Hall, R. Southgate, *J. Chem. Soc.*, C, 721 (1969), and references cited therein.
2) S. Brechbuhler-Bader, C. J. Coscia, P. Loew, Ch. von Szczepanski, D. Arigoni, *Chem. Commun.*, 136 (1968) (for chemistry and stereochemistry of **1**).
3) K. Steth, E. Ramstad, J. Wolinsky, *Tetr. Lett.*, 394 (1961). For previous chemical study, see A. J. Birch, E. Smith, *Aust. J. Chem.*, **9**, 234 (1956); K. W. Merz, H. Lehmann, *Arch. Pharm.*, **290**, 543 (1957).
4) G. Büchi, R. E. Manning, *Tetr.* **18**, 1049 (1962).
5) P. J. Lentz, M. G. Rossmann, *Chem. Commun.*, 1269 (1969).

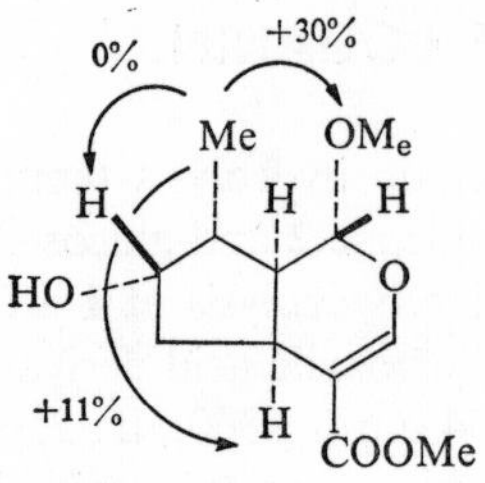

11 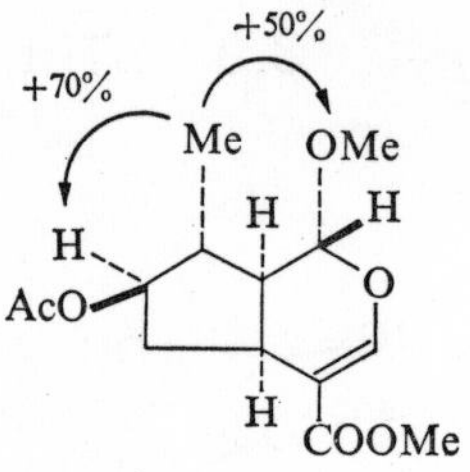   **12**

Remarks

Loganin **1** was first isolated from the fruits of *Strychnos nux-vomica* and other *Strychnos* species. It was also isolated from *Strychnos lucids, Menyanthes trifoliata, Vinca rosea* and others. It has been well established that loganin is a key intermediate in the biosynthesis of three major indole alkaloids. A number of structurally related monoterpene glucosides have been isolated from various plant sources.[8]

Synthesis

Partial synthesis of loganin pentaacetate **4** from asperuloside, whose structure was already known, was performed by the following route.[6]

Total synthesis of **1** has been achieved by Büchi[7] (see "Synthesis of loganin and related compounds").

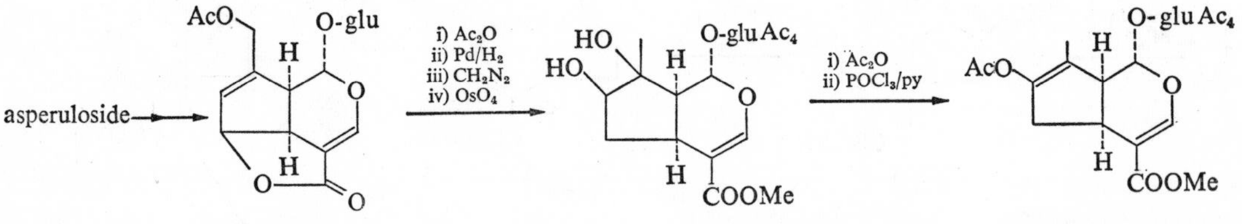

13   **14**   **15**

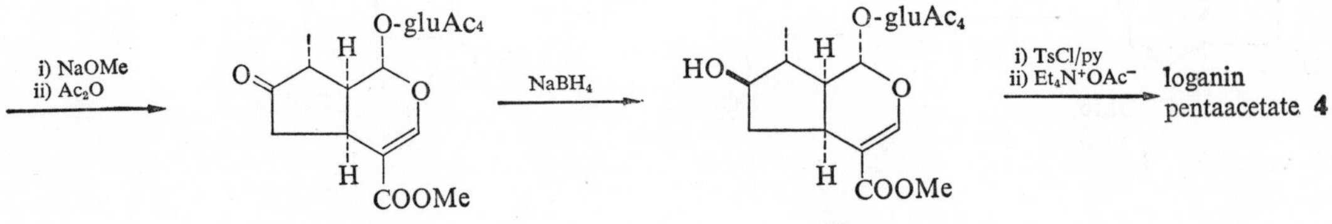

16   **17**

6) H. Inouye, T. Yoshida, S. Tobita, *Tetr.*, **26**, 3905 (1970); *Tetr. Lett.*, 2945 (1968).
7) G. Büchi, J. A. Carlson, J. E. Powell Jr., L. -F. Tietze, *J. Am. Chem. Soc.*, **92**, 2165 (1970).
8) V. Plouvier, J. F-Bonvin, *Phytochemistry*, **10**, 1697 (1971).

3.10 STRUCTURE OF SECOLOGANIN

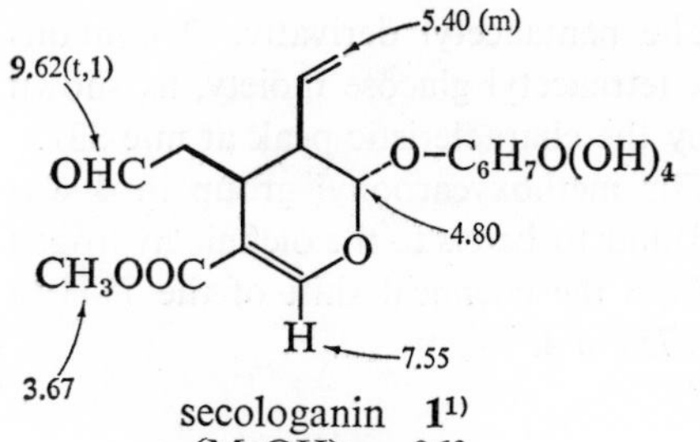

9.62(t,1)

5.40 (m)

OHC

O–$C_6H_7O(OH)_4$

CH_3OOC

O

4.80

H   7.55

3.67

secologanin $1^{1)}$
α_D(MeOH): −96°
uv: 235 (3.96)
ir: 3400, 1700, 1630
nmr in D_2O

1) Emulsin cleaved the glucose residue from **1**, suggesting a β-glucosidic link.
2) Borohydride reduction of **1** affords sweroside **2**.
3) Secologanin **1** reacts with CH_2N_2 to form compound **3**

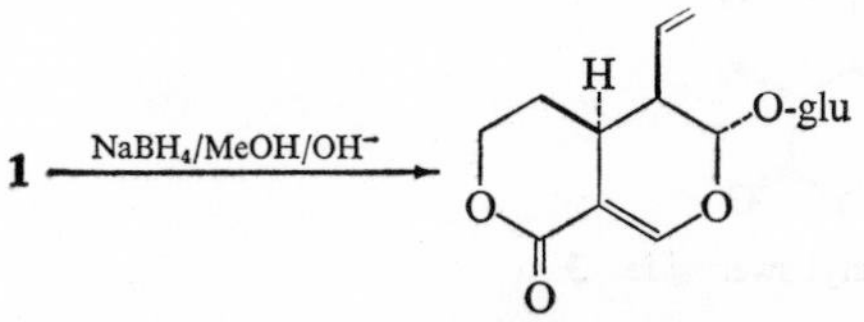

1 $\xrightarrow{\text{NaBH}_4/\text{MeOH/OH}^-}$

H

O-glu

O   O

O

sweroside $2^{2)}$
uv: 246(3.29)

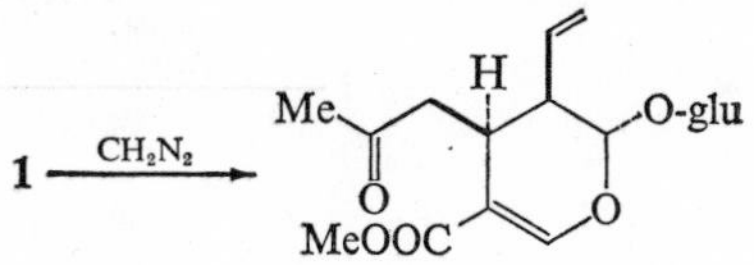

1 $\xrightarrow{\text{CH}_2\text{N}_2}$

Me

H

O-glu

MeOOC   O

3

Remarks

Secologanin **1** has been isolated from *Vinca rosea* as a minor constituent. It is obtainable from menthiafolin, in which secologanin is contained in a masked lactol form, by hydrolysis followed by esterification. Sweroside **2** was obtained[2] from *Swertia japonica*, *S. caroliniensis*, and also from *Vinca rosea*.

Secologanin **1**, which is itself derived from loganin, is a specific precursor of three major types of indole alkaloids, i.e., the families *Corynanthe*, *Aspidosperma*, and *Iboga*.

1) A. R. Battersby, A. R. Burnett, P. G. Parsons, *J. Chem. Soc.* C, 1187 (1969); *Chem. Commun.*, 1280 (1968).
2) H. Inouye, S. Ueda, Y. Nakamura, *Tetr. Lett.*, 3221 (1967); H. A. Linde, M. S. Ragab, *Helv. Chim Acta*, **50**, 991 (1967).

3.11   STRUCTURES OF FOLIAMENTHIN AND MENTHIAFOLIN

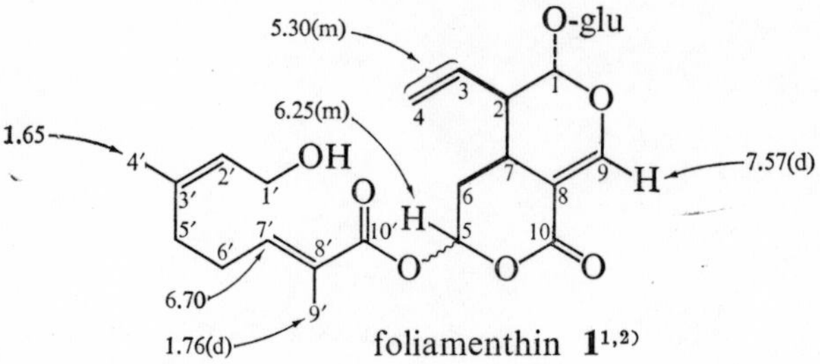

foliamenthin **1**[1,2]

1) The pentaacetyl derivative **2** contains a tetraacetyl glucose moiety, as shown by the characteristic peak at m/e 331.

2) The methoxycarbonyl group in **4** was found to be *cis* to the olefinic hydrogen from the chemical shift of the 7'-H (δ 6.75) in **4**.

mp: 194–196° α_D: $-55°$(for **2**), $-63°$(for **1**)
ms: 750 ($C_{36}H_{46}O_{16}$) for the pentaacetyl deriv. **2**; intense peaks at m/e 525, 403 and 331
uv: 228 (ε 17,200), shoulder at 245
ir: 1622 (α,β-unsat. carbonyl)
nmr in $CDCl_3$[2]

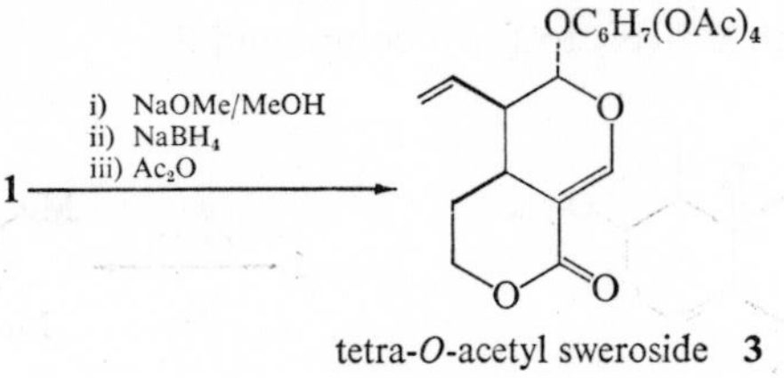

1 $\xrightarrow{\text{i) NaOMe/MeOH}\atop\text{ii) NaBH}_4\atop\text{iii) Ac}_2\text{O}}$

tetra-*O*-acetyl sweroside **3**

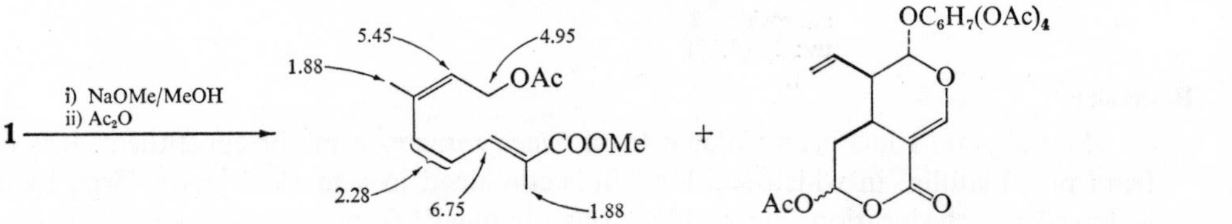

1 $\xrightarrow{\text{i) NaOMe/MeOH}\atop\text{ii) Ac}_2\text{O}}$

4
uv: 216(12000)

+

5
uv: 245(8400)

Remarks

Foliamenthin **1**, dihydrofoliamenthin **6**,[2] and menthiafolin **7**[2] were isolated from the glycosidic fraction of the rhizomes of *Menyanthes trifoliata*, along with loganin and sweroside. Secologanin was prepared from **7**.[3]

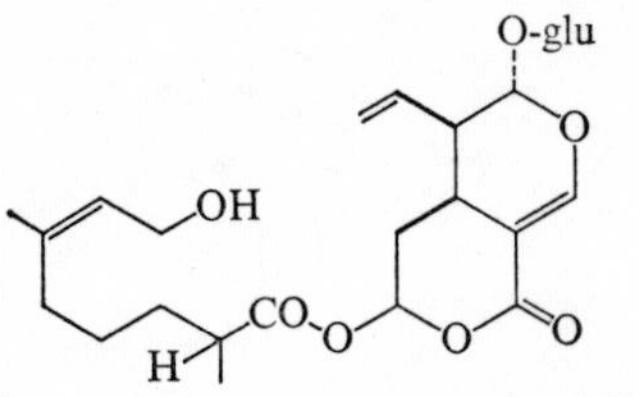

6

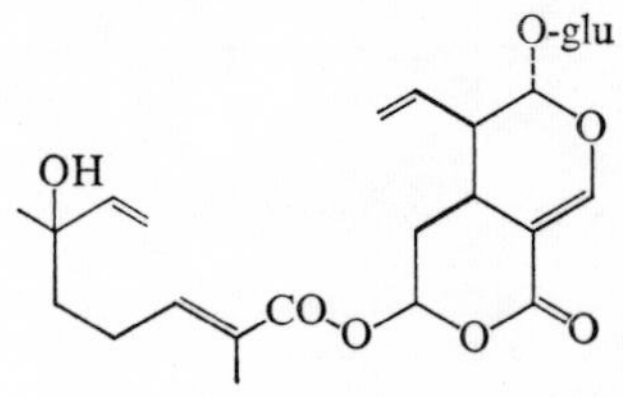

7

1) P. Loew, Ch. v. Sczepanski, C. J. Coscia, D. Arigoni, *Chem. Commun.*, 1276 (1968).
2) A. R. Battersby, A. R. Burnett, G. D. Knowles, P. G. Parsons, *Chem. Commun.*, 1277 (1968).
3) A. R. Battersby, A. R. Burnett, P. G. Parsons, *Chem. Commun.*, 1280 (1968).

3.12   BIOSYNTHESIS OF LOGANIN AND SECOLOGANIN

The non-tryptamine portion of the indole alkaloids has been proved to be derived from an acyclic isoprenoid chain, geraniol 2,[1-3] through the key intermediate loganin 6.

The biosynthetic pathway from geraniol 2 to loganin 6 (or secologanin 8) was postulated as follows, based on considerable accumulated experimental evidence.

1 geranyl pyrophosphate (R=PP)   10-hydroxygeraniol
2 geraniol   (R=H)   3

deoxyloganin 5

4

1) The intermediary role of geraniol 2[1-3] and 10-hydroxygeraniol 3[4,5] was demonstrated by feeding experiments using labeled substrates.

2) Observed randomization of the label at C-9 in 3 between C-9 and C-10 of loganin and the corresponding carbons in alkaloids containing the loganin unit suggests that a 9,10-dioxo intermediate, such as 4 might involved in the pathway, although the exact mechanism of cyclization is unknown.[4]

3) Deoxyloganin 5 has been isolated from natural sources and has been proved to be a biosynthetic intermediate.[6]

4) The cyclopentane ring of loganin may be cleaved to yield 8 via a C-4 hydroxy intermediate, as shown in 7.[3,7,8]

loganin 6   7   secologanin 8

1) A. R. Battersby, R. T. Brown, R. S. Kapil, J. A. Martin, A. O. Plunkett, *Chem. Commun.*, 888 (1966); A. R. Battersby, R. S. Kapil, J. A. Martin, L. Mo, *ibid.*, 133 (1968).
2) P. Loew, D. Arigoni, *Chem. Commun.*, 137 (1968).
3) A. R. Battersby, E. S. Hall, R. Southgate, *J. Chem. Soc. C*, 721 (1969).
4) S. Escher, P. Loew, D. Arigoni, *Chem. Commun.*, 823 (1970).
5) A. R. Battersby, S. H. Brown, T. G. Payne, *Chem. Commun.*, 827 (1970).
6) A. R. Battersby, A. R. Burnett, P. G. Parsons, *Chem. Commun.*, 826 (1970).
7) H. Inouye, S. Ueda, Y. Aoki, Y. Takeda, *Tetr. Lett.*, 2351 (1969).
8) A. R. Battersby, A. R. Burnett, P. G. Parsons, *J. Chem. Soc. C*, 1187 (1969).

3.13   SYNTHESIS OF LOGANIN AND RELATED COMPOUNDS

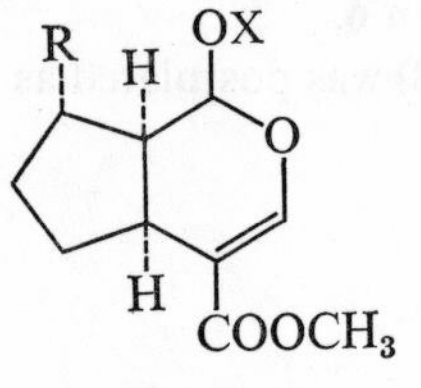

1

A number of monoterpene glucosides possessing the general formula **1** (R = Me, CH$_2$OH,
CHO, alkyl, etc.; X = D-glucose) have been isolated from plant sources.[1] Some of these substances have been shown to be important intermediates in the biosynthesis of the indole alkaloids.
Four different approaches to the synthesis of these
compounds have been developed.

Route I (verbenalol 7)[2]

Verbenalol **7** is the aglucone of verbenalin.

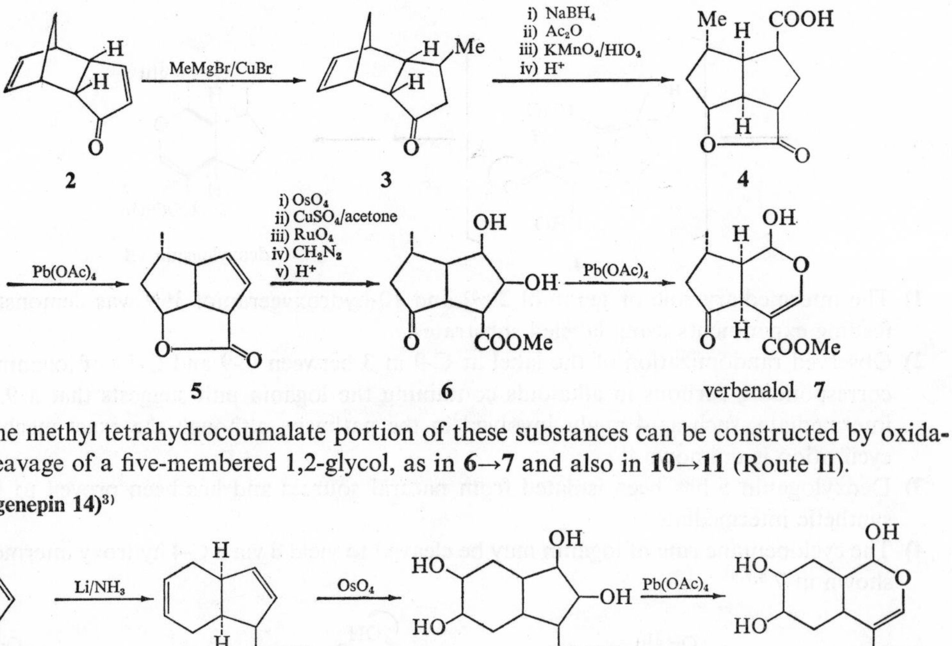

The methyl tetrahydrocoumalate portion of these substances can be constructed by oxidative cleavage of a five-membered 1,2-glycol, as in **6→7** and also in **10→11** (Route II).

Route II (genepin 14)[3]

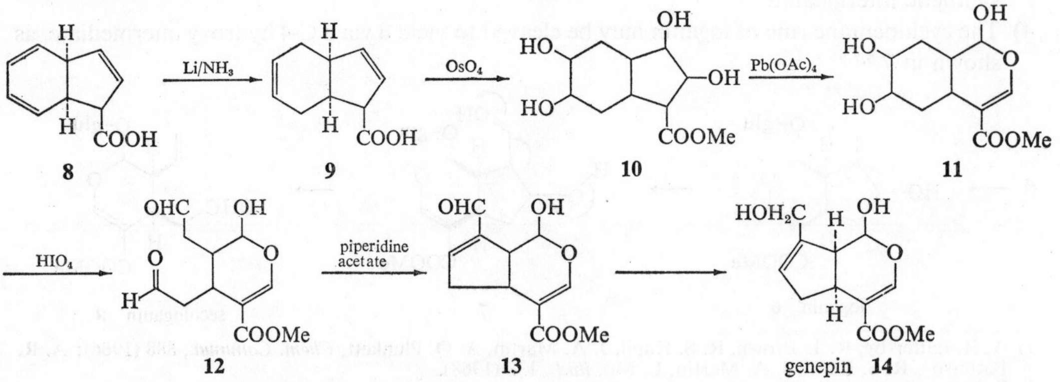

---

1) V. Plouvier, J. F-Bonvin, *Phytochemistry*, **10**, 1697 (1971).
2) T. Sakan, K. Abe, *Tetr. Lett.*, 2471 (1968).
3) G. Büchi, B. Gubler, R. S. Schnieder, J. Wild, *J. Am. Chem. Soc.*, **89**, 2776 (1967).
4) G. Büchi, J. A. Carlson, *J. Am. Chem. Soc.*, **90**, 5336 (1968).
5) G. Büchi, J. A. Carlson, J. E. Powell Jr., L.-F. Tietze, *J. Am. Chem. Soc.*, **92**, 2165 (1970); *ibid.*, **95**, 540 (1973).

A five-membered ring having substituents at C-1 can also be formed by oxidative fission of a six-membered 1,2-glycol and successive base-catalyzed cyclization of the resulting dialdehyde, as in **11→12→13** and also in **18→19→20** (Route III).

Route III (fluvoplumierin 22)[4]

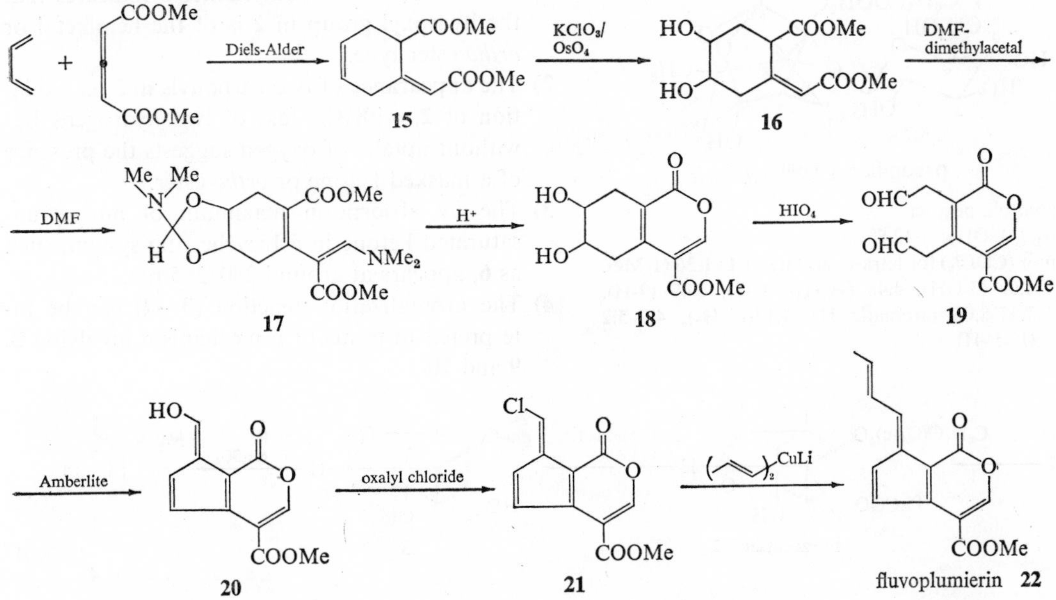

Route IV (loganin 29)[5]

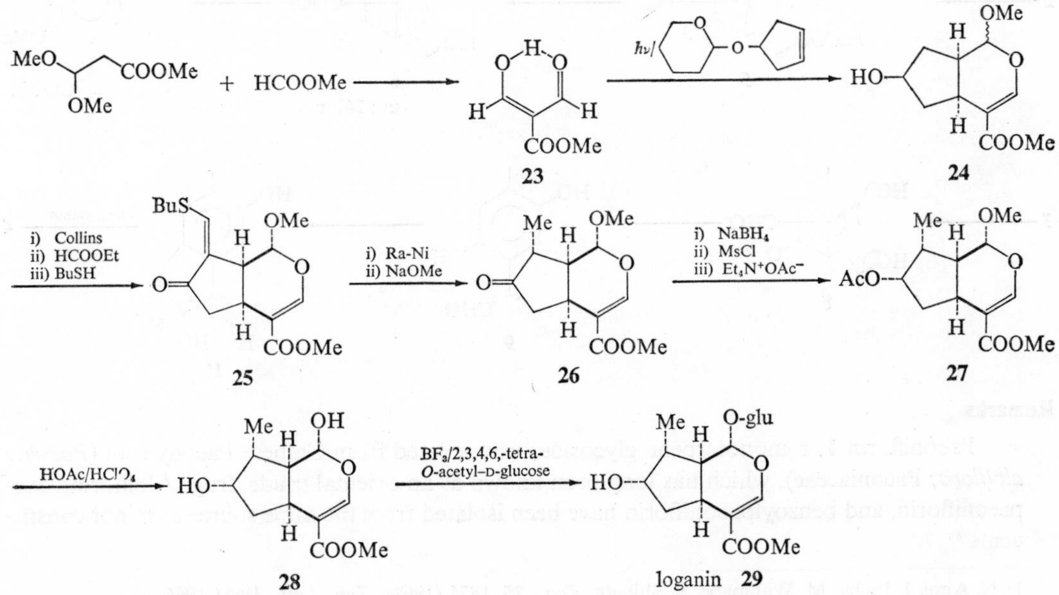

One-step construction of the gross structure was achieved by direct photochemical cycloaddition to a preformed cyclopentene moiety, giving rise to *cis*-fused **24**.

3.14 STRUCTURE OF PAEONIFLORIN

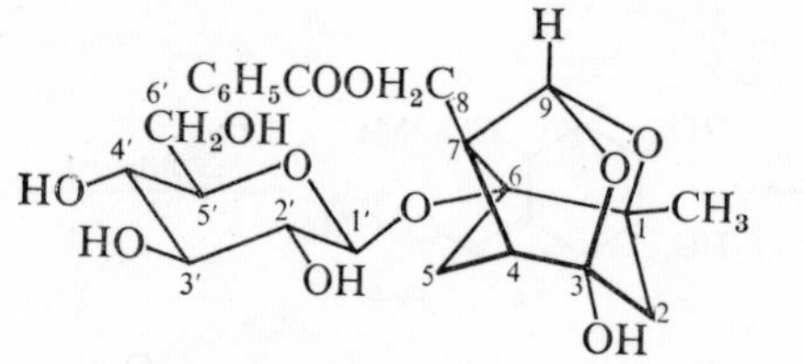

paeoniflorin **1**[1-3]

amorph. powder
α_D (MeOH): $-12.8°$
nmr (CDCl$_3$) for tetra-O-acetate of **1**: 1.36 (1-Me),
4.69 (3-OH), 4.48 (–CH$_2$–, C-8), 5.40 (9-H),
7.35–8.05 (aromatic H), 4.13(6'-H$_2$), 4.7–5.2
(1'–5'-H)

1) Formation of the methyl ether **3** indicates that the hydroxyl group in **2** is of the hemiketal or *ortho* ester type.
2) The appearance of two carbonyls in **5** on oxidation of **2**, with the loss of two hydrogens but without uptake of oxygen suggests the presence of a masked ketone or *ortho* ester.
3) The uv absorption maximum of an α,β-unsaturated ketone in a bicyclic ring system, such as **6**, appears at around 250–265 nm.
4) The aromatization reaction (**3**→**4**) can be interpreted in terms of a mechanism involving **8**, **9** and **10**.

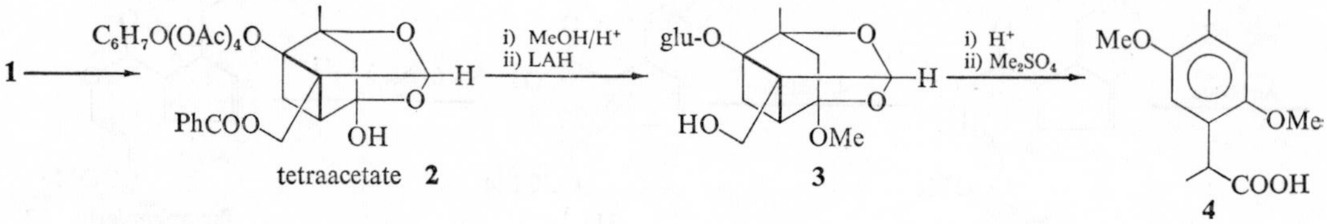

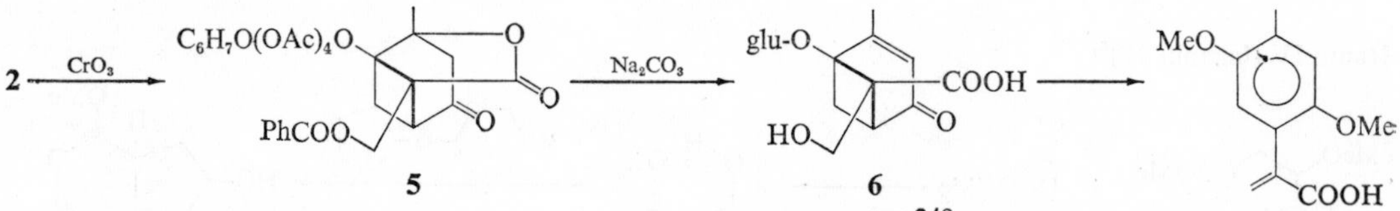

Remarks

Paeoniflorin **1**, a monoterpene glycoside, was isolated from Chinese Paeony root (*Paeonia albiflora;* Paeoniaceae), which has long been known as an oriental crude drug. Albiflorin, oxypaeoniflorin, and benzoylpaeoniflorin have been isolated from the same source as minor constituents.[4]

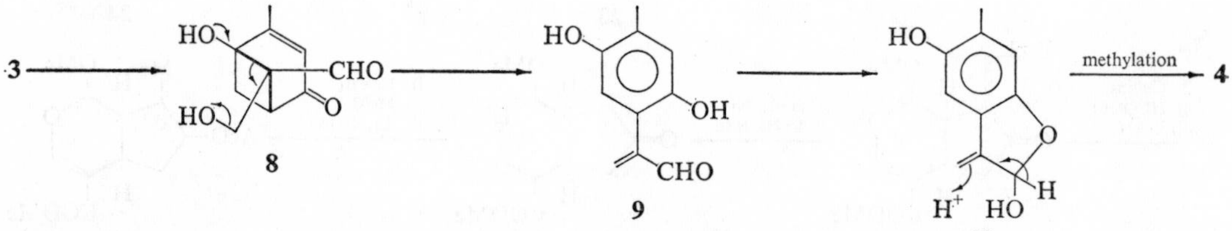

1) N. Aimi, I. Inaba, M. Watanabe, S. Shibata, *Tetr.*, **25**, 1825 (1969); *Tetr. Lett.*, 1991(1964).
2) S. Shibata, M. Nakahara *Chem. Pharm. Bull.*, **11**, 372 (1963).
3) S. Shibata, M. Nakahara, N. Aimi, *Chem. Pharm. Bull.*, **11**, 379 (1963).
4) M. Kaneda, Y. Iitaka, S. Shibata, *Tetr.*, **28**, 4309 (1972).

3.15 BIOSYNTHESIS OF CYCLIC MONOTERPENES IN HIGHER PLANTS

Cyclic monoterpenes are presumed to be formed by the cyclization of geranyl pyrophosphate **3** or its biochemical equivalents, but experimental evidence concerning this process is scanty.

Experiments on the biosynthesis of (−)-thujone **5**,[1] (+)- or (−)-camphor **6**[2] and (+)-pulegone **7**[3] using the leaves of higher plants showed that radioactivity from administered [2-^{14}C]-mevalonic acid was found predominantly in those parts of the skeletons derived from isopentenyl pyrophosphate **1**, i.e. C-6 in **5** and **6**, C-2 and C-6 in **7**. The parts of the skeletons hypothetically derived from dimethylallyl pyrophosphate **2** were essentially unlabeled.

These results have been rationalized in terms of three possibilities: firstly, a large dimethylallyl pyrophosphate pool in the plants; secondly, dimethylallyl pyrophosphate not derived from mevalonic acid is present; thirdly, there may be a compartmentalization effect in higher plants. On the other hand, an equal distribution of radioactivity was found between both the asterisked carbons in linalool **8** biosynthesized by *Cinnamomum camphora* from [2-^{14}C]-mevalonic acid.[4] It has been reported that labeled geraniol was incorporated into (−)-camphor and (−)-borneol in *Salvia officinalis*.[5]

The plants used in the above experiments were *Thuja occidentalis* for (−)-thujone **5**, *Artemisia californica*, *Salvia eucophylla* and *Chrysanthemum balsmita* for camphor **6**, and *Mentha pulegium* for (+)-pulegone **7**.

1) D. V. Banthorpe, J. Mann, K. W. Turnbull, *J. Chem. Soc. C*, 2689 (1970).
2) D. V. Banthorpe, D. Baxendale, *J. Chem. Soc. C*, 2694 (1970).
3) D. V. Banthorpe, B. V. Cahrlwood, M. R. Young, *J. Chem. Soc.*, Perkin I, 1532 (1972).
4) T. Suga, T. Shishibori, M. Bukeo, *Phytochemistry*, **10**. 2725 (1971).
5) A. R. Battersby, D. G. Laing, R. Ramage, *J. Chem. Soc.*, Perkin I, 2743 (1972).

62

3.16 CYCLIC BORATION OF LIMONENE

Hydroboration reactions of mono-olefinic monoterpene and sesquiterpene hydrocarbons such as pinene, carene, thujene, thujopsene, cedrene, etc., have been studied. The steric course of the reactions and the stereochemistry of the alcohols obtained by successive oxidation of alkyl boranes in such compounds have been detailed in the literature.[1]

The addition of thexylborane (2,3-dimethyl-2-butylborane) 2 to limonene 1, which has two double bonds, produces the bicyclic derivative 3, from which the diol 4 and (−)-carvomenthol 6 were derived stereospecifically.[2]

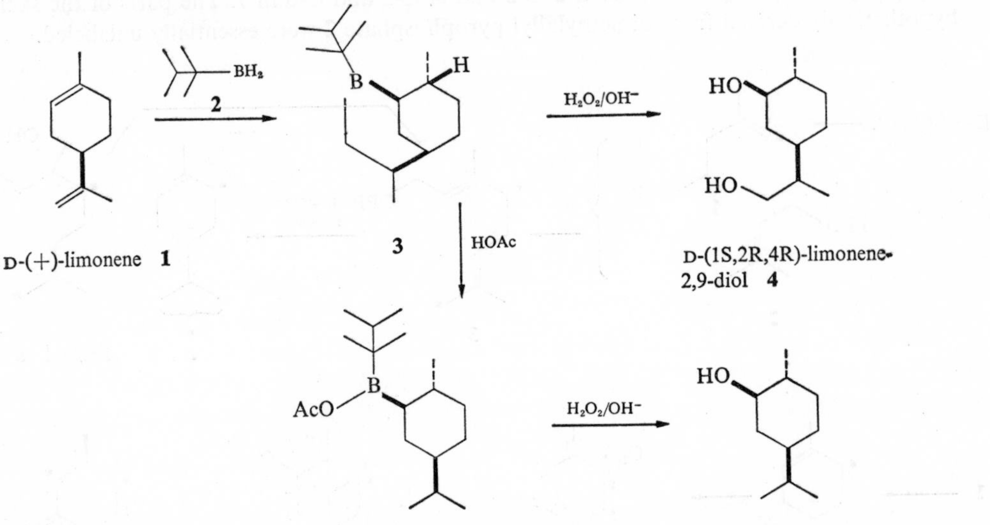

D-(+)-limonene 1

2

3

H_2O_2/OH^-

D-(1S,2R,4R)-limonene-2,9-diol 4

HOAc

5

H_2O_2/OH^-

D-(−)-carvomenthol 6

1) Cyclic boration of 1 with an equimolar amount of thexylborane 2, followed by distillation, yielded 3 in 72% yield.

2) Dihydroboration/oxidation of 1 under the usual conditions results in the formation of a complex mixture of isomeric diols.

3) Hydroboration of 1 under controlled conditions (molar ratio 1:1) followed by oxidation affords 4, but in less satisfactory yield.[3]

4) Essentially pure cis-diol 4 was obtained by oxidation of 3.

5) On treatment with HOAc, 3 was converted to 5, from which essentially pure (−)-carvomenthol 6 was obtained.

1) For leading references, see, S. P. Acharya, H. C. Brown, J. Org. Chem., 35, 196 (1970); ibid., 35, 3874 (1970).
2) H. C. Brown, C. D. Pfaffenberger, J. Am. Chem. Soc., 89, 5475 (1967); H. C. Brown, E. Negishi, ibid., 94, 3567 (1972).
3) H. C. Brown, E. Negishi, P. L. Burke, J. Am. Chem. Soc., 94, 3561 (1972).

3.17   ASYMMETRIC SYNTHESIS USING L-MENTHOL

Asymmetric atrolactic acid synthesis[1-3]

Alkylation of (−)-menthyl phenylglyoxalate **1** with a Grignard reagent, followed by hydrolysis afforded atrolactic acid **3** containing excess (−)-enantiomer. Reduction of **1** with aluminum amalgam followed by hydrolysis gave mandelic acid **4** containing excess (−)-enantiomer.

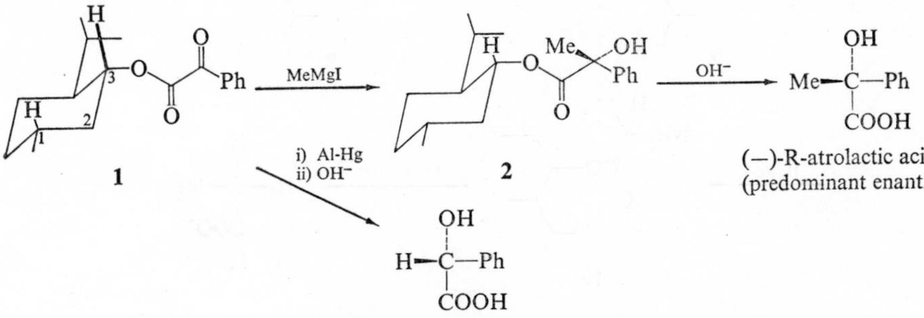

(−)-R-atrolactic acid **3**
(predominant enantiomer)

(−)-mandelic acid **4**
(predominant enantiomer)

1) This result can be explained in terms of attack by the Grignard reagent or reducing agents being more rapid from the less hindered side, i.e. the C-2 methylene side, when the keto ester grouping lies in a transoid conformation in one plane, as depicted in **1**.

2) Since the configuration of the atrolactic acid which is produced in excess is known, the absolute configuration of the hydroxyl-bearing carbon atom in **1** can be determined by this method.[4]

Many examples of asymmetric synthesis using (W)-menthol as an optically active agent have been recorded, and some are shown below.

Enamine alkylation[5]

Condensation of (−)-menthyl *trans*-crotonate and the morpholine enamine of cyclohexanone yields the optically active product **7**, α_D: −0.88°, after hydrolysis. The asymmetric carbon atom was found to possess S configuration.

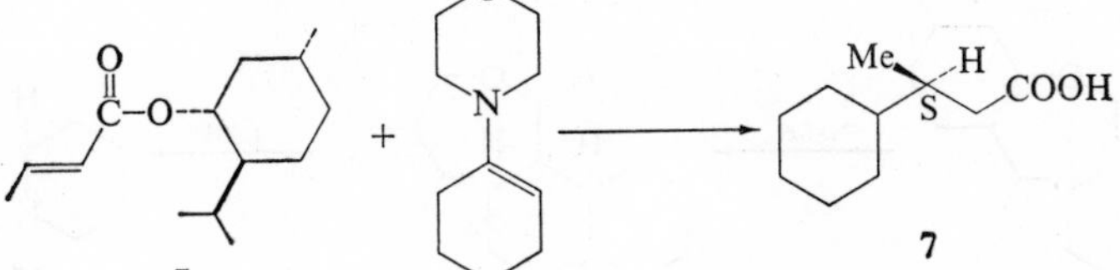

1) For review of asymmetric synthesis, see D. R. Boyd, M. A. McKervey, *Quart. Rev.*, **22**, 95 (1968).
2) A. McKenzie, H. B. Thompson, *J. Chem. Soc.*, 1004 (1905).
3) V. Prelog, *Helv. Chim. Acta*, **36**, 308 (1953).
4) V. Prelog, E. Philbin, E. Watanabe, M. Wilhelm, *Helv. Chim. Acta*, **39**, 1086 (1956); W. G. Dauben, D. F. Dickel, O. Jeger, V. Prelog, *ibid.*, **36**, 325 (1953); V. Prelog, H. Meiser, *ibid.*, **36**, 320 (1953).
5) K. Igarashi, J. Oda, Y. Inouye, M. Ohno, *Agr. Biol. Chem.* (Tokyo), **34**, 811 (1970).

Asymmetric synthesis of alanine[6]

Optically active alanine **11** (50% optical purity) was synthesized in 46% yield from the (−)-menthyl ester **8** by alkylation, followed by reductive removal of the phenyl group and hydrolysis.

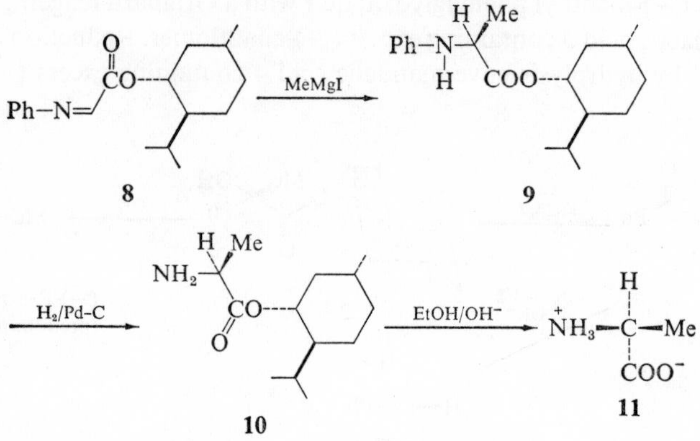

8   **9**

10   **11**

Carbene insertion[7]

The reaction of (−)-menthyl acrylate **12** with diphenyl diazomethane affords the levorotatory diphenyl cyclopropane **14** after hydrolysis of **13**.

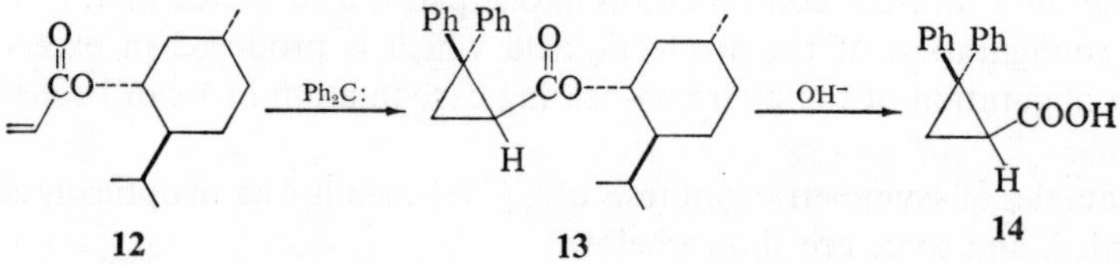

12   **13**   **14**

Synthesis of (+)-lupinene[8]

Partial asymmetric synthesis of optically active lupinene **17** was achieved by NaBH₄ reduction of the double bond in the menthyl ester **15**. LAH reduction of **16** afforded (+)-lupinene **17** in 10% optical yield.

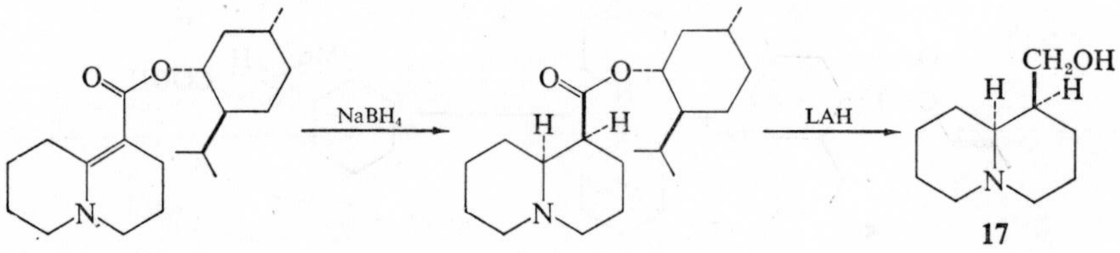

15   **16**   **17**

6) J. C. Fiaud, H. B. Kagan, *Tetr. Lett.*, 1813 (1970); *ibid.*, 1019 (1971).
7) F. J. Impastanp, L. Barash, H. M. Walborsky, *J. Am. Chem. Soc.*, **81**, 1514 (1954).
8) S. I. Goldberg, I. Ragade, *J. Org. Chem.*, **32**, 1046 (1967).

3.18 PHOTOLYSIS OF VERBENONE

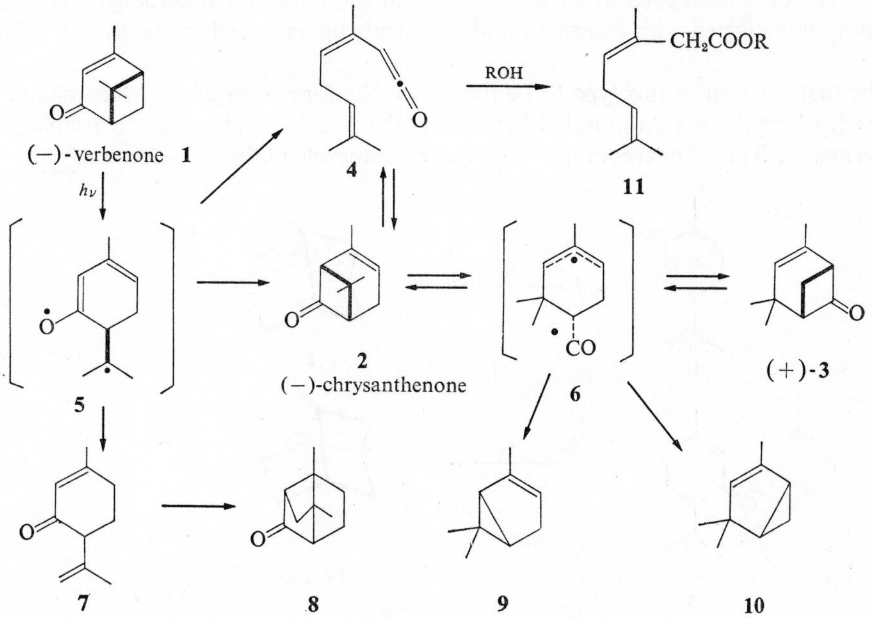

1) Irradiation of **1** with a light over 300 nm affords (−)-chrysanthenone **2** of high optical purity. This might be formed through the intermediate **5** with retention of configuration.

2) With a broad-spectrum mercury arc lamp, **1** yields a mixture of partially racemized **2** and **3** together with several other products (**7–10**). Racemization might occur via the ketenic product **4**.

3) The major products of irradiation of **1** in alcohol are **2** and **11**.

4) Compound **8** is a secondary product of isopiperitenone **7**, which is formed via the intermediate **5**.

5) **9** and **10** are secondary products derived by decarbonylation of intermediate **6**.

1) W. F. Erman, *J. Am. Chem. Soc.*, **89**, 3828 (1967); W. F. Erman, H. C. Kretschmar, *ibid.*, **91**, 779 (1969).

3.19   PHOTOCYCLOADDITION OF ENONES AND OLEFINS

The photochemical cyclization of olefins to an α,β-unsaturated carbonyl system has been extensively investigated with regard to both the synthetic and mechanistic aspects of the reaction.[1]

The first reaction of this type to be found was the formation of carvonecamphor **2** from carvone **1**, where the α,β-unsaturated ketone and the double bond are in a favorable position.[2] Isopiperitenone **3** also undergoes intramolecular photocycloaddition, yielding **4**.[3]

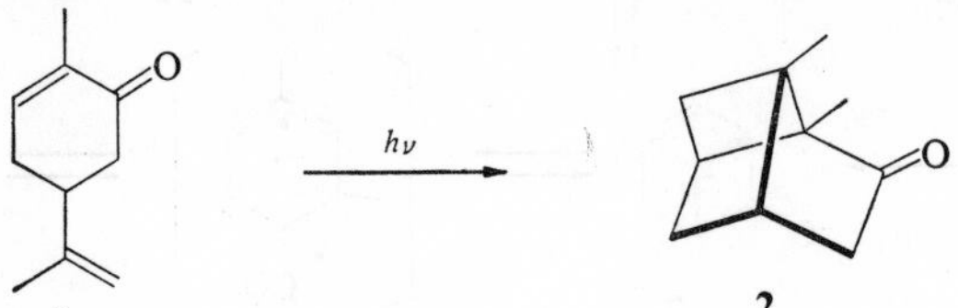

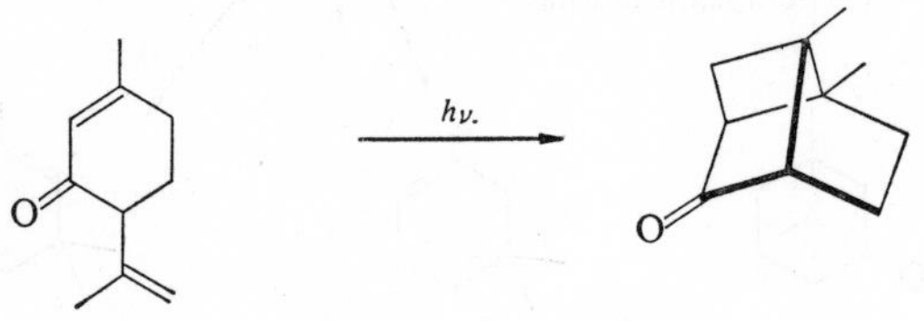

Intermolecular photocycloaddition has considerable potential for organic synthesis, and has frequently been used for the one-step construction of cyclobutane-containing compounds and systems derived by subsequent rearrangement of such rings. Two simple cases are illustrated below.

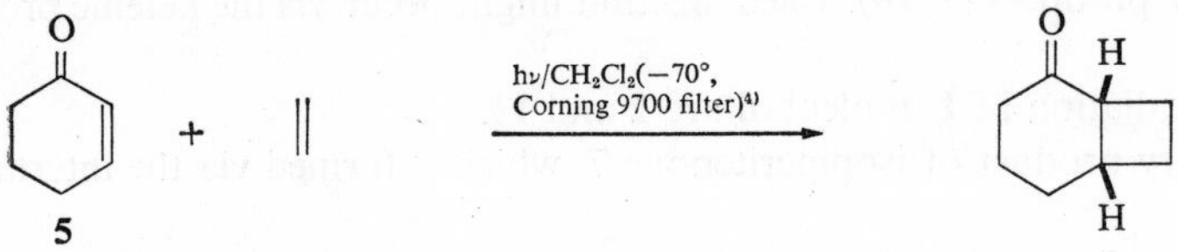

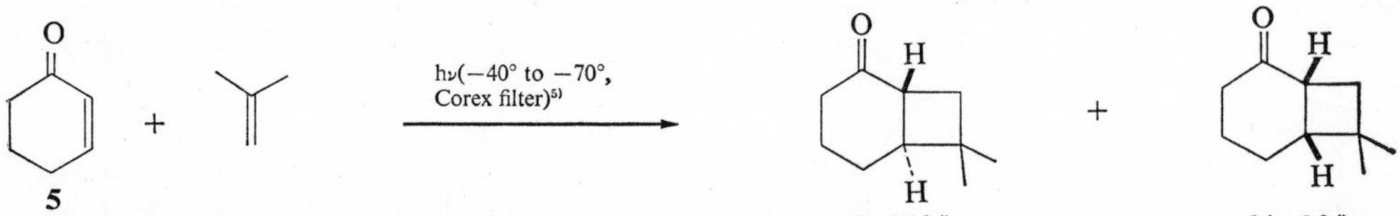

The utility of this reaction has been demonstrated in the synthesis of many terpenes, such as caryophyllene, bourbonene, α-caryophyllene alcohol, himachalene, illudol, loganin, grandisol, etc.[1]

1) For reviews, see P. E. Eaton, *Accounts Chem. Res.*, **1**, 50 (1968); P. G. Bauslaugh, *Synthesis*, 287 (1970); P. de Mayo, *Accounts Chem. Res.*, **4**, 41 (1971).
2) G. Büchi, I. M. Goldman, *J. Am. Chem. Soc.*, **79**, 4741 (1957); J. Meinwald, R. A. Schneider, *ibid.*, **87**, 5218 (1965); T. W. Gibson, W. F. Erman, *J. Org. Chem.*, **31**, 3028 (1966).
3) W. F. Erman, *J. Am. Chem. Soc.*, **89**, 3828 (1967).
4) D. C. Owsley, J. J. Bloomfield, *J. Chem. Soc., C*, 3445 (1971).
5) E. J. Corey, J. D. Bass, R. Lemathieu, R. B. Mitra, *J. Am. Chem. Soc.*, **86**, 5570 (1964).

3.20 BIOSYNTHESIS OF FARNESYL PYROPHOSPHATE

The biosynthesis of farnesyl pyrophosphate **1**, an important common precursor of the sesquiterpenoids, has been studied in detail by means of cell-free systems and partially purified enzymes. The reaction mechanism and stereochemistry of each step are now quite well understood.

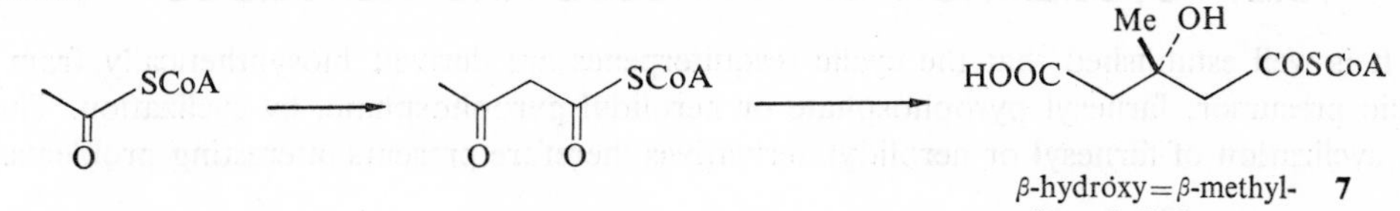

acetyl–CoA **5**   acetoacetyl–CoA **6**   β-hydroxy=β-methyl- **7**
glutaryl–CoA

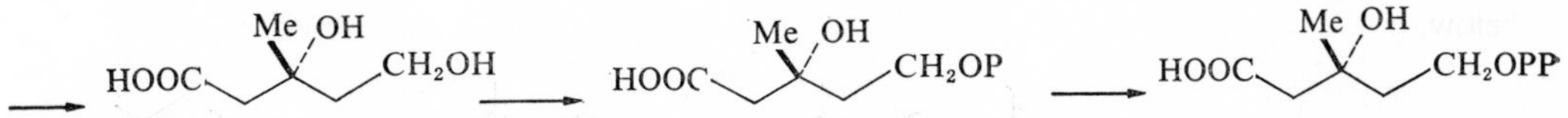

mevalonic acid **8**   **9**   **10**

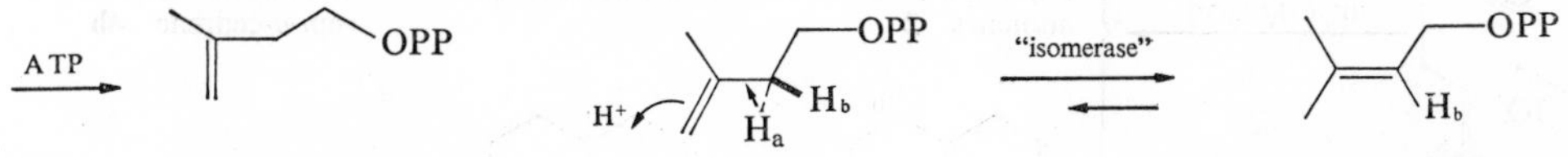

isopentenyl pyrophosphate **2**   **2**   dimethylallyl pyrophosphate **3**

prenyl transferase

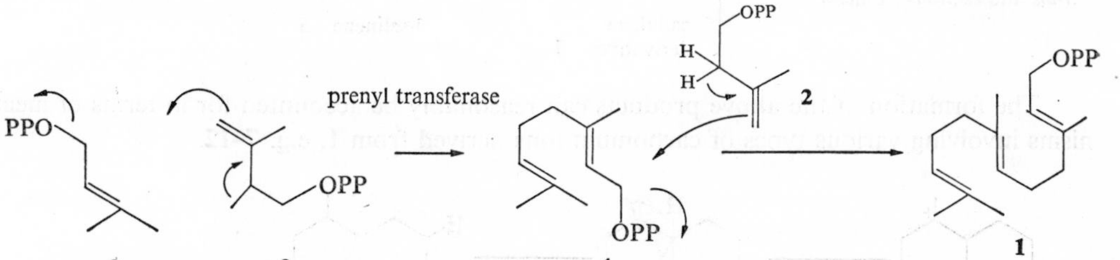

3   **2**   **4**   **1**

1) The biological isoprene unit, isopentenyl pyrophosphate **2**, is biosynthesized from acetyl coenzyme-A via mevalonic acid, as shown above.[1-3]

2) **2** is isomerized to dimethylallyl pyrophosphate **3** by an "isopentenyl pyrophosphate isomerase." **3** is an initiating species for isoprene polymerization.

3) Protonation and deprotonation during this isomerization are stereospecific.[4-6]

4) Two successive condensations of **2** with **3** are catalyzed by prenyl transferase, yielding all-*trans* farnesyl pyrophosphate **1** stereospecifically.[5]

1) F. Lynen, H. Eggerer, U. Henning, I. Kessel, *Angew. Chem.*, **70**, 738 (1958).
2) K. Bloch, S. Chaykin, A. H. Phillips, A. de Waard, *J. Biol. Chem.*, **234**, 2595 (1959).
3) H. Rudney, *The Biosynthesis of Terpenes and Sterols* (ed. G. E. W. Wolstenhome, M. O'Conner), p. 75, Churchill, 1959.
4) J. W. Cornforth, R. H. Cornforth, C. Donninger, G. Popják, *Proc. Roy. Soc.* B, **163**, 492 (1966).
5) J. W. Cornforth, R. H. Cornforth, G. Popják, L. Yengoyan, *J. Biol. Chem.*, **241**, 3970 (1966); J. W. Cornforth, R. H. Cornforth, C. Donninger, G. Popják, G. J. Schroepher Jr., *Proc. Roy. Soc.* B, **163**, 436 (1966); C. Donninger, G. Popják, *ibid.*, **163**, 465 (1966).
6) K. Clifford, J. W. Cornforth, R. Mallaby, G. T. Phillips, *Chem. Commun.*, 1599 (1971).

Remarks

Isopentenyl pyrophosphate isomerase and prenyl transferase used in the experiments were prepared from liver tissues, pumpkin fruit, or yeast cells.

3.21 CYCLIZATION OF FARNESOL AND NEROLIDOL

It is well established that the cyclic sesquiterpenes are derived biosynthetically from an acyclic precursor, farnesyl pyrophosphate or nerolidyl pyrophosphate, by cyclization. The *in vitro* cyclization of farnesyl or nerolidyl derivatives therefore presents interesting problems.[1-3]

Cyclization of farnesol[1]

An example of such cyclization, yielding a number of products including **2–6**, is illustrated below.

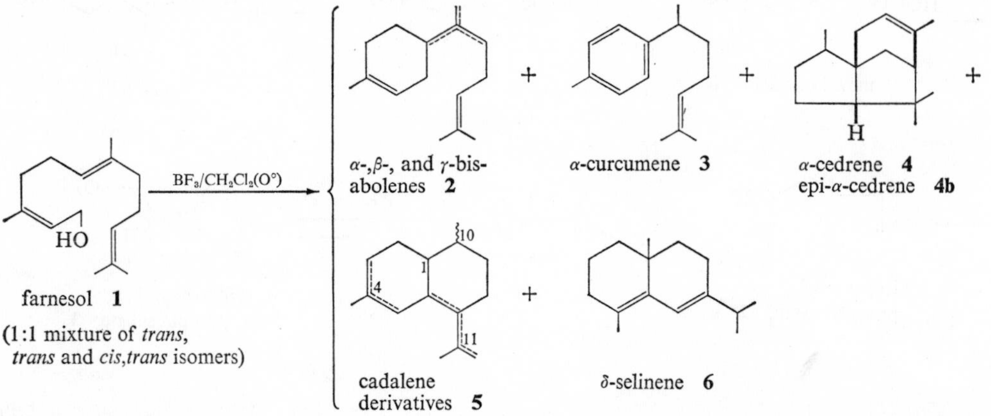

farnesol **1**

(1:1 mixture of *trans, trans* and *cis,trans* isomers)

$BF_3/CH_2Cl_2(0°)$

α-, β-, and γ-bis-abolenes **2**

α-curcumene **3**

α-cedrene **4**
epi-α-cedrene **4b**

cadalene derivatives **5**

δ-selinene **6**

The formation of the above products can reasonably be accounted for in terms of mecha-nisms involving various types of carbonium ions derived from **1**, e.g. **7–12**.

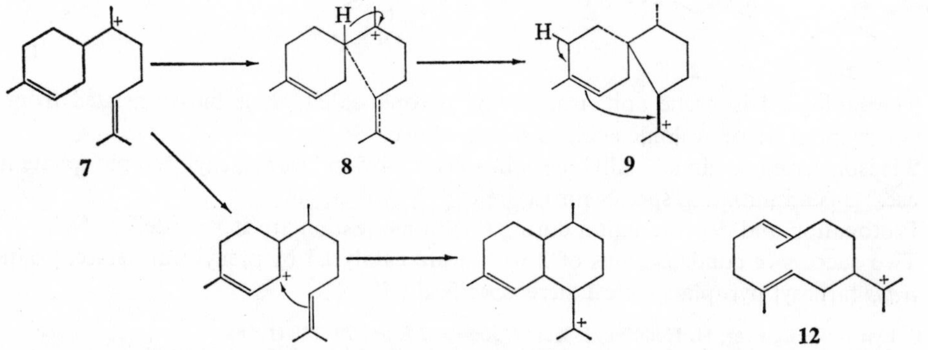

7   **8**   **9**

10   **11**   **12**

1) Y. Ohta, Y. Hirose, *Chem. Lett.*, 263 (1972).
2) C. D. Gutsche, J. R. Maycock, C. T. Chang, *Tetr.*, **24**, 859 (1968).
3) For cyclization of geranyl derivatives, see W. Rittersdorf, F. Cramer, *Tetr.*, **23**, 3015 (1967); *ibid.*, **24**, 43 (1968); R. C. Haley, J. A. Miler, H. C. S. Wood, *J. Chem. Soc. C*, 264 (1969).
4) N. H. Andersen, D. D. Syrdal, *Tetr. Lett.*, 2455 (1972).

1) Deprotonation of the bisabolenium cation **7**, formed by nucleophilic attack of the central double bond on the allylic alcohol in **1**, produces bisabolenes **2**.

2) α-Cedrene **4** is formed through the carbonium ion **9**, which is derived by hydride shift followed by attack of the terminal double bond, as in **8**.

3) Hydride shift or shifts in **7** producing the ion **10** might lead to cadalene derivatives possessing a saturated 10-Me region, but with unsaturation around the isopropyl group, via the ion **11**.

4) δ-Selinene **6** might be produced through the germacrenium ion **12**.

Cyclization of nerolidol[4)]

(+)-nerolidol **13** α- and β-bisabolenes **2**

1) Optically active β- and α-bisabolenes were obtained by cyclization of (+)-**13**. The optical purity of (−)-β-bisabolene obtained is 31% when HCOOH is used, and above 37% when ClSO$_2$NCO in Et$_2$O is used.

2) One-pot synthesis of α-cedrene from nerolidol was achieved by treatment of **13** first with HCOOH and then with CF$_3$COOH.

3.22   CHEMICAL TRANSFORMATION OF NEROLIDOL

A series of transformations leading ultimately from nerolidol (*cis/trans* mixture) to dehydro-α-cedrene through intermediates whose structures are reminiscent of naturally occurring sesquiterpenes, has been studied.[1)]

nerolidol **1**   **2**   **3**

4   **5**   **6**

7   carotol ether **8**

---

1) E. Demole, P. Enggist, C. Borer, *Helv. Chim. Acta*, **54**, 1845 (1971); *ibid.*, **54**, 456 (1971); *Chem. Commun.*, 264 (1969).

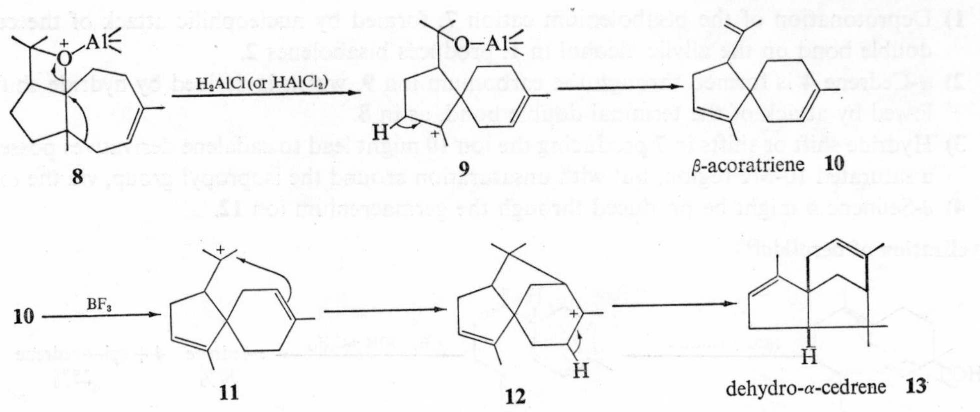

$$ 8 \xrightarrow{\text{H}_2\text{AlCl (or HAlCl}_2)} 9 \longrightarrow \beta\text{-acoratriene } 10 $$

$$ 10 \xrightarrow{\text{BF}_3} 11 \longrightarrow 12 \longrightarrow \text{dehydro-}\alpha\text{-cedrene } 13 $$

1→3: Bromination, in conjunction with the intramolecular hydroxyl group, affords the tetra-hydrofuran derivative **3**.

3→6: Dehydrobromination of **3** yields the allyl vinyl ether **4** (**5**), which immediately undergoes [3,3]-sigmatropic rearrangement to the cycloheptenone **6**.

6→8: On acid treatment, **6** is further cyclized to the oxetane derivative **8** by intramolecular cycloaddition.

8→10: The oxetane ring in **8** is cleaved by Lewis acids with concommitant skeletal rearrangement to the spiro compound **10**.

10→13: The spiro compound **10** further cyclizes to the tricyclic system **13** (c.f. "Acoradienes and alaskenes" and "Synthesis of cedrene").

3.23 CYCLIZATION OF FARNESYL ACETATE, METHYL FARNESOATE AND THEIR EPOXIDES

A small group of bicyclic sesquiterpenes possessing a carbon skeleton reminiscent of the A/B ring system of polycyclic terpenoids is known. These compounds are assumed to be derived biogenetically by protonation of the terminal double bond or terminal epoxide of a farnesyl precursor.[1,2]

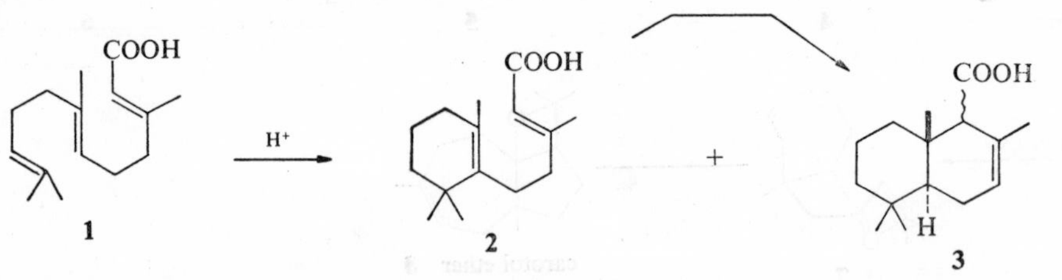

1) A. Caliezi, H. Schinz, *Helv. Chim. Acta*, **35**, 1637 (1952); G. Stork, A. W. Burgstahler. *J. Am. Chem. Soc.*, **77** 5068 (1955).
2) P. A. Stadler, A. Eschenmoser, H. Schinz, G. Stork, *Helv. Chim. Acta*, **40**, 2191 (1957).

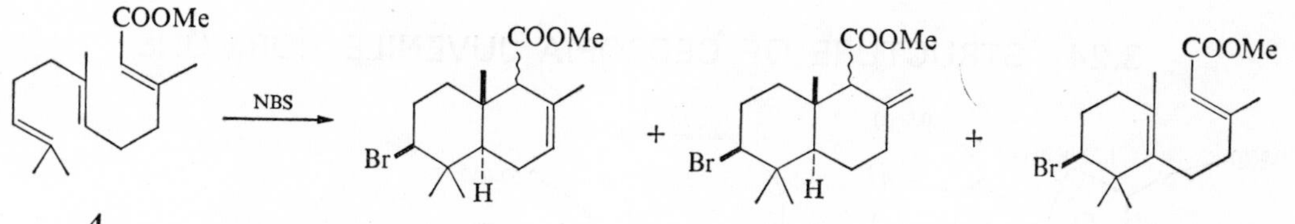

4 → 5 + 6 + 7

Monocyclic farnesoic acid **2** is also converted into the bicyclic product **3** on treatment with BF$_3$.[3] The bicyclic bromo esters **5** and **6** were obtained by direct terminal bromination with NBS.[4]

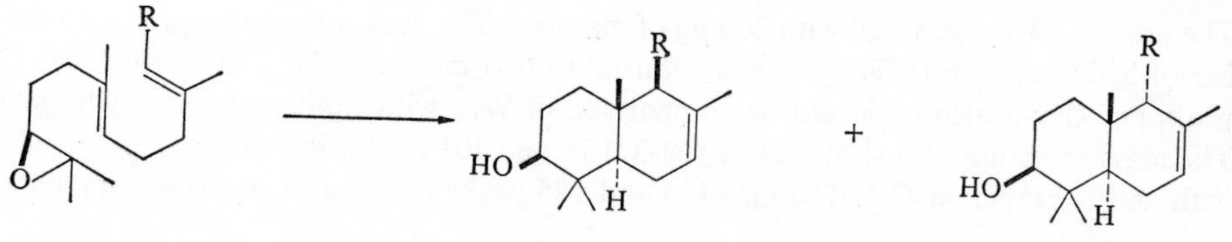

8 R=CH$_2$OAc
11 R=COOMe

9 R=CH$_2$OAc
12 R=COOMe

10 R=CH$_2$OAc
13 R=COOMe

It has been shown that the terminal epoxide **8** derived from farnesyl acetate cyclizes into a mixture of 85% **9** and 15% **10** on catalysis with BF$_3$ etherate. A similar result was obtained with the epoxide **11**. The ratio of formation of **9** and **10** (or **12** and **13**) varied according to the reaction conditions employed.[5-7]

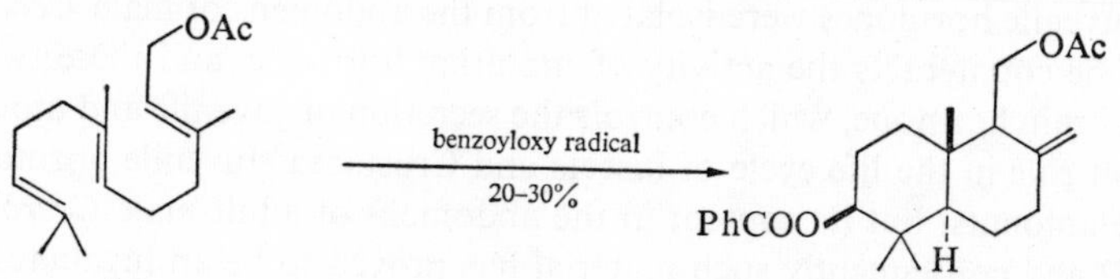

benzoyloxy radical
20–30%

14 → 15

Radical-initiated cyclization of farnesyl acetate under conditions suitable for the copper-catalyzed thermal decomposition of benzoyl peroxide yields the bicyclic ester **15**.[8]

3) Y. Kitahara, T. Kato, T. Suzuki, S. Kanno, M. Tanemura. *Chem. Commun.*, 342 (1969).
4) E. E. van Tamelen, E. J. Hessler, *Chem. Commun.*, 411 (1966).
5) E. E. van Tamelen, A. Storni, E. J. Hessler, M. Schwartz, *J. Am. Chem. Soc.*, **85**, 3295 (1963).
6) E. E. van Tamelen, J. P. McCormick, *J. Am. Chem. Soc.*, **91**, 1847 (1969).
7) For a review, see E. E. van Tamelen, *Accounts Chem. Res.*, **1**, 111 (1968).
8) R. Breslow, S. S. Olin, J. T. Groves, *Tetr. Lett.*, 1837 (1968).

3.24 STRUCTURE OF CECROPIA JUVENILE HORMONE

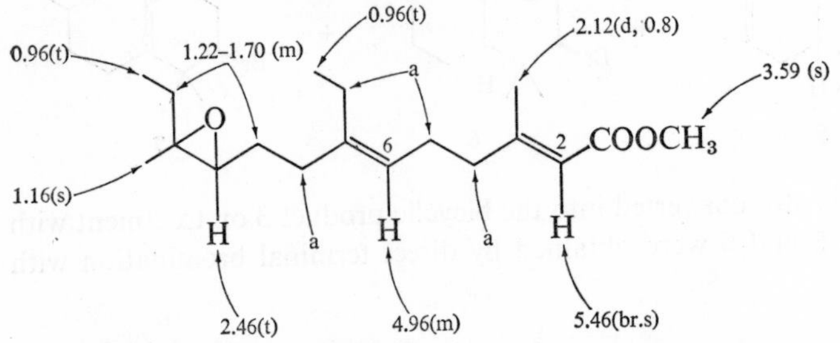

C_{18}-*Cecropia* juvenile hormone **1**
ms: 294 (M$^+$ $C_{18}H_{30}O_3$)
nmr: 1.80–2.30 (a-H)

1) The nmr of **1** was measured with 200 μg of material. The peak intensities agree with the number of hydrogen atoms. The 2-ene is deduced to be *trans* from the *J* value for the 2.12 ppm peak, and this is also indicated by comparison of $W_{1/2}$ with similar terpene hydrogens.

2) The mass spectrum of **2** showed m/e (M-31) 74 and 101 peaks, indicating aliphatic COOMe with an Me branch at C-3. The m/e 143 and 185 peaks indicate Et or di-Me branch at C-7.

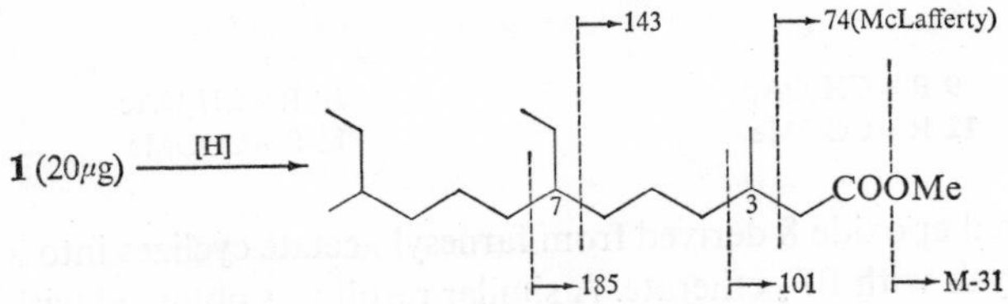

2 ms: 284 (M$^+$$C_{18}H_{36}O_2$)

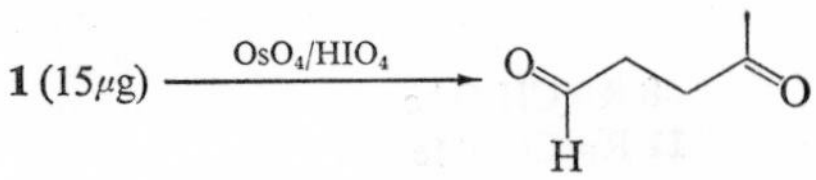

levulinic aldehyde **3** (identified by glc)

Remarks

Cecropia juvenile hormones were isolated from the abdomens of male *Cecropia* silkworms. Juvenile hormone counteracts the activity of moulting hormone, and these two hormones, together with the brain hormone, which controls the secretion of juvenile and moulting hormones, play a dominant role in the life cycle of insects and Crustacea. Juvenile hormone occurs in extremely minute amounts, but its content in the abdomens of adult male *Cecropia* silk moths is relatively high[1] and consequently such material has proved to be an invaluable source.

The structure of hormone **1** has been determined as shown above,[2] with the exception of the stereochemistry of the epoxide and absolute configuration. Less than 300μg was used, this work providing an excellent example of the achievements of modern methodology. The structure, together with the *cis*-arrangement for the oxirane, has been established by independent syntheses.

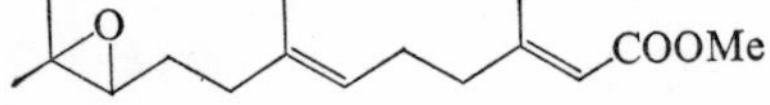

4

A second juvenile hormone **4** has been isolated from adult male abdomens of *Hyalophora cecropia*, the giant silkworm moth. The structure determination[3] was carried out in a manner similar to that described above, the nmr being measured with 55 μg of sample. C_{18}-hormone **1** and C_{17}-hormone **4** are contained in a ratio of about 4:1 in the male silk moth.

1) C. M. Williams, *Nature*, 178, 212 (1956).
2) H. Röller, K. H. Dahm, C. C. Sweely, B. M. Trost, *Angew. Chem. Intern. Ed.*, 6, 179 (1967).
3) A. S. Meyer, H. A. Schneiderman, E. Hanzmann, J. H. Ko, *Proc. Nat. Acad. Sci. U.S.*, 60, 853 (1968).

3.25 ABSOLUTE CONFIGURATION OF (+)-C$_{18}$-JUVENILE HORMONE

The absolute configuration of C$_{18}$-JH **1**, α_D:$+7°$,[1] was established as being 10R, 11S by spectroscopic methods as well as by chemical methods.

Method I (spectroscopic)[2]

The absolute configuration of **1** was deduced by determining the chirality of the glycol **2**, which was obtained by acid-catalyzed fission of the epoxide.

Me   Me   Me
11S
O   10R   COOMe $\xrightarrow{H^+}$ OH   Et 11 10 9 Me H O H$^+$ $\longrightarrow$ OH Et 11 10 9 Me H HO
Me
(+)-JH **1**   **2**

OH OH 9
Et 11 10
Me H
73(17%) ←
(m/e 75; 38% incorp) → m/e 225(100%)

CD$_3$ CD$_3$
O O
Me 9
Et 11 10 H
4

Me
e HO /(e')
HO Et(e')
H
2 cd(Δε$_1$'): -1.11(312 nm)
cd(Δε$_2$): $+1.59$ (286 nm)

1) The mode of epoxide cleavage was clarified by mass fragmentation of the glycol **2** prepared in ^{18}O-enriched medium. Calculations based on the relative intensities of the 73/75 and 225/227 peaks indicated that OH ion attacks C-11 to yield the glycol **2** formed by C-11 inversion and C-10 retention.

2) That the glycol derived from the *cis*-epoxide was *threo*, was also supported by the nmr spectrum of the d$_6$-acetonide **4**. W-type coupling, but no noe, was observed between 11-Me and 10-H. The chemical shift of 11-Me *cis* to C-9 is higher by 0.10–0.15 than that of the *trans* isomer.

3) Chirality of the glycol **2** was determined by a newly devised cd method which utilizes the shift reagent Pr(dpm)$_3$. It has been found that the longer wavelength cd extremum coincides with the chirality, if the glycol chirality is defined as

+ and −, respectively, when a clockwise or counterclockwise twist is seen in a Newman projection. Thus, the glycol **2** has a (10R, 11R) configuration.

Method II (chemical)[3]

Both enantiomeric forms of JH were synthesized from starting materials of known absolute configuration. Dextrorotatory JH was obtained from (S)-(+)-**5**.[4]

HO
+ COOMe $\longrightarrow$ (+)-**1**
MeO OMe OH
(S)-(+)-**5**

---

1) A. S. Meyer, E. Hanzmann, *Biochem. Biophys. Res. Commun.*, **41**, 891 (1970).
2) K. Nakanishi, D. A. Schooley, M. Koreeda, J. Dillon, *Chem. Commun.*, 1235 (1971).
3) D. J. Faukner, M. R. Petersen, *J. Am. Chem. Soc.*, **93**, 3766 (1971).
4) See P. Loew, W. S. Johnson, *J. Am. Chem. Soc.*, **93**, 3765 (1971) for alternative synthesis of optically active JH.

3.26 SYNTHESIS OF CECROPIA JUVENILE HORMONE

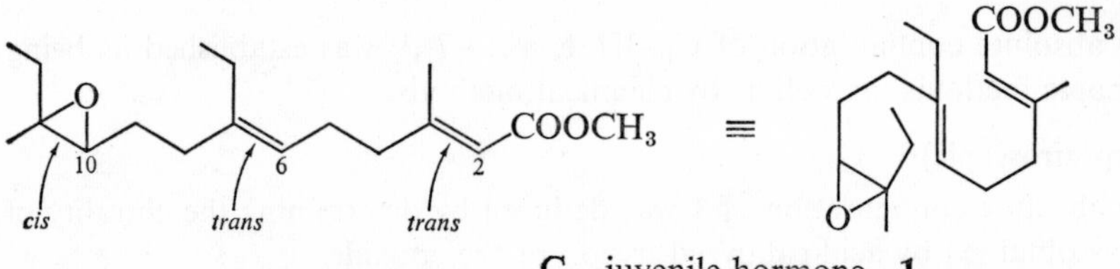

C_{18}-juvenile hormone **1**

C_{18}-juvenile hormone was first synthesized in a non-stereoselective manner (Route I), which consists essentially of two Wittig reactions and an epoxidation reaction with fractionations of the stereoisomers at appropriate stages.[1] Bioassay of the various stereoisomers showed that the 2- and 6-enes had to be *trans* for high activity. The 10-epoxide is *cis* in the natural compound, but the activity of the 10-*trans* epoxide was still half that of the 10-*cis* isomer. The absolute configuration of the natural juvenile hormone was unknown at this stage, but bioassay results indicated that the natural hormone and synthetic racemate exhibit activities of similar magnitude.[1]

Route I[1]

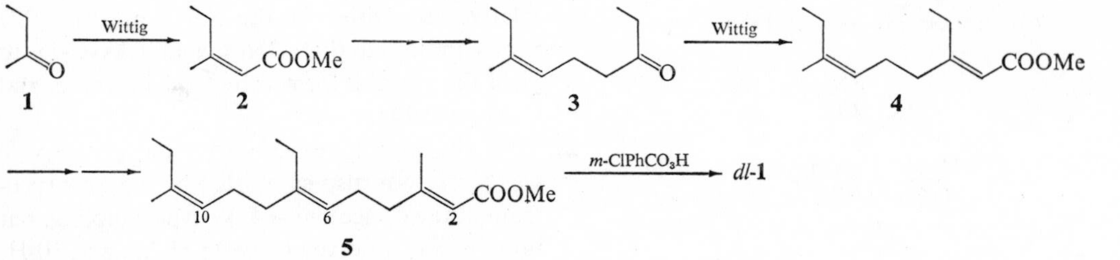

Route II[2]

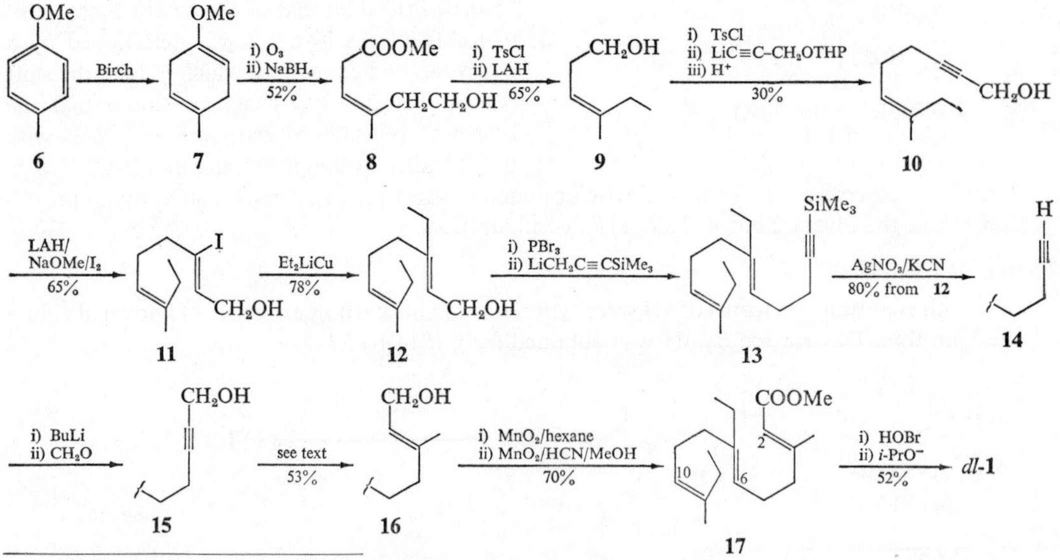

1) K. H. Dahm, B. M. Trost, H. Röller, *J. Am. Chem. Soc.*, **89**, 5292 (1967); see K. H. Dahm, H. Röller, B. M. Trost, *Life Sci.*, **7**, 129 (1968) for a slightly modified synthesis.
2) E. J. Corey, J. A. Katzenellenbogen, N. W. Gilman, S. A. Roman, B. W. Erickson, *J. Am. Chem. Soc.*, **90**, 5618 (1968).

This synthesis is an excellent example of combined reactions employing new reagents.

7→8: Formation of **8** from a benzenoid precursor ensures *cis* stereochemistry of the terminal double bond, and hence the 10-epoxide.

10→11→12, 15→16: Propargylic alcohols can be converted stereospecifically into trisubstituted olefins by LAH reduction in the presence of I_2[3] followed by reaction of the *cis*-iodide with lithium dialkylcopper.[4]

12→13→14: The allylic bromide is treated with 3-lithio-1-trimethylsilylpropyne[5] and subsequently the $SiMe_3$ group is removed;[6] this results in overall selective propynylation.

16→17: The allylic alcohol is oxidized to the conjugated aldehyde and the latter to the corresponding methyl ester as follows in a "one-flask" conversion via the following sequence:[7]

Route III[8]

The stereochemistry of the olefin in **27** is controlled by the stereochemistry of the cyclic compound **24** synthesized stereoselectively as depicted. Optimal geometry for synchronous fragmentation of C-3a-C-4 and C-5-OTs bonds is attained in **24** (or **24a**) and in **26**.[9] The ketone **27** thus synthesized is a key intermediate in the Route I synthesis.[1]

3) E. J. Corey, J. A. Katzenellenbogen, G. H. Posner, *J. Am. Chem. Soc.*, **89**, 4245 (1967).
4) E. J. Corey, G. H. Posner, *J. Am. Chem. Soc.*, **90**, 5615 (1968).
5) E. J. Corey, H. A. Kirst, *Tetr. Lett.*, 5041 (1968).
6) H. M. Schmidt, J. F. Arens, *Rec. Trav. Chim.*, **86**, 1138 (1967).
7) E. J. Corey, N. W. Gilman, B. E. Ganem, *J. Am. Chem. Soc.*, **90**, 5616 (1968).
8) R. Zurflüh, E. N. Wall, J. B. Siddall, J. A. Edwards, *J. Am. Chem. Soc.*, **90**, 6224 (1968).
9) C. A. Grob, H. R. Kiefer, H. Lutz, H. Wilkens, *Tetr. Lett.*, 2901 (1964).

Route IV[10)]

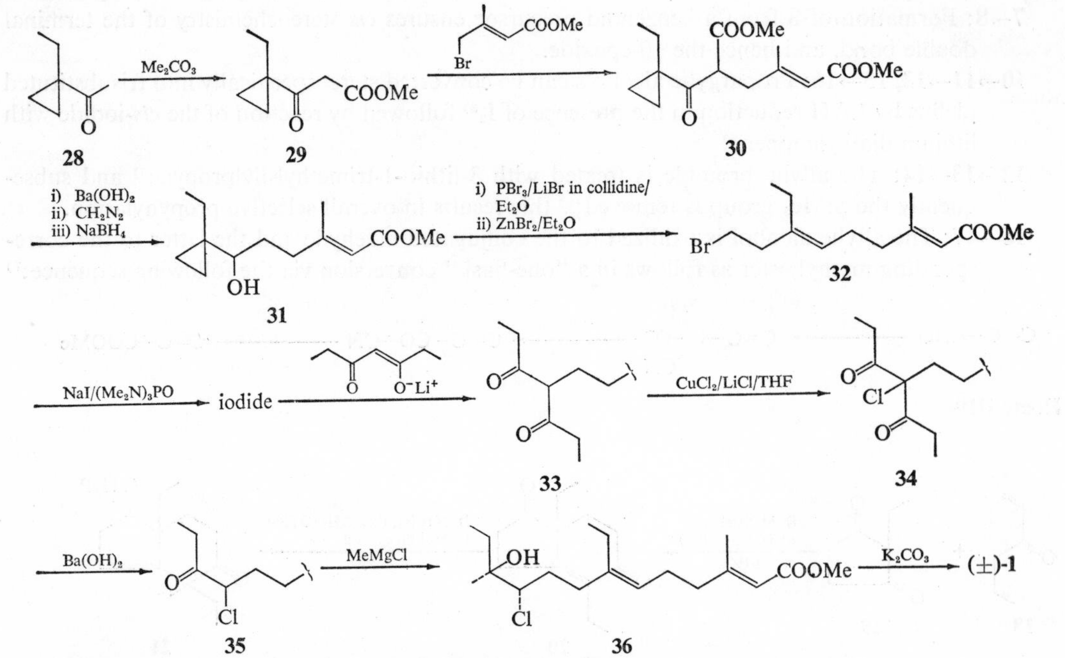

The key step in this synthesis is **31→32**, the highly stereoselective production of a *trans*-trisubstituted double bond[11)] (maximum of 5% of *trans, cis* isomer), which is a modification of the Julia method;[12)] a cyclopropylcarbinyl system is rearranged to a homoallylic system at this step. All but two of the steps in this synthesis proceed in yields of 90% or better.

Route V[13)]

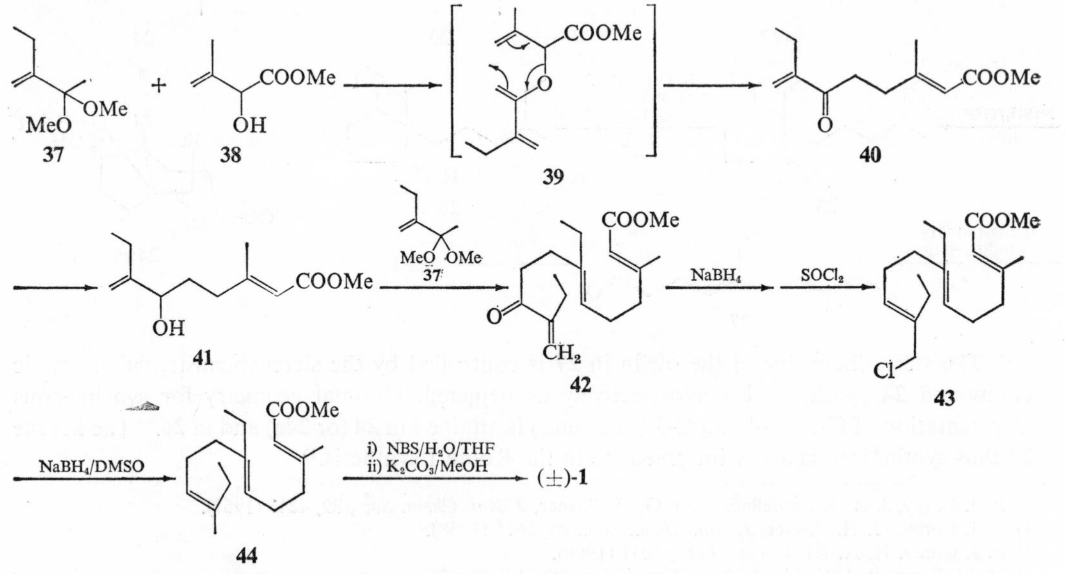

10) W. S. Johnson, T. Li, D. J. Faulkner, S. F. Campbell, *J. Am. Chem. Soc.*, **90**, 6225 (1968).
11) S. F. Brady, M. A. Ilton, W. S. Johnson, *J. Am. Chem. Soc.*, **90**, 2882 (1968).
12) M. Julia, S. Julia, S.-Y. Tchen, *Bull. Soc. Chim. Fr.*, 1849 (1961).

Claisen rearrangement was effective for the stereospecific formation of the *trans* olefinic linkage,[13] as shown in the conversions **37 + 38 → 40** and **37 + 41 → 42**. Rearrangement proceeds through the allylic ether **39**, which is produced *in situ*.

Route VI[14]

45 + 46 → 47 → (sec-BuLi) → 48
β-oxido phosphonium ylide

(paraformaldehyde) → 49 → i) py/SO$_3$ ii) LAH iii) H$^+$ → 50 → (±)C$_{17}$–*Cecropia* JH **51**

49 → i) MnO$_2$ ii) methylene triphenyl phosphorane → 52 → i) diimide ii) H$^+$ → 53 → (±)-C$_{18}$–*Cecropia* JH **1**

A stereospecific synthesis of **49** involving the reaction of β-oxido phosphonium ylide **48** and paraformaldehyde as a key step has been reported. β-Oxido phosphonium ylide generated from the Wittig betaine **47** with *sec*-BuLi reacts with carbonyl compounds to yield the *trans* olefin **49** uncontaminated by stereoisomer, through a β,β′-dioxo phosphonium ion intermediate. Both C$_{17}$- and C$_{18}$-*Cecropia* juvenile hormones (**51** and **1**) were synthesized stereospecifically from the common intermediate **49**.

Route VII[15]

54 → (MeMgBr/MgSO$_4$) → 55
54 → (Me$_2$S=CH$_2$) → 56
55, 56 → i) n-BuLi/DABCO ii) SOCl$_2$/py → 57 → i) n-BuLi/DABCO ii) (epoxide) →

13) W. S. Johnson, T. J. Brocksom, P. Loew, D. H. Rich, L. Werthemann, R. A. Arnold, T. Li, D. J. Faulkner, *J. Am. Chem. Soc.*, **92**, 4463 (1970); see also, P. Loew, J. B. Siddall, V. Spain, L. Werthemann, *Proc. Nat. Acad. Sci. U.S.*, **67**, 1462 (1970); cf. C. A. Henrick, F. Schaub, J. B. Siddall, *J. Am. Chem. Soc.*, **94**, 5374 (1972); R. J. Anderson, C. A. Henrick, J. B. Siddall, R. Zurflüh, *ibid.*, **94**, 5379 (1972); D. J. Faulkner, M. R. Petersen, *ibid.*, **95**, 553 (1973).
14) E. J. Corey, H. Yamamoto, *J. Am. Chem. Soc.*, **92**, 6637 (1970).
15) K. Kondo, A. Negishi, K. Matsui, D. Tsunemoto, S. Masamune, *Chem. Commun.*, 1311 (1972).
16) For reviews see, C. E. Berkoff, *Quart. Rev. Chem. Soc.*, **23**, 372 (1969); B. M. Trost, *Accounts Chem. Res.*, **3**, 120 (1970).

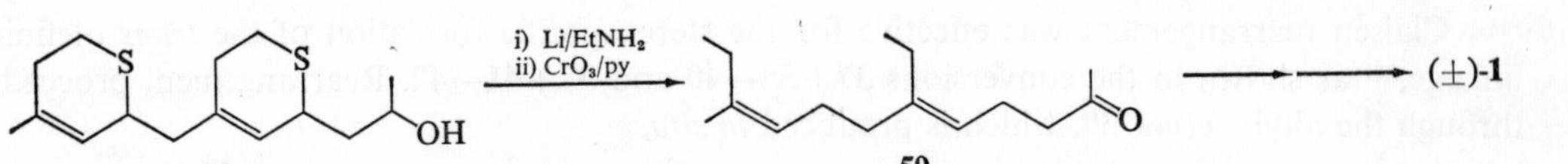

i) Li/EtNH₂
ii) CrO₃/py

$\longrightarrow$ (±)-**1**

58   **59**

It has been reported that C_{18}-JH **1** was synthesized stereospecifically by a route involving the condensation of two dihydrothiopyrans (**55** and **56**) followed by reductive desulfurization.

3.27   SYNTHESIS OF SINENSAL

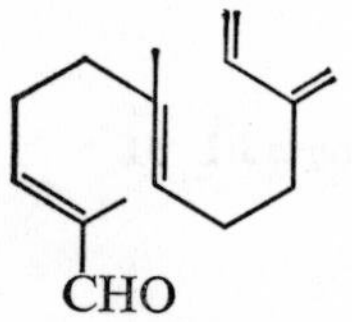

β-sinensal   **1**[1]

bp (1 mm): ~180°
ms: 218
uv (EtOH):   227 (ϵ 31,000)

Route I[2]

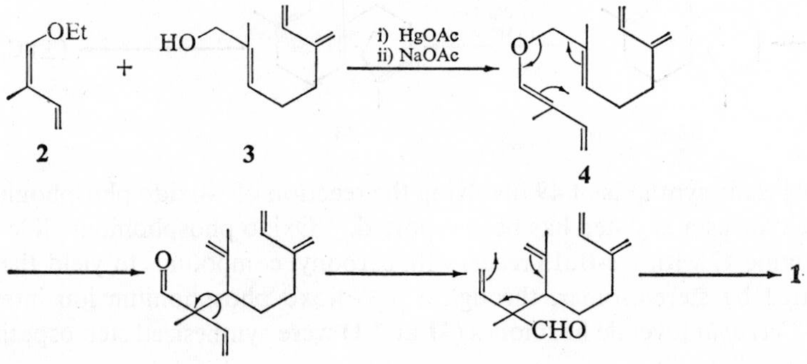

OEt

+

HO

i) HgOAc
ii) NaOAc

O

2   **3**   **4**

O

CHO

$\longrightarrow$ **1**

5   **6**

4→5: The aldehyde **5** is obtained by Claisen rearrangement of **4** on mild pyrolysis.

5→6: Rotation about the single bond in **5** brings the double bond into a suitable position for Cope rearrangement, as shown in **6**, giving rise to *β*-sinensal **1**.

Route II[3]

The lithium salts of the imines **8** and **10** react with the alkyl halide **7** and the aldehyde **9** to yield the aldehyde **9** and sinensal **1**, respectively, after acid hydrolysis.

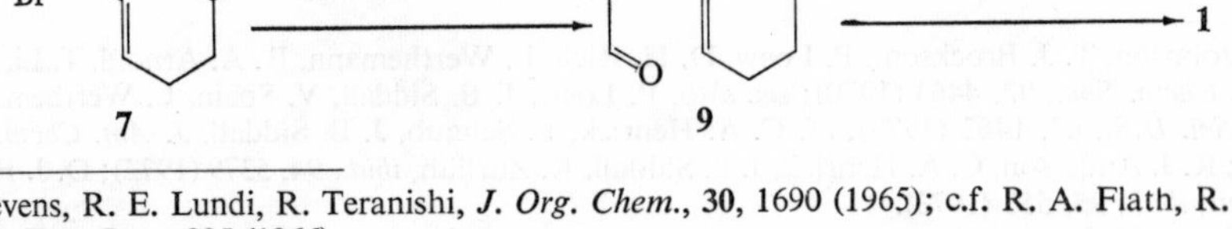

Br

i) MeCH=N–CMe₃ **8**
ii) Li–diisopropyl
iii) oxalic acid

i) MeCH₂CH=N–CMe₃ **10**
ii) Li–diisopropyl
iii) oxalic acid

$\longrightarrow$ **1**

O

7   **9**

1) K. L. Stevens, R. E. Lundi, R. Teranishi, *J. Org. Chem.*, **30**, 1690 (1965); c.f. R. A. Flath, R. E. Lundi, R. Teranishi, *Tetr. Lett.*, 295 (1966).
2) A. F. Thomas, *Chem. Commun.*, 947 (1967).
3) G. Büchi, H. Wüest, *Helv. Chim. Acta*, **50**, 2440 (1967).
4) E. Bertele, P. Schudel, *Helv. Chim. Acta*, **50**, 2445 (1967).

Route III[4]

The non-stereospecific synthesis of sinensals was also achieved by a Wittig reaction.

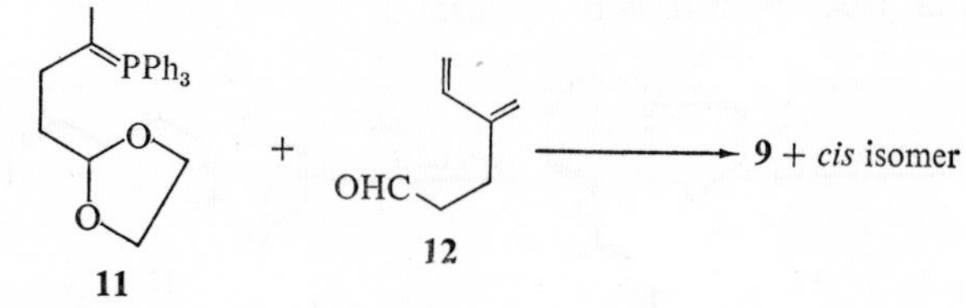

9 + *cis* isomer

11   **12**

Remarks

β-Sinensal **1**, a flavour constituent of Chinese oranges, was isolated from the essential oil of *Citrus sinensis* together with the isomer α-sinensal. β-Sinensal could be isomerized to α-sinensal in the presence of a noble metal catalyst.[4]

3.28   SYNTHESIS OF DENDROLASIN

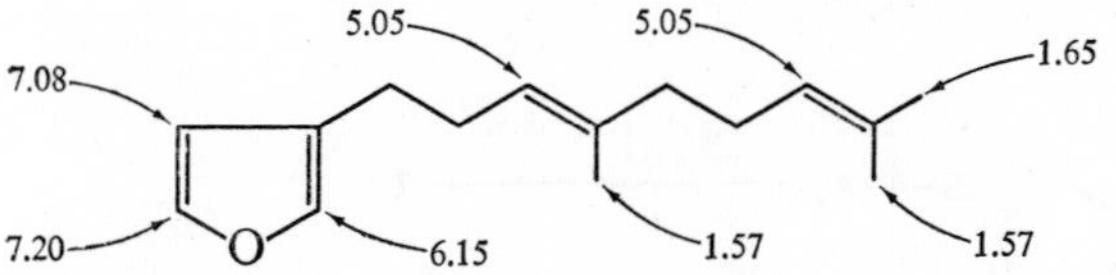

dendrolasin   **1**[1]

Route I[2]

The trisubstituted *trans* double bond in **1** was constructed by a modification of Julia's method.

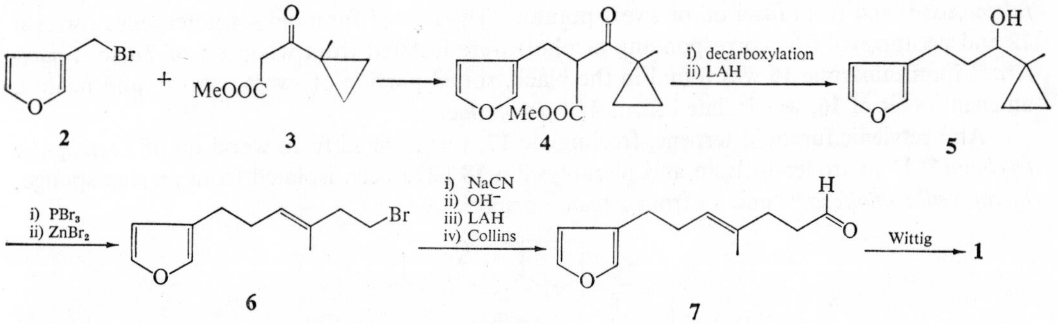

1) A. Quilico, F. Piozzi, M. Pavan, *Tetr.*, **1**, 177 (1957).
2) K. A. Parker, W. S. Johnson, *Tetr. Lett.*, 1329 (1969).
3) A. F. Thomas, *Chem. Commun.*, 1657 (1968).
4) R. Bernardi, C. Cardani, O. Ghiringhelli, A. Selva, A. Baggini, M. Pavan, *Tetr. Lett.*, 3893 (1967).
5) Y. Hirose, M. Abe, Y. Sekiya, *J. Chem. Soc. Japan*, **82**, 725 (1961).
6) T. Sakai, K. Nishimura, Y. Hirose, *Bull. Chem. Soc. Japan*, **38**, 381 (1965).
7) A. J. Birch, R. A. Massy-Westropp, S. E. Wright, T. Kubota, T. Matsuura, M. D. Sutherland, *Chem. Ind.*, 902 (1954); T. Kubota, T. Matsuura, *ibid.*, 521 (1956).
8) A. J. Birch, R. A. Massy-Westropp, S. E. Wright, *Aust. J. Chem.*, **6**, 385 (1953).
9) R. A. Massy-Westropp, G. D. Reynolds, T. M. Spotswood, *Tetr. Lett.*, 1939 (1966).
10) G. Cimino, S. De Stefano, L. Minale, E. Trivellone, *Tetr.*, **28**, 4761 (1972).
11) F. Bohlmann, N. Rao, *Tetr. Lett.*, 1039 (1972).

Route II[3]

1, torreyal 12 and neotorreyol 13 were also synthesized by a route based on the use of a double Claisen-Cope type reaction (see "Synthesis of sinensal").

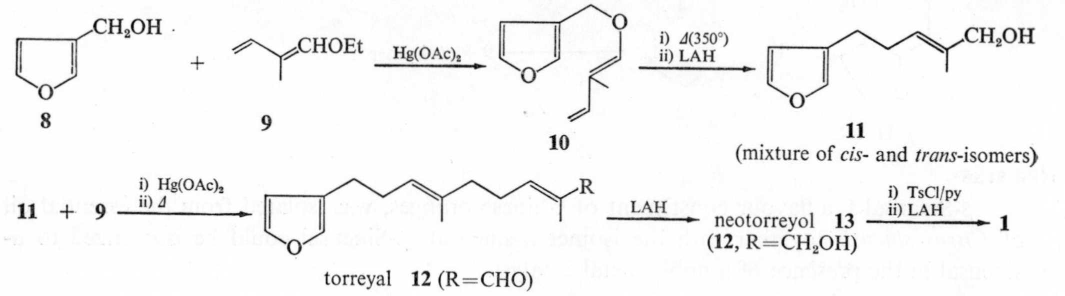

11 (mixture of cis- and trans-isomers)

torreyal 12 (R=CHO)

neotorreyol 13
(12, R=CH₂OH)

Route III[12]

The reaction of the n-butylthiomethylene derivative of the ketone 14 with dimethylsulfonium methylide afforded 1 in good yield. This reaction might be generally be applicable for the synthesis of 3- and 3,4-substituted furans.

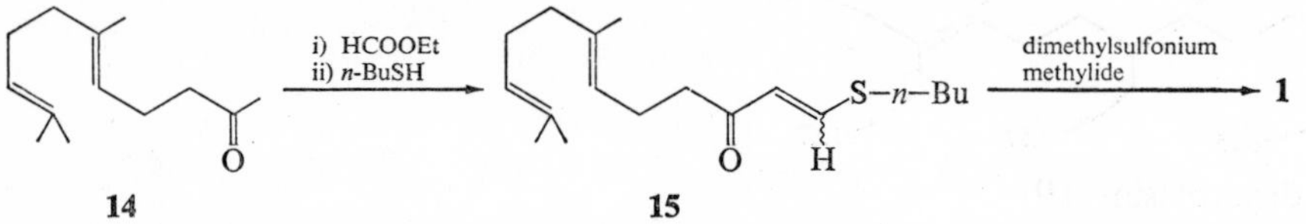

Remarks

Dendrolasin 1 was isolated from the mandibular gland of ants, *Lasius (Dendrolasius) fulginosus*[1,4] and from fusel oil of sweet potato.[5] The related furanoid sesquiterpenes torreyal 12 and neotorreyol (the corresponding alcohol) were isolated from wood oil of *Torreya nucifera*.[6] Ipomeamarone 16 was found in the black, rotted portion of sweet potato[7] and ngaion, an enantiomer of 16, was isolated from *Myoporum laetum*.[8]

An acetylenic furanoid terpene, freelingyne 17, was isolated from wood oil of *Eremophila freelingii*.[9] Dehydrodendrolasin and pleraplysillin 18 have been isolated from marine sponge, *Pleraplysilla spinifera*,[10] and 19 from *Athanasia* species.[11]

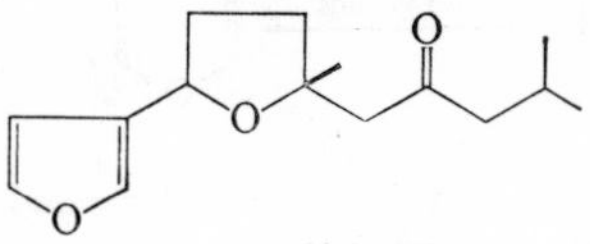

ipomeamarone 16

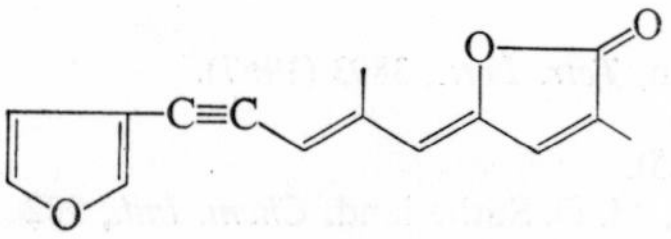

freelingyne 17

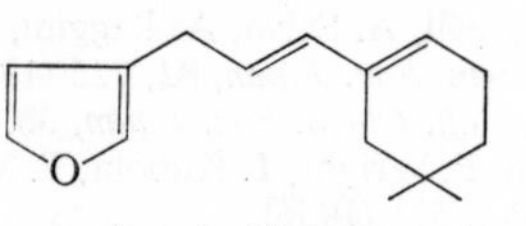

pleraplysillin 18

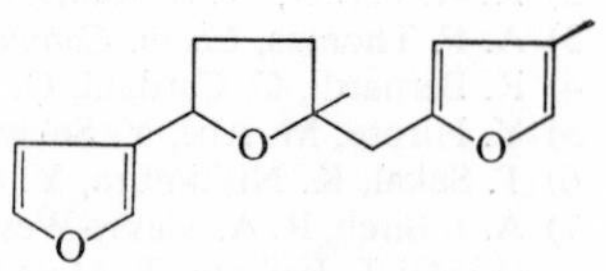

19

12) M. E. Garst, T. A. Spenser, *J. Am. Chem. Soc.*, **95**, 250 (1973).

3.29   STRUCTURE AND SYNTHESIS OF CYCLONERODIOL

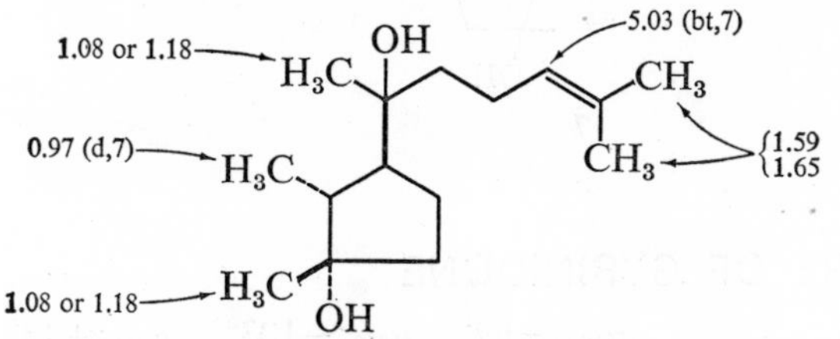

cyclonerodiol **1**[1]

1) The mass fragmentation pattern of **1**, the dehydration product **2** and the hydrogenation product **3** indicate the partial structural units shown.

2) Ozonolysis gave the trisnor compound **4**, showing γ-lactone ir absorption.

3) The appearance of two olefinic methyl signals in **5** reveals the *vic* arrangement of the ring methyls.

4) The relative stereochemistry of the ring substituents as *trans, trans* was established by the synthesis of **1** from **9**, whose stereochemistry was already known.[2]

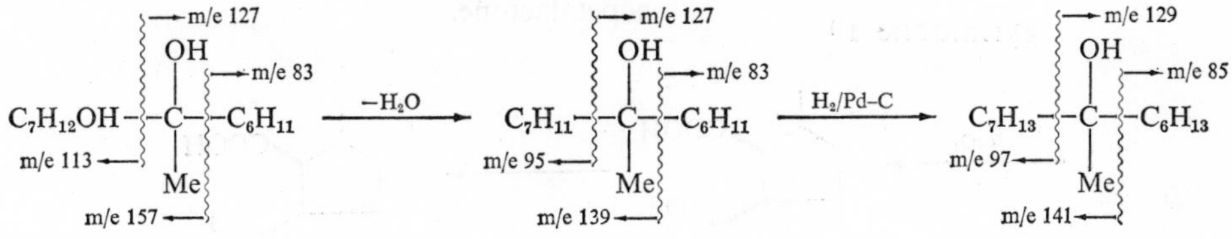

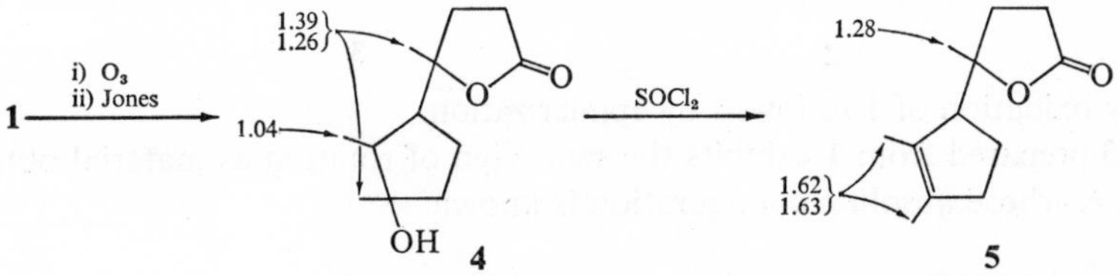

ir: 1766 (γ-lactone)   ir: 1766 (γ-lactone)

Synthesis[2]

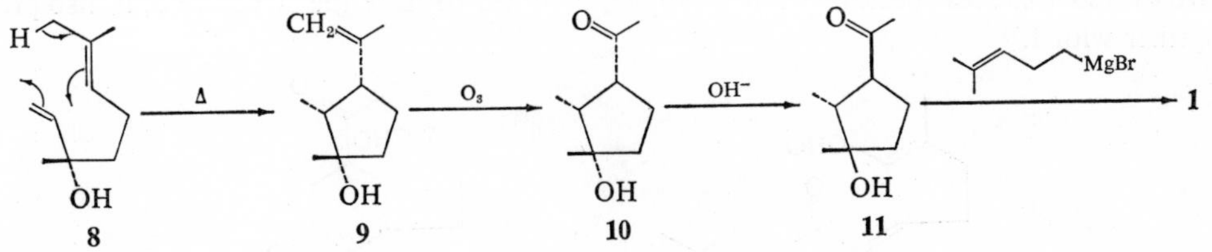

Remarks

Cyclonerodiol **1** was isolated from *Trichothecium roseum* along with the compound **6** and nerolidol **7**.[1] **1** was also found in the metabolites of *Gibberella fujikuroi*.[3]

1) S. Nozoe, M. Goi, N. Morisaki, *Tetr. Lett.*, 1293 (1970).
2) S. Nozoe, N. Morisaki, M. Goi, *Tetr. Lett.*, 3701 (1971).
3) B. E. Cross, R. E. Markwell, J. C. Stewart, *Tetr.*, **27**, 1663 (1971).

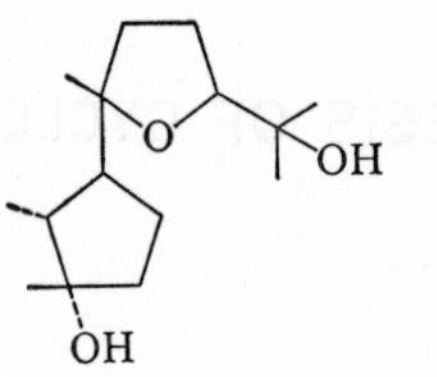

6 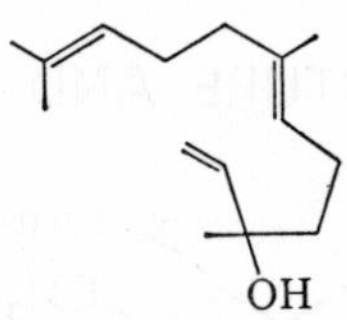   7

3.30   STRUCTURE OF GYRINIDONE

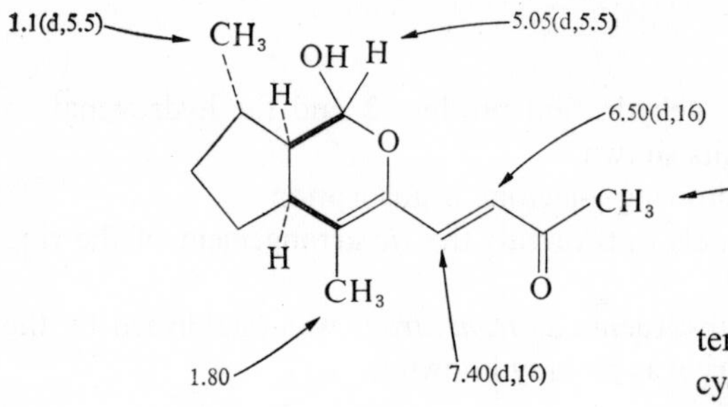

1.1(d,5.5) → CH₃   OH H ← 5.05(d,5.5)

6.50(d,16)

CH₃ ← 2.30(s)

1.80

7.40(d,16)

gyrinidone 1[1]

$\alpha_{400}: -69°$ $\alpha_{350}: -103°$ $\alpha_{255}: +138°$
ms: 236.141($M^+C_{14}H_{20}O_3$)
uv(EtOH): 318(ε 8,700), 234(ε 5,600)
ir: 2.9, 5.98, 6.16, 6.30
nmr in CDCl₃

Gyrinidone 1 is a cyclopentanoid nor-sesquiterpene whose structure is reminiscent of those of cyclopentano-monoterpenes such as iridodial and nepetalactone.

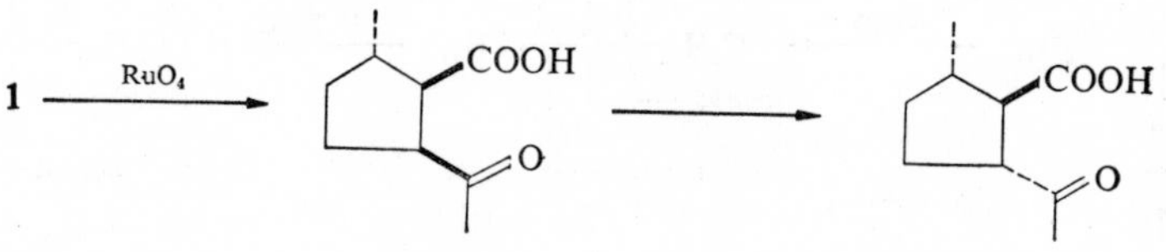

1 →(RuO₄)→ COOH ... O   →   COOH ... O

2   3

1) **3** was obtained by oxidation of **1** followed by epimerization.

2) The 2,4-DNP of **3** prepared from **1** exhibits the same sign of rotation as material obtained from nepetalactone, whose absolute configuration is known.

Remarks

Gyrinidone **1** was isolated from secretions of the whirligig beetle, *Dineutes discolor* Aube (Gyrinidae). The acyclic diketo-aldehyde **4** was found as a major component in the secretions of *Gyrinus* and *Dineutes* species.[2,3] An aldehyde, assumed to have the structure **5**, is also present together with **1**.[1]

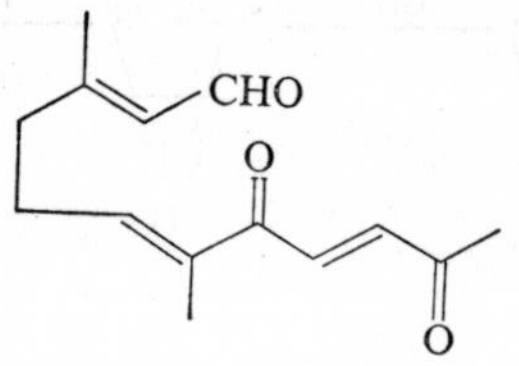

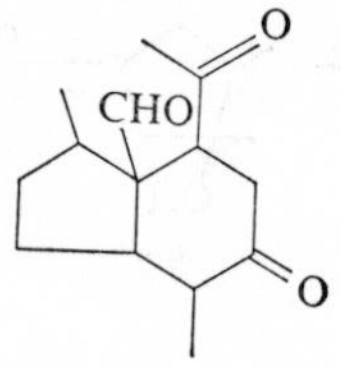

CHO   CHO   O
O   O

4   5

1) J. W. Wheeler, S. K. Oh, E. F. Benfield, S. E. Neff, *J. Am. Chem. Soc.*, **94**, 7589 (1972).
2) H. Schildknecht, H. Neumaier, B. Tauscher, *Ann. Chem.*, **756**, 155 (1972).
3) J. Meinwald, K. Opheim, T. Eisner, *Proc. Nat. Acad. Sci. U. S.*, **69**, 1208 (1972).

3.31 TRANSANNULAR CYCLIZATION OF CARYOPHYLLENE

Caryophyllene **1** is subject to a variety of skeletal rearrangements because of the nonrigidity of the medium-sized ring, as well as the high reactivity of the *trans* double bond in such a ring system. Of particular interest is the acid-catalyzed transannular ring closure to yield a mixture of caryolanol **2**, clovene **3** and neoclovene **4**.[1,3]

Formation of 2 and 3

The genesis of these products can be accounted for on the basis that caryolanol **2** and clovene **3** might arise from two different conformational isomers such as **1a** and **1b**, respectively.

caryophyllene **1**   caryolanol **2**   clovene **3**   neoclovene **4**

1a   **2**   **1b**   **3**

These stereochemical aspects have been verified by deuterium labelling experiments. The position and stereochemistry of the deuterium atom in **2** and **3**, which were prepared from **1** by the action of D_2SO_4, were rigorously established as the 9β position in **2** and the 9α position in **3** (H asterisked in diagrams).[2]

Formation of 4

The formation of neoclovene **4** is explicable in terms of a mechanism involving the ions **6**, **7** and **8**.[3] The feasibility of this was demonstrated by the synthesis[4] of **4** via the cation **6**, which was generated by acid from the corresponding carbinol **11**. In addition, the carbinol **11** was synthesized from the tricyclic ketone **10**, which was prepared by transannular cyclization of norcaryophyllene oxide **9**.

1) A. Abei, D. H. R. Barton, A. W. Burgstahler, A. S. Lindsey, *J. Chem. Soc.*,,4659 (1954).
2) A. Nickon, F. Y. Edamura, T. Iwadera, K. Matsuo, F. G. McGuire, J. S. Roberts, *J. Am. Chem. Soc.*, **90**, 4196 (1968); F. Y. Edamura, A. Nickon, *J. Org. Chem.*, 35, 1509 (1970).
3) W. Parker, R. A. Raphael, J. S. Roberts, *J. Chem. Soc.* C, 2634 (1969); prelim. commun., *Tetr. Lett.*, 2313 (1965).
4) T. F. W. McKillop, J. Martin, W. Parker, J. S. Roberts, J. R. Stevenson, *J. Chem. Soc.* C, 3375 (1971); prelim. commun., *Chem. Commun.*, 162 (1967).

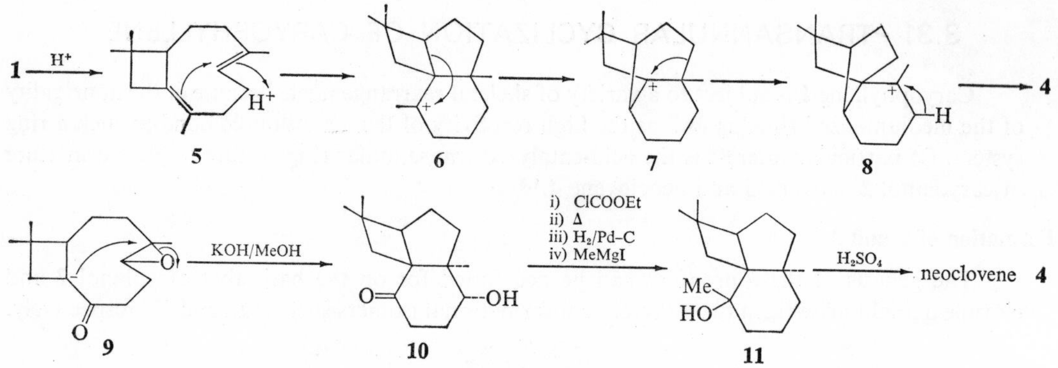

Further rearrangement of caryolanol

Products **13**,[5] **14**[6] and **15**[7] were assumed to be derived from the dienic intermediate **12**.[10]

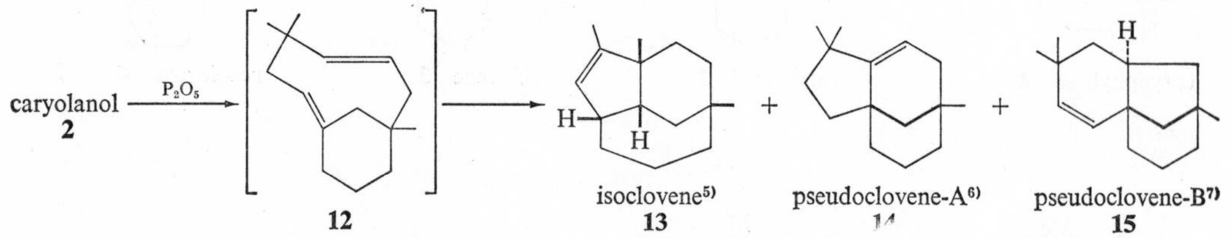

Rearrangement of isocaryophyllene[8,9]

Isocaryophyllene **16** does not rearrange into clovene **3** and caryolanol **2**, but affords neoclovene **4** and the tricyclic product **18** under acid catalysis. Acid treatment of caryophyllene hydrochloride **19** yields another tricyclic compound, **21**.

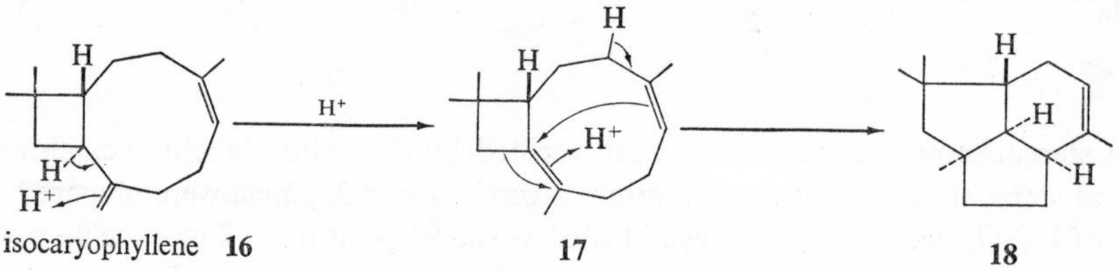

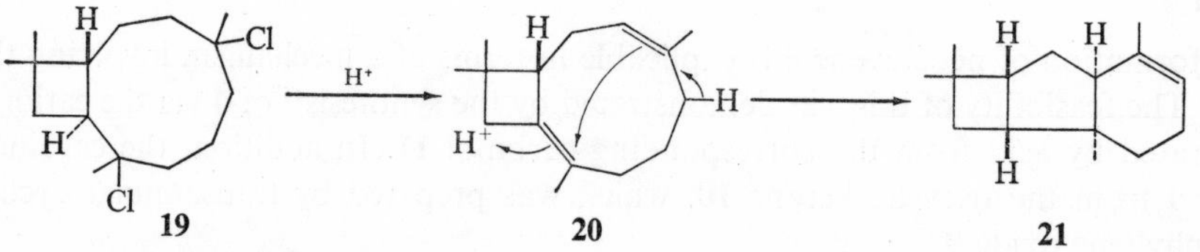

5) J. S. Clunie, J. M. Robertson, *Proc. Chem. Soc.*, 82 (1961); *J. Chem. Soc.*, 4382 (1961).
6) P. Doyle, G. Ferguson, D. M. Hawley, T. F. W. McKillop, J. Martin, W. Parker, *Chem. Commun.*, 1123 (1967).
7) R. I. Crane, C. Eck, W. Parker, A. B. Penrose, T. F. W. McKillop, D. M. Hawley, J. M. Robertson, *Chem. Commun.*, 385 (1972).
8) K. Gollnick, G. Schade, A. F. Camerone, C. Hannaway, J. M. Robertson, *Chem. Commun.*, 46 (1971).
9) K. Gollnick, G. Schade, A. F. Camerone, C. Hannaway, J. S. Roberts, J. M. Robertson, *Chem. Commun.*, 248 (1970).
10) See also D. Baines, C. Eck, W. Parker, *Tetr. Lett.*, 3933 (1973).

3.32 SYNTHESIS OF CARYOPHYLLENE AND ISOCARYOPHYLLENE

Caryophyllene **1** and isocaryophyllene **2** are composed of fused four- and nine-membered rings and contain *trans* and *cis* endocyclic double bonds, respectively. These compounds were synthesized through the tricyclic intermediates **10** and **12**, respectively.[1]

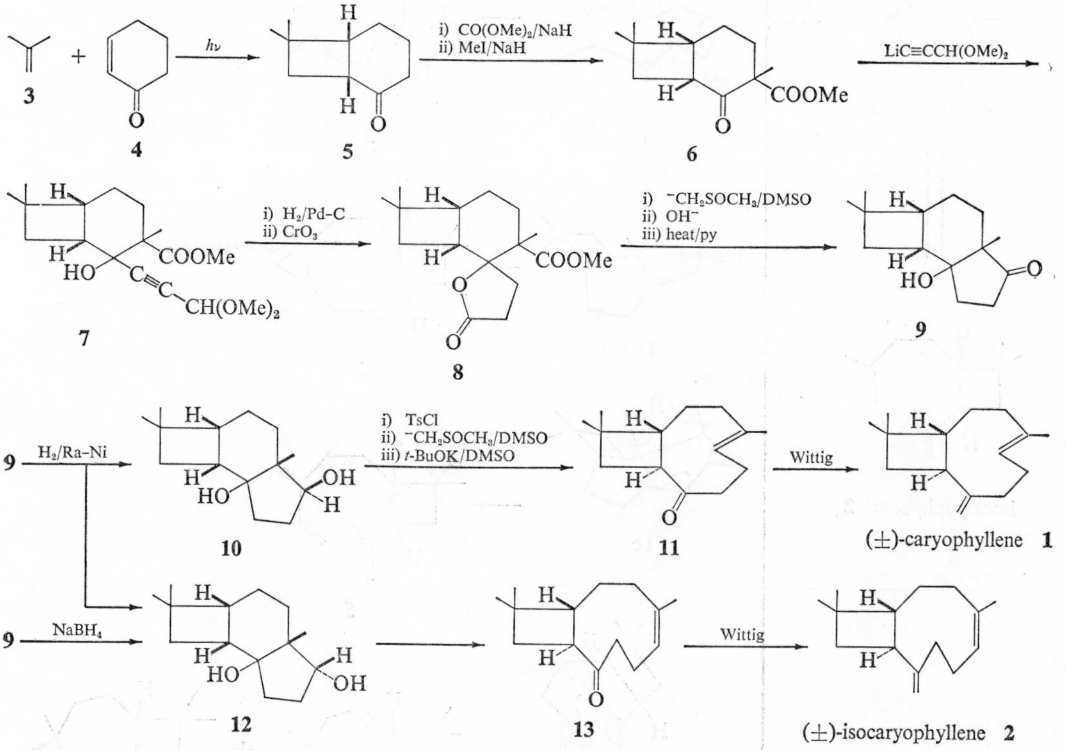

3+4→5: Photochemical cycloaddition of cyclohexenone and isobutylene predominantly produces the unstable *trans*-fused isomer, which readily isomerizes to *cis* on contact with base or activated alumina, or on heating at 200°. Wide applicability of this reaction in organic synthesis has since been demonstrated.

10→11, 12→13: The geometry of the endocyclic double bond can be controlled by means of the stereochemistry of the *sec*-OH groups of the precursors during the fragmentation of 1,3-diol monotosylates. When the angular methyl and the *vic* leaving group are *cis* (as in **10**), a *trans* double bond is formed. The 4,9-*cis* fused ketone initially formed in the above reaction scheme isomerized to the stable *trans*-fused compounds **11** and **13**.

1) E. J. Corey, R. B. Mitra, H. Uda, *J. Am. Chem. Soc.*, **86**, 485 (1964); *ibid.*, **85**, 362 (1963).

3.33 PHOTOREARRANGEMENT OF CARYOPHYLLENE

Irradiation of caryophyllene **1** at its π-π^* transitions affords a variety of isomeric products including the compounds **3–7**.[1]

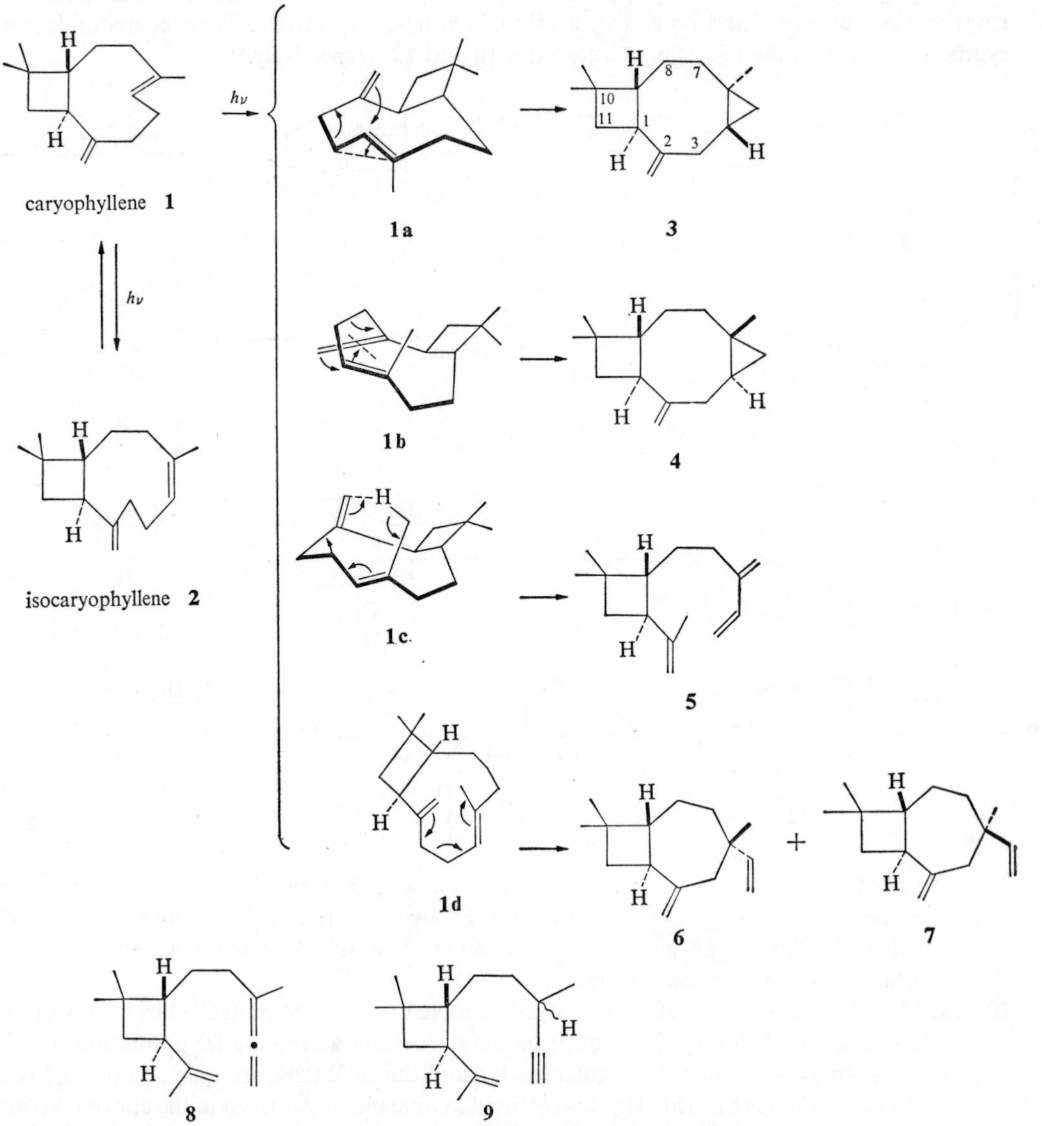

caryophyllene **1**

isocaryophyllene **2**

1a

3

1b

4

1c

5

1d

6 + **7**

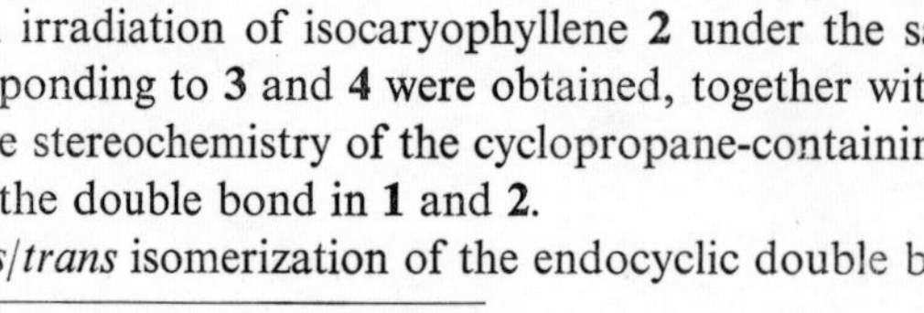

8   **9**

1) The products **3** and **4** are formed from the conformers **1a** and **1b**, respectively.
2) Compounds **6** and **7** are formed by Cope-type rearrangements.
3) On irradiation of isocaryophyllene **2** under the same conditions, C-4,C-6-*cis* isomers corresponding to **3** and **4** were obtained, together with the compounds **6–9**.
4) The stereochemistry of the cyclopropane-containing products is dependent on the geometry of the double bond in **1** and **2**.
5) *Cis/trans* isomerization of the endocyclic double bond also occurs in the reaction.

1) K. H. Schulte-Elte, G. Ohloff, *Helv. Chim. Acta.*, **54**, 379 (1971).

3.34   TRANSANNULAR CYCLIZATION OF HUMULENE

A *trans* double bond in a medium-sized ring is more reactive than an acyclic double bond, due to its steric strain. Humulene has three *trans* olefinic linkages in an 11-membered ring, and undergoes transannular cyclization, giving rise to the bromohydrin **2** on treatment with NBS.[1,2] The dehydration product **4** was converted stereospecifically into caryophyllene **5** and hydrocarbon **6** or back to humulene **1** by hydride reduction of the cyclopropylcarbinyl system in **4**.[1]

humulene **1**   **2**[2]   **3**

2 →(POCl$_3$/py)→ **4** →(LAH)→ caryophyllene **5** + **6** + **1**
                                       (30%)   (50%)   (10%)

Humulene oxide **7** also cyclized in the same way to give the diol **8**,[3] which was proved to be identical with tricyclohumuladiol recently isolated from hop oil.[4] Hydrolysis of the bromohydrin **2** yields the diol **8** with retention of configuration.

7   tricyclohumuladiol **8**

Remarks

Humulene oxide **7** was isolated from *Zingiber zermbet*[5] and tricyclohumuladiol **8** from hop, *Humulus luplus*.[4]

---

1) J. M. Greenwood, M. D. Solomon, J. K. Sutherland, A. Torre, *J. Chem. Soc.* C, 3004 (1968); prelim. form., J. M. Greenwood, J. K. Sutherland, A. Torre, *Chem. Commun.*, 410 (1965).
2) See also, F. H. Allen, D. Rogers, *Chem. Commun.*, 582 (1966); F. H. Allen, E. D. Brown, D. Rogers, J. K. Sutherland, *Chem. Commun.*, 1116 (1967).
3) M. A. McKervey, J. R. Wright, *Chem. Commun.*, 117 (1970).
4) Y. Naya, M. Kotake, *Bull. Chem. Soc. Japan*, **42**, 2405 (1969).
5) N. P. Damodaran, S. Dev. *Tetr.*, 24, 4123, 4133 (1968).

3.35 FORMATION AND SYNTHESIS OF
α-CARYOPHYLLENE ALCOHOL

α-caryophyllene alcohol 1[1,2]

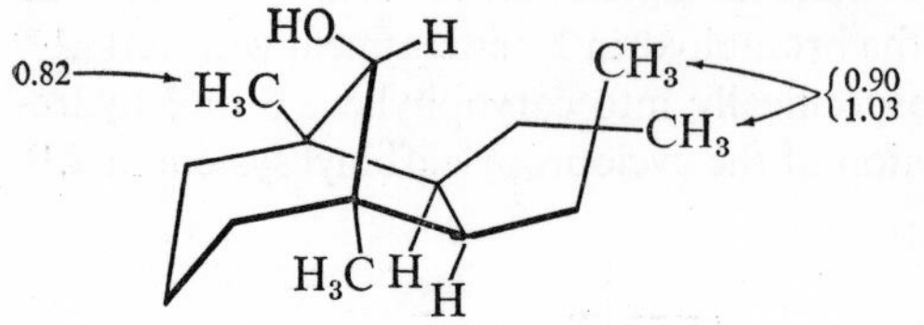

Acid-catalyzed hydration of humulene **2** affords the saturated tricyclic alcohol, α-caryophyllene alcohol **1**.

The genesis[1,3] of **1** has been interpreted in terms of prototropic rearrangement of **2** to the conjugated triene **3**, and protonation of **3** with subsequent formation of the ions **4**, **5** and **6**. The feasibility of this mechanism was demonstrated by the synthesis of **1** from the tertiary alcohol **10**, through the ions **5** and **6** (see below).[3,4]

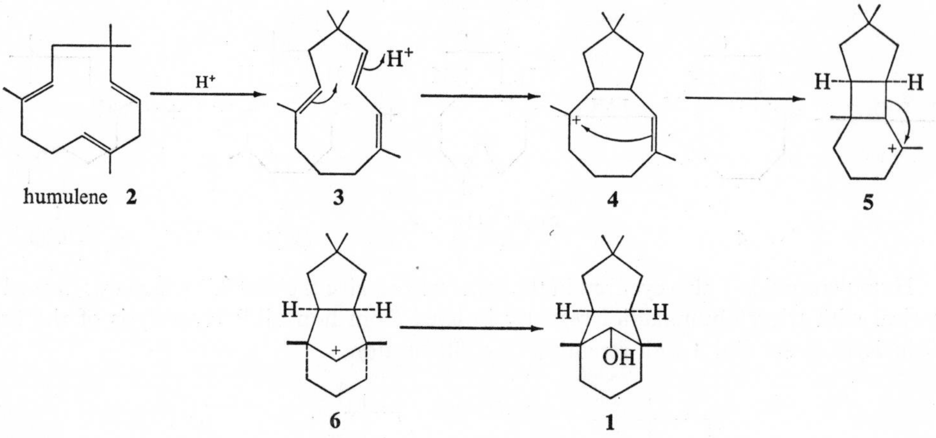

humulene **2**   **3**   **4**   **5**

6   **1**

Synthesis[3]

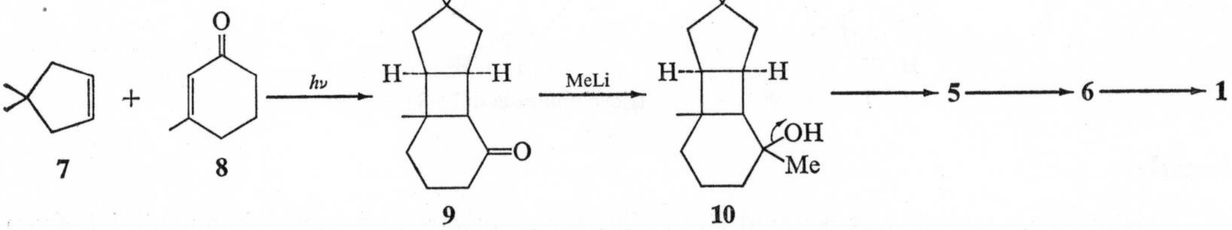

7   **8**   **9**   **10**   **5** ——— **6** ——— **1**

Remarks

The name α-caryophyllene alcohol was given to the alcohol since it was first obtained from commercial "caryophyllene" (which contains humulene). To avoid confusion, the name apollan-11-ol has been proposed for this compound.[1]

1) A. Nickon, T. Iwadare, F. J. McGuire, J. R. Mahajan, S. A. Narang, *J. Am. Chem. Soc.*, **92**, 1688 (1970), and references therein; A Nickon, F. J. McGuire, J. R. Mahajan, B. Umezawa, S. A. Narang, *ibid.*, **86**, 1437 (1964).
2) K. W. Gemmell, W. Parker, J. S. Roberts, G. A. Sim, *J. Am. Chem. Soc.*, **86**, 1438 (1964).
3) E. J. Corey, S. Nozoe, *J. Am. Chem. Soc.*, **87**, 5733 (1965); prelim. commun., *ibid.*, **86**, 1652 (1964).
4) See also Y. Naya, Y. Hirose, *Chem. Lett.*, 727, 133 (1973).

3.36   SYNTHESIS OF HUMULENE

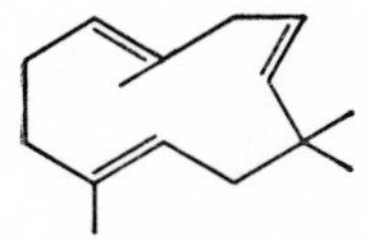

humulene   **1**

Eleven-membered carbocycles containing three *trans* olefinic linkages have been synthesized by the following reaction sequence, which involves a new type of cyclization, using $Ni(CO)_4$ to convert the bisbromide **5** into the macrocyclic compound **6**.[1]

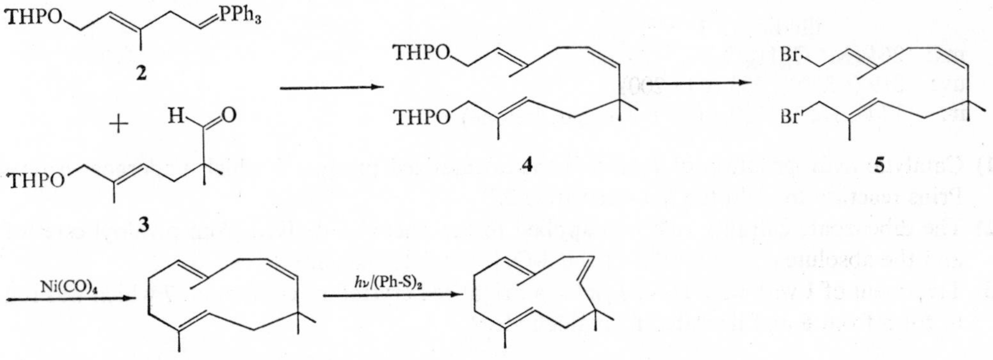

THPO   PPh₃ ... (scheme)

2

+   H   O

THPO

3

THPO
THPO

4

Br
Br

5

$Ni(CO)_4$ $h\nu/(Ph-S)_2$

6   **1**

6→1: The central *cis* double bond in **6** is isomerized to *trans* by irradiation in the presence of diphenyldisulfide.[1]

5→6: The cyclization of a bisallylic α,ω-dibromide using nickel carbonyl, as in this reaction, provides an efficient synthetic route for the construction of C_9 to C_{18} carbocycles containing 1,5-diene moieties,[2] as shown in the examples below.

13→14: This reagent can also be used for the synthesis of macrocyclic lactones such as **14**.[3]

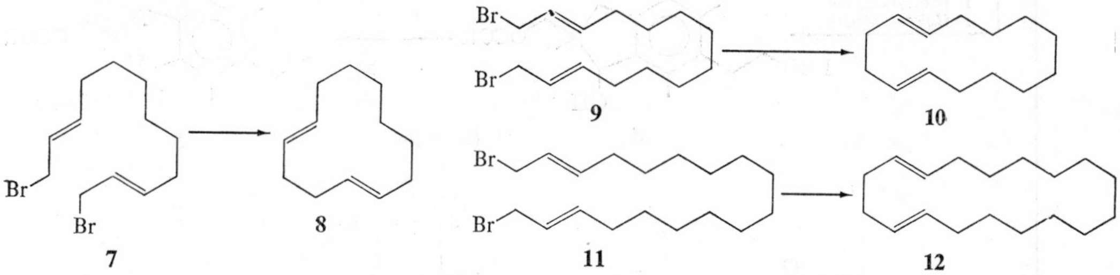

Br
Br
7

8

Br
Br
9

10

Br
Br
11

12

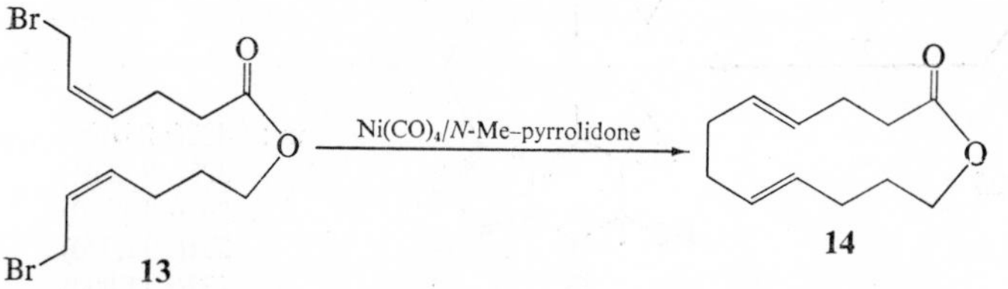

Br   O
O
13

$Ni(CO)_4/N$-Me-pyrrolidone

O
O
14

1) E. J. Corey, E. Hamanaka, *J. Am. Chem. Soc.*, **89**, 2758 (1967); *ibid.*, **86**, 1641 (1964).
2) E. J. Corey, E. K. W. Wat, *J. Am. Chem. Soc.*, **89**, 2757 (1967); E. J. Corey, M. F. Semmelhack, *Tetr. Lett.*, 6237 (1966).
3) E. J. Corey, H. A. Kirst, *J. Am. Chem. Soc.*, **94**, 668 (1972).

3.37 STRUCTURE OF ILLUDIN-S

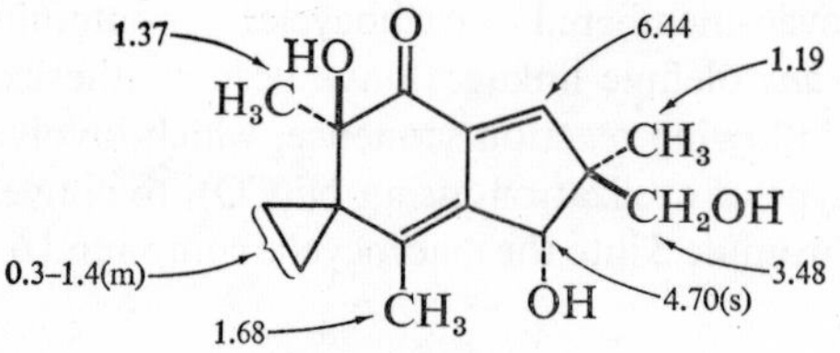

illudin-S **1**[1-3,12]

ms: 264 (M⁺ C₁₅H₂₀O₄)
ms: 264 (M^+ $C_{15}H_{20}O_4$)
uv: 319 (ϵ 3600), 233 (ϵ 13,200)
ir: 1706, 1652, 1610, 1600 (α,β-unsat. ketone)

1) Catalytic hydrogenation of **1** affords the aromatized product **2** which undergoes a reverse Prins reaction to yield the nor-derivative **3**.[1]

2) The dibenzoate chirality rule was applied to the phenol **5** derived from pivaloyl ester of **1**, and the absolute configuration of the 3-OH was determined.[4]

3) Treatment of **1** with cold H_2SO_4 yields a bright red crystalline compound **7** which was found to form from **6** and liberated formaldehyde.[5]

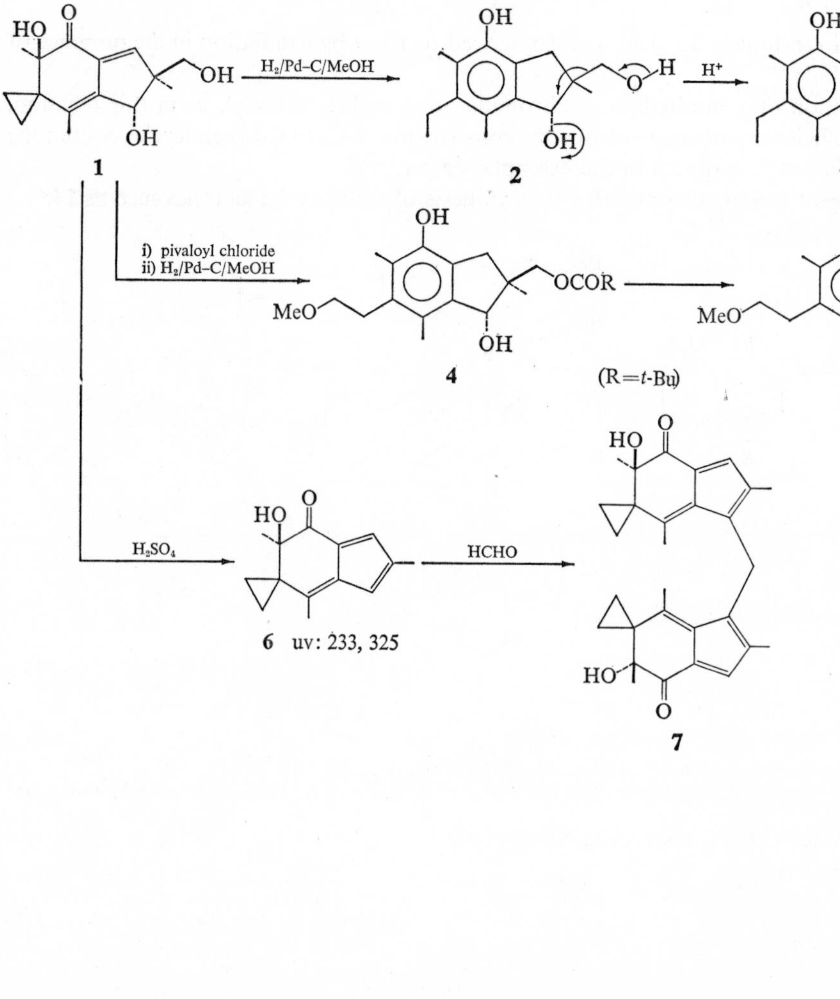

uv: 422(ϵ 2640)
     328(ϵ 7400)
     269(ϵ 10,500)
     251(ϵ 11,700)
     237(ϵ 11,800)

Remarks

Illudin-S (lampterol) **1** is a toxic compound isolated from the bioluminescent mushroom *Clitocybe illudens*[1] (or *Lampteromyces japonicus*)[2,3] and shows antibacterial activity. Illudin-M **8**[1], dihydroilludin-S **9**[6] and -M[7], illudol, illudalic acid **10**[8] and illudinin **11** were also isolated from these mushrooms.

A number of 1-indanone derivatives named pterosins (**12**, R = OH, O-glu, Cl or acyl) have been isolated from brackens, *Pteridium aquilinum* and *Hypolepis punctata*.[9] The secoilludoid hypacrone **13** was obtained from *Hypolepis punctata*[10]

 1) T. C. McMorris, M. Anchel, *J. Am. Chem. Soc.*, **87**, 1594 (1965); *ibid.*, **85**, 831 (1963).
 2) K. Nakanishi, M. Ohashi, M. Tada, Y. Yamada, *Tetr.*, **21**, 1231 (1965).
 3) T. Matsumoto, H. Shirahama, A. Ichihara, Y. Fukuoka, Y. Takahashi, Y. Mori, M. Watanabe, *Tetr.*, **21**, 2671 (1965).
 4) N. Harada, K. Nakanishi, *Chem. Commun.*, 310 (1970).
 5) S. M. Weinreb. T. C. McMorris, M. Anchel, *Tetr. Lett.*, 3489 (1971).
 6) A. Ichihara, H. Shirahama, T. Matsumoto, *Tetr. Lett.*, 3965 (1969).
 7) P. Singh, M. S. R. Nair, T. C. McMorris, M. Anchel, *Phytochemistry.*, **10**, 2229 (1971).
 8) M. S. R. Nair, H. Takeshita, T. C. McMorris, M. Anchel, *J. Org. Chem.*, **34**, 240 (1969).
 9) K. Yoshihara, M. Fukuoka, M. Kuroyanagi, S. Natori, *Chem. Pharm. Bull.*, **19**, 1491 (1971); *ibid.*, **20**, 426 (1972); H. Hikino, T. Takahashi, T. Takemoto, *ibid.*, **19**, 2424 (1971); *ibid.*, **20**, 210 (1972); Y. Hayashi, M. Nishizawa, S. Harita, T. Sakan, *Chem. Lett.*, 375 (1972).
10) Y. Hayashi, M. Nishizawa, T. Sakan, *Chem. Lett.*, 63 (1973).
11) See J. R. Hanson, T. Marten, *Chem. Commun.*, 171 (1973). (biosynthesis of illudin-M).
12) A. Furusaki, H. Shirahama, T. Matsumoto, *Chem. Lett.*, 1293 (1973). (x-ray studies of **1**)

3.38   SYNTHESIS OF ILLUDIN-M

The total stereospecific synthesis of (±)-illudin-M has been achieved via the following reaction sequence.[1-3]

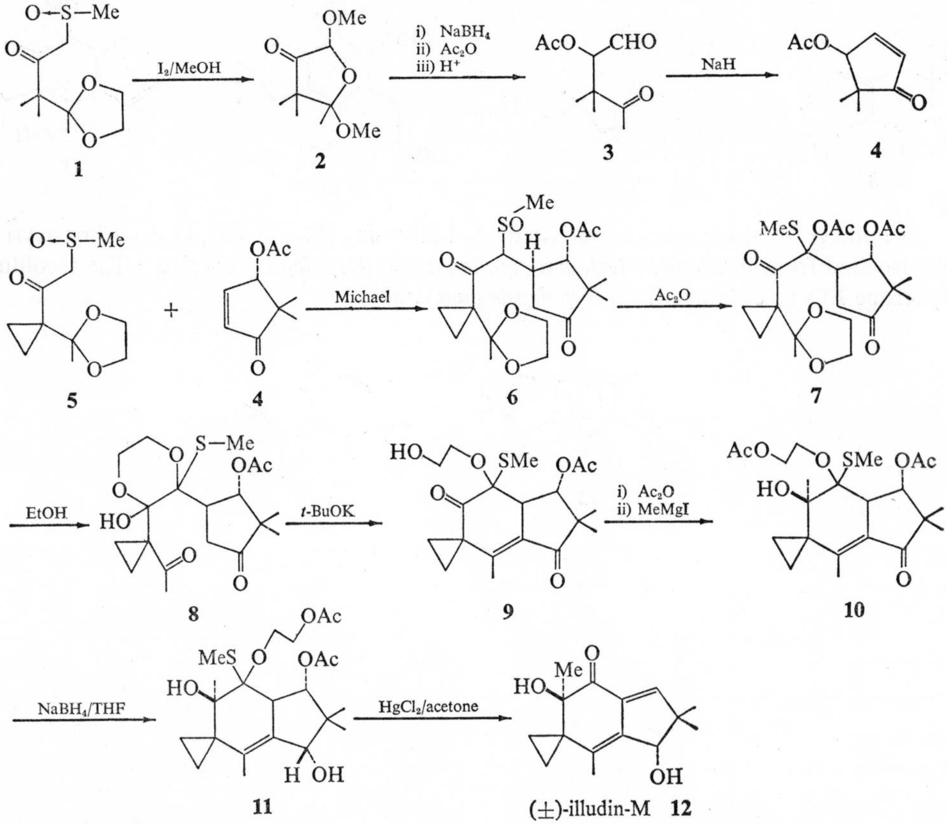

1→2: β-Ketosulfoxides give, in general, α-ketoacetals on treatment with I_2 in MeOH.

5,4→6: Michael addition of the carbanion derived from the β-ketosulfoxide **5** proceeds stereoselectively[2] to afford *trans* **6**.

6→7: The sulfoxide rearranges, on treatment with acid anhydride, into an acyloxy-sulfide. This is known as the Pummerer rearrangement.

7→8: Transketalization.

9→10: The Grignard reagent attacks the six-membered ring ketone stereospecifically from the α-side.

10→11: Hydride attack occurs from the β-side due to the large steric hindrance on the α-side.

1) T. Matsumoto, H. Shirahama, A. Ichihara, H. Shin, S. Kagawa, F. Sakan, S. Matsumoto, S. Nishida, *J. Am. Chem. Soc.*, **90**, 3280 (1968); cf. T. Matsumoto *et al.*, *Tetr. Lett.*, 1171 (1970).
2) A. Ichihara, J. Morita, K. Kobayashi, S. Kagawa, H. Shirahama. T. Matsumoto, *Tetr.*, **26**, 1331 (1970).
3) For analogous synthesis of illudin-S, see T. Matsumoto, H. Shirahama, A. Ichihara, H. Shin, S. Kagawa, F. Sakan, K. Miyano, *Tetr. Lett.*, 2049 (1971).

3.39 SYNTHESIS OF ILLUDOL

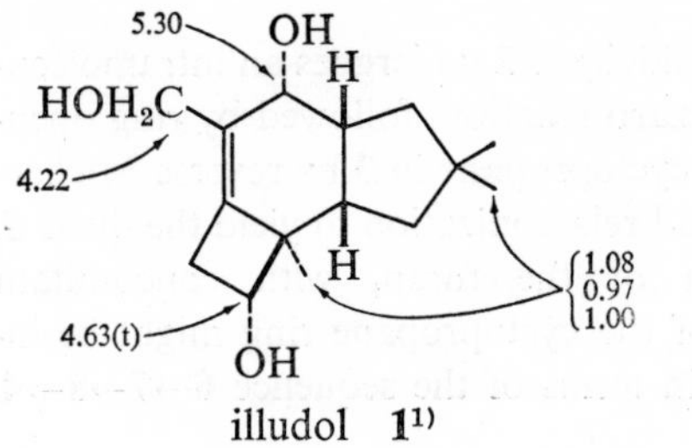

5.30 — OH H
HOH₂C
4.22 —
{1.08
0.97
1.00
4.63(t) — OH
H

illudol 1[1]

Illudol **1** was synthesized by a reaction sequence involving a stereo- and position-specific photochemical cycloaddition reaction as the key step, as shown below.[2]

mp: 130–132° α_D(AcOH): −116°
uv: 207 (ϵ 9,400)
ms: 234 (M-H₂O), 216 (M-2H₂O),
208 (M-44, CH₂CHOH)
nmr in CDCl₃

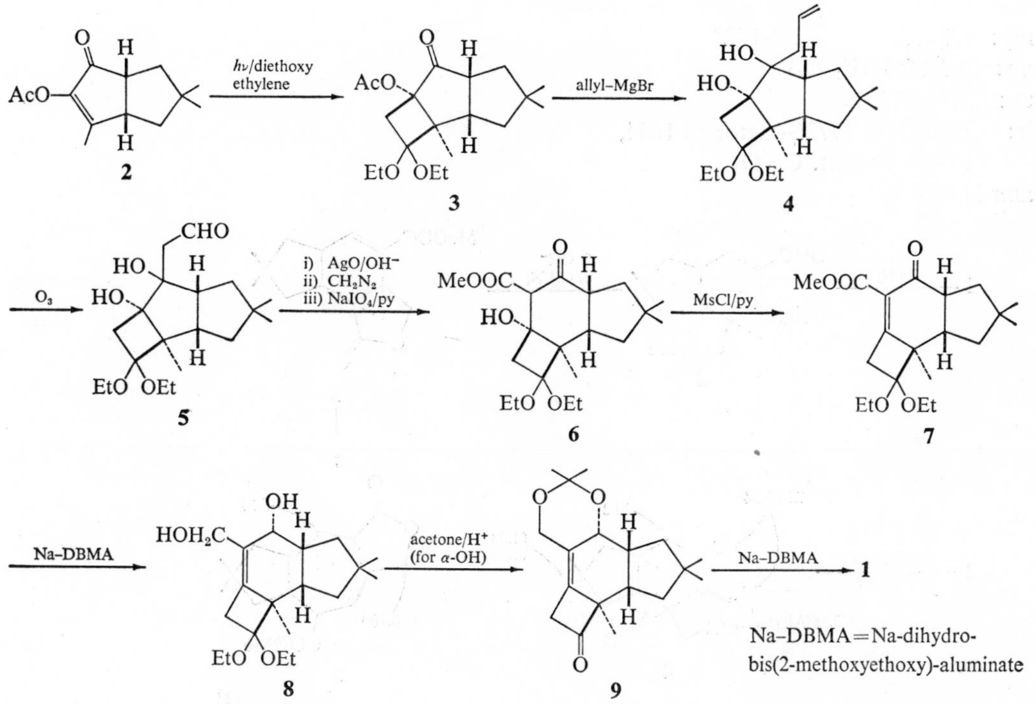

Remarks

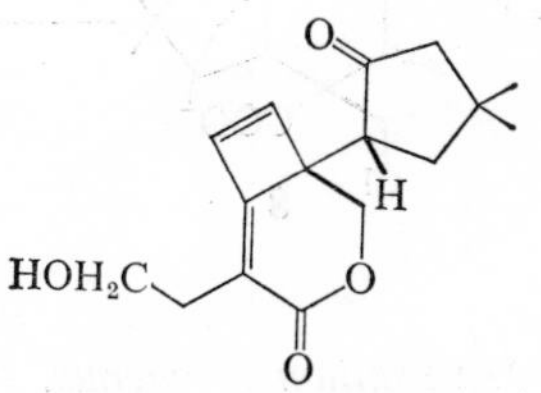

10

Illudol **1** is a physiologically inactive metabolite of the Basidiomycete *Clitocybe illudens*. Its carbon skeleton, named protoilludane, suggests a biogenetic relationship with illudin-M, marasmic acid and other similar sesquiterpenes. The phytotoxic substance fomannosin **10**,[3] isolated from *Fomes annosus*, can be considered to be derived from a compound having the protoilludane skeleton.

1) T. C. McMorris, M. S. R. Nair, M. Anchel, *J. Am. Chem. Soc.*, **89**, 4562 (1967); T. C. McMorris, M. S. R. Nair, I. Singh, M. Anchel, *Phytochemistry*, **10**, 1611 (1967); P. D. Cradwick, G. A. Sim, *Chem. Commun.*, 431 (1971).
2) T. Matsumoto, K. Miyano, S. Kagawa, S. Yu, J. Ogawa, A. Ichihara, *Tetr. Lett.*, 3521 (1971).
3) J. A. Kepler, M. E. Wall, J. E. Mason, C. Basset, A. T. McPhail, G. A. Sim, *J. Am. Chem. Soc.*, **89**, 1260 (1967).

3.40   STRUCTURE OF MARASMIC ACID

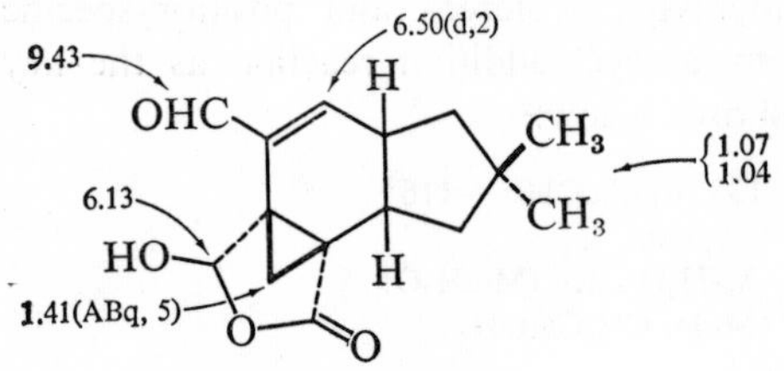

OHC 9.43

6.50(d,2)

H

CH₃ {1.07
      {1.04

HO 6.13

H

1.41(ABq, 5)

O   O

marasmic acid   **1**[1]

1) The 1,2-dialdehyde **2** undergoes an intramolecular Cannizzaro reaction, followed by ring opening of the cyclopropane in **3** by reverse Michael reaction and relactonization to yield the diene **5**.
2) Formation of the furan, with concomitant cleavage of the cyclopropane ring might be interpreted in terms of the sequence **6→7→8→4**.

mp: 173–174° α_D: +182°
ms: 262(M⁺C₁₅H₂₀O₄)
uv: 241(ε 9,700)
ir: 3380(OH), 1773(γ-lactone), 1684,
    1631(α,β-unsat. C=O)
nmr in CDCl₃

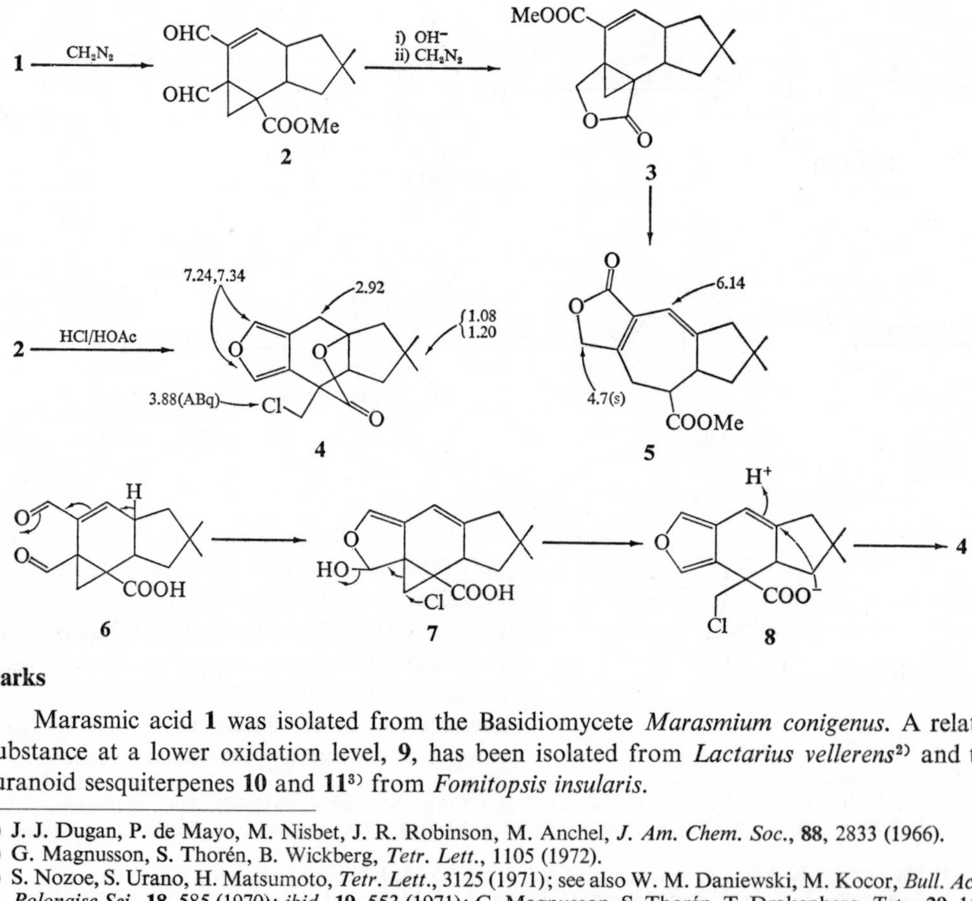

Remarks

Marasmic acid **1** was isolated from the Basidiomycete *Marasmium conigenus*. A related substance at a lower oxidation level, **9**, has been isolated from *Lactarius vellerens*[2] and the furanoid sesquiterpenes **10** and **11**[3] from *Fomitopsis insularis*.

1) J. J. Dugan, P. de Mayo, M. Nisbet, J. R. Robinson, M. Anchel, *J. Am. Chem. Soc.*, **88**, 2833 (1966).
2) G. Magnusson, S. Thorén, B. Wickberg, *Tetr. Lett.*, 1105 (1972).
3) S. Nozoe, S. Urano, H. Matsumoto, *Tetr. Lett.*, 3125 (1971); see also W. M. Daniewski, M. Kocor, *Bull. Acad. Polonaise Sci.*, **18**, 585 (1970); *ibid.*, **19**, 553 (1971); G. Magnusson, S. Thorén, T. Drakenberg, *Tetr.*, **29**, 1621 (1973).

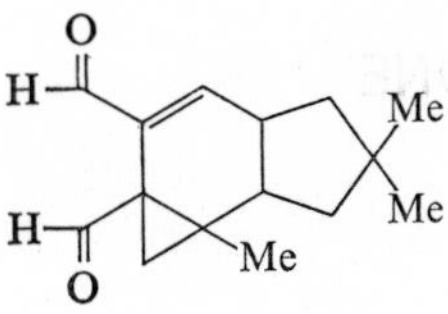

9 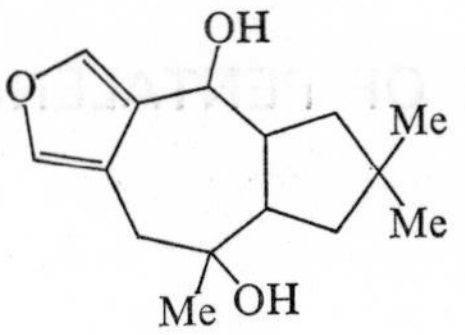   10 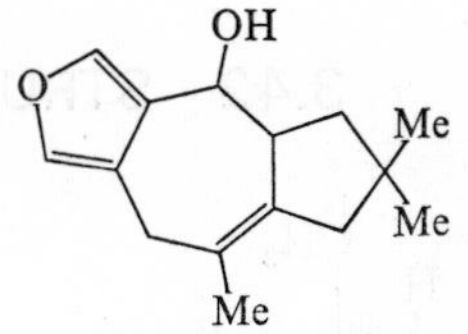   11

3.41 HIRSUTIC ACID AND CORIOLIN

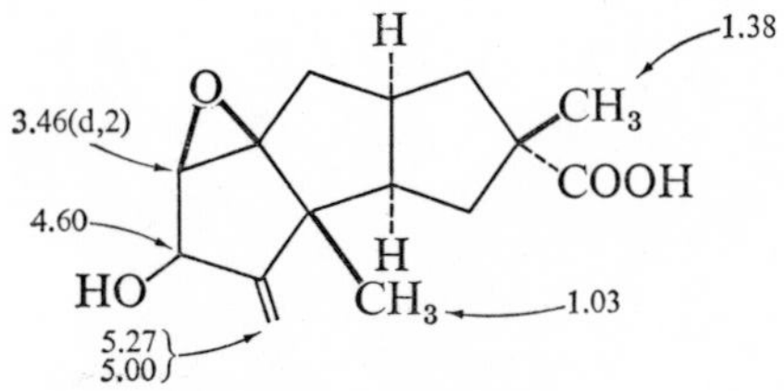

hirsutic acid **1**[1]

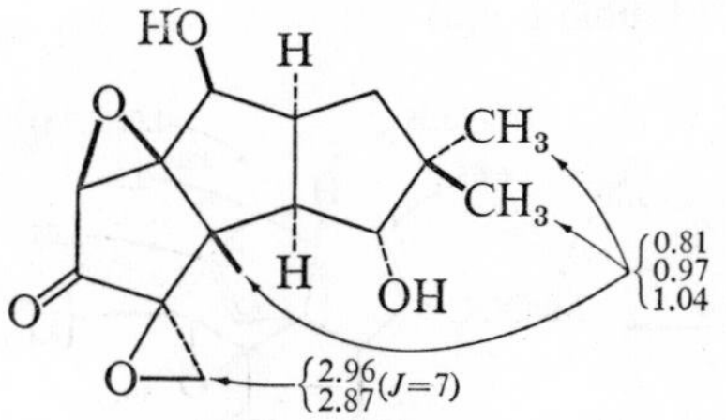

mp: 179–180° α_D: +116°
ms: 278 (M⁺ for methyl ester)
uv: end absorption
ir: 3520(OH), 3200, 1700(COOH), 1655,
     890(C=CH₂)

coriolin **2**[2]

mp: 175°
ms: 280(M⁺C₁₅H₂₀O₅)
ir: 3450–3300(OH), 1382, 1369 (*gem*-
     di-Me)
nmr in d₆-DMSO

Remarks

Hirsutic acid **1** was isolated from the Basidiomycete *Stereum hirsutum* and coriolin **2** from *Coriolus consor*, together with the congeners[3] coriolin B and C.

1 and **2** possess the same carbon skeleton, consisting of three fused five-membered rings, which were assumed to be derived biogenetically from farnesyl pyrophosphate by the following skeletal rearrangement.

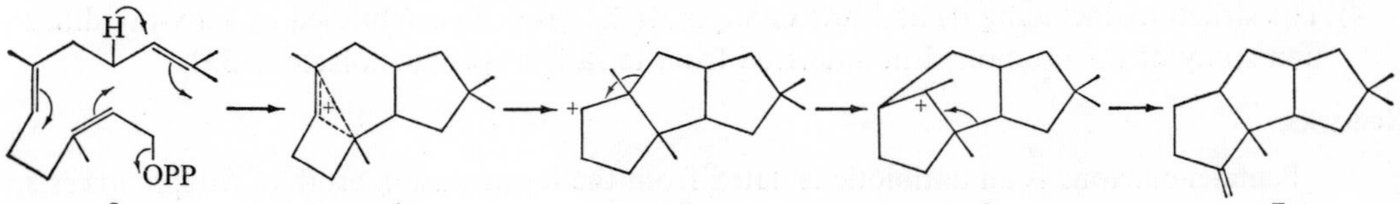

3   4   5   6   7

1) F. W. Comer, F. McCapra, I. H. Qureshi, A. I. Scott, *Tetr.*, **23**, 4761 (1967).
2) S. Takahashi, H. Naganawa, H. Iinuma, T. Takita, K. Maeda, H. Umezawa, *Tetr. Lett.*, 1955 (1971).
3) S. Takahashi, H. Iinuma, T. Takita, K. Maeda, H. Umezawa, *Tetr. Lett.*, 4663 (1969); *ibid.*, 1637 (1970).

3.42 STRUCTURE OF PENTALENOLACTONE

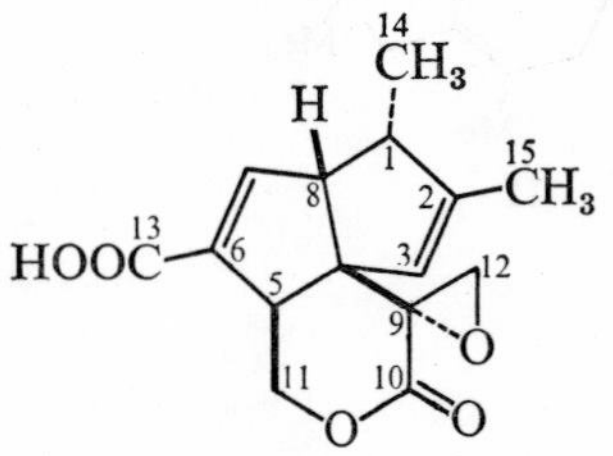

pentalenolactone **1**[1,2]

mp: 61–2° α_D(MeOH): −172°
ms: 276($M^+C_{15}H_{16}O_5$)
uv: 218.5 (ε 8625)
ir: 1765(lactone), 1695(COOH),
     1635(double bond)

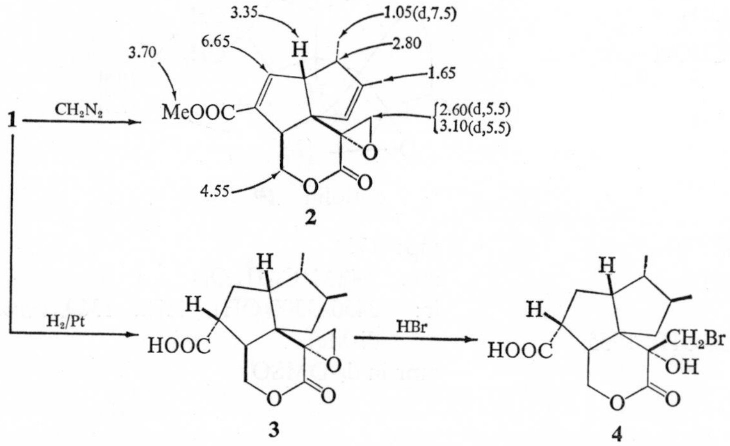

1) In the nmr spectrum of **2**, noe was observed between 5-H and 3-H, 14-H and 7-H, and 1-H and 7-H.[1]

2) The doublets observed at 2.60 and 3.10 ($J = 5.5$) in **2**, previously assigned to *vic* protons on the endocyclic epoxide, can now be assigned to *gem* protons on the exocyclic oxirane ring.[1,2]

3) A new methyl group appeared on LAH reduction of **3**, supporting the presence of an exocyclic oxirane ring.[2]

4) The structure, including its absolute configuration, has been established by an x-ray diffraction study of the bromohydrin **4** derived from tetrahydropentalenolactone **3**.[2]

Remarks

Pentalenolactone is an antibiotic isolated from the fermentation broth of *Streptomyces* sp.

1) S. Takeuchi, Y. Ogawa, H. Yonehara, *Tetr. Lett.*, 2737 (1969).
2) D. G. Martin, G. Slomp, S. Mizsak, D. J. Duchamp, C. G. Chidester, *Tetr. Lett.*, 4901 (1970).

3.43 STRUCTURE OF GERMACRENES

Several kinds of sesquiterpenes having a ten-membered ring structure have recently been isolated. This group consist of cyclodecadienes (or cyclodecatriene), such as **1–6**, and more highly oxygenated and/or esterified compounds.

germacrene C **1**   hedycaryol **2**   bicyclogermacrene **3**

costunolide **4**   furanodiene **5**   laurenobiolide **6**

1) In the nmr of this group of compounds, one of the olefinic methyl signals is shielded by the transannular π-electrons of the other double bond.

2) The conformation of compounds of this type in solution has been well studied by application of the nuclear Overhauser effect.

3) The acid-catalyzed cyclization of compounds of this type and the corresponding oxides yields bicyclic compounds having eudesmane and guaiane skeletons; this may have a biogenetic implication.

4) A Cope rearrangement of these compounds gives elemane-type sesquiterpenes.

Remarks

Compounds **1–6** were isolated from *Kadzura japonica*,[1] *Hedycarya angustifolia*,[2] *Citrus junos*,[3] *Saussurea lappa*,[4] *Curcuma zedoaria*[5] and *Laurus nobilis*,[6] respectively.

1) K. Morikawa, Y. Hirose, *Tetr. Lett.*, 1799 (1969); K. Yoshihara, Y. Ohta, T. Sakai, Y. Hirose, *ibid.*, 2263 (1969).
2) R. V. H. Jones, M. D. Sutherland, *Chem. Commun.*, 1229 (1968).
3) K. Nishimura, N. Shinoda, Y. Hirose, *Tetr. Lett.*, 3097 (1969).
4) A. S. Rao, G. R. Kelkar, S. C. Bhattacharyya, *Tetr.*, **9**, 275 (1960); cf. M. Suchy, V. Herout, F. Sorm, *Coll. Czech. Chem. Commun.*, **31**, 2899 (1966).
5) K. Takeda, H. Hikino, K. Agatsuma, T. Takemoto, *Tetr. Lett.*, 931 (1968).
6) H. Tada, K. Takeda, *Chem. Commun.* 1391, (1971).

3.44 CONFORMATION OF GERMACRADIENES
AND RELATED COMPOUNDS

The conformation of germacradiene compounds has been investigated by several methods.

X-ray diffraction studies

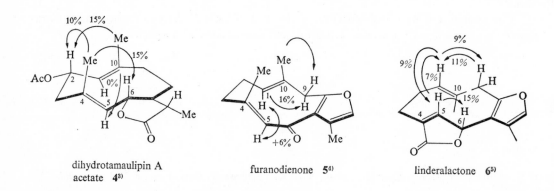

germacrene B **1**   shiromodiol **2**   elephantol **3**

The crystal structures of the silver nitrate adduct of germacrene B **1**[1] and heavy atom derivatives of shiromodiol **2**[2] and elephantol **3**[2] were determined by x-ray diffraction studies. It has been shown that the planes of both *endo* double bonds in **1** are approximately perpendicular to that of the 10-membered ring, in which the 14-Me and 15-Me are in a *syn* relationship.

Shiromodiol **2** and elephantol **3** also possess *syn* substituents at C-4 and C-10, though all substituents except at C-7 are antipodal to each other in the two compounds.

Nmr studies

The conformation of 10-membered rings in solution can be determined by means of measurements of the nuclear Overhauser effect of substituents in the nmr spectra. The compounds dihydrotamaulipin A **4**,[3] furanodienone **5**,[4] and linderalactone **6**[5] were shown to have the depicted conformations by noe techniques, for example.

dihydrotamaulipin A
acetate **4**[3]   furanodienone **5**[4]   linderalactone **6**[5]

1) F. H. Allen, D. Rogers, *Chem. Commun.*, 588 (1967).
2) R. J. McClure, G. A. Sim, P. Goggon, A. T. McPhail, *Chem. Commun.*, 128 (1970).
3) N. S. Bhacca, N. H. Fischer, *Chem. Commun.*, 68 (1969).
4) H. Hikino, C. Konno, T. Takamoto, K. Tori, M. Ohtsuru, I. Horibe, *Chem. Commun.*, 662 (1969).
5) K. Takeda, K. Tori, I. Horibe, M. Ohtsuru, H. Minato, *J. Chem. Soc.* C, 2697 (1970).

An increase in the integrated intensity of the 2-H and 6-H signals caused by irradiation of the 4-Me signal, as well as enhancement of the 2-H signal by irradiation of the 10-Me signal, indicate that the 10-membered ring in **4** adopts a conformation where the 4-Me, 10-Me, 2-H and 6-H are in the same direction, as illustrated.

The conformations of furanodienone[4] and linderalactone[5] were also determined, as illustrated in **5** and **6**, by the same technique.[10]

Existence of two conformers and ring inversion

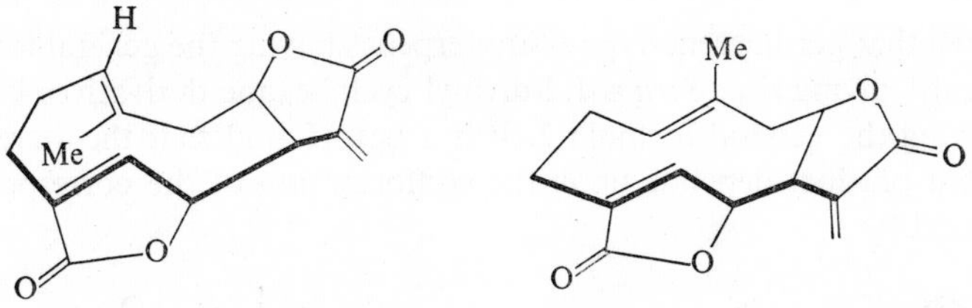

isabelin **7**

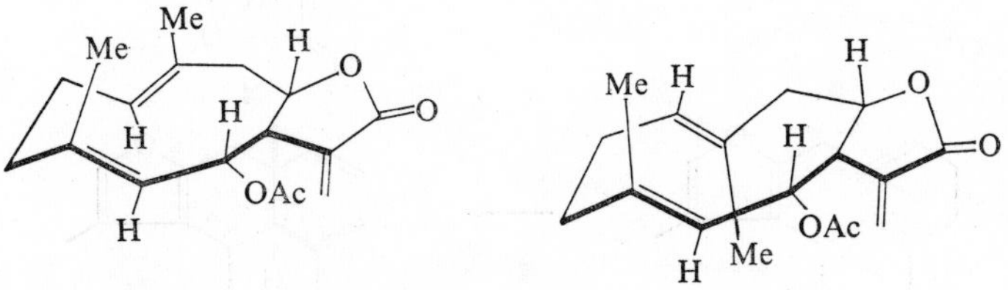

laurenobiolide **8**

Isabelin **7**[6] and laurenobiolide **8**[7] were found to exist in two conformational isomers at low temperature, as illustrated above. In the spectra of **8**, rapid inversion of the 10-membered ring was observed at higher temperature (distinguishable by variable-temperature nmr studies).

The conformations of neolideralactone having *cis* (C-1:C-10) and *trans* (C-4:C-5) double bonds, and sericenine having *trans* (C-1:C-10) and *cis* (C-4:C-5) double bonds were also studied by intramolecular noe and variable-temperature investigations.[8]

Urospermal A and B, which are conformational isomers, have been isolated as crystalline compounds and shown to be interconvertible in solution.[9] Each conformer is considered to be stabilized by intramolecular hydrogen bonding.

6) K. Tori, I. Horibe, Y. Yoshioka, T. J. Mabry, *J. Chem. Soc.* B, 1084 (1971).
7) K. Tori, I. Horibe, K. Kuriyama, H. Tada, K. Takeda, *Chem. Commun.*, 1393 (1971).
8) K. Tori, I. Horibe, H. Minato, K. Takeda, *Tetr. Lett.*, 4355 (1971).
9) R. K. Bentley, J. G. St. C. Buchanan, T. G. Halsall, V. Thaller, *Chem. Commun.*, 435 (1970).
10) For recent studies on the conformation of the germacradiene system, see K. Nishimura, I. Horibe, K. Tori, *Tetr.*, **29**, 271 (1973); I. Horibe, K. Tori, K. Takeda, T. Ogino, *Tetr. Lett.*, 735 (1973).

3.45   COPE REARRANGEMENT OF LINDERALACTONE
AND RELATED COMPOUNDS

1   **2**

It is well-known that germacrene-type sesquiterpenes having the general formula **1** undergo Cope rearrangement,[1] giving rise to *trans*-1,2-divinyl cyclohexane derivatives known as elemane sesquiterpenes, having the general formula **2**. It has been found that the stereochemistry of a Cope rearrangement product depends upon the conformation of the corresponding ten-membered ring compound.[2]

3   **4**

5   **6**

7   **8**

1) On heating linderalactone **3** at 160° a mixture of **3** and isolinderalactone **4** was obtained in the ratio 2:3. When isolinderalactone **4** was heated under the same conditions, the same equilibrated mixture was obtained.[2]

2) The absolute configuration of **4** was found to be antipodal to that of common elemane-type sesquiterpenes.[3]

3) On the other hand, the diol **5**, prepared by reductive cleavage of the lactone ring of **3**, and the ether **7** yield products **6** and **8**, respectively, on Cope rearrangement, with absolute configurations of the elemane type.[2]

4) These results were explained in terms of conformational differences in the original ten-membered ring compounds. Conformational studies of **3**, **5** and **7** by nmr techniques support this view (see "conformation of germacradienes").

1) For mechanistic explanation of the rearrangement, see R. B. Woodward, R. Hoffmann, *J. Am. Chem. Soc.*, **87**, 395 (1965); G. B. Gill, *Quart. Rev.*, **22**, 386 (1968).

2) K. Takeda, I. Horibe, H. Minato, *J. Chem. Soc.* C, 1142 (1970); K. Takeda, K. Tori, I. Horibe, M. Ohtsuru, H. Minato, *ibid.*, 2697 (1970); K. Takeda, *Pure Appl. Chem.*, **21**, 181 (1970).

3) K. Takeda, I. Horibe, M. Teraoka, H. Minato, *J. Chem. Soc.* C, 1491 (1969); see also K. Takeda, I. Horibe, H. Minato, *ibid.*, 2212 (1973).

3.46 TRANSANNULAR CYCLIZATION OF THE GERMACRADIENE SYSTEM

Transannular cyclization of the 1,5-cyclodecadiene system was postulated in the biogenesis of the eudesmane- and guaiane-type sesquiterpenoids.[1] In fact, germacradiene (or -triene) derivatives and the corresponding epoxides are readily converted into bicyclic compounds under acid-catalyzed or electrophile-induced conditions. A number of examples of such cyclization have been reported. Transannular cyclization of germacrene-B **1** is described below.

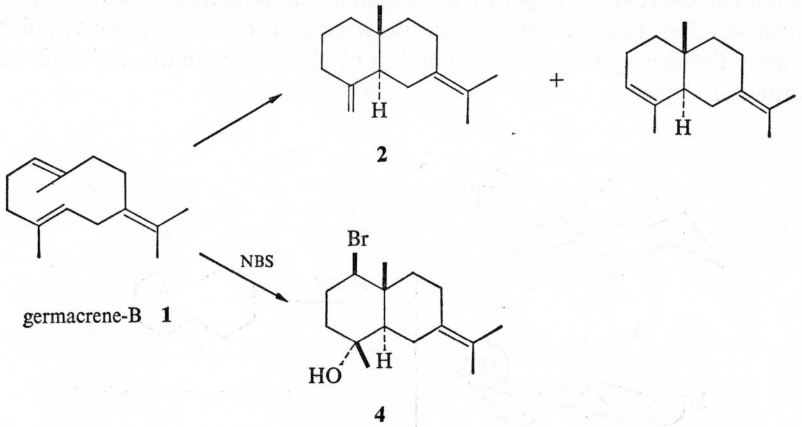

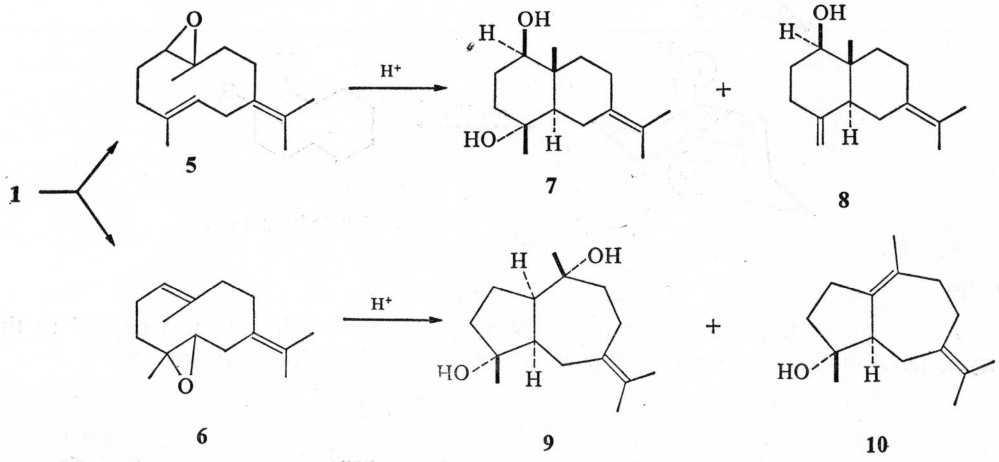

1) Compounds **2** and **3** were obtained on percolation of **1** through alumina.[2]
2) Bromination of **1** with NBS yielded the eudesmane derivative **4**.[2]
3) Acid-catalyzed cyclization of the monoepoxide **5** yielded **7** and **8** as major products.[3]
4) On the other hand, the epoxide **6** was converted into **9** and **10**, both of which possess the guaiane skeleton.[3]

1) D. H. R. Barton, P. de Mayo, *J. Chem. Soc.*, 150 (1957); J. B. Hendrickson, *Tetr.*, **7**, 82 (1959).
2) E. D. Brown, M. D. Solomon, J. K. Sutherland, A. Torre, *Chem. Commun.*, 111 (1967).
3) E. D. Brown, J. K. Sutherland, *Chem. Commun.*, 1060 (1968).

3.47 SYNTHESIS OF CYCLODECENE AND CYCLODECADIENE BY FRAGMENTATION

A growing number of sesquiterpenoids with ten-membered ring structures have been iso-lated, and this has stimulated interest in the synthesis of cyclodecane derivatives. Some synthetic routes are shown below.

Route I[1]

Fragmentation of a cyclic 1,3-diol monosulfonate ester provides an efficient route to cyclodecenes. The C-9:C-10 linkage of 1,10-decalindiol monotosylate **1, 2** and **3** is readily cleaved on treatment with base to yield stereospecifically *trans* and *cis* cyclodecenone in high yield. In all these instances, the leaving group and the bond to be cleaved are in the anti-coplanar position in the conformation depicted.

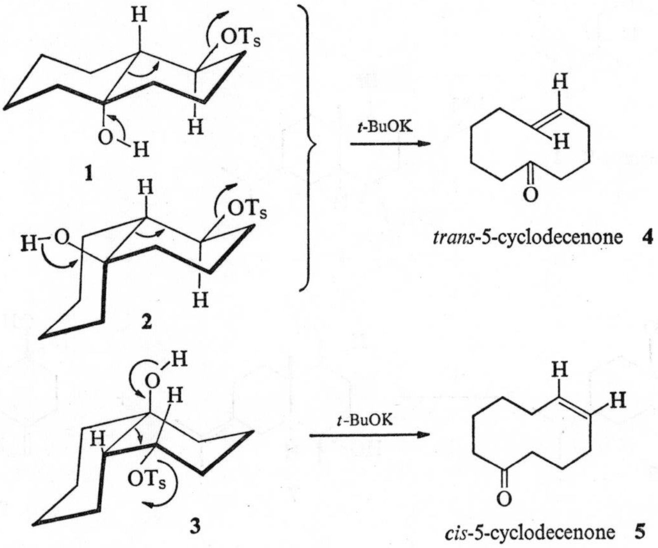

trans-5-cyclodecenone **4**

cis-5-cyclodecenone **5**

Route II[2]

An analagous fragmentation reaction of a 1,3-diol monosulfonate was applied to the bicyclo [5.3.1]-undecane system to prepare the substituted *cis*-cyclodecene derivative 9.

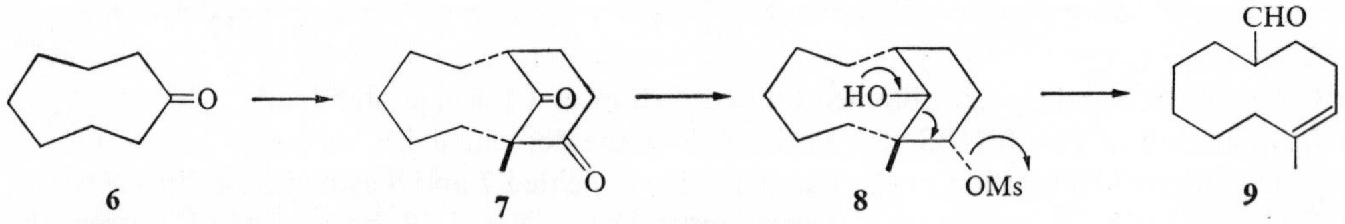

1) P. S. Wharton, G. A. Hiegel, *J. Org. Chem.*, **30**, 3254 (1965).
2) J. A. Marshall, C. J. V. Scanio, *J. Org. Chem.*, **30**, 3019 (1965).
3) J. A. Marshall, G. L. Bundy, *J. Am. Chem. Soc.*, **88**, 4291 (1966).
4) J. A. Marshall, G. L. Bundy, *Chem. Commun.*, 854 (1967).
5) J. A. Marshall, J. H. Babler, *Chem. Commun.*, 993 (1968).

Route III[3,5)]

A new type of fragmentation reaction of alkyl borane derivatives such as **11** and **14** affords
1,5-[4)] and 1,6-cyclodecadiene[3,5)] **12** and **15**, respectively, in high yield.

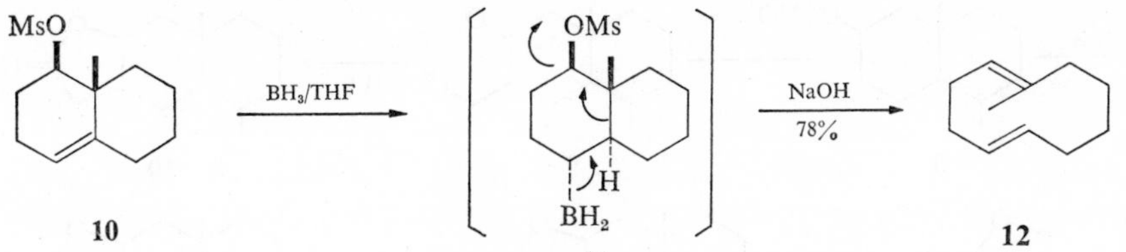

3.48 SYNTHESIS OF EUDESMANES

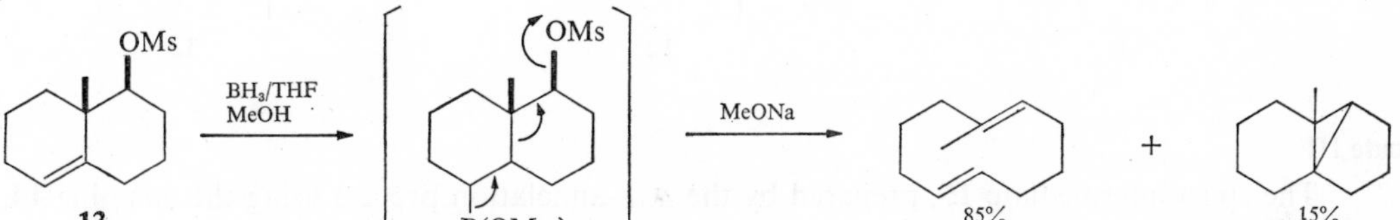

Route I

Robinson annelation reactions are most frequently used for the construction of the bicyclic
skeleton characteristic of this group of sesquiterpenes. If a monoterpene ketone such as dihydro-
carvone **2** is used as the starting material, the desired skeleton, such as **3**, is obtained.[1)]

Many sesquiterpenes of this class have been synthesized from the starting compounds **4**, **5**
or **6**, which are all obtainable via the Robinson reaction.[2)] The compound **6** has also been synthesized by non-annelation processes, e.g. Birch reduction of tetralone derivatives.[3)]

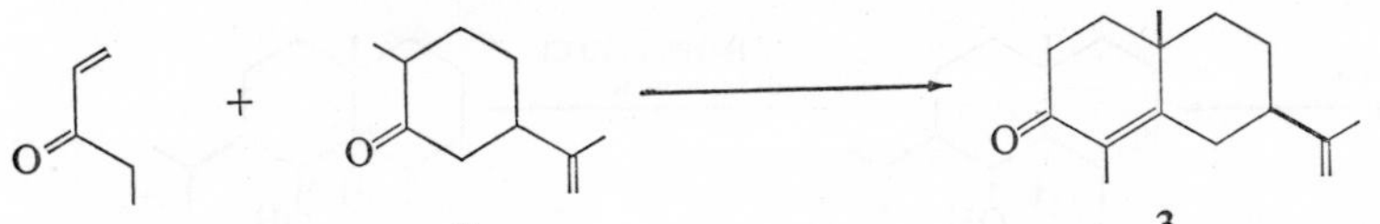

---

1) For examples, see D. C. Humber, A. R. Pinder, R. A. Williams, *J. Org. Chem.*, **32**, 2335 (1967); J. A. Marshall,
   M. T. Pike, *ibid.*, **33**, 435 (1968).
2) For examples, see C. H. Heathcock, T. R. Kelly, *Tetr.*, **24**, 1801 (1968); J. A. Marshall, M. T. Pike, R. D.
   Carroll, *J. Org. Chem.*, **31**, 2933 (1966).
3) J. W. Huffman, M. L. Mole, *J. Org. Chem.*, **37**, 13 (1972).

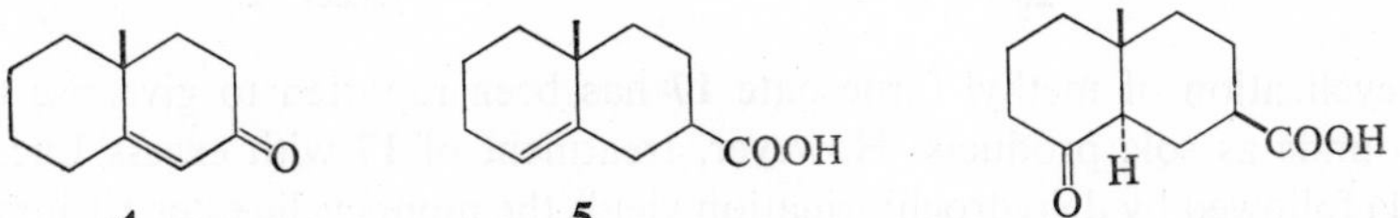

Route II[4)]

Alantolactone **13** was synthesized by solvolytic cyclization of the diene alcohol **7**, leading to the bicyclic compound **8**, followed by functionalization at C-7.

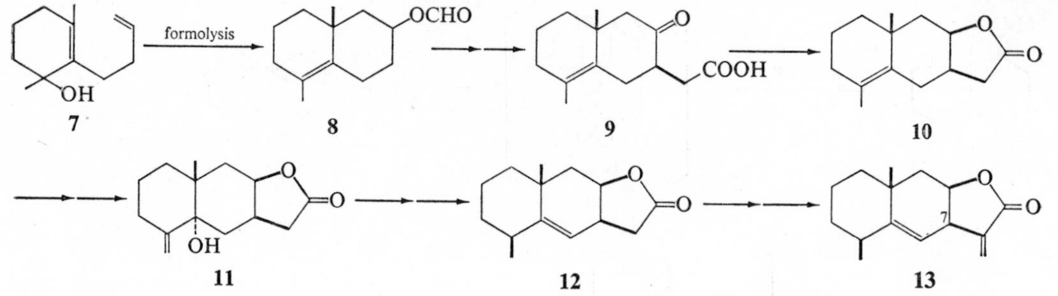

Route III[5)]

The spiro intermediate **15**, prepared by the α,α'-annelation process using the enamine **14**, can be used for the synthesis of the eudesmane skeleton.

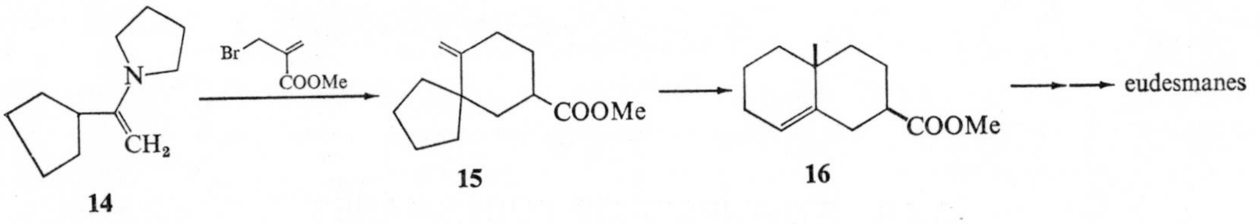

Route IV[6)]

The stepwise cyclization of methyl farnesoate can also be utilized.

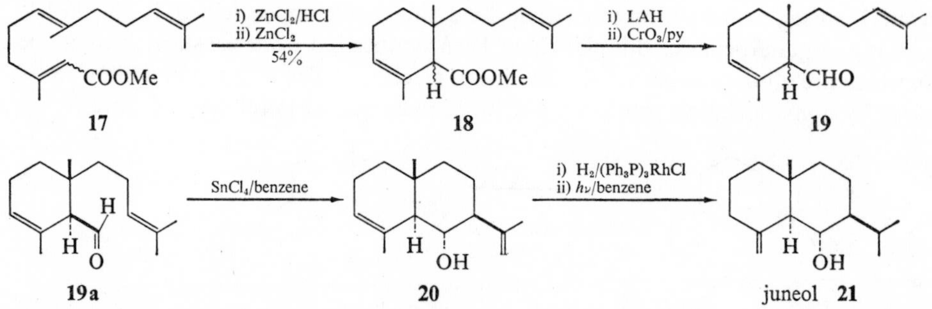

17→18: Acid-catalyzed cyclization of methyl farnesoate **17** has been reported to give rise to drimane-type compounds as sole products. However, treatment of **17** with excess Lucas reagent ($ZnCl_2/HCl$) followed by dehydrochlorination yields the monocyclic ester **18** along with other products.

4) J. A. Marshall, N. Cohen, A. R. Hochstetler, *J. Am. Chem. Soc.*, **88**, 3409 (1966).
5) D. J. Dunham, R. G. Lawton, *J. Am. Chem. Soc.*, **93**, 2075 (1971).
6) M. A. Schwartz, J. D. Crowell, J. H. Musser, *J. Am. Chem. Soc.*, **94**, 4361 (1972).
7) R. G. Carlson, E. G. Zey, *J. Org. Chem.*, **37**, 2468 (1972).

19→20: The epimeric aldehydes **19** could be separated by preparative tlc. Cyclization of **19a** proceeded non-stereospecifically, yielding the *trans*-eudesmane derivative, from which the desired isomer **20** was isolated in 18% yield.

Route V[7)]

An alternative synthetic approach to eudesmol has been reported.

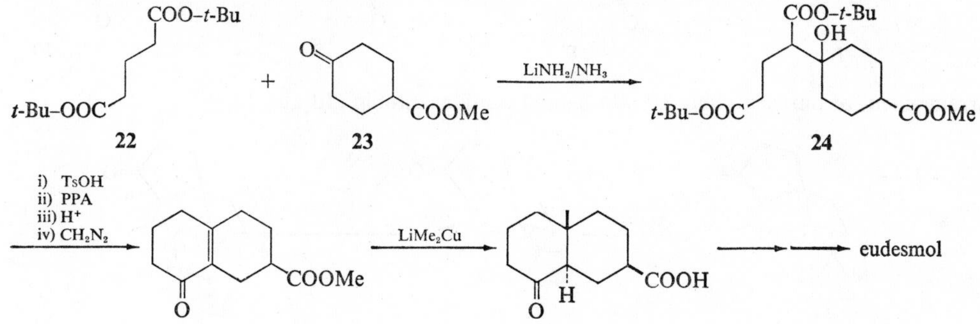

22 **23** **24** **25** **6**

3.49 REARRANGEMENT OF SANTONIC ACID

Santonic acid **3**[1)] is a product of the action of strong alkali on santonin **1**. The formation of **3** can be explained in terms of an intramolecular Michael addition of the carbanion, as in **2**. The tricyclodecane structure of santonic acid **3** can undergo a variety of skeletal rearrangements, including those shown below.[2,6)]

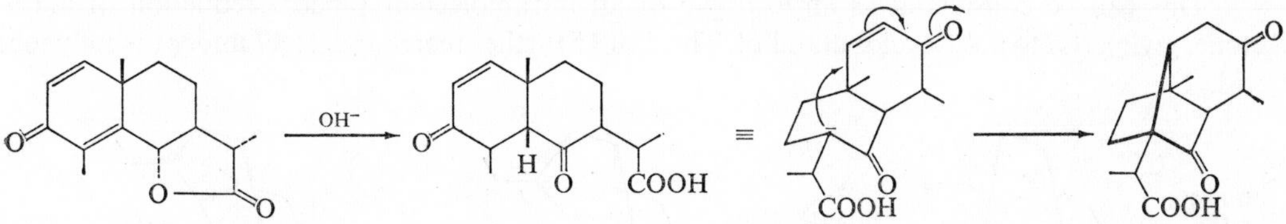

santonin **1** **2** santonic acid **3**

Formation of γ-metasantonin 6[3)]

γ-Metasantonin **6** is formed by a reverse Michael reaction.

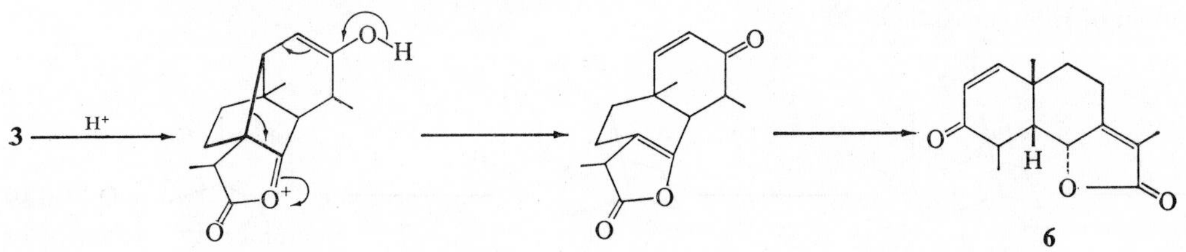

4 **5** **6**

Formation of parasantonide 10 and its C-11 epimer[3)]

Acid treatment followed by pyrolysis yields a ketolactone **10**. This rearrangement is interpreted as follows.

1) R. B. Woodward, F. J. Brutschy, H. Baer, *J. Am. Chem. Soc.*, **70**, 4216 (1948).
2) R. B. Woodward, E. G. Kovach, *J. Am. Chem. Soc.*, **72**, 1009 (1950).

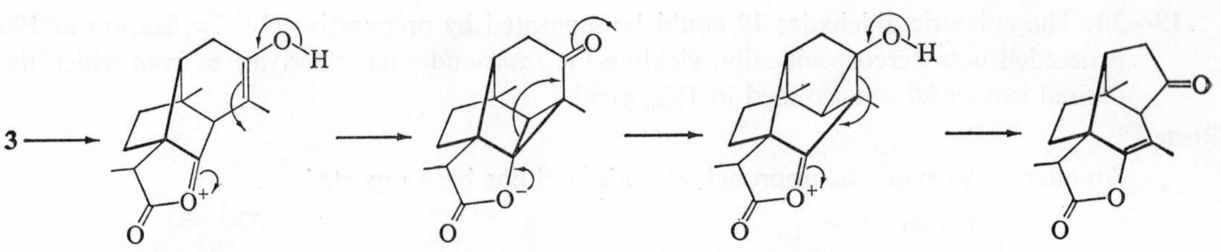

$3 \longrightarrow$

7   **8**   **9**   **10**

Rearrangement of parasantonide 10 into α- and β-metasantonin (13 and 14)[3]

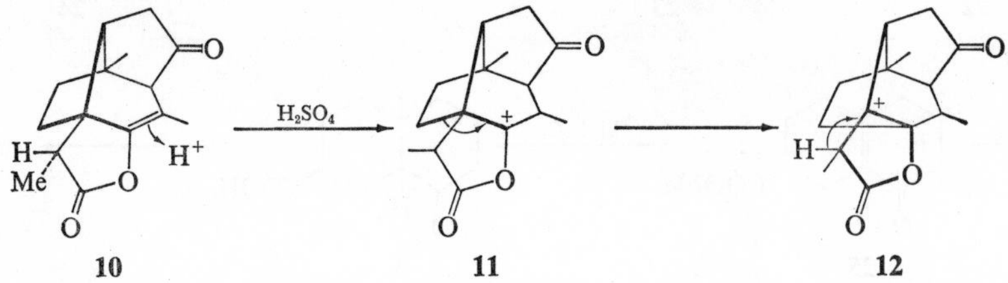

10   **11**   **12**

13 $+$ **14**

Pinacol reduction[4]

This can be considered as an example of an intramolecular pinacol reduction of a 1,4-diketone, giving rise to a 1,2-cyclobutandiol. The diol **15** further rearranges to **17** under the influence of acid.

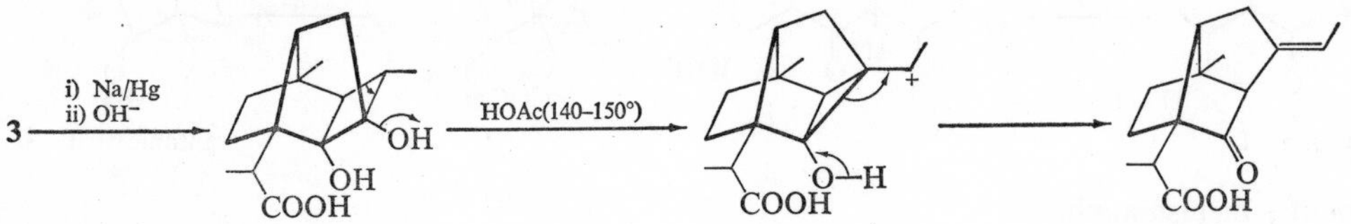

$3 \longrightarrow$ i) Na/Hg ii) OH⁻

15   **16**   **17**

Formation of tribromosantonin 20[5]

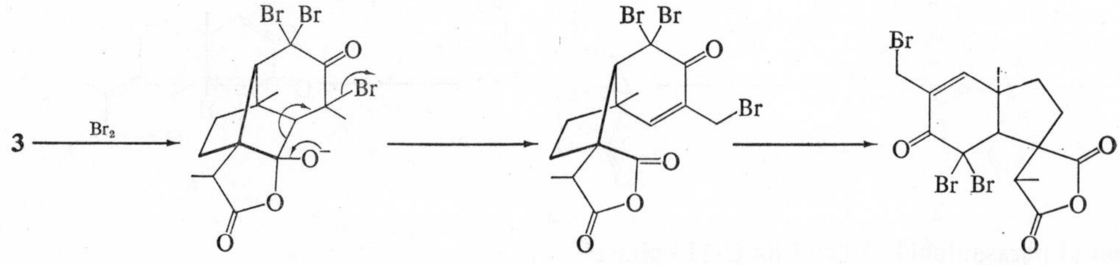

$3 \xrightarrow{Br_2}$

18   **19**   **20**

3) R. B. Woodward, P. Yates, *J. Am. Chem. Soc.*, **85**, 551, 553 (1963).
4) A. G. Hortmann, *Tetr. Lett.*, **52**, 99 (1970).
5) R. B. Woodward, S. G. Levine, P. Yates, *J. Am. Chem. Soc.*, **85**, 557 (1963).
6) See also A. G. Hortmann, D. Daniel, *J. Org. Chem.*, **37**, 4446 (1972) for chemistry of santonic acid.

3.50 SYNTHESIS OF AGAROFURANS

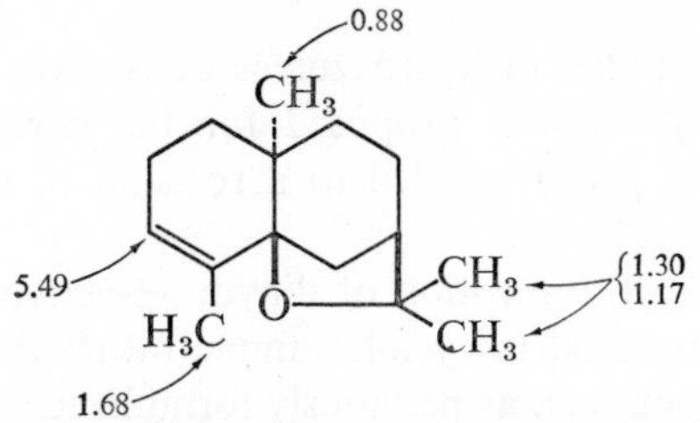

α-agarofuran **1**[1)]

bp (6 mm):134° α_D($CHCl_3$):37.09°

ir:1653, 838 (trisubst. C=C),
1389, 1370 (*gem*-dimethyl)

Route I[2)]

The synthesis of **1** from (−)-epi-α-cyperene **4** of known absolute configuration established the correct structure of **1**.

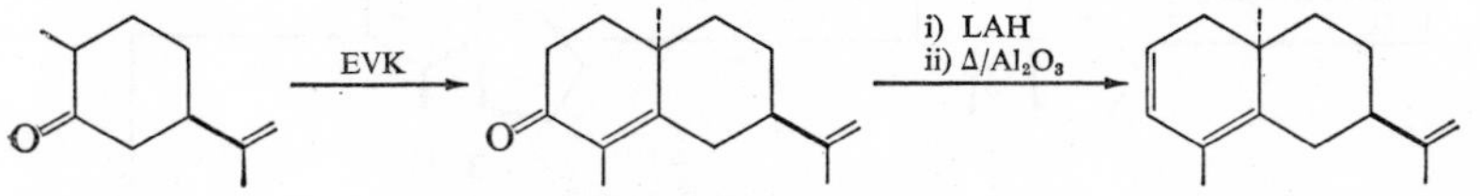

3 (−)-epi-α-cyperene **4** **5**

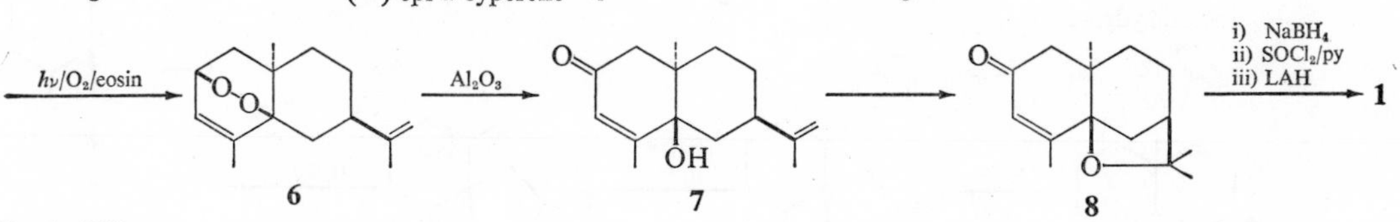

6 **7** **8**

Route II[3)]

An alternative synthesis of α- and β-agarofurans is shown below.

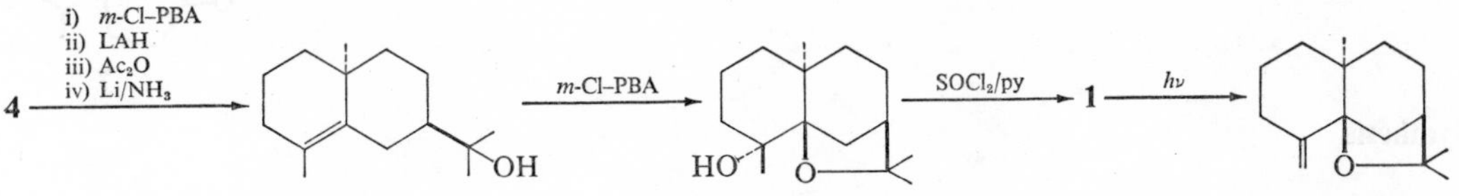

9 **10** β-agarofuran **2**

9→10: Epoxidation of **9** did not yield the corresponding epoxide, but gave instead the naturally occurring compound **10**.

Remarks

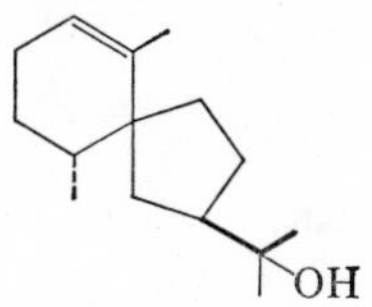

11

α- and β-agarofuran (**1** and **2**) were isolated from fungus-infected agarwood, *Aquillaria agallocha*. Co-occurrence of the closely related spiro compound agarospirol **11** was reported in the same source.[4)]

1) M. L. Maheshwari, T. C. Jain, R. B. Bates, S. C. Bhattacharyya, *Tetr.*, **19**, 1079 (1963).
2) H. C. Barrett, G. Büchi, *J. Am. Chem. Soc.*, **89**, 5665 (1967).
3) J. A. Marshall, M. T. Pike, *J. Org. Chem.*, **33**, 435 (1968).
4) K. R. Varma, M. L. Maheshwari, S. C. Bhattacharyya, *Tetr.*, **21**, 115 (1965).

3.51 STRUCTURE AND SYNTHESIS OF β-VETIVONE

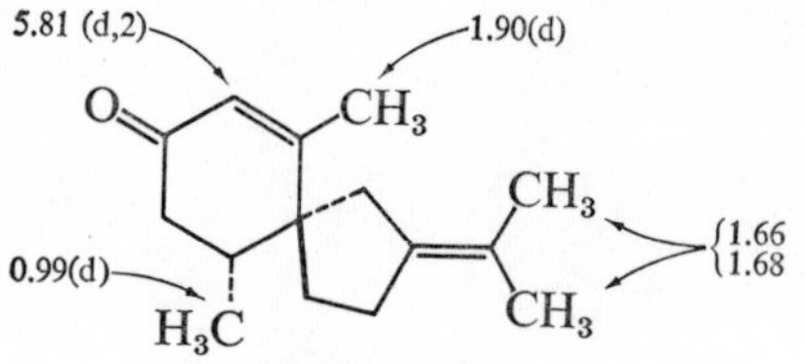

5.81 (d,2)   1.90(d)

β-vetivone **1**[1)]
mp: 42–44°
uv: 240 (4.11)
ir: 2950, 2910
1670, 1614 (α,β-unsat. CO)

1) Synthetic studies of hydroazulene derivatives of the type previously proposed for the carbon skeleton of β-vetivone led to a revision of the structure.

2) The carbonyl absorption of dihydro-β-vetivone must be assigned to cyclohexanone rather than a cycloheptanone, as previously formulated.

3) Synthesis of the hydrocarbon **3**, a degradation product of **1**, verified the spiro[4.5]decane ring system of **1**.

4) The absolute configuration of **1** was established by chemical correlation with hinesol, which was degraded to yield (+)-methylglutaric acid.

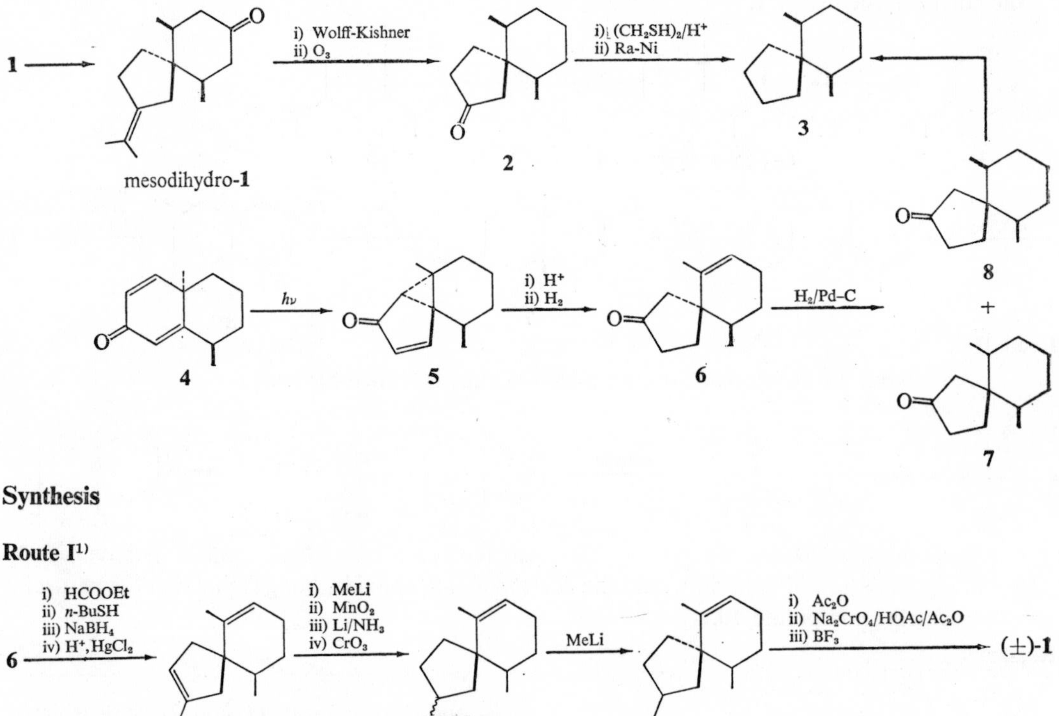

Synthesis

Route I[1)]

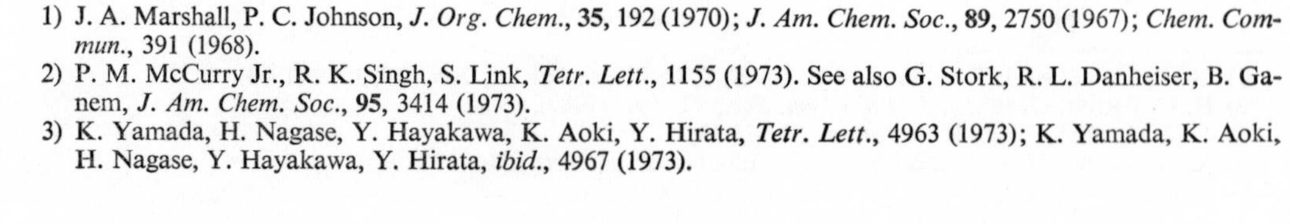

---

1) J. A. Marshall, P. C. Johnson, *J. Org. Chem.*, **35**, 192 (1970); *J. Am. Chem. Soc.*, **89**, 2750 (1967); *Chem. Commun.*, 391 (1968).
2) P. M. McCurry Jr., R. K. Singh, S. Link, *Tetr. Lett.*, 1155 (1973). See also G. Stork, R. L. Danheiser, B. Ganem, *J. Am. Chem. Soc.*, **95**, 3414 (1973).
3) K. Yamada, H. Nagase, Y. Hayakawa, K. Aoki, Y. Hirata, *Tetr. Lett.*, 4963 (1973); K. Yamada, K. Aoki, H. Nagase, Y. Hayakawa, Y. Hirata, *ibid.*, 4967 (1973).

Route II[2]

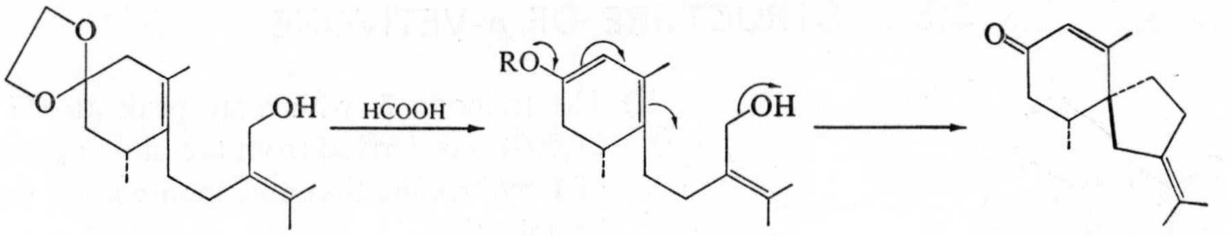

12	**13**	10-epi-β-vetivone **14**

Acid treatment of **12** yielded 10-epi-β-vetivone **14**.

Route III[3]

An alternative synthesis of **1**, hinesol, α-vetispirene by the spirocondensation of 4(3'-formylpropyl)-3-cyclohexenone derivative have been reported.

Remarks

β-Vetivone was isolated from essential oil of vetiver along with α-vetivone. Several closely related compounds, e.g. hinesol, agarospirol, bicyclovetivenol and isovetivenene were also obtained.

3.52   STRUCTURE OF α-VETIVONE

α-vetivone **1**[1]

mp: 30–35° α_D(CHCl$_3$): +248°
uv(EtOH); 238.5 (ε 13,900), 240 (ε 16,200)
ir(film): 5.99 (CO), 6.17 (C=C)
nmr in CCl$_4$/py (1:1)

1) The trienone **2**, with a uv peak at 347 nm (ε 21,000) was derived from the dihydro derivative of **1** by bromination and spontaneous dehydrobromination.

2) The trienone **2** was chemically correlated with a known compound **3**.

3) Absolute configuration was assigned as in **1** from the ord curve, which showed a positive Cotton effect in comparison with that of Δ^4-3-keto steroids.

4) The structure and absolute configuration of **1** were also confirmed by interrelation[2] of **4** with eremophilone **5**.

i) Li/NH$_3$/EtOH
ii) CrO$_3$
iii) Br$_2$

1 ⟶   **2**   ←   nootkatone **3**

1 O$_2$/OH$^-$ ⟶   **4**

eremophilone **5**

i) enol acetylation
ii) NaCr$_2$O$_7$
iii) OH$^-$ ⟶   enantiomer of **4**

Remarks

α-Vetivone **1** was isolated from essential oil of vetiver, and the structure was revised in 1967.[1,2] The name "isonootkatone" was proposed[1] from the close structural relationship of **1** to nootkatone **3**.

Nootkatone **3** and the corresponding hydrocarbons (nootkatene, valencene) were obtained from grapefruit (or orange) peel and Alaskan yellow cedar.[3] A series of sesquiterpenes having methyl groups at C-4 and C-5 of opposite configuration, such as eremophilone **5**, petasin, eremophilenolide, furanopetasin, etc., are also known.[4,5]

1) J. A. Marshall, N. H. Andersen, *Tetr. Lett.*, 1611 (1967).
2) K. Endo, P. de Mayo, *Chem. Commun.*, 89 (1967).
3) W. D. MacLeod Jr., *Tetr. Lett.*, 4779 (1965), and references therein.
4) L. H. A. Zalkow, F. X. Markley, C. Djerassi, *J. Am. Chem. Soc.*, **82**, 6354 (1960); C. J. W. Brooks, G. H. Draffen, *Chem. Commun.*, 701 (1966).
5) For review, see A. R. Pinder, *Perfumery Essent. Oil Record*, **59**, 280, 645 (1968).

3.53 SYNTHESIS OF EREMOPHILANES AND VALENCANES

Route I

This route utilizes the Robinson annelation technique.

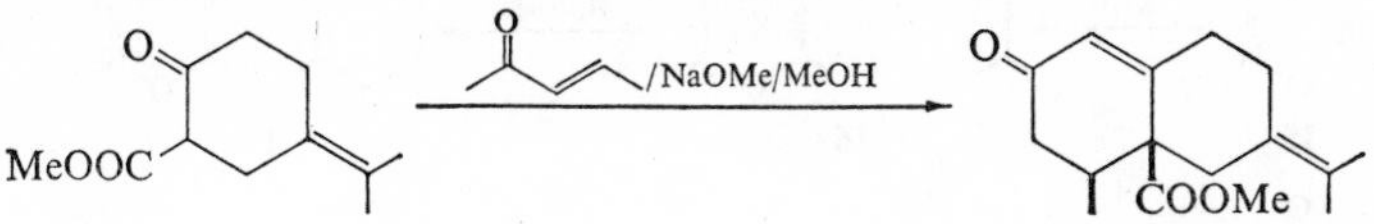

1 → **2**: Isonootkatone **4** was synthesized[1] by stereoselective condensation of the keto-ester **1** and *trans*-3-penten-2-one, yielding **2**, in which there is a *cis* relationship between the methoxycarbonyl and methyl groups.

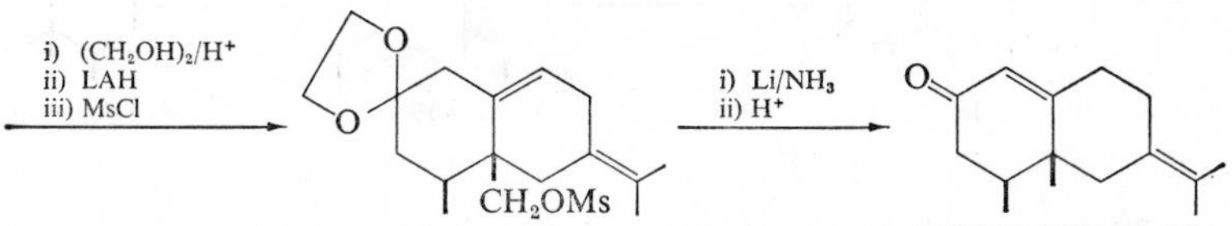

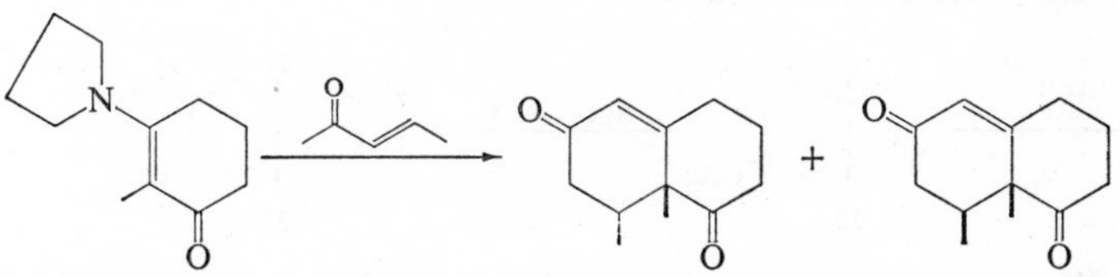

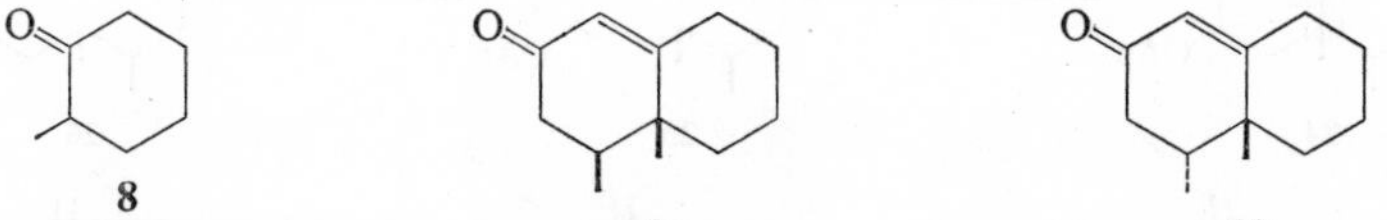

5 → **6,7**: When the enamine **5** was used as a starting material, a mixture of isomeric octalones with *cis* and *trans* vicinal methyl groups was obtained, e.g. **6** and **7**, which lead ultimately to eremophilane-type and valencane-type sesquiterpenes.[2]

8 → **9**; **8** → **10**: It has been found that treatment of a suspension of the sodium enolate of **8** with *trans*-pentenone in dioxane afforded exclusively *cis* dimethyl octalone **9**, while in DMSO solution the *trans* isomer **10** is obtained.[3]

Route II[4,5]

A stereoselective route to fukinone and related sesquiterpenes has been reported using octalone **13** as a key intermediate. The octalone was synthesized from α-protected 2,3-dimethyl cyclohexanone.

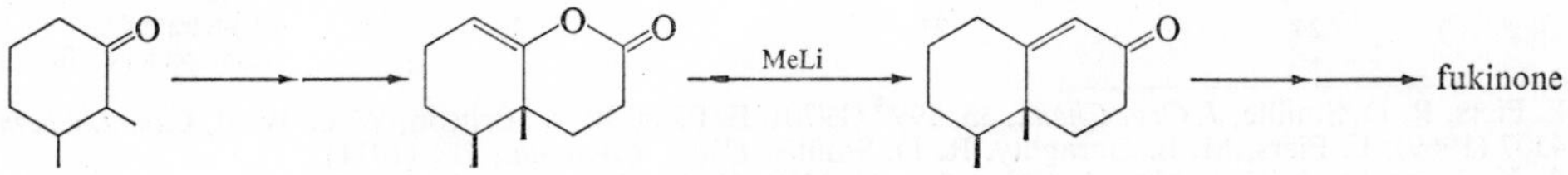

1) J. A. Marshall, T. M. Warne Jr., *J. Org. Chem.*, **36**, 178 (1971); J. A. Marshall, R. A. Ruden, *ibid.*, **36**, 594 (1971).
2) R. M. Coates, J. E. Shaw, *Chem. Commun.*, 47, 515 (1968); *J. Org. Chem.*, **35**, 2597 (1970).
3 C. J. V. Scanio, R. M. Starrett, *J. Am. Chem. Soc.*, **93**, 1539 (1971).

Route III[6]

An isoxazole annelation has also been utilized for reactions of this type.

14   15 $\xrightarrow{\text{KOH}}$ 16 $\xrightarrow{\text{H}_2/\text{Pd–C}}$ 17

i) isopropoxy
   methylenation
ii) MeI/NaH
iii) deprotection

i) triethyloxonium
   fluoroborate
ii) KOH

18   19   (±)-dehydrofukinone 20

Route IV[7]

Fukinone **29** was synthesized by a stereoselective route involving conjugate methylation of the enone **23** with Me₂CuLi as a key step.

i) HCOOH
ii) LAH
iii) Ac₂O

21   22 $\xrightarrow{\text{Na}_2\text{CrO}_4/\text{HOAc}}$ 23

$\xrightarrow{\text{Me}_2\text{CuLi}}$

i) H₂NNH₂/KOH
ii) CrO₃
iii) Ph₃CLi/Ac₂O

i) ArCOOOH
ii) heat

24   25   26

i) DMSO/SO₃
ii) Ac₂O/H⁺

i) Ca/NH₃
ii) Al₂O₃

27   28   (±)-fukinone 29

Route V[8]

30   31 $\xrightarrow{\text{MeLi}}$ 32 $\xrightarrow{\text{POCl}_3}$ 33

i) HCOOH
ii) LAH

i) Jones
ii) Wolff-Kishner
iii) hν/O₂
iv) LAH (for 34)

i) O₃
ii) Ac₂O
iii) Ca/NH₃

34   +   35   36   (±)-tetrahydro-
eremophilone 37

4) E. Piers, R. D. Smillie, *J. Org. Chem.*, **35**, 3997 (1970); E. Piers, R. W. Britton, W. de Waal, *Can. J. Chem.*, **47**, 4307 (1969); E. Piers, M. B. Geraghty, R. D. Smillie, *Chem. Commun.*, 614 (1971).
5) A. K. Torrence, A. R. Pinder, *J. Chem. Soc.* C, 3410 (1971); *Tetr. Lett.*, 745 (1971).
6) M. Ohashi, *Chem. Common.*, 893 (1969).
7) J. A. Marshall, G. M. Cohen, *J. Org. Chem.*, **36**, 877 (1971).
8) S. Murayama, D. Chan, M. Brown, *Tetr. Lett.*, 3715 (1968).
9) For other syntheses, see M. Peraro, G. Bozzato, P. Schudel, *Chem. Commun.*, 1152 (1968); C. Berger, M. F. Newmann, G. Ourisson, *Tetr. Lett.*, 3451 (1968); H. C. Odom, A. R. Pinder, *J. Chem. Soc.*, Perkin I, 2193 (1972); K. P. Dastur, *J. Am. Chem. Soc.*, **95**, 6509 (1973); E. Piers, M. B. Geraghty, *Can. J. Chem.*, **51**, 2166 (1973).

3.54 STRUCTURE OF ISHWARONE AND SYNTHESIS OF ISHWARANE

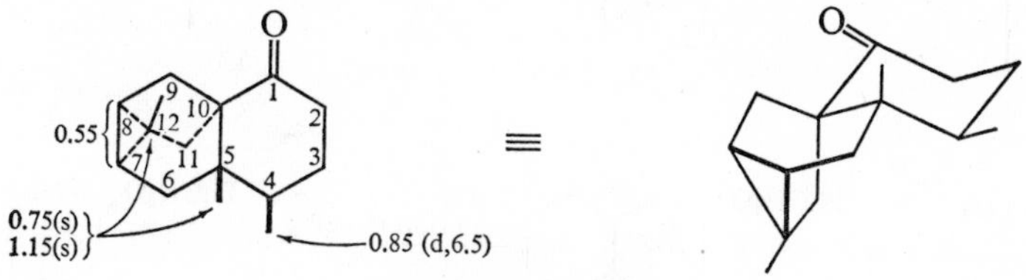

ishwarone $1^{1)}$

ms: 218 (M^+ $C_{15}H_{22}O$)

uv: 211 (ϵ, 275)

288–290 (ϵ, 30)

ir: 1706 (cyclohexanone)

1418 (–CO–CH$_2$)

1) Acid treatment of **1** yields isoishwarone **2** by cleavage of the cyclopropane ring.

2) A C-C linkage of the bicyclo [2.2.2]-octane ring in **3** is cleaved by a retroaldol-type reaction, yielding the bicyclic compound **4**.

3) The absolute configuration was established by correlation of **4** with (+)-nootkatane.[2] Synthesis of **15** and **16** confirmed the structure of **1**.[3]

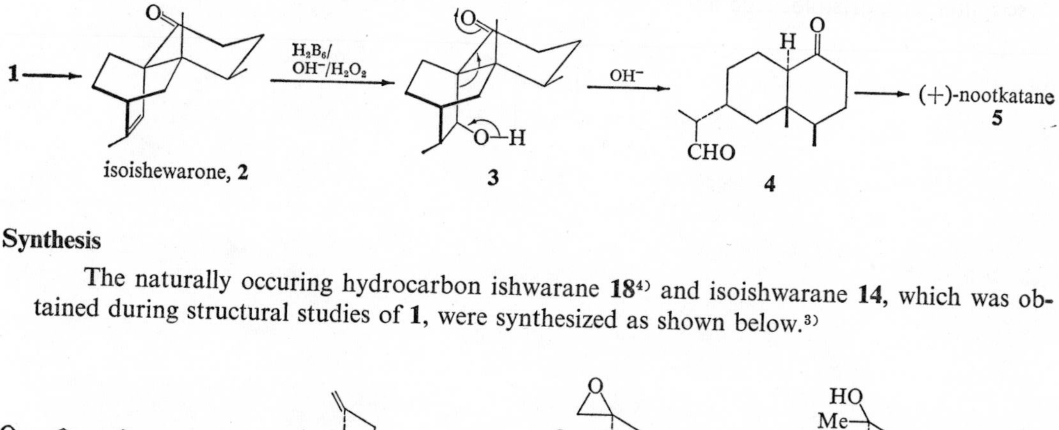

Synthesis

The naturally occuring hydrocarbon ishwarane **18**[4] and isoishwarane **14**, which was obtained during structural studies of **1**, were synthesized as shown below.[3]

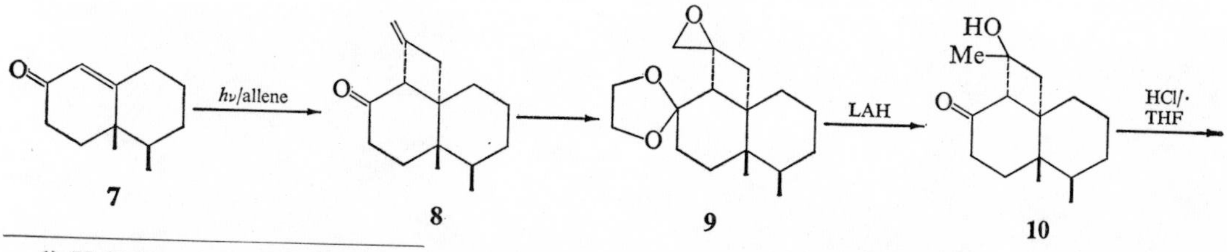

1) H. Fuhrer, A. K. Ganguly, K. W. Copinath, T. R. Govindachari, K. Nagarajan, G. R. Pai, P. C. Parthasarathy, *Tetr.*, **26**, 2371 (1970); A. K. Ganguly, K. W. Copinath, T. R. Govindachari, K. Nagarajan, G. R. Pai, P. C. Parthasarathy, *Tetr. Lett.*, 133 (1969)

2) T. R. Govindachari, K. Nagarajan, P. C. Parthasarathy, *Chem. Commun.*, 823 (1969).

3) R. B. Kelly, J. Zamecnik, *Chem. Commun.*, 1102 (1970); R. B. Kelly, J. Zamecnik, B. A. Beckett, *ibid.*, 479 (1971).

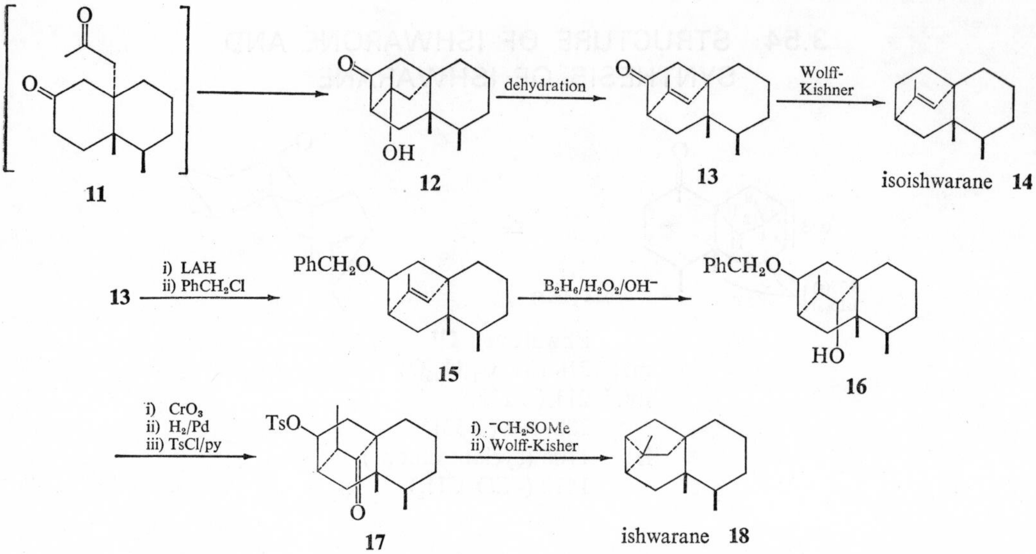

$7 \rightarrow 8$: This is the photochemical cycloaddition of an allene and an $\alpha\beta$-unsaturated ketone.

$10 \rightarrow 12$: Retro-aldol opening of cyclobutane followed by aldol cyclization.

Remarks

Ishwarone **1** and ishwarane **18** were isolated from *Aristolochia indica* along with a bicyclic sesquiterpene, aristolochene **6**.[4]

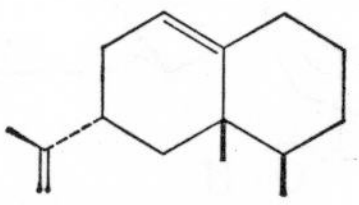

6

4) T. R. Govindachari, P. A. Mohamed, P. C. Parthasarathy, *Tetr.*, **26**, 615 (1970).

3.55   SYNTHESIS OF TRICYCLIC SESQUITERPENES WITH A FUSED DIMETHYLCYCLOPROPANE RING

A small group of tricyclic sesquiterpenes (e.g. those with maaliane, aristolane and aroma-dendrane skeletons) possessing a *gem*-dimethylcyclopropane ring in the molecule are known. Various methods have been used for the construction of the fused *gem*-dimethylcyclopropane ring in the synthesis of some of these compounds.

Route I[1]

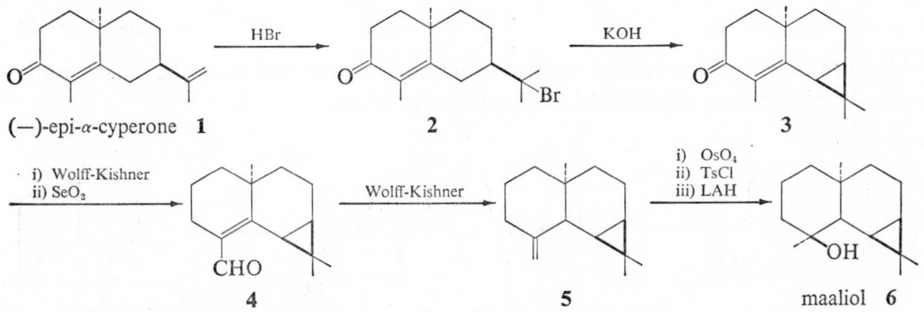

(−)-epi-α-cyperone **1**   **2**   **3**

maaliol **6**

2→3: Dehydrobromination of **2** yielded the cyclopropane derivative **3**, from which maaliol **6** was synthesized.

Route II[2]

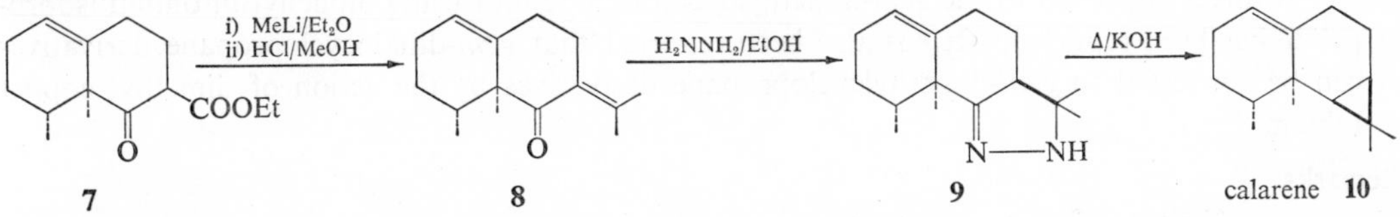

7   **8**   **9**   calarene **10**

9→10: Pyrolysis of **9** yielded calarene **10** by isomerization followed by expulsion of nitrogen.

Route III[3]

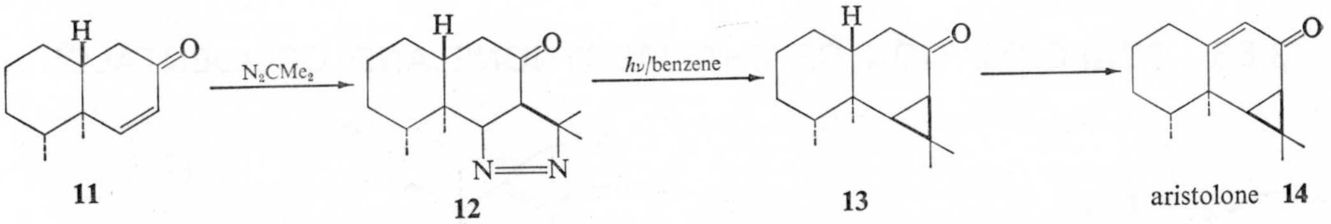

11   **12**   **13**   aristolone **14**

12→13: Photolysis of the pyrazoline derivative **12**, prepared by 1,3-dipolar addition of diazo-2-propane to the enone **11**, yielded **13**, from which aristolone **14** was derived.

1) R. B. Bates, G. Büchi, T. Matsuura, B. R. Schaffer, *J. Am. Chem. Soc.*, **82**, 2327 (1960).
2) R. M. Coates, J. E. Shaw, *J. Am. Chem. Soc.*, **92**, 5657 (1970).
3) C. Berger, M. Franck-Neumann, G. Ourisson, *Tetr. Lett.*, 3451 (1968).
4) E. Piers, R. W. Britton, W. de Waal, *Can. J. Chem.*, **47**, 831 (1969).
5) A. E. Greene, J. C. Muller, G. Ourisson, *Tetr. Lett.*, 4147 (1971).
6) E. J. Corey, G. H. Posner, *J. Am. Chem. Soc.*, **89**, 3911 (1967).
7) E. J. Corey, M. Jautelat, *J. Am. Chem. Soc.*, **89**, 3912 (1967).

Route IV[4)]

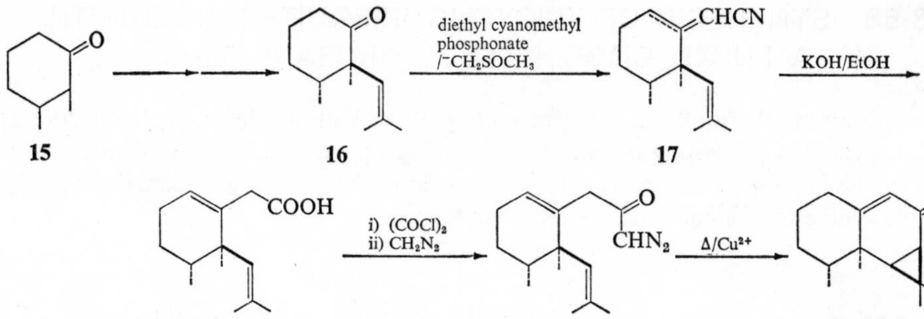

15   **16**   **17**

18   **19**   aristolone **14**

19→14: Intramolecular cyclization of the keto carbene generated from the diazoketone **19** onto the olefinic linkage afforded **14** as the major product.

Route V[5)]

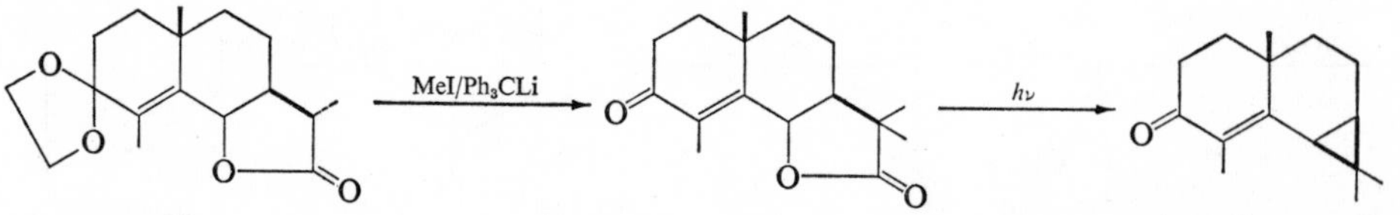

20   **21**   epi-maalienone **22**

21→22: Photochemical decarboxylation provides a new route for the synthesis of these compounds.

Other routes

A direct approach to the *gem*-dimethylcyclopropane unit using diphenylsulfonium isopropylide has been reported.[6)] It has also been reported that *gem*-dihalocyclopropane derivatives can be converted to *gem*-dimethylcyclopropane derivatives by the action of dimethyl copper lithium.[7)]

Remarks

Aromadendrene and related sesquiterpenes also possess a carbon skeleton in which a *gem*-dimethylcyclopropane ring is present, fused to a hydroazulenic nucleus. The enantiomer of 9-aristolene, α-ferulene, has been isolated.

3.56   CONFORMATION OF CHAMAECYNONE AND OCCIDENTALOL

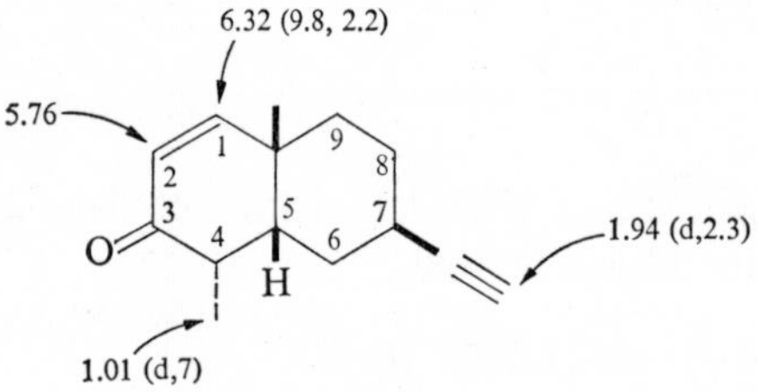

(−)-chamaecynone[1)] **1**

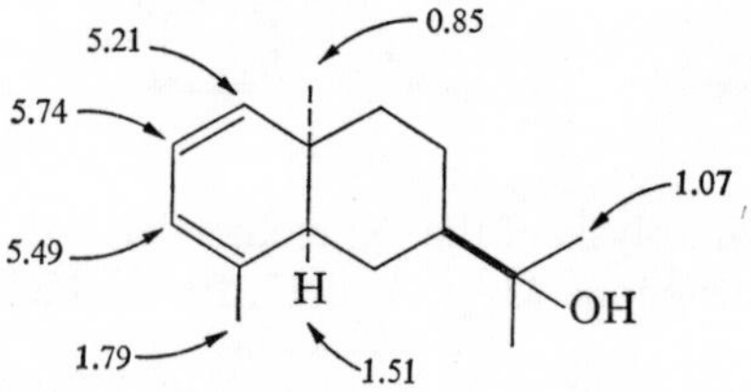

(+)- occidentalol[2)] **2**

1) Ord curves for **1, 3** and **5** show positive Cotton effect, indicating that this series compounds exists in solution in the non-steroidal conformation (e.g. **1b**).

2) On the other hand, the ord curves of isochamaecynone (C-4 epimer of **1**), **4** and **6** exhibit negative Cotton effects, showing that isochamaecynone-series compounds exist in a steroidal conformation such as **1a.**

3) Occidentalol **2** also exists in a non-steroidal conformation such as **2b.**

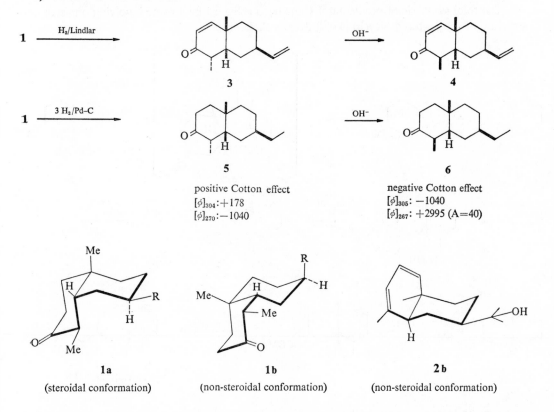

positive Cotton effect
$[\phi]_{304}: +178$
$[\phi]_{270}: -1040$

negative Cotton effect
$[\phi]_{305}: -1040$
$[\phi]_{267}: +2995 \ (A=40)$

1a
(steroidal conformation)

1b
(non-steroidal conformation)

2b
(non-steroidal conformation)

Remarks

Chamaecynone **1** and related substances were isolated from *Chamaecyparis formosensis*. (+)-Occidentalol **2** was obtained from *Thuja occidentalis*.

A unique biosynthetic pathway involving valence tautomerism of the 1,3,5-cyclodecatriene system **8** was suggested for the formation of these *cis*-fused eudesmane-type sesquiterpenes.[3]

7   **8**   **9**

1) T. Toda, Y. S. Chang, T. Nozoe, *Tetr. Lett.*, 3663 (1966); T. Nozoe, T. Asao, M. Ando, K. Takase, *ibid.*, 2821 (1967); T. Asao, S. Ide, K. Takase, Y. S. Chang T. Nozoe, *ibid.*, 3639 (1968).
2) A. G. Hortmann, J. B. De Roos, *J. Org. Chem.*, **34**, 736 (1969).
3) A. G. Hortmann, *Tetr. Lett.*, 5785 (1968).

3.57 SYNTHESIS OF OCCIDENTALOL

Route I[1]

The total synthesis of occidentalol **1** has been achieved by a route involving one-step construction of the *cis*-fused bicyclo [4.4.0]-decane skeleton.

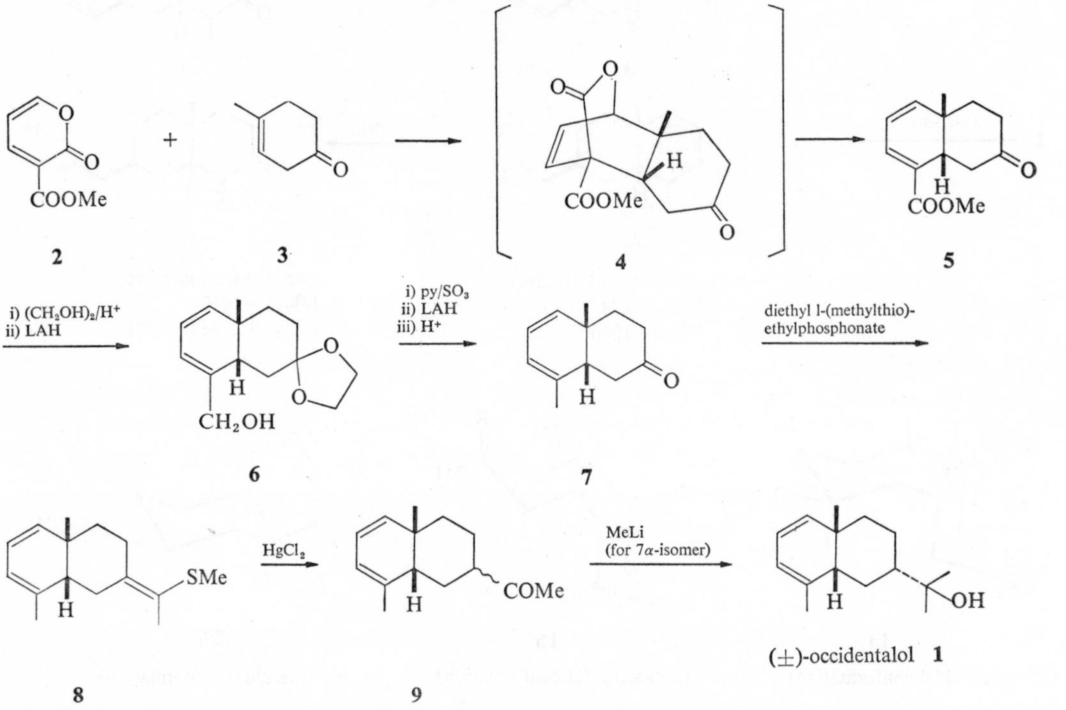

2+3→5: Diels-Alder addition of **2** and **3** gave stereoselectively *cis*-fused **5**.
8→9: The formation ratio of the two isomers is $\alpha:\beta = 2.9:1$.

Route II[2,3]

Two syntheses of (+)-**1** have been reported starting from (+)-dihydrocarvone **10** (**11→12→1** and **11→14→1**). These syntheses confirmed the abolute stereochemistry of **1**.

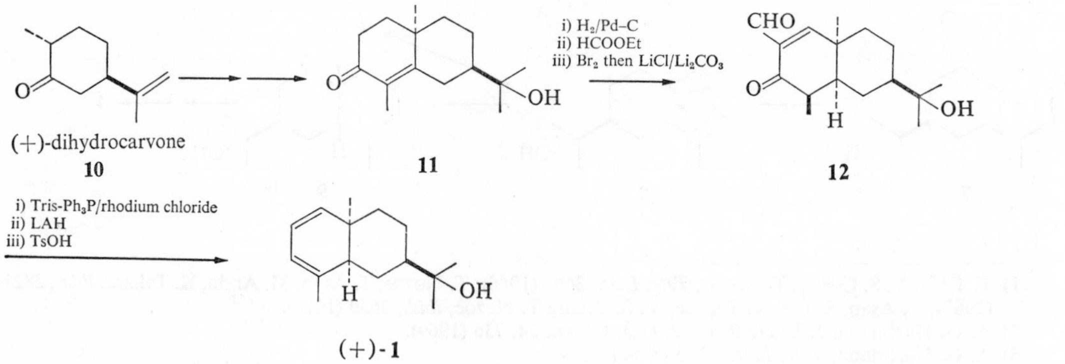

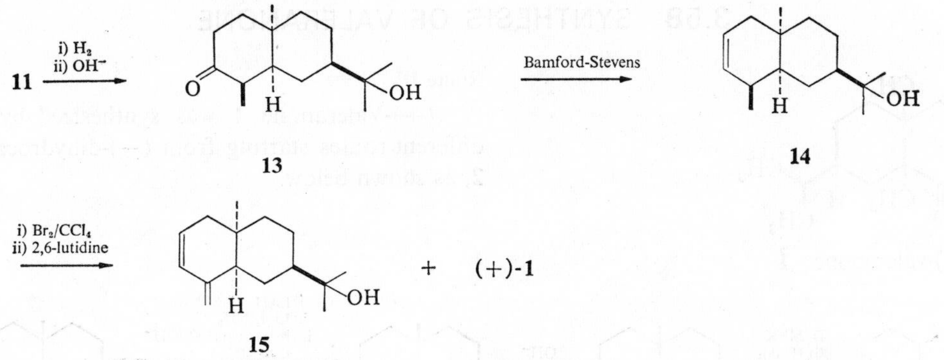

11 $\xrightarrow[\text{ii) OH}^-]{\text{i) H}_2}$ **13** $\xrightarrow{\text{Bamford-Stevens}}$ **14**

i) Br$_2$/CCl$_4$
ii) 2,6-lutidine
$\longrightarrow$ **15** $+$ $(+)$-**1**

Route III[4)]

$(-)$-Occidentalol **1** was also synthesized from **16**, which has been prepared from α-santonin.

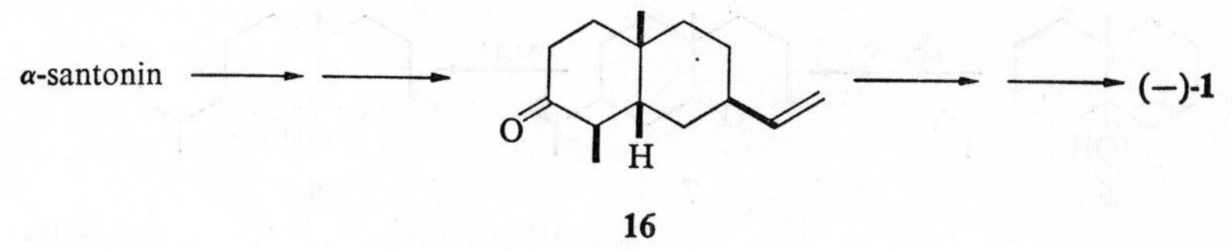

α-santonin $\longrightarrow$ $\longrightarrow$ **16** $\longrightarrow$ $\longrightarrow$ $(-)$-**1**

Route IV[5)]

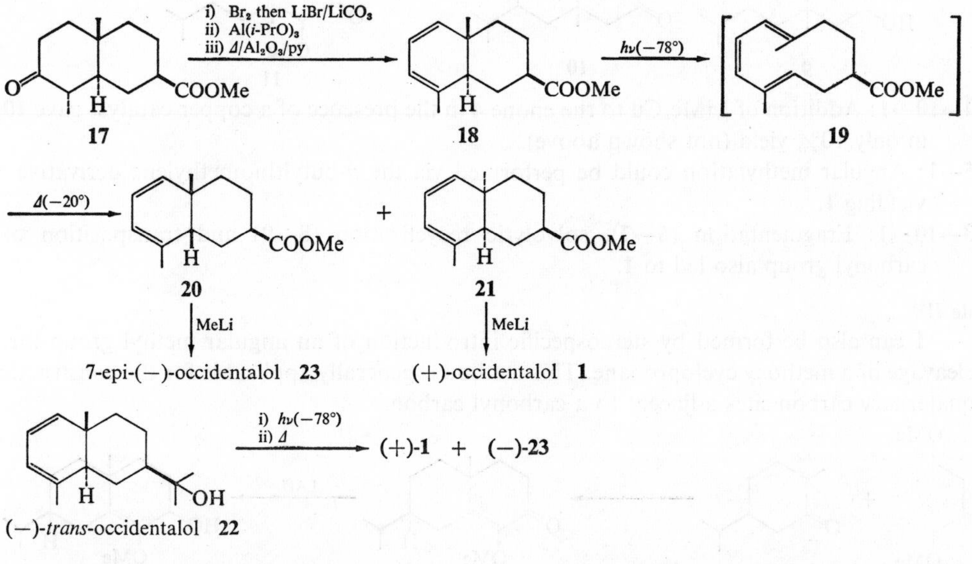

17 $\xrightarrow[\substack{\text{i) Br}_2 \text{ then LiBr/LiCO}_3 \\ \text{ii) Al}(i\text{-PrO})_3 \\ \text{iii) } \Delta/\text{Al}_2\text{O}_3/\text{py}}]{}$ **18** $\xrightarrow{h\nu(-78°)}$ **19**

$\xrightarrow{\Delta(-20°)}$ **20** $+$ **21**

7-epi-$(-)$-occidentalol **23** $\xleftarrow{\text{MeLi}}$ **20**

$(+)$-occidentalol **1** $\xleftarrow{\text{MeLi}}$ **21**

$(-)$-*trans*-occidentalol **22** $\xrightarrow[\text{ii) } \Delta]{\text{i) } h\nu(-78°)}$ $(+)$-**1** $+$ $(-)$-**23**

Biogenetically patterned syntheses of $(+)$-**1** and $(-)$-**23** including a photolysis-recyclization sequence (e.g., **18**→**19**→**20**+**21**) have been reported. Photolysis-recyclization of **22** also gave $(+)$-**1** and $(-)$-**23** in a ratio of 2:3.

1) D. S. Watt, E. J. Corey, *Tetr. Lett.*, 4651 (1972).
2) M. S. Sergent, M. Mongrain, P. Deslongchamps, *Can. J. Chem.*, **50**, 336 (1972).
3) Y. Amano, C. H. Heathcock, *Can. J. Chem.*, **50**, 340 (1972).
4) M. Ando, K. Nanaumi, T. Nakazawa, T. Asano, K. Takase, *Tetr. Lett.*, 3891 (1970).
5) A. G. Hortmann, D. S. Daniel, J. E. Martinelli, *J. Org. Chem.*, **38**, 728 (1973).

3.58   SYNTHESIS OF VALERANONE

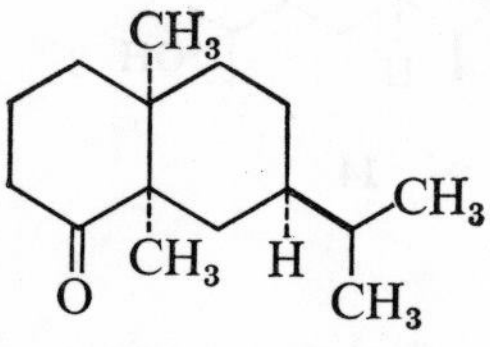

(−)-valeranone **1**

Route I[1)]

(+)-Valeranone **1** was synthesized by three different routes starting from (−)-dihydrocarvone **2**, as shown below.

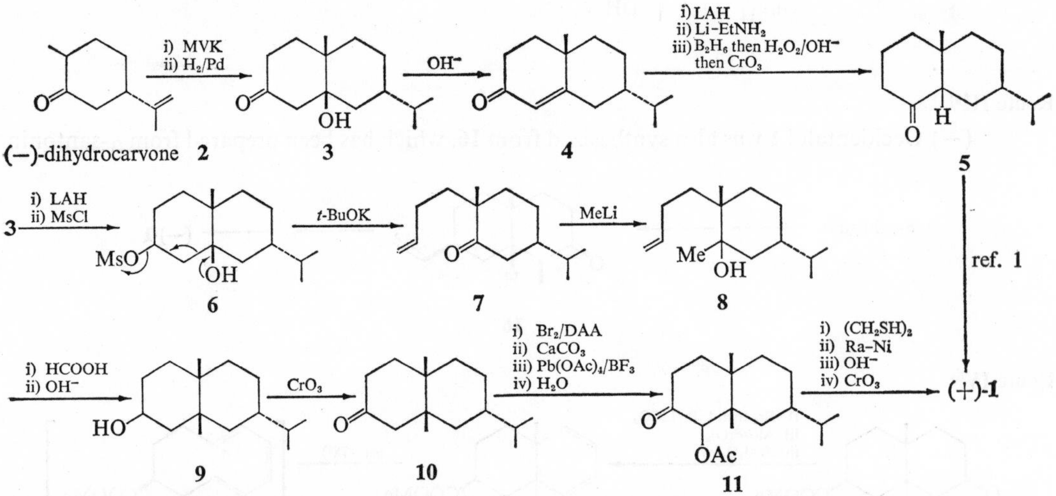

(−)-dihydrocarvone 2

i) MVK
ii) H₂/Pd → **3**

OH⁻ → **4**

i) LAH
ii) Li-EtNH₂
iii) B₂H₆ then H₂O₂/OH⁻
then CrO₃ → **5**

3 → i) LAH
ii) MsCl → **6**

t-BuOK → **7**

MeLi → **8**

ref. 1

i) HCOOH
ii) OH⁻ → **9**

CrO₃ → **10**

i) Br₂/DAA
ii) CaCO₃
iii) Pb(OAc)₄/BF₃
iv) H₂O → **11**

i) (CH₂SH)₂
ii) Ra-Ni
iii) OH⁻
iv) CrO₃ → (+)-**1**

4→10→1: Addition of LiMe/Cu to the enone **4** in the presence of a copper catalyst gave **10**, but in only 10% yield (not shown above).

5→1: Angular methylation could be performed via the *n*-butylthiomethylene derivative of **5**, yielding **1**.

3→10→1: Fragmentation (**6→7**), solvolytic recyclization (**8→9**) and transposition of the carbonyl group also led to **1**.

Route II[2)]

1 can also be formed by stereospecific introduction of an angular methyl group through cleavage of a methoxy cyclopropane. This method is generally applicable for the construction of quaternary carbon sites adjacent to a carbonyl carbon.

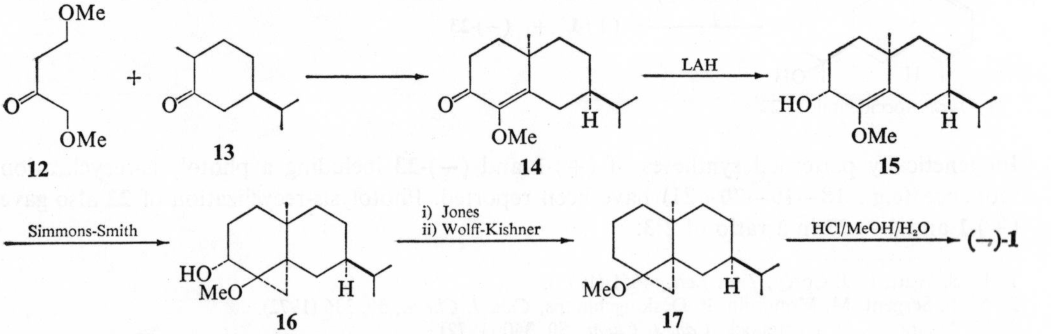

12 + **13** → **14**

LAH → **15**

Simmons-Smith → **16**

i) Jones
ii) Wolff-Kishner → **17**

HCl/MeOH/H₂O → (−)-**1**

1) J. A. Marshall, G. L. Bundy, W. I. Fanta, *J. Org. Chem.*, **33**, 3913 (1968).
2) E. Wenkert, D. A. Berges, *J. Am. Chem. Soc.*, **89**, 2507 (1967).

15→16: Methylenation of **15** by the Simmons-Smith reaction (CH_2Cl_2/Zn) occurs stereo-specifically from the same side (α) as the adjacent hydroxyl group.

3.59 GUAIANE DERIVATIVES AND GUAIANOLIDES

Many sesquiterpene hydrocarbons, alcohols, ethers and lactones possessing a guaiane skeleton have been isolated from plant sources. The carbon skeleton of compounds of this type is assumed to be derived from a germacradiene-type precursor. Representative compounds of this class are shown below.[1]

Hydrocarbons, alcohols, esters and ethers

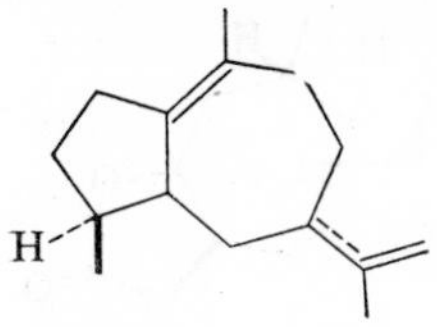

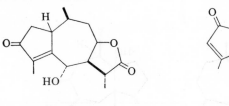

bulnesenes **1**	guaiol **2**	torilin **3**	liguloxide **4**

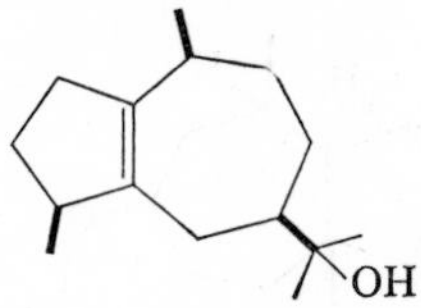

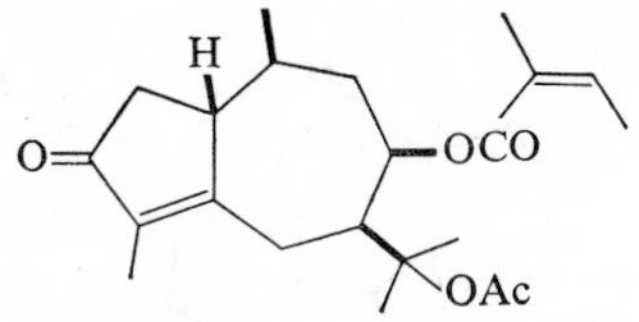

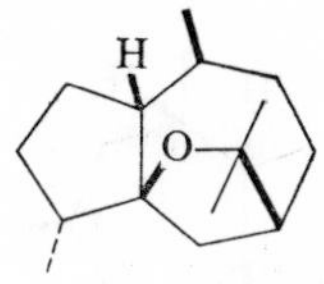

Guaianolides having saturated lactones

geigerin **5** 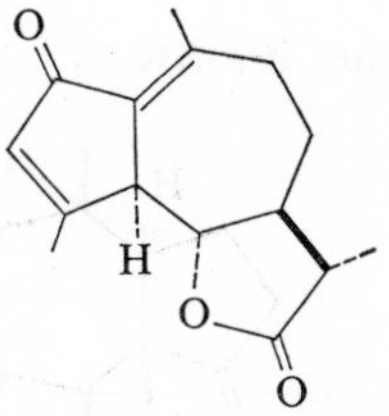   desacetoxymatricarin **6**

Guaianolides having unsaturated lactones

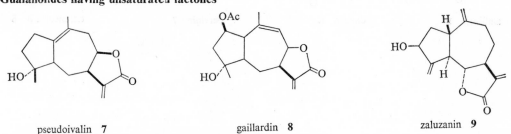

pseudoivalin **7**   gaillardin **8**   zaluzanin **9**

Guaianolides containing an additional ring

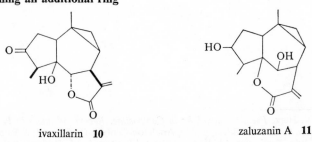

ivaxillarin **10**   zaluzanin A **11**

1) For reviews, see F. Sorm, L. Dolejs, *Guaianolides and Germacranolides*, Hermann (Paris), 1965.

3.60 PSEUDOGUAIANOLIDES

A series of hydroazulenic sesquiterpene lactones having an angular methyl group at the C-5β position, instead of at C-4, as in normal guaianolides, are known and are classified as pseudoguaianolides.[1]

Some members of this group have an α-methyl substituent at C-10, as in **1–4**, and some have a β-oriented methyl group at this position, as in **5–8**

Pseudoguaianolides having 10α-Me

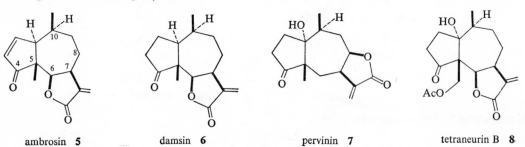

helenalin **1**   mexicanin I **2**   pullchelin **3**   mexicanin H **4**

Pseudoguaianolides having 10β-Me

ambrosin **5**   damsin **6**   pervinin **7**   tetraneurin B **8**

1) For reviews, see W. Herz, *Pseudoguaianolides in Compositae, Recent Advances in Phytochemistry* (ed. T. J. Mabry, R. E. Alston, V. C. Runeckles), p. 220, Appleton-Century-Croft, 1968; J. Romo, A. Romo de Vivar, *The Pseudoguaianolides, Progress in Chemistry of Natural Products* (ed. L. Zechmeister), vol. 25, p. 90, Springer-Verlag, 1967; T. J. Mabry, *Phytochemical Phylogeny* (ed. J. B. Harborne), p. 269, Academic Press ,1970.

3.61 SECOGUAIANOLIDES

Many sesquiterpene lactones are known whose carbon skeletons are derived by oxidative fission of a carbon-carbon linkage of a guaianolide or pseudoguaianolide.[1-3]

4,5-Secoguaianolides[1]

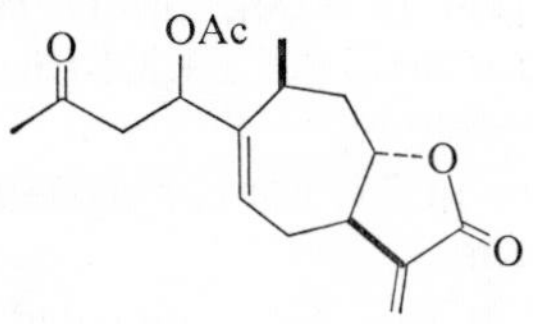

xanthinin **1** 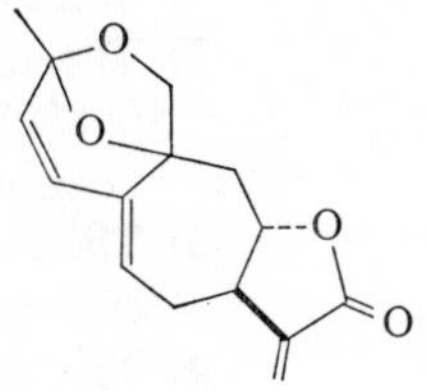　　　　griesenin **2** 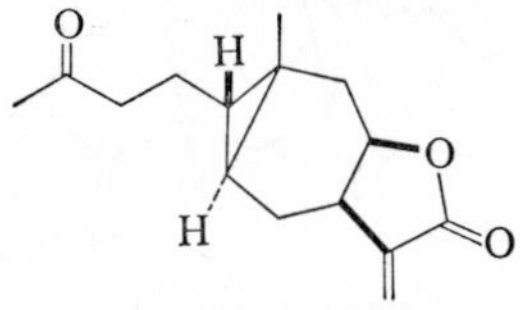　　　　carabrone **3**

Many related compounds, known generally as xanthanolides, have been isolated, mainly from *Xanthium* and related species, as well as from *Inuleae* species.

4,5-Secopseudoguaianolides[2]

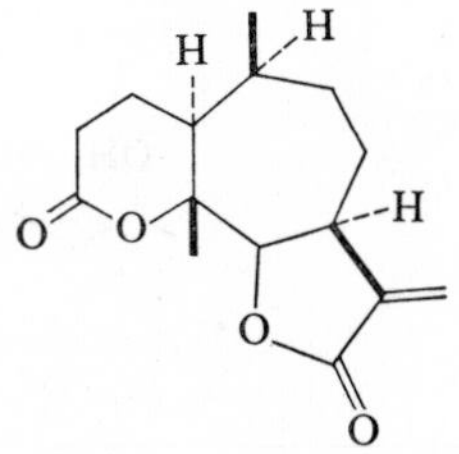

psilostachyin C **4** 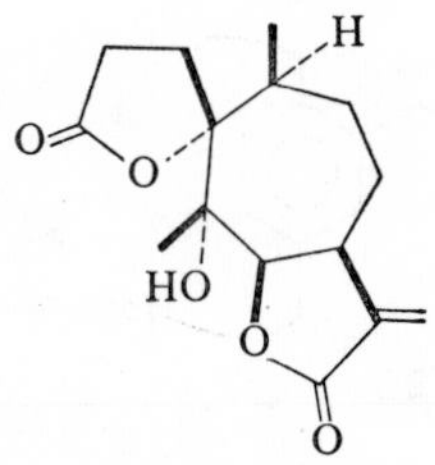　　　　psilostachin **5**

In vitro conversions of the pseudoguaianolides damsin and coronopilin into psilostachyin C and psilostachyin, respectively, were performed by peracetic acid oxidation.

Psilostachyins have been isolated from *Ambrosia puloclachya* and related species.

3,4-Secopseudoguaianolides[3]

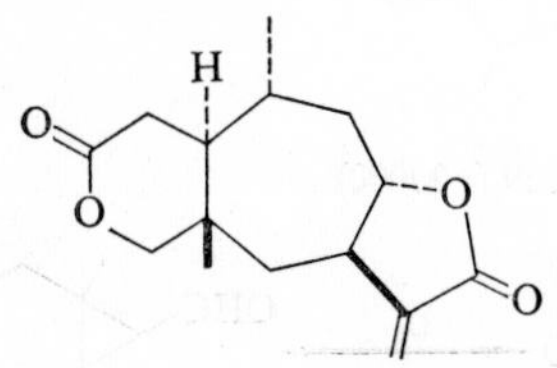

vermeerin **6**

Vermeerin and floribundin (C-8 epimer) were isolated from *Hymenoxys* species.

1) T. A. Geissman, *J. Org. Chem.*, **27**, 2692 (1962); For leading references, see W. Herz, S. V. Bhat, A. L. Hall *J. Org. Chem.*, **35**, 1110 (1970).
2) T. J. Mabry, W. Renold, H. E. Miller, H. B. Kagan, *J. Org. Chem.*, **31**, 681 (1966); *ibid.*, **31**, 1629 (1966); H. B. Kagan, H. E. Miller, W. Renold, M. V. Lakshmikantham, L. R. Tether, W. Herz, T. J. Mabry, *ibid.*, **31**, 1629 (1966).
3) For leading references, see W. Herz, K. Aota, A. L. Hall, *J. Org. Chem.*, **35**, 4117 (1970); W. Herz, K. Aota, M. Holub, Z. Samek, *ibid.*, **35**, 2611 (1970).

3.62 STRUCTURE AND SYNTHESIS OF CARABRONE

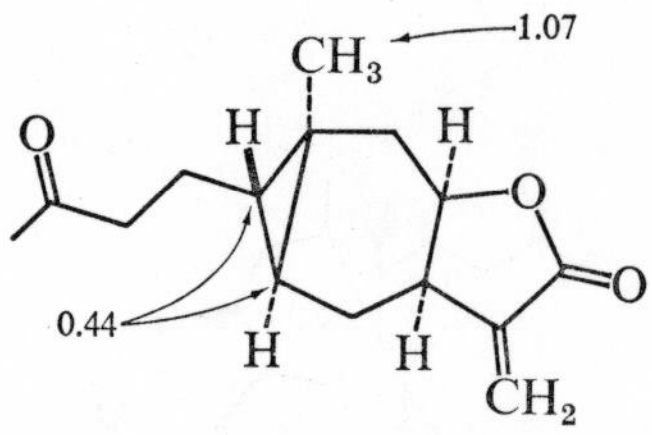
CH_3
1.07
H
H
O
O
O
0.44
H
H
CH_2

carabrone **1**[1)]

mp: 90–91° α_D(EtOH): +116.9°
uv(EtOH): 213 (ϵ 8150)
ir($CHCl_3$): 1712 (C=O), 1758 (unsat. γ-lactone), 3100, 1665, 1266, 822 (double bond)

1) The uv and ir spectra of **1** indicate the presence of an α,β-unsaturated γ-lactone.
2) The uv of **2** [217 (ε 9700), 267 (ε 777), 275.5 (ε 830)] is characteristic of a 1,3,4-trisubstituted benzene chromophore.
3) The ring ketone in **5** is not conjugated with the cyclopropane ring.
4) The aldehyde in **6** is not conjugated with the cyclopropane ring.
5) The double bond in **8** and the aldehyde in **9** are conjugated with the cyclopropane ring.
6) The structure was confirmed by total synthesis[2)] of (±)-**1**.

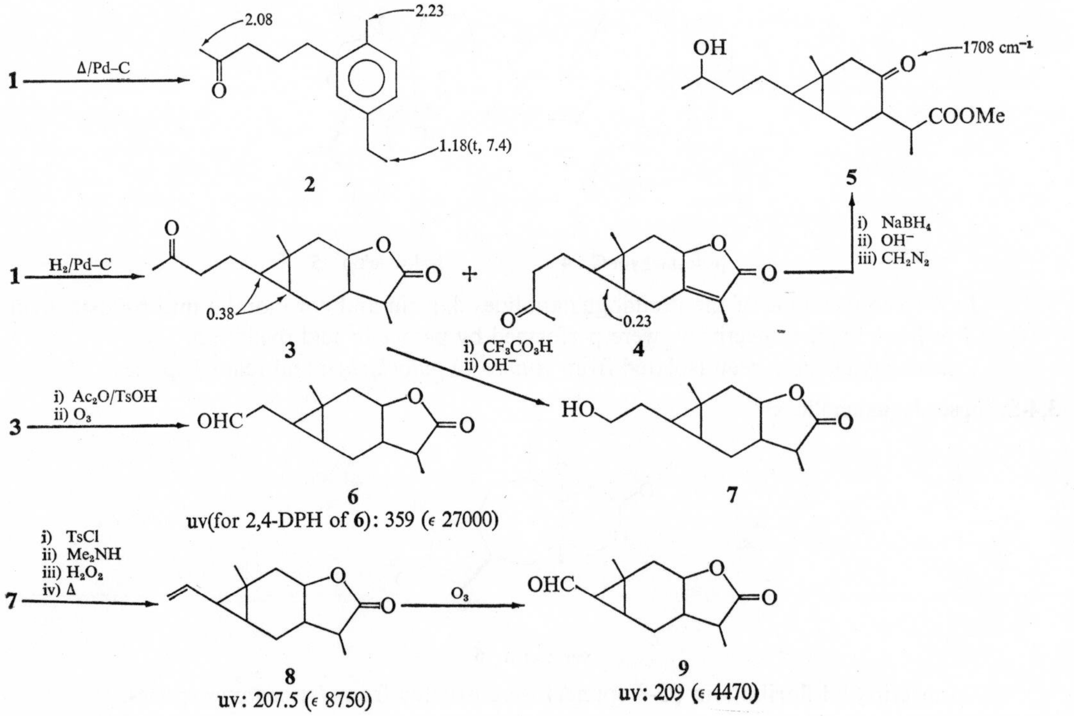
1
Δ/Pd–C
2.08
2.23
O
1.18(t, 7.4)
2
OH
O
1708 cm⁻¹
COOMe
5
i) NaBH₄
ii) OH⁻
iii) CH₂N₂
1
H₂/Pd–C
O
O
O
0.38
3
+
O
O
O
0.23
4
i) CF₃CO₃H
ii) OH⁻
3
i) Ac₂O/TsOH
ii) O₃
OHC
O
O
6
HO
O
O
7
uv(for 2,4-DPH of 6): 359 (ε 27000)
7
i) TsCl
ii) Me₂NH
iii) H₂O₂
iv) Δ
O
O
O₃
OHC
O
O
8
uv: 207.5 (ε 8750)
9
uv: 209 (ε 4470)

1) H. Minato, S. Nosaka, I. Horibe, *J. Chem. Soc.*, 5503 (1964).
2) H. Minato, I. Horibe, *J. Chem. Soc.* C, 2131 (1968).

Synthesis[2]

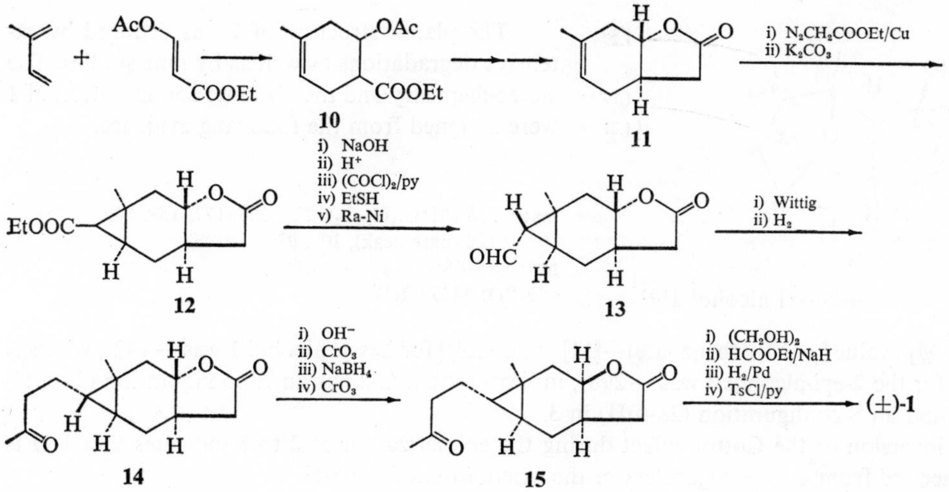

Remarks

The secoguaianolide carabrone **1** was isolated from fruits of *Carpesium abrotanoides*.

3.63 STRUCTURE OF KESSYL ALCOHOL

α-kessyl alcohol **1**[1,2)]

The planar structure of **1** was deduced by extensive degradations as well as by nmr studies. The stereochemistry and the absolute configuration of **1** were assigned from the following evidence.

ms: 238 ($M^+C_{15}H_{26}O_2$), 223, 205, 177, 159, 149, 126 (base peak), 108, 97
mp: 85–86° α_D: −38.4°
ir(KBr): 3425 (OH)

1) The $[M]_D$ value $[=[M]_D$ (benzoate)$-[M]_D$ (alcohol)] for kessyl alcohol **1** was −192°, whereas that for the 2-epi-alcohol **3** was +298°, indicating that C-2 has an R configuration (2β-OH) in **1** and an S configuration (2α-OH) in **3**.

2) The inversion of the Cotton effect during the epimerization of **2** to **5** indicates that 1-H is epimerized from α to β, regardless of the orientation of the 5-H.

3) An appreciable downfield shift of the 4-Me signal is observed in **3**, indicating that 2-OH and 4-Me are *cis* in the epi series.

4) The coupling constant of the 1-H signal in the lactone **7** is 10.2 Hz, whereas in the lactone **8** it is 4.0 Hz, and therefore, it appears from the Karplus equation that 1-H and 5-H are *trans* in **7** and *cis* in **8**. Since Baeyer-Villiger oxidation of ketones is known to retain the stereochemistry, ring fusions in **2** and **5** are *trans* and *cis* respectively.

5) The oxide bridge was shown to be in the β orientation by observation of a concentration-independent band at 3570 cm⁻¹ in **1** (intramolecular hydrogen bonding).

Remarks

α-Kessyl alcohol **1** and kessyl glycol were isolated from Japanese valerian root (*Valeriana officinalis*).

1) S. Ito, M. Kodama, T. Nozoe, H. Hikino, Y. Hikino, Y. Takeshita, T. Takemoto, *Tetr.*, **23**, 553 (1967); *Tetr. Lett.*, 1787 (1963).
2) H. Hikino, Y. Takeshita, Y. Hikino, T. Takemoto, S. Ito, *Chem. Pharm. Bull.*, **13**, 520 (1965); *ibid.*, **15**, 485 (1967).

3.64 SYNTHESIS OF HYDROAZULENE DERIVATIVES

There are a number of naturally occurring sesquiterpenoids possessing hydroazulene skeltons. In recent years, many synthetic approaches to this class of compounds have been devised.

Route I

Rearrangement of a preformed hydronaphthalene derivative into a hydroazulene is frequently used for construction of the carbon skeleton.[1-4] Thus, for example, solvolytic rearrangement of the sulfonates **1**[1] and **6**[2] and the bromide **10**[3] yield the hydroazulenes **2, 7** and **11**, respectively. **2** is easily converted into **3** by a Wittig reaction.

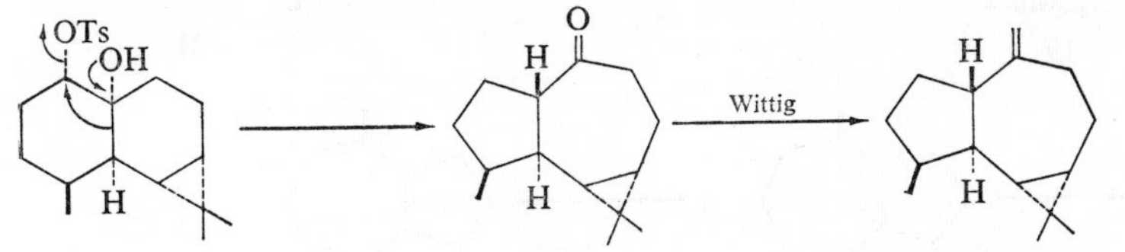

1   2   (−)-aromadendrene   3

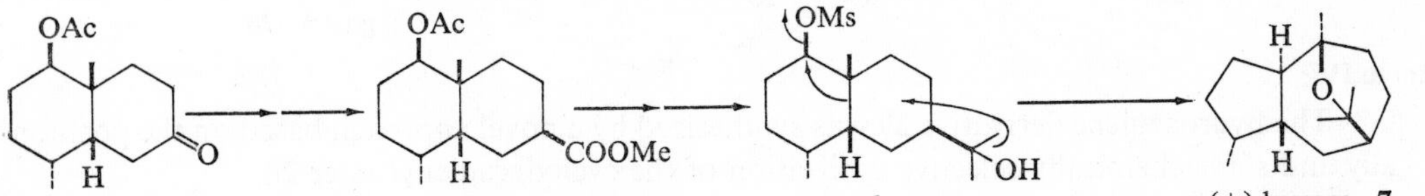

4   5   6   (+)-kessane   7

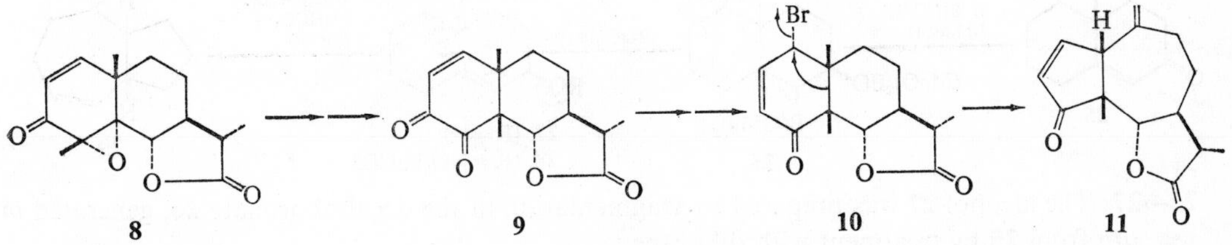

8   9   10   11

Route II[5]

Hydroazulene derivatives are also obtainable by solvolytic rearrangement of the bicyclo-[4.3.1]dec-10-yl system **15**, which is stereospecifically derived from **12**.

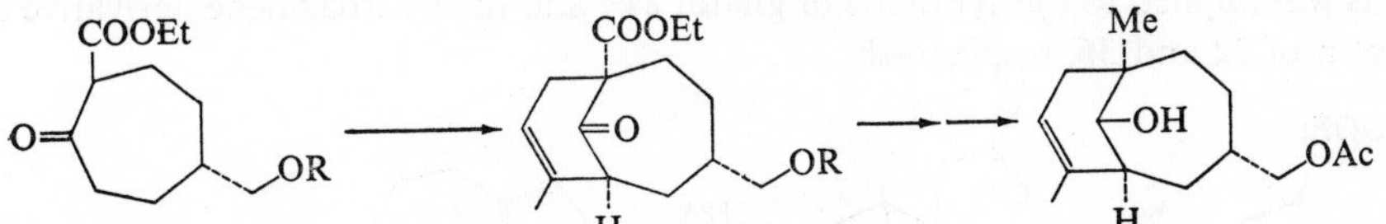

12   13   14

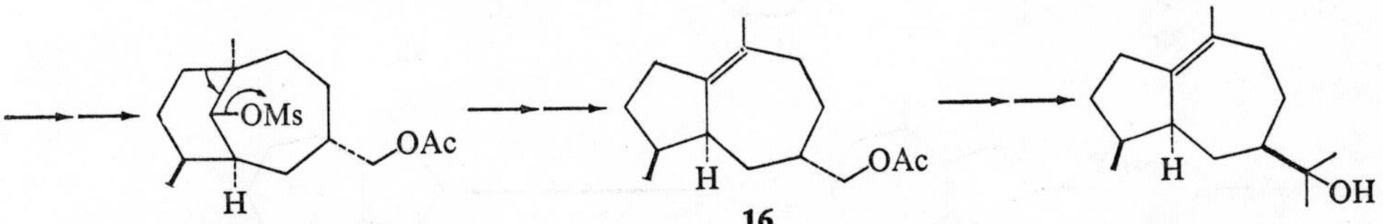

15   16   bulnesol   17

1) G. Büchi, W. Hofheinz, J. V. Pauksteils, *J. Am. Chem. Soc.*, **91**, 6473 (1969).
2) M. Kato, H. Kosugi, A. Yoshikoshi, *Chem. Commun.*, 934 (1970).
3) J. B. Hendrickson, C. Ganter, D. Dorman, H. Link, *Tetr. Lett.*, 2235 (1968).
4) For analagous syntheses, see: C. H. Heathcock, R. Ratcliffe, *J. Am. Chem. Soc.*, **93**, 1746 (1971); M. Kato, H. Kosugi, A. Yoshikoshi, *Chem. Commun.*, 185 (1970); C. H. Heathcock, R. Ratcliffe, J. Van, *J. Org. Chem.*, **37**, 1796 (1972).

Route III[6]

An alternative stereoselective route to guaiol **24**, involving the solvolytic rearrangement of hydrindanyl mesylate **20** to the hydroazulenic acetate **23** as a key step, has been reported. The rearrangement must proceed through the ions **21** and **22**.

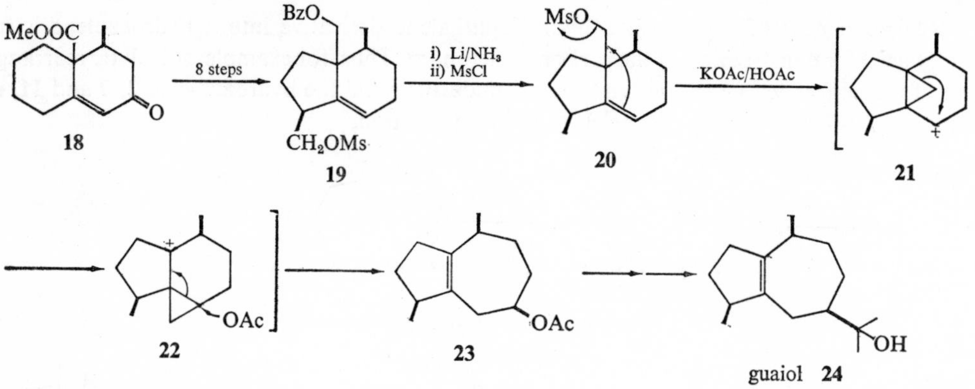

Route IV[7]

The hydroazulene derivative **29** was synthesized by a novel approach based on the positionally and stereochemically selective cyclization of the cyclodecadienyl ester **28**.

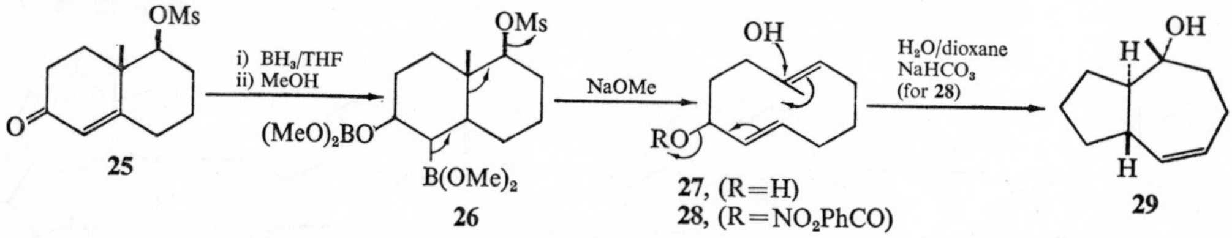

25→27: The alcohol **27** was prepared by fragmentation of the decalinboronate **26**, generated *in situ* from **25** by treatment with diborane.

Route V[8,9]

Bond cleavage of bridged bicyclic ketones provides a useful route to substituted medium-sized rings. This was applied to the synthesis of guaiol **24**[8] and the hydroazulene derivative **37**[9] by bridge scission of **32** and **36**, respectively.

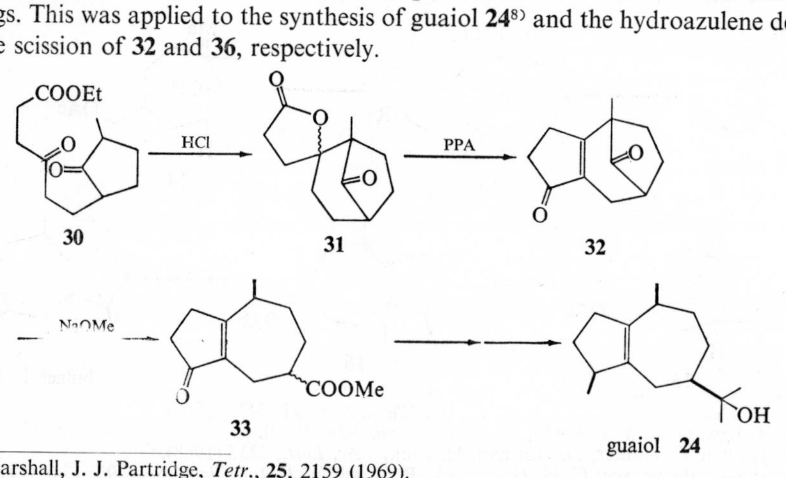

5) J. A. Marshall, J. J. Partridge, *Tetr.*, **25**, 2159 (1969).
6) J. A. Marshall, A. E. Greene, *J. Org. Chem.*, **37**, 982 (1972); cf. C. J. V. Scanio, L. P. Hill, *Chem. Commun.*, 242 (1971).

34 35 36 37

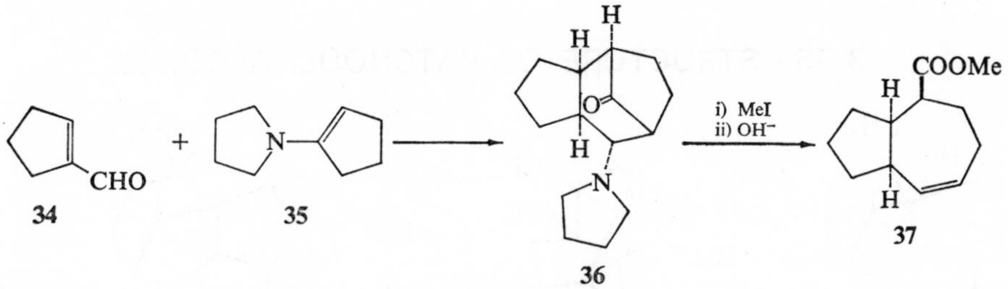

Route VI

Photochemical rearrangement products such as isophotosantonic acid lactone **39**, obtained by irradiation of santonin **38**, are potentially useful intermediates for the partial synthesis of hydroazulene derivatives.[10] For example, desacetoxymatricarin **40** was synthesized by a four-step reaction from **39**, which determines the stereochemistry of the product.[11] α-Bulnesene **44** was also derived from **41** by a photochemical approach.[12]

38 39 40

41 42 43 α-bulnesene 44

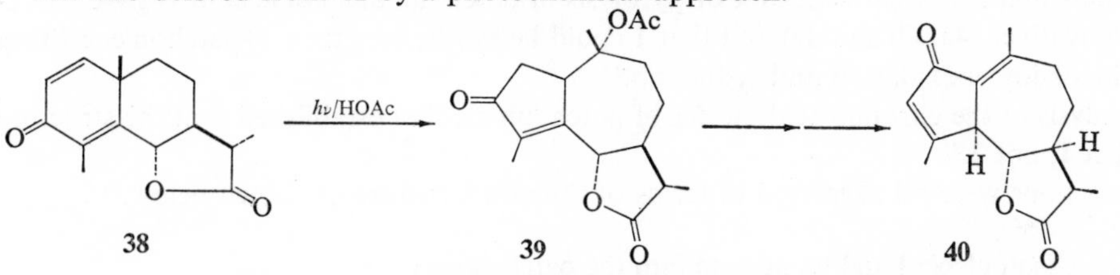

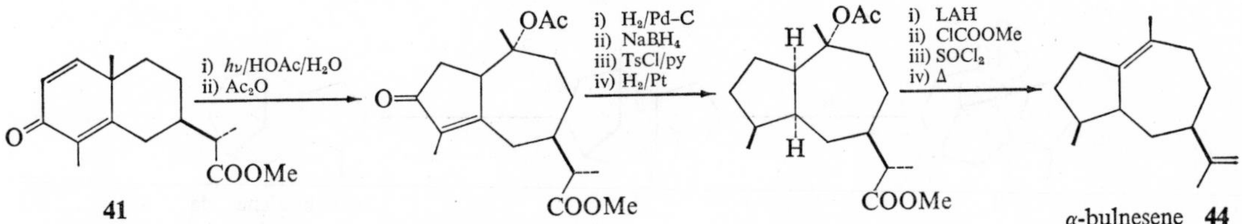

Route VII[13]

A stereoselective synthesis of hydroazulenes based on aldehyde-olefin cyclization (e.g. **49**→**50**) has been reported.[14] This synthesis led to a revision of the previously formulated structure of vetivone.

45 46 47 48 49 50

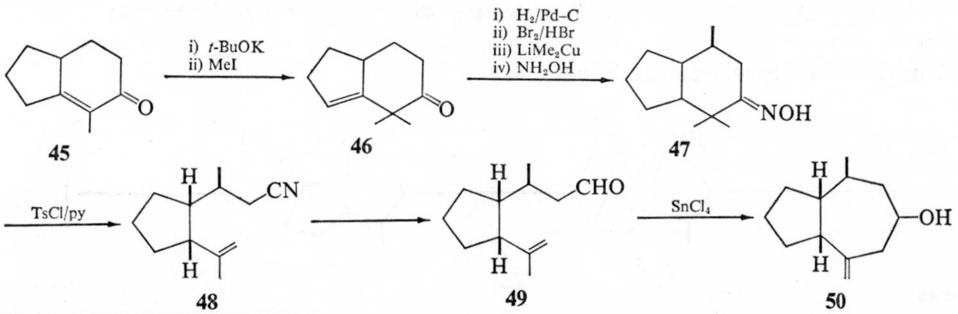

7) J. A. Marshall, W. F. Huffman, *J. Am. Chem. Soc.*, **92**, 6358 (1970); J. A. Marshall, W. F. Huffman, J. A. Ruth *ibid.*, **94**, 4691 (1972).
8) G. L. Buchanan, G. A. R. Young, *Chem. Commun.*, 643 (1971).
9) J. B. Hendrickson, R. K. Boeckman Jr., *J. Am. Chem. Soc.*, **93**, 1307 (1971).
10) D. H. R. Barton, P. Levsalles, J. T. Pinhey, *J. Chem. Soc.*, 3472 (1962); D. H. R. Barton, J. T. Pinhey, R. J. Wells, *ibid.*, 2518 (1964).
11) J. N. Marx, E. H. White, *Tetr.*, **25**, 2117 (1969); E. H. White, S. Eguchi, J. N. Marx, *ibid.*, **25**, 2099 (1969); c.f. M. Suchy, V. Herout, F. Sorm, *Coll. Czech. Chem. Commun.*, **29**, 1829 (1964).
12) E. Piers, K. F. Cheng, *Chem. Commun.*, 562 (1969).
13) J. A. Marshall, N. H. Andersen, P. C. Johnson, *J. Org. Chem.*, **35**, 186 (1970).
14) For recent synthesis of **24**, see N. H. Andersen, H. Uh, *Tetr. Lett.*, 2079 (1973).

3.65   STRUCTURE OF PATCHOULI ALCOHOL

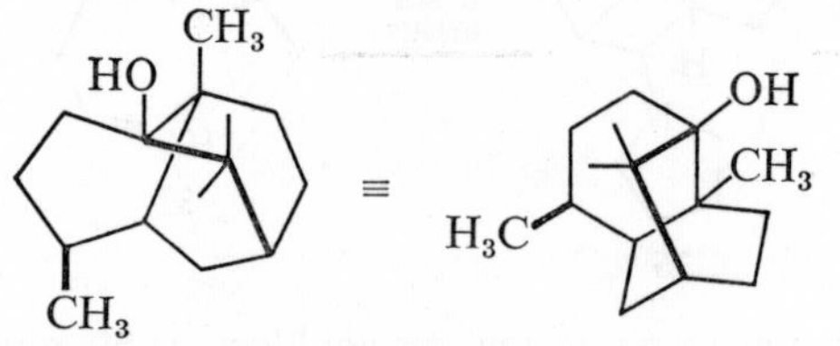

patchouli alcohol **1** 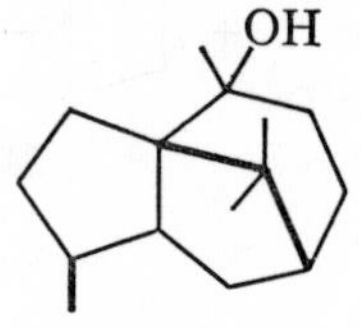   **2**   (incorrect structure for **1**)

1) The structure of patchouli alcohol had erroneously been considered to be **2**, based on the observations that acid treatment of **1** or pyrolysis of patchouli alcohol acetate **5** gave patchoulene derivatives (**4a**, **4b** and **4c**) and that **1** could be synthesized from α-patchoulene **4b** via a route involving epoxidation and reduction.[1]

2) X-ray analysis of the chromic acid diester of patchouli alcohol established that the structure was in fact **1**, not **2**.[2]

3) The conversions were reinterpreted in terms of the new structure as shown below.[3]

Conversion of patchouli alcohol and its acetate into the patchoulenes

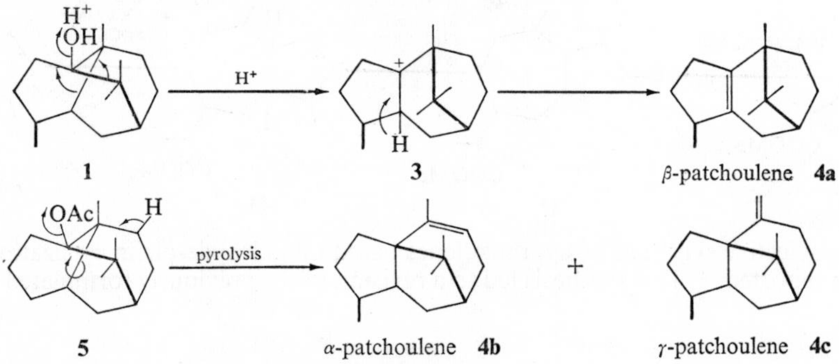

1   **3**   β-patchoulene **4a**

5   α-patchoulene **4b**   γ-patchoulene **4c**

Partial synthesis of 1 from α-patchoulene 4

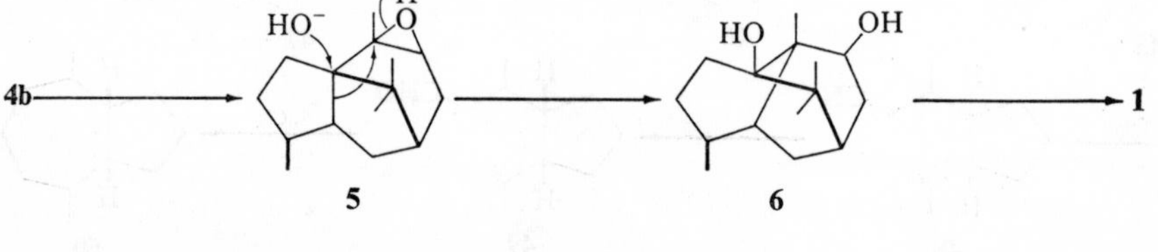

4b   **5**   **6**   **1**

Remarks

Patchouli alcohol **1** is a constituent of the East Indian shrub *Pogostemon patchouli* and a major component of commercial patchouli oil.

1) G. Büchi, W. D. MacLeod Jr, *J. Am. Chem. Soc*, **84**, 3205 (1962); G. Büchi, R. E. Erickson, N. Wakabayashi, *ibid*, **83**, 927 (1961).
2) M. Dobler, J. D. Dunitz, B. Gubler, H. P. Weber, G. Büchi, J. Padilla O., *Proc. Chem. Soc.*, 383 (1963).
3) G. Büchi, W. D. MacLeod Jr., J. Padilla, *J. Am. Chem. Soc.*, **86**, 4438 (1964).
4) For total synthesis of **1**, see R. N. Mirrigton, K. J. Schmalzl, *J. Org. Chem.*, **37**, 2871 (1972) and references therein.

3.66 SYNTHESIS OF SEYCHELLENE

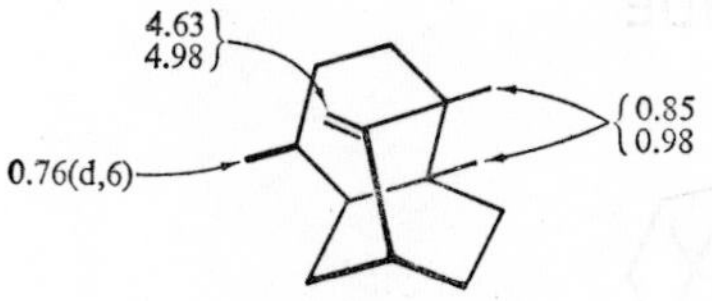

seychellene **1**[1]

ms: 204 (M$^+$ C$_{15}$H$_{24}$)
ir: 1640, 880 (*exo*-methylene)

Three different approaches for construction of the tricyclic carbon framework of seychellene have been successfully employed, and are shown below.

Route I[2]

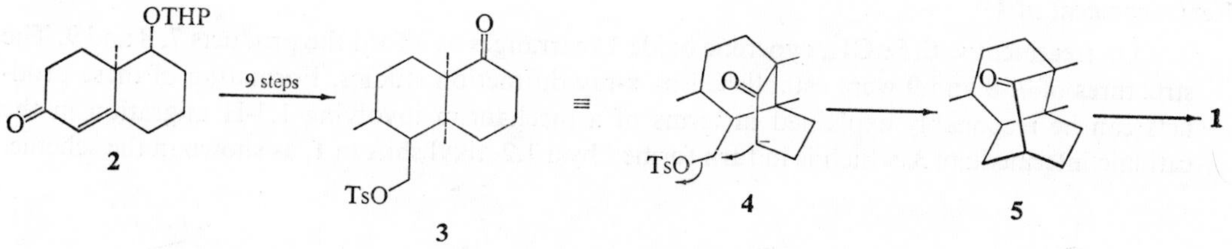

Route II[3]

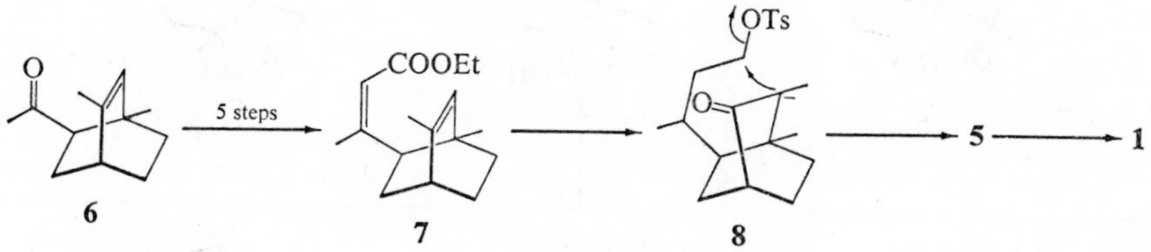

Route III[4]

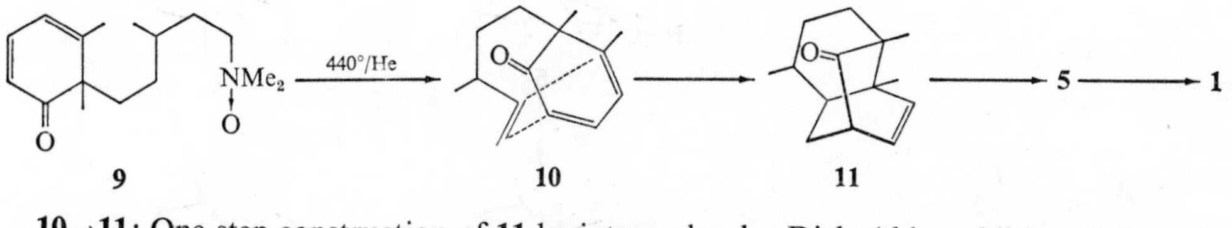

10→11: One-step construction of **11** by intramolecular Diels-Alder addition of the monocyclic intermediate **10** is a key step of this synthesis.

Remarks

Seychellene **1** was isolated from the essential oil of *Pogostemon patchouli* (patchouli oil).[1,5]

Patchouli alcohol **12**, obtained from the same source, has a structurally and biogenetically close relationship with seychellene **1**.

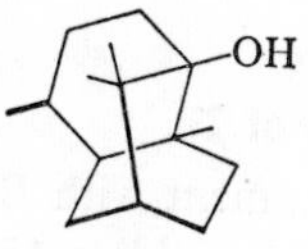

1) G. Wolff, G. Ourisson, *Tetr. Lett.*, 3849 (1968); *Tetr.*, **25**, 4903 (1969).
2) E. Piers, R. W. Britton, W. de Waal, *Chem. Commun.*, 1069 (1969).
3) K. J. Schmalzl, R. N. Mirrington, *Tetr. Lett.*, 3219 (1970); *J. Org. Chem.*, **37**, 2877 (1972).
4) N. Fukamiya, M. Kato, A. Yoshikoshi, *Chem. Commun.*, 1120 (1971); *J. Chem. Soc.*, Perkin I, 1843 (1973).
5) See also S. J. Terhune, J. W. Hogg, B. M. Lawrence, *Tetr. Lett.*, 4705 (1973).

3.67 REARRANGEMENT OF CYPERENE OXIDE AND β-PATCHOULENE OXIDE

cyperene oxide **1**   β-patchoulene oxide **2**

Rearrangement of 1[1]

On treatment with SnCl$_4$, cyperene oxide **1** rearranges to afford the products **7**, **8** and **9**. The structures of **7**, **8** and **9** were established by x-ray diffraction studies. Formation of these products can be reasonably explained in terms of a mechanism involving 1,3-H migration in the cationic intermediate **3**, which is in turn formed by a 1,2-alkyl shift in **1**, as shown in the scheme.

cyperene oxide **1**   **3**   **4**   **7**

5   **8**

6   **9**

Rearrangement of 2[2]

On treatment with BF$_3$ or SnCl$_4$, patchoulene oxide **2** undergoes a deep-seated skeletal rearrangement yielding **15**, **16**, and **17**. Formation of these products was interpreted in terms of the mechanisms outlined below.

---

1) L. Bang, G. Ourisson, *Tetr. Lett.*, 3761 (1969); *Tetr.*, **29**, 2097 (1973); cf. L. Bang, M. A. Diaz-Parra, G. Ourisson, *ibid.*, **29**, 2087 (1973).
2) L. Bang, I. G. Guest, G. Ourisson, *Tetr. Lett.*, 2089 (1972).

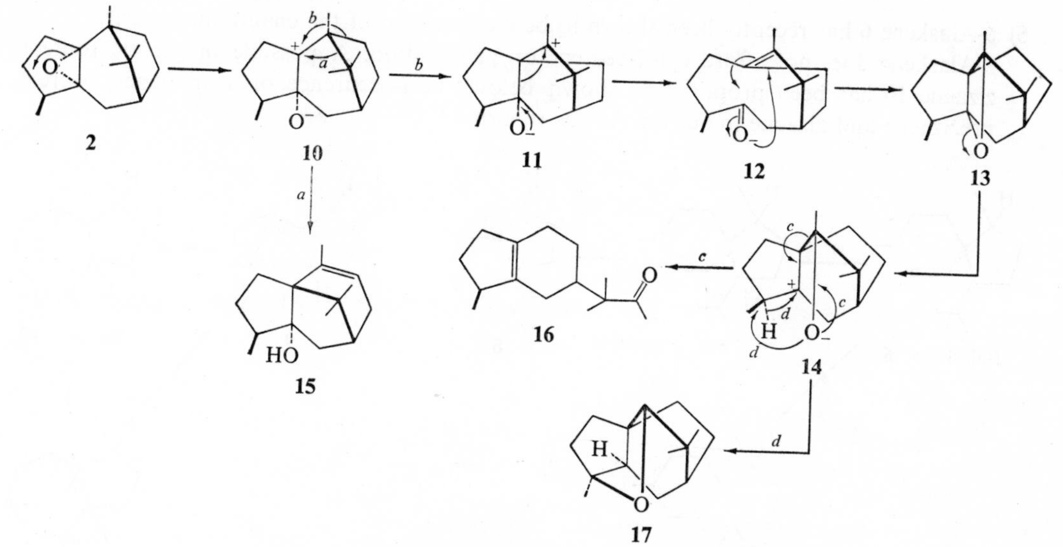

3.68 ACORADIENES AND ALASKENES

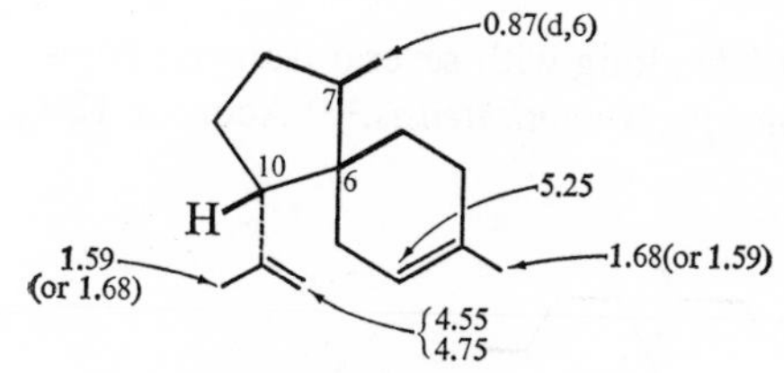

α-acoradiene **1**[1]

ms: 204 (M⁺ C₁₅H₂₄)

α_D: —20°

ir(film): 3070, 1654, 890 (=CH₂)
805(trisubst. double bond)

1) Tetrahydroacoradiene was found to be identical with acorane derived from acorone **12**.

2) Formation of (−)-α-cedrene **5** established the absolute configuration of C-6, C-7 and C-10 of **1**.

3) α-Alaskene **2**[2] and α-acorenol **3**[1] both give rise to α-cedrene **5** on acid treatment.

4) Synthesis of (±)-α-cedrene by a route involving cyclization of the cationic intermediate **4** has been accomplished.[3]

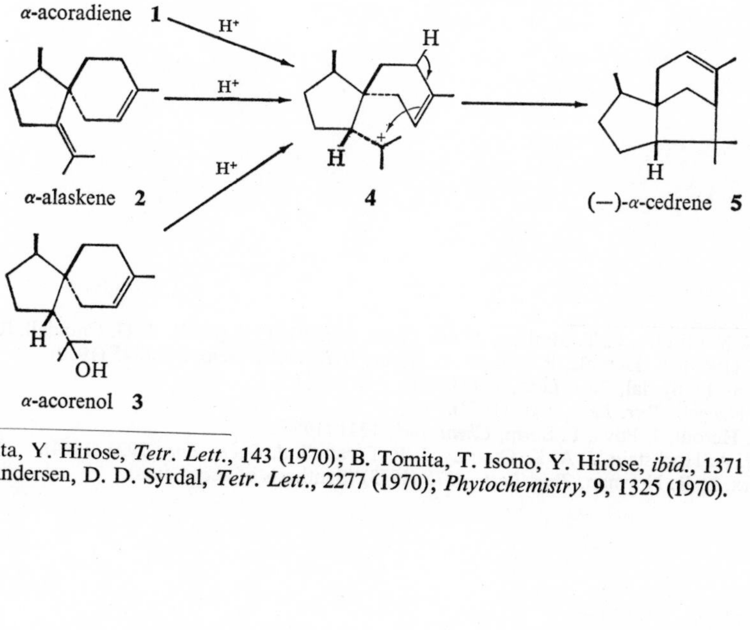

1) B. Tomita, Y. Hirose, *Tetr. Lett.*, 143 (1970); B. Tomita, T. Isono, Y. Hirose, *ibid.*, 1371 (1970).
2) N. H. Andersen, D. D. Syrdal, *Tetr. Lett.*, 2277 (1970); *Phytochemistry*, **9**, 1325 (1970).

5) β-Alaskene **6** has recently been shown to be a compound of the enantiomeric series.[4]
β-Alaskene does not afford epi-α-cedrene on protonation. A possible biogenetic route to
zizaene **10** has been proposed as shown below.[4] Co-occurence of acoradienes, cedrene,
prezizaene and zizaene in vetiver oil was reported.[5]

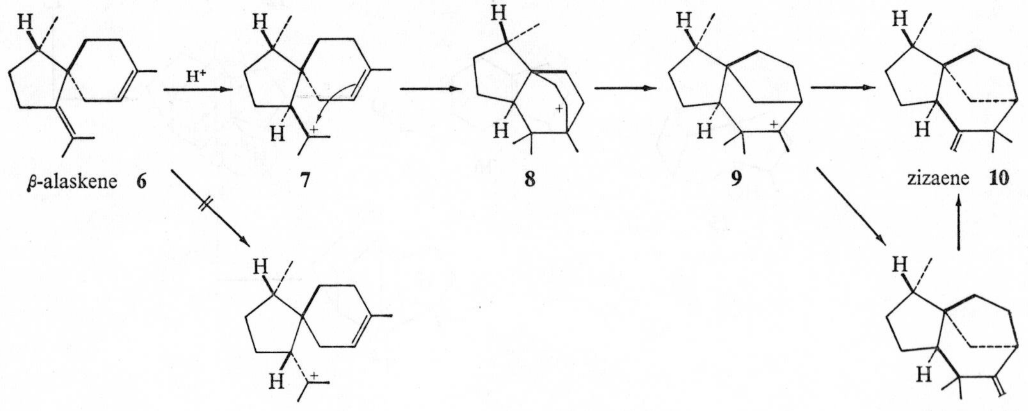

β-alaskene **6**   **7**   **8**   **9**   zizaene **10**

14   prezizaene **11**

Remarks

1 and **3** were isolated from the wood of *Juniperus ridida* along with several isomeric compounds.[1] **2** and **6** were obtained from leaf oil of *Chamaecyparis nootkatensis*.[2,4] Acorone **12**[6] and acoric acid **13**[7] were isolated from *Acorus calamus*.

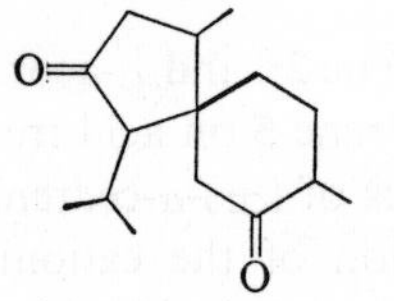

12 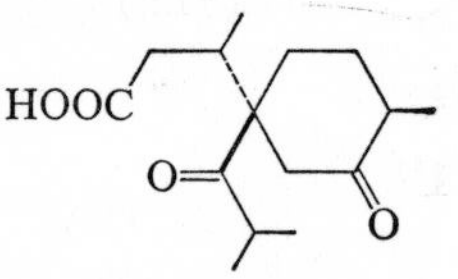   **13**

---

3) E. J. Corey, N. N. Girotra, C. T. Mathew, *J. Am. Chem. Soc.*, **91**, 1557 (1969); T. G. Crandall, R. G. Lawton, *ibid.*, **91**, 2127 (1969); E. Demole, P. Enggist, C. Borer, *Helv. Chim. Acta*, **54**, 1845 (1970).
4) N. H. Andersen, D. Syrdal, *Tetr. Lett.*, 899 (1972).
5) R. Kaiser, P. Naegeli, *Tetr. Lett.*, 2009 (1972).
6) V. Sykora, V. Herout, J. Pliva, F. Sorm, *Chem. Ind.*, 1231 (1956).
7) A. J. Birch, F. A. Hochstein, J. A. K. Quartey, J. P. Turnbull, *J. Chem. Soc.*, 2923 (1964).
8) See J. N. Marx, L. R. Norman, *Tetr. Lett.*, 4375 (1973) (Synthesis of acordienes).

3.69 SYNTHESIS OF CEDROL AND CEDRENE

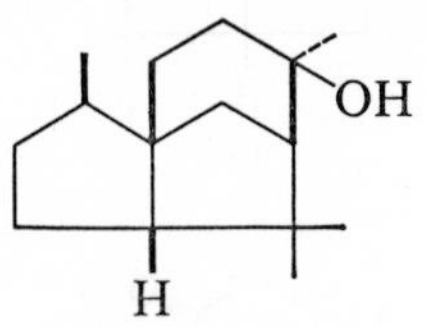

cedrol **1**

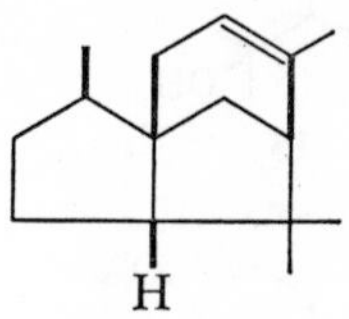

cedrene **2**

Route I[1]

This route to cedrol **1** involves a stepwise construction of the tricyclic ring system. The synthesis is characterized by the high stereospecificity in each reaction step.

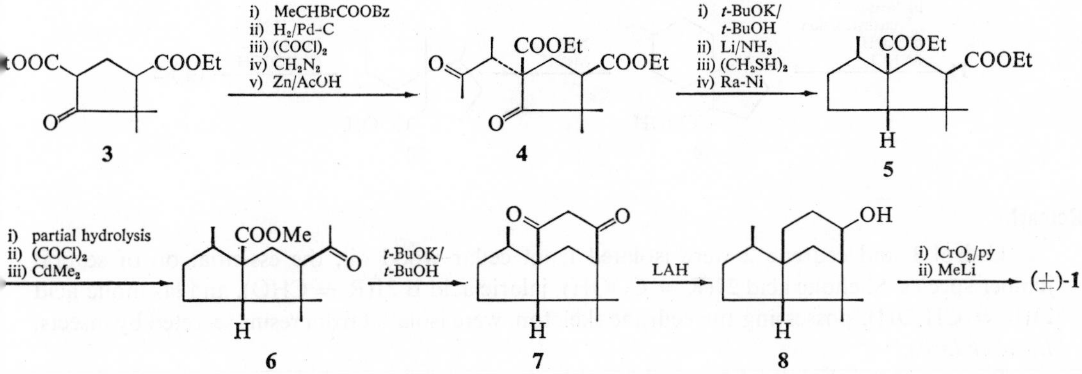
i) MeCHBrCOOBz
ii) H_2/Pd–C
iii) $(COCl)_2$
iv) CH_2N_2
v) Zn/AcOH
COOEt
i) t-BuOK/ t-BuOH
ii) Li/NH_3
iii) $(CH_2SH)_2$
iv) Ra-Ni
3
4
5
i) partial hydrolysis
ii) $(COCl)_2$
iii) $CdMe_2$
COOMe
t-BuOK/ t-BuOH
LAH
OH
i) CrO_3/py
ii) MeLi
(±)-1
6
7
8

Route II[2,3]

Cedrene **2** was synthesized by a route involving cyclization via cations **13** and **14**. The sequence **13**→**2** parallels the proposed biosynthetic pathway for cedrene.[2]

The conversions, **15**→**16**,[2] **17**→**18**,[2] and **20**→**2**[3] also proceed through corresponding cations.[4]

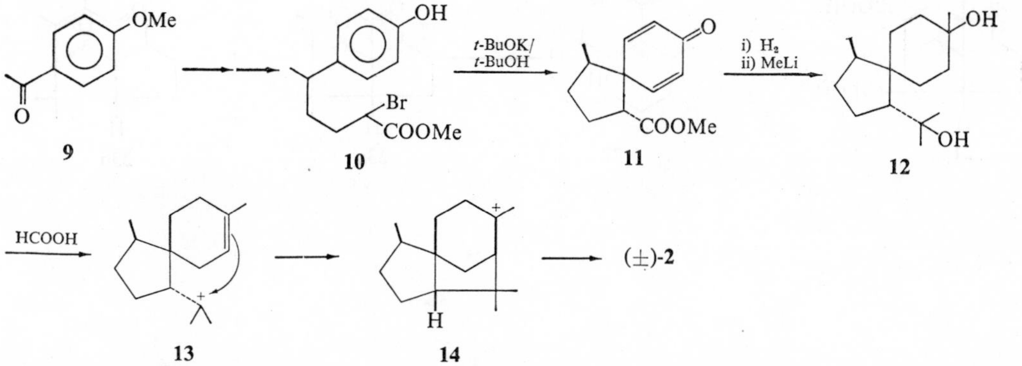
OMe
9
OH
Br
COOMe
10
t-BuOK/ t-BuOH
COOMe
11
i) H_2
ii) MeLi
OH
OH
12
HCOOH
13
14
(±)-2

1) G. Stork, F. H. Clarke Jr., *J. Am. Chem. Soc.*, **83**, 3114 (1961).
2) E. J. Corey, N. N. Girotra, C. T. Mathew, *J. Am. Chem. Soc.*, **91**, 1557 (1969).
3) T. G. Crandall, R. G. Lawton, *J. Am. Chem. Soc.*, **91**, 2127 (1969).
4) See also "Acoradienes and alaskenes" and "Cyclization of nerolidol"; E. J. Corey, R. D. Balanson, *Tetr. Lett.*, 3153 (1973).

15 16

17 18

19 20

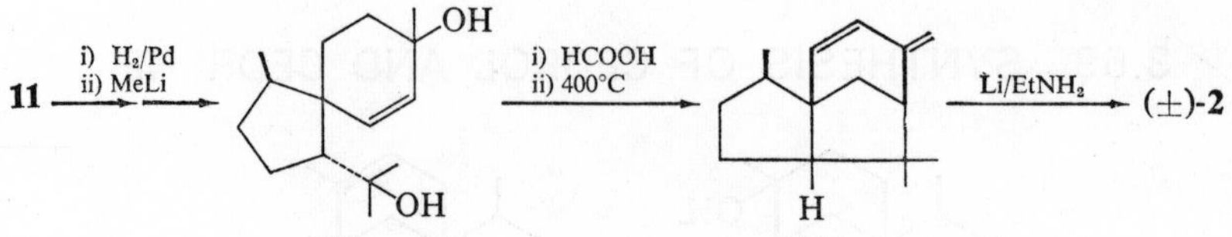

Remarks

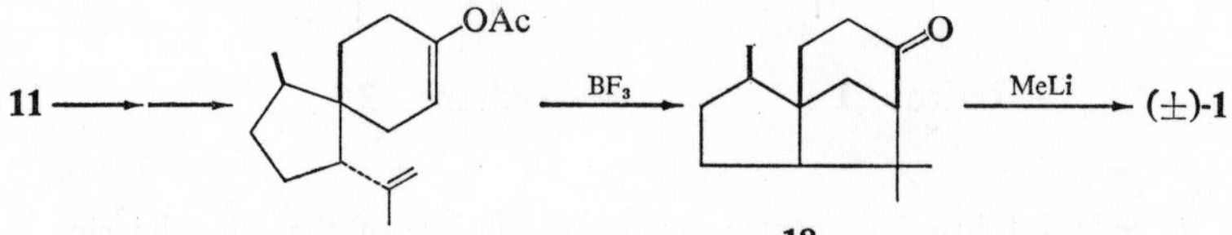

Cedrol **1** and cedrene **2** were isolated from cedar wood oil, the essential oil of several juniper species. Shellolic acid **21**(R = COOH), jalaric acid B **21**(R = CHO), and laksholic acid **21**(R = CH_2OH), possessing the cedrane skeleton, were isolated from resins secreted by insects, *Laccifer lacca.*[5]

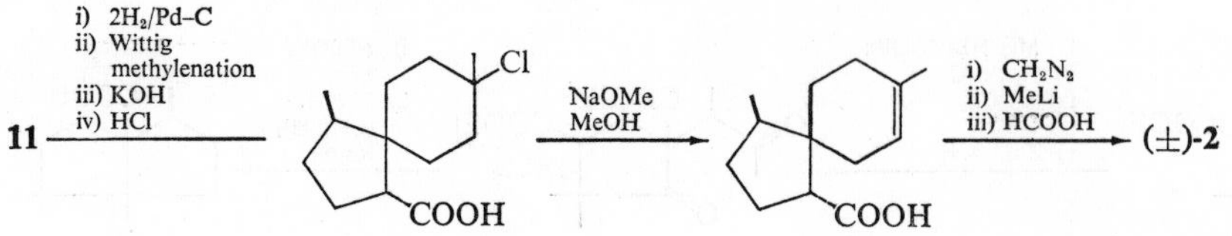

Perezone **22** and pipitzol **23** were found in the root of *Perrezia curenvaco.*[6] **23** was formed from **22** by thermal reaction. One of the isomers, α-pipitzol **23a**, possesses a carbon skeleton, including absolute stereochemistry, identical with that of **1**.

In addition, (+)-2,5-diepi-β-cedrene has been isolated from *Sciadopitys verticillata.*[7]

21 22 23a 23b

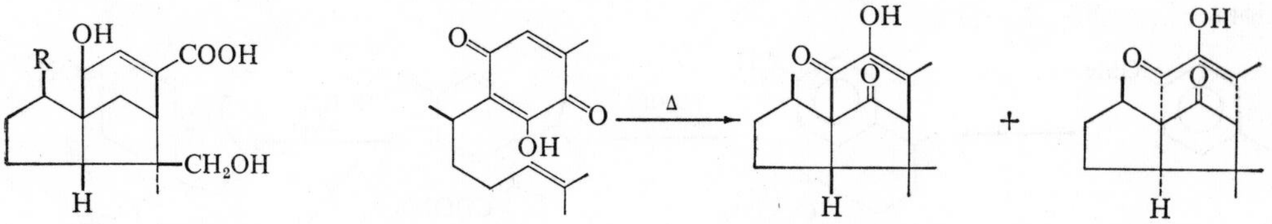

5) M. S. Wadia, R. G. Khurana, V. V. Mhaskar, S. Dev. *Tetr.*, **25**, 3841 (1969).
6) F. Walls, J. Padilla, P. J. Nathan, F. Giral, M. Escober, J. Romo, *Tetr.*, **22**, 2387 (1966); *Tetr. Lett.*, 1577 (1965); E. R. Wagner, R. D. Moss, R. M. Brooker, *ibid.*, 4233 (1965).
7) T. Norin, S. Sundin, B. Carlsson, P. Kierkegard, A-M. Pollotti, A-C. Wiehager, *Tetr. Lett.*, 17 (1973).

3.70 STRUCTURE OF ZIZANOIC ACID AND RELATED COMPOUNDS

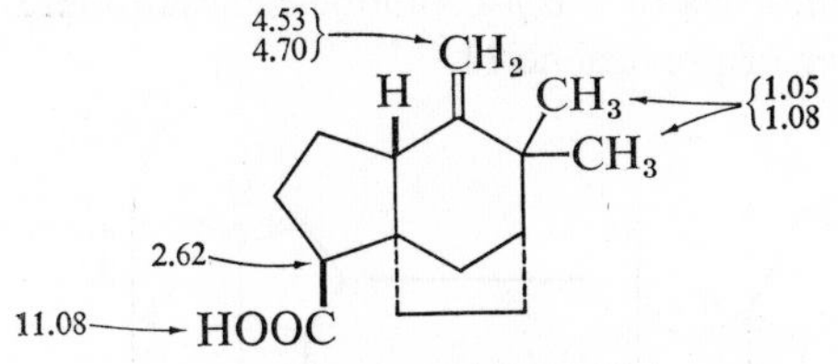

4.53 ⎱
4.70 ⎰ → CH$_2$

CH$_3$ ← ⎰ 1.05
CH$_3$ ⎱ 1.08

H

2.62 —

11.08 — → HOOC

zizanoic acid **1**[1)]
mp(cyclohexylamine salt): 146°
ir(film): 2600, 1700 (–COOH),
1640, 892(=CH$_2$)
nmr in CCl$_4$

1) The ord curve of **6** exhibits a strong negative Cotton effect, $[\phi]_{312} -5200$, $[\phi]_{278} +5200$, a = -104. Comparing these values with those of C/D *trans* steroidal compounds indicates both the presence of a *trans* indanone ring junction and the absolute configuration.

2) The ord curve of **2** shows a strong positive Cotton effect, $[\phi]_{312} +7200$, $[\phi]_{273} -6100$, a = $+133$. The octant rule predicts a *trans* ring junction of the indanone system in **2**.

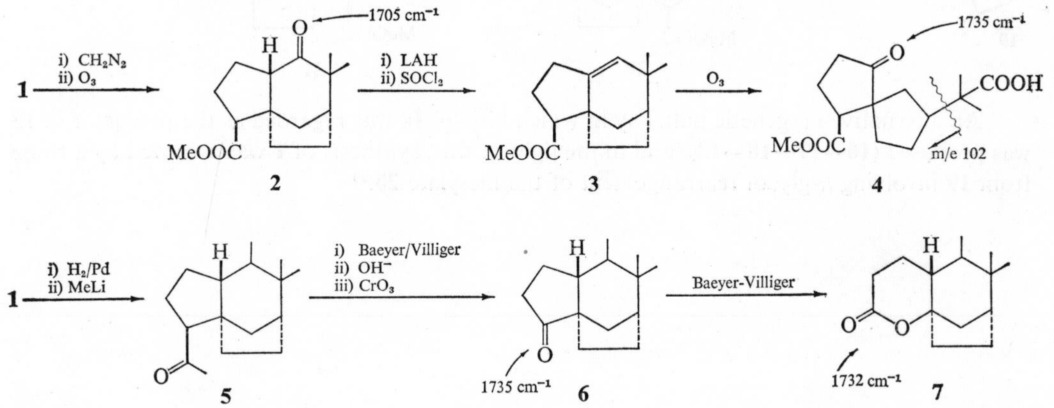

1 $\xrightarrow{\text{i) CH}_2\text{N}_2 \atop \text{ii) O}_3}$

←1705 cm^{-1}

MeOOC **2**

$\xrightarrow{\text{i) LAH} \atop \text{ii) SOCl}_2}$

MeOOC **3**

$\xrightarrow{\text{O}_3}$

←1735 cm^{-1}

COOH

MeOOC *m/e* 102 **4**

1 $\xrightarrow{\text{i) H}_2/\text{Pd} \atop \text{ii) MeLi}}$

O **5**

$\xrightarrow{\text{i) Baeyer/Villiger} \atop \text{ii) OH}^- \atop \text{iii) CrO}_3}$

O
1735 cm^{-1} **6**

$\xrightarrow{\text{Baeyer-Villiger}}$

O O
1732 cm^{-1} **7**

Remarks

FPP → →

8

→

9

→

10

→

prezizaene **11**

↓

12

↓

H H
zizaene **13**

β-acoradiene **14**

$\xrightarrow{\text{i) HCOOH} \atop \text{ii) KOH}}$

OH
allocedrol **15**

$\xrightarrow{\text{i) BsCl/py} \atop \text{ii) HCOOH}}$

enantiomer of **11**

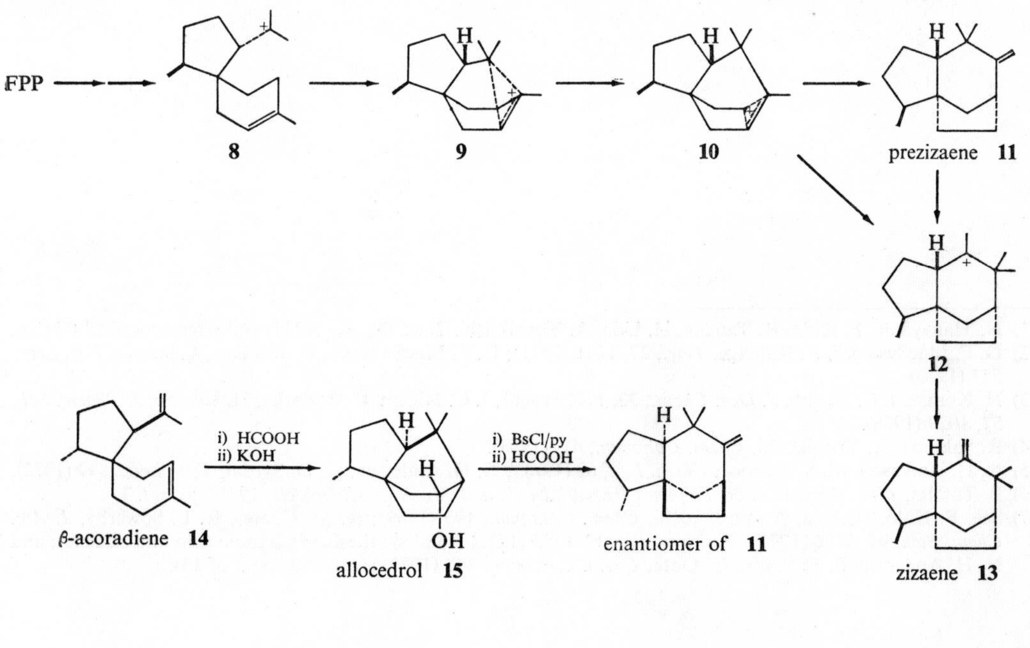

Zizanoic acid **1** (khusenic acid)[3], khusimol[3] and zizaene **13**[4] (tricyclovetivene) and epizizanoic acid[1] were isolated from vetiver oil. A possible biogenetic pathway for zizaene **13** and prezizaene **11** through the ions **9, 10** and **12** has been proposed[5,1] (see "Acoradienes and alaskenes").

In connection with this, it has been found that formolysis of the *p*-bromobenzenesulfonate of allocedrol **15**, obtained from **14** gave the enantiomer of prezizaene **11**.[6]

An alternative biogenetic pathway in which hinesol **16** was regarded as the precursor of **13** was proposed (**16→17→18→13**), and in the light of this, synthesis of **1** was achieved by a route from **19** involving α-glycol rearrangement of the mesylate **20**.[2]

1) N. Hanayama, F. Kido, R. Tanaka, H. Uda, A. Yoshikoshi, *Tetr.*, **29**, 945 (1973) and references cited therein.
2) D. F. MacSweeney, R. Ramage, *Tetr.*, **27**, 1481 (1971); D. F. MacSweeney, R. Ramage, A. Statter, *Tetr. Lett.*, 557 (1970).
3) H. Komae, I. C. Nigam, *J. Org. Chem.*, **33**, 1771 (1968); I. C. Nigam, C. Radecka, H. Komae, *J. Pharm. Sci.*, **57**, 1029 (1968).
4) R. Sakuma, A. Yoshikoshi, *Chem. Commun.*, 41, (1968).
5) N. H. Andersen, M. S. Falcone, *Chem. Ind.*, 62 (1971); N. H. Andersen, D. D. Syrdal, *Tetr. Lett.*, 899 (1972).
6) B. Tomita, *15th Symposium on Terpenes, Essential Oils and Aromatic Chemicals*, 1971; cf., *ref.* 1.
7) See, F. Kido, H. Uda, A. Yoshikoshi, *Chem. Commun.*, 195 (1969); R. M. Coates, R. L. Sowerby, *J. Am. Chem. Soc.*, **94**, 5386 (1972); *J. Chem. Soc. C*, 1755 (1972) (total synthesis of zizanoic acid and zizaene); and N. H. Andersen, S. E. Smith, Y. Ohta, *Chem. Commun.*, 447 (1973) (rearrangement of **13**).

3.71 STRUCTURE OF SIRENIN

sirenin **1**[1,2)]
viscous oil $\alpha_D(CHCl_3)$: $-45°$
ms: 236 (M$^+$ C$_{15}$H$_{24}$O$_2$)
ir: 3600 (OH)
nmr in CDCl$_3$

1) Upon esterification, sirenin affords bis-4-(4-nitrophenylazo)benzoate (NABz) **2**.

2) MnO$_2$ oxidation of **1** yields a dialdehyde, nmr: 9.41 (2H), ir: 2710, 1686, 1672 and 1630 cm^{-1}, uv: 232 and 260 nm.

3) Ozonolysis of the ester **2** yields three products, **3**, **4** and **5**, all of which contain the azo chromophore.

4) The downfield shift of the *tert*-Me signal (0.95→1.33) indicates a *cis* relationship between the Me group and the newly formed aldehyde in **4**.

5) The *J* value of the aldehydic H in **4** indicates that it must be on a cyclopropane ring.

6) Oxidation of **2** affords a tricarboxylic acid **6**.

7) The trimethyl ester of **6** and three of its stereoisomers were synthesized from compound **7** and diazoacetate.

8) Comparison of the solvent shift in the nmr spectra of the synthesized isomers established the stereochemistry of the natural triester as in **6**.

3 (R=NABz)   **4**   **5**

6 (R=H)   **7**

Remarks

Sirenin **1** is a sperm attractant produced by female gametes of water mold, *Allomyces sabuiscula* or *A. javanicus*, and is biologically active at concentrations of 10^{-10} M.[1)]

A compound possessing the structure of the parent hydrocarbon of sirenin (sesquicarene[3)]) has been isolated from essential oil of *Schysandra chreum*. The compounds have been chemically correlated.

1) L. Machlis, W. H. Nutting, M. W. Williams, H. Rapoport, *Biochemistry*, **5**, 2147 (1966).
2) L. Machlis, W. H. Nutting, H. Rapoport, *J. Am. Chem. Soc.*, **90**, 6434 (1968); *ibid.*, **90**, 1674 (1968).
3) Y. Ohta, Y. Hirose, *Tetr. Lett.*, 1251 (1968).

3.72 SYNTHESIS OF SIRENIN

Route I

Sirenin **1** was synthesized by the following reaction sequence, involving intramolecular addition of an α-keto-carbene to the *trans* olefinic linkage as the key step.[1]

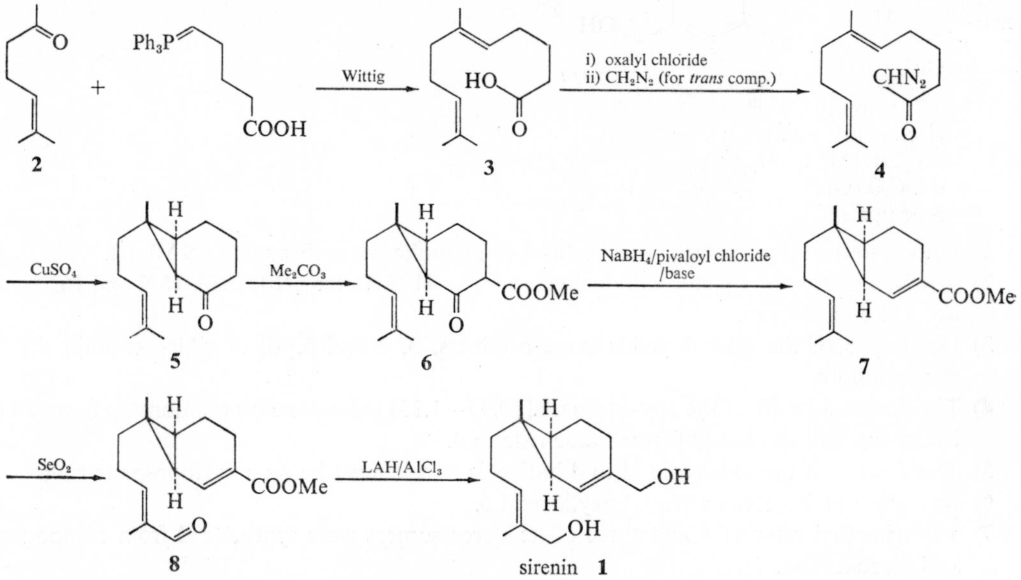

The synthetic ketone **5** was converted to diastereomeric ketals **9** and **10**, which were separated into *d* and *l* forms by gas-liquid chromatography. The absolute configuration of natural sirenin was assigned[2] from the cd spectrum of **11** and by direct conversion of monodeoxysirenin **13** into the parent hydrocarbon, sesquicarene **14**.[3]

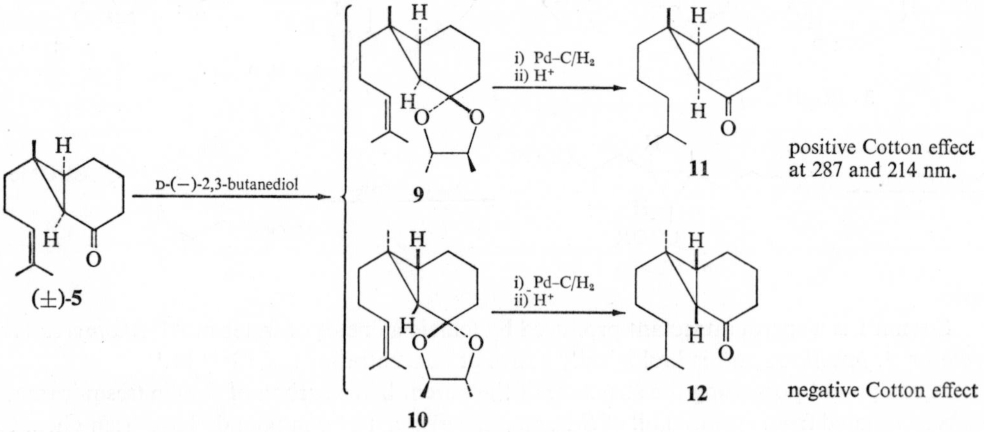

positive Cotton effect at 287 and 214 nm.

negative Cotton effect

1) U. T. Bhalerao, J. J. Plattner, H. Rapoport, *J. Am. Chem. Soc.*, **92**, 3429 (1970); *ibid.*, **91**, 4933 (1969).
2) J. J. Plattner, H. Rapoport, *J. Am. Chem. Soc.*, **93**, 1758 (1971).
3) Y. Ohta, Y. Hirose, *Tetr. Lett.*, 1251 (1968).
4) E. J. Corey, K. Achiwa, J. A. Katzenellenbogen, *J. Am. Chem. Soc.*, **91**, 4318 (1969).

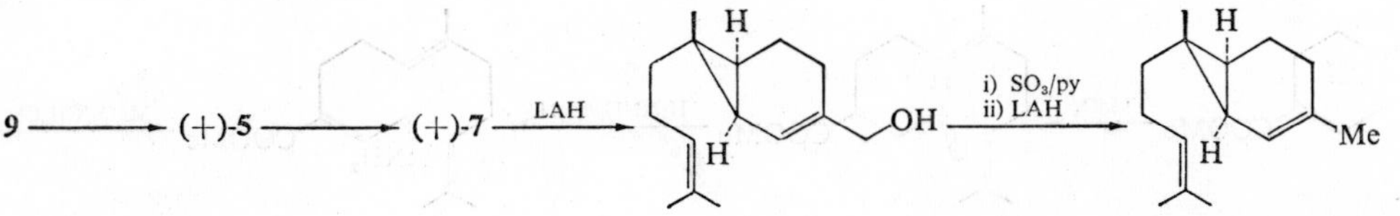
9
(+)-5
(+)-7
LAH
H
H
OH
i) SO_3/py
ii) LAH
H
H
Me

(−)-13 (−)-sesquicarene **14**

Route II[4)]

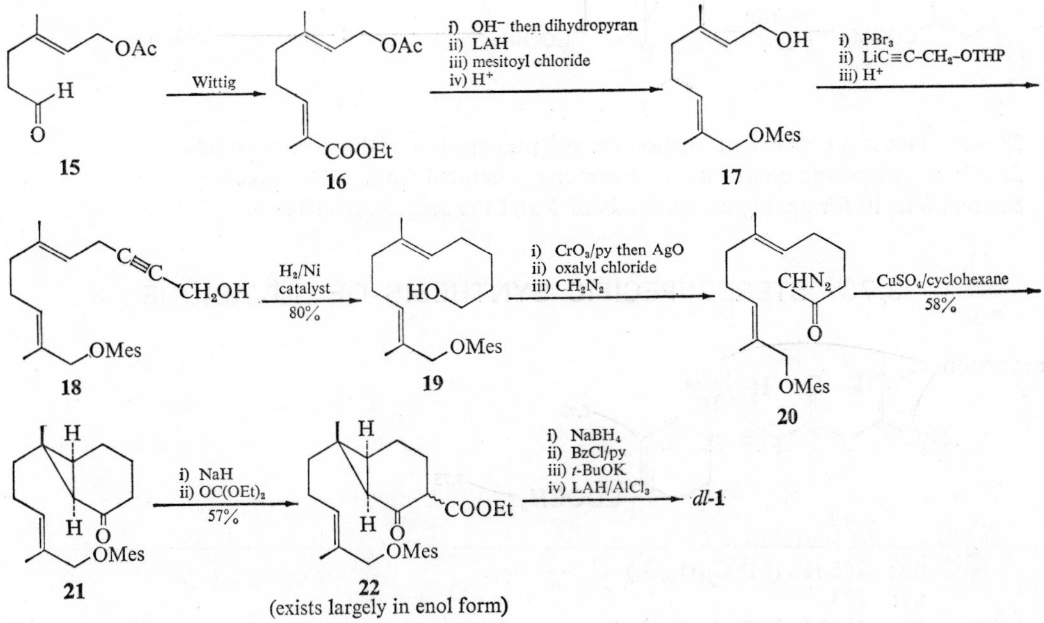
OAc
H
O
15
Wittig
OAc
COOEt
16
i) OH⁻ then dihydropyran
ii) LAH
iii) mesitoyl chloride
iv) H⁺
OH
OMes
17
i) PBr₃
ii) LiC≡C-CH₂-OTHP
iii) H⁺
C≡C
CH₂OH
OMes
18
H₂/Ni
catalyst
80%
HO
OMes
19
i) CrO₃/py then AgO
ii) oxalyl chloride
iii) CH₂N₂
CHN₂
O
OMes
20
CuSO₄/cyclohexane
58%
H
H
O
OMes
21
i) NaH
ii) OC(OEt)₂
57%
H
H
O
COOEt
OMes
22
(exists largely in enol form)
i) NaBH₄
ii) BzCl/py
iii) t-BuOK
iv) LAH/AlCl₃
dl-1

18→19: Selective saturation of the triple bond in **18** was achieved by hydrogenation over a nickel boride catalyst.

20→21: This gives stereospecific addition of the α-keto-carbene to the *trans* double bond.

Route III[5)]

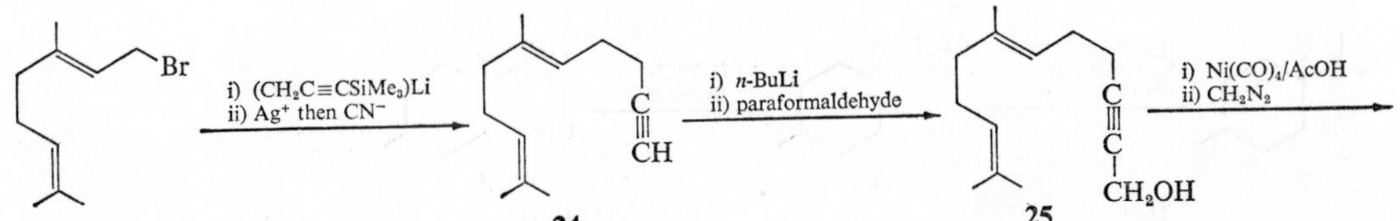
Br
23
i) (CH₂C≡CSiMe₃)Li
ii) Ag⁺ then CN⁻
C
CH
24
i) n-BuLi
ii) paraformaldehyde
C
C
CH₂OH
25
i) Ni(CO)₄/AcOH
ii) CH₂N₂

5) E. J. Corey, K. Achiwa, *Tetr. Lett.*, 2245 (1970).
6) P. A. Grieco, *J. Am. Chem. Soc.*, **91**, 5660 (1969).
7) K. Mori, M. Matsui, *Tetr. Lett.*, 4435 (1969); *Tetr.*, **26**, 2801 (1970).
8) R. M. Coates, R. M. Freidinger, *Chem. Commun.*, 871 (1969); *Tetr.*, **26**, 3487 (1970).
9) E. J. Corey, K. Achiwa, *Tetr. Lett.*, 1837, 3257 (1969).
10) K. Mori, M. Matsui, *Tetr. Lett.*, 2729 (1969).

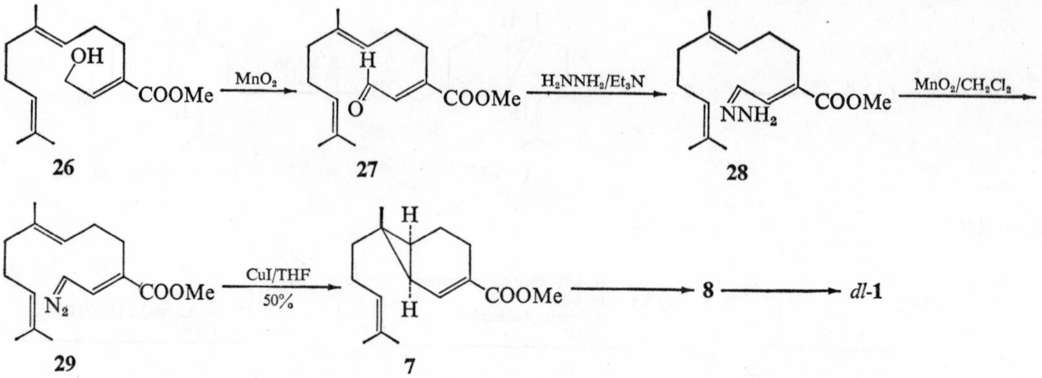

28→29: This is a convenient method for the preparation of diazo compounds.
29→7: Stereospecific cyclization of a carbene generated from a diazoalkene.
See refs 6 to 10 for analogous synthesis of **1** and the sesquicarene **14**.

3.73 STEREOSPECIFIC SYNTHESIS OF JUVABIONE

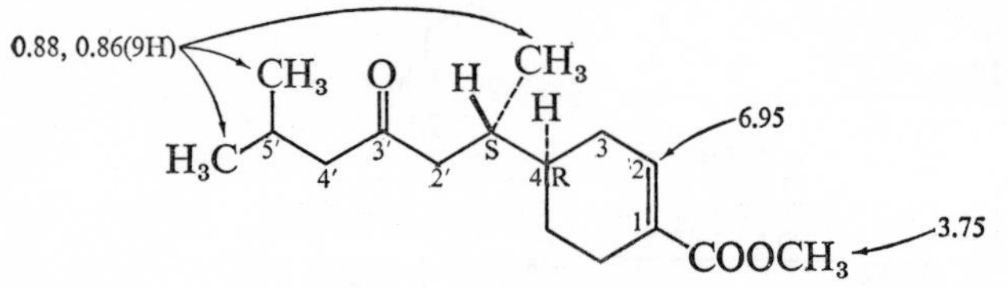

(+)-juvabione **1**[1]

ms: 266.188 (M^+ $C_{16}H_{26}O_3$)
uv: 223
ir: 1722 (α,β-unsat. ester),
     1645 (sat. ketone)

Route I[2]

Juvabione **1** has been synthesized in an optically active form from R-(+)-limonene. This synthesis established the absolute configuration as in **1** (4R, 1'S), since the absolute configuration of the alcohol **3** has been determined as 4R, 8R by x-ray analysis.[3]

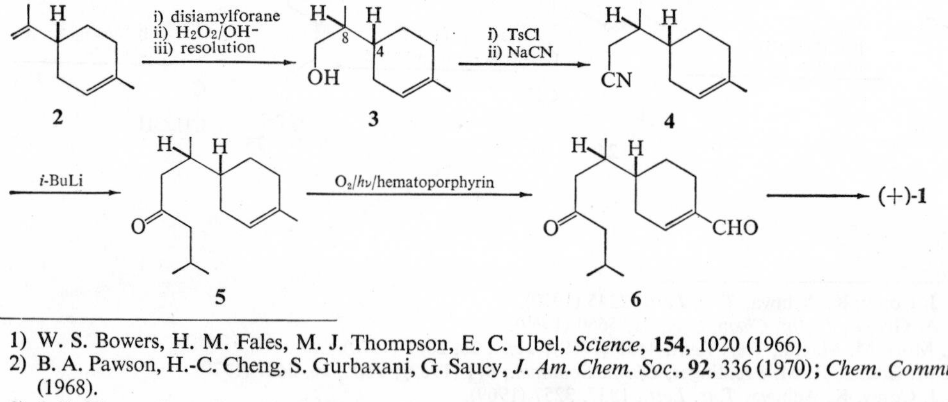

1) W. S. Bowers, H. M. Fales, M. J. Thompson, E. C. Ubel, *Science*, **154**, 1020 (1966).
2) B. A. Pawson, H.-C. Cheng, S. Gurbaxani, G. Saucy, *J. Am. Chem. Soc.*, **92**, 336 (1970); *Chem. Commun.*, 1057 (1968).
3) J. F. Blount, B. A. Pawson, G. Saucy, *Chem. Commun.*, 715 (1969).

Route II[4]

An alternative stereospecific synthesis of **1** starting from the Diels-Alder adduct **9**, which undergoes acid-catalyzed ring opening to a 4-substituted cyclohexanone, has been reported.

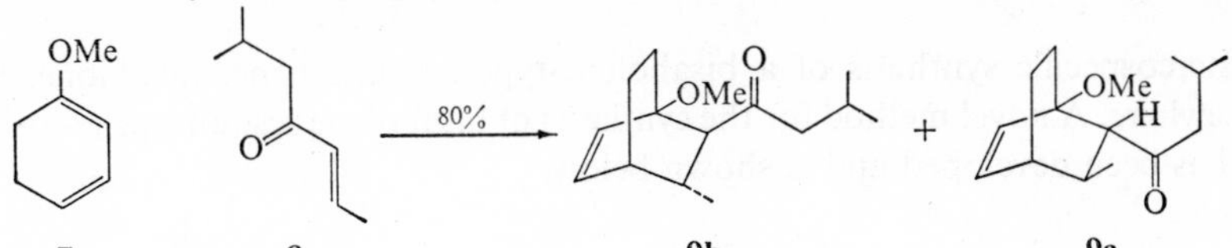

7,8→9: The equal mixture of **9a** and **9b** can be separated by distillation.
11→12: The ratio of **12a** to **12b** is about 1:2.

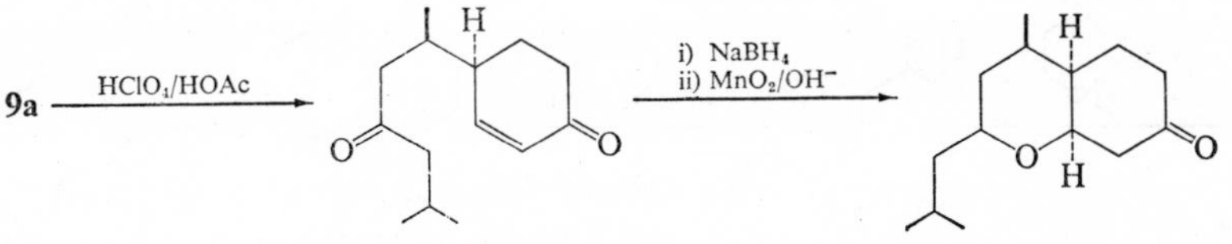

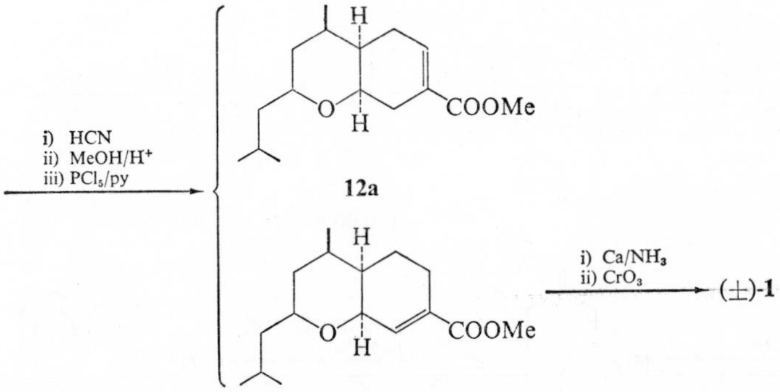

Remarks

(+)-Juvabione was isolated from balsam fir. It has been shown to be present in pulp and various paper products, and has been called "paper factor", since it shows strong juvenile hormone activity for the hemipteran bug, *Pyrrhocoris apterus*.[1] The corresponding free acid is known as todomatuic acid **13**, and has been isolated from *Abies sachalinensis*.[5]

(+)-Methyl todomatuate **14**,[6] pseudotsugonal **15**,[6] and the keto-acid **16**[7] isolated from Douglas fir wood, as well as todomatuic acid **13** have been found to have the same 4R, 1'R configuration.[6] Dehydrojuvabione, possessing juvenile hormone activity, was also isolated from balsam fir.[8]

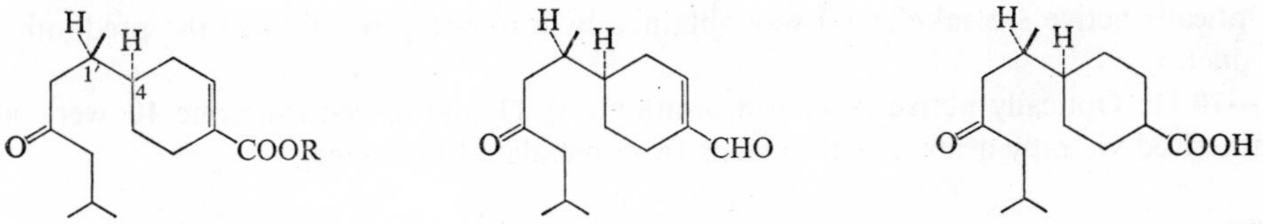

13 (R=H)
14 (R=Me)
15
16

4) A. J. Birch, P. L. Macdonald, V. H. Powell, *J. Chem. Soc., C,* 1469 (1970); *Tetr. Lett.,* 351 (1969).
5) T. Momose, *J. Pharm. Soc. Japan,* **61**, 288 (1941); T. Nakazaki, S. Isoe, *Bull. Chem. Soc. Japan,* **36**, 1198 (1963).
6) T. Sakai, Y. Hirose, *Chem. Letr.,* 491 (1973); *ibid.,* 825 (1973).
7) I. H. Rogers, J. F. Manville, *Can. J. Chem.,* **50**, 2380 (1972).
8) V. Cerny, L. Dolejs, L. Lábler, F. Sorm, K. Sláma, *Tetr. Lett.,* 1053 (1967); also see K. Sláma et al., *Science.* **162**, 582 (1968).

3.74   SYNTHESIS OF BISABOLENE-TYPE SESQUITERPENES

The stereospecific synthesis of a bisabolene-type sesquiterpene, juvabione, has been described elsewhere. A novel method for the synthesis of bisabolene sesquiterpenes using metalated limonene has been developed and is shown below.

Route I[1]

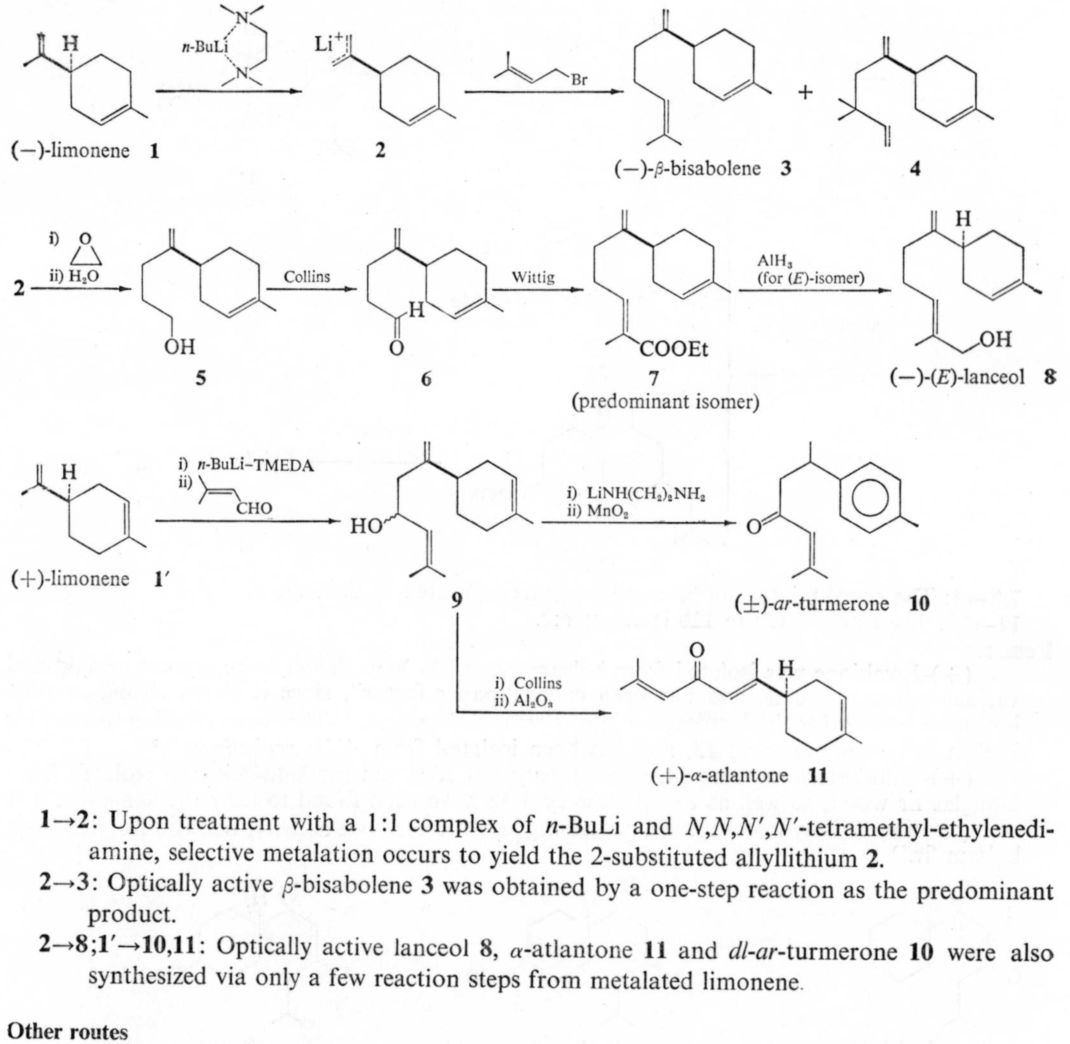

(−)-limonene **1**   **2**   (−)-β-bisabolene **3**   **4**

5   **6**   **7**
                                                                       (predominant isomer)   (−)-(E)-lanceol **8**

(+)-limonene **1′**   (±)-ar-turmerone **10**

9

(+)-α-atlantone **11**

1→2: Upon treatment with a 1:1 complex of n-BuLi and N,N,N′,N′-tetramethyl-ethylenediamine, selective metalation occurs to yield the 2-substituted allyllithium **2**.

2→3: Optically active β-bisabolene **3** was obtained by a one-step reaction as the predominant product.

2→8;1′→10,11: Optically active lanceol **8**, α-atlantone **11** and dl-ar-turmerone **10** were also synthesized via only a few reaction steps from metalated limonene.

Other routes

Several sesquiterpenes of this group have been synthesized by stepwise construction of the side-chain starting from acetophenone derivatives, or by other classical methods.[2]

1) R. J. Crawford, W. F. Erman, C. D. Broaddus, *J. Am. Chem. Soc.*, **94**, 4298 (1972).
2) For example: G. Büchi, H. Wüest, *J. Org. Chem.*, **34**, 1122 (1969); K. Mori, M. Matsui, *Tetr.*, **24**, 3127 (1968); C. D. Gutsche, J. R. Maycock, C. T. Chang, *ibid.*, **24**, 859 (1968); K. S. Ayyar, G. S. K. Rao, *Can. J. Chem.*, **46**, 1467 (1968); A. Manjarrez, A. Guzman, *J. Org. Chem.*, **31**, 348 (1966).

3.75 STRUCTURE AND SYNTHESIS OF CAMPHERENONE

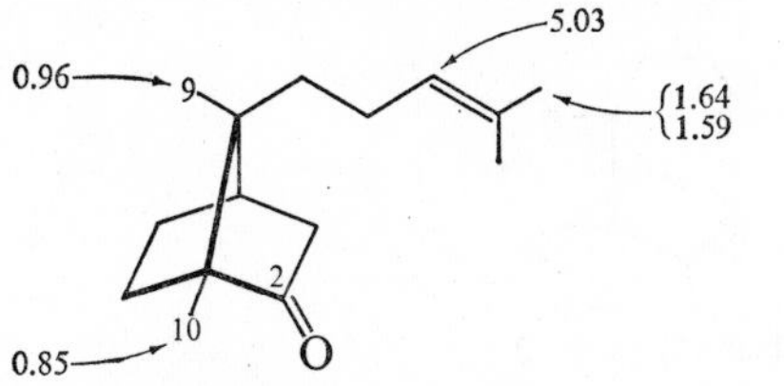

0.96 ← 9
5.03
{1.64
{1.59
2
10
0.85
O

campherenone **1**[1]

α_D: −33.6°

ir(CCl$_4$): 1745, 1415 (five-membered ring
ketone flanked by methylene)

nmr in CCl$_4$

1) The chemical shifts of the C-9 and C-10 methyls in **1** coincide with those of camphor.
2) The mass fragmentation of **1** is well explained in terms of an isopentenyl camphor.
3) Reduction of **1** yields 2-epicampherenol **3**.
4) Naturally occurring campherenol **2** can be oxidized to **1**.
5) The chemical shifts and splitting patterns of the 2-H in **2** and **3** are in accord with those of borneol and isoborneol, respectively.
6) The absolute configuration of **1** was deduced from the ord spectrum.

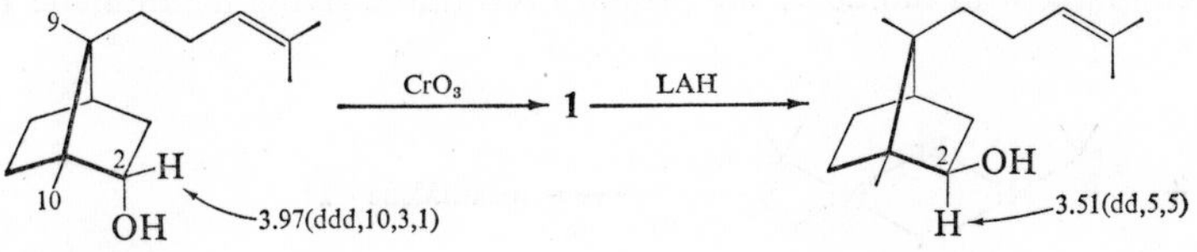

$\xrightarrow{CrO_3}$ **1** $\xrightarrow{LAH}$

9
10
2 H
OH ← 3.97(ddd,10,3,1)

2

2 OH
H ← 3.51(dd,5,5)

3

Remarks

Campherenone **1** and campherenol **2** were isolated from the high-boiling fraction of essential oil of the camphor tree, *Cinnamomum camphora,* which also contains camphor and santalene.

Synthesis

Route I[2]

1 has been synthesized by the route shown below, which is generally applicable for construction of the [2.2.1]-bicyclic ketone system.[2]

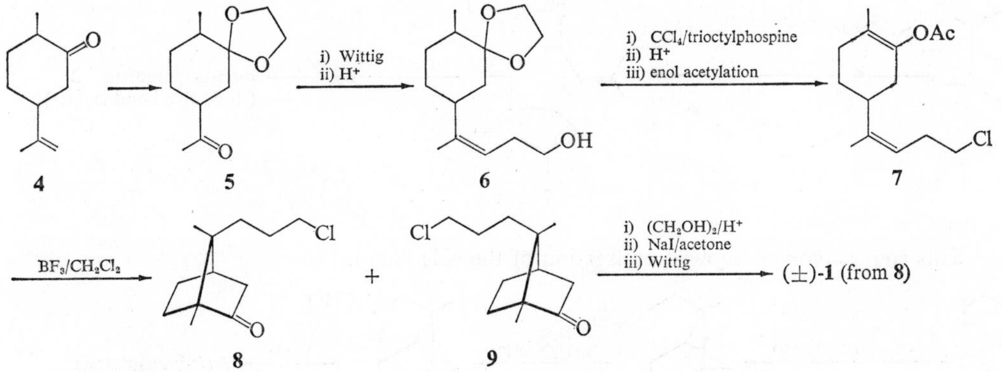

i) Wittig
ii) H$^+$

i) CCl$_4$/trioctylphospine
ii) H$^+$
iii) enol acetylation

OAc

4   **5**   **6**   **7**

BF$_3$/CH$_2$Cl$_2$

Cl

+

Cl

i) (CH$_3$OH)$_2$/H$^+$
ii) NaI/acetone
iii) Wittig

$\xrightarrow{}$ (±)-**1** (from **8**)

8   **9**

Route II[3]

Optically active campherenone and epicampherenone were synthesized by condensation of iodocamphor and a π-allylnickel complex.

1) H. Hikino, N. Suzuki, T. Takemoto, *Tetr. Lett.*, 5069 (1967).
2) G. L. Hodgson, D. F. MacSweeney, T. Money, *Chem. Commun.*, 766 (1971); *Tetr. Lett.*, 3683 (1972); *J. Chem. Soc.*, Perkin I, 2113 (1973).
3) G. L. Hodgson, D. F. MacSweeney, R. W. Mills, T. Money, *Chem. Commun.*, 235 (1973).

3.76 SYNTHESIS OF SANTALENES AND BERGAMOTENES

α-santalene **1**   α-*cis*-bergamotene **2**

Santalenes and bergamotenes belong to the group of sesquiterpenes which are isoprene homologs of known monoterpenes. These compounds can therefore be synthesized from the corresponding monoterpene halides, e.g. by an isoprene homologation reaction.

Route I[1)]

The nickel reagent **4**, derived from α,α-dimethylallyl bromide and Ni(CO)₄ in DMF solution, undergoes preferential coupling with halide at the primary terminus, giving α-santalene **1** in excellent yield.

3   +   **4** $\xrightarrow{88\%}$ α-santalene **1**

Route II[2)]

Reaction of the Mg derivative of **3** with α,α-dimethylallyl mesitoate also affords **1** as the sole product, but in less satisfactory yield.

Route III[3)]

An alternative isoprene homologation has been achieved by alkylation of the Li salt of allylic 2-pyridyl sulfide **6** with the halide, followed by reductive desulfurization.

5   **6** + *n*-BuLi   **7** $\xrightarrow{\text{CuCl}_2/\text{LAH/MeOLi}}$ α-*cis*-bergamotene **2** (+ double bond isomer)

Route IV[4)]

This route involves stepwise elongation of the side chain.

8 $\xrightarrow[\text{ethylenediamine/DMSO}]{\text{LiC≡CH/}}$ **9** $\xrightarrow[\text{ii) H}_2\text{O}_2/\text{OH}^-]{\text{i) H}_2\text{B}_6}$ **10** $\xrightarrow{\text{Wittig}}$ β-*cis*-bergamotene (β-*cis*-**2**)

1) E. J. Corey, M. F. Semmelhack, *J. Am. Chem. Soc.*, **89**, 2756 (1967); cf. G. L. Hodgson, D. F. MacSweeney, R. W. Mills, T. Money, *Chem. Commun.*, 235 (1973).
2) E. J. Corey, S. W. Chow, R. A. Scherrer, *J. Am. Chem. Soc.*, **79**, 5773 (1957).
3) K. Narasaka, M. Hayashi, T. Mukaiyama, *Chem. Lett.*, 259 (1972).
4) T. W. Gibson, W. F. Erman, *J. Am. Chem. Soc.*, **91**, 4771 (1969); *Tetr. Lett.*, 905 (1967).
5) E. J. Corey, D. E. Cane, L. Libit, *J. Am. Chem. Soc.*, **93**, 7016 (1971).

Other route[5)]

Both α- and β-*trans*-bergamotenes have been synthesized by a route involving the photolytic formation of the bicyclo [2.1.1]-hexane ring system followed by ring expansion.

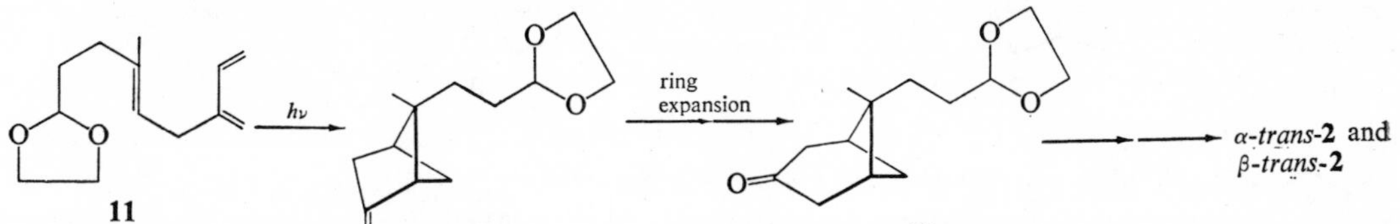

11 **12** **13**

3.77 SYNTHESIS OF α-SANTALOL

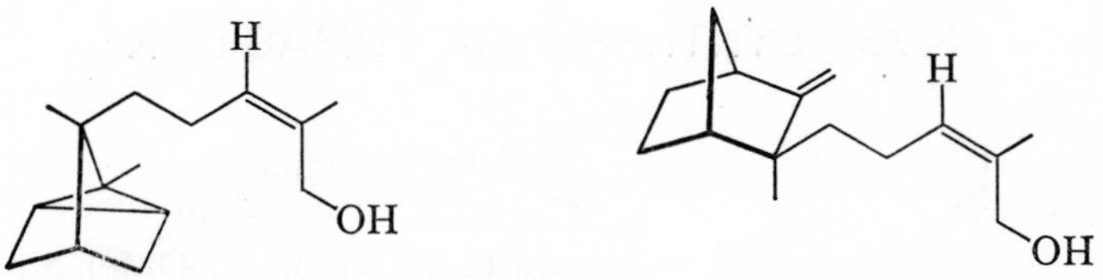

α-santalol **1** β-santalol **11**

A trisubstituted double bond system R–CH_2–CH=C(Me)CH_2OH having a methyl group and an olefinic hydrogen *cis*, such as that present in the side chain of α- or β-santalol, can be constructed stereospecifically by two different methods.

Route I[1)]

The β-oxido phosphonium ylide **4**, derived from the Wittig betaine **3** by reaction with BuLi, reacts with formaldehyde to yield the β,β′-dioxido phosphonium derivative **5**, an immediate precursor of **1**.

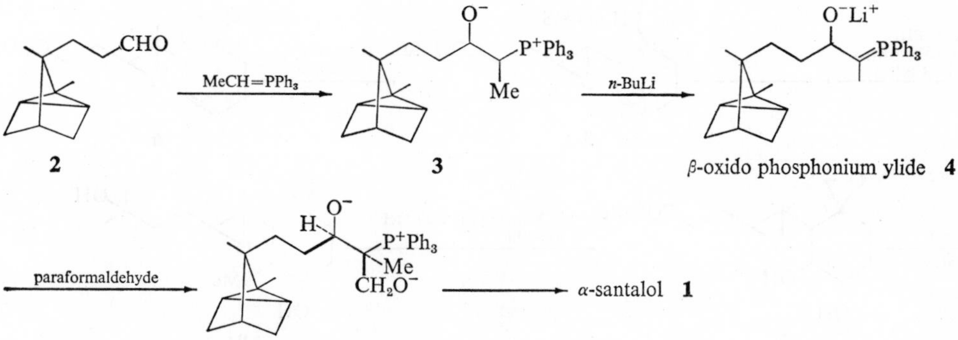

2 **3** β-oxido phosphonium ylide **4**

β,β-dioxido phosphonium derivative **5**

---

1) E. J. Corey, H. Yamamoto, *J. Am. Chem. Soc.*, **92**, 226 (1970).
2) E. J. Corey, H. A. Kirst, J. A. Katzenellenbogen, *J. Am. Chem. Soc.*, **92**, 6314 (1970).
3) For other syntheses, see: R. G. Lewis, D. H. Gustafson, W. F. Erman, *Tetr. Lett.*, 401 (1967); S. Y. Kamat, K. K. Chakravarti, S. C. Bhattacharyya, *Tetr.*, **23**, 4487 (1967); J. Colonge, G. Descotes, Y. Bahurel, A. Menet, *Bull. Soc. Chim. Fr.*, 374 (1966); H. C. Kretschmar, W. F. Erman, *Tetr. Lett.*, 41 (1970).

Route II[2]

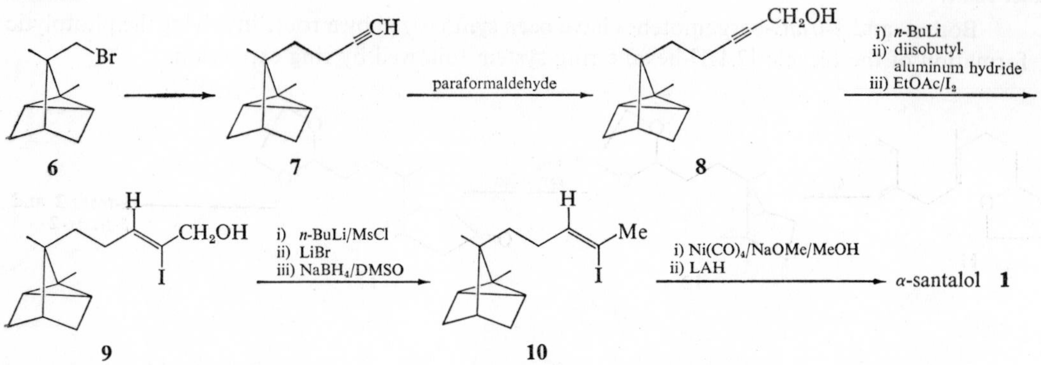

CH

Br

6   7

paraformaldehyde

CH₂OH

8

i) n-BuLi
ii) diisobutyl-
   aluminum hydride
iii) EtOAc/I₂

H   CH₂OH

I

9

i) n-BuLi/MsCl
ii) LiBr
iii) NaBH₄/DMSO

H   Me

I

10

i) Ni(CO)₄/NaOMe/MeOH
ii) LAH

α-santalol **1**

8→9: This reaction permits the stereospecific conversion of the propargylic alcohol into the iodo derivative **9**.

3.78 SYNTHESIS OF FUMAGILLIN

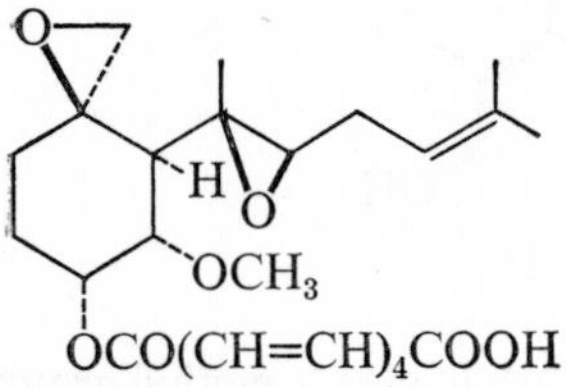

O

H
O

OCH₃

OCO(CH=CH)₄COOH

fumagillin **1**[1-3]

Data for fumagillol:[1]
$C_{16}H_{20}O_4$ $\alpha_D: -68°$
nmr (CDCl₃): 3.4 (s,–OMe), 5.1 (olefinic H),
            1.74 and 1.65 (C=CMe₂),
            1.12 (–O–C–Me)

Route I[4]

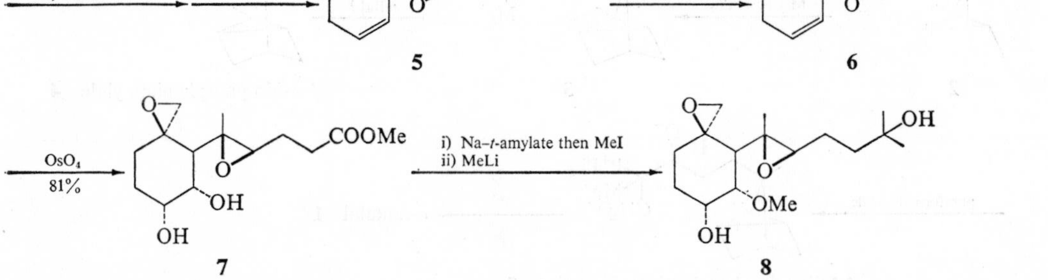

OHC   COOMe

2

Wittig
84%

COOMe

3

Br
CHO

Br CHO
COOMe

4

i) NaBH₄
ii) TMSCl/Et₃N
iii) m-Cl-PBA

Br CH₂OTMS
COOMe
O

5

i) Bu₄N⁺F⁻
ii) NaOMe

O
COOMe
O

6

OsO₄
81%

O
COOMe
O
OH
OH

7

i) Na–t-amylate then MeI
ii) MeLi

O
OH
O
OMe
OH

8

1) D. S. Tarbell *et al.*, *J. Am. Chem. Soc.*, **83**, 3096 (1961).
2) N. J. McCorkindale, J. G. Sime, *Proc. Chem. Soc.*, 331 (1961) (x-ray analysis).
3) J. R. Turner, D. S. Tarbell, *Proc. Nat. Acad. Sci. U.S.*, **48**, 733 (1962) (stereochemistry).
4) E. J. Corey, B. B. Snider, *J. Am. Chem. Soc.*, **94**, 2549 (1972).
5) H. P. Sigg, H. P. Weber, *Helv. Chim. Acta.*, **51**, 1395 (1968); P. Bollinger, H. P. Sigg, H. P. Weber, *ibid.*, **56**, 819 (1973).

i) Ac₂O/py
ii) MsCl/Et₃N
iii) Bu₄N⁺Br⁻   i) K₂CO₃/MeOH
⎯⎯⎯⎯⎯⎯⎯⎯⎯⎯⎯→ (±)-fumagillyl acetate ⎯ii) decatetraenedioyl chloride⎯→ (±)-1
                                          9

2→3: The central double bond in the triene **3** was a mixture of *cis* and *trans* isomers (1:1), which isomerized at 80° to the *trans* isomer.

3→4: The desired product **4** was predominantly formed by Diels-Alder addition.

4→5: Peracid attacks the trisubstituted double bond in **4** from one side only, to yield the epoxide **5**, due to the large steric shielding on the other side by the ring substituents.

Remarks

Fumagillin **1**, obtained from *Aspergillus fumigatus*, is an antibiotic exhibiting antiparasitic and carcinolytic activities.

The analogous substance ovalicin **10** was isolated from *Pseudeurotium ovalis*.[5]

10

3.79 ACID-CATALYZED REARRANGEMENT OF THUJOPSENE

thujopsene **1**

Thujopsene **1**, containing an olefinic linkage conjugated with a cyclopropane ring, has been shown to undergo a variety of rearrangements under acid conditions.[1-4]

Rearrangements in cyclopropylcarbinyl-homoallyl system[1]

1) Under mild acid conditions (0.02M HClO₄ in dioxane) thujopsene **1** rearranges into the naturally occuring widdrol **4** and the homoallylic alcohol **5**. It has been shown that the rearrangement goes through the cyclopropylcarbinyl cation **3** by an experiment using the 6,6-dideutero derivative of **1**.

2) On prolonged treatment under mild acid conditions, **1** yields mainly the bicyclic diene **6**. It was demonstrated that the diene **6** is formed through intermediate **2** directly and not from widdrol **4**.

1 ⎯→   **2**   **3**   **4**

                                                                +

                                                       5   **+**   **6**

1) W. G. Dauben, L. E. Friedrich, P. Oberhansli, E. I. Aoyagi, *J. Org. Chem.*, **37**, 9 (1972), and references cited therein.

2) W. G. Dauben, L. E. Friedrich, *J. Org. Chem.* **37**, 241 (1972).

3) S. Ito, M. Yatagai, K. Endo, *Tetr. Lett.*, 1149 (1971); S. Ito, M. Yatagai, K. Endo, M. Kodama, *ibid.*, 1153 (1971).

4) W. G. Dauben, E. I. Aoyagi, *J. Org. Chem.*, **37**, 251 (1972); see also H. U. Daeniker, A. R. Hochstetler, K. Kaiser, G. C. Kitchens, *J. Org. Chem.*, **37**, 1 (1972).

Further rearrangement of bicyclic diene 6 into tricyclic compound 9[2)]

1) The diene **6** rearranges to the tricyclic compound **9** under more vigorous conditions (0.02M HClO₄ in refluxing HOAc) through the strained bridgehead cation **7**.

2) This rearrangement (**7→9**) is reminiscent of the reaction pathway in the conversion of caryophyllene into neoclovene.

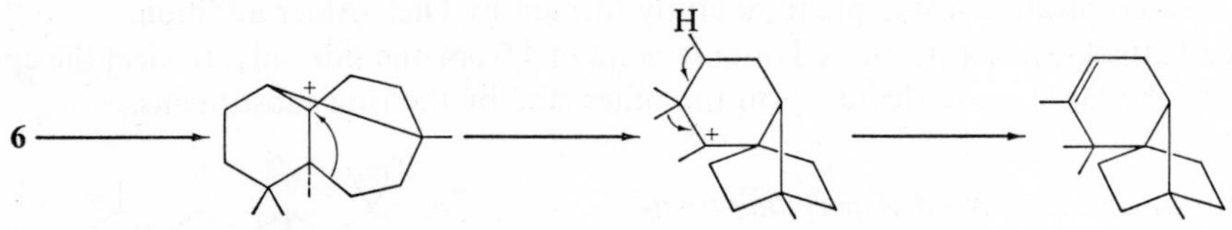

Formation of chamigrenes and subsequent conversion into tricyclic isomers[4)]

1) When **1** was heated under reflux in HOAc, **11** (30%), **12** (30%) and **6** (40%) were formed.

2) Upon treatment with 0.02M HClO₄ in HOAc at 25°, β-chamigrene **11** further rearranges into the tricyclic olefins **21** and **20**.

3) In refluxing 0.02M HClO₄ in HOAc, β-chamigrene **11** rearranges to the tricyclic olefins **18** and **17**.

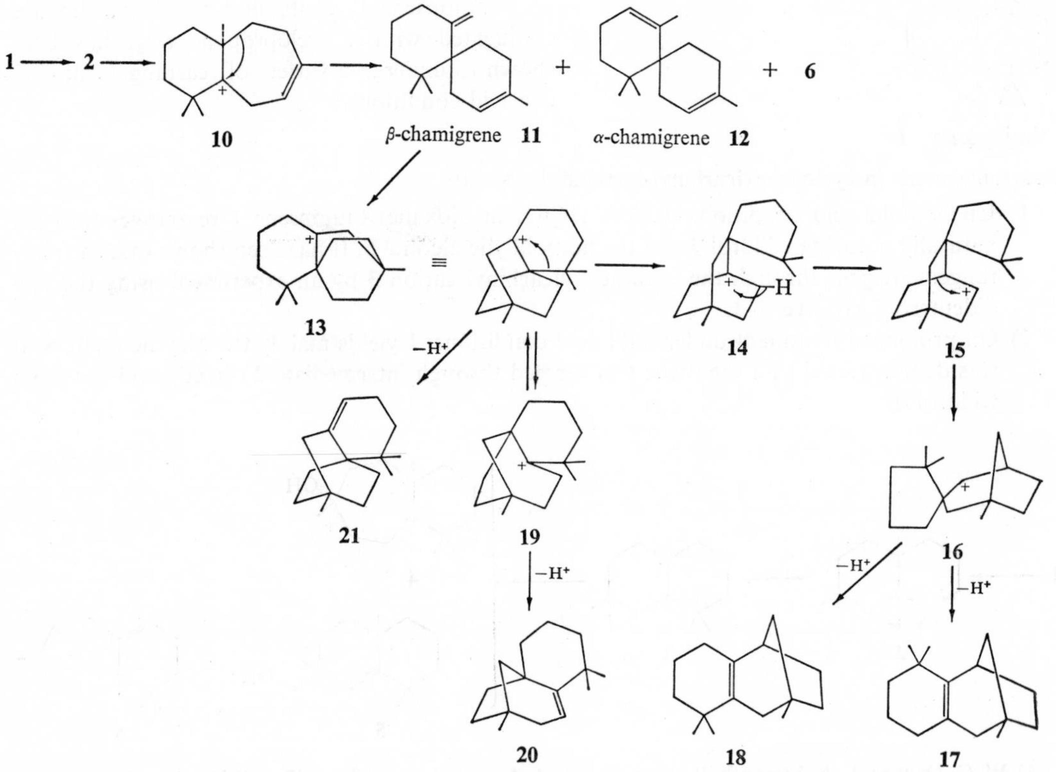

Remarks

(−)-Thujopsene **1** was isolated from wood oil of *Thujopsis dolabrata* and was later found in cedarwood oil. It also occurs widely in the oils of Cupressales.[5)]

---

5) T. Norin, *Acta. Chem. Scand.*, **15**, 1676 (1961); **17**, 738 (1963).

3.80 SENSITIZED PHOTOOXIDATION OF THUJOPSENE

The sensitized photooxidation of thujopsene **1**, followed by reduction, yielded the products **2–7**.[1,3]

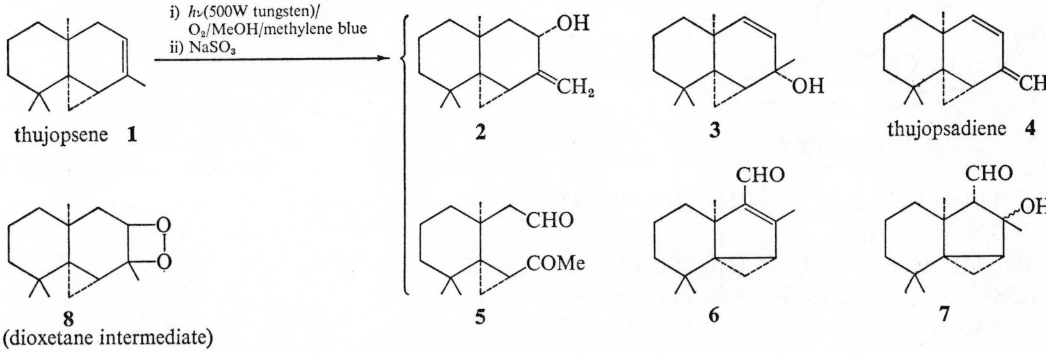

thujopsene **1**

i) $h\nu$(500W tungsten)/
O₂/MeOH/methylene blue
ii) NaSO₃

2   **3**   thujopsadiene **4**

8
(dioxetane intermediate)

5   **6**   **7**

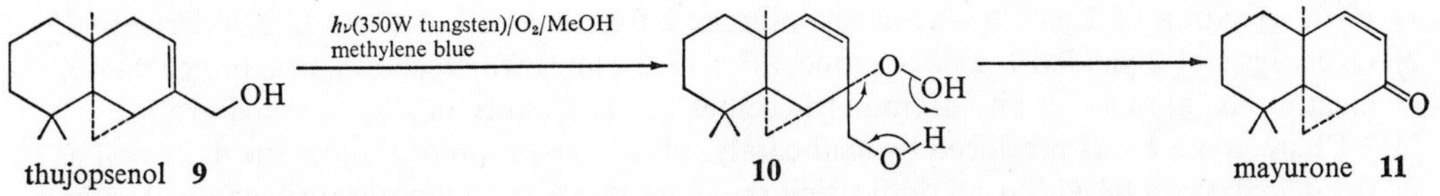

thujopsenol **9**

$h\nu$(350W tungsten)/O₂/MeOH
methylene blue

10   mayurone **11**

1) The formation of hydroperoxides corresponding to **2** and **3** indicates that the α-side of **1** is less hindered, in spite of the presence of the cyclopropane ring.

2) The small product ratio **2/3** demonstrates the preferred abstraction by O₂ of the 9α-H (which is oriented axially in a stable conformation) rather than the methyl hydrogens.

3) The products **6** and **7** might be formed from **5** during the isolation process.

4) Production of the keto aldehyde **5** through the dioxetane intermediate **8** has been verified by deuterium labeling experiments.[1]

5) Photooxygenation of **9** yielded the nor compound **11**, implying the biogenesis of mayurone by such oxygenation.[2]

Remarks

Thujopsene **1**, thujopsenol **9** and mayurone **11** were isolated from *Thujopsis dolabrata*, and thujopsadiene **4** from *Biota orientalis*.

1) S. Ito, H. Takeshita, M. Hirama, *Tetr. Lett.*, 1181 (1971); S. Ito, H. Takeshita, T. Muroi, *ibid.*, 3091 (1969).
2) H. Takeshita, T. Sato, T. Muroi, S. Ito, *Tetr. Lett.*, 3095 (1969).
3) See also G. Ohloff H. Strickler, B. Willhalm, C. Borer, M. Hinder, *Helv. Chim. Acta*, **53**, 623 (1970).

3.81   STRUCTURE AND SYNTHESIS OF CHAMIGRENE

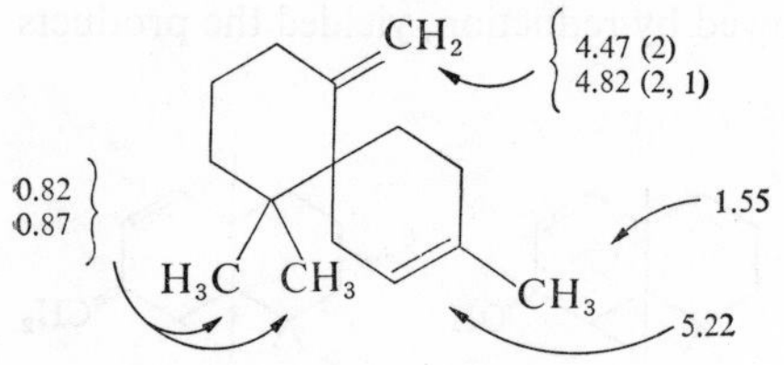

β-chamigrene $\mathbf{1}^{1)}$

bp: 110–113°(13 mm) α_D(CHCl$_3$): $-52.7°$
ms: 204(M$^+$C$_{15}$H$_{24}$)
ir (liquid): 1638, 890 (C=CH$_2$), 800–820
  (CH=C<), 1388, 1368 (*gem*-dimethyl)
nmr in CCl$_4$

1) Hydrogenation of **1** yields dihydrochamigrene **2** together with the tetrahydro derivative.
2) Ozonolysis of **2** yielded the nor-ketone **3** (ir: 1711 cm^{-1}) which shows a plain ord curve, indicating the absence of any asymmetric center in the vicinity of the carbonyl group.
3) β-Chamigrene **1** was produced by acid-catalyzed isomerization of thujopsene **4**, as well as by the dehydration of widdrol **5**, indicating the close structural relationship between **1**, **4** and **5** (see "Rearrangement of thujopsene").

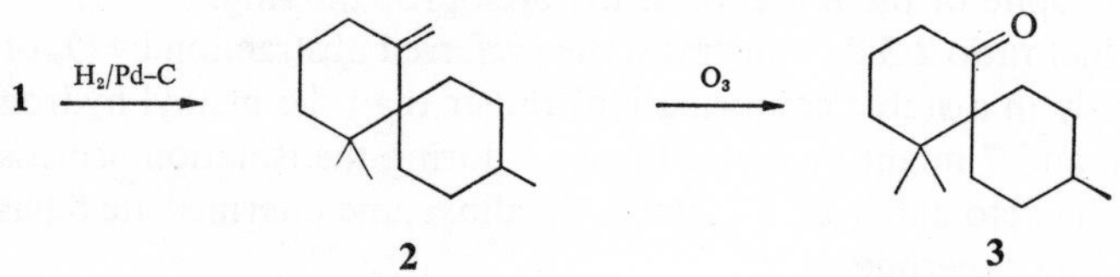

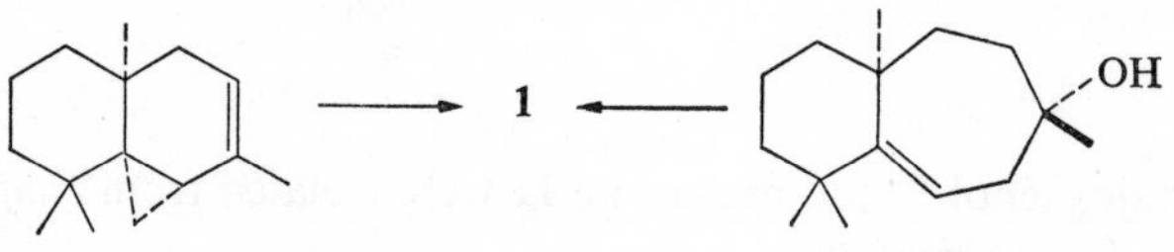

Synthesis
Route I$^{2)}$

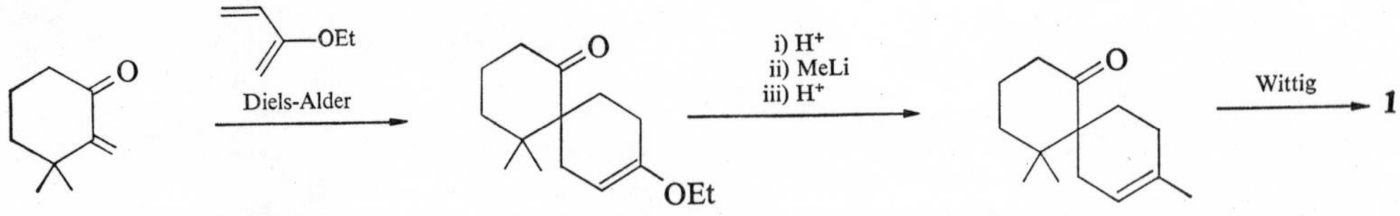

---

1) S. Ito, K. Endo, T. Yoshida, M. Yatagai, M. Kodama, *Chem. Commun.*, 186 (1967).
2) A. Tanaka, H. Uda, A. Yoshikoshi, *Chem. Commun.*, 188 (1967).

Route II[3]

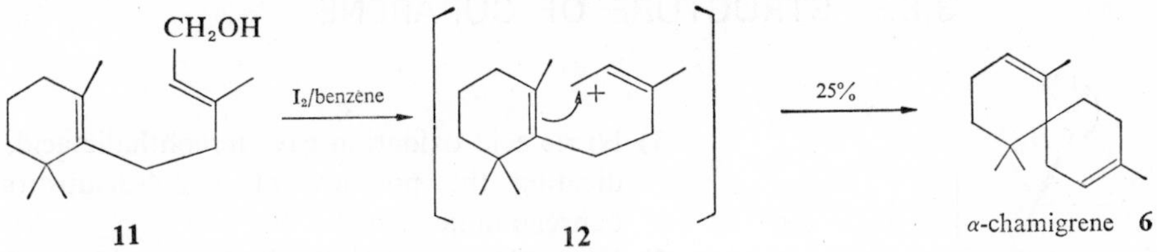

11   12   α-chamigrene **6**

Remarks

β-Chamigrene **1** was isolated from essential oils of leaves of *Chamaecyparis taiwanensis*, together with thujopsene **4**, widdrol **5**, cuparene and many other sesquiterpenes.[1] From oils of the fruits of *Schizandra chinensis*, **1**, α-chamigrene **6**, and chamigrenal **7** have been isolated.[4] A bromine- and chlorine-containing marine product, pacifenol **13**, having a chamigrene skeleton has been found in the red alga *Laurencia pacifica*.[5] Prepacifenol **14**, a possible precursor of pacifenol **13** was isolated from the red alga, *Laurencia filiformis*.[6]

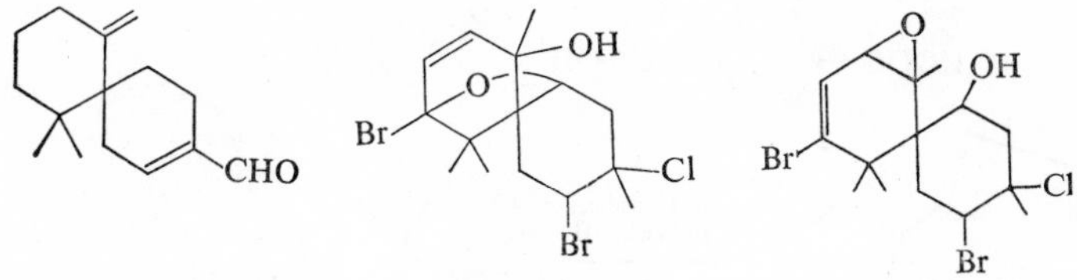

7   13   14

3) S. Kanno, T. Kato, Y. Kitahara, *Chem. Commun.*, 1257 (1967).
4) Y. Ohta, Y. Hirose, *Tetr. Lett.*, 2483 (1968).
5) J. J. Sims, W. Fenical, R. M. Wing, P. Radlick, *J. Am. Chem. Soc.*, **93**, 3774 (1971); *Tetr. Lett.*, 195 (1972).
6) J. J. Sims, W. Fenical, R. M. Wing, P. Radlick, *J. Am. Chem. Soc.*, **95**, 972 (1973). cf. A. G. González, J. Darias, J. D. Martin, *Tetr. Lett.*, 2381 (1973).

3.82   STRUCTURE OF CUPARENE

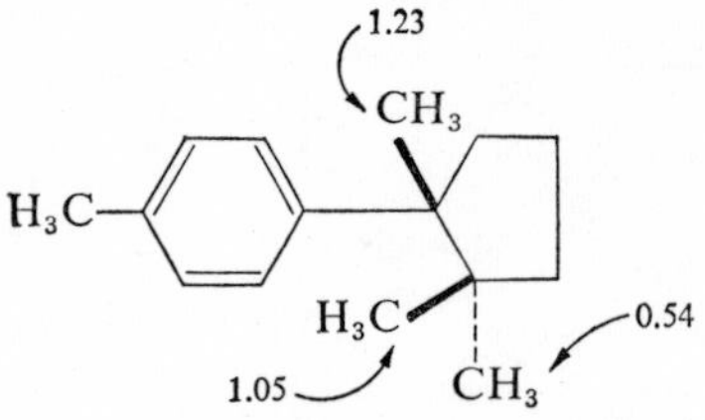

(+)-cuparene 1[1]

bp (19 mm): 138° α_D (CHCl$_3$): +65°
ms: 202 (M$^+$C$_{15}$H$_{22}$)
uv: 259 (ε 260), 265 (ε 360), 273 (ε 350)

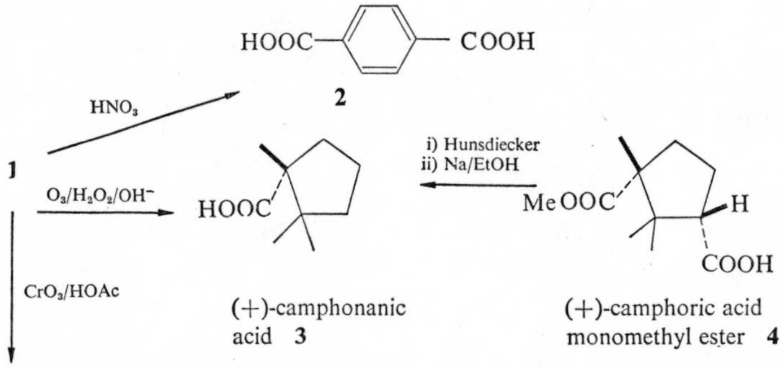

1) Nitric acid oxidation gave terephthalic acid, indicating the presence of a 1,4-disubstituted benzene moiety.
2) Formation of (+)-camphonanic acid 3, which has been correlated with (+)-methyl camphorate 4, established the absolute configuration of 1.

(+)-camphonanic acid 3

(+)-camphoric acid monomethyl ester 4

cuparenic acid 5

Remarks

(+)-Cuparene 1 was isolated from *Chamaecyparis thyoides, Biota orientalis* and various species of the Cupressales.[1] Cuparic acid 5,[1] cuprenenes 6,[2] cuparenones[3] and cuparenols[4] have also been found in nature.

The fungal pigment helicobasidin 7[5] was isolated from *Helicobasidium mompa*, together with the phenolic compound 8,[6] cuparene, cuprenene and cuparenol.

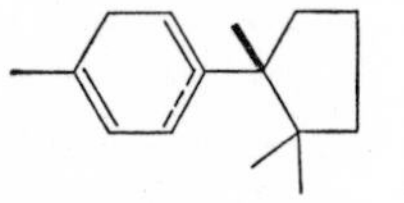

cuprenene 6

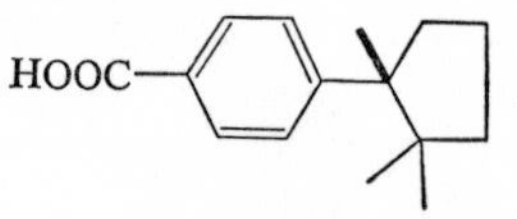

cuparenic acid 5

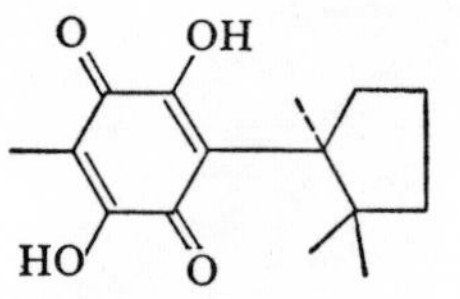

helicobasidin 7

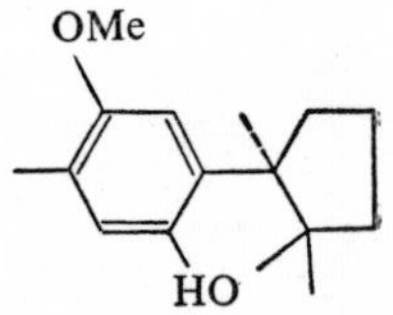

8

1) C. Enzell, H. Erdtman, *Tetr.*, 4, 361 (1958).
2) T. Nozoe, H. Takeshita, *Tetr. Lett.*, 14 (1960); W. G. Dauben, P. Oberhänsli, *J. Org. Chem.*, 31, 315 (1966).
3) B. Tomita, Y. Hirose, T. Nakatsuka, *Tetr. Lett.*, 843 (1968).
4) G. L. Chety, S. Dev. *Tetr. Lett.*, 73 (1964); A. Matsuo, M. Nakayama, S. Hayashi, *Chem. Lett.*, 341 (1972).
5) S. Natori, H. Nishikawa, H. Ogawa, *Chem. Pharm. Bull.*, 12, 236 (1964).
6) S. Nozoe, M. Morisaki, H. Matsumoto, *Chem. Commun.*, 926 (1970).

3.83 LAURENE, LAURINTEROL AND APLYSIN

2.31 6.99 1.29
CH₃
CH₃
1
2 3
4.81
6.99
0.68 CH₃ CH₂

laurene 1[1]

bp: 131–133°(21 mm) α_D:+48.7°
uv(EtOH):253(ε 280), 259(ε 280),
265(ε 280), 274(ε 240)
ir(film): 1512, 875(aromatic);
1653, 875(exo-methylene)
nmr in CCl₄

7.05
2.32 Br
CH₃ 1.25
CH₃
6.52
O
1.30 CH₃ CH₃ 1.09(d)

aplysin 2[4,5]

mp: 85–86° α_D:−85.4°
uv: 234(3.95), 294(3.66)
ir: 1520, 1586, 1486, 1277, 1240
ms: 296, 281(base peak), 279

1) Upfield shift of the 2-Me signal indicates that this is *cis*-oriented to the aromatic group.
2) 1 is readily isomerized to the endocyclic double bond isomer, isolaurene, by passing through a silica gel column.
3) The absolute configuration of 1 was establised by correlation with (+)-cuparene 5 by the following route. The ketone 3 was synthesized from *p*-tolyl-cyclopentanone.

1 i) OsO₄/py
ii) HIO₄ →

O

i) HCOOEt
ii) *n*-BuSH/H⁺
iii) MeI/KO*t*-Bu →

CHS-*n*-Bu
O

3
ir: 1737

4

i) KOH
ii) EDT/H⁺
iii) Ra-Ni →

(+)-cuparene 5

Remarks

Laurene 1 was isolated from the seaweeds *Laurencia glandulifera* and *L. nipponica*.[1] Laurinterol 6[2,5] and isolaurinterol 7,[2] close relatives of 1 have been isolated from *L. intermeda* and *L. glandulifera*. The compound 6 was shown to be converted to aplysin 2 by treatment with *p*-TsOH. Aplysin 2 was isolated from the sea hare, *Aplysia kurodai* and this compound

1) T. Irie, Y. Yasunari, T. Suzuki, N. Imai, E. Kurosawa, T. Masamune, *Tetr. Lett.*, 3619 (1965); *Tetr.*, **25**, 459 (1969).
2) T. Irie, M. Suzuki, E. Kurosawa, T. Masamune, *Tetr.*, **26**, 3271 (1970); *Tetr. Lett.*, 1837 (1966).
3) T. Irie, A. Fukuzawa, M. Izawa, E. Kurosawa, *Tetr. Lett.*, 1343 (1969).
4) S. Yamamura, Y. Hirata, *Tetr.*, **19**, 1485 (1963).
5) F. Cameron, G. Ferguson, J. M. Robertson, *Chem. Commun.*, 271 (1967).

was considered to be derived from **6** *in vivo*, since the sea hares feed on *L. glandulifera*.[4,5) Laurenisol **8** was obtained from *L. nipponica*.[3)

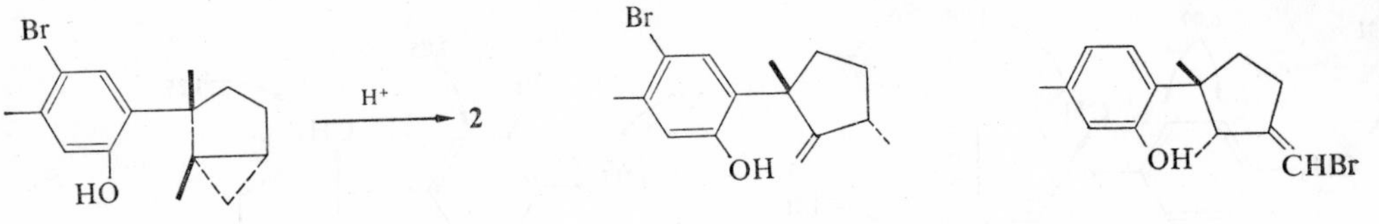

laurinterol **6**[2)

isolaurinterol **7**[2)

laurenisol **8**[3)

3.84 REARRANGEMENT OF TRICHOTHECOLONE

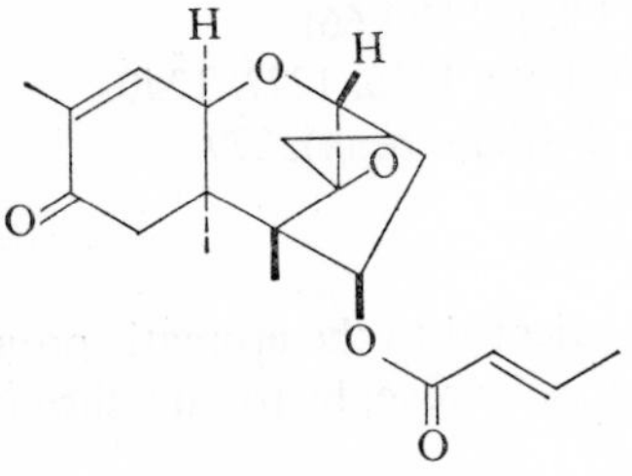

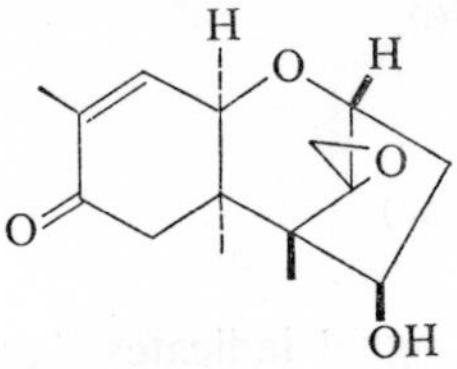

trichthecin   **1**[1,2)

uv:230 (ε 10,000), 325 (ε 37)

trichothecolone   **2**

Trichothecolone **2**, a hydrolysis product of **1**, undergoes a variety of rearrangements, as shown below.

Acid-catalyzed rearrangement of 2[2,5)

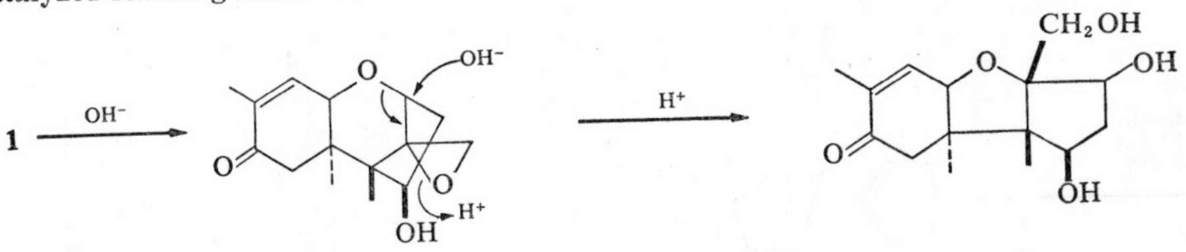

1 $\xrightarrow{OH^-}$ $\xrightarrow{H^+}$

trichothecolone **2**

3

Base-catalyzed rearrangement of 2[4)

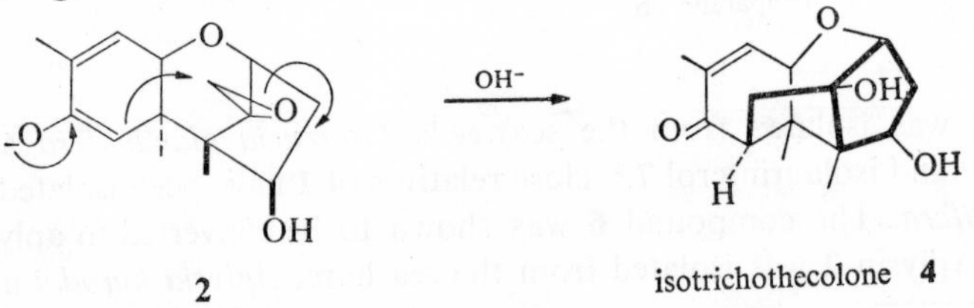

$\xrightarrow{OH^-}$

2

isotrichothecolone   **4**

1) G. G. Freemen, J. E. Gill, W. S. Waring, *J. Chem. Soc.*, 1105 (1959); J. Fishman, E. R. H. Jones, G. Lowe, M. C. Whiting, *ibid.*, 3948 (1960).
2) W. O. Godtfredsen, S. Vangedal, *Proc. Chem. Soc.*, 188 (1964); *Acta. Chem. Scand.*, **19**, 1088 (1965).
3) W. O. Godtfredsen, J. F. Grove, Ch. Tamm., *Helv. Chim. Acta.*, **50**, 1666 (1967).
4) J. Gutzwiller, Ch. Tamm, H. P. Sigg, *Tetr. Lett.*, 4495 (1965).
5) P. M. Adams, J. R. Hanson, *J. Chem. Soc.* C, 2283 (1972).

Reductive rearrangement of 2[4]

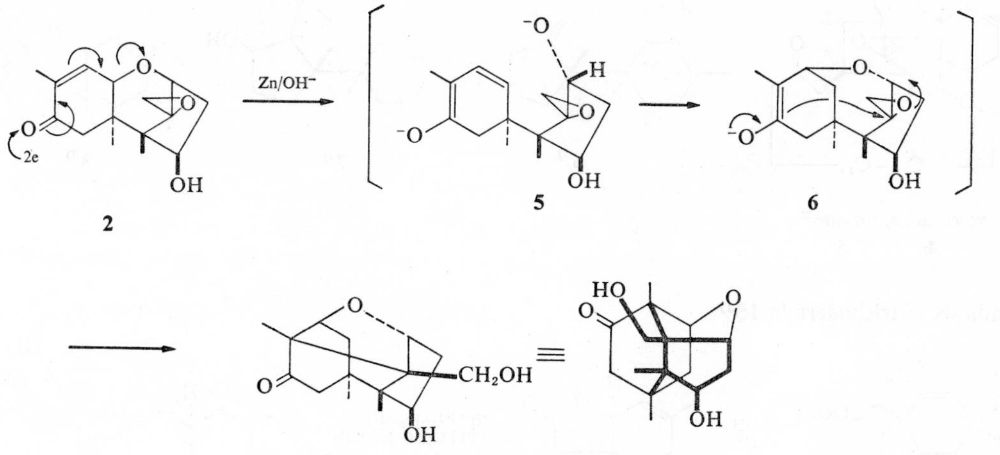

allodehydrotrichothecolone **7**

Remarks

Trichothecin **1** was isolated from the fungi *Trichothecium roseum*. The original structure of trichothecin was revised in 1964 by interrelating it with trichodermol whose structure was established by an x-ray diffraction study.[2] The nomenclature of these compounds is discussed in ref. 3.

3.85   TRICHOTHECIN AND RELATED COMPOUNDS

Naturally occuring esters of the group of sesquiterpenes possessing the trichothecane skeleton, e.g. trichothecin, trichodermin **15**,[1] verrucarins **4**[2] and roridins **5**[2] (antibiotics isolated from fungi *Trichothecium, Trichoderma* and *Myrothecium* spp.), the compounds **1**,[3] **2**[4] and **3**[5] (phyto- or cytotoxic substances produced by *Fusarium* spp.), constitute an important class of physiologically active terpenoids.[6]

Trichodiene **6**, trichodiol **7**, and **8** have been isolated as minor metabolites of *Trichothecium roseum*.[7] The intermediary role of **6** in the biosynthesis of trichothecin was verified by feeding experiments.[8]

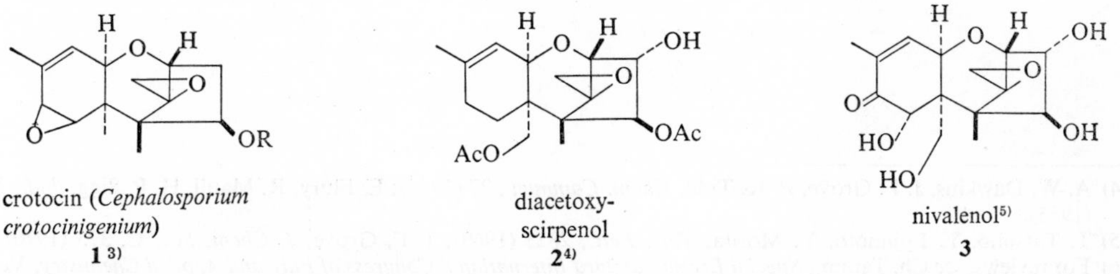

crotocin (*Cephalosporium crotocinigenium*)
1 [3]

diacetoxy-
scirpenol
2 [4]

nivalenol[5]
3

1) W. O. Godtfredsen, S. Vangedal, *Proc. Chem. Soc.*, 188 (1964).
2) B. Böhner, E. Fetz, E. Härri, H. P. Sigg, Ch. Stoll, Ch. Tamm, *Helv. Chim. Acta.*, 48, 1079 (1965); W. Zürcher, Ch. Tamm. *ibid.*, 49, 2594 (1966) and references therein.
3) J. Gyimesi, A. Melera, *Tetr. Lett.*, 1665 (1967).

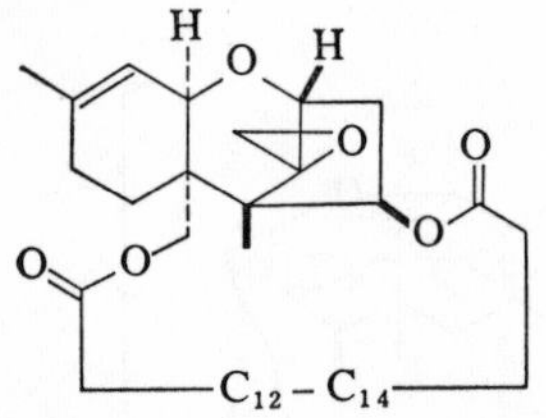

verrucarins, roridins[2]

4   5

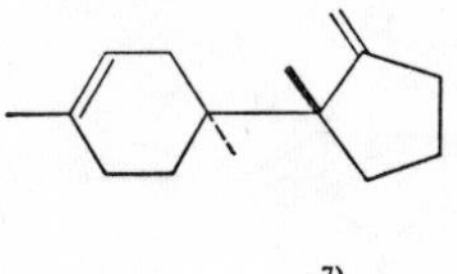

$6^{7)}$ 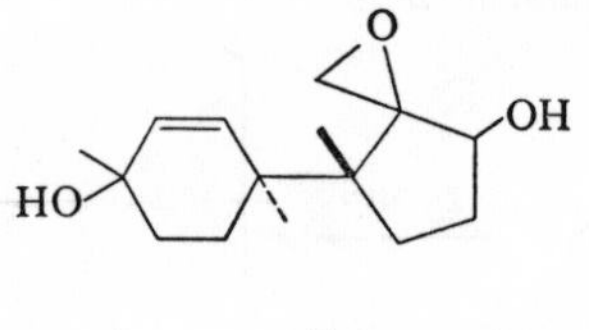 $7^{7)}$ 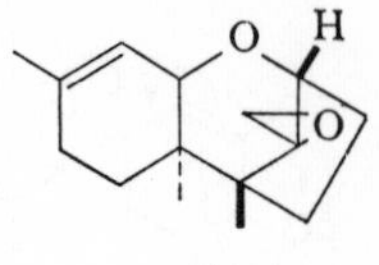 $8^{7)}$

Synthesis of trichodermin 15[9]

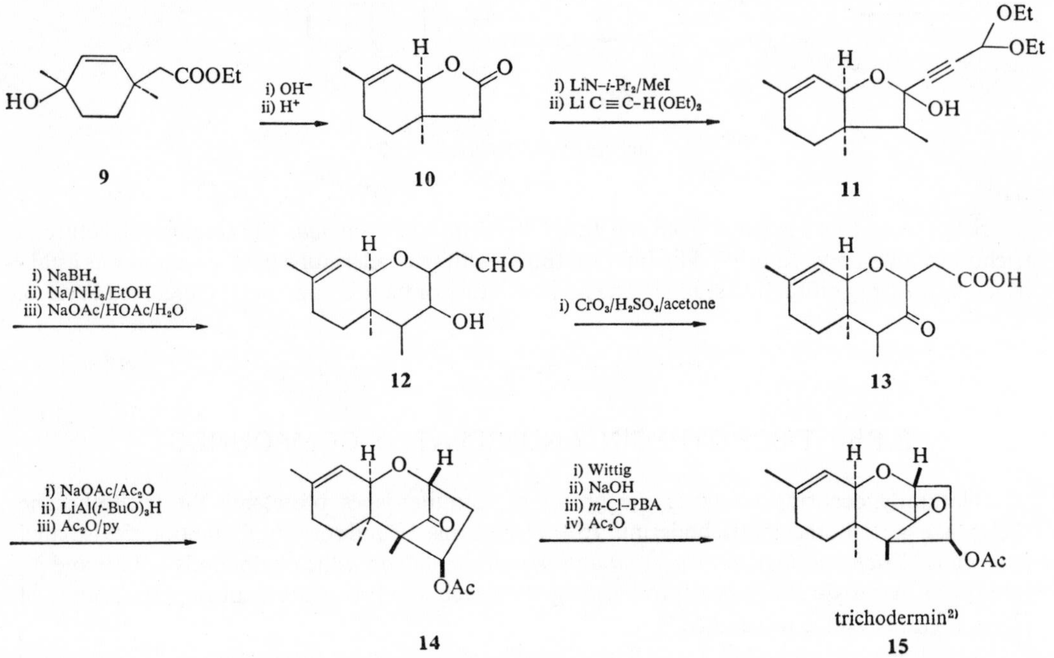

9

i) OH⁻
ii) H⁺

10

i) LiN-i-Pr₂/MeI
ii) Li C≡C–H(OEt)₂

11

i) NaBH₄
ii) Na/NH₃/EtOH
iii) NaOAc/HOAc/H₂O

12

i) CrO₃/H₂SO₄/acetone

13

i) NaOAc/Ac₂O
ii) LiAl(t-BuO)₃H
iii) Ac₂O/py

14

i) Wittig
ii) NaOH
iii) m-Cl–PBA
iv) Ac₂O

trichodermin[2]
15

4) A. W. Dawkins, J. F. Grove, B. K. Tidd, *Chem. Commun.*, 27 (1965); E. Flury, R. Mauli, H. P. Sigg, *ibid.*, 26 (1965).
5) T. Tatsuno, Y. Fujimoto, Y. Morita, *Tetr. Lett.*, 2823 (1969); J. F. Grove, *J. Chem. Soc.* C, 375 (1970).
6) For reviews, see Ch. Tamm., *Special Lecture at 23rd International Congress of Pure and Applied Chemistry*, Vol. 5, p. 49, Butterworth, 1971.
7) S. Nozoe, Y. Machida, *Tetr.* **28**, 5105, 5113 (1972); *Tetr. Lett.*, 1177, 2671 (1970).
8) Y. Machida, S. Nozoe, *Tetr. Lett.*, 1969 (1972).
9) E. W. Colvin, R. A. Raphael, J. S. Roberts, *Chem. Commun.*, 858 (1971); E. W. Colvin, S. Malchenko, R. A. Raphael, J. S. Roberts, *J. Chem. Soc.*, Perkin I, 1989 (1973).

3.86 GYMNOMITROL AND TRICHODIENE

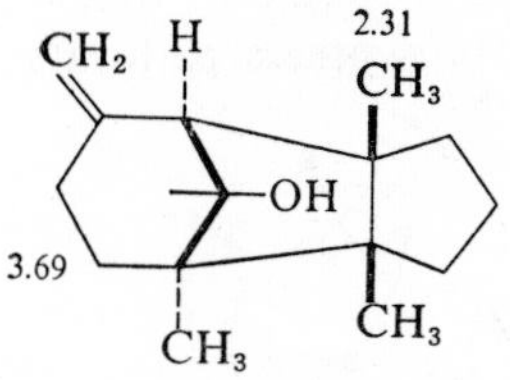

CH$_2$ H 2.31
CH$_3$

—OH

3.69

CH$_3$ CH$_3$

gymnomitrol 1[1]
mp:114–6° $\alpha_D + 7°$

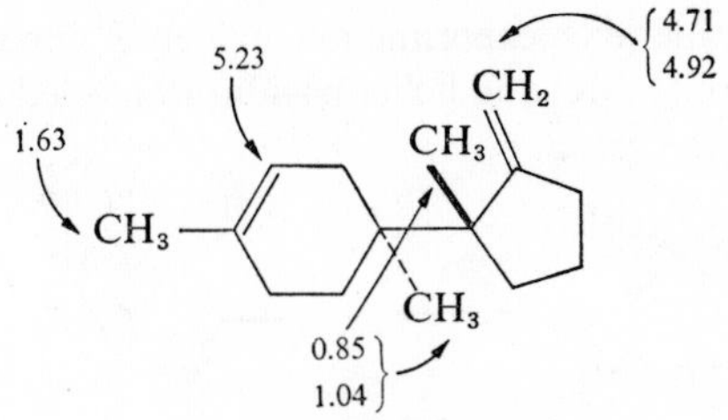

5.23   4.71
            4.92
CH$_2$
1.63   CH$_3$
CH$_3$—

CH$_3$
0.85
1.04

trichodiene 2[2,3]
oily substance $\alpha_D + 21°$
ms: 204 (M$^+$, C$_{15}$H$_{24}$)

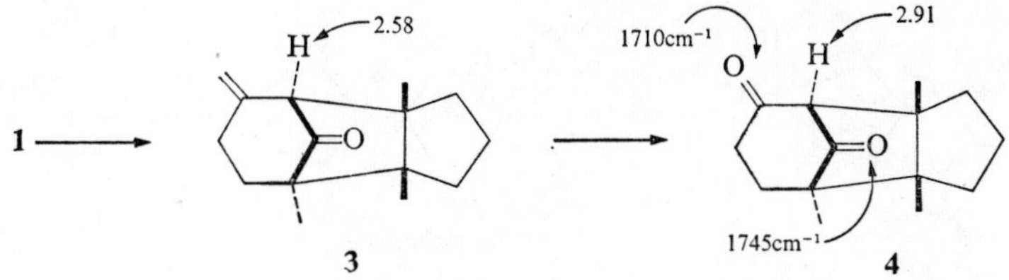

H 2.58

O

1 —→

3

1710cm^{-1} 2.91
O   H

O

1745cm^{-1}
4

Remarks

Gymnomitrol **1** was isolated from the liverwort, *Gymnomitrion obtusum* together with a hydrocarbon **5**, epoxide-acetate **6**, and epoxide-diacetate **7**.[1] Trichodiene **2** and trichodiol **8** were isolated from *Trichothecium roseum* as minor metabolites.[3]

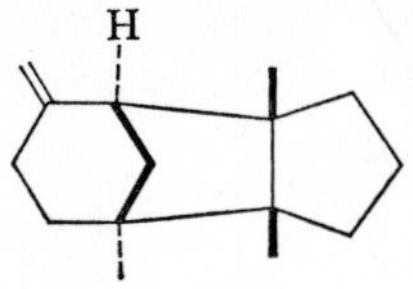

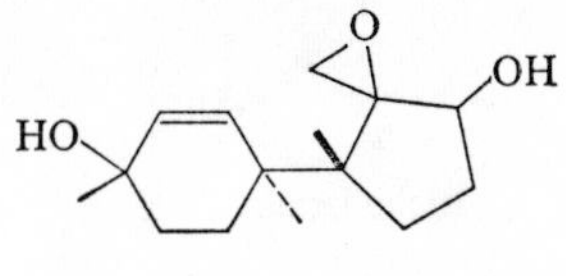

H

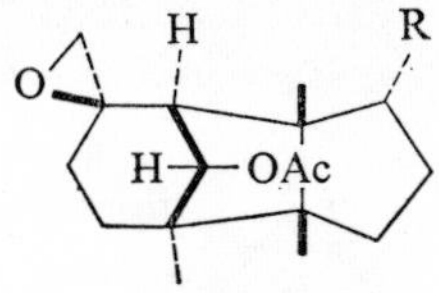

5

H R

O

H—OAc

6 (R=H)
7 (R=OAc)

O

HO   OH

HO—

8

5 and **2** are assumed to be derived biogenetically from the common cationic intermediate **10**, which in turn is generated from farnesyl pyrophosphate through cyclization followed by hydride and methyl shifts.[4]

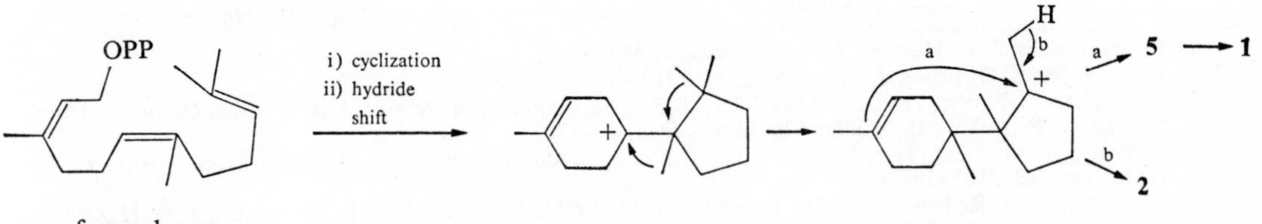

OPP

i) cyclization
ii) hydride
    shift

+

10

H
a   b
    +   a ⟶ 5 ⟶ 1

b
        2

11

farnesyl pyro-
phosphate **9**

1) J. D. Connolly, A. E. Harding, I. M. S. Thornton, *Chem. Commun.*, 1320 (1972).
2) S. Nozoe, Y. Machida, *Tetr. Lett.*, 2671 (1970).
3) S. Nozoe, Y. Machida, *Tetr.*, **28**, 5105 (1972).
4) see "Biosynthesis of trichothecin and helicobasidin"

3.87   BIOSYNTHESIS OF TRICHOTHECIN AND HELICOBASIDIN

Considerable experimental evidence concerning the biosynthesis of trichothecin **8** (R = COCH = CH–Me) and helicobasidin **10** has led to the following postulated biosynthetic pathway.

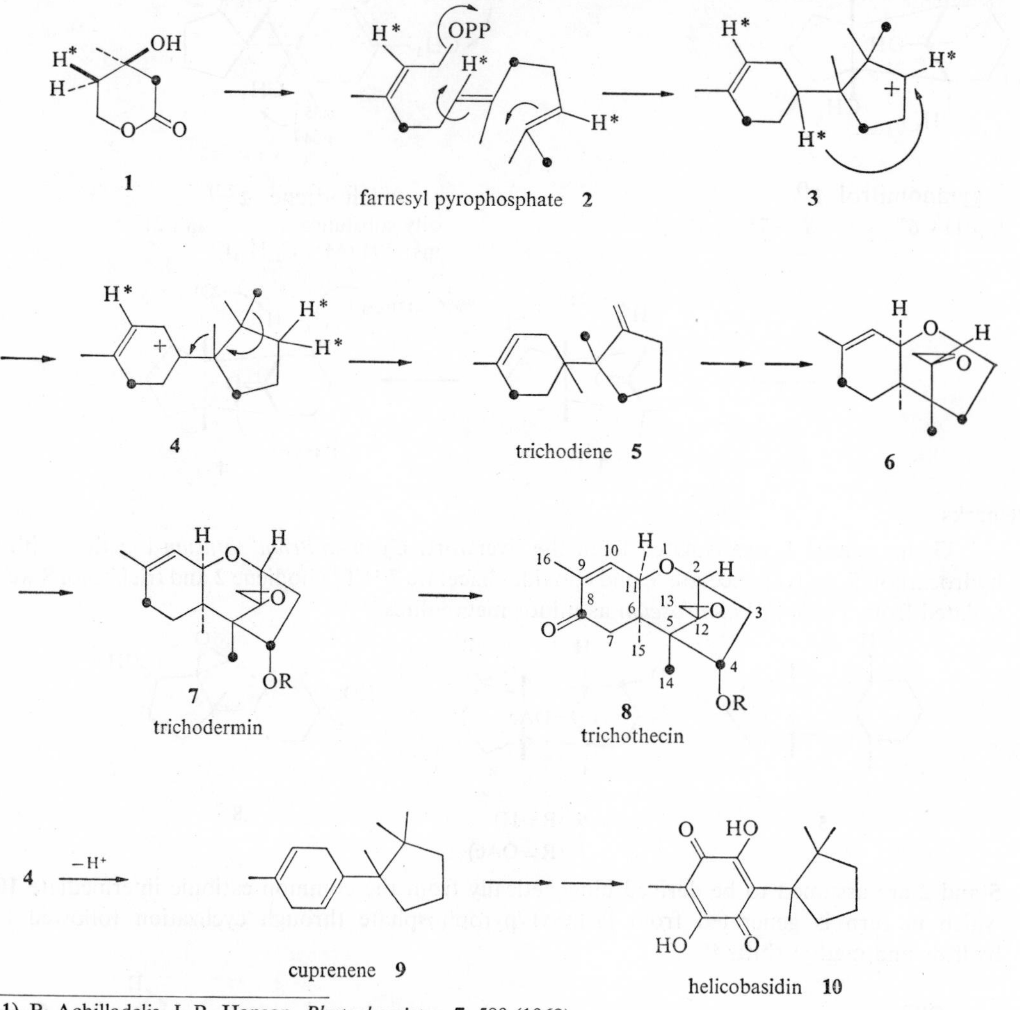

1

farnesyl pyrophosphate **2**

3

4

trichodiene **5**

6

7

trichodermin

8

trichothecin

4 $\xrightarrow{-H^+}$

cuprenene **9**

helicobasidin **10**

1) B. Achilladelis, J. R. Hanson, *Phytochemistry*, **7**, 589 (1968).
2) E. R. H. Jones, G. Lowe, *J. Chem. Soc.*, 3959 (1960).
3) Y. Machida, S. Nozoe, *Tetr.* **28**, 5113 (1972); cf. B. Achilladelis, P. M. Adams, J. R. Hanson, *Chem. Commun.*, 511 (1970); R. Achini, B. Müller, Ch. Tamm, *ibid.*, 404 (1971).
4) S. Nozoe, Y. Machida, *Tetr. Lett.*, 2671 (1970); Y. Machida, S. Nozoe, *ibid.*, 1969 (1972); S. Nozoe, Y. Machida, *Tetr.* **28**, 5105 (1972).
5) P. M. Adams, J. R. Hanson, *Chem. Commun.*, 1414 (1971); B. Achilladelis, P. M. Adams, J. R. Hanson, *J. Chem. Soc.*, Perkin I, 1425 (1972).
6) S. Nozoe, M. Morisaki, H. Matsumoto, *Chem. Commun*, 926 (1970). P. M. Adams, J. R. Hanson, *J. Chem. Soc.*, Perkin I, 586 (1972).
7) P. M. Adams, J. R. Hanson, *Chem. Commun.*, 1569 (1970).
8) See also J. M. Forrester, T. Money, *Can. J. Chem.*, **50**, 3310 (1972) for biosynthesis of **8**.
9) R. E. Evans, A. M. Holton, J. R. Hanson, *Chem. Commun.*, 465 (1973) (biosynthesis of **5** from **2** using cell-free systems).

1) [2-^{14}C]-MVA **1** and ^{14}C-FPP **2** were shown to be incorporated into trichothecin **8** by feeding experiments.[1]

2) Localization of ^{14}C atoms at the C-4, C-8, and C-14 positions was demonstrated by degradation of **8** biosynthesized from [2-^{14}C]-MVA.[2,3] This result was interpreted in terms of successive double methyl migration through the ion **4** to give the skeleton of trichodiene **5**.[9]

3) Trichodiene has been isolated from *Trichothecium roseum* and shown to be incorporated into **8**.[4]

4) That long-range hydride shifts occur (e.g. **3**→**4**) during the cyclization of **2** into **5** was demonstrated by experiments using 4R-[4-^{3}H, 2-^{14}C]-MVA and [2-^{3}H, 2-^{14}C]-geranyl pyrophosphate as substrates.[5]

5) The same hydride shifts were observed in the biosynthesis of helicobasidin **10**, which might be derived from the same ion **4** through cuprenene **9**.[6]

6) Trace amounts of **6** were found in cells of *Trichothecium roseum*.[3]

7) It was demonstrated that the epoxide, crotocin, may be an intermediate of trichothecin biosynthesis.[7,8]

3.88   PREISOCALAMENDIOL AND ISOCALAMENDIOL

preisocalamendiol **1**[1,2]   isocalamendiol **2**[3]

1) The ten-membered ring ketone **1** cyclizes to **2** under acidic conditions.[1]

2) Pyrolysis of **1** gave the cyclic compound **3**.

3) Upon heating shobunone **6** in a sealed tube, **1**, **3** and **9** were produced. Formation of the products was explained in term of intermediates **7** and **8**, which are formed by Cope rearrangement of **6**.[4]

3

1) M. Iguchi, M. Niwa, S. Yamamura, *Chem. Commun.*, 974 (1971).
2) M. Iguchi, A. Nishiyama, S. Yamamura, Y. Hirata, *Tetr. Lett.*, 855 (1970).
3) M. Iguchi, A. Nishiyama, H. Koyama, S. Yamamura, Y. Hirata, *Tetr. Lett.*, 3729 (1969).
4) M. Iguchi, A. Nishiyama, S. Yamamura, Y. Hirata, *Tetr. Lett.*, 4295 (1969).
5) M. Iguchi, M. Niwa, S. Yamamura, *Tetr. Lett.*, 1687 (1973).
6) See also M. Iguchi, M. Niwa, A. Nishiyama, S. Yamamura, *Tetr. Lett.*, 2759 (1973); M. Iguchi, M. Niwa, S. Yamamura, *ibid.*, 4367 (1973).

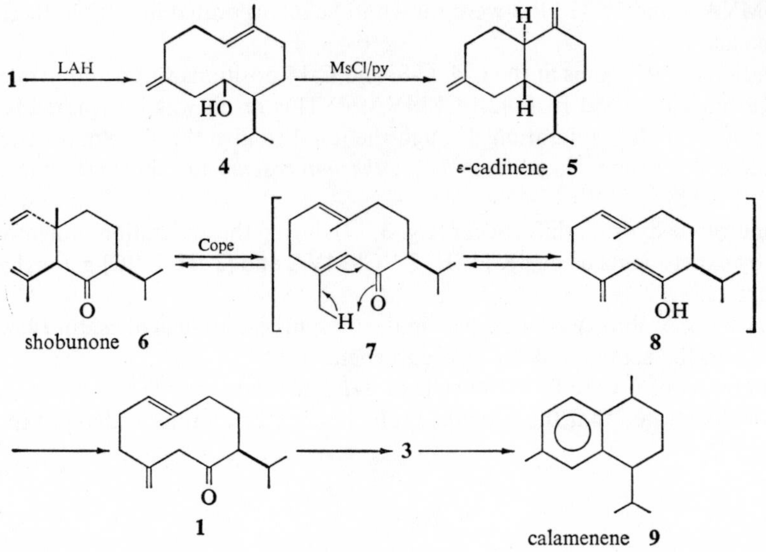

ε-cadinene **5**

shobunone **6**   **7**   **8**

1   calamenene **9**

4) Transannular cyclization of **10** in 80% AcOH afforded products possessing only the cadinane skeleton. However, when the cyclization reaction is conducted in thiophenol, a product with the guaiane skeleton **11** is obtained in 50% yield.[5,6]

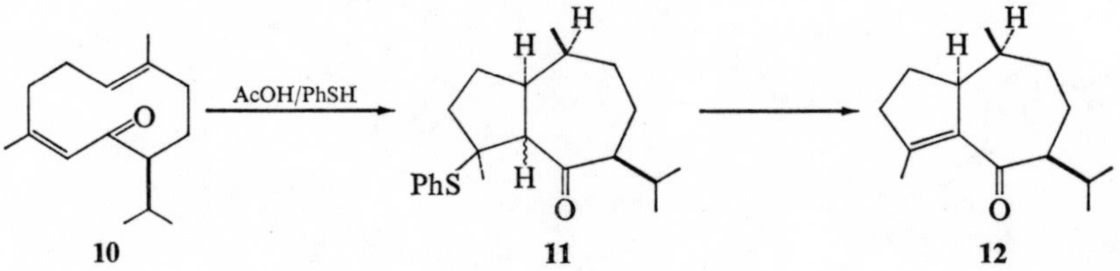

10   **11**   **12**

Remarks

Preisocalamendiol **1**, isocalamendiol **2**, calamendiol (C-9 epimer of **2**), and shobunone **6** were all isolated from the rhizomes of *Acorus calamus*.

3.89 SESQUITERPENES HAVING A CADALANE SKELETON

By extensive separation of the high-boiling hydrocarbon fraction of essential oils from *Mentha piperita* or[1] *Pinus silvestris*[2] the bicyclic sesquiterpenes **1** through **6** were obtained and characterized. All these substances have the same carbon skeletons but there are differences in the relative stereochemistry at C-1, C-6 and C-7.

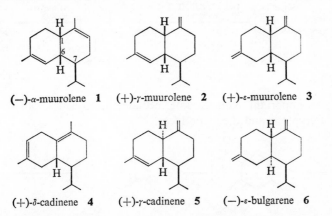

(−)-α-muurolene **1**   (+)-γ-muurolene **2**   (+)-ε-muurolene **3**

(+)-δ-cadinene **4**   (+)-γ-cadinene **5**   (−)-ε-bulgarene **6**

To avoid confusion in the nomenclature of these diastereomeric compounds, systematic names which divided compounds of this type into four subgroups were proposed. These were cadinane **7**, muurolane **8**, amorphane **9** and bulgarane **10**. All compounds having these skeletons or their mirror images can be named on the basis of these four subgroups.

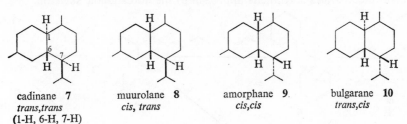

cadinane **7**   muurolane **8**   amorphane **9**   bulgarane **10**
trans,trans   *cis, trans*   *cis,cis*   *trans,cis*
(1-H, 6-H, 7-H)

1) O. Motl, M. Romanuk, V. Herout, *Coll. Czech. Chem. Commun.*, **31**, 2025 (1966); A. Zabza, M. Romanuk, V. Herout, *ibid.*, **31**, 3373 (1966); R. Vlahov, M. Holub, I. Ognjanov, V. Herout, *ibid.*, **32**, 808 (1967); R. Vlahov, M. Holub, V. Herout, *ibid.*, **32**, 822 (1967).

2) L. Westfelt, *Acta. Chem. Scand.*, **20**, 2829, 2841, 2852 (1966).

3.90 SYNTHESIS OF CADINANES

Route I[1]

The Robinson annelation reaction of menthone derivatives with methyl vinyl ketone provides a simple way to construct the cadinane skeleton.

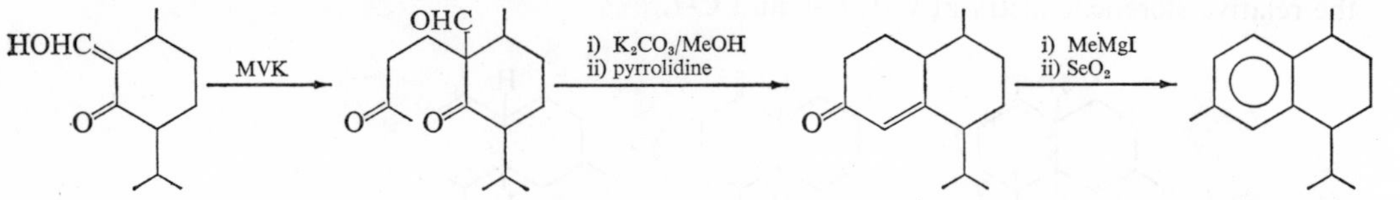

1   **2**   **3**   (+)-calamenene **4**

Route II[2,3]

The diketone **17**, an important synthetic intermediate to cadinanes, can be prepared by Diels-Alder reaction of cryptone and the diene **14**.

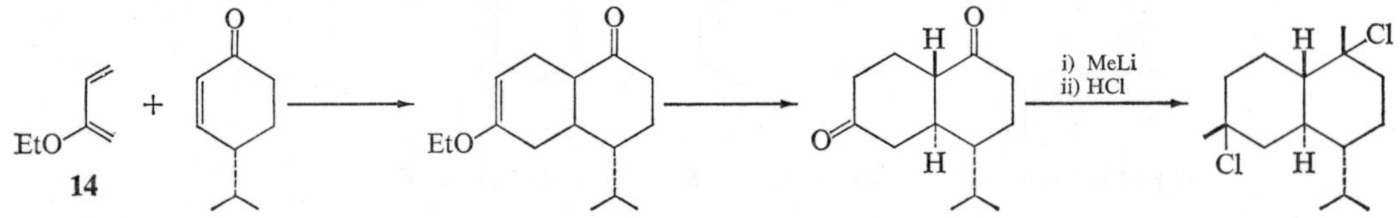

14   cryptone **15**   **16**   **17**   **18**

Route III[4,5]

A tetrahydropyranyloxy group activates a double bond in cyclohexadiene, and it reacts with the enone **7** to yield the diketone **9**. This can be cyclized to **10** and **11** by base and acid, respectively. This route provides a synthetic approach to the norcadinane skeleton.

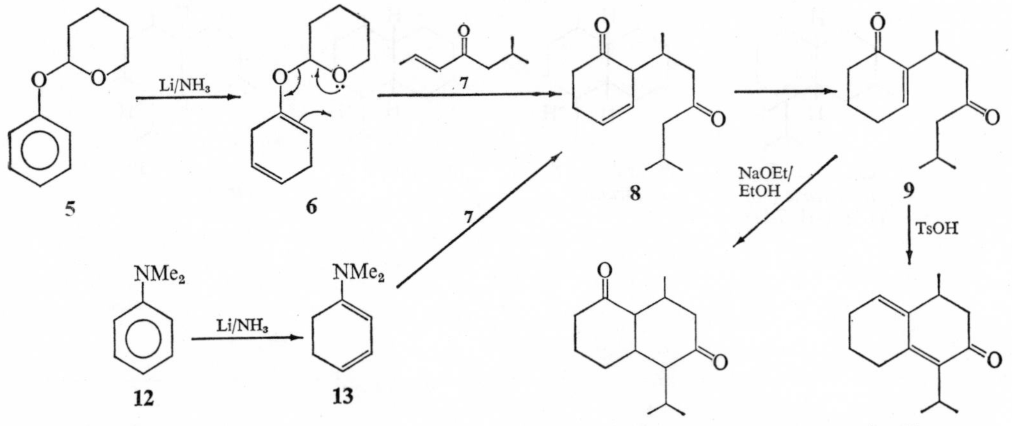

5   **6**   **8**   **9**

12   **13**   **10**   **11**

Route IV[6]

(±)-α-Amorphene has been synthesized by a route involving the [3.3]-sigmatropic reaction of bicyclo[2.2.2.]octenone derivative as the key step.

1) For example, see P. H. Ladwa, G. D. Joshi, S. N. Kulkarni, *Chem. Ind.*, 1601 (1968).
2) M. D. Soffer, G. E. Gunay, O. Korman, M. B. Adams, *Tetr. Lett.*, 389 (1963); *ibid.*, 1355 (1965).
3) M. D. Soffer, L. A. Burk, *Tetr. Lett.*, 211 (1970).
4) A. J. Birch, J. Diekman, P. L. MacDonald, *Chem. Commun.*, 52 (1970).
5) A. J. Birch, E. G. Hutchinson, G. S. Rao, *Chem. Commun.*, 657 (1970).
6) R. P. Gregson, R. N. Mirrington, *Chem. Commun.*, 598 (1973).

3.91 STRUCTURE AND SYNTHESIS OF COPAENE AND YLANGENE

copaene **1**[1,2]

bp (7.5 mm): 112–113° α_D(CHCl$_3$): −6.5
ir (film): 3010, 1660, 788
nmr in CCl$_4$

1) On treatment of **1** with HCl, a C-C linkage is cleaved to give rise to (−)-cadinene dihydrochloride **2**.

2) That the acid-labile ring is four-membered is shown by ir absorption at 1780 cm^{-1} due to cyclobutanone in **4**.

3) Stereochemistry at C-7 and C-8 in **1** must be the same as that of **2**.

4) Only a *cis* junction is possible for two six-membered rings.

Synthesis[3]

Copaene **1** and ylangene **10** were synthesized by a route involving base-catalysed cyclization of the keto-tosylate as a key reaction. This is generally applicable to the synthesis of the tricyclo [4.4.0.0²·⁷]-decane ring system.

Remarks

Copaene **1** was first isolated from African copaiba balsam oil, and was later found in essential oil of *Cedrela toona, Cyperus rotundus,* etc. An isomeric hydrocarbon, ylangene **10**, which differs from **1** only in the configuration of the isopropyl group, was obtained from *Schizandra chinensis,*[4] and the double bond isomers β-copaene and β-ylangene were isolated from *Pinus silvestris.*[5]

1) V. H. Kapadia, B. A. Nagasanpagi, V. G. Naik, S. Dev, *Tetr.*, **21**, 607 (1965).
2) P. de Mayo, R. E. Williams, G. Büchi, S. H. Feairheller, *Tetr.*, **21**, 619 (1965).
3) C. H. Heathcock, R. A. Badger, J. W. Patterson Jr., *J. Am. Chem. Soc.*, **89**, 4133 (1967); *ibid.*, **88**, 4111 (1966).
4) O. Motl, V. Herout, F. Sorm, *Tetr. Lett.*, 451 (1965).
5) L. Westfelt, *Acta. Chem. Scand.*, **21**, 152 (1967).
6) C. J. W. Brooks, M. M. Campbell, *Chem. Commun.*, 630 (1969).

Mustakone **9**[1] and brachylaenalone **11**[6] were found in *Cedrela toona* and *Brachylaena hutchinsii*, respectively.

| 9 | 10 | 11 |

3.92   REARRANGEMENT OF SATIVENE

1.05

{ 4.40
{ 4.72

0.87 (d,5)
0.90 (d,5)

sativene **1**[1]

ms: 204 (M⁺ $C_{15}H_{26}$)
ir: 3060, 1660, 885(*exo*-methylene)
nmr in CCl_4

1) On treatment with refluxing $Cu(OAc)_2$/HOAc, sativene **1** yields a mixture of the isomers **2** and **3**. After 50 hr, an equilibrium consisting of 7% **1**, 32% **2** and 61% **3** was established.[2]

2) Copacamphene **4** differs from **1** in being epimeric at the isopropyl group, but rearranges into (+)-sativene, an enantiomer of **1**, under acid catalysis.[2]

cyclosativene **2**   isosativene **3**

copacamphene **4**

sativene
(enantiomer of **1**)

Remarks

Sativene **1** was isolated from the plant pathogenic fungus, *Helminthosporium sativum* as a minor constituent.[1]

Cyclosativene **2** also occurs in nature, and is obtained from the cortical turpentine of California red fir, *Abies magnifica*.[3]

Copacamphene **4**, a rearrangement product of copaborneol, has not itself been found in nature, although related compounds, such as cyclocopacamphenic acid and cyclocopacamphene have been isolated from vetiver oil.[4]

1) P. de Mayo, R. E. Williams, *J. Am. Chem. Soc.* **87**, 3275 (1965); *ibid.*, **86**, 4438 (1964).
2) J. E. McMurry, *J. Org. Chem.*, **36**, 2826 (1971); *Tetr. Lett.*, **55** (1969); *ibid.*, 3735, 3731 (1970).
3) L. Smedman, E. Zavarin, *Tetr. Lett.*, 3833 (1968); L. Smedman, E. Zavarin, R. Teranishi, *Phytochemistry*, 1457 (1969)
4) F. Kido, R. Sakuma, H. Uda, A. Yoshikoshi, *Tetr. Lett.*, 3169 (1969).

3.93 SYNTHESIS OF HELMINTHOSPORAL

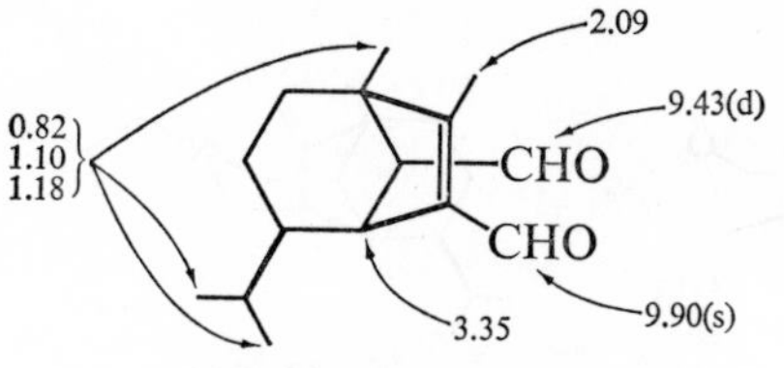

helminthosporal **1**[1]

2.09
9.43(d)
0.82
1.10
1.18
—CHO
CHO
3.35
9.90(s)

ir: 2720 (aldehyde), 1715, 1685
(unsat. aldehyde), 1618 (–C=C–)
uv: 266 (ϵ 11,000)

The total synthesis of natural, levorotatory helminthosporal **1** from (−)-carvomenthone **2**[2] established the absolute configuration of the natural compound as **1**. This synthesis is shown below.

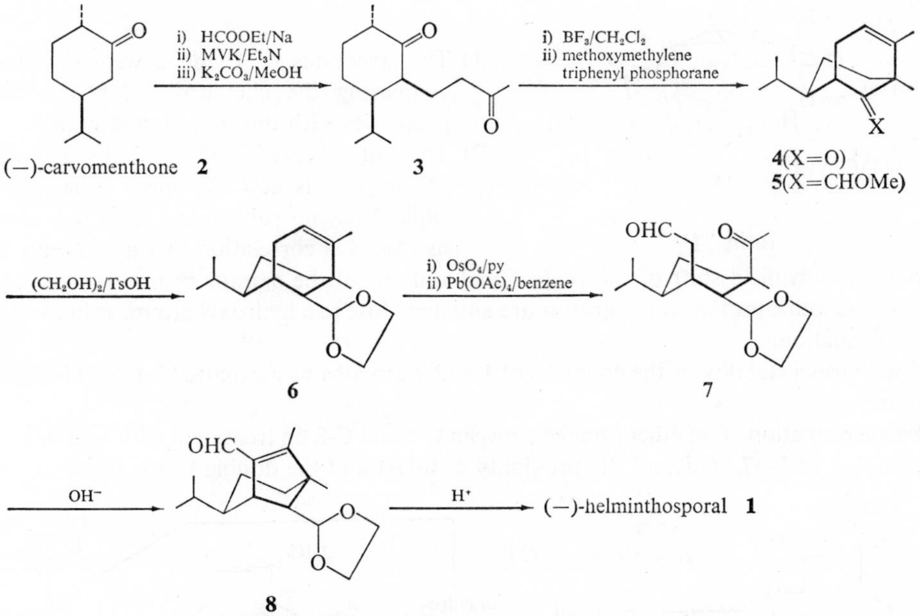

i) HCOOEt/Na
ii) MVK/Et$_3$N
iii) K$_2$CO$_3$/MeOH

(−)-carvomenthone **2**

3

i) BF$_3$/CH$_2$Cl$_2$
ii) methoxymethylene triphenyl phosphorane

4(X=O)
5(X=CHOMe)

(CH$_2$OH)$_2$/TsOH

6

i) OsO$_4$/py
ii) Pb(OAc)$_4$/benzene

OHC   O

7

OHC

OH⁻

H⁺   (−)-helminthosporal **1**

8

Remarks

Helminthosporal **1** is a crop-destroying toxin isolated from *Helminthosporium sativum* (*Bipolaris sorokiniana*), a plant-pathogenic fungus which produces a seedling blight, foot-and-root rot, and head blight of cereals. It was reported that helminthosporal **1** does not exist in the dialdehyde form in the culture. The immediate precursor of **1**, prehelminthosporal[4] has been isolated as an acetal **10**, from which **1** was generated on acid, base or heat treatment. Prehelminthosporol **11**,[4] the precursor of helminthosporol,[5] has also been isolated. Helminthosporal **1**

1) P. de Mayo, E. Y. Spenser, R. W. White, *Can. J. Chem.*, **39**, 1608 (1961); *ibid.*, **41**, 2996 (1963); *J. Am. Chem. Soc.*, **84**, 494 (1962).
2) E. J. Corey, S. Nozoe, *J. Am. Chem. Soc.*, **87**, 5728 (1965); *ibid.*, **85**, 3527 (1963).
3) P. de Mayo, R. E. Williams, *J. Am. Chem. Soc.*, **87**, 3275 (1965).
4) P. de Mayo, R. E. Williams, E. Y. Spenser, *Can. J. Chem.*, **43**, 1357 (1965).
5) S. Tamura, A. Sakurai, K. Kainuma, M. Takai, *Agr. Biol. Chem.* (Tokyo), **27**, 738 (1963).
6) see P. de Mayo, J. R. Robinson, E. Y. Spenser, R. W. White, *Experientia*, **18**, 359 (1962) for the biosynthesis of **1**.
7) F. Dorn, D. Arigoni, *Chem. Commun.*, 1342 (1972).

was shown to be derived from the hydrocarbon sativene **9**, which was obtained from the same fungus as a minor constituent.[3] The toxic base victoxinine **12** has recently been isolated from *Helminthosporium victoriae*.[7]

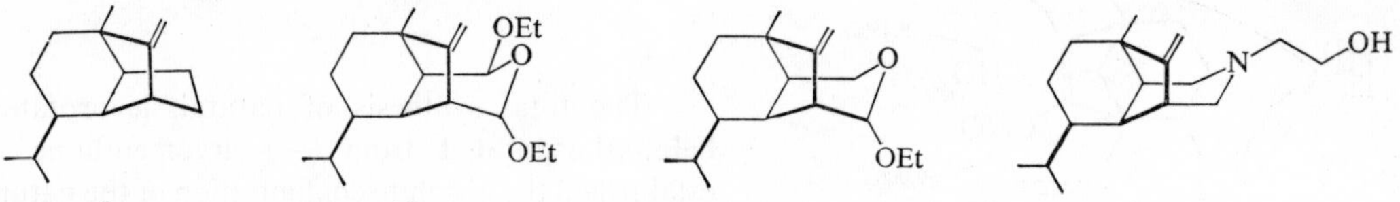

| 9 | 10 | 11 | 12 |

3.94   STRUCTURE OF TUTIN AND RELATED COMPOUNDS

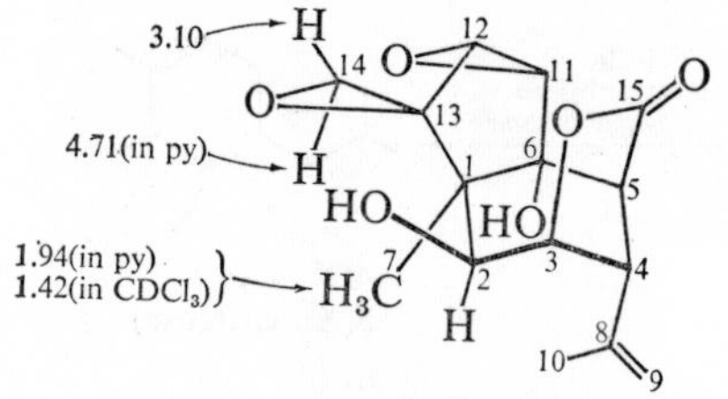

tutin   **1**[1]

1) The structures of **1** and **2** were elucidated by comparing the chemical and physicochemical properties with those of picrotoxinin **9**.

2) The difference of structure between tutin and coriamyrtin is at C-2, where **1** has a *sec*-OH while **2** has no substituent. This was confirmed by chemical correlation of **1** and **2** with **3**.

3) A large downfield shift of the 1-Me signal and one of the oxirane proton signals (14-H) in pyridine indicates that these groups are situated close to a hydroxyl group, as in the depicted conformation.

4) The unusual stability of the epoxides of **1** and **2** can also be accounted for by this conformation.

5) Facile formation of an ether linkage between C-6 and C-8 on treatment with acid or bromine, e.g. **2→6** or **1→7**, indicates the proximity of 6-OH and the double bond.

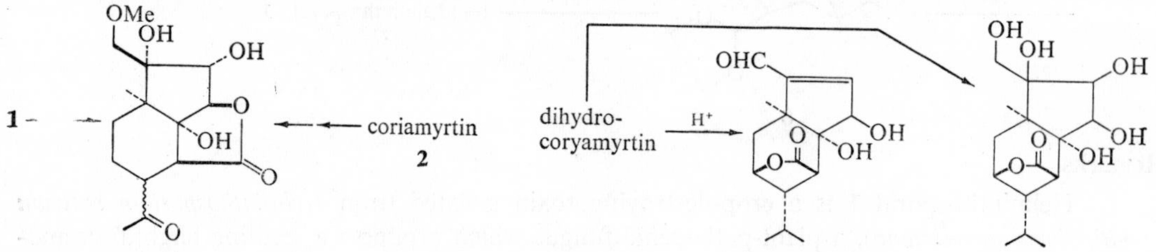

| 3 | 4 | 5 |

| 6 | *α*-Br-tutin **7** | *α*-Br-isotutin **8** |

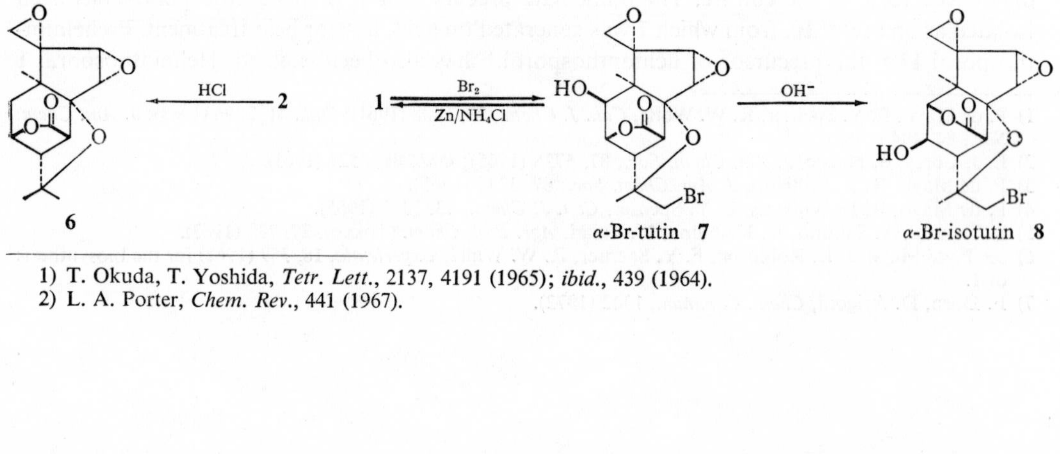

1) T. Okuda, T. Yoshida, *Tetr. Lett.*, 2137, 4191 (1965); *ibid.*, 439 (1964).
2) L. A. Porter, *Chem. Rev.*, 441 (1967).

Remarks

Tutin **1** and coriamyrtin **2**[1] are bitter, toxic lactones isolated from the genus *Coriaria*. Picrotoxinin **9**,[2] and other related substances at a higher oxidation level, nitrogen-containing compounds such as dendrobine **10**, and substances having additional carbon atoms, such as capenicin **11**, have been found in nature.[2]

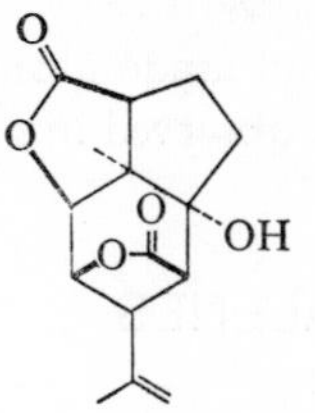

picrotoxinin **9** 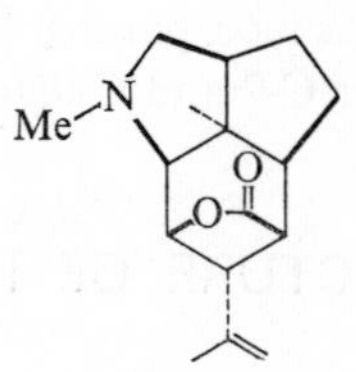   dendrobine **10** 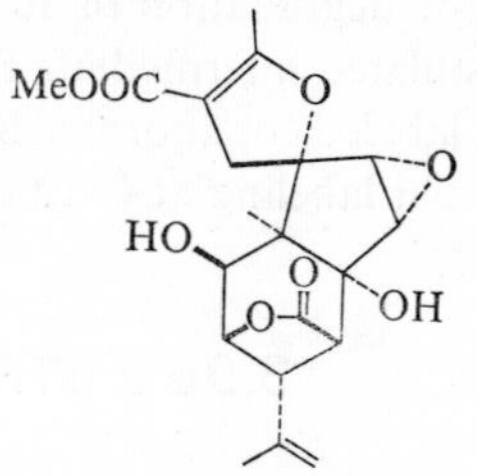   capenicin **11**

3.95 BIOSYNTHESIS OF TUTIN AND HELMINTHOSPORAL

The biosynthetic route to helminthosporal **7** and tutin **10**, both of which have an unusual carbon skeleton was postulated as follows, supported by various experimental evidence.

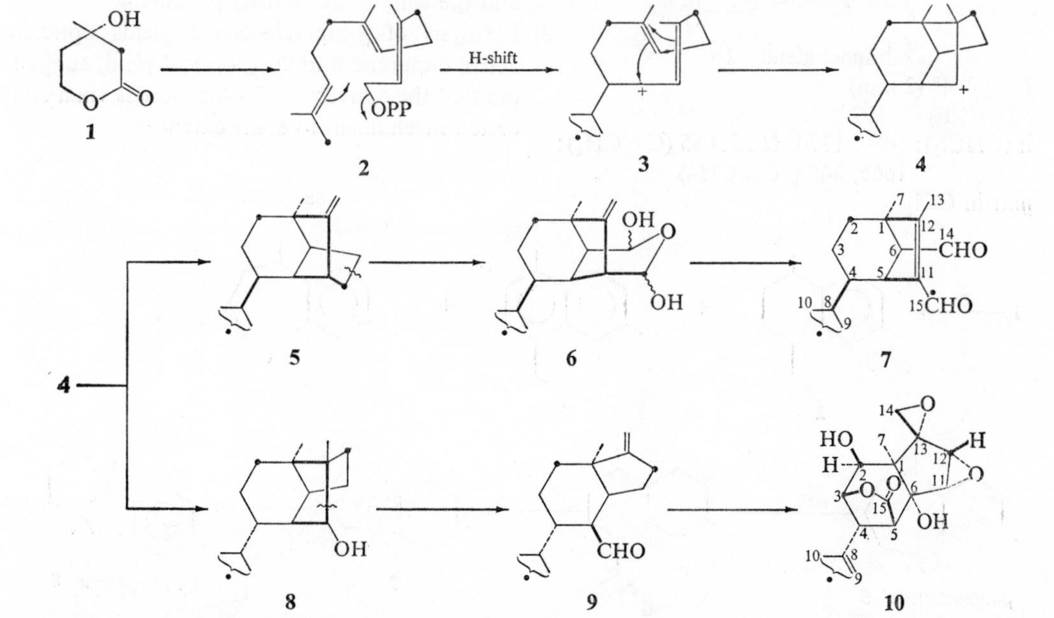

1) One-third of the radioactivity was shown to be located at C-15 by degradation of **7** biosynthesized from [2-14C]-MVA **1**.[1]

1) P. de Mayo, J. R. Robinson, E. Y. Spenser, R. W. White, *Experientia*, **18**, 359 (1962).
2) P. de Mayo, R. E. Williams, *J. Am. Chem. Soc.*, **87**, 3275 (1965).
3) M. Biollaz, D. Arigoni, *Chem. Commun.*, 633 (1969).
4) A. Corbella, P. Garifoldi, G. Jommi, S. Scolastico, *Chem. Commun.*, 634 (1969).
5) M. Kolbe, L. Westfelt, *Acta Chem. Scand.*, **21**, 585 (1967).
6) K. W. Turnbull, W. Acklin, D. Arigoni, A. Corbella, P. Gariboldi, G. Jommi, *Chem. Commun.*, 598 (1972).
7) cf. A. Corbella, P. Gariboldi, G. Tommi, *Chem. Commun.*, 729 (1973).

2) The precursor hydrocarbon, sativene **5**, and prehelminthosporal **6** were isolated from *Helminthosporium sativum,* a fungus producing **7**.[2]

3) Degradation studies with tutin **10** and coriamyrtin (deoxytutin) revealed radioactivity localized at C-12, C-9, and C-10, with equal distribution observed between the latter two carbons.[3]

4) That tritium atoms were localized at C-2, C-12, C-9, and C-10 positions was also demonstrated, by degradation of **10** biosynthesized from [2-^{3}H,2-^{14}C]-MVA.[4,7]

5) The postulated intermediate **8** has been isolated from *Pinus silvestris*.[5]

6) Tritium labeled copaborneol **8** was incorporated into **10** without randomization of the label.[6]

7) Preferential labeling at C-12 over C-2 and C-9(10) in **10** was observed in some experiments.

3.96   STRUCTURE OF HIMACHALENES

α-himachalene **1**[1]

bp: 93–4° (2 mm)

α_D(CHCl$_3$): −192.3°

ir (CHCl$_3$): 3060, 1770, 1625, 855 (C=CH$_2$);
                   1665, 865 (–C=CH–)

nmr in CCl$_4$

1) α- and β-himachalenes (**1** and **5**) afford the same dihydrochloride.

2) Upon dehydrogenation with Se, **1** yields the bond-cleaved product **2**, rebonded product **3**, and the simply aromatized product **4**.

3) Pyrolysis of β-himachalene **5** yields optically active cuparene **8** in 20% optical yield, suggesting that the conversion **5**→**8** proceeds via a concerted mechanism to some extent.[4]

2   **3**

β-himachalene **5**   **6**   **7**   (+)-cuparene **8**

Remarks

α- and β-himachalenes (**1** and **5**), himachalol **9**,[2] and allohimachalol **10**[3] were isolated from

1) T. C. Joseph, S. Dev. *Tetr.,* **24**, 3809 (1968).
2) S. C. Bisarya, S. Dev. *Tetr.,* **24**, 3861 (1968).
3) S. C. Bisarya, S. Dev, *Tetr.,* **24**, 3869 (1968).
4) H. N. Subba Rao, N. P. Damodaran, S. Dev. *Tetr. Lett.,* 2213 (1968).
5) S. C. Bisarya, S. Dev. *Tetr. Lett.,* 3761 (1964).

essential oils of Himalayan cedar, *Cedrus deodara*. The structure of allohimachalol **10** was revised[3] in 1968 from a previous structure with a bicyclo-[4.4.1]-undecane ring system.[5] Solvolysis[3] of the tosylate ester of allohimachalol **10** produces a mixture of **1**, **5**, **9**, and **10**.

9   **10**

3.97   REARRANGEMENT OF LONGIFOLENE

Treatment of (+)-longifolene **1** with $Cu(OAc)_2$/HOAc yields a mixture of longicyclene **2** (29%), isolongifolene **3** (15%) and racemic longifolene **1a** and **1b** (51%).[1,2]

(+)-longifolene **1a**   longicyclene **2**   isolongifolene **3**

4   **5**   (−)-longifolene **1b**

6   **7**   **8**   **9**   **10**

1) Formation of **2** is accounted for by deprotonation of the ion **4**.

2) Racemization of **1a** can be interpreted in terms of a mechanism involving the ions **4** and **5**.

3) The extensive rearrangement to isolongifolene **3** is rationalized[3,4] in terms of a pathway through the ion **6**, which undergoes an *exo*-2,3-methyl shift to give the ion **7** with successive rearrangement via the ions **8**, **9** and **10**.

4) An alternative mechanism involving an *endo*-2,3-methyl shift in **4**, yielding **3**, has been proposed.[2]

1) J. R. Prahlad, R. Ranganathan, U. R. Nayak, T. S. Santhanakrishnan, S. Dev, *Tetr. Lett.*, 417 (1964).
2) G. Ourisson, *Proc. Chem. Soc.*, 274 (1964).
3) J. A. Berson, J. H. Hammons, A. W. McRowe, R. G. Bergman, A. Remanick, D. Houston, *J. Am. Chem. Soc.*, 89, 2590 (1967).
4) J. E. McMurry, *J. Org. Chem.*, 36, 2826 (1971).

3.98 TRANSANNULAR HYDROGEN SHIFT IN THE LONGIFOLENE SERIES

longifolene **1**

The proximity of the C-3 and C-7 positions in the longifolene molecule is responsible for facile transannular 1,5-hydrogen shift or transannular cyclization.[1-4]

1) Halomethylation of **1**[2] yields the 3α-bromo derivative **4**. The reaction involves free radical hydrogen transfer from C-3 to C-7, as in **2→3**

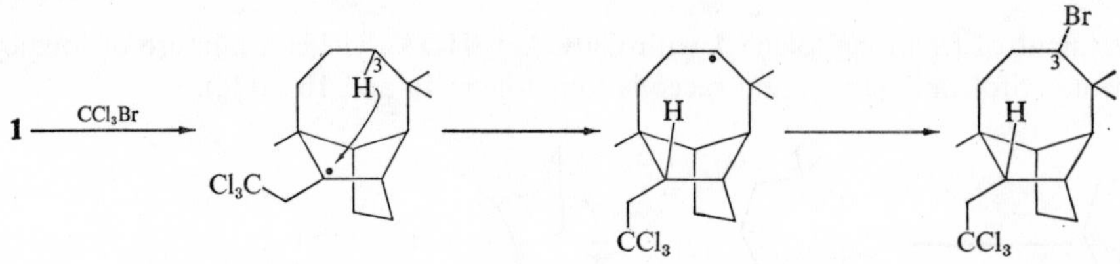

2) Solvolysis of the 3α-bromo derivative **5** affords longifolene **1** in 80% yield by a transannular 1,5-hydride shift from C-7 to C-3.[1]

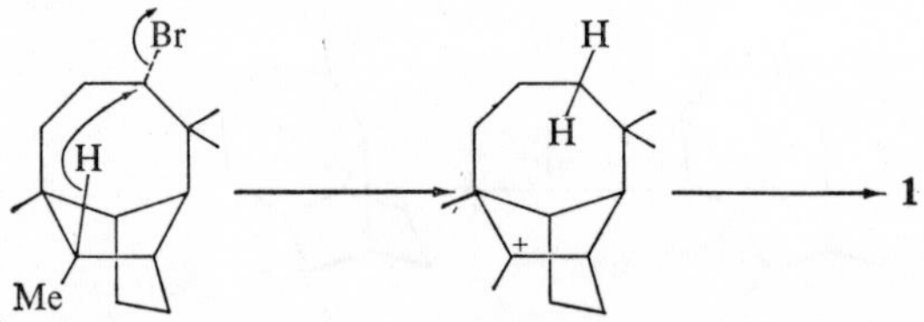

3) When alkylborane **7** was oxidized by AgO, the 3α-hydroxyl derivative **8** and longifolol **9** were formed.[3,5] Formation of **8** can also be interpreted in terms of a transannular H shift. Oxidation with alkaline H_2O_2 gives **9**.

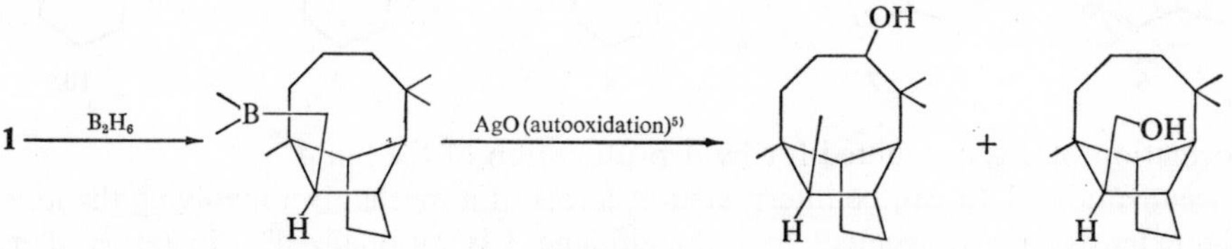

4) Solvolysis of the 3α-bromo-longifolene **10** gives the transannular cyclization products **11** and **12**.[4] This cyclization also provides a chemical proof of the proximity of the C-3 and C-7 positions in **1**.

1) L. Stéhelin, J. Lhomme, G. Ourisson, *J. Am. Chem. Soc.*, **93**, 1650 (1971).
2) D. Helmlinger, G. Ourisson, *Ann. Chem.*, **686**, 19 (1965).
3) J. Lhomme, G. Ourisson, *Tetr.*, **24**, 3167 (1968).
4) D. Helmlinger, G. Ourisson, *Tetr.*, **25**, 4895 (1969).
5) Y. Tanahashi, J. Lhomme, G. Ourisson, *Tetr.*, **28**, 2655 (1972).

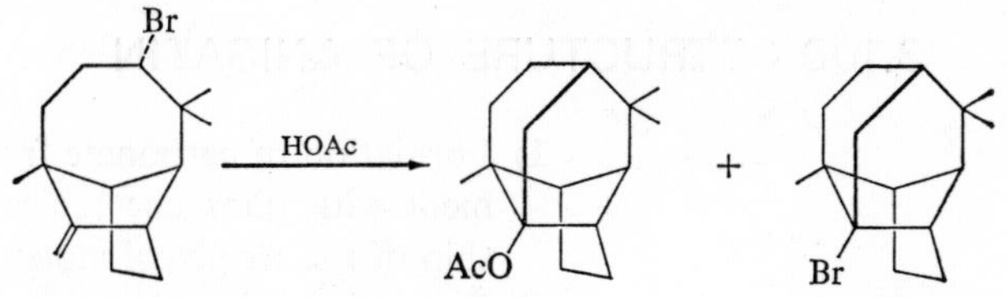

10   **11**   **12**

3.99   SYNTHESIS OF LONGIFOLENE

The tricyclic sesquiterpene longifolene **1** was synthesized by a route involving intramolecular Michael cyclization of the homodecalin derivative **6** as the key step.[1]

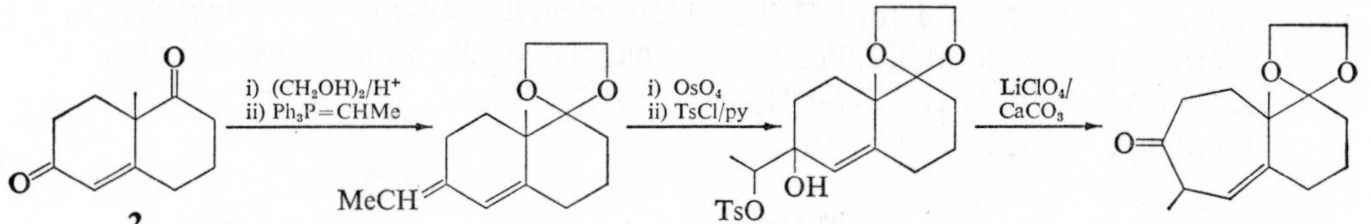

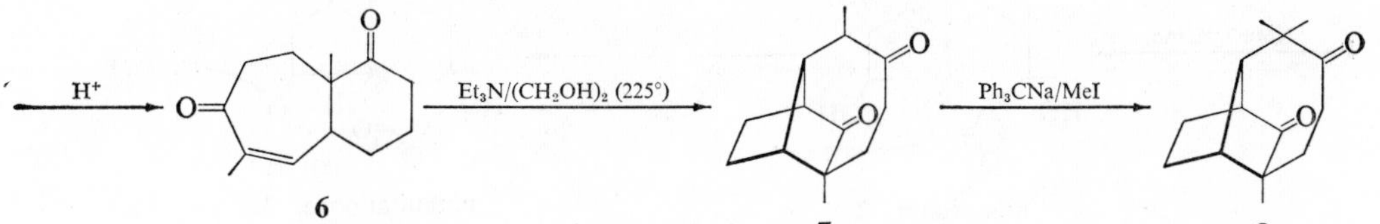

4→5: Vinyl migration predominates over alkyl migration, giving the desired ring-expanded product **5** from the pinacol rearrangement of **4**.

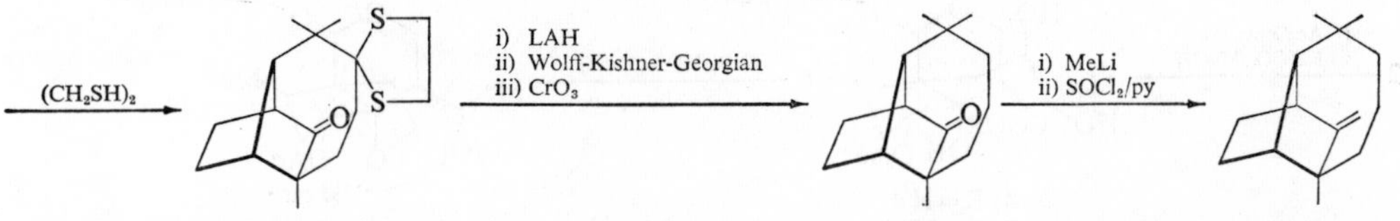

6→7 (5→7): The same type of internal Michael addition is seen in the facile rearrangement of santonin to santonic acid, though the cyclization of **6** is much less facile than that of santonin.

8→9: (+)-Longifolene was obtained from the optically active monothioketal prepared from **8** by treatment with L-(+)-2,3-butanedithiol, followed by chromatographic separation of the resulting diastereomeric mixture of monothioketals.

9→10: Direct Wolff-Kishner-Georgian reduction of the hydroxy thioketal derived from **9** was successfully applied for selective reduction of the carbonyls.

1) E. J. Corey, M. Ohno, R. B. Mitra, P. A. Vatakencherry, *J. Am. Chem. Soc.*, **86**, 478 (1964); *ibid.*, **83**, 1251 (1961).

3.100   STRUCTURE OF ANISATIN

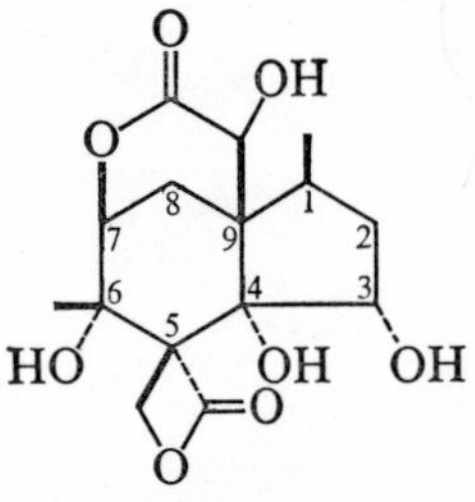

anisatin **1**[1]

mp: 227–228° α_D (dioxane): −27°
ms: 328 (M⁺ $C_{15}H_{20}O_8$)
ir (CHCl₃): 1826 (β-lactone), 1739 (δ-lactone)
nmr: data for triacetate of **1** (in CDCl₃)
0.87 (sec-Me), 1.87 (6-Me), 2.06, 2.09, 2.26 (3 × Ac), 5.40 (–OH), 4.17 (–CH₂–O–) (ABq), 5.88 (q, 6.5 and 8.0)

1) Formation of carbonate from **1** and **2** on treatment with phosgene indicates the *cis* relationship of the *vic* glycol moiety.
2) Formation of δ-lactone in **4** indicates that the methylene group of the β-lactone at C-5 and the carboxyl group at C-9 are in a 1,3-diaxial position.
3) Another lactonization, to **6**, reveals that the 4-OH and the lactone carbonyl at C-9 are in a *trans* relationship.
4) The ester **5** consumed 1 mole of Pb(OAc)₄, indicating the *trans* relationship of the *vic* glycol in the six membered ring.
5) The complete structure was established by x-ray analysis[2] of the bromo derivative of **3**.

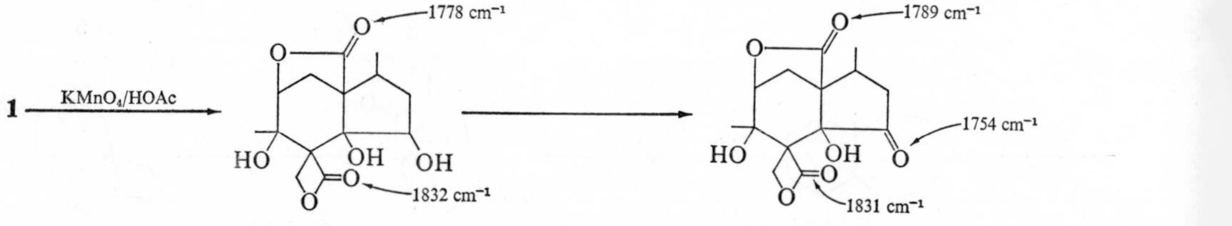

noranisatin **2**   noranisatinone **3**

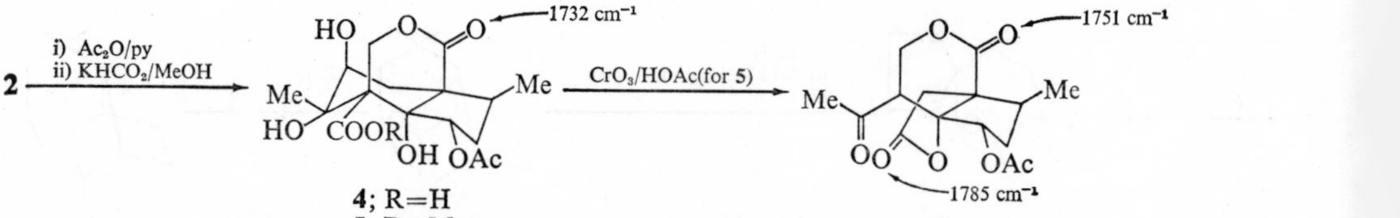

4; R=H
5; R=Me

6

Remarks

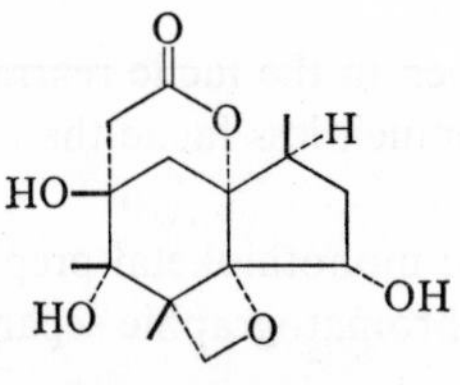

pseudoanisatin **7**

Anisatin **1** is a toxic component of the seeds of Japanese star anise, *Illicium religiosum* (*Illicium anisatum*)[1,4]. From the same source, neoanisatin (OH at C-3 replaces H in **1**), and pseudoanisatin **7** have been isolated.[3]

1) K. Yamada, S. Takada, S. Nakamura, Y. Hirata, *Tetr. Lett.*, 4785, 4797 (1965).
2) N. Sakabe, Y. Hirata, A. Furusaki, Y. Tomie, I. Nitta, *Tetr. Lett.*, 4795 (1965).
3) M. Okigawa, N. Kawano, *Tetr. Lett.*, 75 (1971).
4) J. F. Lane, W. T. Koch, N. S. Leeds, G. Gorin, *J. Am. Chem. Soc.*, **24**, 3211 (1952).

3.101 STRUCTURE OF LASERPITINE

laserpitine **1**[1]

1) Dehydrogenation of **1** gave **2** with skeletal rearrangement.
2) Periodate oxidation of **4** yielded **7** and formic acid, indicating the presence of a *vic* triol in **4**.
3) Production of the formate **8** in this reaction can be accounted for via **5** and **6**.
4) The β-hydroxyketone **7** is readily dehydrated yielding **9**. The bathochromic shift in the uv of **9** might be due to the presence of a hydroxyl group at C-4. The high value of the *vic* coupling constant ($J=16.5$) of the olefinic proton signals in **9** indicates a *trans* configuration of the double bond.

2   laseol **3**   **4**

5   **6**   **7** R=H
                         8 R=CHO

9
uv: 239 (4.17)

Remarks

Laserpitine **1** was isolated from *Laserpitium latifolium*.[1] The related sesquiterpenes carotol **11**, daucol **10** and daucene **12** were isolated from common carrot, *Daucus carota*.[2,3] Jaeschkea-nadiol **13** has been isolated from the roots of *Ferula jaeschkeana*,[7] and its stereochemistry was established by direct correlation with laseol **3**.[7]

Recently, (+)-daucene **12**, (+)-carotol **11** and (−)-daucol **10** have been synthesized starting from (−)-dihydrocarvone.[4] (−)-Daucene has also been synthesized from (+)-limonene.[5,6]

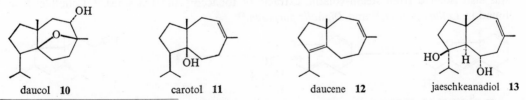

daucol **10**   carotol **11**   daucene **12**   jaeschkeanadiol **13**

1) M. Holub, Z. Samek, V. Herout, F. Sorm, *Coll. Czech. Chem. Commun.*, **32**, 591 (1967); M. Holub, V. Herout, F. Sorm, A. Linek, *Tetr. Lett.*, 1441 (1965).
2) M. Soucek, *Coll. Czech. Chem. Commun.*, **27**, 2929 (1962); J. Levisalles, H. Rudler, *Bull. Soc. Chim. Fr.*, **8**, 2020 (1964).
3) R. B. Bates, C. D. Green, T. C. Sneath, *Tetr. Lett.*, 3461 (1969).
4) H. De Broissa, J. Levisalles, H. Rudler, *Chem. Commun.*, 855 (1972).
5) M. Yamasaki, *Chem. Commun.*, 606 (1972).
6) For an alternative approach to the daucane skeleton, see P. Naegeli, R. Kaiser, *Tetr. Lett.*, 2013 (1972).
7) M. C. Sriraman, B. A. Nagasampagi, R. C. Pandey, S. Dev, *Tetr.*, **29**, 985 (1973).

3.102   STRUCTURE OF ACTINIDIOLIDE

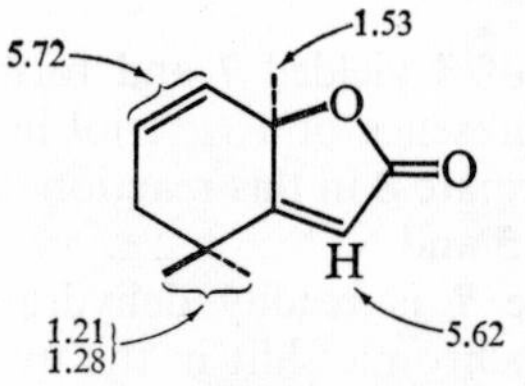

actinidiolide **1**[1]
ms:178 (M$^+$ C$_{11}$H$_{14}$O$_2$)
uv: 241 (ϵ 10,000)
ir: 1745 (α,β-unsat. γ-lactone)

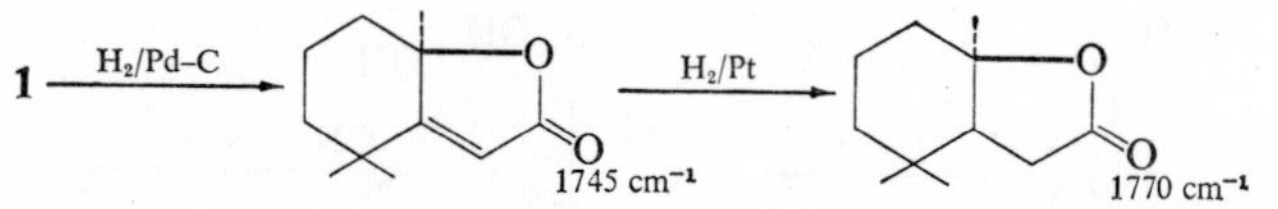

2H-actinidiolide **2**   4H-actinidiolide **3**

1) The structure was confirmed by the chemical synthesis of **2** and **3**[6] as follows.[1]

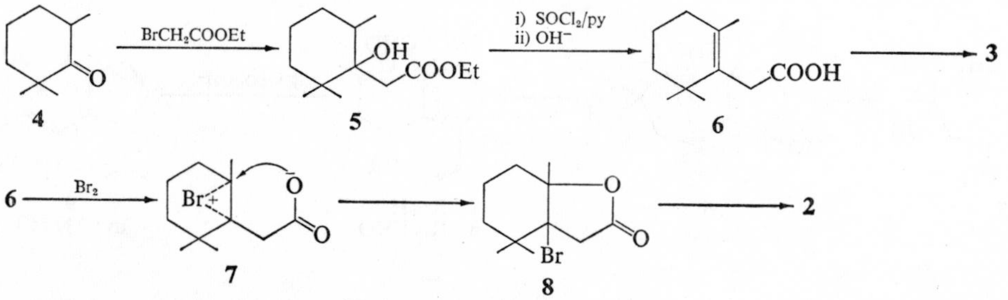

Remarks

Actinidiolide **1** and dihydroactinidiolide **2** were isolated from essential oil of the leaves of *Actinidia polygama*. These compounds were found to be active, albeit at a low level, towards Felidae animals. A related substance, actinidol **9**, was isolated from the same source.[1] Actinidol **9** is a very unstable liquid readily subject to air oxidation to yield actinidiolide **1**. Compound **2** was also isolated from steam-volatile extracts of tobacco[2] and black tea.[3] Loliolide **1** was found in *Lolium perenne*[4] and *Digitalis purpurea*.[5]

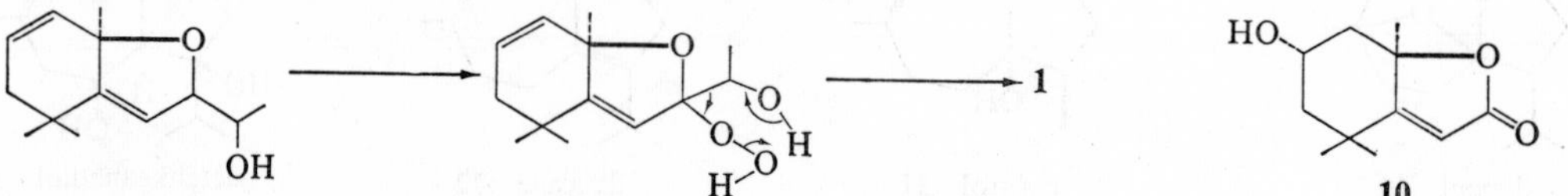

1) T. Sakan, S. Isoe, S. B. Hyeon, *Tetr. Lett.*, 1623 (1967).
2) W. C. Bailey Jr., A. K. Bose, R. M. Ikeda, R. H. Newman, H. V. Secor, C. Varsel, *J. Org. Chem.*, **33**, 2819 (1968); H. Kaneko, K. Ijichi, *Agr. Biol. Chem.* (Tokyo), **32**, 1337 (1968).
3) K. Ina, Y. Sakato, H. Fukami, *Tetr. Lett.*, 2777 (1968); J. Bricout, R. Viani, F. Muggler-Chavan, J. P. Marion, D. Raymond, R. H. Egli, *Helv. Chim. Acta*, **50**, 1517 (1967).
4) R. Hodges, A. L. Porte, *Tetr.*, **20**, 1463 (1964).
5) T. Wada, D. Satoh, *Chem. Pharm. Bull.*, **12**, 752 (1964); T. Wada, *ibid.*, **12**, 1117 (1964); *ibid.*, **13**, 43 (1965).

3.103 SYNTHESIS OF ALLENIC TERPENE FROM GRASSHOPPER

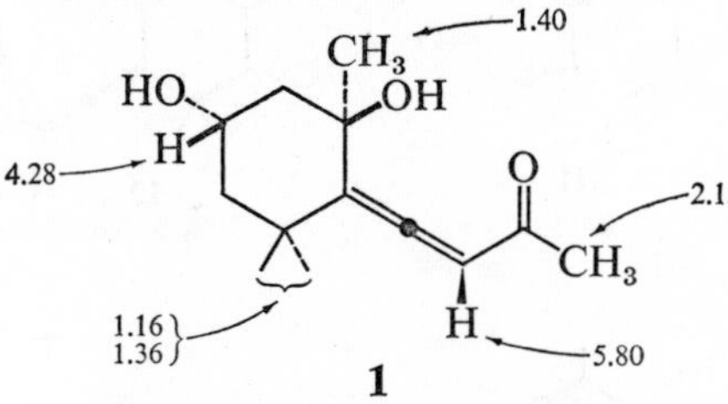

1

allenic ketone from *R. microptera*[1]

ms: 368.2201 for trimethylsilyl deriv., corresponding to the bis-trimethyl-silyl ester of the diol $C_{13}H_{20}O_3$.

ir: 6.00 (C=O), 5.15 (allene), 2.9 (–OH)

uv: 232 (ε 12,500)

nmr in $CDCl_3$: 2H signals at 1.40–1.55 are exchangable with D_2O

The structure of **1** was confirmed by the stereo-specific synthesis of the racemic compound, and the stereochemistry of the allenic double bond was established by an x-ray study of the synthetic ketone **1**.

Route I[2,3]

Two independent syntheses, both of which utilize reduction of acetylenic compounds (**3** and **8**, respectively) for formation of the allenic linkage, have been reported. An x-ray diffraction study of synthetic **1** showed that the 5-OH is *cis* to the allenic hydrogen.[4] S_N2' attack by a hydride proceeds with *cis* stereochemistry in the reduction of acetylenic compounds.

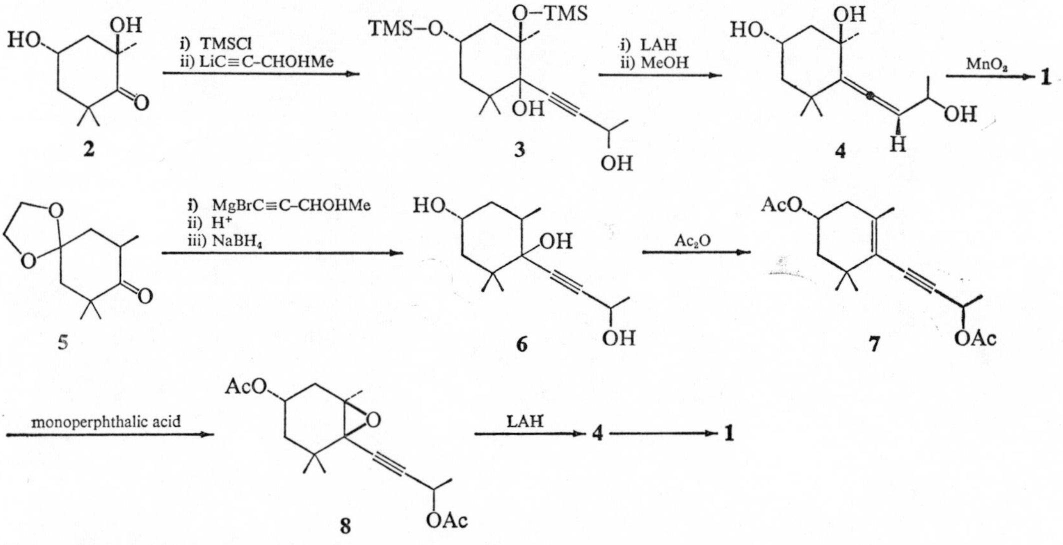

1) J. Meinwald, K. Erickson, M. Hartshorn, Y. C. Meinwald, T. Eisner, *Tetr. Lett.*, 2959 (1968).
2) J. Meinwald, L. Hendry, *Tetr. Lett.*, 1657 (1969).
3) S. W. Russell, B. C. L. Weedon, *Chem. Commun.*, 85 (1969).
4) T. E. DeVille, M. B. Hursthouse, S. W. Russell, B. C. L. Weedon, *Chem. Commun.*, 754 (1969).

Route II[5]

Allenic ketone **1** of natural stereochemistry was also synthesized via another route involving photooxygenation and photostereoisomerization.

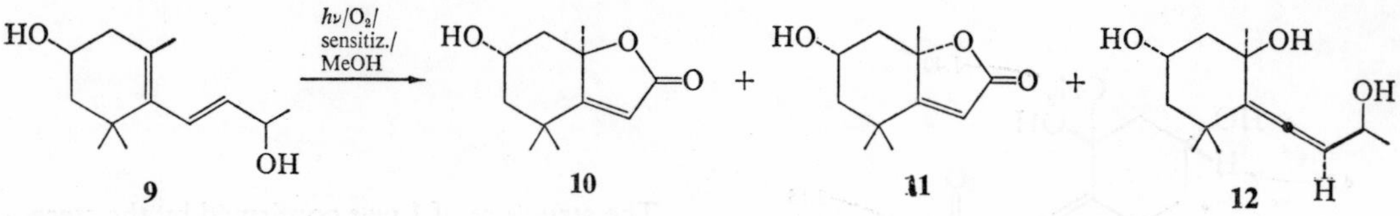

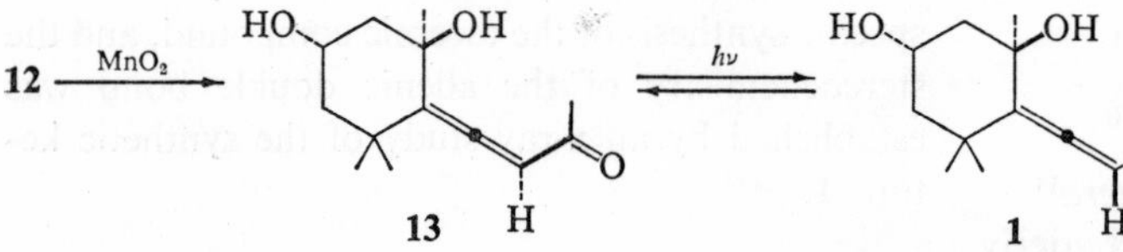

1) The reaction of β-ionol **9** with singlet oxygen affords the allenic triol **12**, together with **10** and **11**.
2) It was confirmed by x-ray studies that the ketone **13** derived from **12** is different from the natural compound with regard to the stereochemistry of the allenic moiety.[6]
3) Irradiation of **13** results in a 1:1 photoequilibrium between **13** and **1**. This is the first example of photochemically induced stereoisomerization of an allenic ketone.

Remarks

The allenic C_{13} terpene **1** was isolated from the secretion of the large flightless grasshopper *Romea microptera*. This compound shows repellent effects against ants and other natural enemies of the insect. It is assumed to be derived *in vivo* from a leaf pigment of carotenoid origin.

5) S. Isoe, S. Katsumura, S. B. Hyeon, T. Sakan, *Tetr. Lett.*, 1089 (1971).
6) T. E. DeVille, J. Hora, M. B. Hursthouse, T. P. Toube, B. C. L. Weedon, *Chem. Commun.*, 1231 (1970).

3.104 SYNTHESIS OF LATIA LUCIFERIN

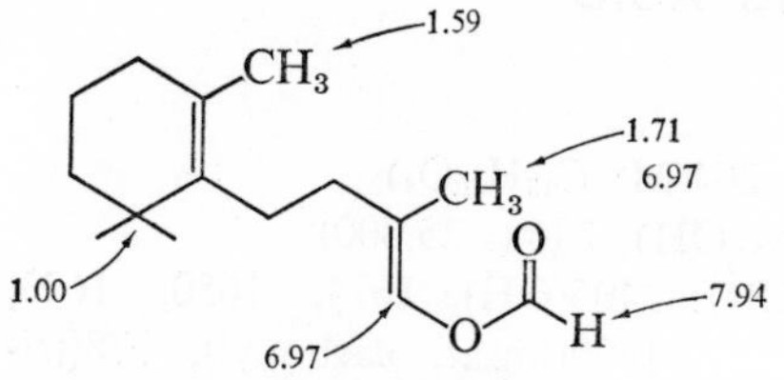

CH$_3$ ——1.59

CH$_3$ ——1.71
    6.97

1.00

O
    —7.94
H

6.97

6.97

Latia luciferin **1**[1)]
ms: 236 (M$^+$ C$_{15}$H$_{24}$O$_2$)
ir: 1742,1170(ester),1375(*gem*-dimethyl)
uv: 207 (ϵ 13,700)

Two syntheses, starting from β-ionone, have been recorded.

Route I[2]

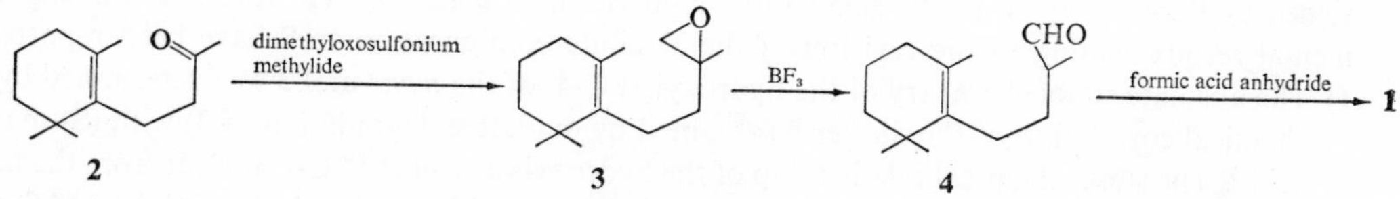

dimethyloxosulfonium methylide

BF$_3$

CHO

formic acid anhydride

1

2   **3**   **4**

Route II[3)]

Baeyer-Villiger rearrangement of the α,β-unsaturated aldehyde **5a** and **5b** occurs stereospecifically, giving rise to *trans* and *cis* formate esters, respectively.

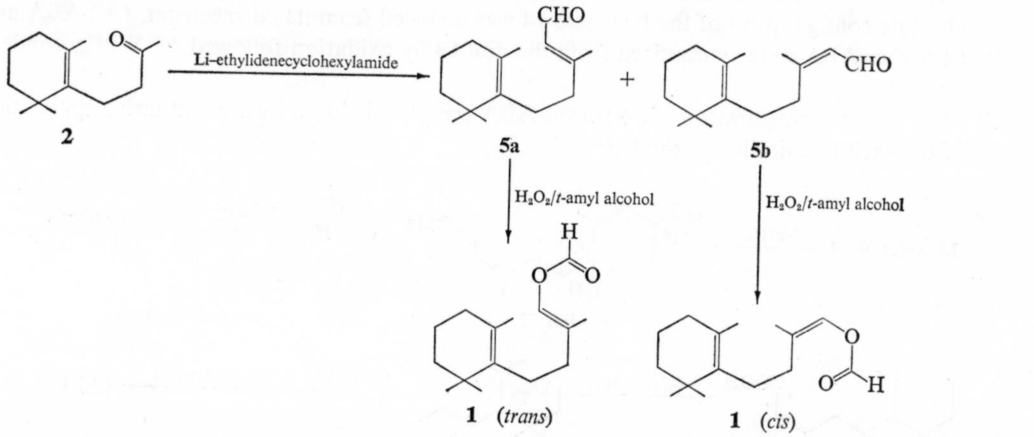

Li-ethylidenecyclohexylamide

CHO

+

CHO

2   **5a**   **5b**

H$_2$O$_2$/*t*-amyl alcohol   H$_2$O$_2$/*t*-amyl alcohol

H
O——O

O
O——H

1 (*trans*)   **1** (*cis*)

Remarks

1 is a specific substrate of the bioluminescence enzyme system of the fresh water limpet *Latia neriloides*. Synthetic *trans* isomer shows the same activity as natural luciferin against *Latia* luciferase, while the *cis* isomer shows 50–60% activity.[3)] A possible mechanism for the bioluminescence of *Latia* luciferin was proposed involving a dioxetane intermediate derived from the Schiff base of the aldehyde.[4)]

1) O. Shimomura, F. H. Johnson, *Biochemistry*, **7**, 1734 (1968).
2) M. G. Fracheboud, O. Shimomura, R. K. Hill, F. H. Johnson, *Tetr. Lett.*, 3951 (1969).
3) F. Nakatsubo, Y. Kishi, T. Goto, *Tetr. Lett.*, 381 (1970).
4) F. McCapra, R. Wrigglesworth, *Chem. Commun.*, 91 (1969).

3.105 ABSOLUTE CONFIGURATION AND SYNTHESIS OF ABSCISIC ACID

5.99(s)
2.06
7.82(d,16)
CH_3
COOH
5.78
OH
4'
1'
H
1.04
1.12
H CH_3 ← 1.94(d,1.5)
6.16(d,1.6)

(+)-abscisic acid **1**[1-3]

ms: 264(M+ $C_{15}H_{20}O_4$)
uv (MeOH): 246 (ε 25,000)
ir (KBr): 3405(OH), 1674, 1650, 1623, 1600(unsat. carbonyl), 978(tri-subst. double bond)
nmr in $CDCl_3$[12]

The absolute configuration of natural (+)-abscisic acid (ABA) had been erroneously assigned as R by applying Mill's rule to the epimeric diols **2** and **3**.[4] However, several experimental results leading to the revision of the absolute configuration to S have been reported.

1) The absolute stereochemistry of the hydroxyl at C-4' of the *trans*-diol **3** was determined by a chemical correlation of the diester **6** (obtained by oxidative degradation of **3**) with (S)-malic acid.[5] The stereochemical relationship of the hydroxyls at C-4' and C-1' is clear from the fact that one diol, which was assigned as *cis*, was shown to be identical with the diol derived form the *dl*-epidioxide **5** by hydrogenation with Lindlar's catalyst followed by esterification.[4]

2) *Cis*-diol derived from the peroxide **8** was optically resolved into the enantiomers **10** and **12** through the (+)-α-methoxy-α-trifluoromethylphenyl acetate (MTP) derivatives **9**. The absolute configuration of the benzoate **14** was deduced from its cd spectrum. (+)-ABA and (+)-*trans*-ABA were synthesized from the diol **13** by oxidation followed by Wittig condensation.[6]

3) The absolute configuration of (+)-*trans*-ABA was also deduced by a quantitative application of the exciton chirality methods.[7]

Me-ester of **1** —NaBH₄→

HO
H
4'
1'
OH
COOMe

cis-diol **2**

+

HO
H
4'
1'
OH
COOMe

trans-diol **3**

COOH —O₂/hν/eosin→

4

O–O
COOH —0.01N NaOH(100°)→ (±)-**1**

5

1) K. Ohkuma, J. L. Lyon, F. T. Addicott, O. E. Smith, *Science*, **142**, 1952 (1964).
2) K. Ohkuma, F. T. Addicott, O. E. Smith, W. E. Thiessen, *Tetr. Lett.*, 2529 (1956).
3) J. W. Cornforth, B. V. Milborrow, G. Ryback, *Nature*, **206**, 715 (1965); *ibid.*, **210**, 627 (1966).
4) J. W. Cornforth, W. Draber, B. V. Milborrow, G. Ryback, *Chem. Commun.*, 114 (1967).
5) G. Ryback, *Chem. Commun.*, 1190 (1972).
6) M. Koreeda, G. Weiss, K. Nakanishi, *J. Am. Chem. Soc.*, **95**, 239 (1973).
7) N. Harada, *J. Am. Chem. Soc.*, **95**, 241 (1973).

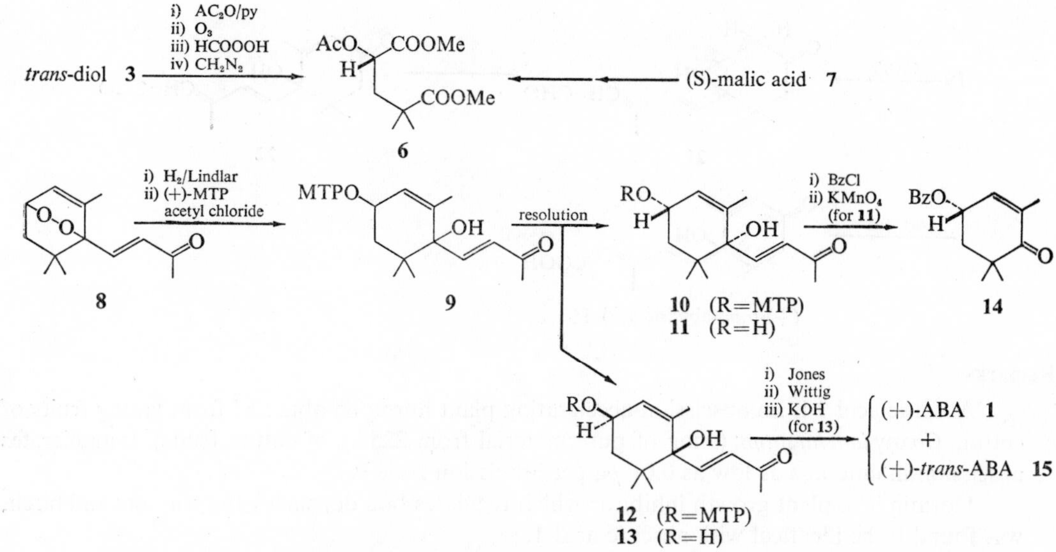

4) It was also shown recently that xanthoxin **19**, obtained from violaxanthin **16** by photooxygenation, could be converted into (+)-*trans*-abscisic acid **15**. This also indicates that the absolute configuration of (+)-abscisic acid should be revised to S.[8,9] Zeaxanthin **17** also gave **18** and **20** on photooxygenation.[9]

5) (−)-Abscisic acid ester has been synthesized from (−)-R-α-ionone. This also supports an S configuration for (+)-**1**.[10]

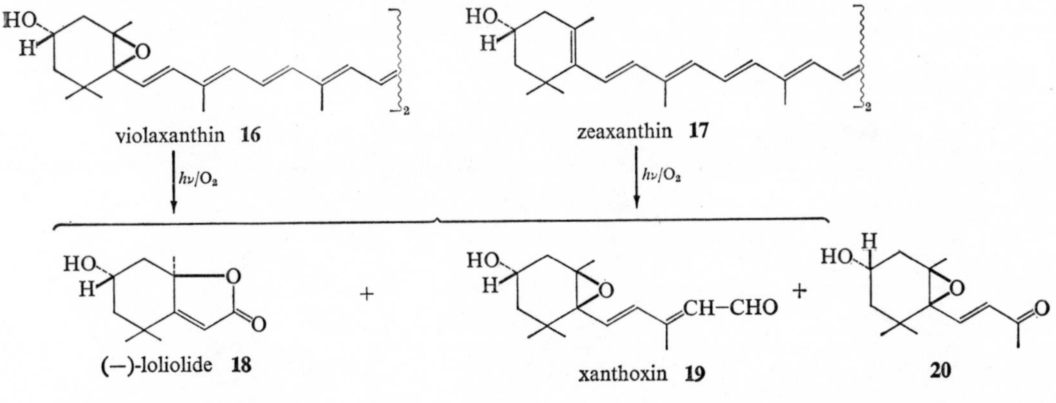

8) R. S. Burden, H. F. Taylor, *Tetr. Lett.*, 4071 (1970).

9) S. Isoe, S. B. Hyeon, S. Katsumura, T. Sakan, *Tetr .Lett.*, 2517 (1972).

10) T. Oritani, K. Yamashita, *Tetr. Lett.*, 2521 (1972).

11) J. W. Cornforth, B. V. Milborrow, G. Ryback, P. F. Wareing, *Nature*, **205**, 1269 (1965).

12) Y. Isogai, T. Okamoto, Y. Kodama, *Chem. Pharm. Bull.*, **15**, 1256 (1967).

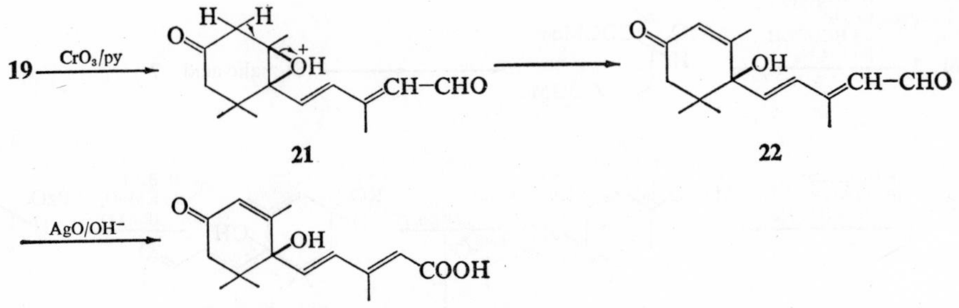

$$19 \xrightarrow{\text{CrO}_3/\text{py}} \mathbf{21} \longrightarrow \mathbf{22}$$

$$\xrightarrow{\text{AgO/OH}^-}$$

(+)-*trans*-abscisic acid **15**

Remarks

Abscisic acid **1** is an abscission-accelerating plant hormone obtained from young fruits of cotton, *Gossypium hirsutum* (9 mg of pure material from 225 kg of cotton fruits). It accelerates abscission in amounts as low as 0.01 μg per abscission zone.[1]

Dormin,[11] a plant growth inhibitor which regulates bud dormancy in sycamore and birch, was found to be identical with abscisic acid **1**.

1 occurs in buds, twigs, leaves, tubers, seeds or fruits of a wide variety of plants.

Synthesis

Racemic ABA was synthesized through the epidioxide **5**, which was obtained from **4** by photooxygenation.[3] (+)-ABA was synthesized from **8** via **12**,[6] and (−)-ABA from (−)-R-α-ionone[10] as mentioned above. Various approaches for the synthesis of **1** have been reported.[13-15]

13) M. Mousseron-Canet, J. C. Mani, J. P. Dalle, J. L. Olive, *Bull. Soc. Chim. Fr.*, 3874 (1966).
14) D. L. Roberts, R. A. Heckman, B. P. Hege, S. A. Bellin, *J. Org. Chem.* **33**, 3566 (1968).
15) J. A. Findlay, W. D. Mackay, *Can. J. Chem.*, **49**, 2369 (1971).

3.106 STRUCTURE OF STRIGOL

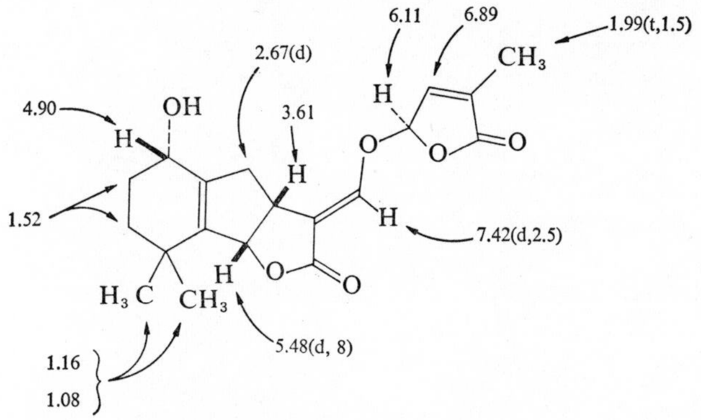

strigol 1[1)]

ms: 346.1408 ($M^+ C_{19}H_{22}O_6$)
mp: 200–202°
uv: 234 (ε 17,700)
ir: 3590 (OH), 1787, 1745(butenolide),
1682 (unsat. ketone)
nmr in $CDCl_3$

1) The mass spectrum of **1** indicated rapid cleavage to $C_5H_5O_2$ and $C_{14}H_{17}O_4$ fragments.
2) Hydrogenation of **1** yielded a hexahydro derivative which gave a $C_5H_7O_2$ fragment as an ms base peak.
3) MnO_2 (or DDQ) oxidation of **1** affords an α,β-unsaturated ketone, strigone **2**, indicating that the *sec*-OH group is allylic.
4) X-ray crystallographic analysis of **1** established the structure.

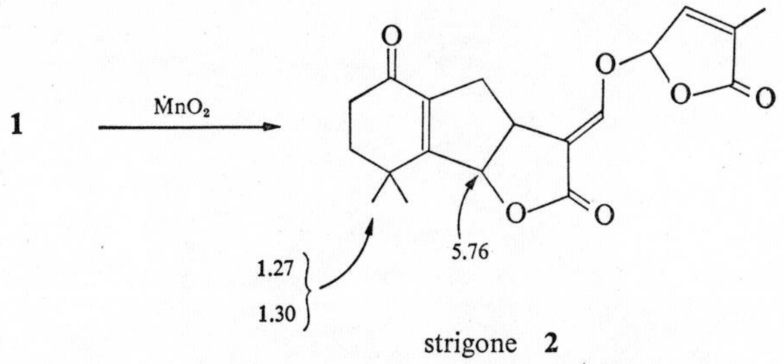

strigone **2**
uv: 234 (ε 27,000)

Remarks

Strigol **1** was isolated from root exudates of cotton, *Gossypium hirsutum*, and is a potent seed germination stimulant for witchweed, *Striga lutea*.

1) C. E. Cook, L. P. Whichard, M. E. Wall, G. H. Egley, P. Coggon, P. A. Luhan, A. T. McPhail, *J. Am. Chem. Soc.*, **94**, 6198 (1972).

CHAPTER 4

Diterpenes

4.1 INTRODUCTION

The diterpenes are C_{20} compounds biogenetically derived from geranylgeranyl pyrophosphate. The notable feature of diterpene structures is the fascinating variation encountered in their skeletons and the occurrence in nature of both normal and antipodal stereochemical series.

They are mainly of plant and fungal origin and usually occur as mixtures of closely related compounds. They include the resin acids and gibberellins. The following correlation chart shows the main diterpene skeletons according to the classification recommended by Rowe et al.[1] In this chapter, diterpenes possessing these skeletons are described first, following approximately the sequence of the chart; the less common diterpenes are collected at the end. A convenient reference source for diterpenes reported up to 1970 (totalling *ca.* 650) is ref. 2.

Schematic Correlation of the Main Diterpene Skeletons

The skeletons marked with * occur exclusively or almost exclusively in their antipodal forms in nature.

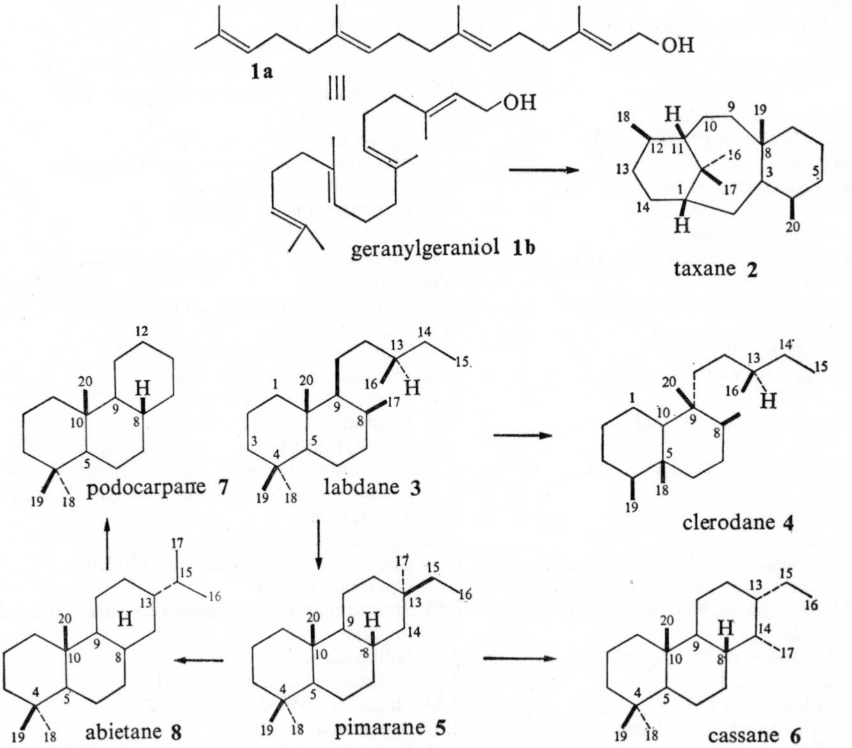

geranylgeraniol **1b**

taxane **2**

podocarpane **7**   labdane **3**   clerodane **4**

abietane **8**   pimarane **5**   cassane **6**

---

1) *The Common and Systematic Nomenclature of Cyclic Diterpenes*, prepared by J. W. Rowe, Forest Products Laboratories, USDA, 1968, co-sponsored by J. ApSimon, M. Fetizon, E. Fujita, L. Gough, W. Herz, P. R. Jefferies, D. Mangoni, T. Norin, K. Overton, S. W. Pelletier, E. Wenkert; a proposal submitted to the International Union of Pure and Applied Chemistry, Commission on Nomenclature.

2) T. K. Devon, A. I. Scott, *Handbook of Naturally Occuring Compounds*, Vol. II (Terpenes), Academic Press, 1972.

from **8**

from **5**

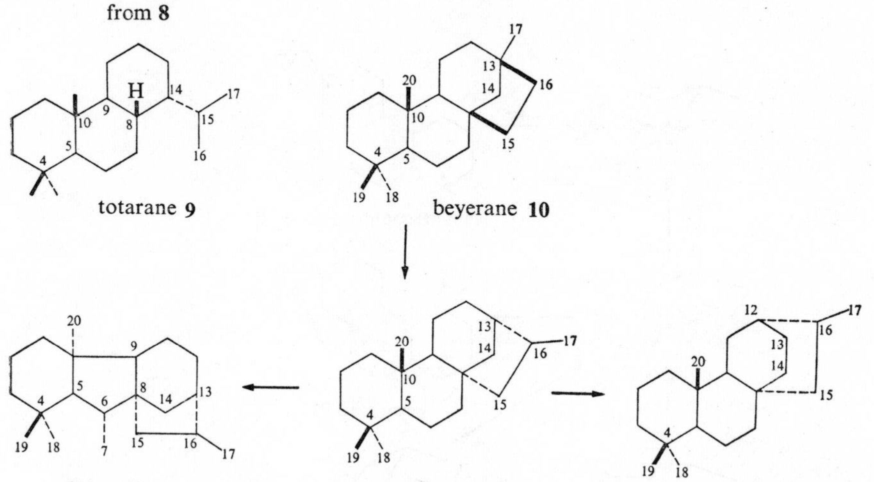

totarane **9**

beyerane **10**

gibberellane* **13**   kaurane* **11**   atisane* **12**

4.2   BIOGENESIS OF SOME CYCLIC DITERPENES

Geranylgeraniol was postulated as a general precursor of cyclic diterpenes in 1953.[1] It was demonstrated that geranylgeraniol pyrophosphate **1** was converted into kaurene (wild cucumber homogenate).[2] Also, in cell-free extracts of castor beans, radioactive **1** was incorporated into **7** through **11**[3] (expected positions of ^{14}C are indicated by *, although actual characterizations of labeled carbons were not performed).

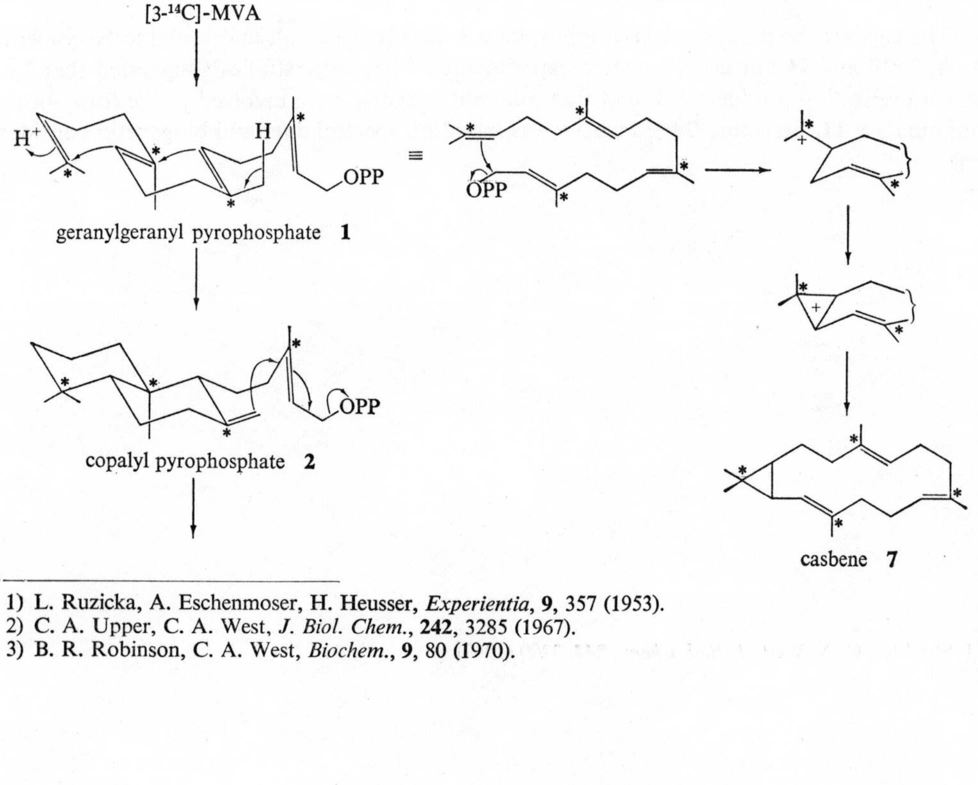

[3-^{14}C]-MVA

geranylgeranyl pyrophosphate **1**

copalyl pyrophosphate **2**

casbene **7**

1) L. Ruzicka, A. Eschenmoser, H. Heusser, *Experientia*, **9**, 357 (1953).
2) C. A. Upper, C. A. West, *J. Biol. Chem.*, **242**, 3285 (1967).
3) B. R. Robinson, C. A. West, *Biochem.*, **9**, 80 (1970).

from **2**

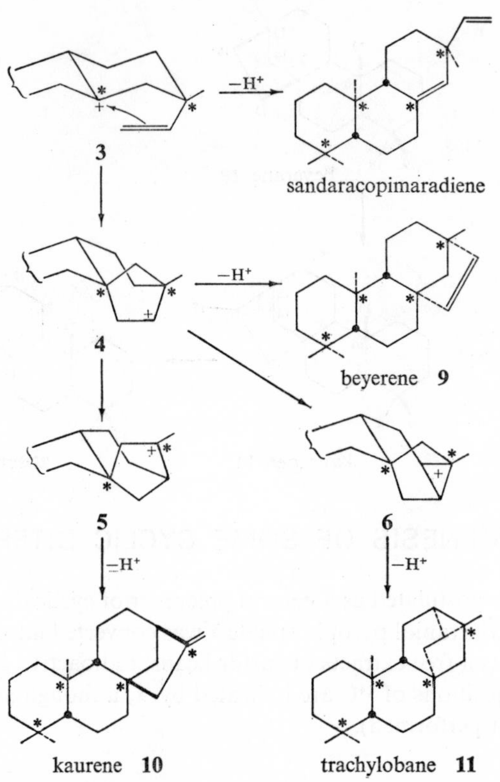

sandaracopimaradiene **8**

3

4

beyerene **9**

5   **6**

kaurene **10**   trachylobane **11**

The key intermediate copalyl pyrophosphate **2** was also isolated, and found to be converted into **8**, **9**, **10** and **11** but not **7**[4] (tracer experiments). Enzymatic studies[3] suggested that **7** was a direct cyclization product of **1**, and that different enzymes were involved in the formation of compounds **8–11**. Structure **7** for casbene was based on spectral data and biogenetic considerations.

4) I. Shechter, C. A. West, *J. Biol. Chem.*, **244**, 3200 (1969).

4.3   BIOGENESIS OF VIRESCENOSIDES BY ^{13}C NMR

Biosynthetic studies utilizing ^{13}C-labeled precursors offer several advantages over radioactive precursors.[1,2] The most obvious is that, provided the pertinent ^{13}C peaks can be assigned in the ^{13}C nmr of the nonlabeled compound, locations of carbons enriched from the ^{13}C precursor can be recognized directly without relying on systematic degradations. The major disadvantages, on the other hand, are the requirements for a much larger amount of the enriched substrate and a high incorporation yield. As the natural abundance of ^{13}C is *ca.* 1%, an incorporation of 1% of ^{13}C from the precursor into the substrate would lead to a two-fold intensification of the enriched ^{13}C peaks; a 0.1% incorporation would not be practical.

Virescenosides A **1** and B **2** are metabolites of the mushroom *Oospora virescens*. The entire ^{13}C nmr spectra of the aglycones, virescenols A **3** and B **4** and their hydrolysis artifacts, **5** and **6**, were clarified by comparisons with ^{13}C spectra of pimaradienes (e.g. **5**). Addition of sodium [1-^{13}C] acetate to the culture medium, and isolation of **5** after hydrolysis of virescenoside A **1** led to enhancements of carbon signals as indicated in **7**. Similarly, feeding with [2-^{13}C] acetate led to the enhanced signals depicted in **8**. These distributions are in accord with the accepted biosynthetic pathway.

1 R =β-D-altropyranosyl;
   R′=OH
2 R =β-D-altropyranosyl;
   R′=H
3 R =H; R′=OH
4 R =R′=H

5 nmr in CHCl₃, ppm from TMS
6 no OH at C-2

7 ■ [1-^{13}C] acetate
8 ● [2-^{13}C] acetate

4.4   ACYCLIC DITERPENES

Acyclic diterpenes have been isolated from various sources ranging from bacteria to human sera.[1] Geranylgeraniol, the biogenetic precursor for cyclic diterpenes accounts for the pleasant aroma of some trees.[2,3] The structure determination and synthesis of two members of this class will be described.

Solanone

Solanone **2**, and its precursor **1**, are tobacco terpenoids which apparently violate the bio-

4.3
1) E. Wenkert, B. L. Buckwalter, *J. Am. Chem. Soc.*, **94**, 4367 (1972).
2) J. Polonsky, Z. Baskevitch, N. Cagoloni-Bellarita, P. Ceccherelli, B. L. Buckwalter, E. Wenkert, *J. Am. Chem. Soc.*, **94**, 4369 (1972), and references therein for other biosynthetic studies with ^{13}C-enriched acetates.

genetic isoprene rule. The structures (excluding stereochemistry) of its two thermally unstable precursors have been determined in the following manner.[4]

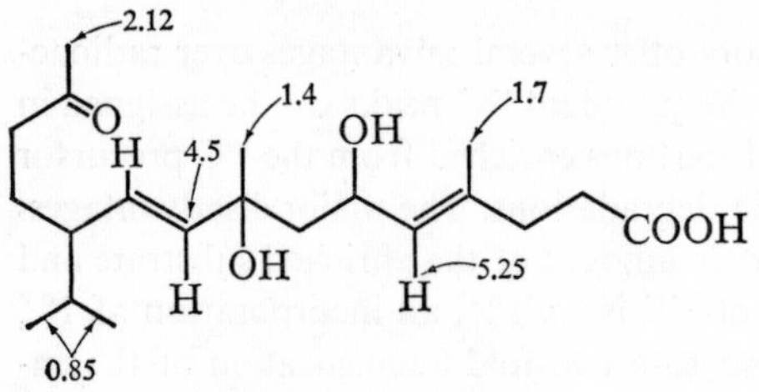

solanone precursor **1**

oil α_D^{25}: $+14.0°$
ir (film): 3410 (OH); 1740, 1720 (2C=O),
970 (*trans* C=C)
ms: 368 (M$^+$), 350 (M$^+$−H$_2$O), ?32

1) The hydroxyls are not vicinal, since **1** is inert to NaIO$_4$.
2) Formation of keto ester **3** indicated **1** to be acyclic.
3) The production of **4** and **5** determined the two ends of the molecule.
4) Reverse aldolization of **6** yielded **7** and **8** containing all the carbon atoms in **1**, which in conjunction with the formation of solanone **2** upon pyrolysis, defined the structure of the solanone precursor **1**.

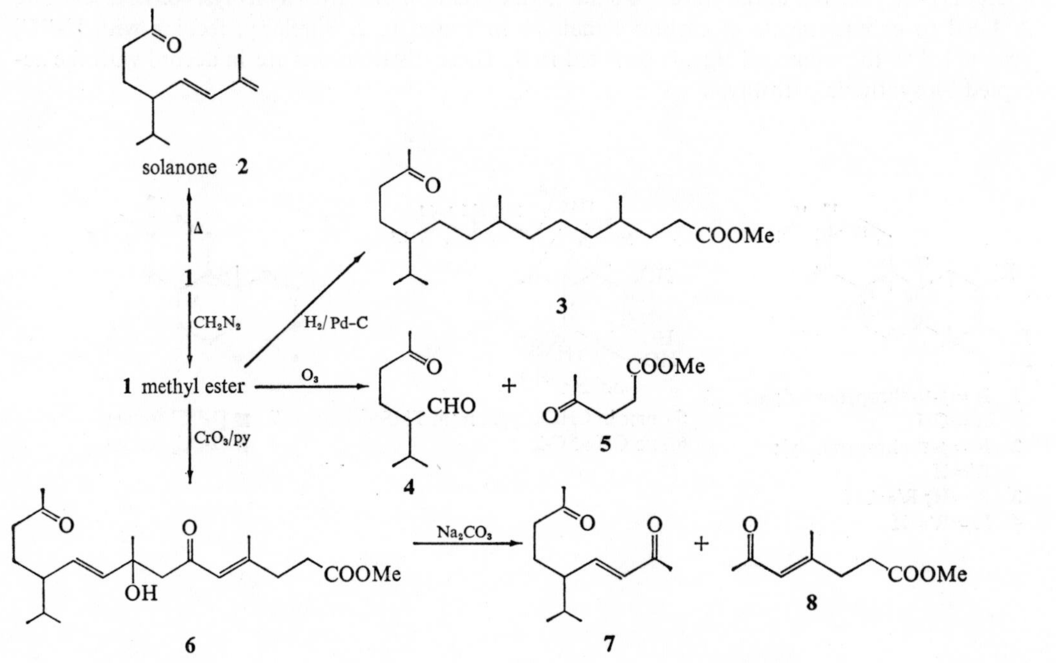

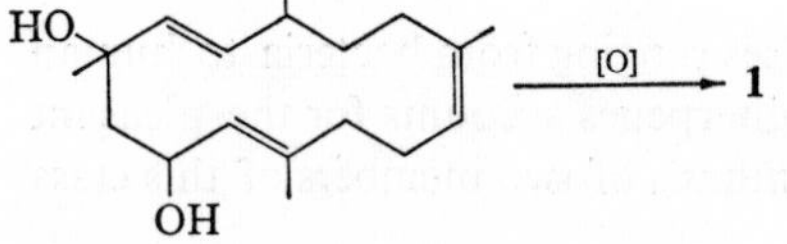

solanone **2**

It has been suggested that **1** is biogenetically derived from the cyclic tobacco terpenoid **9**, which is consistent with the isoprene rule.

1) R. P. Hansen, *Nature*, **210**, 841 (1966), and references therein.
2) T. W. Goodwin, *Biochem. J.*, **123**, 293 (1971); E. Heftman, *Steroid Biochemistry*, Academic Press, 1970.
3) B. A. Nagasampagi, L. Yankov, S. Dev, *Tetr. Lett.*, 189 (1967).
4) G. W. Kinzer, T. F. Page, R. R. Johnson, *J. Org. Chem.*, **31**, 1797 (1966).
5) J. W. K. Burrell, R. F. Garwood, L. M. Jackman, E. Oskay, B. C. L. Weedon, *J. Chem. Soc.* C, 2144 (1966).
6) E. Fedeli, P. Capella, M. Cirimele, G. Jacini, *J. Lipid Res.*, **7**, 437 (1966).
7) M. Blumer, A. W. Thomas, *Science*, **147**, 1148 (1965).

Phytol

Phytol **15** occurs abundantly in plants as an ester of the propionic acid side chain of chlorophyll. Its synthesis from optically active citronellol **10** and the half ester **12** has elucidated the 7(R), 11(R) configurations for the compound.[5]

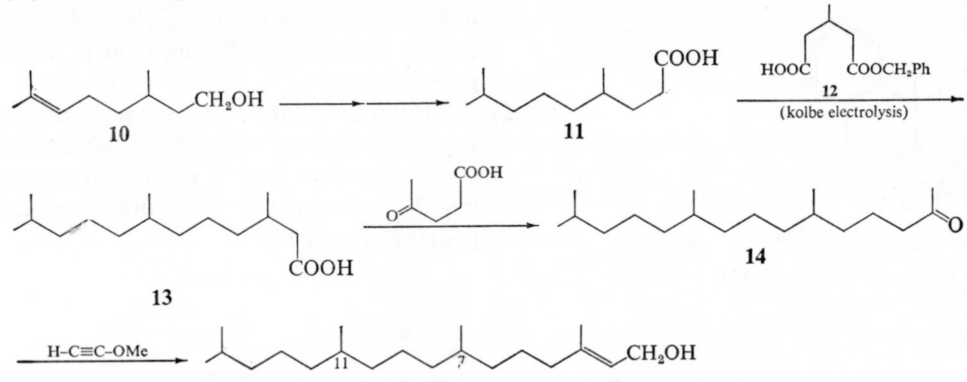

phytol **15**

Remarks

It is noteworthy that although geranylgeraniol has been postulated as the biological precursor of diterpenes for some time, its isolation from natural sources has been quite recent. The structures and sources of some related compounds are shown below.

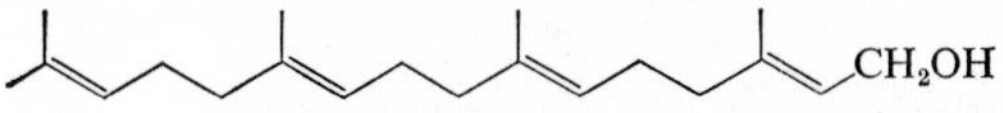

geranylgeraniol **16**   *Cederela toona*[3]
                               linseed oil[6]

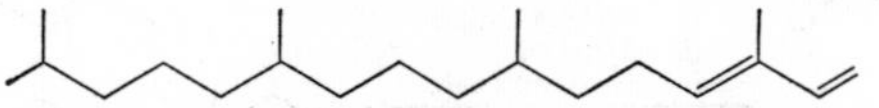

phytadiene **17**   zooplankton[7]

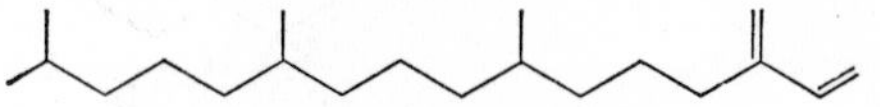

neophytadiene **18**   tobacco[8],
                               zooplankton[7]

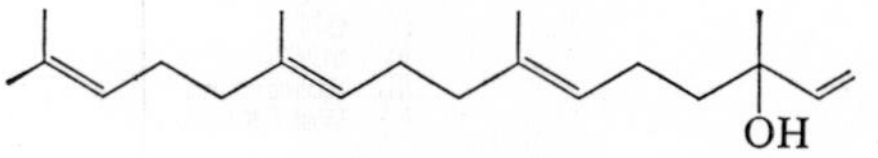

geranyllinalool (3S, 3R) **19**   jasmin oil[9],
                               Norwegian spruce[10]

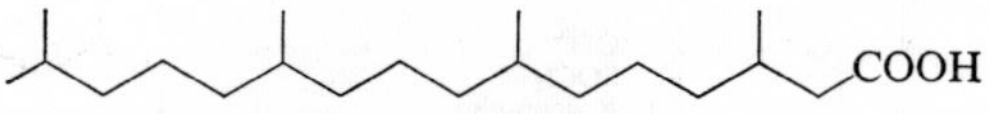

phytanic acid **20**   animal fat[11],
                               bacteria[11]

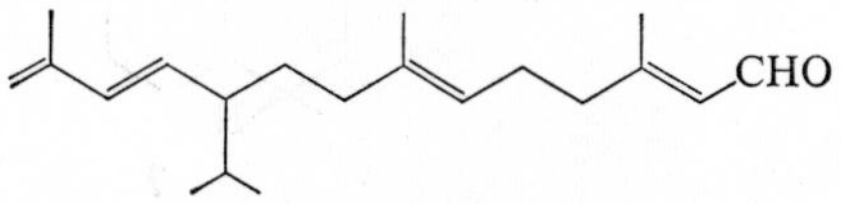

tetradecatetraenal **21**   tobacco[11]

8) R. L. Rowland, *J. Am. Chem. Soc.*, **79**, 5007 (1957).
9) E. Demole, E. Lederer, *Bull. Soc. Chim. France*, 1128 (1958).
10) B. Kimland, T. Norin, *Acta Chem. Scand.*, **21**, 825 (1967).
11) J. L. Courtney, S. McDonald *Tetr. Lett.*, 459 (1967).

4.5 STRUCTURE AND SYNTHESIS OF PODOCARPIC ACID
[Podocarpane group]

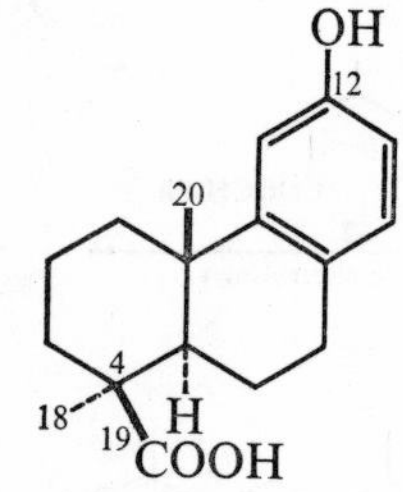

podocarpic acid **1**[1,2]

mp: 193.5° α_{578} (EtOH): $+144°$

1) The dehydrogenation product **3**, whose structure was confirmed by synthesis, clarified both the carbon skeleton and the location of carboxyl group at C-4.[1]

2) The isopropyl derivative **6** from podocarpic acid **1** was not identical with **8** from dehydroabietic acid **9**. However, both **6** and **8** were converted into **7**, establishing the configuration at C-4 of podocarpic acid **1**.[2]

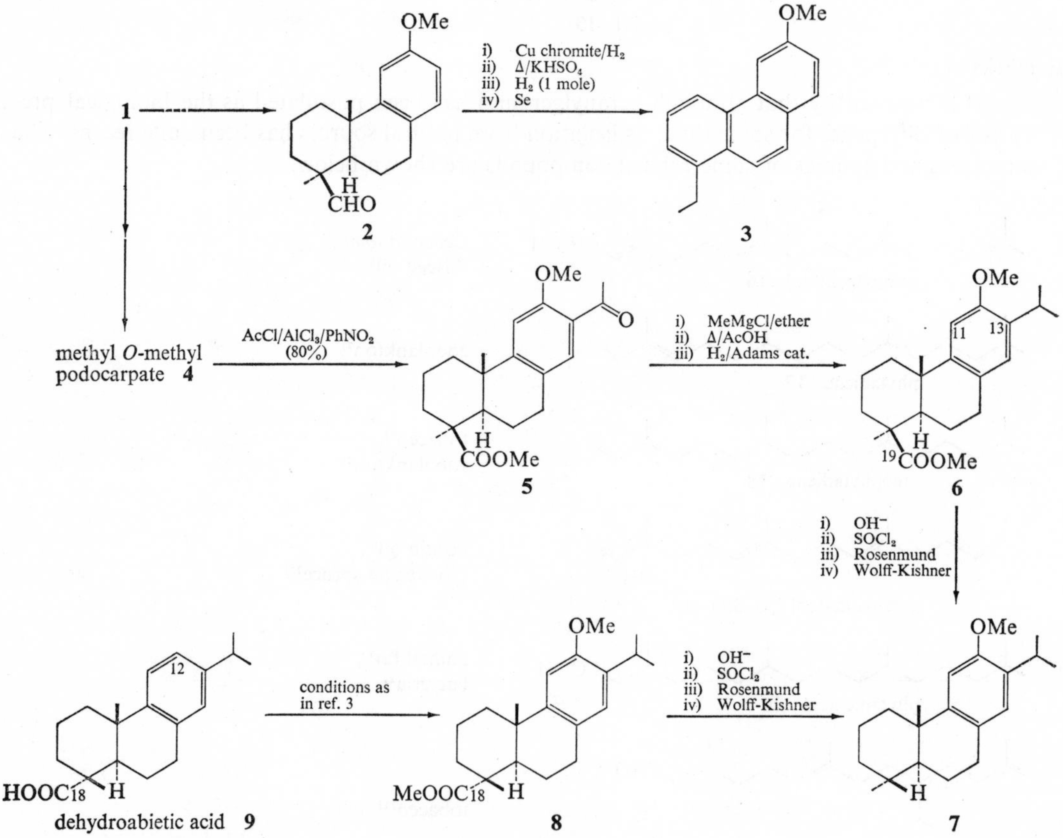

1) W. P. Campbell, D. Todd, *J. Am. Chem. Soc.*, **62**, 1287 (1940).
2) W. P. Campbell, D. Todd, *J. Am. Chem. Soc.*, **64**, 928 (1942).
3) L. F. Fieser, W. P. Campbell, *J. Am. Chem. Soc.*, **61**, 2528 (1939).

Synthesis

The first synthesis was carried out by King *et al.*,[4] and involved acid-catalyzed cyclization of the intermediate **10**; it was essentially the same as that used in the synthesis of ferruginol (p. 222). Wenkert *et al.* synthesized *d*-podocarpic acid employing a new annelation step to **12**, and optical resolution of **14** through the cinchonine salt.[5-7]

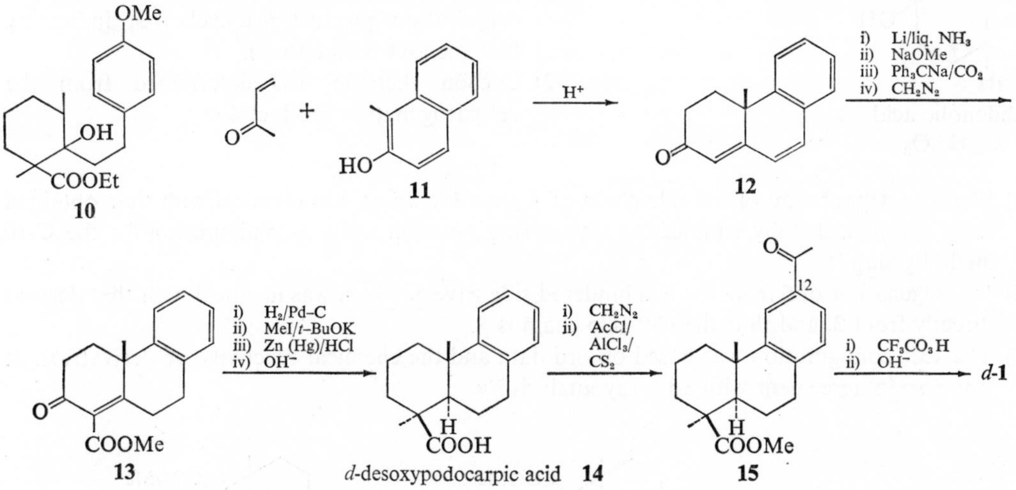

Remarks

Podocarpic acid **1** was first isolated in 1873 from the resin of *Podocarpus cupressium* (Podocarpaceae), and was later found in various other resins.

The related compound nimbiol **16** (shown below) was isolated from the tree *Melia azadirachta* (syn. *Azadirachta indica* Juss) (Meliaceae), commonly known as "neem" or Indian Lilac.

Nimbiol

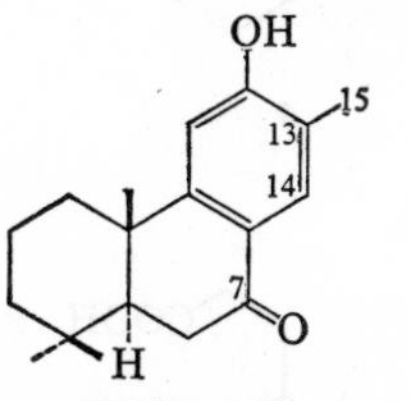

nimbiol **16**[8]

mp: 250–251°
α_D +32.3°

1) The structure was based on spectroscopic similarities (especially, uv, ord) to those of sugiol (isopropyl instead of methyl at C-13 in **16**).

2) Co-occurrence of nimbiol **16** and sugiol in the same tree suggested a common biogenetic intermediate of the pimaric acid type, from which both could be derived.

4) F. E. King, T. J. King, J. G. Topliss, *Chem. Ind.*, 113 (1956).
5) E. Wenkert, B. G. Jackson, *J. Am. Chem. Soc.*, **80**, 217 (1958).
6) E. Wenkert, B. G. Jackson, *J. Am. Chem. Soc.*, **81**, 5601 (1959).
7) E. Wenkert, A. Tahara, *J. Am. Chem. Soc.*, **82**, 3229 (1960).
8) P. Sengupta, S. N. Choudhuri, H. N. Khastgir, *Tetr.*, **10**, 45 (1960).

4.6   STRUCTURE AND SYNTHESIS OF LABDANOLIC ACID
[Labdane group]

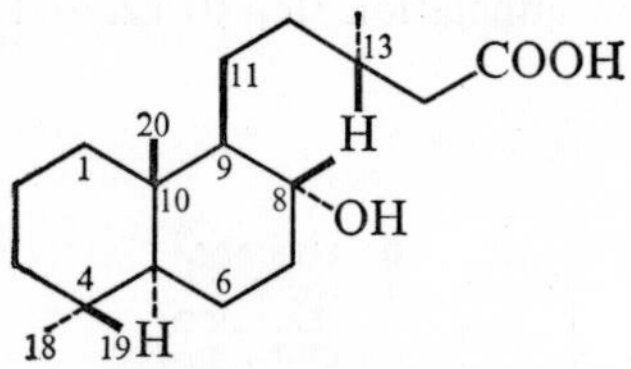

labdanolic acid **1**
$C_{20}H_{36}O_3$

1) Position of the *tert*-OH followed from ozonolysis of the dehydration product **3**. There is only one dehydration product (exocyclic) **3**, indicating that the OH is equatorial.

2) Carbon skeleton was determined from the dehydrogenation product **4**.

3) Stepwise degradation of the side chain of **6** gave **10**, which was identical with that obtained from ambrein, thereby establishing a *trans* ring junction and a β-configuration for the C-10 methyl group.[1]

4) Hydrogenation of **5** from the less hindered side gave **6**, which was identical with that derived directly from **3**, and thus the C-9 side chain is β.

5) The 13R configuration was based on ord data and the chemical reactivity of derivatives. It was also in agreement with an x-ray analysis.[2]

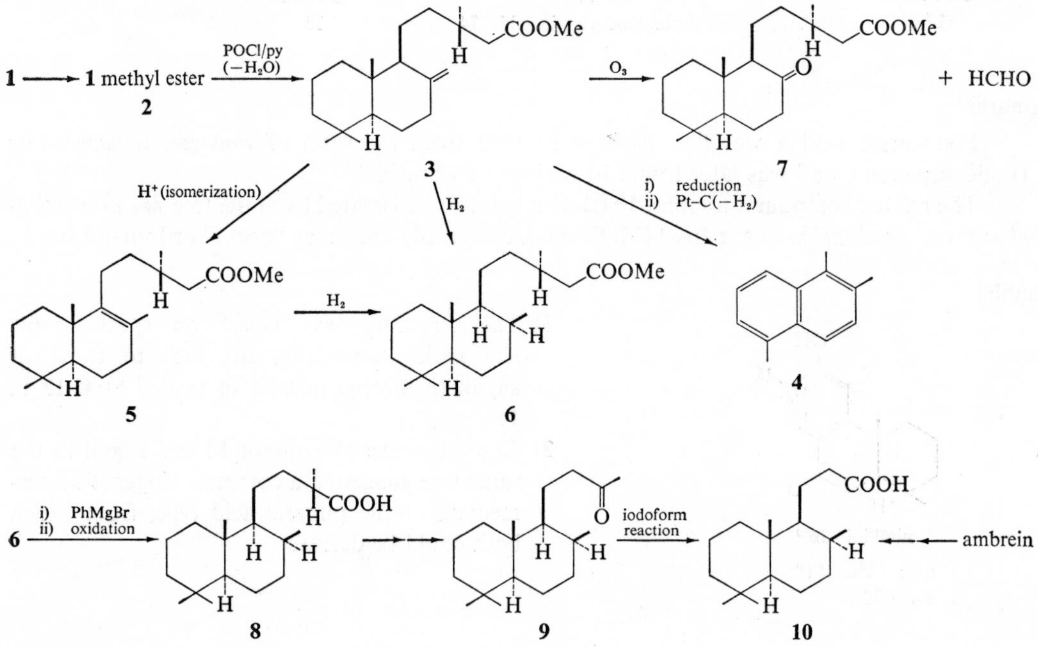

1 ⟶ **1** methyl ester $\xrightarrow[(-H_2O)]{POCl/py}$

2

$\xrightarrow{O_3}$ + HCHO

7

$\xrightarrow{H^+ \text{ (isomerization)}}$

$\Bigg\downarrow H_2$

i) reduction
ii) Pt–C(−H$_2$)

5 $\xrightarrow{H_2}$ **6**   **4**

3

6 $\xrightarrow[\text{ii) oxidation}]{\text{i) PhMgBr}}$ **8**   ⟶   **9** $\xrightarrow[\text{reaction}]{\text{iodoform}}$ **10** ⟵ ambrein

Remarks

Labdanolic acid **1** was obtained from gum labdanum, from *Cistus labdaniferus*.

1) J. D. Cocker, T. G. Halsall, *J. Chem. Soc.*, 4262 (1956).
2) K. H. Overton, A. J. Renfrew, *J. Chem. Soc.* C, 931 (1967).
3) D. B. Bigley, N. A. Rogers, J. A. Barltrop, *J. Chem. Soc.*, 4613 (1960).

Synthesis[3]

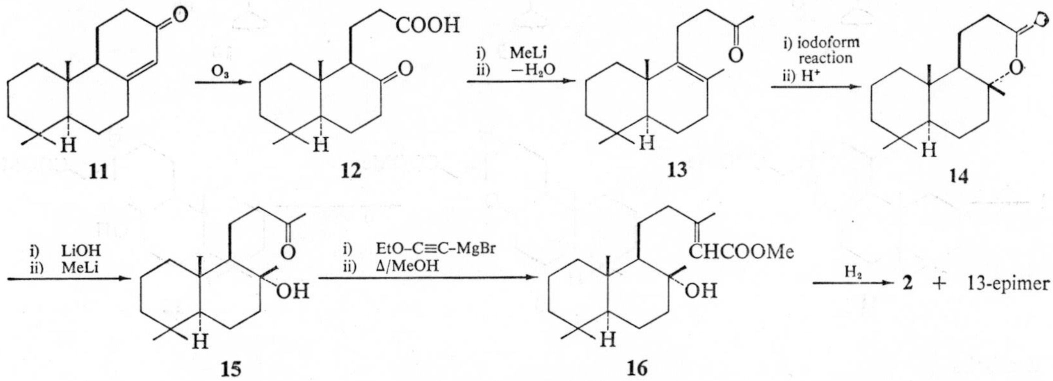

11 → (O₃) → **12** → (i) MeLi, ii) −H₂O) → **13** → (i) iodoform reaction, ii) H⁺) → **14**

15 (i) LiOH, ii) MeLi) → **16** (i) EtO–C≡C–MgBr, ii) Δ/MeOH) → **16** → (H₂) → **2** + 13-epimer

The versatile starting material **11** was transformed to ambreinolide **14** which could be further converted to a mixture of methyl labdanolate and its 13-epimer.

4.7   STRUCTURE OF EPERUIC ACID
[Labdane group]

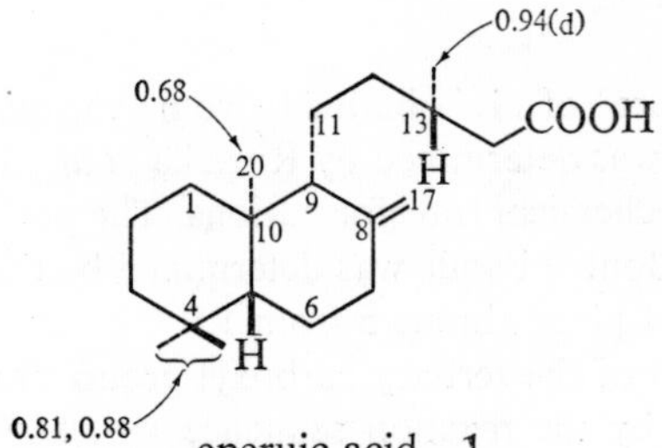

eperuic acid **1**
$C_{20}H_{34}O_2$

bp: 190° (0.7 mm)
uv: none above 210

1) Hydrogenation of **1** yielded dihydroeperuic acid.
2) The dehydrogenation products **2** and **3** revealed the structure of the carbon skeleton.
3) Stepwise degradation of the acidic side chain confirmed the side chain structure.[1]
4) Comparisons of ord and ir showed the keto ester **11** to be the antipode of **13**, which in turn is derived from the stereochemically known 13-epilabdanolic acid **12**. This confirmed the stereochemistry of **1**.[2,3]

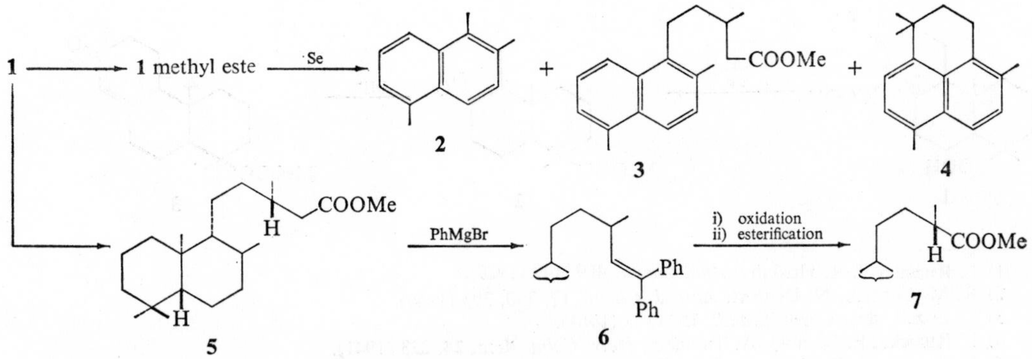

1 ⟶ 1 methyl ester ⟶ (Se) → **2** + **3** + **4**

5 → (PhMgBr) → **6** → (i) oxidation, ii) esterification) → **7**

1) F. E. King, G. Jones, *J. Chem. Soc.*, 658 (1955).
2) E. M. Graham, K. H. Overton, *J. Chem. Soc.*, 126 (1965).
3) C. A. Henrick, P. R. Jefferies, *Tetr.*, **21**, 1175 (1965).

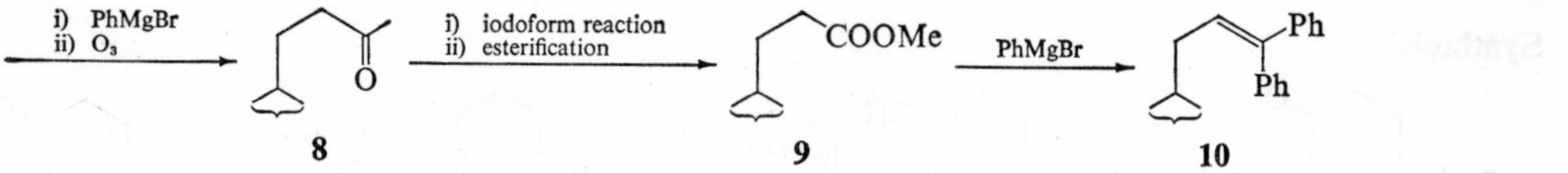

8 → **9** → **10**

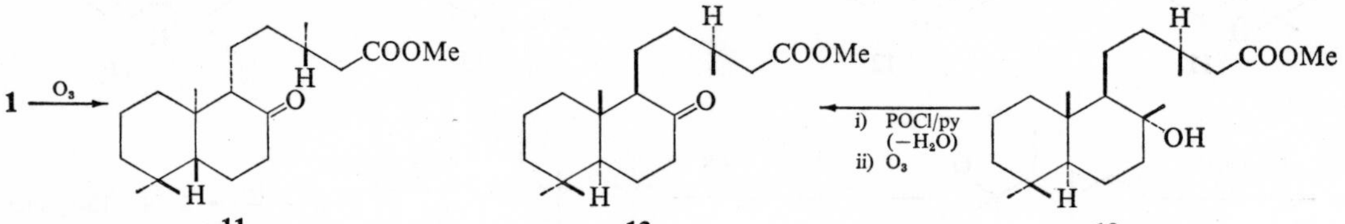

i) PhMgBr
ii) O₃

i) iodoform reaction
ii) esterification

PhMgBr

1 $\xrightarrow{O_3}$ **11** → **13** $\xleftarrow[\text{ii) } O_3]{\text{i) POCl/py (}-H_2O)}$ **12**

Remarks

The source of eperuic acid **1** is *Eperua falcata* (wallaba tree).

4.8   STRUCTURE OF AGATHENEDICARBOXYLIC ACID (AGATHIC ACID)
[Labdane group]

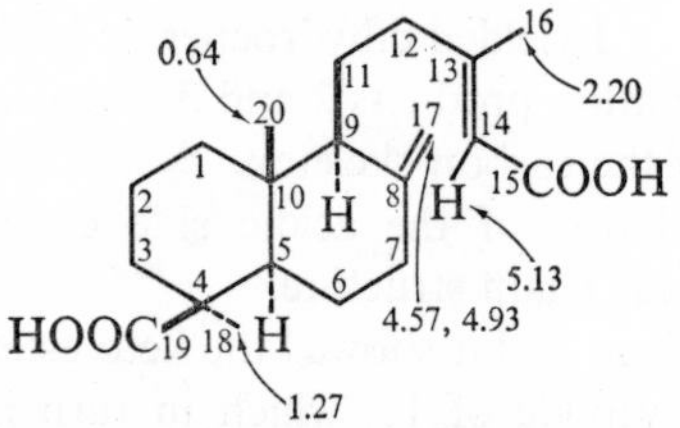

agathenedicarboxylic acid   **1**[1]

mp: 203° α_D (EtOH): +65°
uv (EtOH): 218 (4)
ir (KBr): 1690, 1680 (unsat. acid),
     929, 894, 867, 792 (C=CH₂)

1) The structure of **1**, obtained from various sources,[1-8] was determined by Ruzicka *et al.* on the basis of chemical transformations. The position of the double bonds was determined by the formation of pimanthrene **5** from **1**.[4]

2) The location of the tertiary carboxyl group was determined by the reaction sequence **1**→**10**.[5-7]

3) The configuration of the tertiary carboxyl group was obtained by the formation of **12**.[8]

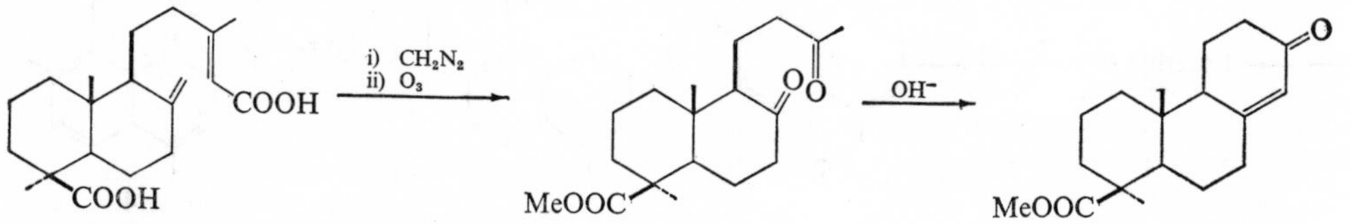

1 $\xrightarrow[\text{ii) } O_3]{\text{i) } CH_2N_2}$ **2** $\xrightarrow{OH^-}$ **3**

1) L. Ruzicka, J. R. Hosking, *Ann. Chem.*, **469**, 147 (1929).
2) R. M. Carman, N. Dennist, *Aust. J. Chem.*, **17**, 390, 393 (1964).
3) C. Enzell, *Acta Chem. Scand.*, **15**, 1303 (1961).
4) L. Ruzicka, E. Bernold, A. Tallichet, *Helv. Chim. Acta*, **24**, 223 (1941).
5) L. Ruzicka, J. R. Hosking, *Helv. Chim. Acta*, **13**, 1402 (1930).
6) L. Ruzicka, H. Jacobs, *Rec. Trav. Chim.*, **57**, 509 (1938).
7) L. Ruzicka, E. Bernold, *Helv. Chim. Acta*, **24**, 1167 (1941).
8) L. Ruzicka, E. Bernold, *Helv. Chim. Acta*, **24**, 931 (1941).

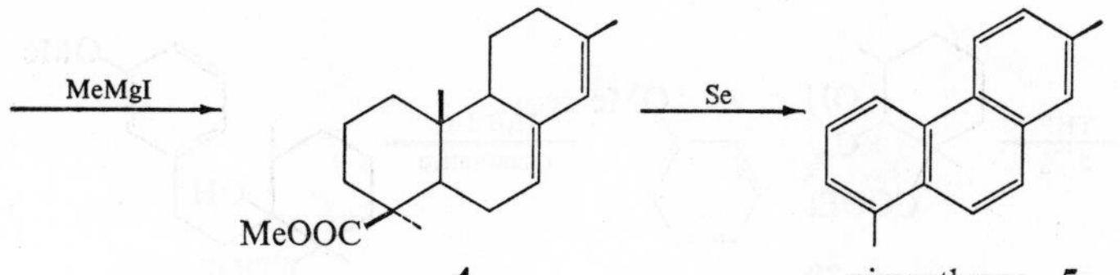

pimanthrene **5**

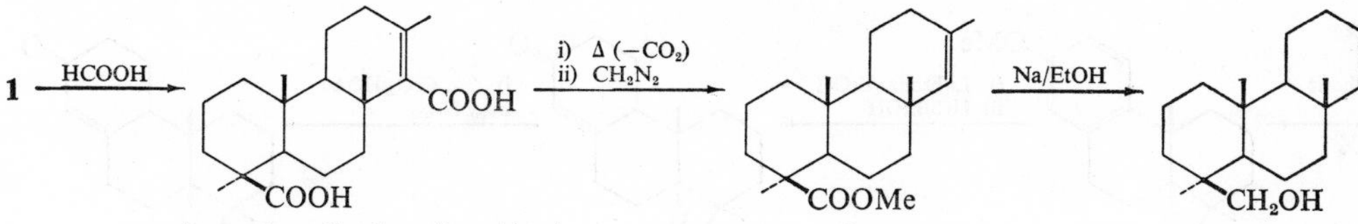

isoagathenedicarboxylic acid **6**

7

8

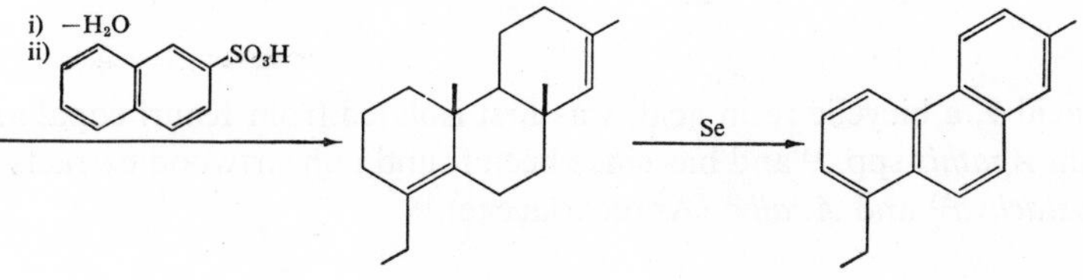

9

10

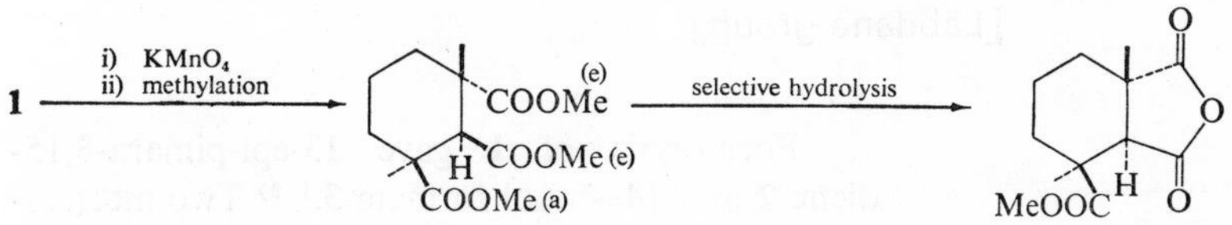

11

12

$\alpha_D : + 57.9°$

4) The A/B ring junction was shown to be *trans* by the formation of **15** from **1**. Compound **15** is identical to that obtained from manool, which is known to have a *trans* A/B ring junction.[9]

5) Further confirmation of the structure of **1** was obtained by synthesis of the degradation product **3** by the sequence **18→3**.[10]

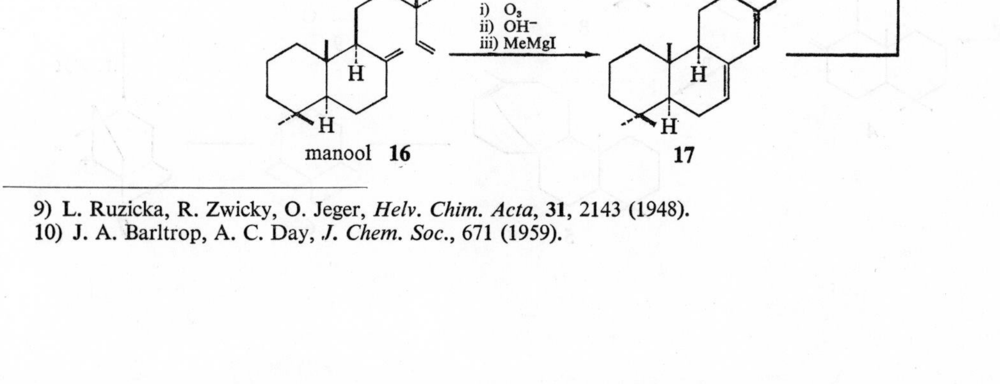

13

14

15

manool **16**

17

9) L. Ruzicka, R. Zwicky, O. Jeger, *Helv. Chim. Acta*, **31**, 2143 (1948).
10) J. A. Barltrop, A. C. Day, *J. Chem. Soc.*, 671 (1959).

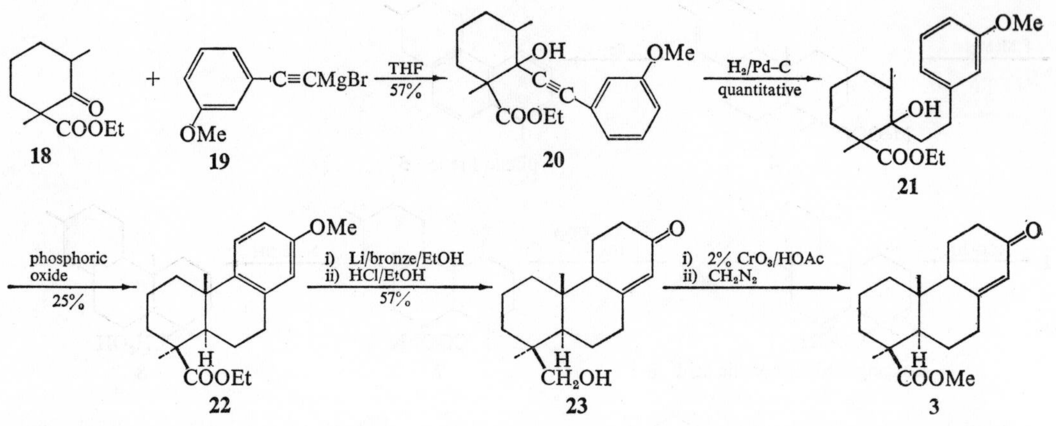

Remarks

Agathenedicarboxylic acid **1**, a bicyclic resin acid, was first isolated from Kauri copal and Manila copal, the resins from *Agathis* spp.,[1] and has since been found in heartwood extracts of *Agathis australis*,[2] *A. microstachya*[2] and *A. alba* (Araucariaceae).[3]

4.9   ACID-CATALYZED CYCLIZATION OF MANOOL
[Labdane group]

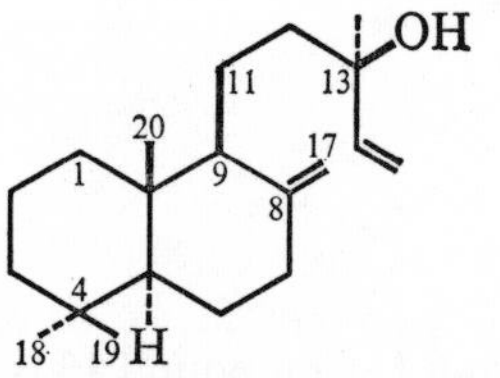

manool **1**

Formolysis of **1** gave 13-epi-pimara-8,15-diene **2** and 14α-hibol formate **3**.[1-3] Two mechanisms were considered for the formation of **3**, namely, path A, based on biogenetic considerations of the formation of tetracyclic diterpenes, and the alternative path B.[2]

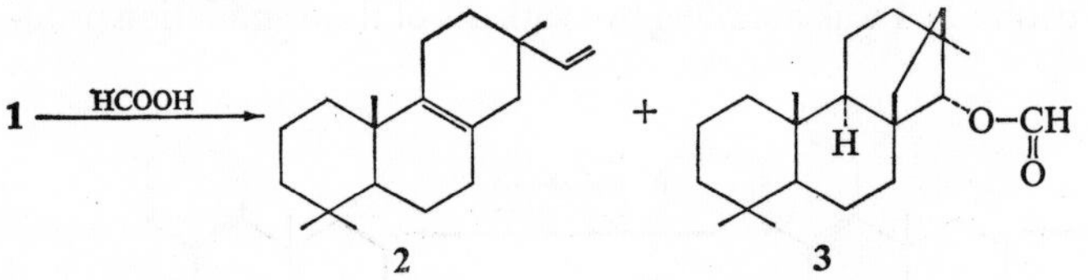

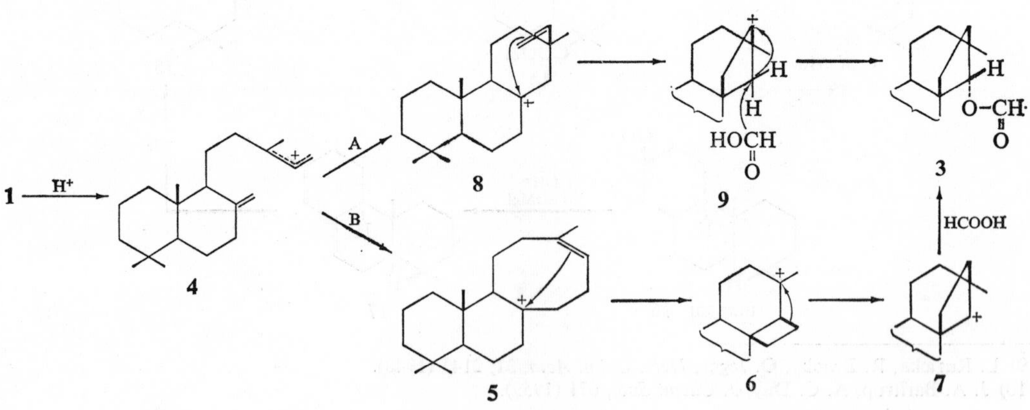

Labelling experiments with ^{14}C and ^{2}H supported path B, involving the unusual formation of a cyclooctenyl cation **5**. It was shown that formolysis of [14-^{14}C]isomanool **10** yielded **11** with the radioactivity exclusively at C-14,[4] and that formolysis of **12** gave **13**, the deuterium atoms of which were located by nmr and ms.[5] The transition state **5** for path B stereochemically favors the formation of **3**.

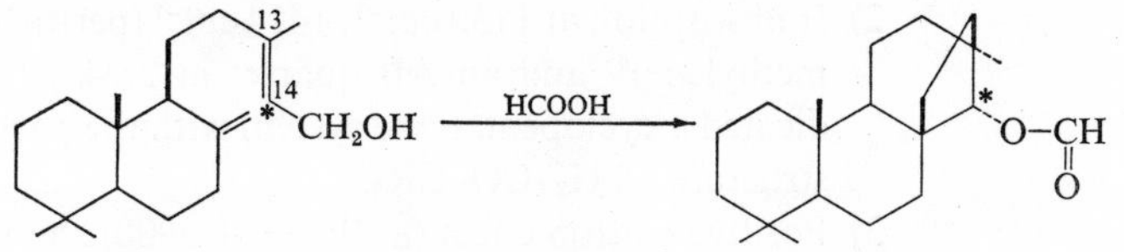

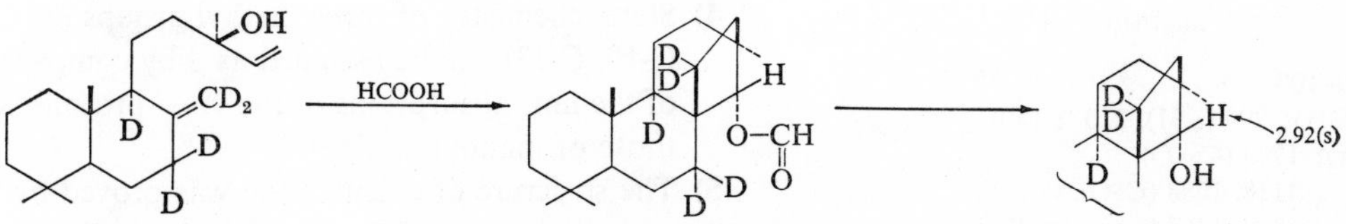

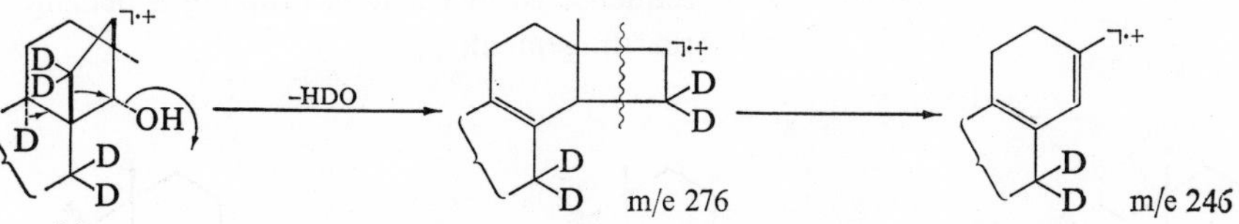

1) O. E. Edwards, R. S. Rosich, *Can. J. Chem.*, **46**, 1113 (1968).
2) E. Wenkert, Z. Kumazawa, *Chem. Commun.*, 140 (1968).
3) T. McCreadie, K. H. Overton, *Chem. Commun.*, 288 (1968).
4) J. Fourrey, J. Polonsky, E. Wenkert, *Chem. Commun.*, 714 (1969).
5) O. E. Edwards, B. S. Mootoo, *Can. J. Chem.*, **47**, 1189 (1969).

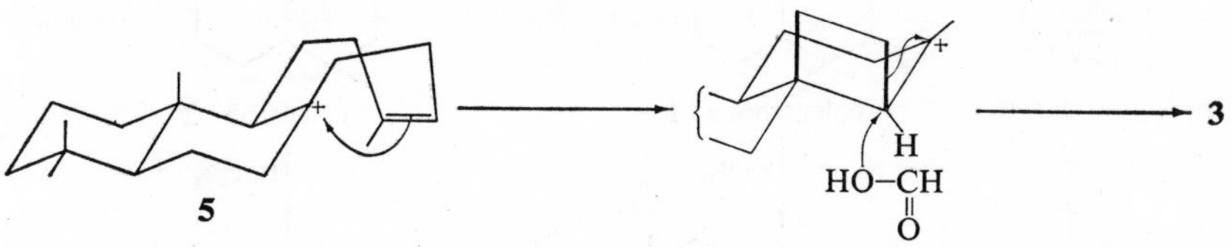

4.10   STRUCTURE OF COLENSENONE
[Labdane group]

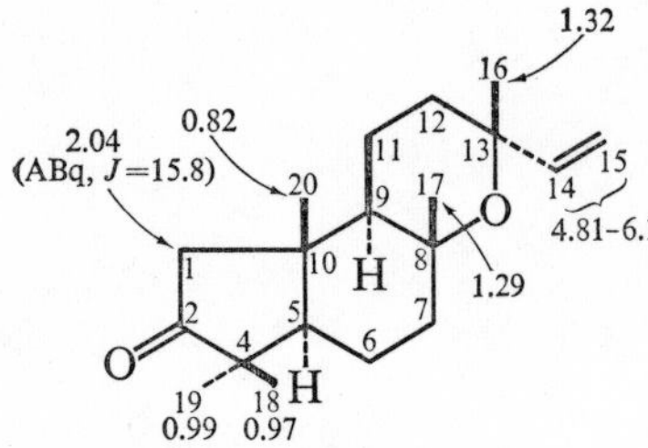

colensenone   **1**[1]

mp: 99–100°
uv (EtOH): 299 (ε 31), 210 (ε 140)
ir (KBr): 1733 (C=O),
            1118, 1088 (C–O),
            1387, 1367 (*gem*-dimethyl),
            3190, 1643, 991, 911, (–CH=CH$_2$)
ms: 290 (M$^+$)

1) Deuteration followed by mass spectroscopy showed two exchangeable hydrogens.
2) Ir absorption at 1730 cm^{-1}, 1414 cm^{-1} (perturbed methylene)[2] and an AB quartet at 2.04 δ[3] indicated a cyclopentanone system with the partial structure –CH$_2$–CO–CR$_3$.
3) Positive Cotton effect ($\alpha_D{}^{317.5} = +3900$) showed a *trans* A/B ring junction.
4) Stereochemistry of three methyl groups (at C-8, C-10, C-13) can be assigned as β by comparison of the nmr absorptions of the vinyl protons with those of manoyl oxide.[3]
5) The structure of colensenone was proved by the transformation of **6** to colensanone **4** via the sequence shown below, utilizing a benzilic acid rearrangement.

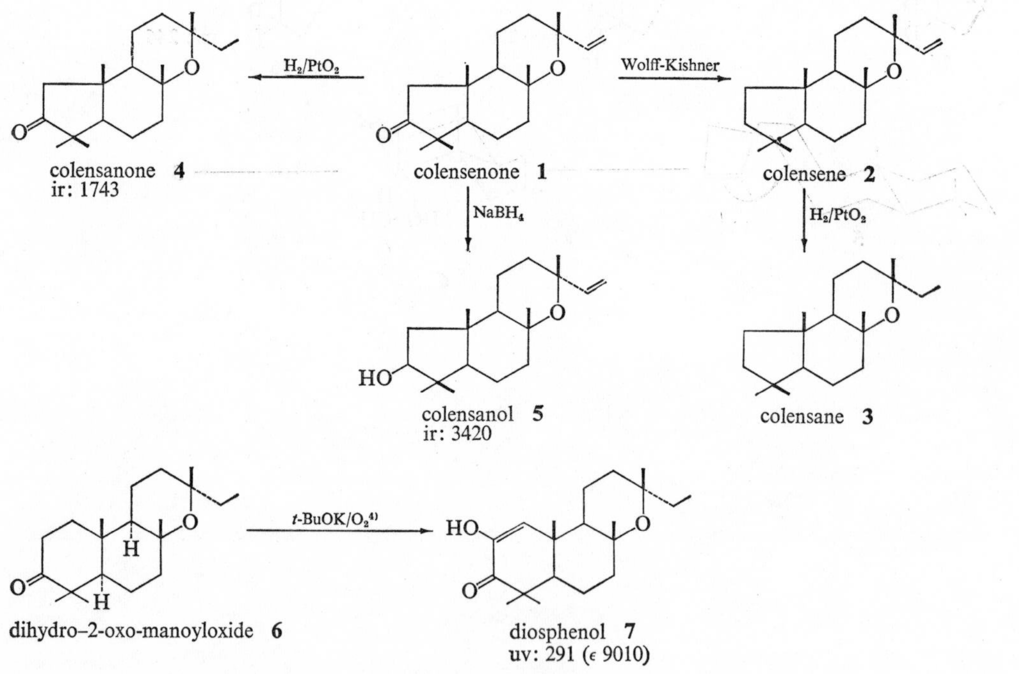

colensanone **4**
ir: 1743

colensenone **1**

colensene **2**

colensanol **5**
ir: 3420

colensane **3**

dihydro–2-oxo-manoyloxide **6**

diosphenol **7**
uv: 291 (ε 9010)

1) P. K. Grant, R. M. Carman, *J. Chem. Soc.*, 3740 (1962).
2) A. R. H. Cole, *Fortsch. Chem. Org. Naturst.*, **13**, 1 (1956).
3) E. Wenkert, P. Beak, P. K. Grant, *Chem. Ind.*, 1574 (1961).
4) D. H. R. Barton, S. K. Pradhan, S. Sternhell, J. F. Templeton, *J. Chem. Soc.*, 255 (1961).
5) P. K. Grant, M. J. A. McGrath, *Tetr.*, **26**, 1619, 1631 (1970).

$$7 \xrightarrow[\text{(benzylic rearr.)}]{8\% \text{ aq. NaOH}} \quad 8 \xrightarrow[\text{ii) HIO}_4]{\text{i) LAH}} \quad \text{colensanone } 4$$

Remarks

Colensenone **1** is a nor-diterpenoid from *Dacrydium colensoi*. The compound **9** was isolated from the same source.[5]

colensan-2β-ol-1α-hydroxymethyl-
2α-carboxylactone **9**

4.11 STRUCTURE OF ANDROGRAPHOLIDE
[Labdane group]

andrographolide **1**

mp: 230–231° α_D: −127°
uv (EtOH): 223 (ε 12,300) (unsat. γ-lactone)
ir (KBr): 3448, 3390–3299(OH),1726, 1672
          (unsat. γ-lactone), 1647, 906 (exocyclic
          C=CH$_2$)

1) Presence of three OH groups was shown by the formation of the triacetate.[1]
2) Easy hydrogenolysis (PtO$_2$/HOAc or Al/Hg) indicated the presence of a labile OH group.[1-3]
3) The bicarbocyclic skeleton and substitution pattern were suggested by the Se dehydrogenation product **4**.[2]
4) The exocyclic α,β-unsaturated γ-lactone was assigned by nmr analysis of **2**[4] (comparison of the vinyl 12-H absorption with that of 2-butenolide).
5) Stereochemistry at C-9 and C-10 was suggested by the strongly positive ord curve of **3**[5-7] and was proved by chemical evidence.
6) The A/B ring junction and stereochemistry at C-3 and C-4 were determined by both chemical and spectral evidence.

1) M. K. Gorter, *Rec. Trav. Chim.*, **30**, 151 (1911).
2) R. Schwyzer, H. G. Biswas, P. Karrer, *Helv. Chim. Acta.*, **34**, 652 (1952).
3) C. Chakravarti, R. N. Chakravarti, *J. Chem. Soc.*, 1697 (1952).
4) M. P. Cava, B. Weinstein, W. R. Chan, L. J. Haynes, L. F. Johnson, *Chem. Ind.*, 167 (1963).
5) M. P. Cava, W. R. Chan, L. J. Haynes, C. F. Johnson, *Tetr.*, **18**, 397 (1962).
6) C. D. Djerassi, R. Rikiner, B. Rikiner, *J. Am. Chem. Soc.*, **78**, 6362 (1956).
7) C. D. Djerassi, M. Cais, L. A. Mitscher, *J. Am. Chem. Soc.*, **81**, 2386 (1959).
8) M. P. Cava, W. R. Chan, R. P. Sein, C. R. Willis, *Tetr.*, **21**, 2619 (1965).

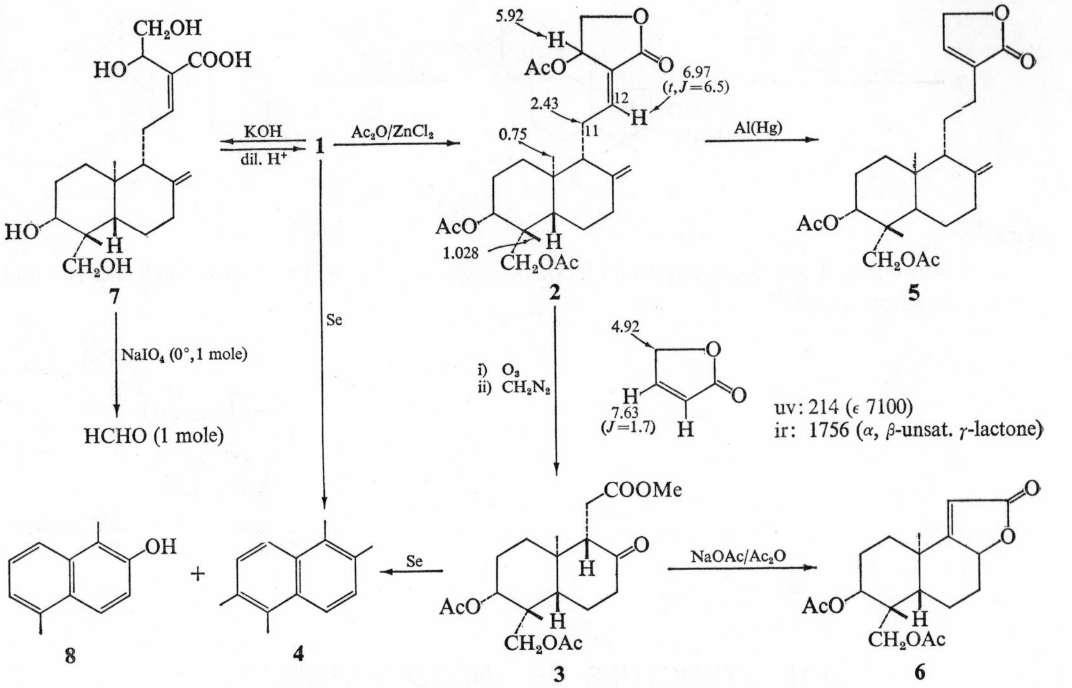

ord: $\alpha_{305} \pm 1718°$, $\alpha_{267.5} -1512°$
cd: 289 $(\Delta\epsilon +2.67)^{10)}$

7) The structure of the degradation product **6** was confirmed by synthesis, as shown below.[9]

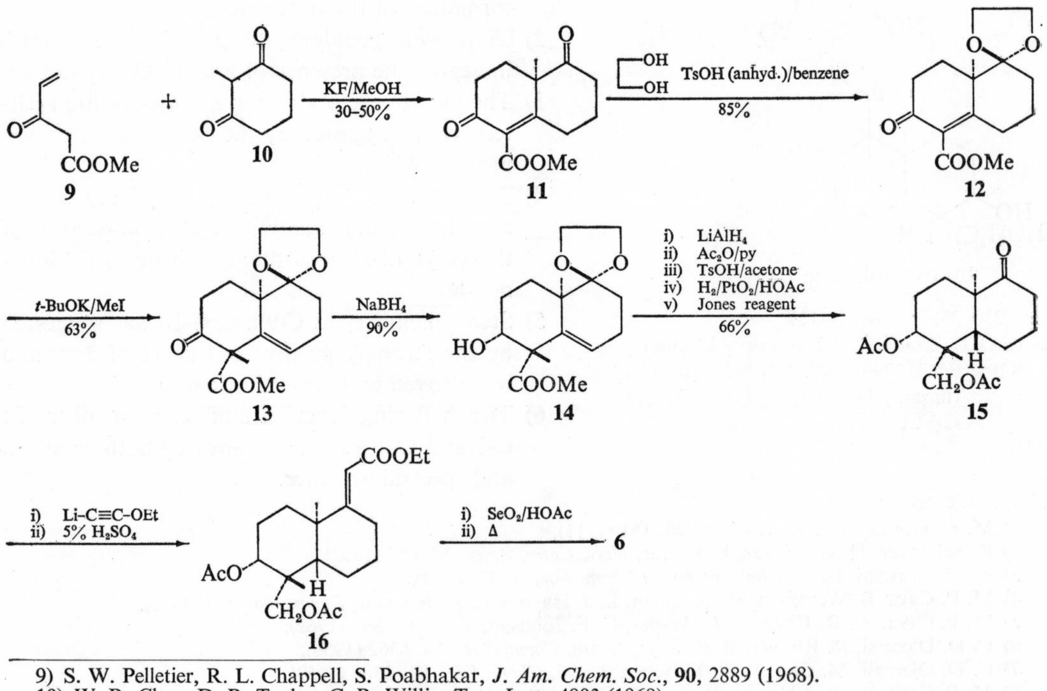

9) S. W. Pelletier, R. L. Chappell, S. Poabhakar, *J. Am. Chem. Soc.*, **90**, 2889 (1968).
10) W. R. Chan, D. R. Taylor, C. R. Willis, *Tetr. Lett.*, 4803 (1968).

Remarks

neoandrographolide **17**[10]

Andrographolide **1** was obtained from *Andrographis paniculata* Nees (Acanthaceae). The related compound neoandrographolide **17** was obtained from the same source.

4.12   STRUCTURE AND SYNTHESIS OF MARRUBIIN
[Labdane group]

marrubiin **1**
$C_{20}H_{28}O_4$

mp: 160° α_D^{20} (CHCl$_3$): +35.8°
uv: 216, 212, 208 (3.70–3.75)
ir (CS$_2$): 1765 (γ-lactone)

1) Besides the characteristic uv, ir and color test, the oily adduct with diethylacetylenedicarboxylate which yielded **2** strongly supports the presence of a β-substituted furan ring.[1]

2) The carbon skeleton was found from the dehydrogenation product **3**.[2]

3) Ozonolysis of **4** gave **5** and **6**, establishing the position of the *tert*-OH group.[3]

4) The oxidation product **7** was converted to the known compounds isoambreinolide **9** and an ambrein derivative **10**,[4] thus establishing the *trans* ring junction.

5) $W_{1/2}$ of 6-H of **1** and the δ value at 19-H of **13** suggested a *β-cis* lactone.[5]

6) 8α-Me was correlated to that of **5**, which was assumed to be α from the ord.[6,7] The fact that no epimerization occurred at C-8 during ozonolysis of **4** was shown by carrying out the reaction in MeOD.[8]

7) The 9α-OH was correlated to that of **7**, which can be synthesized stereoselectively from **5**.[6]

1) D. P. Moody, *Chem. Ind.*, 75 (1960).
2) A. Lawson, E. D. Eustice, *J. Chem. Soc.*, 587 (1939).
3) D. G. Hardy, W. Rigby, D. P. Moody, *J. Chem. Soc.*, 2955 (1957).
4) D. Burn, W. Rigby, *J. Chem. Soc.*, 2964 (1957).
5) R. A. Appleton, T. W. B. Fulke, M. S. Henderson, R. McCrindle, *J. Chem. Soc.* C, 1943 (1967).
6) L. Mangoni, M. Adinolfi, *Tetr. Lett.*, 269 (1968).
7) C. Djerassi, W. Klyne, *J. Chem. Soc.*, 4929 (1962).
8) L. J. Stephens, D. M. S. Wheeler, *Tetr.*, **26**, 1561 (1970).
9) M. S. Henderson, R. McCrindle, *J. Chem. Soc.*, 2014 (1969).
10) L. Mangoni, M. Adinolfi, G. Laonigro, R. Caputo, *Tetr.*, **28**, 611 (1972).

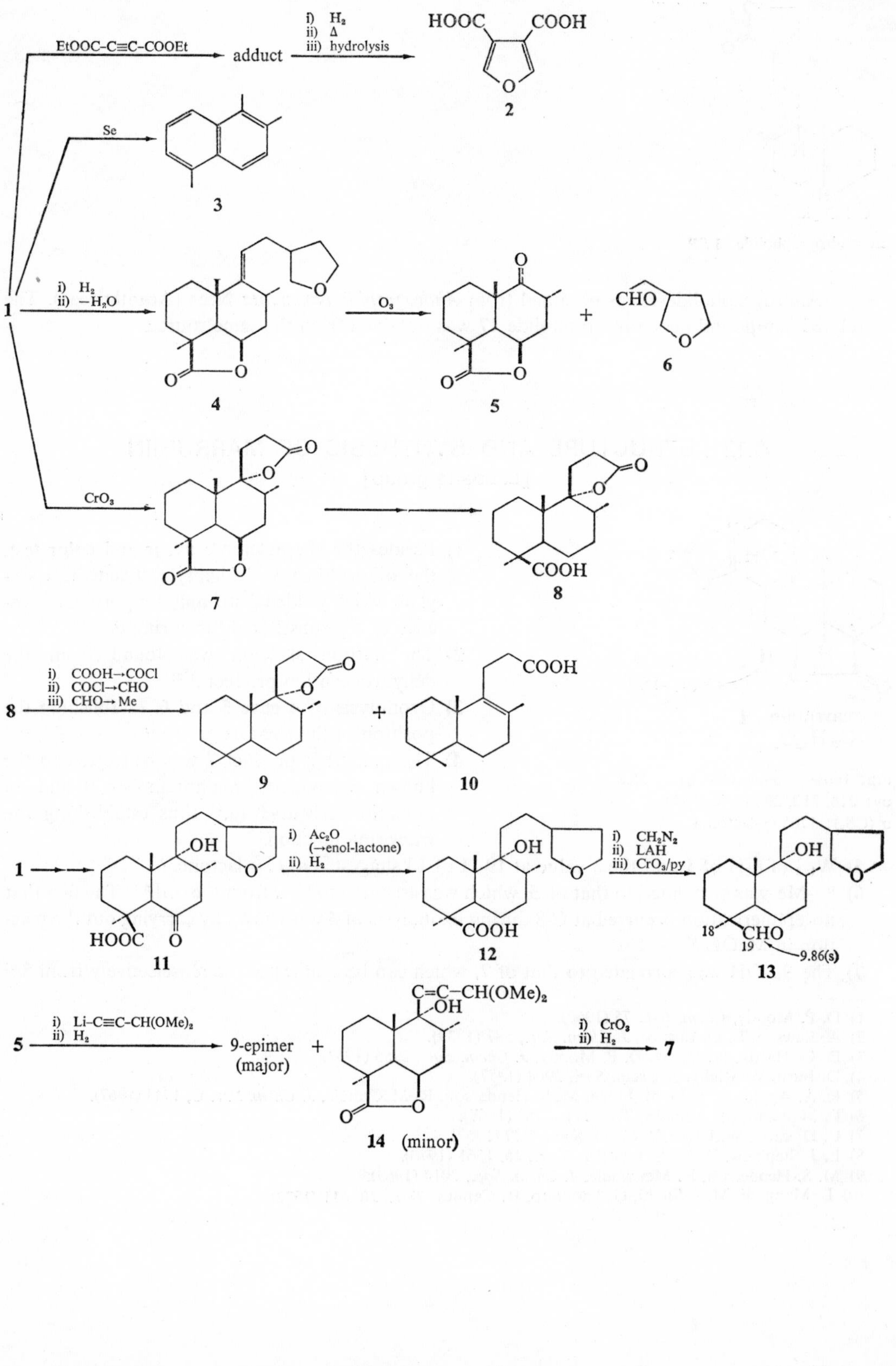

EtOOC-C≡C-COOEt
adduct
i) H₂
ii) Δ
iii) hydrolysis
HOOC
COOH
O
2
Se
3
1
i) H₂
ii) −H₂O
4
O₃
5
CHO
6
CrO₃
7
COOH
8
8
i) COOH→COCl
ii) COCl→CHO
iii) CHO→Me
9
10
1
OH
HOOC
11
i) Ac₂O
(→enol-lactone)
ii) H₂
12
i) CH₂N₂
ii) LAH
iii) CrO₃/py
18
CHO
19
9.86(s)
13
5
i) Li-C≡C-CH(OMe)₂
ii) H₂
9-epimer
(major)
C=C−CH(OMe)₂
OH
14 (minor)
i) CrO₃
ii) H₂
7

Remarks

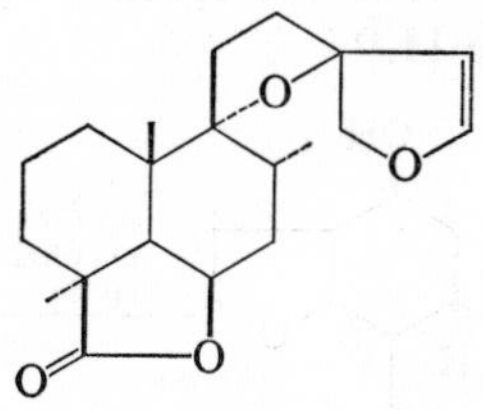

Marrubiin **1** was isolated from *Marrubium vulgare* L. (hoarhound) (Labiatae).

Premarrubiin **21**, isolated from the same source, appeared to be the major substrate from which marrubiin was generated as an artefact during the extraction process.[9]

21

Synthesis[10]

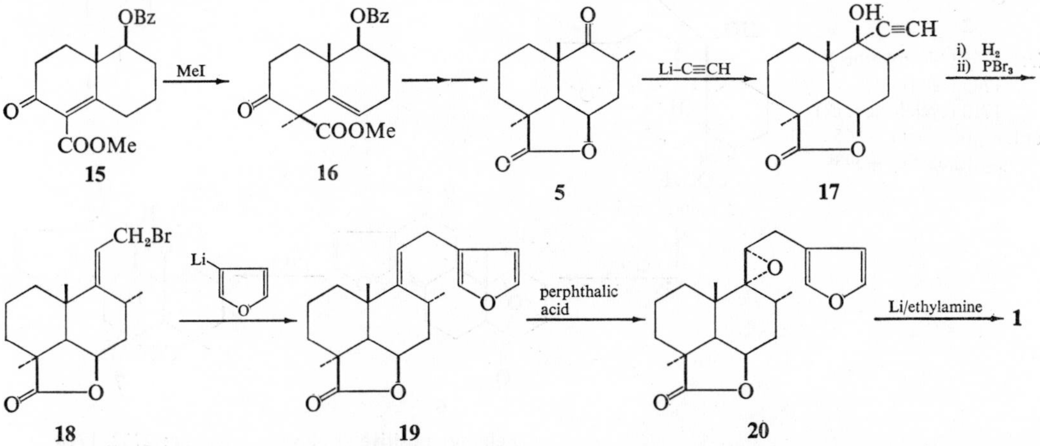

The key intermediate was **5**, which could be prepared stereospecifically from **15**. **19**→**20**: epoxidation from the less hindered side controlled the stereochemistry at C-9 of **1**.

4.13   STRUCTURE OF SCIADIN
[Labdane group]

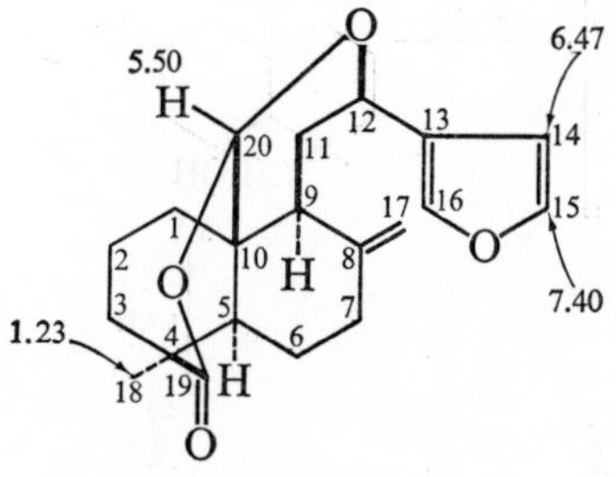

sciadin **1**[1]

1) A positive Ehrlich test and characteristic ir and nmr absorptions showed the presence of a β-substituted furan.[3]

2) The ir and ozonolysis results (formaldehyde and a C_6 cyclic ketone) indicate a double bond *exo* to cyclohexane.

3) The configurations at C-9 and C-10 were determined by the negative Cotton effect of **2**.

4) The bicarbocyclic ring system was proved by the Se dehydrogenation product **4**.

5) The presence of a δ-lactol ether was shown by chemical and spectral evidence.

mp: 160° α_D: +10.3°
uv: 206 (ϵ 9100)
ir (nujol): 1736 (δ-lactol ether), 1636, 894 (C=CH$_2$), 1636, 1607, 1505, 870 (furan)[2]

1) M. Sumimoto, *Tetr.*, **19**, 643 (1963).
2) T. Kubota, *Tetr.*, **4**, 68 (1958).
3) E. J. Corey, G. Slomp, Sukh Dev, S. Tobinaga, R. Glazier, *J. Am. Chem. Soc.*, **80**, 1204 (1958).

6) The structure was determined by chemical transformations and spectral evidence. **1** was correlated with dimethyl sciadinonate **12** and sciadinone **11**.[4]

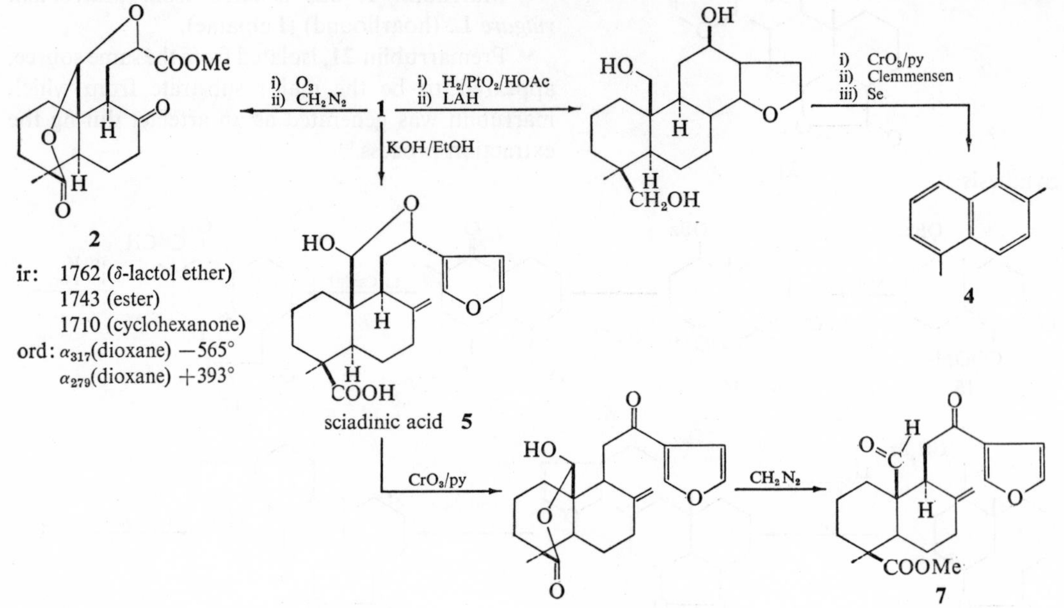

2

ir: 1762 (δ-lactol ether)
1743 (ester)
1710 (cyclohexanone)
ord: α_{317}(dioxane) −565°
α_{279}(dioxane) +393°

sciadinic acid **5**

4

6

Fehling: positive
uv: 254 (ε 3600)
(β-furyl ketone).

7

ir: 2752, 1716
(aldehyde)
DNP derivative

Remarks

Sciadin **1** was isolated from *Sciadopitys verticillata* Sieb. et Zucc. The related compounds dimethyl sciadinonate **12** and sciadinone **11** were isolated from the same source.[4] The relationship between **1**, **11** and **12** is illustrated by the scheme below.[4]

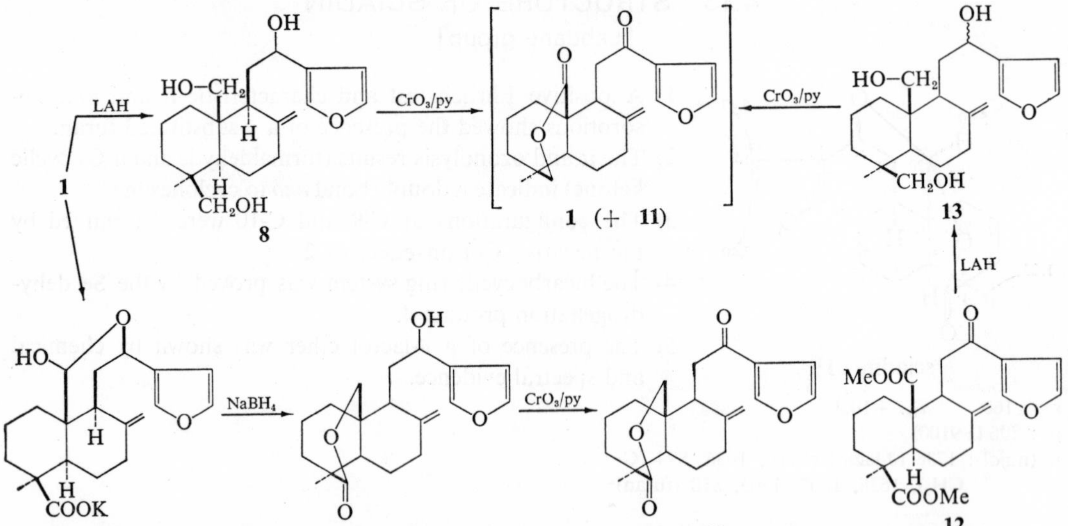

8

1 (+ 11)

13

9

10

11

12

4) C. Kaneko, T. Tsuchiya, M. Ishikawa, *Chem. Pharm. Bull.*, **11**, 1346 (1963).

4.14 STRUCTURE OF NEPETAEFOLIN, NEPETAEFURAN AND NEPETAEFURANOL

[Labdane group]

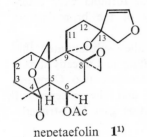

nepetaefolin **1**[1]

mp: 260°
α_D (25°, *c* 0.90, CHCl$_3$): −14.6°

1) The relationship between nepetaefolin **1**, nepetaefuran **2**, and nepetaefuranol **6**, is shown by the transformations **1→3** and **2→6**.

2) Presence of the terminal epoxide and δ-lactone follows from the transformation of **2** to **4**, whose structure was derived by nmr.

3) The relationship of functionalities in ring B is shown by the sequence **2→6→7**.

4) Unambiguous proof of the structure of **1** was obtained by correlation with leonotin **9** of known structure and relative stereochemistry.[2]

5) Only the stereochemistry at C-13 in **1** is still unknown.

nepetaefolin **1** Δ/CHCl$_3$ nepetaefuran **2**   LAH/THF   **4**

i) TsCl/py
ii) LAH

OH⁻   HClO$_4$ / i) TsCl / ii) OH⁻   OH⁻

5   nepetaefuranol **6**   **3**   leonotol **8**

i) NaIO$_4$
ii) alumina

LAH

ir(nujol): 1670
7   leonotin **9**

Remarks

Compounds **1, 2** and **6** are present in the leaves of the plant *Leonotis nepetaefolia* (Labiatae), which is widespread throughout the West Indies, South America, and the African continent.

1) J. D. White, P. S. Manchand, *J. Am. Chem. Soc.*, **92**, 5527 (1970).
2) J. D. White, P. S. Manchand, W. B. Whalley, *Chem. Commun.*, 1315 (1969).

4.15 STRUCTURE OF CLERODIN
[Clerodane group]

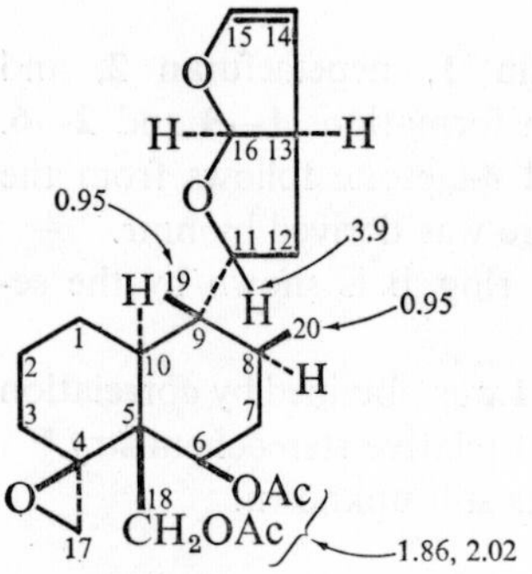

clerodin **1**
$C_{24}H_{34}O_7$

mp: 164–165° α_D: $-47°$
ir (nujol): 1727, 1252 (OAc), 1615,
        738 (–O–C=C)
ir (CCl₄): 3025 (H–C=C), 3045 $\left(\bigvee\hspace{-0.3em}\diagup\hspace{-0.5em}\diagdown \right)$

1) Structure and molecular formula were determined by x-ray study of the bromolactone derivative **3**.[1]
2) Chemical and nmr studies substantiated this formulation;[2] investigation had already shown the presence of two acetates, two C-methyl groups (Kuhn-Roth) and the absence of methoxyl; dehydrogenation gave 1,2,5-trimethylnaphthalene.
3) Solution of **1** in acetic acid gives the hemiacetal ester **6**. The vinyl ether moiety was shown to belong to a 5-membered ring by formation of the lactone **5** via **4**.

$$\mathbf{1} \xrightarrow[\text{H}_2\text{O/HOAc}]{\text{Br}_2/\text{H}_2\text{O}}$$

2 R₁=OH, R₂=Br
4 R₁=OH, R₂=H
6 R₁=OAc, R₂=H

$$\xrightarrow{\text{CrO}_3}$$

3 R=Br
5 R=H

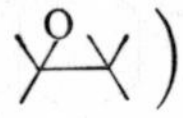

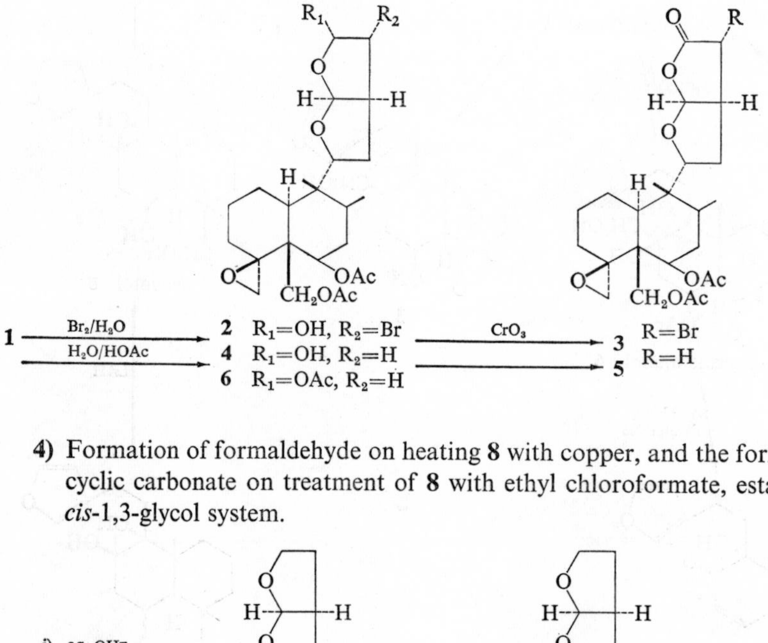

4) Formation of formaldehyde on heating **8** with copper, and the formation of a six-membered cyclic carbonate on treatment of **8** with ethyl chloroformate, established the presence of a *cis*-1,3-glycol system.

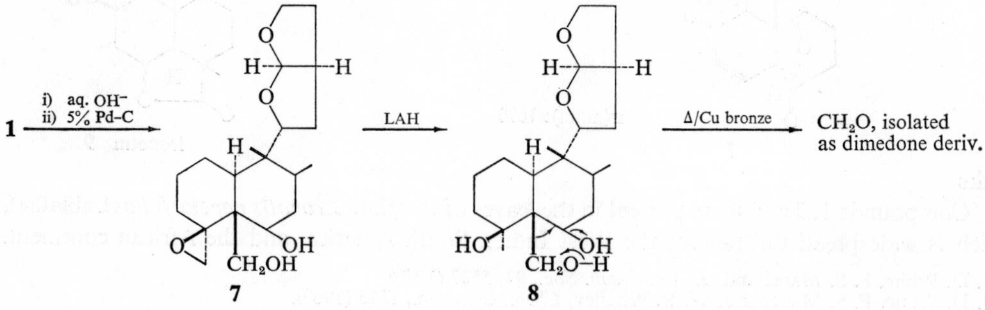

$$\mathbf{1} \xrightarrow[\text{ii) 5\% Pd-C}]{\text{i) aq. OH}^-}$$

7

$$\xrightarrow{\text{LAH}}$$

8

$$\xrightarrow{\Delta/\text{Cu bronze}}$$

CH₂O, isolated
as dimedone deriv.

5) The nature of the epoxide and the partial structure shown in **7** are indicated by the transformation of **8** to **11** via the highly unstable **10**.

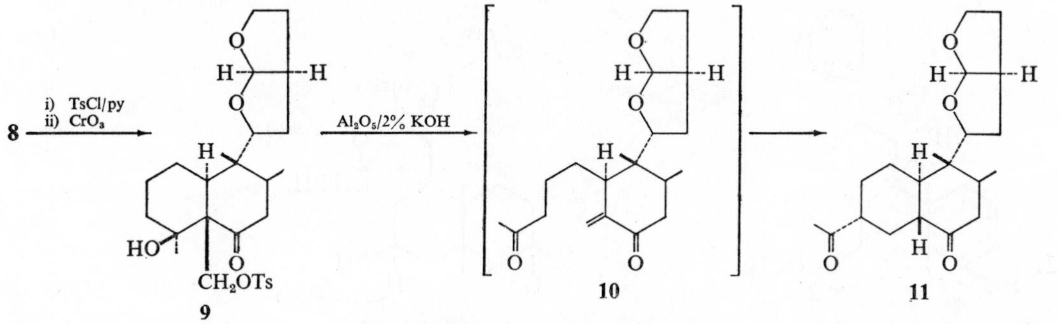

6) The position of the remaining ethereal oxygen atom was indicated by the facile formation of the lactam **13** from the amide **12** due to the "acetalic" nature of C-16.

7) The absolute configuration was tentatively assigned by comparison of the ord curve of **14** with that of a 6-keto *trans* A/B steroid.

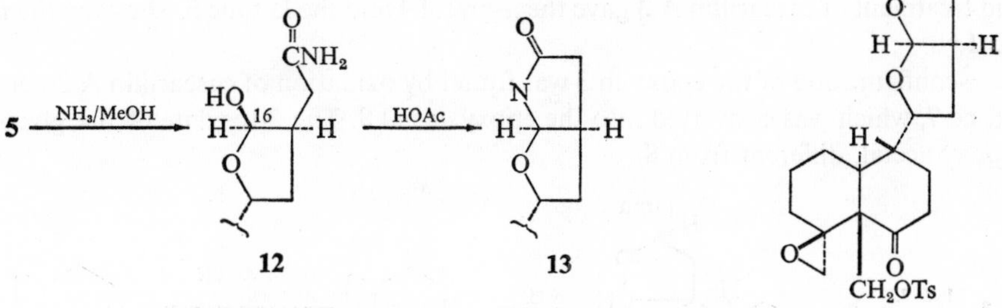

Remarks[3)]

Clerodin is found in the ether extract of leaves and twigs of the Indian bhat tree, *Clerodendron infortunatum* (Verbenaceae), and a crude preparation is used locally as a vermifuge.

1) G. A. Sim, T. A. Hamor, J. C. Paul, J. M. Robertson, *Proc. Chem. Soc.*, 75, (1961).
2) D. H. R. Barton, H. T. Cheung, A. D. Cross, L. M. Jackman, M. Martin-Smith, *J. Chem. Soc.*, 5061 (1961); *Proc. Chem. Soc.*, 76 (1961).
3) H. N. Banerjee, *J. Indian Chem. Soc.*, **14**, 51 (1937).

4.16 STRUCTURE OF CASCARILLIN AND CASCARILLIN A
[Clerodane group]

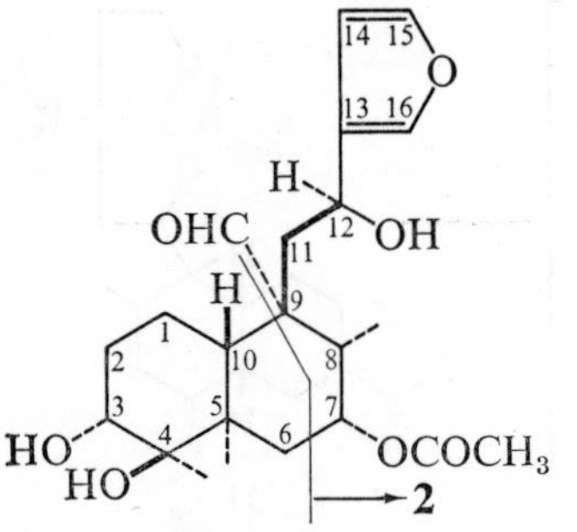

cascarillin **1**
mp: 203.5°

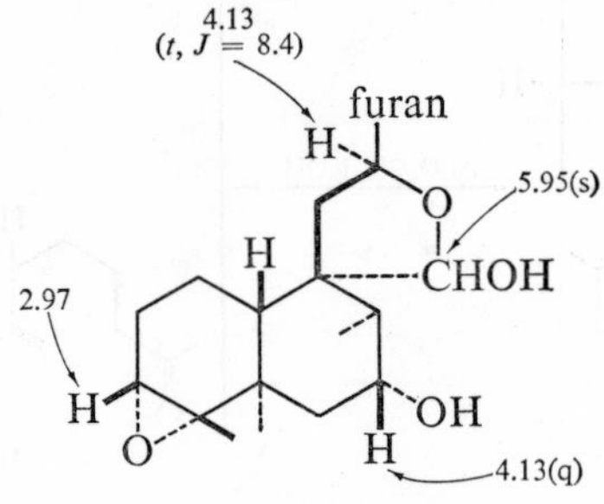

cascarillin A **3**
mp: 187–203° (formation of acetal upon heating)

1) The partial structure **2** of cascarillin was obtained by chemical and spectroscopic methods.[1]
2) Hydrolysis of **1** gave the acetal **4**; x-ray studies of **4** iodoacetate elucidated the structure **1** (without absolute configuration).[2]
3) Absolute configuration of **1** was obtained from a comparison of the ord of ketones **5** and **6** with D-homosteroids.[3] This was confirmed by Bijvoet's anomalous dispersion method.[4]
4) Acid treatment of cascarillin A **3** gave the α-glycol **4** and the ketone **5**, showing the relation of **3** to **1**.[5]
5) The α-configuration of the epoxy in **3** was found by oxidation of cascarillin A 3-acetate to the lactone **7**, which was converted into the epoxy acetal **8**. The 3-tosylate of the glycol **4** gave a β-epoxy acetal, different from **8**.

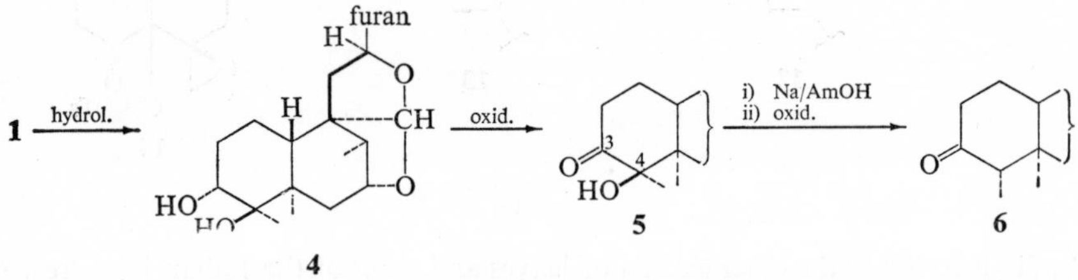

4

5

6

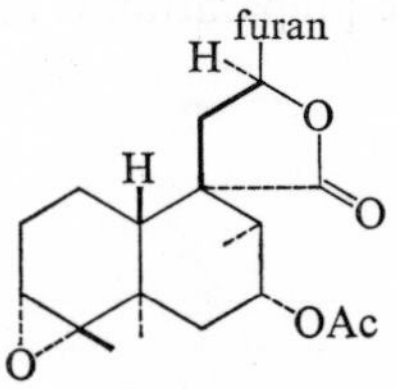

7

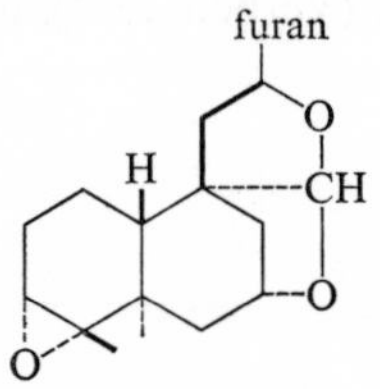

8

Remarks

The source of **1** and **3** is *Croton eleuteria* (Cascarilla bark).

1) J. S. Birtwistle, D. E. Case, P. C. Dutta, T. G. Halsall, G. Mattewo, H. D. Sabel, V. Thaller, *Proc. Chem. Soc.,* 329 (1962).
2) J. M. Robertson, *Proc. Chem. Soc.*, 235 (1963).
3) D. E. Case, T. G. Halsall, A. W. Oxford, Abstracts, *IUPAC Symposium on the Chemistry of Natural Products,* 54, 1964.
4) C. E. McEachan, A. T. McPhail, G. A. Sim, *J. Chem. Soc.* (B), 633 (1966).
5) T. G. Halsall, A. W. Oxford, W. Rigby, *Chem. Commun.*, 218 (1965).

4.17   STRUCTURE OF COLUMBIN AND RELATED COMPOUNDS
[Clerodane group]

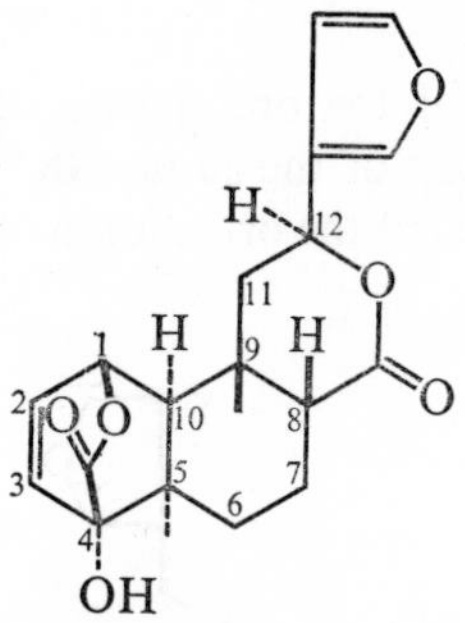

columbin   **1**[2)]
$C_{20}H_{22}O_6$

ir (CCl₄): 3620 (OH)
1762, 1743 (δ-lactone)

isocolumbin **2**: C-8 epi
jateorin   **3**: β-2,3-oxido
chasmanthin **4**: β-2,3-oxido, 12-epi
palmarin   **5**: β-2,3-oxido, 8-epi, 12-epi

1) Columbin **1** is easily isomerized to isocolumbin **2** by mild alkali; both **1** and **2** behave as dilactones towards alkali.

2) Loss of CO_2 and *tert*-OH on melting **1** and **2** yields monolactones **6** and **7**, suggesting a β,γ-unsaturated lactone structure.

3) Hydrogenolysis of the second lactone in the transformation of **6** to **8** defines the relationship of one lactone to the furan ring, whose presence was shown by ozonolysis of **9** to **10** and **11**.

4) Dehydrogenation of **6** to **12**, involving migration of the 5-Me helped to establish the points of attachment of the lactone carbonyls.

5) Quantitative ir (1380 cm⁻¹) on **8** and Kuhn-Roth determination of **1** indicated two C-methyl groups.

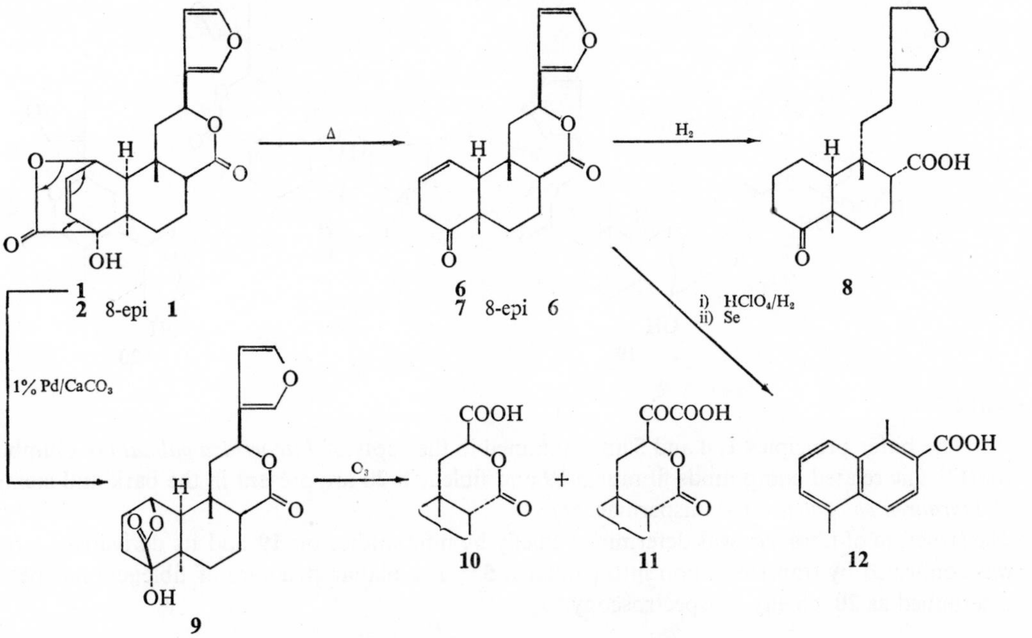

6) Based on the examination of models, the transformation of octahydroisocolumbin **14** to **16** leads to the conclusion that **2** has a *cis*-fused decalin system with *cis* C-1 and C-8 substituents

1) F. Wessely, K. Dinjaski, W. Isemann, G. Singer, *Monatsh.*, 65, 87 (1935).
2) D. H. R. Barton, D. Elad, *J. Chem. Soc.*, 2085 (1956).

(1 has *trans* C-1, C-8). Similarly **13** to **17** suggests a *trans* relationship of the C-5 and C-9 methyls.

7) The absolute configuration at C-12 was derived by application of the Hudson-Klyne[3] and Bose[4] rules to the molecular rotation values for octahydrocolumbic acid **15** and related compounds.

8) The absolute configuration *in toto* was determined by measurement of the ord spectra of several ketonic columbin derivatives and was confirmed by x-ray analysis of the adduct **18**.[5]

9) The interrelationships of **1, 2, 3, 4** and **5** had been previously determined;[6] the orientation of the epoxides was determined by nmr.

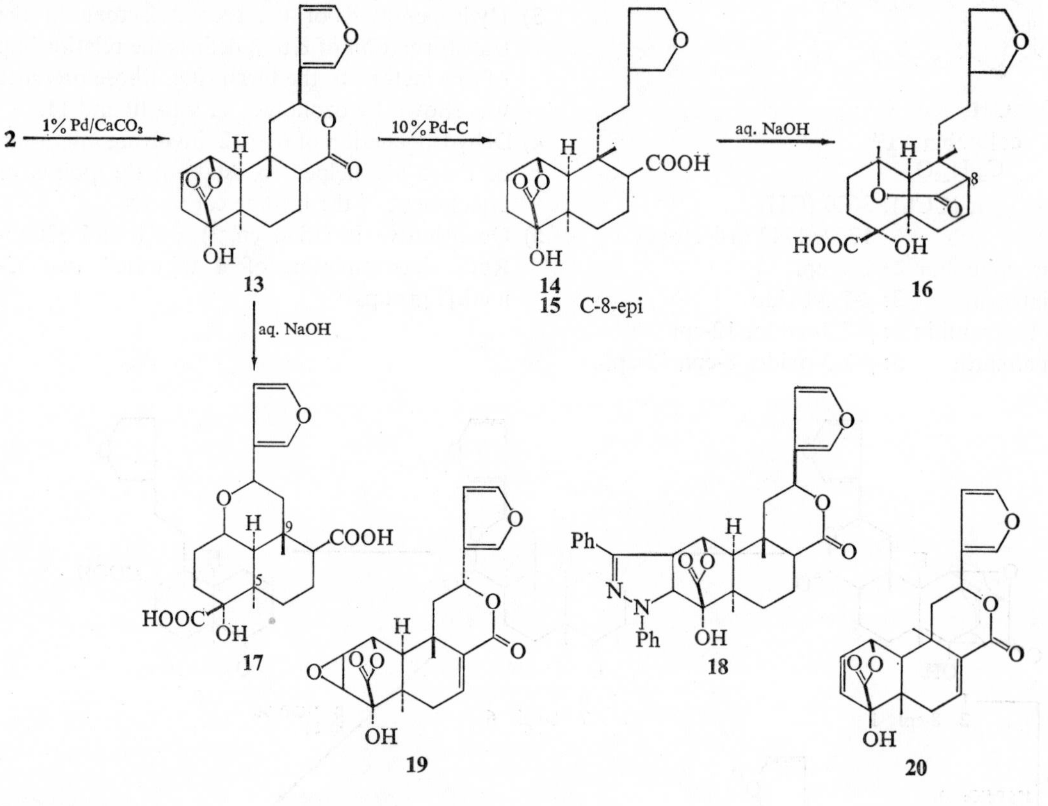

Remarks

The bitter principles **1, 4** and **5** are contained in the roots of *Jateorrhiza palmata* (Columbo root).[1] The related compounds fibraurin **19** and fibleucin **20** are present in the bark and stems of *Fibraurea chloroleuca* (Menispermaceae).[7,8]

The structure of fibraurin was determined chiefly by nmr studies on **19** and its derivatives, and was confirmed by transformation into palmarin **5**.[7] The planar structure of fibleucin has been determined as **20**, chiefly by spectroscopy.[8]

3) W. Klyne, *Chem. Ind.*, 1198 (1954).
4) A. K. Bose, B. G. Chatterjee, *J. Org. Chem.*, **23**, 1425 (1958).
5) K. H. Overton, H. G. Weir, A. Wylie, *J. Chem. Soc.*, 1482 (1966).
6) D. H. R. Barton, K. H. Overton, A. Wylie, *J. Chem. Soc.*, 4809 (1962).
7) T. Hori, A. K. Kiang, K. Nakanishi, S. Sasaki, M. C. Woods, *Tetr.*, 2649 (1967).
8) K. Ito, H. Furukawa, *Chem. Commun.*, 653 (1969).

4.18   STRUCTURE AND SYNTHESIS OF RIMUENE
[Pimarane group]

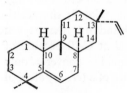

rimuene **1**
mp: 55–56°
α_D: +56°

1) Tricyclic skeleton was established by dehydrogenation to the known compound **2**.[1]
2) Vinyl group and trisubstituted double bond were determined by nmr and ir studies.[1]
3) C-9 methyl and the position of the 5,6-double bond were confirmed by degradation to the keto acid **4**.[2]
4) The stereochemistry at C-13 was established by comparison with the diene from erythroxydiol Y **5**,[3] which had different glc behavior but the same degradation product **6**. This conclusion proved previously proposed structures to be in error.

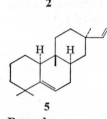

2

5

i) H₂
ii) osmic acid

1

H H

OH
OH

3

CrO₃

H H

O

COOH

4

i) oxidation
ii) H₂O₂
iii) reduction

H H

6

Remarks

A total synthesis has been achieved of rimuene,[4,5] which was first isolated from essential oil of *Dacrydium cupressium*.[6]

Synthesis

The total synthesis[4] from the tricyclic ketone **7**[5] confirmed the C-13 stereochemistry.

O

CH₃O

7

i) Li/NH₃
ii) LAH

OH

H

CH₃O

8

i) Li/NH₃/t-BuOH
ii) t-C₄H₉OK/MeI

OH

H H

O

9

i) (HSCH₂)₂
ii) Ra–Ni
iii) CrO₃/H⁺
iv) Me₂S=CH₂

H H

O

10

i) BF₃·Et₂O
ii) MeI/t-C₄H₉OK

CHO

H H

11

Ph₃P=CH₂

dl-**1**

1) J. D. Connolly, R. McCrindle, R.D.H. Murray, K. H. Overton, *J. Chem. Soc.* C, 273, (1966).
2) R. E. Corbett, S. G. Wyllie, *Tetr. Lett.*, 1903 (1964).
3) J. D. Connolly, R. McCrindle, R. D. H. Murray, A. J. Renfrew, K. H. Overton, H. Melesa, *J. Chem. Soc.* C, 268 (1966).
4) R. E. Ireland, L. N. Mander, *Tetr. Lett.*, 3453 (1964).
5) F. H. Howell, D. A. H. Taylor, *J. Chem. Soc.*, 1248 (1958).
6) F. H. McDowall, H. J. Finlay, *J. S. C. I.*, 44, 42T (1925).

4.19   ROSENONOLACTONE
[Pimarane group]

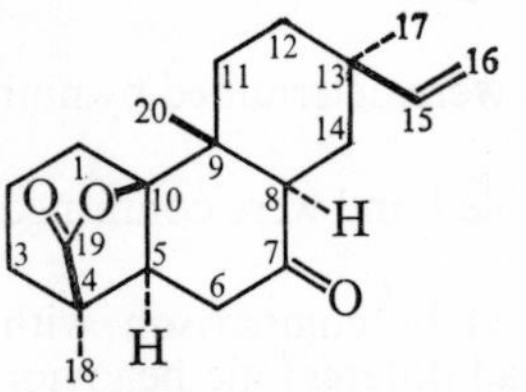

rosenonolactone **1**

mp: 208° $\alpha_D{}^{18}$: $-116°$
uv (CHCl₃): 289 (1.6) (C=O)
ir (CCl₄): 1786 (γ-lactone),
                1724 (C=O)[1]

1) The planar structure of **1** was based on the following: formation of 1,7-dimethyl-9-phenanthiol **2**; characterization of degradation products **5** and **6**;[1] the oxygen terminus of the γ-lactone was fixed at C-10 because of the formation of the triol **7**. This could be converted only to a diacetate, indicating the presence of a tertiary alcohol group.

2) The lactone and the C-9 methyl were assigned as *cis* since 7-deoxo **3**, prepared by condensation of ethanedithiol with **3** followed by desulfurization with Raney nickel, was not identical to the corresponding pimeric acid lactone **8**. The *trans-syn-trans* arrangement of rings was suggested by the less hindered carbonyl group in **1** as compared to its C-8 *iso* compound.[2]

3) The absolute stereochemistry was deduced by comparison of the ord and cd spectra of **4** with similar known compounds[3] and by x-ray analysis of the dibromo derivative of **1**.[4]

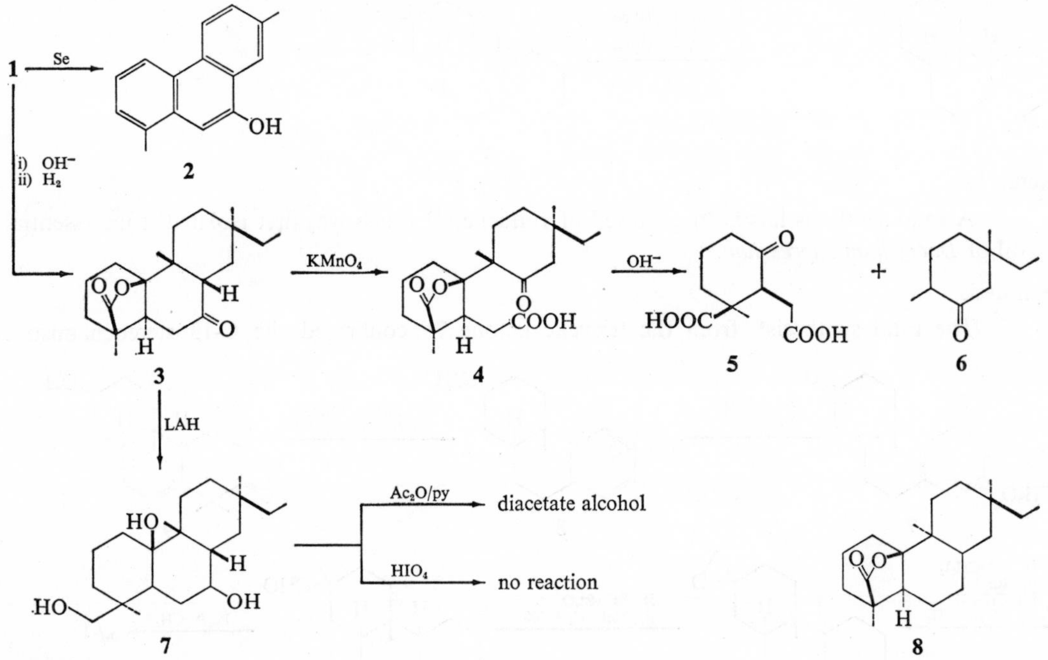

Remarks

Rosenonolactone was isolated from *Trichothecium roseum* (apple mold). The related compounds rosololactone **9** and 6β-hydroxyrosenonolactone **10** are known.

1) A. Harris, A. Robertson, W. B. Whalley, *J. Chem. Soc.*, 1799, 1807 (1955).
2) B. Green, A. Harris, W. B. Whalley, *Chem. Ind.*, 1369 (1958).
3) W. B. Whalley, B. Green, D. Arigoni, J. J. Britt, C. Djerassi, *J. Am. Chem. Soc.*, **81**, 5520 (1959).
4) A. I. Scott, S. A. Southerland, O. W. Young, L. Gugliemetti, D. Arigoni G. A. Sim, *Proc. Chem. Soc.*, 19 (1964).

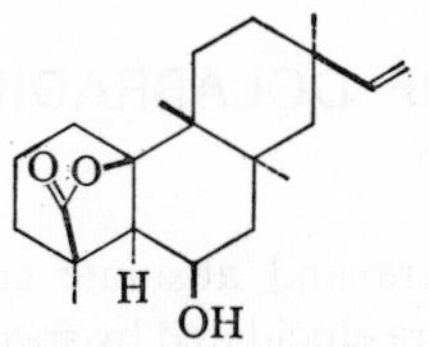

rosololactone[1] **9** 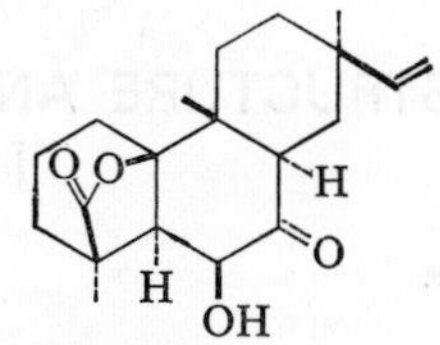   6β-hydroxyrosenonolactone[5] **10**

The rosenonolactone skeleton does not follow the isoprene rule. It has been shown by biosynthetic experiments that it is a diterpene which has undergone a C-10→C-9 methyl migration[6,7] which appears to be concerted with closure of the C ring, and a C-9→C-8 hydride shift.[9] The precise stage at which oxidation occurs remains to be elucidated.

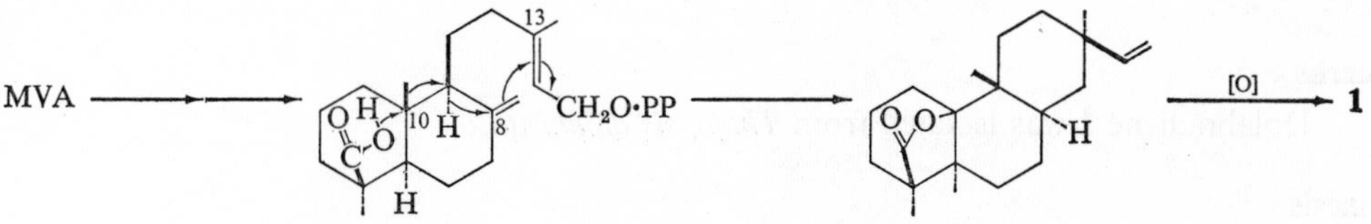

MVA ⟶ ... ⟶ **11** ... **12** ⟶[O]⟶ **1**

Synthesis

The methyl ester of naturally occurring isocupresic acid,[9] **13**, was used as starting material for the synthesis, which is modeled on the biogenesis of **1**.[10]

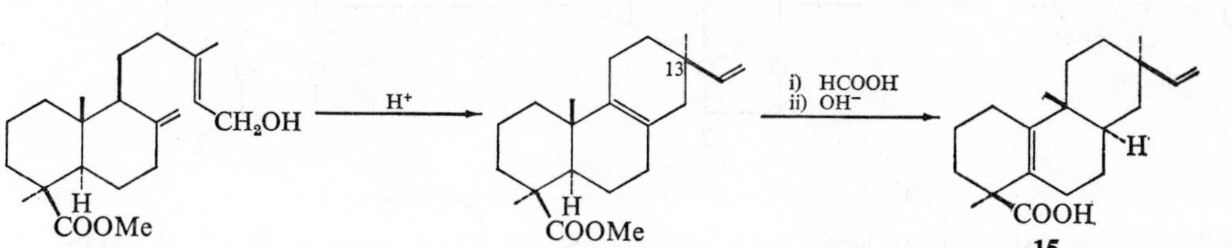

13 →H+→ **14 + 13β-Me** →i) HCOOH ii) OH−→ **15**

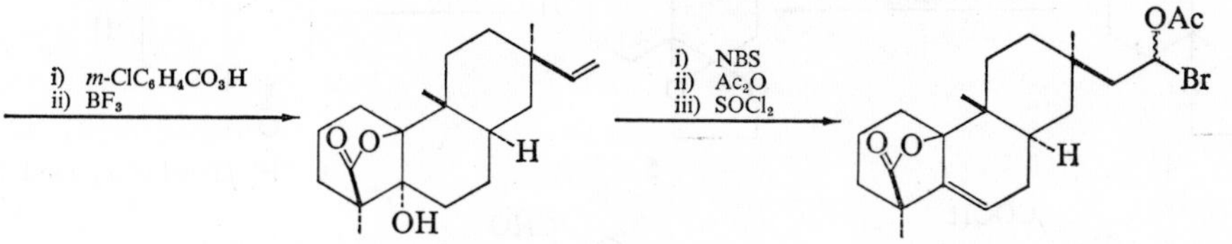

i) m-ClC₆H₄CO₃H ii) BF₃ → **16** →i) NBS ii) Ac₂O iii) SOCl₂→ **17**

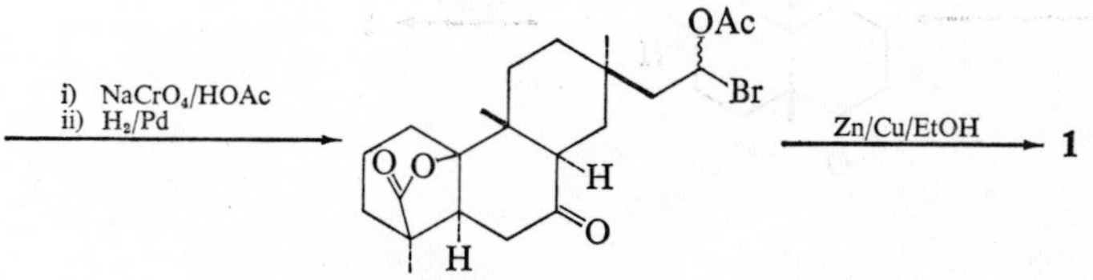

i) NaCrO₄/HOAc ii) H₂/Pd → **18** →Zn/Cu/EtOH→ **1**

5) C. W. Holzapfel, P. S. Steyn, *Tetr.*, **24**, 3321 (1968); A. J. Allison, J. D. Connolly, K. H. Overton, *J. Chem. Soc.* C, 2122 (1968).
6) J. J. Britt, D. Arigoni, *Proc. Chem. Soc.*, 224 (1958).
7) A. J. Birch, R. W. Richards, H. Smith, A. Harris, W. B. Whalley, *Tetr.*, **7**, 241 (1959).
8) B. Achilladeis, J. R. Hanson, *J. Chem. Soc.* C, 2010, (1969).
9) L. Mangoni, M. Bellaridne, *Gazz. Chim. Ital.*, **94**, 1108 (1969).
10) T. McCreadie, K. H. Overton, A. J. Allison, *Chem. Commun.*, 959 (1969).

4.20   STRUCTURE AND SYNTHESIS OF DOLABRADIENE
[Pimarane group]

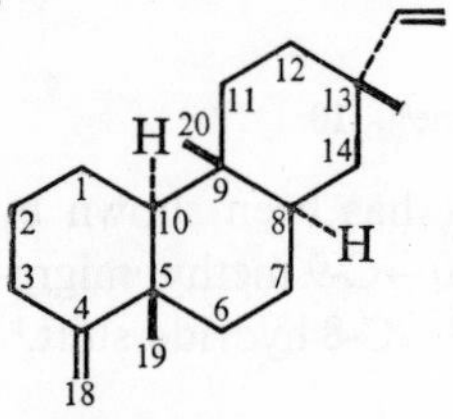

The structure and absolute configuration of dolabradiene were elucidated by spectroscopic analysis of **1** and its various chemical degradation products.[1]

dolabradiene **1**

bp: 169° (7 mm) $\alpha_D{}^{25}$: −70°
ir: 1783, 890 (–C=CH$_2$),
    1820, 988, 908 (–CH=CH$_2$)

Remarks

Dolabradiene **1** was isolated from *Thujopsis dolabrata*.

Synthesis

The synthesis of this compound was accomplished in the following manner.[2]

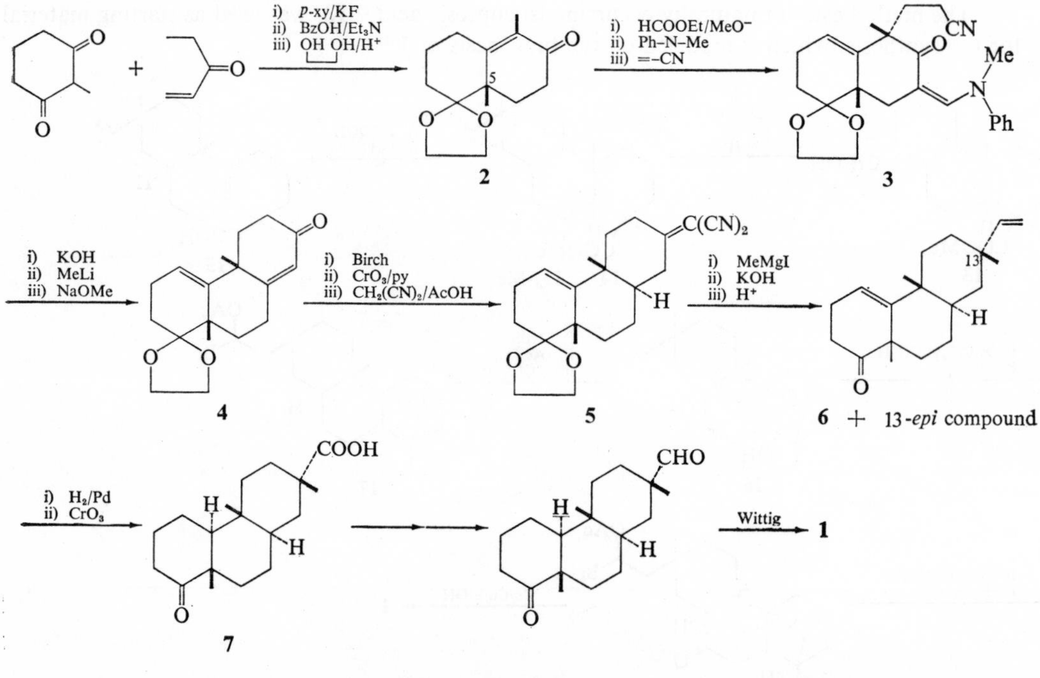

i) *p*-xy/KF
ii) BzOH/Et$_3$N
iii) OH OH/H$^+$

i) HCOOEt/MeO$^-$
ii) Ph–N–Me
iii) =–CN

2

3

i) KOH
ii) MeLi
iii) NaOMe

i) Birch
ii) CrO$_3$/py
iii) CH$_2$(CN)$_2$/AcOH

i) MeMgI
ii) KOH
iii) H$^+$

4

5

6 + 13-*epi* compound

i) H$_2$/Pd
ii) CrO$_3$

Wittig ⟶ **1**

7

The stereospecific addition of acrylonitrile to **2** can be rationalized by the blocking effect of the 5α methyl group. The (±)-keto acid **7** was resolved as its quinidine methohydroxide salt and the synthesis was completed by relay with this compound derived from the natural product. This accounts for the circuitous introduction of the vinyl function.

1) Y. Kitahara, A. Yoshikoshi, *Tetr.*, **26**, 1753 (1964).
2) Y. Kitahara, A. Yoshikoshi, S. Oida, *Tetr.*, **26**, 1763 (1964).

4.21 SYNTHESIS OF (-)-SANDARACOPIMARIC ACID [Pimarane group]

This synthesis is a conversion of a steroid, testosterone, into $(-)$-sandaracopimaric acid,[1,2] a diterpenoid resin acid isolated from *Callitris quadrivalvis*.

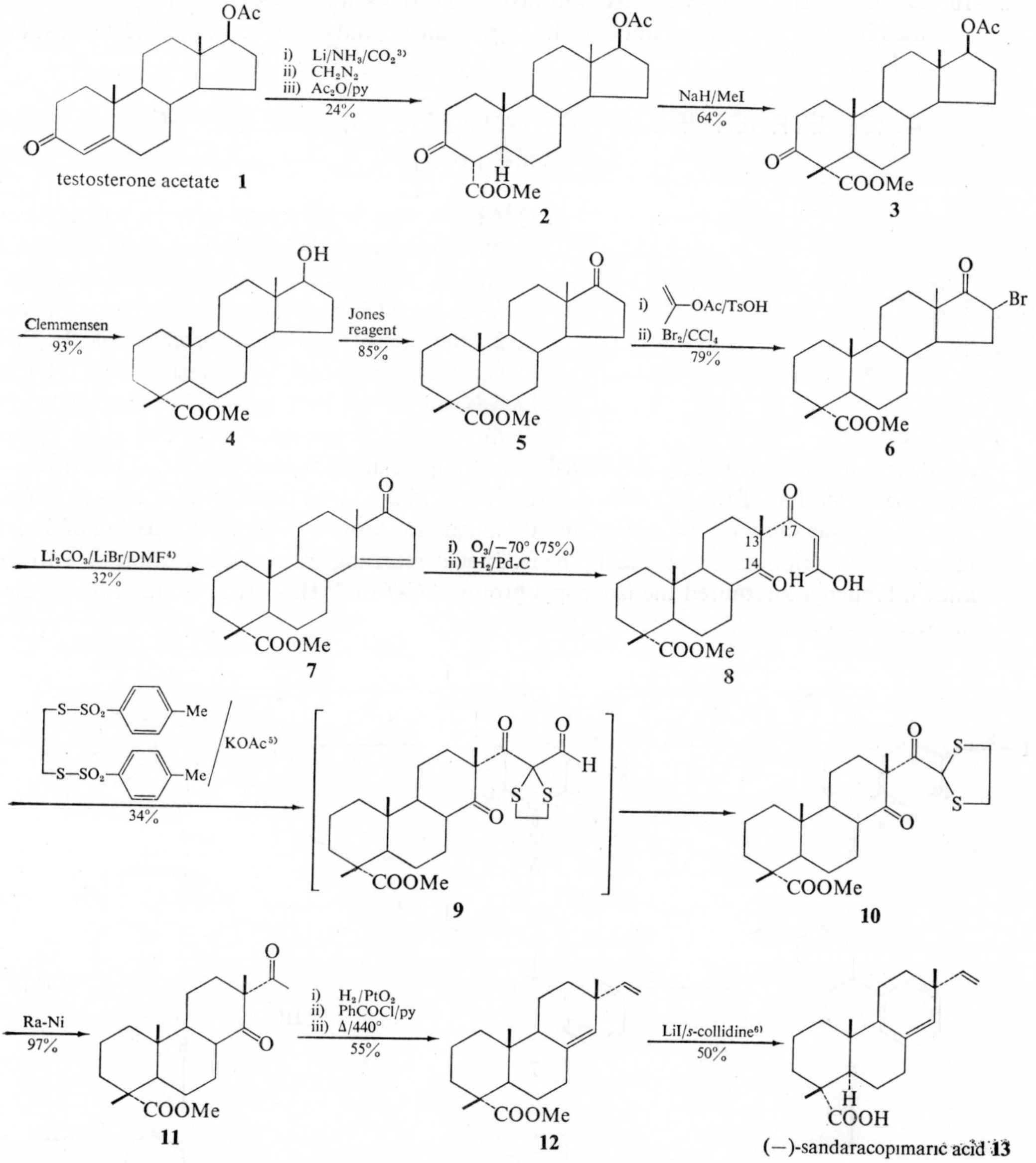

testosterone acetate **1**

i) Li/NH₃/CO₂ [3]
ii) CH₂N₂
iii) Ac₂O/py
24%

2

NaH/MeI
64%

3

Clemmensen
93%

4

Jones reagent
85%

5

i) ⟩—OAc/TsOH
ii) Br₂/CCl₄
79%

6

Li₂CO₃/LiBr/DMF [4]
32%

7

i) O₃/−70° (75%)
ii) H₂/Pd-C

8

—S—SO₂—⟨ ⟩—Me / —S—SO₂—⟨ ⟩—Me   KOAc [5]
34%

9

10

Ra-Ni
97%

11

i) H₂/PtO₂
ii) PhCOCl/py
iii) Δ/440°
55%

12

LiI/s-collidine [6]
50%

$(-)$-sandaracopimaric acid **13**

1) A. Afonso, *J. Org. Chem.*, **35**, 1949 (1970); *J. Am. Chem. Soc.*, **90**, 7375 (1968).
2) O. E. Edwards, A. Nicholson, M. N. Roger, *Can. J. Chem.*, **38**, 663 (1960); V. Galik, J Kalhan, F. Petru, *Chem. Ind.*, 722 (1960); A. K. Bose, *ibid.*, 1104 (1960).
3) G. Stork, P. Rosen, N Goldman, R. V. Coombs, J. Tsuji, *J. Am. Chem. Soc.*, **87**, 275 (1965).
4) A. Afonso, *Can. J. Chem.*, **47**, 3693 (1969).
5) R. B. Woodward, A. A. Patchett, D. H. R. Barton, D. A. J. Ives, R. B. Kelly, *J. Chem. Soc.*, 1131 (1957).
6) F. Elsinger, J. Schreiber, A. Eschenmoser, *Helv. Chim. Acta.*, **43**, 113 (1960).

1→2:   The Stork procedure for reductive carboxymethylation was employed. **3** is the expected
        product of stereoelectronically controlled axial alkylation.

6→7:   The deconjugated ketone **7** is the major product of dehydrobromination of **6**.

7→8:   This conversion involves the formation of a stable crystalline ozonide and its catalytic
        hydrogenolysis to the hydroxymethylene ketone **8** (positive ferric chloride test).

8→10: Decarbonylation of **8** was accomplished by the Woodward method. This procedure was
        used in order to circumvent bond cleavages which could occur between C-13 and C-14 or
        C-13 and C-17.

4.22   STRUCTURE AND SYNTHESIS OF ABIETIC ACID
[Abietane group]

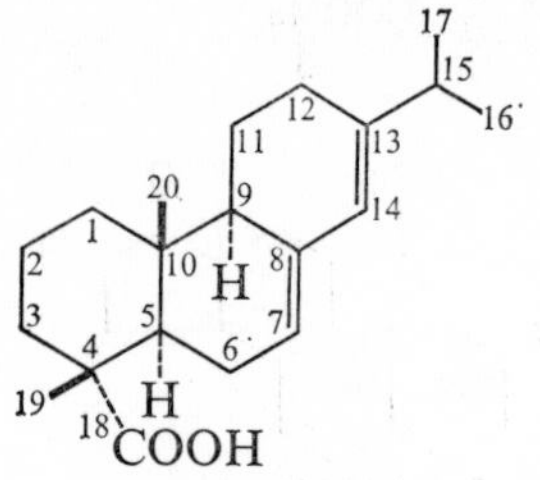

abietic acid **1**

1) Abietic acid is extremely difficult to obtain in
pure form and this hampered early structural
studies. After confirmation of the molecular
formula as $C_{20}H_{30}O_4$, the identification of retene
2, the dehydrogenation product of **1**, then
became the focus of investigations. Retene
was shown to be 1-methyl-7-isopropylphenan-
threne.[2,3] The substitution of two alkyl groups
on the terminal ring(s) and the location of one group adjacent to the central ring was deduced
from the formation of the fluorenone carboxylic acid **5** and its base cleavage to acid **6**. One
alkyl group was shown to be methyl and the other isopropyl by the transformation of **3** to **4**,
involving oxidation accompanied by benzilic acid rearrangement. Degradation of **4** to the
known biphenyl **9** required the isopropyl group at C-7 in **2**. Haworth's synthesis confirmed
the structure of retene.[4]

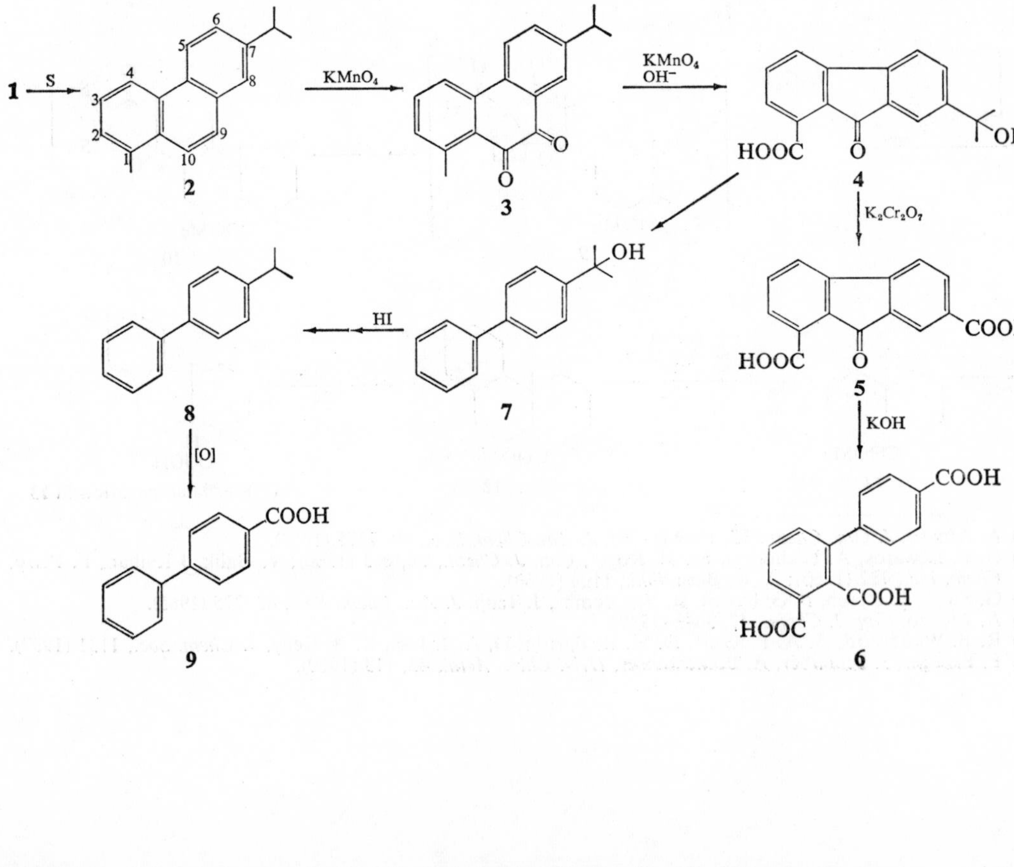

2) Dehydrogenation to retene implies the presence in abietic acid of three rings and two double bonds. The loss of two carbon atoms in this process can be accounted for by the elimination of carboxyl and quarternary methyl groups. The presence of two double bonds was substantiated by catalytic hydrogenation to tetrahydroabietic acid and formation of a dihydrobromide,[5] while their conjugation was first indicated by reaction with diazotized p-nitroaniline (a qualitative test for conjugated dienes) and later confirmed by observation of a uv maximum at 240 nm (4.15).[6]

3) Permanganate oxidation of abietic acid[7] led to two tribasic acids **10** and **11**, which were dehydrogenated to homomellitene and m-xylene respectively. This required a 1,3 relationship of the ring A methyl groups and suggested the presence of carboxyl at C-4 and a double bond between C-7 and C-8. Vigorous oxidation of abietic acid to yield isobutyric acid indicated attachment of the isopropyl group to a double bond.

4) The location of the conjugated diene system was established by transformation of **1** to 9-keto-8-azo-dehydroabietic acid **15** and thence to 8-azaretene **16**.[8]

5) The location of the carboxyl group, finally determined to be C-4, was obscured for some time by the fact that while abietenol **18** was dehydrogenated to retene **2**, dehydration of **18** to **19** followed by dehydrogenation gave homoretene **20**, due to rearrangement during the dehydration.[9]

6) In assigning relative stereochemistry, the lack of optical activity in **10** and **11** indicated a cis-1,3 relationship of the two methyl groups and two of the carboxyl groups. The thermodynam-

1) J. Simonsen, D. H. R. Barton, *The Terpenes*, vol. III, p. 374–382, Cambridge University Press, 1952.
2) E. Bamberger, S. Hooker, *Ann. Chem.*, **229**, 114 (1885); *Chem. Ber.*, **18**, 1750 (1885).
3) J. E. Bucher, *J. Am. Chem. Soc.*, **32**, 374 (1910).
4) R. D. Haworth, *J. Chem. Soc.*, 1784 (1932).
5) P. Levy, *Ann. Chem.*, **81**, 148 (1913).
6) L. F. Fieser, W. P. Campbell, *J. Am. Chem. Soc.*, **60**, 159 (1938).
7) L. Ruzicka, J. Meyer, M. Pfeiffer, *Helv. Chim. Acta*, **8**, 637 (1925).
8) L. Ruzicka, L. Sternbach, *Helv. Chim. Acta*, **21**, 565 (1938); *ibid.*, **23**, 333, 341, 355 (1940); *ibid.*, **24**, 492 (1941); *ibid.*, **25**, 1036 (1942).
9) L. Ruzicka, J. Meyer, *Helv. Chim. Acta*, **5**, 581 (1922); L. Ruzika, G. B. R. de Groff, J. H. Muller, *ibid.*, **5**, 1300 (1922); R. D. Haworth, *J. Chem. Soc.*, 2717 (1932).

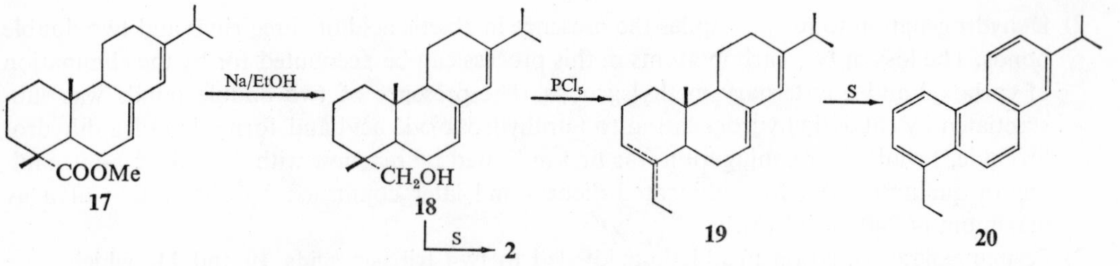

17   → Na/EtOH → 18   → PCl₅ → 19   → S → 20

COOMe (17)   CH₂OH (18)   S → 2

ic dissociation constants of **11** showed the central carboxyl to be *trans* to the other two, a situation demanding a *trans* A/B ring junction.[10] The remaining asymmetric center at C-9 was determined through correlation with cholesta-3,5-diene by the method of molecular rotation differences.[11]

Remarks

Abietic acid **1** occurs widely as the major constituent of commercial resins, which are prepared from the crude oleoresins of conifers by heat and mineral acid treatment. The primary resin acids, levopimaric, neoabietic, pimaric and isopimaric acids, undergo facile isomerisation to abietic acid under these conditions.[1] **1** is also a primary constituent of many conifer resins.

Synthesis

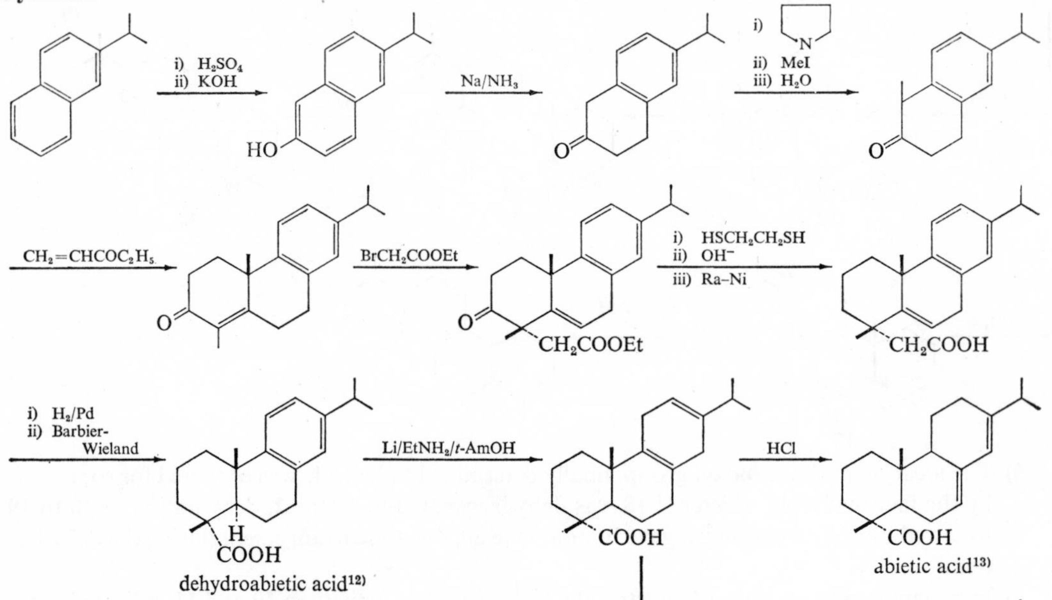

i) H₂SO₄   ii) KOH

Na/NH₃

i) (pyrrolidine) ii) MeI iii) H₂O

CH₂=CHCOC₂H₅

BrCH₂COOEt

i) HSCH₂CH₂SH ii) OH⁻ iii) Ra–Ni

CH₂COOEt   CH₂COOH

i) H₂/Pd ii) Barbier–Wieland

H / COOH
dehydroabietic acid[12]

Li/EtNH₂/t-AmOH

COOH

HCl

COOH
abietic acid[13]

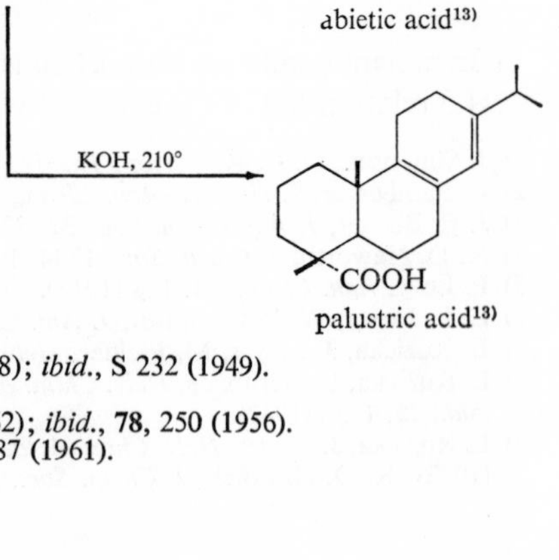

KOH, 210°

COOH
palustric acid[13]

Studies by Stork[12] and Burgstahler[13] constitute a total synthesis of abietic acid (above). An alternative synthesis has been executed.[14]

10) D. H. R. Barton, G. A. Schmeidler, *J. Chem. Soc.*, 1197 (1948); *ibid.*, S 232 (1949).
11) W. Klyne, *J. Chem. Soc.*, 3072 (1953).
12) G. Stork, J. W. Schulenberg, *J. Am. Chem. Soc.*, **84**, 284 (1962); *ibid.*, **78**, 250 (1956).
13) A. W. Burgstahler, L. R. Worden, *J. Am. Chem. Soc.*, **83**, 2587 (1961).
14) R. Ireland, R. C. Kierstead, *J. Org. Chem.*, **27**, 703 (1962).

4.23 STRUCTURE OF LEVOPIMARIC, NEOABIETIC AND PALUSTRIC ACIDS
[Abietane group]

levopimaric acid **1**

1) Levopimaric acid **1** affords retene on dehydrogenation.[1] The presence of the same carbocyclic skeleton as in abietic acid was deduced from its facile isomerization to abietic acid upon treatment with heat or acid.[2]

2) Hydrogenation[3] and peracid titration[4] showed that **1** contained two double bonds which constituted a homoannular diene as judged from the uv at 272.5 nm.[5] An earlier observation[6] that **1** reacts quantitatively at room temperature with maleic anhydride (see **2**) provided another indication of the presence of a homocyclic diene system.

3) Production of isobutyric acid upon ozonolysis[7] showed that one of the double bonds bears the isopropyl group. The positions of the two double bonds were established by converting the Diels-Alder adduct **2** to anthraquinone-1,3-dicarboxylic acid **4**.[8]

4) As both abietic acid and **1** yield the same Diels-Alder adduct, the stereochemistry at ring junctions was deduced to be identical. The 9-H orientation was proven chemically.[9]

5) The strongly negative ord Cotton effect showed that the two double bonds constitute a counterclockwise helix.[10]

1) L. Ruzicka, Fr. Balas, Fr. Willim, *Helv. Chim. Acta*, **7**, 458 (1924).
2) G. Dupon, *Bull. Soc. Chim. Fr.*, **29**, 718, 727 (1921).
3) L. Ruzicka, R. C. R. Bacon, *Helv. Chim. Acta*, **20**, 1542 (1937).
4) K. Kraft, *Ann. Chem.*, **524**, 1 (1936).
5) G. C. Harris, T. F. Sanderson, *J. Am. Chem. Soc.*, **70**, 334 (1948).
6) R. G. R. Bacon, L. Ruzicka, *Chem. Ind.*, **55**, 546 (1936).
7) L. Ruzicka, R. G. R. Bacon, R. Lukes, J. S. Rose, *Helv. Chim. Acta*, **21**, 583 (1938).
8) B. A. Arbuzov, *Bull. Acad. Sci. U. R. S. S., Classe Sci. Chim.*, 95 (1940); cf. *Chem. Abst.*, **35**, 2898 (1941).
9) W. H. Schuller, R. V. Lawrence, *J. Am. Chem. Soc.*, **83**, 2563 (1961).
10) A. W. Burgstahler, H. Ziffer, U. Weiss, *J. Am. Chem. Soc.*, **83**, 4660 (1961).

Remarks

The primary resin **1** was first isolated in a pure state from French galipot.[11] All three acids (**1, 5** and **6**) are contained in oleoresin of *Pinus palustris* (Pinaceae) as primary constituents, together with pimaric acid and isopimaric acid.

Neoabietic acid is conveniently prepared by heating abietic acid at 300°. Palustric acid has been synthesized from dehydroabietic acid.[14]

neoabietic acid **5**   palustric acid **6**

4.24   STRUCTURE AND SYNTHESIS OF FERRUGINOL
[Abietane group]

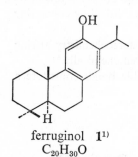

ferruginol **1**[1]
$C_{20}H_{30}O$

1) Ferruginol forms a formate, acetate and benzoate. It is slightly soluble in aq. NaOH and gives a deep red color with diazotized sulfanilic acid. The phenolic nature of **1** is further corroborated by the 277.5 uv maximum of **1** acetate.

2) Catalytic hydrogenation results in the loss of a hydroxyl group and formation of a saturated hydrocarbon[2] with molecular formula $C_{20}H_{36}$, indicating a tricyclic structure.

3) The carbocyclic skeleton was identified by Se dehydrogenation of **2** to retene **3**; dehydrogenation of **1** gave 6-hydroxyretene **4** (erroneously considered to be 8-hydroxyretene in initial investigations).

4 $\xleftarrow{\text{Se}}$ **1** $\xrightarrow{\text{H}_2/\text{Pt}}$ **2** $\xrightarrow{\text{Se}}$ retene **3**

Remarks

The Miro tree *Podocarpus ferrugineus*, endemic to New Zealand, bleeds a soft, sticky resin, the main constituent of which is ferruginol.[1]

4.23

11) A. Vesterberg, *Chem. Ber.*, **20**, 3248 (1887).
12) G. C. Harris, T. F. Sanderson, *J. Am. Chem. Soc.*, **70**, 339 (1948).
13) W. H. Schuller, R. W. Moore, R. V. Lawrence, *J. Am. Chem. Soc.*, **82**, 1734 (1960); V. M. Loeblich, D. E. Baldwin, R. V. Lawrence, *J. Am. Chem. Soc.*, **77**, 2823 (1955).
14) A. W. Burgstahler, L. R. Worden, *J. Am. Chem. Soc.*, **83**, 2587 (1961).

4.24

1) C. W. Brandt, L. G. Neubauer, *J. Chem. Soc.*, 1031 (1939).
2) W. P. Campbell, D. Todd, *J. Am. Chem. Soc.*, **62**, 1287 (1940).

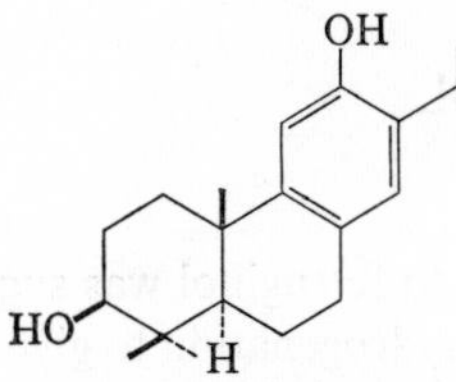

hinokiol **5**

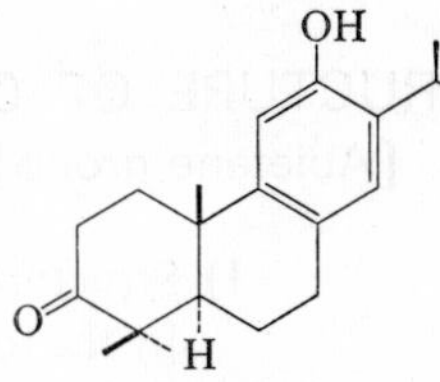

hinokione **6**

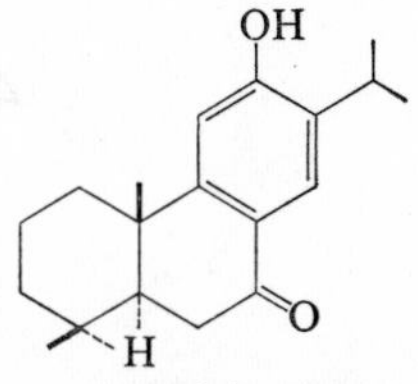

sugiol **7**

Related compounds include hinokiol **5** and hinokione **6** which are derived from *Chamaecyparis obtusa* ("hinoki" in Japanese) and other conifer species.[3,4] The structure[4] and synthesis[5] have been investigated. Sugiol **7** has been isolated from *Cryptomeria japonica* D. Don ("sugi" in Japanese).[6]

Synthesis

Route I

The partial synthesis of **1** from 6-hydroxydehydroabietic acid shown below confirmed the structure **1**.[7,8]

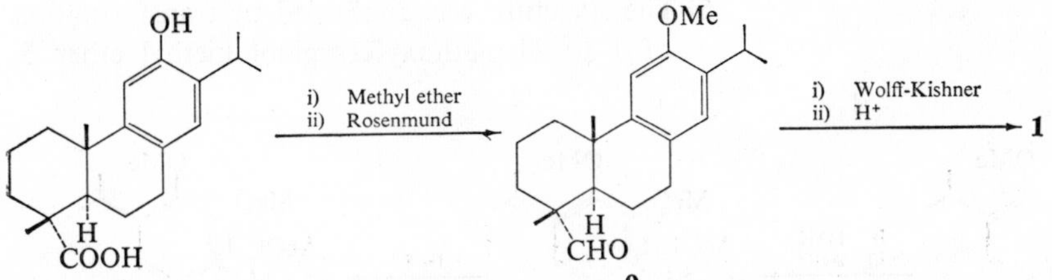
OH
COOH
i) Methyl ether
ii) Rosenmund
OMe
CHO
i) Wolff-Kishner
ii) H⁺
1

8 **9**

Route II

Total synthesis of **1** has been achieved.[9] Route II shows a recent synthesis.[10]

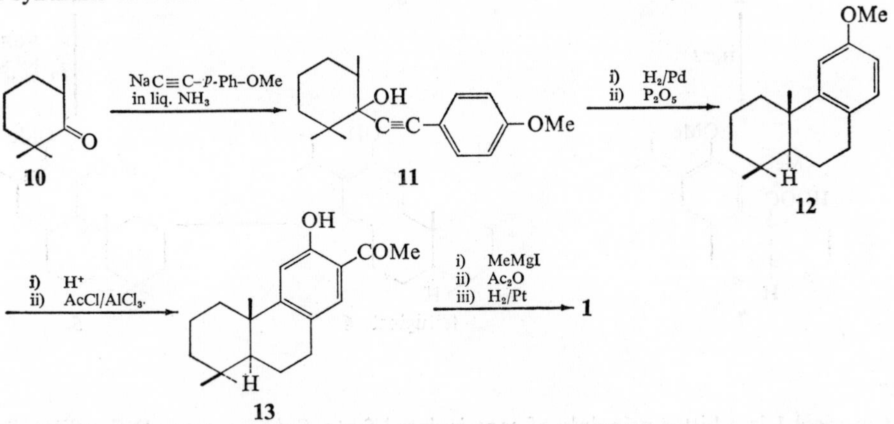
NaC≡C-p-Ph-OMe
in liq. NH₃
OH
C≡C
OMe
i) H₂/Pd
ii) P₂O₅
OMe
i) H⁺
ii) AcCl/AlCl₃
OH
COMe
i) MeMgI
ii) Ac₂O
iii) H₂/Pt
1

10 **11** **12** **13**

---

3) Y. Yoshiki, T. Ishiguro, *Yakugaku Zasshi*, **53**, 73 (1933).
4) Y. Chow, H. Erdtman, *Acta Chem. Scand.*, **16**, 1296 (1962).
5) Y. Chow, *Acta Chem. Scand.*, **16**, 1301 (1962).
6) K. Keimatsu, T. Ishiguro, G. Fukui, *Yakugaku Zasshi*, **57**, 92 (1937).
7) W. P. Campbell, D. Todd, *J. Am. Chem. Soc.*, **64**, 928 (1942).
8) L. F. Fieser, W. P. Campbell, *J. Am. Chem. Soc.*, **61**, 2528 (1939).
9) F. E. King, T. J. King, J. G. Topliss, *J. Chem. Soc.*, 573 (1957).
10) M. Ohashi, T. Maruishi, H. Kakisawa, *Tetr. Lett.*, 719 (1968).

4.25 STRUCTURE OF CARNOSOL
[Abietane group]

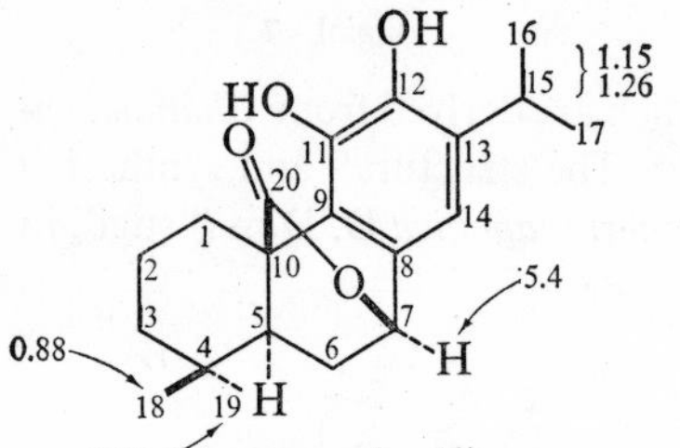

0.88
18 19
0.93
carnosol **1**[1]
$C_{20}H_{26}O_4$
mp: 221–226° (AcOH)

OH 16 } 1.15 / 1.26
15
17
;5.4

1) Structural similarity to ferruginol was suggested by the following: dehydrogenation to give retene and alkaline peroxide oxidation to isobutyric acid; presence in the nmr of aromatic, isopropyl and two additional methyl signals.

2) The ir absorption at 1745 cm⁻¹ in **1** and its diacetate and dimethyl ether suggests γ or δ lactone.

3) The benzylic *O*-terminus of the lactone is indicated by the nmr signal at 5.4 and by the hydrogenolysis of **2** to **7**.

4) Decarboxylation at 230° shows the presence of benzylic carboxyl.[2]

5) The structure was confirmed by transformation of **1** to 11-methoxyferruginol methyl ether **5**.

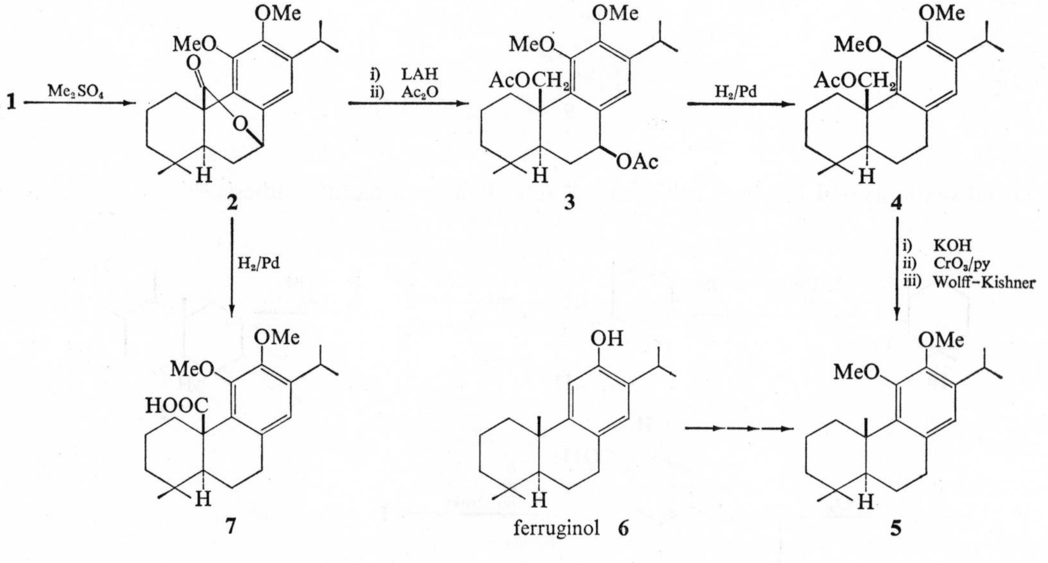

Remarks

Carnosol **1** is a bitter principle of sage isolated from *Salvia carnosa*,[1] *S. officinalis* and *S. triloba*,[2] and from *Rosmarinus officinalis* (Labiatae).[3]

A total synthesis of carnosol has been reported.[4]

1) A. I. White, G. L. Jenkins, *J. Am. Pharm. Assoc., Sci. Ed.*, **31**, 33, 37 (1942).
2) C. H. Brieskorn, A. Fuchs, *Chem. Ber.*, **95**, 3034 (1962).
3) C. H. Brieskorn, A. Fuchs, J. B. Bredenberg, J. D. McChesney, E. Wenkert, *J. Org. Chem.*, **29**, 2293 (1964).
4) D. C. Shew, W. L. Meyer, *Tetr. Lett.*, 2963 (1968).

4.26 STRUCTURE OF CALLICARPONE
[Abietane group]

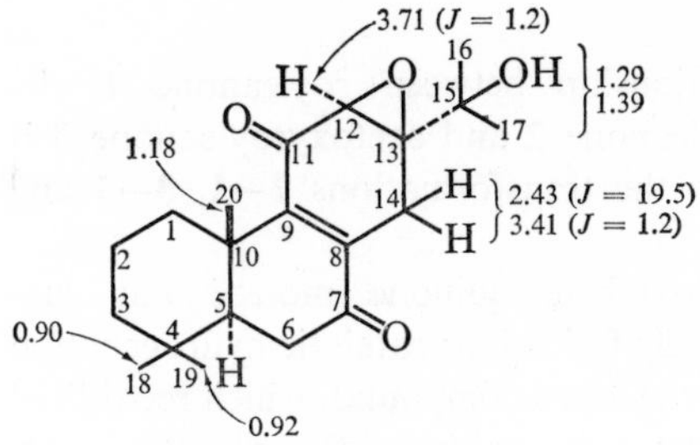

callicarpone **1**[1]

1) The presence of five tertiary methyl groups was shown by nmr.

2) Tertiary hydroxyl was shown by ir (CCl₄) at 3375 cm⁻¹: dehydration gives a tetrasubstituted olefin, ir (CCl₄) at 1640 cm⁻¹, but with no olefinic signals in the nmr.

3) Epoxide was suggested by the 1-H doublet at 3.71 ($J = 1.2$) and lack of alkoxyl or *sec* hydroxyl group; treatment with pyridine hydrochloride gives the chlorohydrin **2**.

4) Ir (CCl₄) at 1700–1675 cm⁻¹ in **1** changes to 1707 and 1679 cm⁻¹ in **2**, suggesting the presence of ene-1,4-dione (uv: 266.5 nm (ε 6.689)) adjacent to the epoxide.

5) The tetraol **3** has two *sec* and two *tert* hydroxyls, since it gives the diacetate **4**. Lead tetraacetate cleavage of **3** gives acetone, thus placing the hydroxy isopropyl group on the epoxide ring.

6) **1** was transformed to the known methoxyferruginol methyl ether **8**[2] via cryptojaponol **7**.[3]

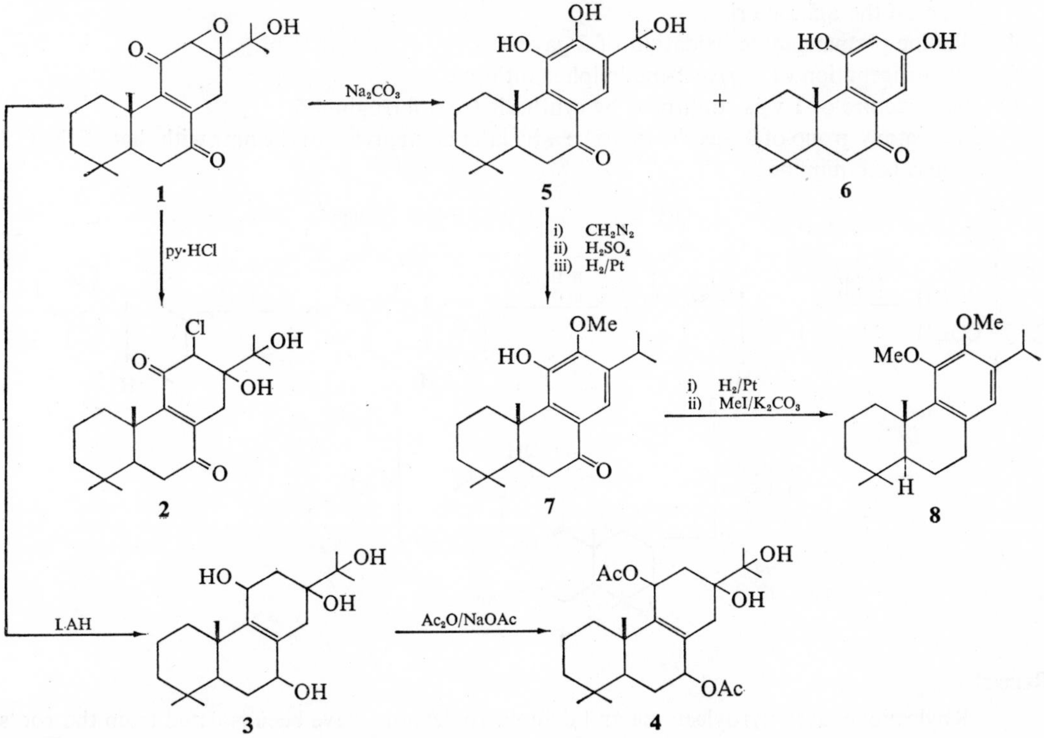

Remarks

The potent fish poison **1** was isolated from *Callicarpa candicans*.

1) K. Kawazu, T. Mitsui, *Tetr. Lett.*, 3519 (1966); *cf. C. A.* **67**, 9065 (1967); *C. A.* **67**, 54275 (1967).
2) C. H. Brieskorn, A. Fuchs, J. B. Bredenberg, J. D. McChesney, E. Wenkert, *J. Org. Chem.*, **29**, 2293 (1964).
3) T. Kondo, M. Suda, M. Teshima, *J. pharm. Soc., Japan*, **82**, 1252 (1962); *cf. C.A.* **59**, 1685 (1962).

4.27 STRUCTURE OF ROYLEANONES, TAXODIONE AND FUERSTION

[Abietane group]

royleanone **1**[1]

mp: 179–181° α_D: +134°
uv: 403 (ε 510), 283 (ε 15,000), 277 (ε 15,900)
ir (CHCl₃): 3350, 1672, 1632

1) The relationship between royleanone **1**, dehydroroyleanone **2** and acetoxyroyleanone **3** is shown by the transformations **2→1**, **3→1** and **3→2**.

2) The hydroxybenzoquinone moiety was suggested by the following: catalytic hydrogenation of **1** gave the leuco-compound, which reoxidized to **1** in air; **1** is weakly acidic (pK_a 8.5 in aq. MeOH); treatment with diazomethane gave a monomethyl ether;   shows hydrogen-bonded hydroxyl (3350 cm⁻¹), α,β-unsat. carbonyl (1672 cm⁻¹) and hydrogen-bonded α,β-unsat. carbonyl (1632 cm⁻¹); the uv is similar to that of other hydroxy-p-benzoquinones.

3) Alkaline peroxide treatment of **1** gave isobutyric acid, showing the presence of an isopropyl group on the quinone ring.

4) Alkaline permanganate oxidation of **2** gave **4**.

5) Dehydrogenation of **1** gave 1-methylphenanthrene.

6) The structure of **1** was confirmed by synthesis from ferruginol **5**.

7) The acetoxy group of **3** was shown to be α by later comparison of the nmr with that of the C-7 epimer, taxoquinone.[2]

Remarks

Royleanone, acetoxyroyleanone and dehydroroyleanone have been isolated from the roots of *Inula royleana* D. C. which grows in the western Himalayas.[1] Royleanone has also been found in the seeds of *Taxodium distichium* from the south-eastern United States. These seeds also contain taxodione, taxodone and taxoquinone (7β-hydroxyroyleanone).[2]

1) O. E. Edwards, G. Feniak, M. Los, *Can. J. Chem.*, **40**, 1540 (1962).
2) S. M. Kupchan, A. Karim, C. M. Marcks, *J. Org. Chem.*, **34**, 3912 (1969).

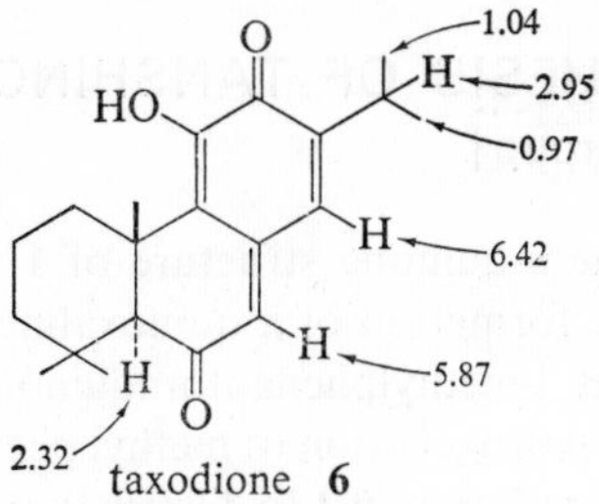

taxodione **6**
mp : 115–116°

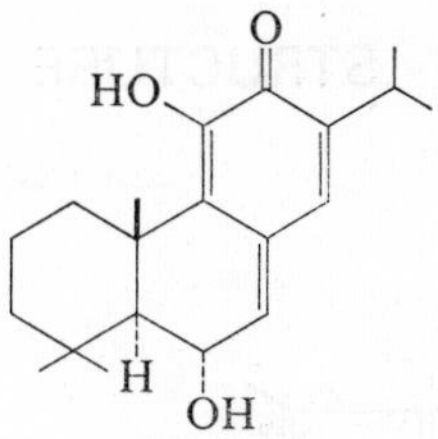

taxodone **7**
mp : 164–165°

The structures of the tumor inhibitory quinone methides, taxodione **6** and taxodone **7**, were determined chiefly by spectroscopic methods.[2] The structures were confirmed by the transformation of **8**, derived from either **6** or **7**, to 11,12-dimethoxyabieta-8,11,13-triene **11**, previously derived from sugiol **12**.[3] The α-configuration of the C-5 hydrogen in **6** and **7** was indicated by the ease of dehydration of **10** and the failure of **9** to epimerize under acidic conditions.

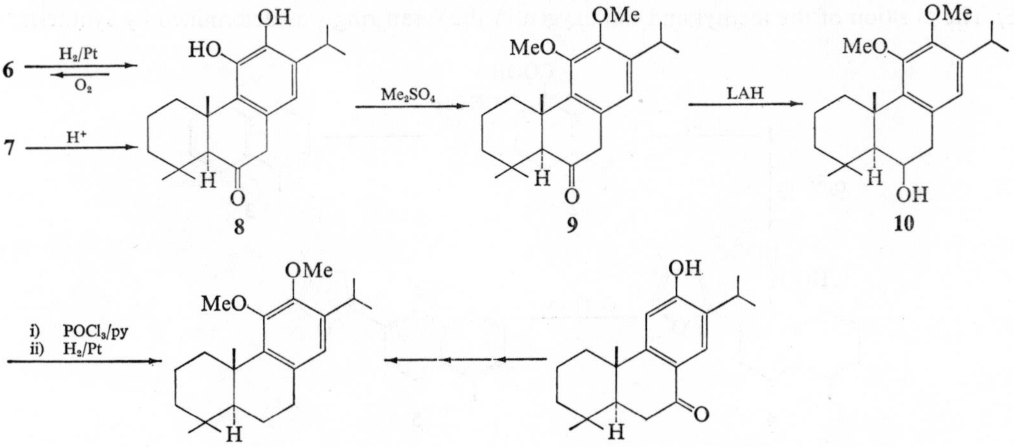

8   **9**   **10**

11   **12**

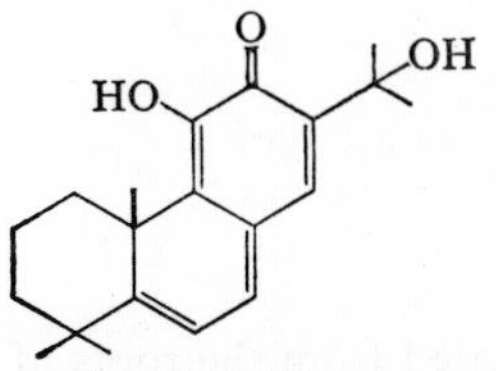

fuerstion **13**

A formal total synthesis of taxodione **6** from podocarpic aci¹ has been carried out.[4] The structure of the closely related quinone methide, fuerstion **13**, has also been determined.[5]

3) C. H. Brieskorn, A. Fuchs, J. B. Bredenberg, J. D. McChesney, E. Wenkert, *J. Org. Chem.*, **29**, 2293 (1964)
4) K. Mori, M. Matsui, *Tetr.*, **26**, 3467 (1970).
5) D. Karanatsios, J. S. Scarpa, C. H. Eugster, *Helv. Chim. Acta.*, **49**, 1151 (1966).

4.28   STRUCTURE AND SYNTHESIS OF TANSHINONES
[Abietane group]

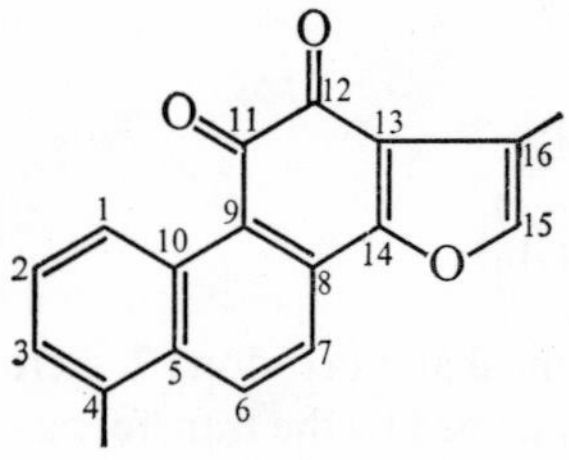

tanshinone I **1**

1) The *o*-quinone structure of **1** was indicated by the formation of a quinoxaline.[1-3]
2) The l-methylphenanthraquinone was shown by de..ydrogenation to methyl phenanthrene and by degradation of **1** to **2** (structure by synthesis) and **3**.
3) The position of the *o*-quinone moiety was determined by degradation of **1** to **6**.
4) The methyl furan moiety was deduced from the unreactivity of the remaining oxygen function, the resistance of **1** to hydrogenation after uptake of 1 mole of hydrogen (quinone reduction), and the presence of two methyl groups (Kuhn-Roth determination).
5) The position of the methyl and the oxygen in the furan ring was determined by synthesis.[4]

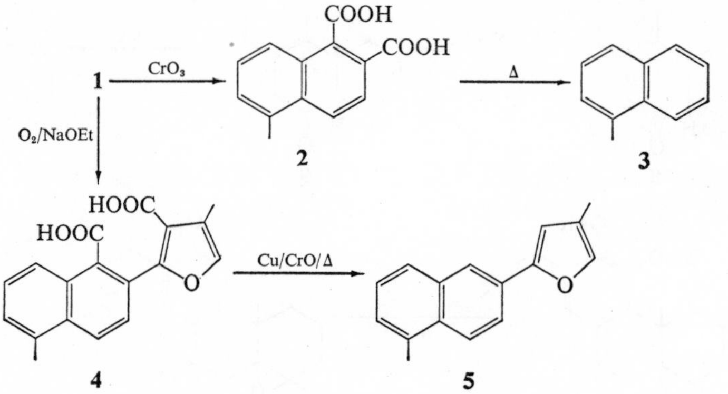

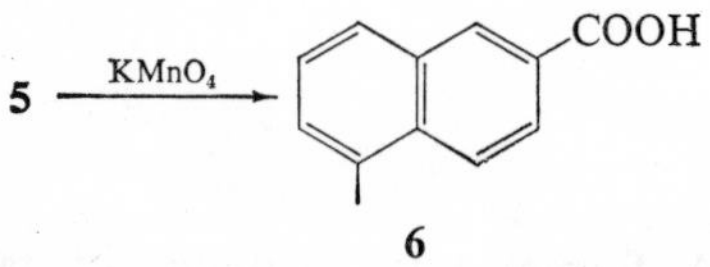

Remarks

The tanshinones comprise a number of pigments isolated from the roots of *Salvia miltiorrhiza* (Labiatae),[1] having a common norditerpenoid *o*-quinone structure.

Synthesis

The synthesis of cryptotanshinone **13** is illustrative of a general synthetic approach to tanshinones.[5]

1) M. Nakao, T. Fukushinia, *Nippon Kagaku Zasshi*, **54**, 154 (1934).
2) Y. Wessely, S. Wang, *Chem. Ber.*, **73-B**, 19 (1940).
3) K. Takuira, *Nippon Kagaku Zasshi*, **63**, 41 (1943), in C. A. **44**, 7280 (1924).
4) G. C. Baillie, R. H. Thompson, *J. Chem. Soc.* C, 48 (1968).
5) Y. Inouye, H. Kakisawa, *Bull. Chem. Soc. Japan*, **42**, 3318 (1969); *cf. Chem. Commun.*, 1327 (1967).

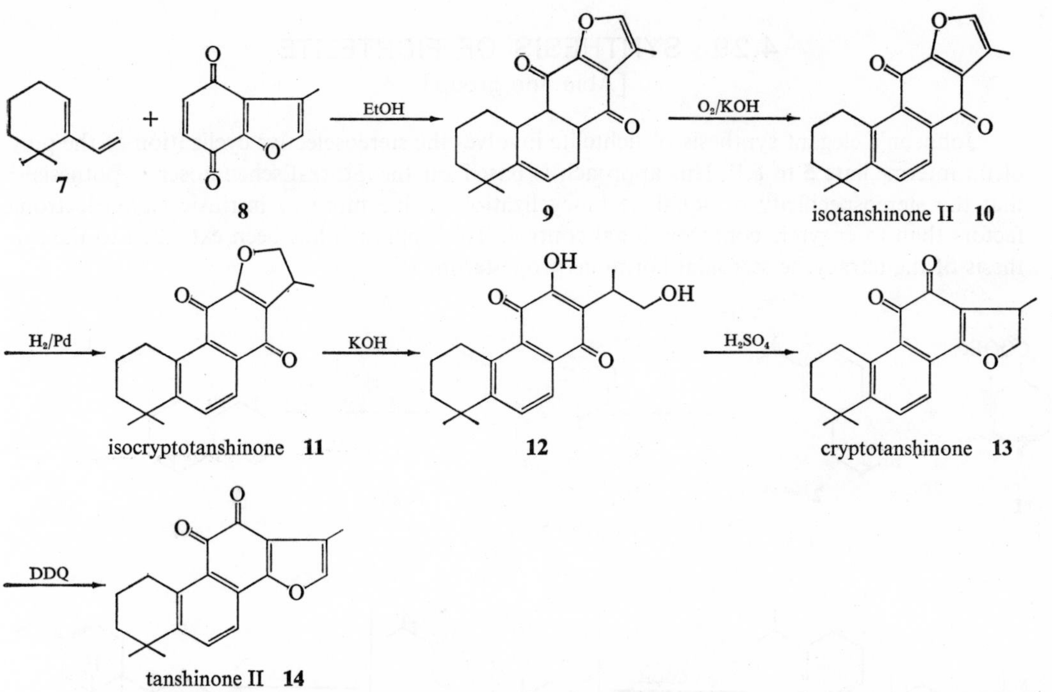

isotanshinone II 10

isocryptotanshinone 11   12   cryptotanshinone 13

tanshinone II 14

The diene **7** adds readily to the more electrophilic double bond of **8**. The orientation in **9** is indicated by the identity of **10**, **11**, **13** and **14** with the natural products, and is that predicted on the basis both of steric requirements and the relative stabilities of the diradical intermediates.

Tanshinone I **1** was synthesized in an analogous manner using *o*-methylstyrene in place of **7**.

4.29   SYNTHESIS OF FICHTELITE
[Abietane group]

Johnson's elegant synthesis of fichtelite involves the stereoselective cyclization of the poly-olefin intermediate **5** to **8**.[1] This approach is based on the Stork-Eschenmoser hypothesis[2, 3] that the stereospecificity of squalene biocyclization is due more to intrinsic stereoelectronic factors than to enzymic conformational control. This approach has been extended to the synthesis of the tetracyclic steroidal hormone progesterone.[4]

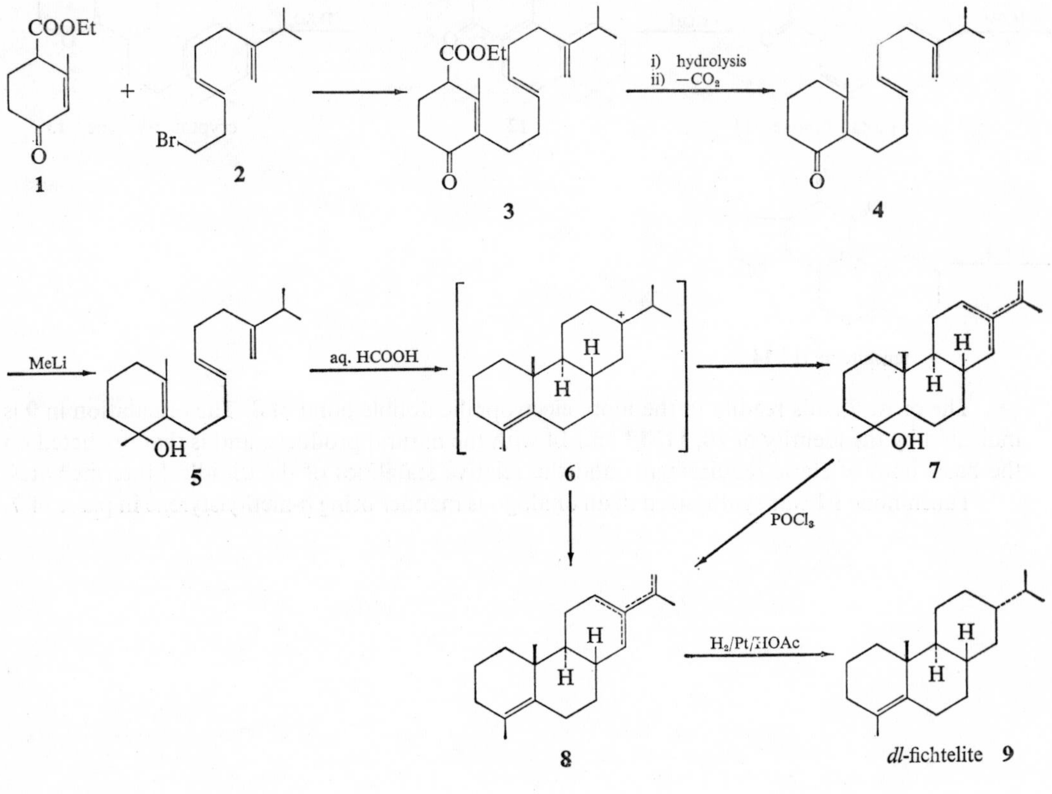

1) W. S. Johnson, N. P. Jensen, J. Hooz, *J. Am. Chem. Soc.*, **88**, 3859 (1966).
2) G. Stork, G. W. Burgstahler, *J. Am. Chem. Soc.*, **77**, 5068 (1955).
3) P. Eschemoser, L. Ruzicka, O. Jeger, D. Arigoni, *Helv. Chim. Acta.*, **38**, 1890 (1955).
4) W. S. Johnson, A. G. Gravestock, B. E. McCarry, *J. Am. Chem. Soc.*, **93**, 433 (1971).

4.30 STRUCTURE AND SYNTHESIS OF TOTAROL
[Totarane group]

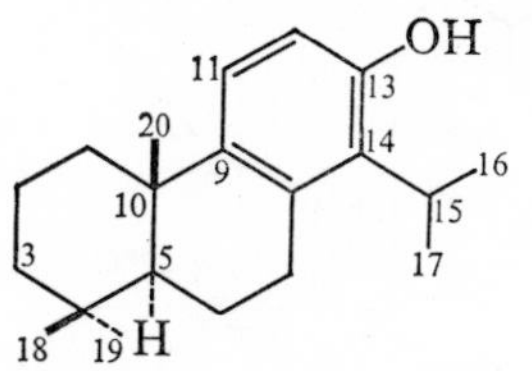

totarol **1**

mp: 132° α_D^{20} (EtOH):+41.3°
uv: 280–285 (3.29)

1) The structure of **1** is based on the following:[1]
presence of phenolic OH, indicated by the formation of substitution products (bromototarol, formyltotarol, etc.); synthetic characterization of the dehydrogenation products (**3**, **4**, etc); no significant amount of HCHO or acetone upon ozonolysis.

2) 5α· , 10β-Me in **1**[2,3] were indicated because totarolone **2**, which shows a large positive Cotton effect similar to that shown by lanost-8-en-3-one, was converted into **1**.

3) Chemical proof for 5α-H, 10β-Me was obtained from comparison of the degradation product **6** with that obtained from dehydroabietic acid.[2,3]

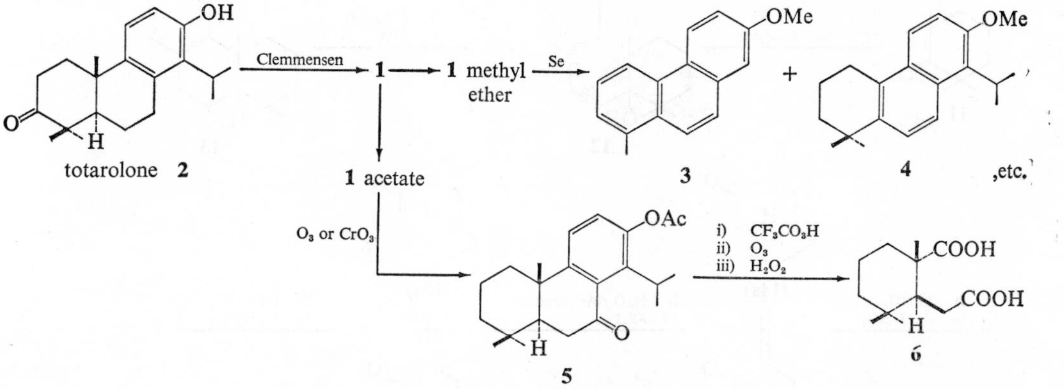

totarolone **2**   **3**   **4**   ,etc.

1 acetate

5   **6**

Remarks

The source of **1** is *Podocarpus totara* (Podocarpaceae). The related compounds **7** and **8** are known.[4,5] **7** was obtained from *Tetraclinus articulata* and **8** from *Cupressus sempervirens*. Total synthesis of **1** has been achieved.[6-8]

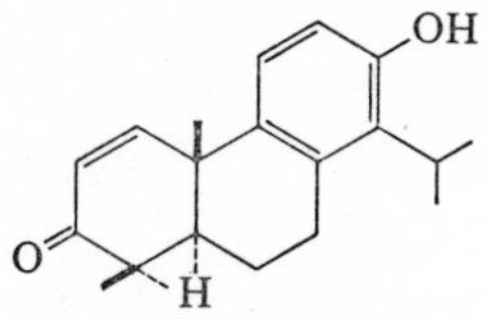

totarelenone **7**[4] 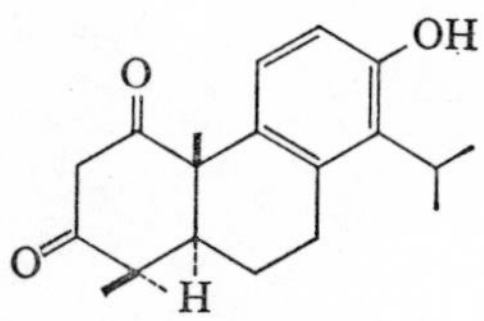   **8**[5]

1) W. F. Short, H. Wang, *J. Chem. Soc.*, 2979 (1951).
2) Y. -L. Chow, H. Erdtman, *Acta Chem. Scand.*, **14**, 1852 (1960).
3) Y. -L. Chow, H. Erdtman, *Acta Chem. Scand.*, **16**, 1305 (1962).
4) Y. -L. Chow, H. Erdtman, *Acta Chem. Scand.*, **14**, 1852 (1960).
5) L. Mangoni, M. Belardini, *Tetr. Lett.*, 2643 (1964).

The totarol structure does not follow the isoprene rule. On the basis of its co-occurrence with ferruginol 9, it has been suggested that the totarol skeleton is formed through intermediate 10[9] or by 1,2-migration of the isopropyl group upon breakdown of an abietane type (e.g. ferruginol 9) dimer.[10]

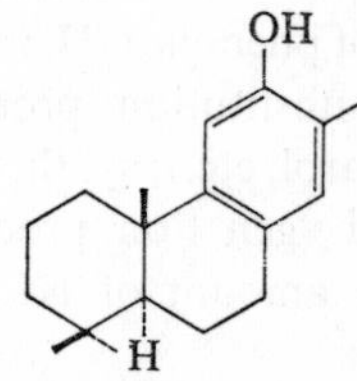

ferruginol 9 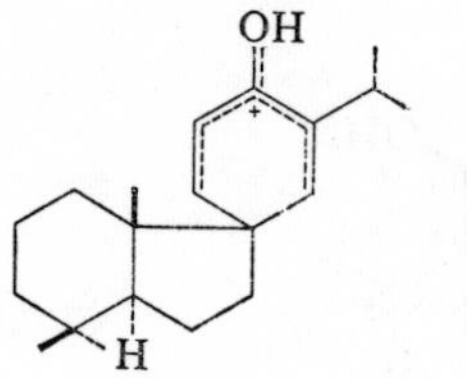   10

Synthesis

The versatile synthetic intermediate 14[6] was employed[7] in the above synthesis. 18-Oxygenated totarols (CH$_2$OH, CHO, and COOH) were synthesized from optically active 18-oxygenated compounds corresponding to 14a.[8]

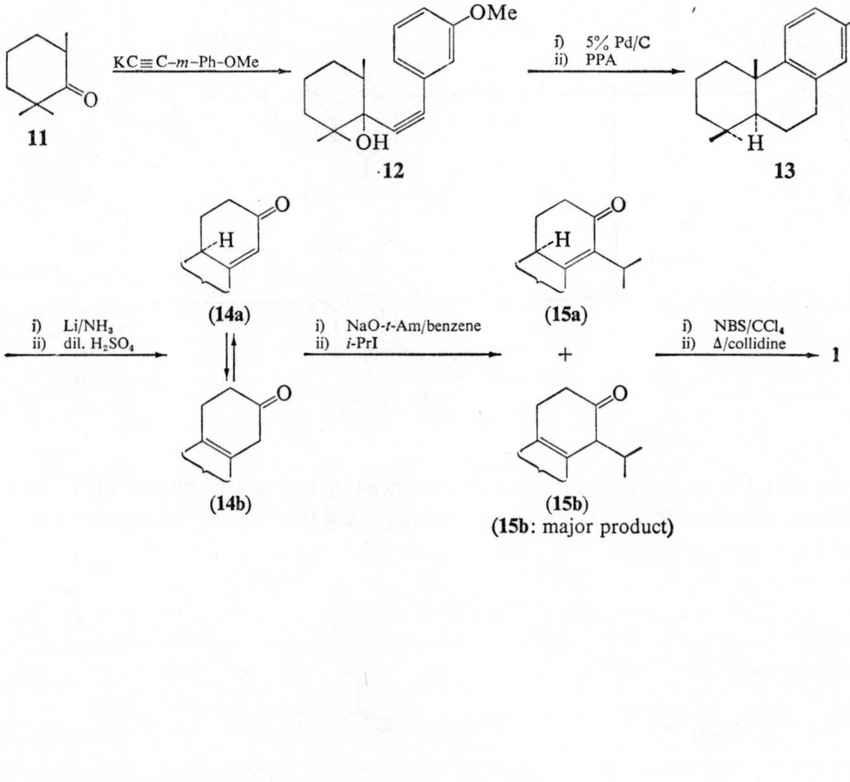

(15b: major product)

6) J. A. Barltrop, N. A. J. Rogers, *Chem. Ind.*, 20 (1957).
7) J. A. Barltrop, N. A. J. Rogers, *J. Chem. Soc.*, 2566 (1958).
8) A. C. Day, *J. Chem. Soc.*, 3001 (1964).
9) E. Wenkert, B. C. Jackson, *J. Am. Chem. Soc.*, **80**, 211 (1958).
10) A. W. Johnson, T. J. King, R. J. Martin, *J. Chem. Soc.*, 4420 (1961).

4.31 STRUCTURE OF INUMAKILACTONES AND RELATED LACTONES
[Totarane group]

Inumakilactone A[1]

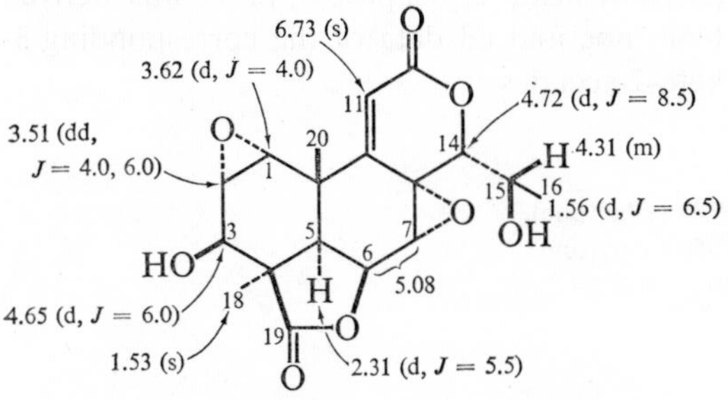

6.73 (s)
3.62 (d, $J = 4.0$)
3.51 (dd, $J = 4.0, 6.0$)
4.72 (d, $J = 8.5$)
4.31 (m)
1.56 (d, $J = 6.5$)
HO
5.08
4.65 (d, $J = 6.0$)
1.53 (s)
2.31 (d, $J = 5.5$)

inumakilactone A **1**[1]

mp: 251–253° (dec.) α_D: ±0°
ir (KBr): 3460 (OH), 1760 (γ-lactone), 1705, 1640
($\Delta^{\alpha,\beta}$-lactone)
uv (EtOH): 220 (ε 11,000)

1) $J_{5,6} = 4.8$ and $J_{6,7} = 1.3$ in **2** gave the 5,6-*cis*-6,7-*trans* structure.
2) The *cis* relation between 18-H and 3-H was deduced from the positive noe (18-H→3-H).
3) The α-7,8-epoxide structure was found from the negative Cotton effect of the α,β-epoxy ketone **4**.
4) 14β-H was indicated by the large upfield shifts of 7-H upon acetylation (0.64) and oxidation (0.81ppm) of 15-OH.
5) 15S configuration was determined from the benzoate rule and Horeau's asymmetric synthesis.

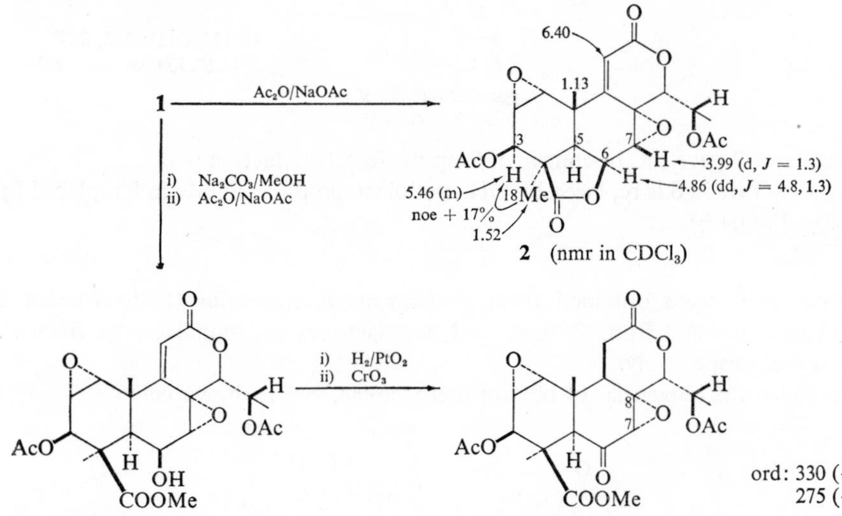

6.40
1.13
AcO
3.99 (d, $J = 1.3$)
4.86 (dd, $J = 4.8, 1.3$)
5.46 (m)
noe + 17%
1.52
2 (nmr in CDCl₃)

1 — Ac₂O/NaOAc →

i) Na₂CO₃/MeOH
ii) Ac₂O/NaOAc

AcO
OH
COOMe
3

i) H₂/PtO₂
ii) CrO₃

AcO
COOMe
4

ord: 330 (−1180)
275 (+3060)

Other inumakilactones[2]

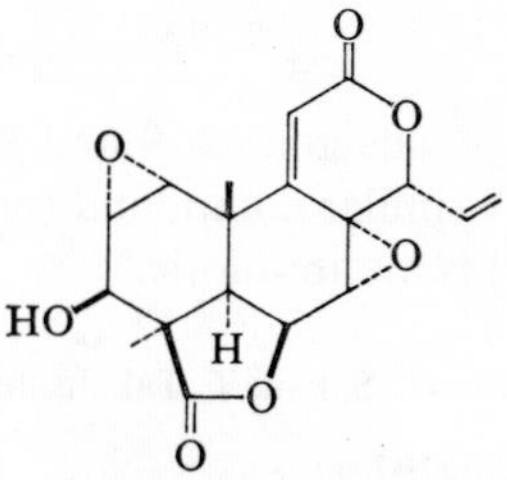

HO
inumakilactone B **5**

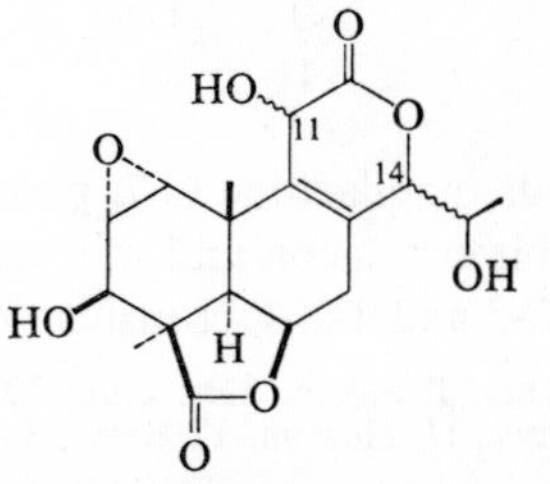

HO
11
14
OH
HO
inumakilactone C **6**
(11,14-substituents: α-*cis* or β-*cis*)

1) S. Ito, M. Kodama, M. Sunagawa, T. Takahashi, H. Imamura, O. Honda, *Tetr. Lett.*, 2065 (1968).
2) S. Ito, M. Sunagawa, M. Kodama, H. Honma, T. Takahashi, *Chem. Commun.*, 91 (1971).

Nagilactones A—D[3,4]

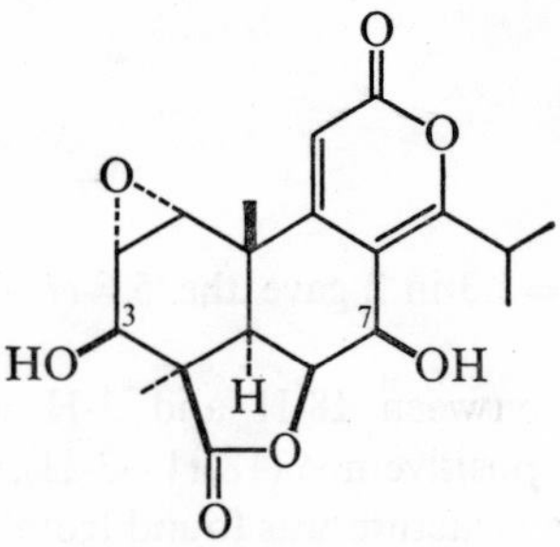

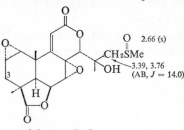

nagilactone C **7**

1) All the nagilactones (A–D) possess an α-pyrone structure.
2) The stereochemistry of nagilactones C **7** and D (ethyl instead of isopropyl in **7**) was derived from noe and cd data of the corresponding 3-keto-7-acetates.

ir (nujol): 3400 (OH), 1770 (γ-lactone),
1710, 1640, 1550 (α-pyrone)
uv (EtOH): 300 (ε 5,200)

Podolactones A–D[5,6] and ponalactones[7]

podolactone C **8**
(nmr in C₅D₅N)

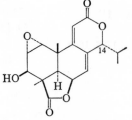

ponalactone A   **9**
(also β-glucoside)

uv (MeOH): 262, 227
(ε 12,500, sh. *ca.* 7000)

3) The 7-monoacetate of **7** gave the 14-epimer of **9** upon NaBH₄ reduction.[7]
4) In podolactone A, SOMe in **8** is replaced by OH. Podolactone B has SOMe in **8** replaced by OH, and also has 3β-OH.[5,6]

Remarks

Inumakilactones A–C were obtained from *Podocarpus macrophyllus* (Podocarpaceae). Nagilactones A–D were isolated from *P. nagi*, and podolactones and ponalactones from *P. neriifolius* and *P. nakaii*, respectively.

A possible route for the biogenesis of the norditerpene-lactones is shown below.[3]

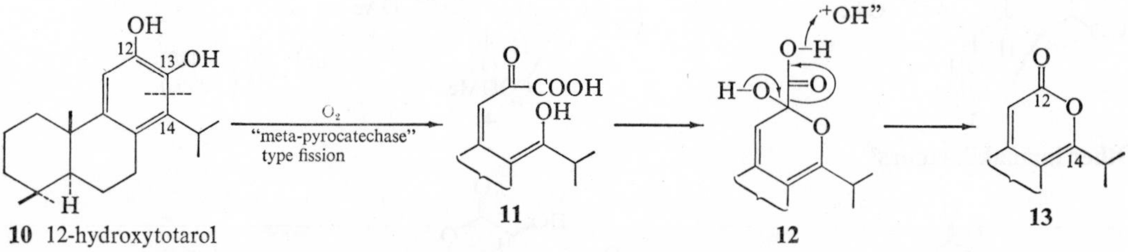

10 12-hydroxytotarol   **11**   **12**   **13**

Some of these compounds (nagilactone C **7**, podolactones A,B,[5] ponalactone A **9** and β-glucoside of **9**[7]) strongly inhibit expansion and mitosis of plant cells. A similar compound (no substituents at C-1 through C-3 and 14α-OH instead of isopropyl in **9**) is an antibiotic.[8]

3) Y. Hayashi, S. Takahashi, H. Ona, T. Sakan, *Tetr. Lett.*, 2071 (1968).
4) S. Ito, M. Kodama, M. Sunagawa, H. Honma, Y. Hayashi, S. Takahashi, H. Ona, T. Sakan, T. Takahashi, *Tetr. Lett.*, 2951 (1969).
5) M. N. Galbraith, D. H. S. Horn, J. M. Sasse, D. Adamson, *Chem. Commun.*, 170 (1970).
6) M. N. Galbraith, D. H. S. Horn, J. M. Sasse, *Chem. Commun.*, 1362 (1971).
7) S. Ito, M. Kodama, M. Sunagawa, M. Koreeda, K. Nakanishi, *Chem. Commun.*, 855 (1971).
8) G. E. Ellestad, R. H. Evans Jr., M. P. Kunstman, J. E. Lancaster, G. O. Morfin, *J. Am. Chem. Soc.*, **92**, 5483 (1970).

4.32   STRUCTURE AND SYNTHESIS OF PODOTOTARIN
(BITOTAROL)
[Totarane group]

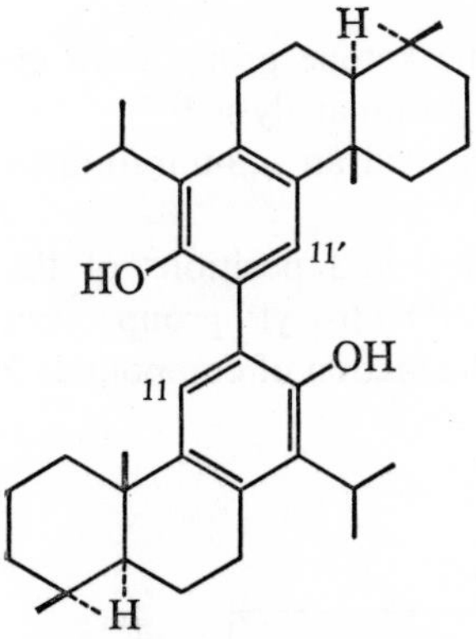

podototarin **1**[1,2]
mp: 225–226° α_D^{24} +76°
ir (CS$_2$): 3472 (OH)

1) The 3472 cm^{-1} ir absorption was assigned to nonbonded and sterically hindered phenolic groups.

2) A weak uv band at 254 nm (ε 12,700) which is absent in totarol suggested a biphenyl system with restricted rotation.

3) Chemical shifts of 11,11'-H at 7.00 in nmr suggested a biphenyl dihedral angle of about 70° (based on a curve calculated for biphenyls by Jackman).

Remarks

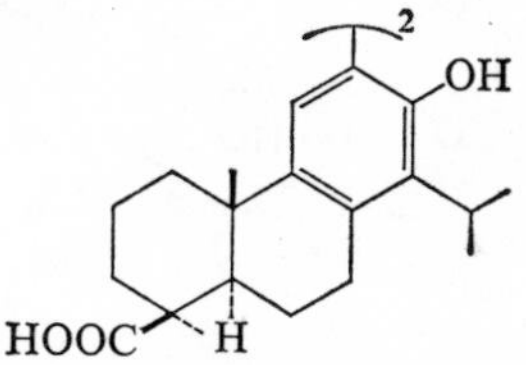

macrophyllic acid **2**[3]

1 was obtained from *Podocarpus totara*.

A similar compound, macrophyllic acid **2**, has been obtained from *Podocarpus macrophyllus*. **2** was synthesized from the monomeric methyl ester by treatment with KOH/K$_3$[Fe(CN)$_6$] followed by hydrolysis.[3,4]

Synthesis

Route I[1]

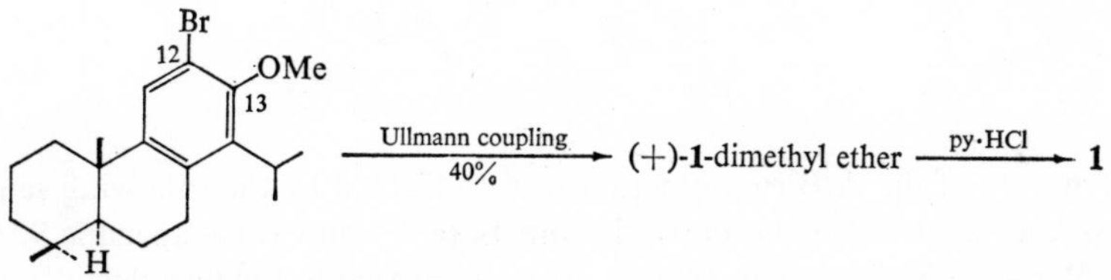

3

$\xrightarrow[\text{40\%}]{\text{Ullmann coupling}}$ (+)-**1**-dimethyl ether $\xrightarrow{\text{py·HCl}}$ **1**

Route II[2]

totarol $\xrightarrow[\text{43\%}]{\text{KOH/K}_3\text{[Fe(CN)}_6\text{]}}$ **1**

1) R. C. Cambie, W. R. J. Simpson, L. D. Colebrook, *Chem. Ind.*, 1757 (1962).
2) C. P. Falshaw, A. W. Johnson, T. J. King, *J. Chem. Soc.*, 2422 (1963).
3) S. M. Bocks, R. C. Cambie, T. Takahashi, *Tetr.*, **19**, 1109 (1963).
4) A. C. Day, *Chem. Ind.*, 1760 (1963).

4.33   STRUCTURE OF CASSAIC ACID
[Cassane group]

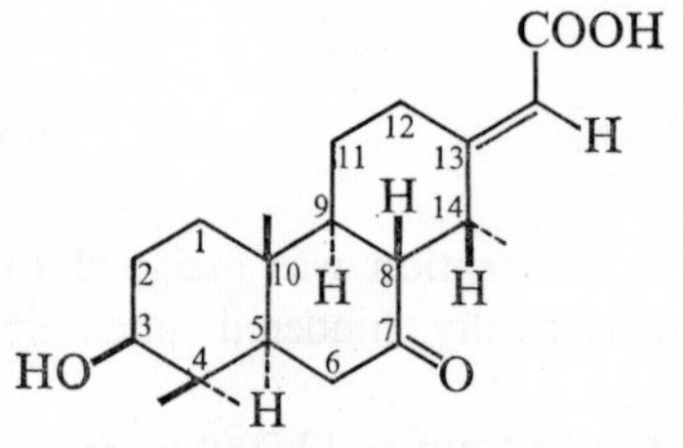

cassaic acid   **1**[1-3]
mp: 203° α_D^{20} (MeOH): $-126.3°$
uv: 215 (4.3)

1) One hydroxyl and one ketone group were indicated by functional group analysis.[2]

2) The uv spectrum suggested an α,β-unsaturated acid.[3]

3) The nuclear structure,[3] and positions of the carboxyl,[4] carbonyl and hydroxyl[5] groups were determined from the formation of compounds **2** to **7** from **1**.

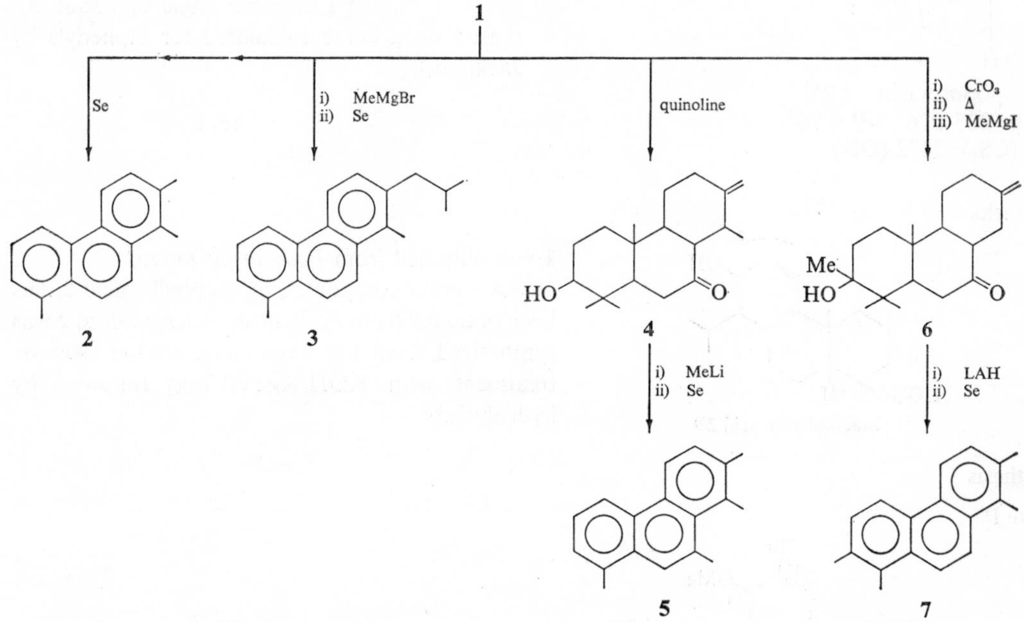

4) The *trans* nature of the A/B ring junction was confirmed by the following sequence of reactions. Cassanic acid **8** can be derived from both **1**[6] and vouacapenane **9**.[7] A compound related to **9**, methyl vinhaticoate **10**, was dehydrogenated to the tetradehydro derivative **12**.[8] This in turn was oxidized to the tricarboxylic acid **13** whose absolute configuration was already known through chemical and optical correlations with abietic acid.[9]

1) G. Dalma, *Ann. Chim. Appl.*, **25**, 569 (1935).
2) G. Dalma, *Helv. Chem. Acta*, **22**, 1497 (1939).
3) L. Ruzicka, G. Dalma, *Helv. Chim. Acta*, **22**, 1516 (1939).
4) L. Ruzicka, B. G. Engel, A. Ronco, K. Beise, *Helv. Chim. Acta*, **28**, 1038 (1945).
5) W. J. Gensler, G. M. Sherman, *J. Am. Chem. Soc.*, **81**, 5217 (1959).
6) L. Ruzicka, *Helv. Chim. Acta*, **23**, 757 (1940).
7) F. E. King, T. J. King, J. M. Uprichard, *J. Chem. Soc.*, 3428 (1958)
8) F. E. King, T. J. King, *J. Chem. Soc.*, 4158 (1953).
9) D. H. R. Barton, G. A. Schmeidler, *J. Chem. Soc.*, 1197 (1948).

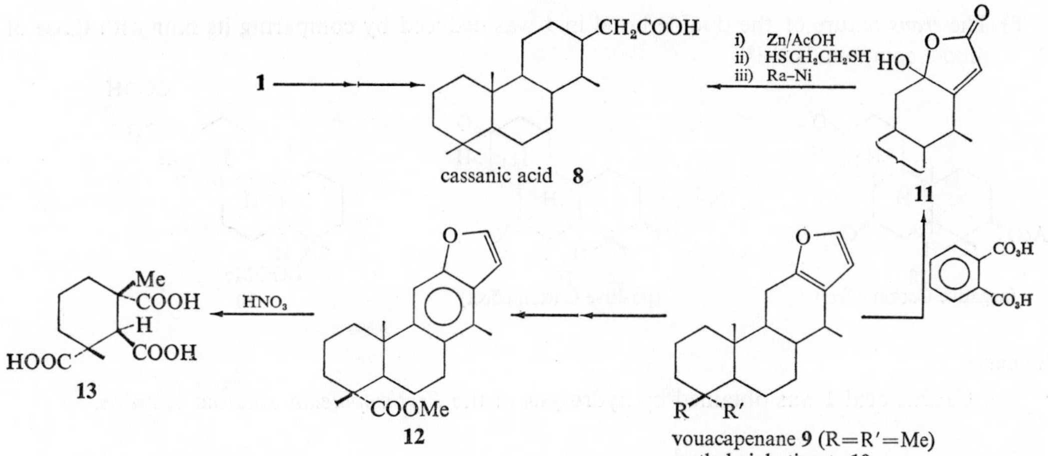

i) Zn/AcOH
ii) HSCH₂CH₂SH
iii) Ra–Ni

cassanic acid **8**

11

HNO₃

13

12

vouacapenane **9** (R=R′=Me)
methyl vinhaticoate **10**
(R=Me, R′=COOMe)

5) An equatorial orientation was deduced[11] for the 3-OH on the basis of a retropinacol re-arrangement of ring A in **1**.[10] This was supported by the following reasoning: The LAH reduction of synthetic **14** results in equatorial 3-OH. Hydrogenation of **15**, by analogy to the transformation of **17a** to **17**[12] should yield a *trans* A/B ring junction and the equatorial 3-OH must be β. Then, the transformation of **15** to **16**, which can also be derived from **1**, suggests the 3-OH in cassaic acid to be β-oriented.[11,12]

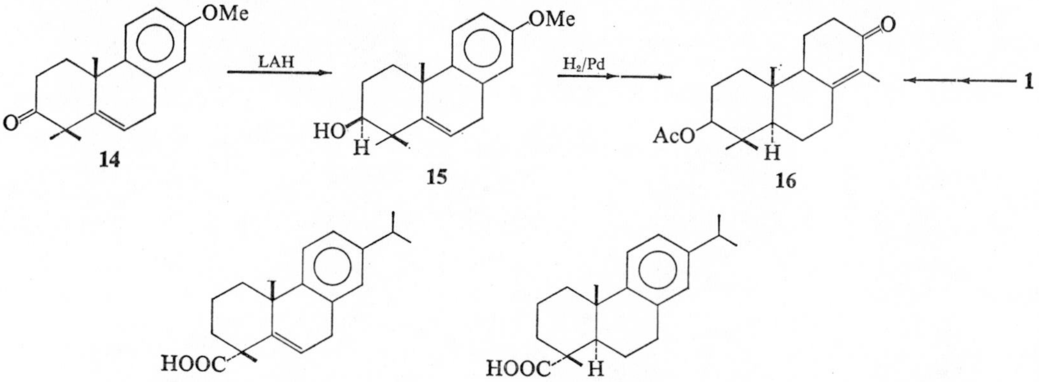

14

LAH

15

H₂/Pd

16

← **1**

17a

17

6) The *trans* nature of the B/C ring junction was suggested[11] by the greater stability of the *trans-anti-trans* system and the conclusion of ord studies[13] that the majority of 9-H in diter-penoids have an α orientation.

7) The α configuration of the 14-Me was suggested[11] from the Cotton effect of **18** and **19** (octant rule). This was supported by nmr studies[14,16,17] and the total synthesis of **1**[11] and of ($\pm$)-14-epi-7-desoxocassamic acid **20**.[15]

10) E. R. H. Jones, T. G. Halsall, *Progress in the Chemistry of Organic Natural Products* (ed. L. Zechmeister) vol. 12, ch. 2, Springer-Verlag, 1955.
11) R. B. Turner, O. Buchardt, F. Herzog, R. B. Morin, A. Riebel, J. M. Sanders, *J. Am. Chem. Soc.*, **88**, 1766 (1965).
12) G. Stork, J. W. Schulenberg, *J. Am. Chem. Soc.*, **78**, 250 (1956).
13) W. Klyne, *J. Chem. Soc.*, 3072 (1953).

8) The *trans* nature of the double bond in **1** was deduced by comparing its nmr with those of model compounds.[14, 16]

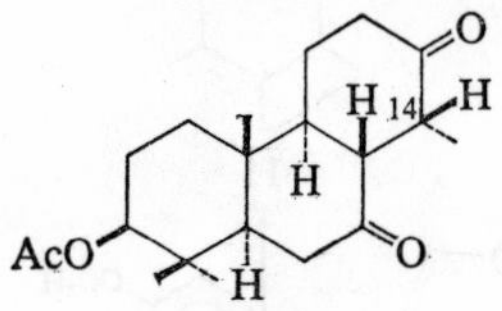

18
(negative Cotton effect)

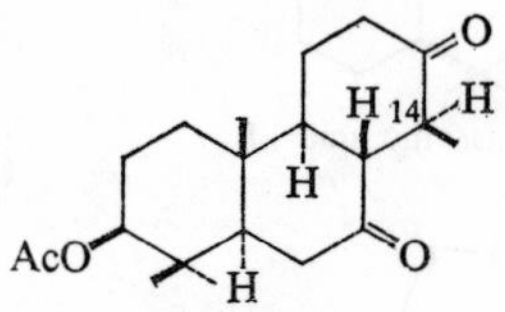

19
(positive Cotton effect)

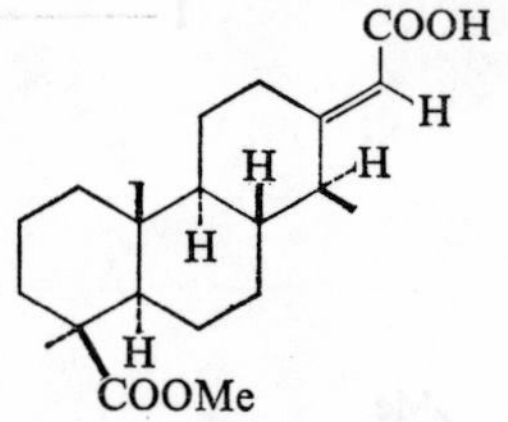

20

Remarks

Cassaic acid **1** was obtained by hydrolysis of the *Erythrophleum* alkaloid cassaine.

14) H. Hauth, D. Stauffacher, P. N. Klaus, A. Melera, *Helv. Chem. Acta.*, **48**. 1087 (1965).
15) K. Mori, M. Matsui, *Tetr.*, **22**. 2883 (1966).
16) R. L. Clark, S. J. Daum, P. E. Shaw, R. K. Kulling, *J. Am. Chem. Soc.*, **88**, 5865 (1966).
17) D. R. Mathieson, A. Kaim. *J. Chem. Soc.*. 1705 (1970).

4.34   STRUCTURE OF α-, β- AND δ-CAESALPINS
[Cassane group]

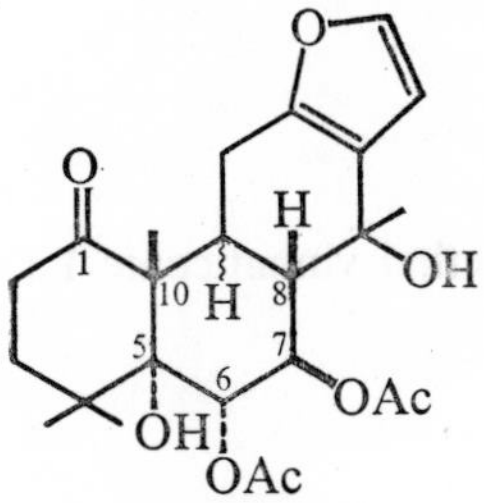

α-caesalpin **1** 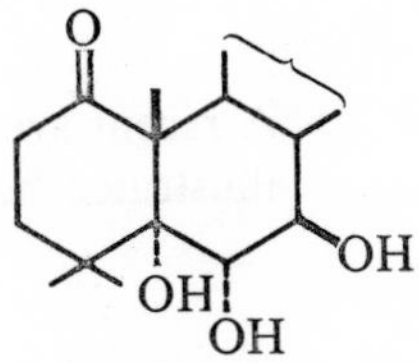   **2** β-caesalpin   **3** δ-caesalpin

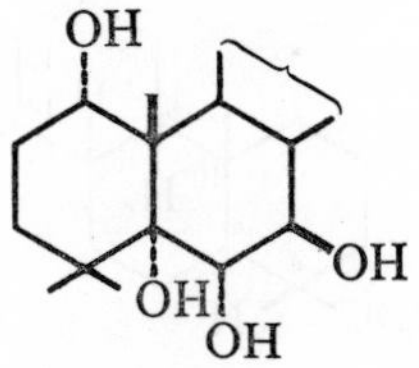

1) The relative configurations at C-5, C-6 and C-10 for **1**, **2** and **3** were determined by the conversions shown below.[1] The Serini transformation **4→6** results in an inversion of configuration at C-5.

2) The high field shift of the 4α-Me in **7**, attributable to the aromatic C ring, suggests an A/B *cis* ring junction; the junction of **1** is thus *trans*.

3) The hemiketal **11** was isolated from a pinacol-type rearrangement. This suggests that the 6-OH is equatorial.

4) In addition, $J_{6,7}$ (10 Hz) and $J_{7,8}$ (10 Hz) in 6-acetates of **2** and **3** support the C-7 and C-8 configurations.

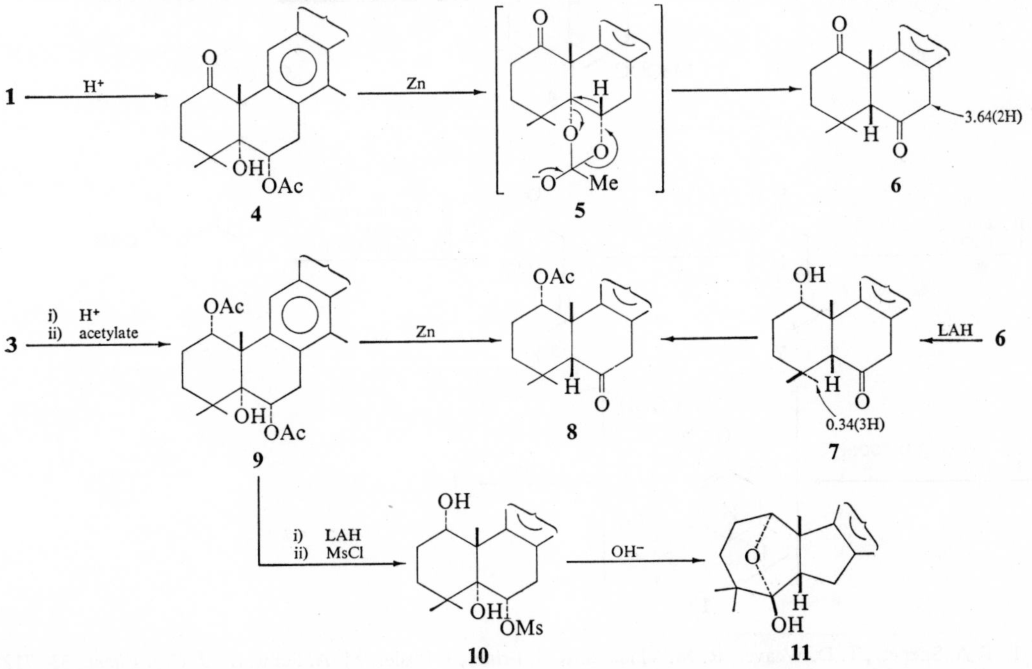

1) A. Balmain, J. D. Connolly, M. Ferrari, E. L. Chisalberti, V. M. Pagnoni, F. Pelezzoni, *Chem. Commun.*, 1244 (1970).

4.35 SYNTHESIS OF METHYL VINHATICOATE
[Cassane group]

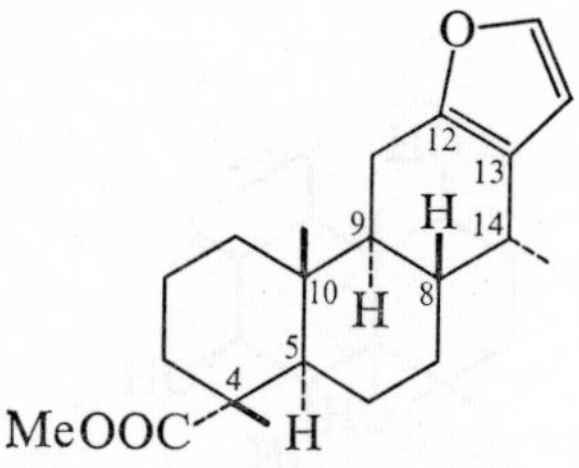

(±)-methyl vinhaticoate **1**

A recent synthesis of (±)-methyl vinhaticoate **1** is illustrated below.[1-3]

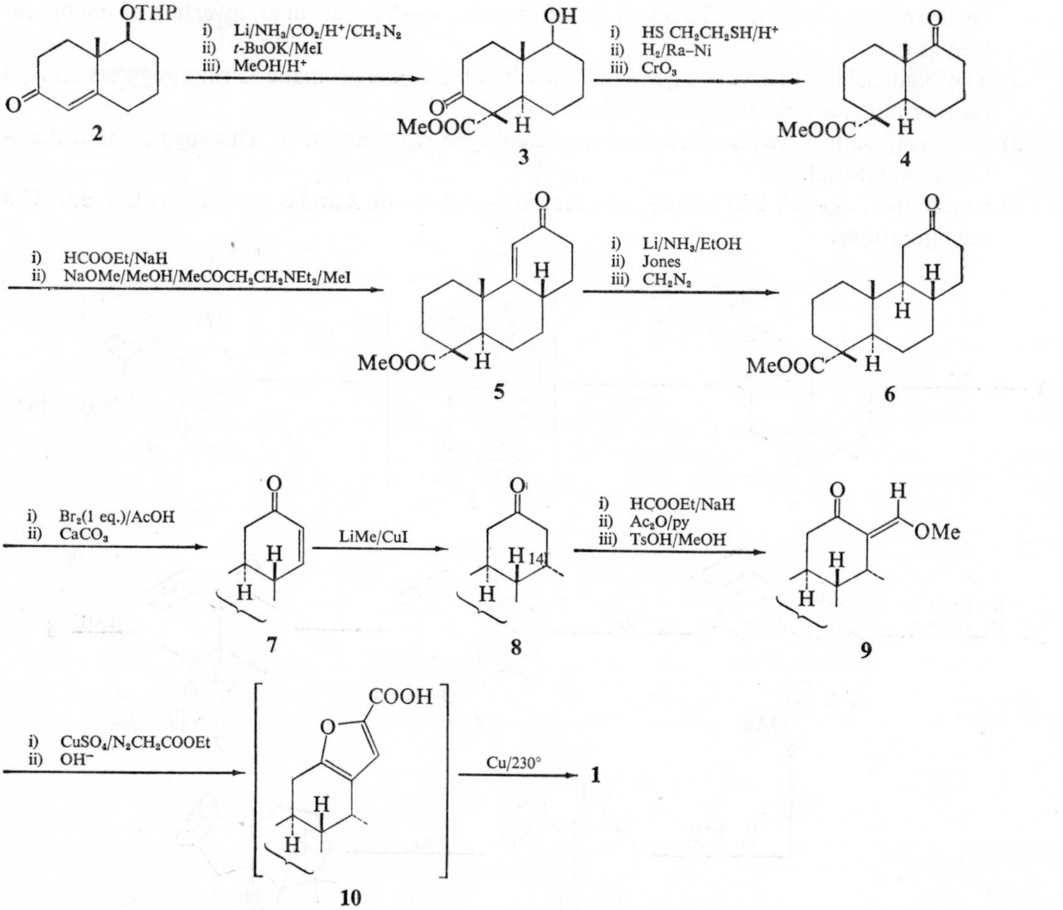

1) T. A. Spencer, T. D. Weaver, R. M. Villarica, R. J. Friary, J. Posler, M. A. Schwarts, *J. Org. Chem.*, **33**, 712 (1968).
2) T. A. Spencer, R. A. J. Smith, D. L. Storm, R. M. Villarica, *J. Am. Chem. Soc.*, **93**, 4856 (1971).
3) T. A. Spencer, R. J. Friary, W. W. Schmiegel, J. F. Sieone, D. S. Watt, *J. Org. Chem.*, **33**, 719 (1968).

2→3: The first step involves the Stork reductive carbomethoxylation procedure.

4→5: This involves a Michael addition followed by cyclization in which the conditions employed (NaOMe in refluxing MeOH) should result in formation of the more stable β-epimer at 8-H.

5→6: The Li/NH_3 reduction of **5** afforded the *trans* B/C ring junction.

7→8: Methylation at C-14 gave a 1 . 1 mixture of epimers, but fortuitously the desired α-derivative crystallized free of its isomer.

9→10: The furan moiety is formed by a 1,4-methylene addition with spontaneous elimination of MeOH.

Remarks

Methyl vinhaticoate **1** was isolated from the heartwood of *Plathymania reticulata*.

4.36 STRUCTURE OF CLEISTANTHOL
[Cassane group]

cleistanthol **1**
mp 193–194° α_D (EtOH): —40°
nmr for triacetate

The structure of this novel cassane derivative (with C-ring ethyl and methyl groups reversed) was deduced as follows.[1]

1) The triol **3** reacts with HIO_4, suggesting a glycol.

2) Spectral data (ir, uv, nmr, ms) for **2** were indistinguishable from those of **4** obtained from methyl vinhaticoate, but the compounds differed in chromatographic properties. This was ascribed to differences in the positions of the methyl and ethyl groups in the phenolic ring.

3) The uv spectrum of **3** suggested a *p*-hydroxyphenyl ketone moiety.

4) When the 2,3-dimesylate was reacted with NaI in refluxing acetone, compounds **5** and **6** were isolated. This, with nmr evidence, suggested a 2-axial, 3-equatorial glycol.

5) Comparison of α_D values of **9** with those of the methyl ethers of hinokione **7** and totarolone **8** indicates an antipodal relationship, and thus the complete structure **1** was derived.

5

6

7: R_1=OMe, R_2=CHMe$_2$, R_3=H; α_D + 119°
8: R_1=H, R_2=OMe, R_3=CHMe$_2$; α_D + 99°

9 α_D —96°

Remarks

Cleistanthol **1** was isolated from *Cleistanthus schlechteri*.

1) E. J. McGarry, K. H. Pegel, L. Phillips, E. S. Waight, *Chem. Commun.*, 1074 (1969).

4.37   STRUCTURE OF BEYEROL
[Beyerane group]

beyerol **1**[1]

mp: 242–243°
α_D (c, 1.1 py):   +61°
uv: end absorption
ir(KBr): 3051, 1650, 750.

1) Beyerol **1** is formed from alkaline hydrolysis of the natural product, its cinnamate **2**.

2) Of the three OH's, two form a 1,3-glycol; this was deduced from the formation of chlorophosphochloridate and ethylidene derivatives, and inertness to $Pb(OAc)_4$ and $NaIO_4$.

3) The C-17 hydroxymethylene is attached to *tert* carbon since it resists Barbier-Wieland degradation (no dehydration of *tert*-OH).

4) Uptake of one mole of H_2 plus formation of a monoepoxide shows the presence of one double bond (*cis* by ir).

5) Dihydrobeyerol is saturated and hence is tetracyclic (negative tetranitromethane test).

6) Formation of a 6-membered anhydride, **11**, shows ring D to be five-membered.

7) The structure of ring A follows from **1→7** and **1→3→4**.

8) These partial structures, plus the formation of **10**, allow structure **1** to be formulated based on isoprene biogenesis.

9) The possibility of **12** can be excluded by **13→14**, i.e., decarboxylation of the malonic acid monoester.

10) Regeneration of **1** from **3** shows the 3-OH to be equatorial.

11) The C-19 is axial from the nmr of the 19-CHO and 19-CH_2OAc functions.

12) Ring B must be *trans* since bromination of *cis* A/B steroid-3-ketones favors substitution at C-4[3] (*cf*. **5**).

13) Shielding of the 10-Me by the double bond and comparison of this shift with phyllocladene and kaurene derivatives show a *trans-anti-trans* backbone.

14) This arrangement is verified by the negative Cotton effect of the ketone **8**, and comparisons with 16-ketones in phyllocladene and related compounds.[2]

15) This structure was corroborated by conversion of **1** to isostevane[4] of known stereochemistry.

Remarks

Beyerol **1** is found in the leaves and stems of *Beyeria leschenaulyii* var. *drummondi*, found in Western Australia.

1) P. R. Jefferies, R. S. Rosich, D. E. White, M. C. Woods, *Aust. J. Chem.*, **15**, 521 (1962).
2) P. R. Jefferies, R. S. Rosich D. E. White, *Tetr. Lett.*, 1793 (1963).
3) H. Butenandt, J. Schmidt, *Chem. Ber.*, **67**, 1901 (1934).
4) C. Djerassi, P. Quitt, E. Mosettig, R. C. Cambie, P. S. Rutledge, L. H. Briggs, *J. Am. Chem. Soc.*, **83**, 3720 (1961).

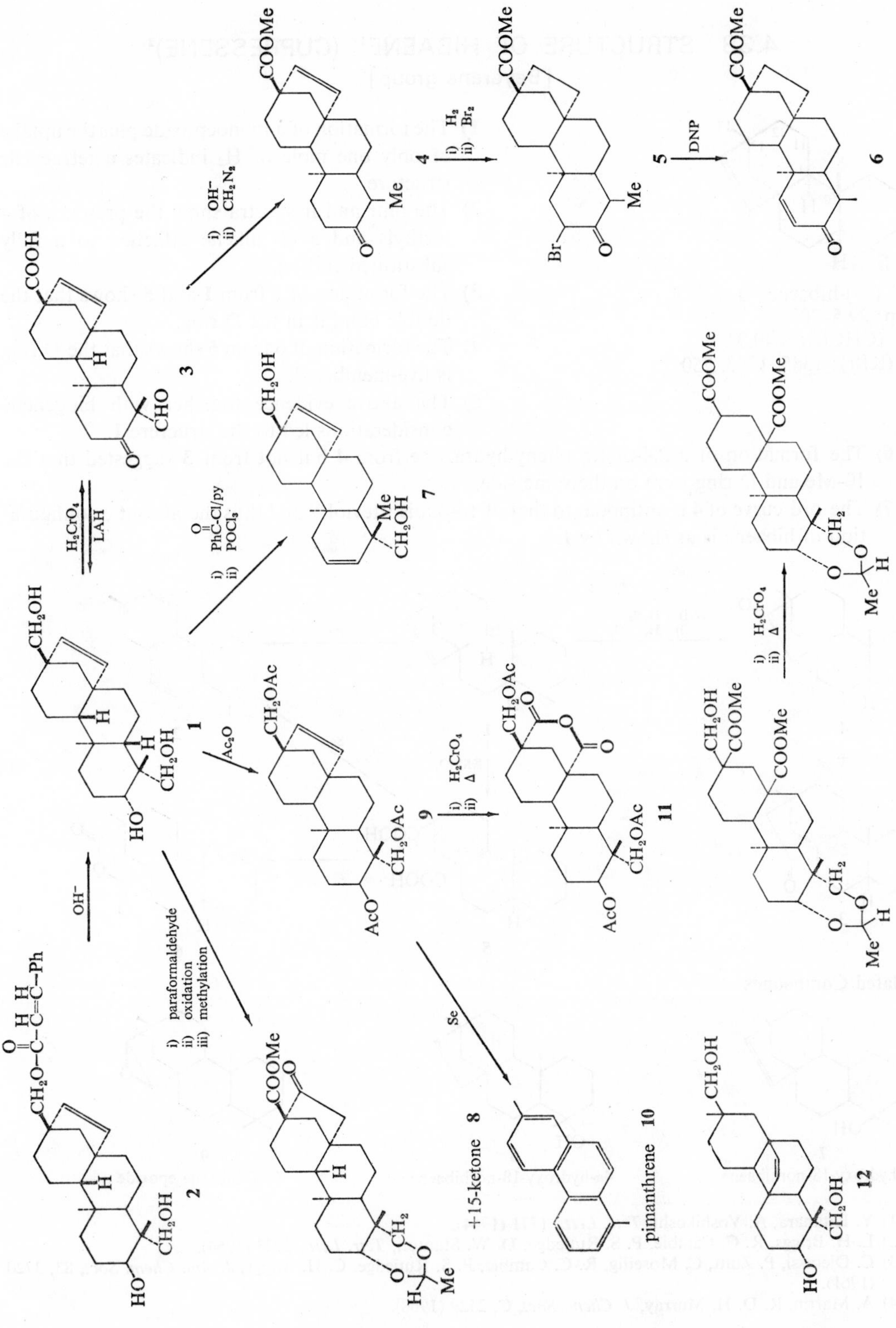
OH⁻
H2CrO4
LAH
i) OH⁻
ii) CH2N2
i) PhC-Cl/py
ii) POCl3
Ac2O
i) H2CrO4
ii) Δ
i) paraformaldehyde
ii) oxidation
iii) methylation
Se
+15-ketone 8
pimanthrene 10
i) H2
ii) Br2
DNP
1
2
3
4
5
6
7
9
11
12
13
14

4.38   STRUCTURE OF HIBAENE[1] (CUPRESSENE)[2]
[Beyerane group]

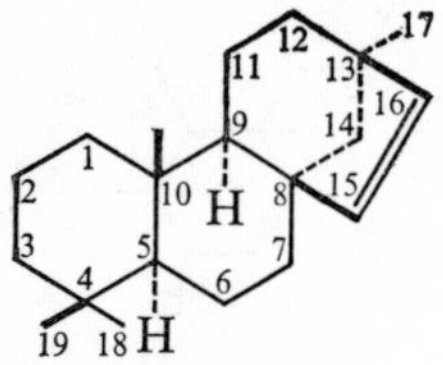

(−)-hibaene **1**
mp: 29.5–30°
α_D^{23} (CHCl$_3$): −49.9°
ir (KBr): 1385, 1363, 750

1) The formation of a monoepoxide plus the uptake of only one mole of H$_2$ indicates a tetracyclic structure.

2) The nmr and ir spectra show the presence of 4 methyls and a *cis* alkene attached to a fully substituted carbon.

3) The formation of **2** from **1** and **5** shows that the double bond is in the D ring.

4) The formation of **6** from **5** shows that the D ring is five-membered.

5) The above evidence together with biogenetic considerations led to the structure **1**.

6) The formation of a 2,4-dinitrophenylhydrazone from **4** but not from **3** suggested that the 10-Me and D ring were on the same side.

7) The ord curve of **4** is antipodal to that of 16-oxoisosteviol,[3] and thus the absolute configuration of hibaene is as shown by **1**.

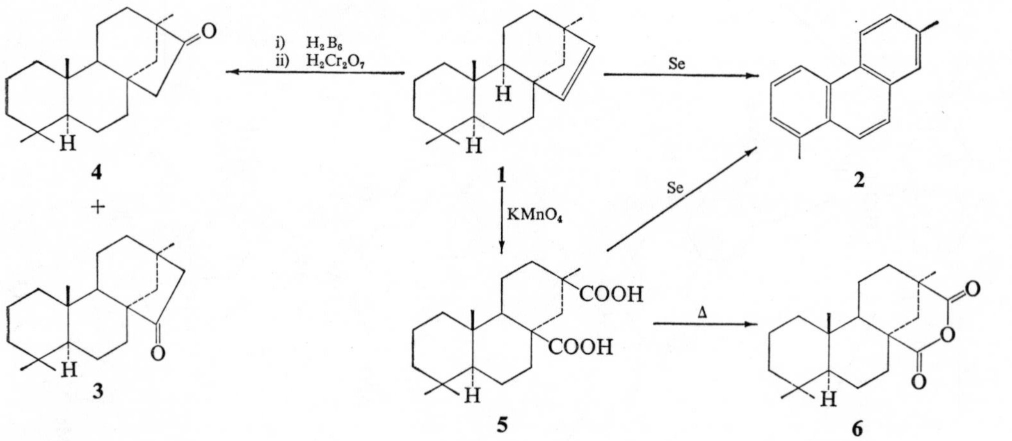

Related Compounds

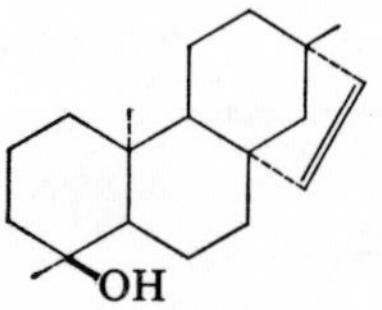

7
4β-hydroxy-18-norhibaene

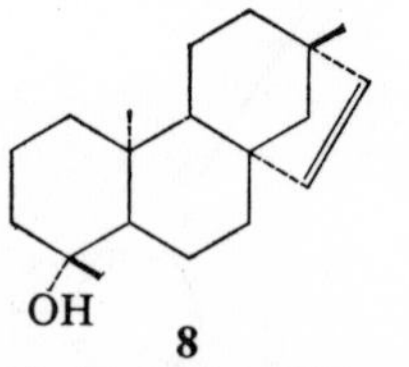

8
4α-hydroxy-18-norhibaene

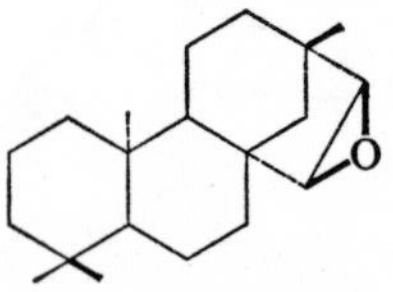

9
(+)-hibaene epoxide

1) Y. Kitahara, A. Yoshikoshi, *Tetr. Lett.*, 1771 (1964).
2) L. H. Briggs, R. C. Cambie, P. S. Rutledge, D. W. Stanton, *Tetr. Lett.*, 2223 (1964).
3) C. Djerassi, P. Zuitt, C. Mosetlig, R. C. Cambie, P. S. Rutledge, C. H. Briggs, *J. Am. Chem. Soc.*, 83, 3720 (1961).
4) A. Martin, R. D. H. Murray, *J. Chem. Soc.*, C, 2529 (1968).

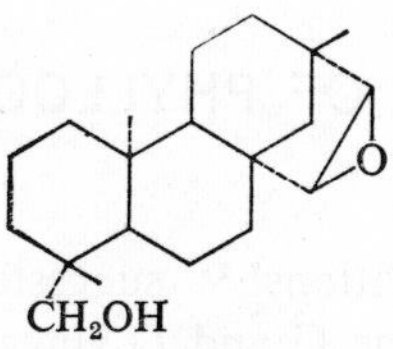

10
erythroxylol A epoxide

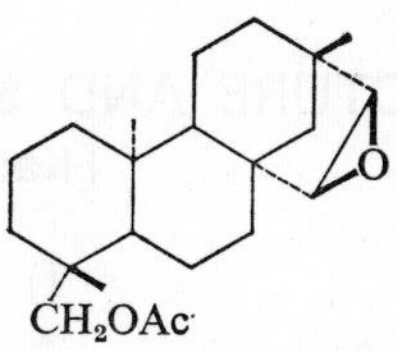

11
erythroxylol A acetate epoxide

The equatorial and axial nature of 4-OH in **7** and **8**, respectively, were based on dehydration studies leading to a 4,18-ene and 4,5-ene.[4]

Remarks

Hibaene **1** is obtained from essential oil of the leaves of *Thujopsis dolabrata*. **1** is also found in *Cupressus macrocarpa*. The related compounds **7** to **11** were isolated from trunkwood of *Erythroxylon monogynum*.

4.39   SYNTHESIS OF HIBAENE
[Beyerane group]

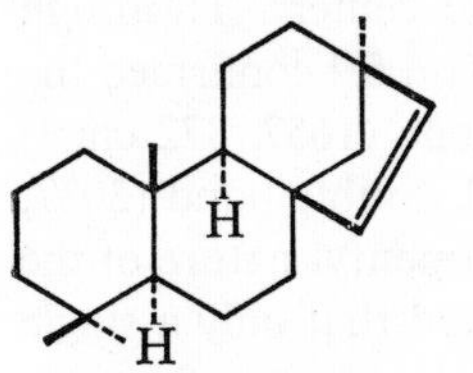

hibaene **1**

Ireland's synthesis of (±)-hibaene **1** from the intermediate **2** employs as a key step the acid-catalyzed rearrangement (**8→9**) of the bridgehead hydroxyolefin.[1] This same intermediate, synthesized in ten steps from *m*-methoxybenzaldehyde,[2] was also used in the synthesis of (±)-kaurene.[3]

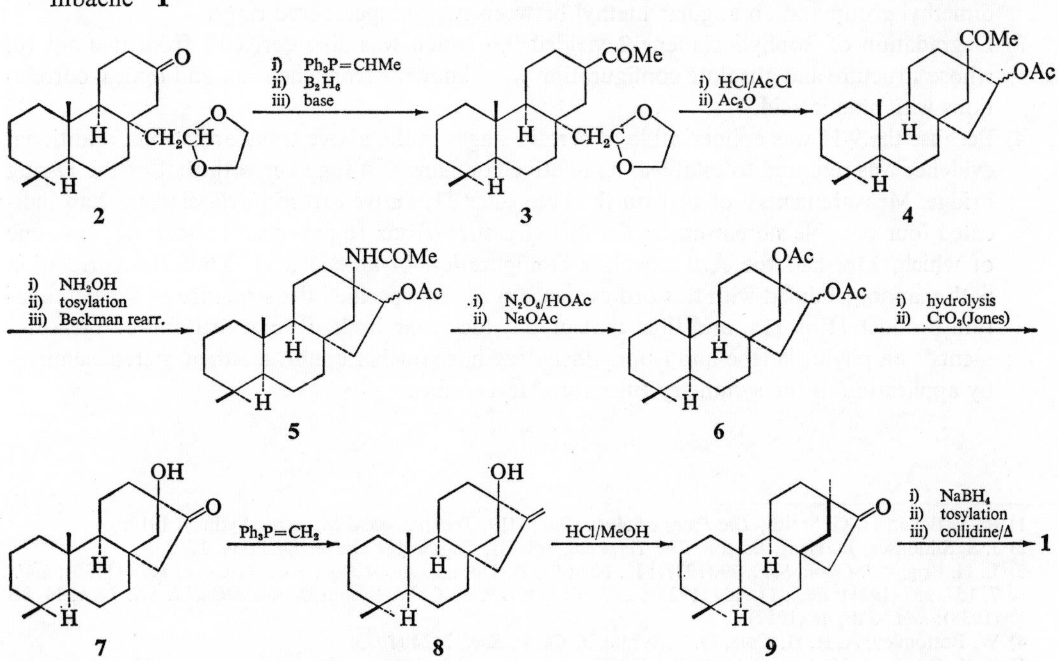

1) R. E. Ireland, L. H. Mander, *Tetr. Lett.*, 2627 (1965).
2) R. F. Church, R. E. Ireland, J. A. Marshall, *Tetr. Lett.*, 1 (1960).
3) R. A. Bell, R. E. Ireland, R. A. Pactyka, *J. Org. Chem.*, **27**, 374 (1962).

4 40 STRUCTURE AND SYNTHESIS OF PHYLLOCLADENE
[Kaurane group]

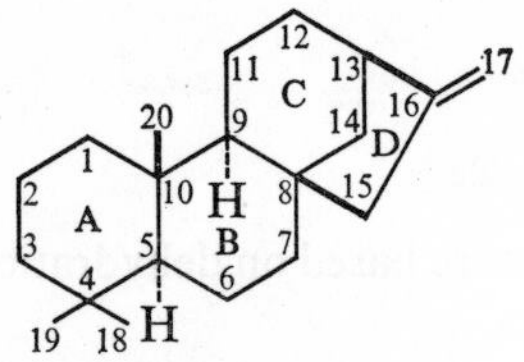

phyllocladene **1**

mp: 98°

$\alpha_D^{25}: +15.8°^{2)}$

ir$(CS_2, CCl_4)^{4)}: 1657(C=CH_2), 872$

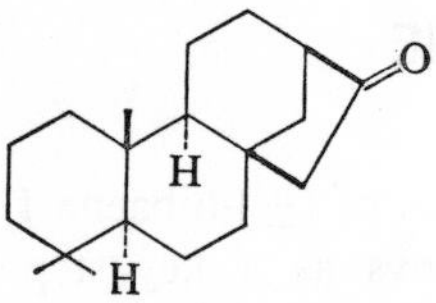

5

1) Early investigations[1,2] suggested the general configuration of the C and D rings in the $C_{20}H_{32}$ compound. The following chart summarizes the data[3] which led Brandt to suggest the structure **1** (without stereochemical detail) for phyllocladene and **2** for isophyllocladene. This evidence left unspecified the location of two methyl groups and the terminus of the D-ring bridge (whether at C-8 or C-9). Brandt assigned the Me's by analogy with similar terpenes and the D-ring bridge by the constraint of the biogenetic isoprene rule.

2) Later evidence confirmed that the above intuited structure was correct and elucidated the relative stereochemistry at the five asymmetric centers. Thorough analysis of the ir spectra of **1**, **2** and **5**[4] confirmed the presence of the exocyclic methylene (1657, 872 cm⁻¹) in **1**, as well as the trisubstituted double bond (3037, 1635, 825 cm⁻¹) in **2** and the cyclopentyl nature of the ketone **5** (1742 cm⁻¹) and suggested that only a single methylene flanked this carbonyl. In addition, other data suggested the presence of a *gem*-dimethyl group and an angular methyl between two six-membered rings.

3) Degradation of isophyllocladene **2** yielded **9**,[5] which was also derived[6] from manool **10**, whose structure and absolute configuration were known[7] from chemical and optical correlations with abietic acid.

4) Because the 9-H was epimerizable at certain stages in the above transformations, additional evidence was required to establish its relative configuration together with that of the D-ring bridge. Measurements[8] of ord on the keto ester **11**, derived from phyllocladene, had indicated four possible stereoisomers for this structure (from 16 conceivable isomers), only one of which, **11a**, had the A/B absolute configuration of abietic acid. Thus the correlation with manool, coupled with this ord evidence uniquely specified the structure of the degradation product **11** as **11a**, and thus that of phyllocladene itself. Recent ord and cd measurements[9] on phyllocladene and isophyllocladene have further confirmed their stereochemistry by application of the symmetry rules for chiral olefins.

1) R. T. Baker, H. G. Smith, *The Pines of Australia*, p.419, Technological Museum, Sydney, 1910.
2) J. S. Simonsen, D. H. R. Barton, *The Terpenes*, vol. III, Cambridge University Press, 1952.
3) L. H. Briggs, *J. Chem. Soc.*, 79 (1937); K. Nishida, H. Uota, *J. Agr. Chem. Soc., Japan*, 6, 1078 (1930); *ibid.*, **7**, 157, 957 (1931); *ibid.*, **11**, 489 (1935); *ibid.*, **12**, 308 (1936); C. W. Brandt, *New Zealand J. Sci. Tech*, **20**, 8B (1938): *ibid.*, **34B**, 46 (1952).
4) W. Bottomley, A. R. H. Cole, D. E. White, *J. Chem. Soc.*, 2624 (1955).
5) P. K. Grant, R. Hodges, *Tetr.*, **8**, 261 (1960).
6) J. R. Hosking, *Chem. Ber.*, **69**, 780 (1936).
7) W. Klyne, *J. Chem. Soc.*, 3072 (1953), and references therein.
8) C. Djerassi, M. Cais, L. A. Mitscher, *J. Am. Chem. Soc.*, **81**, 2386 (1959): see especially note 61.

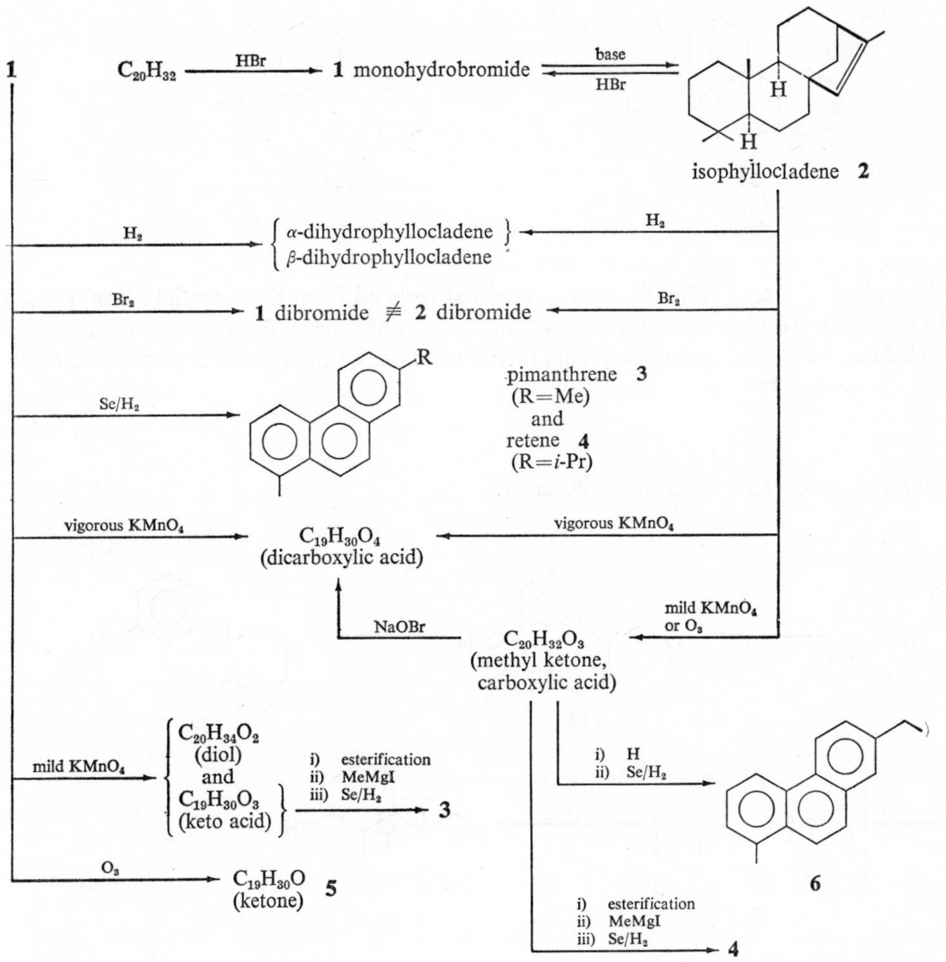

$$1 \quad C_{20}H_{32} \xrightarrow{\text{HBr}} 1 \text{ monohydrobromide} \underset{\text{HBr}}{\overset{\text{base}}{\rightleftarrows}}$$

isophyllocladene **2**

$$\xrightarrow{H_2} \left\{ \begin{array}{l} \alpha\text{-dihydrophyllocladene} \\ \beta\text{-dihydrophyllocladene} \end{array} \right\} \xleftarrow{H_2}$$

$$\xrightarrow{Br_2} 1 \text{ dibromide} \neq 2 \text{ dibromide} \xleftarrow{Br_2}$$

$$\xrightarrow{Se/H_2}$$

pimanthrene **3**
(R=Me)
and
retene **4**
(R=*i*-Pr)

$$\xrightarrow{\text{vigorous KMnO}_4} \underset{\text{(dicarboxylic acid)}}{C_{19}H_{30}O_4} \xleftarrow{\text{vigorous KMnO}_4}$$

$$\xrightarrow{\text{NaOBr}} \underset{\substack{\text{(methyl ketone,} \\ \text{carboxylic acid)}}}{C_{20}H_{32}O_3} \xleftarrow[\text{or O}_3]{\text{mild KMnO}_4}$$

$$\xrightarrow{\text{mild KMnO}_4} \left\{ \begin{array}{l} C_{20}H_{34}O_2 \\ \text{(diol)} \\ \text{and} \\ C_{19}H_{30}O_3 \\ \text{(keto acid)} \end{array} \right\} \begin{array}{l} \text{i) esterification} \\ \text{ii) MeMgI} \\ \text{iii) Se/H}_2 \end{array} \longrightarrow \textbf{3}$$

i) H
ii) Se/H₂

6

$$\xrightarrow{O_3} \underset{\text{(ketone)}}{C_{19}H_{30}O} \quad \textbf{5}$$

i) esterification
ii) MeMgI
iii) Se/H₂ ⟶ **4**

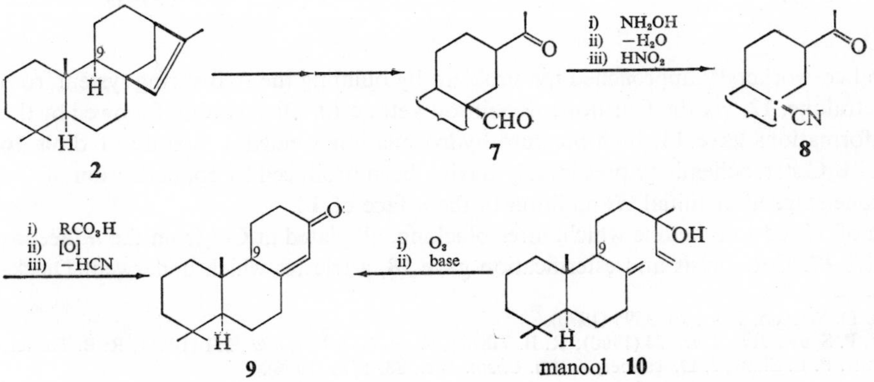

2

i) NH₂OH
ii) −H₂O
iii) HNO₂

7

8

i) RCO₃H
ii) [O]
iii) −HCN

9

i) O₃
ii) base

manool **10**

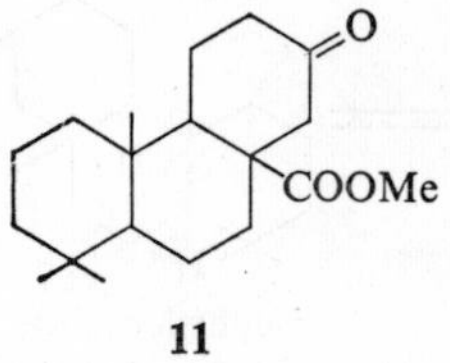

11

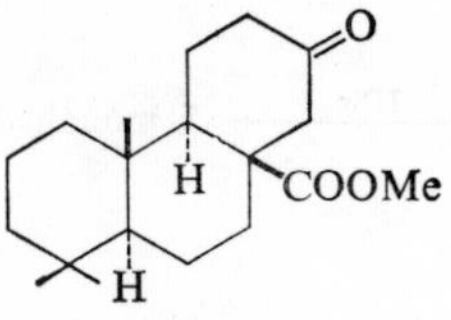

11a

Remarks

Phyllocladene **1**, a tetracyclic diterpene, was first isolated from the essential oil of *Phyllocladus rhomboidalis* (Podocarpaceae) leaves[1] and has since been found, occasionally with isophyllocladene, in a wide variety of plants including those of the genera *Phyllocladus, Podocarpus* and *Dacrydium*.[2]

Synthesis

Total synthesis corroborated the structure **1**.

Route I

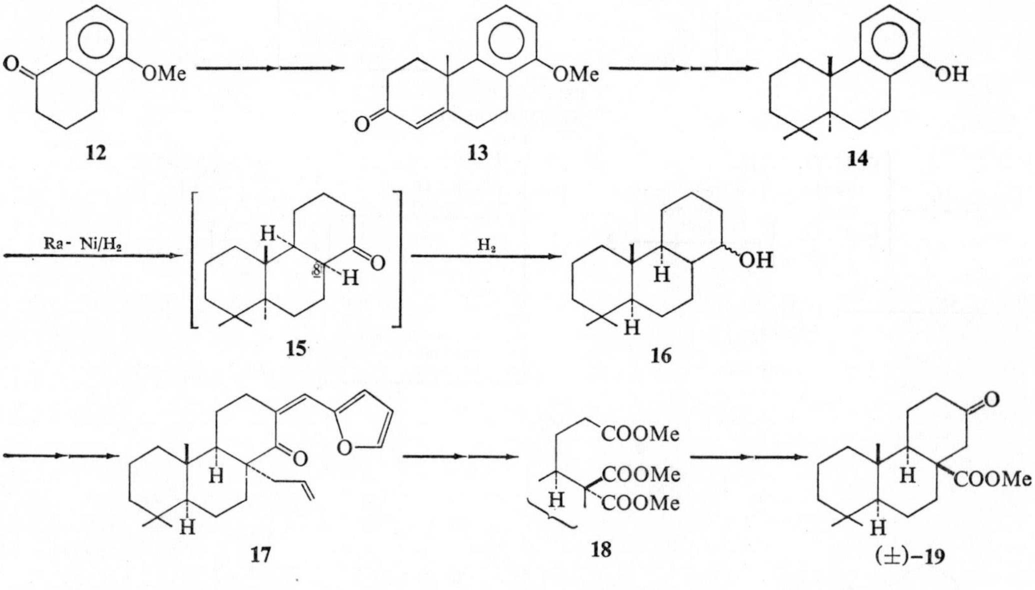

12 **13** **14**

15 **16**

17 **18** (±)–**19**

Turner and co-workers[10] approached the problem by building the A/B/C ring system from 5-methoxy-2-tetralone, **12**, via the Cornforth-Robinson ketone **13**. Dialkylation followed by the required transformations gave **14**; high-pressure hydrogenation yielded a mixture of diols **16** with the crucial B/C stereochemistry presumably having been produced by epimerization of the 8-H at the ketone stage after initial *cis* addition to the α face of **14**.

Oxidation of **16** led to a ketone which, after blocking, alkylated at C-8 from the α face, as expected, to give **17**. Ozonolysis and esterification gave **18**, a triester which underwent Dieck-

9) A. I. Scott, A. D. Wrixon, *Tetr.*, **26**, 3695 (1970).
10) R. B. Turner, P. Shaw, *Tetr. Lett.*, 24 (1960); R. B. Turner, K. H. Gänshirt, *ibid.*, 231 (1961); R. B. Turner, K. H. Gänshirt. P. E. Shaw, J. D. Tauber, *J. Am. Chem. Soc.*, **88**, 1776 (1966).

mann cyclization followed by decarboxylation to yield racemic **19** whose solution ir spectrum was identical with **11** from phyllocladene. The synthesis was completed using naturally derived **11** as a relay compound.

Keto ester **11** was elaborated to diacid **20**, the barium salt of which was then pyrolyzed to give ketone **21**. The acetal of the hydroxymethylene derivative of **21** was reduced with LAH and treated with mineral acid; Wolff-Kishner reduction of the resulting α,β-unsaturated aldehyde, **22**, had been shown to give **1**.

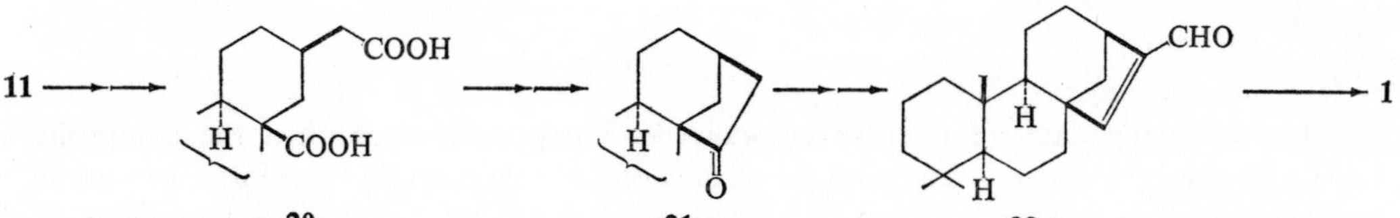

11 ⟶⟶   **20**   ⟶⟶   **21**   ⟶⟶   **22**   ⟶ **1**

Route II

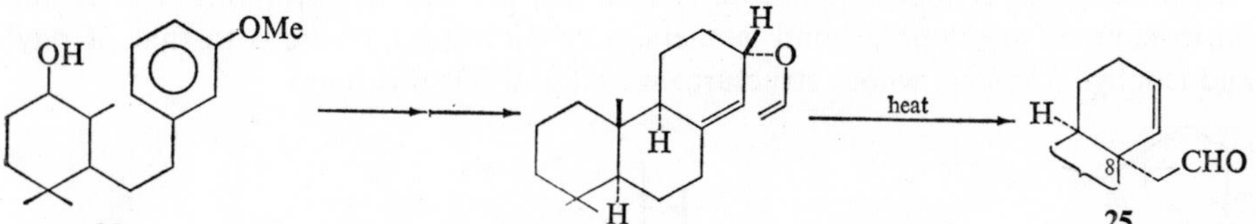

23   ⟶   **24**   —heat→   **25**

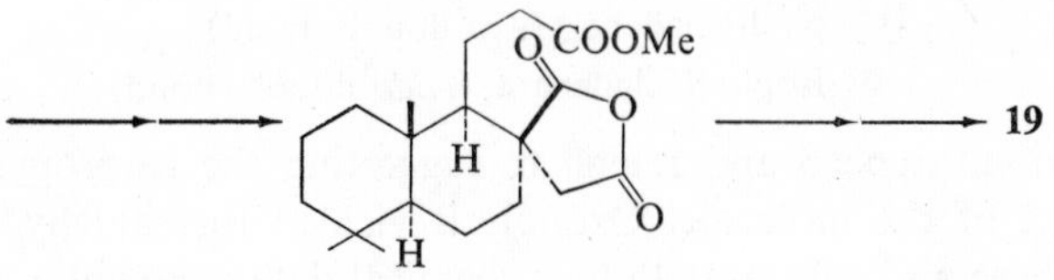

⟶   **26**   ⟶   **19**

The synthesis of **19** has also been reported[11] by a route involving acid-catalyzed cyclization of **23** followed by reductive transformations and ether formation to yield **24**. Pyrolysis gave a product **25**, with undesired stereochemistry at C-8; this orientation was reversed, as in the previous synthesis,[10] by oxidative cleavage to the tricarboxylate stage and Dieckmann reclosure followed by decarboxylation to give **19**.

11) R. F. Church, R. E. Ireland, J. A. Marshall, *Tetr. Lett.*, 1 (1960).

4.41   STRUCTURE AND SYNTHESIS OF KAURENE
[Kaurane group]

0.82, 0.87, 1.03

ent-kaurene **1**

mp : 51° α_D^{11} : −72°

kaurene:[1]
mp: 49–50° α_D^{17}:74°
ir (KBr): 876 (CH=CH$_2$), 1387, 1368 (CMe$_2$)

The following data all refer to *ent*-kaurene. Early work established the composition C$_{20}$H$_{32}$ for this hydrocarbon. The presence of only one double bond (hydrochloride formation,[2] hydrogenation[3]) suggested a tetracyclic structure. Mineral acid treatment of kaurene yielded isokaurene **2** (identical with α-cryptomerene in the earlier literature), which yielded the same hydrogenation products and hydrochloride as kaurene.[1] This information together with the ir spectrum: 820 (trisubst. double bond), and nmr: 1.71 (d), 3H (allylic methyl), indicated that kaurene and isokaurene are double bond isomers, a relationship analogous to that of phyllocladene and isophyllocladene, whose structures were known at this time.

isokaurene **2**

phyllocladene **3** (*exo* double bond)
isophyllocladene **4** (*endo* double bond)

Selenium dehydrogenation gave pimanthrene **5** and retene **6**, suggesting the isopropyl-substituted perhydrophenanthrene nature of the molecule. Ozonolysis yielded formaldehyde and a cyclopentanone (ir: 1745 cm^{-1}). These and substantial other chemical data were entirely parallel with the results obtained with phyllocladene and isophyllocladene, implying that kaurene differed only in stereoisomeric detail, the centers in question being C-8, C-9 and C-13.[1]

5 (R=Me)
6 (R=*i*-Pr)

steviol **7**

8

Steviol **7**, whose D-ring configuration was known through correlations of a related ketone with gibberellic acid derivatives, was converted into β-dihydro-*ent*-kaurene,[4] one of the hydrogenation products of **1**, thus establishing the configuration at C-8 and C-13. Substantial doubt remained concerning the configuration at C-9 because epimerization at this center during the conversion had not been ruled out. The problem was resolved[5] by ord measurements on ketones **8** derived from *ent*-kaurene and steviol **7**, with application of the octant rule and correlation with steroid models.

1) L. H. Briggs, B. F. Cain, R. C. Cambie, B. R. Davis, P. S. Rutledge, J. K. Wilmshurst, *J. Chem. Soc.*, 1345 (1963), and references therein.
2) J. R. Hosking, *Rec. Trav. Chim.*, **47**, 578 (1928).
3) J. R. Hosking, *Rec. Trav. Chim.*, **49**, 1036 (1930).

The absolute configuration of *ent*-kaurene **1** was first established[6] unequivocally when podocarpic acid **9** of known configuration was converted into **10**, which was enantiomeric with the phenol derived from the alkaloid atisine. Atisine had been related to another alkaloid, garryfoline, and the latter to *ent*-kaurene, thus demonstrating the absolute configuration of *ent*-kaurene.

podocarpic acid **9**   **10**

Phyllocladene **3** and *ent*-kaurene **1** have been converted into enantiomeric ketoesters **11** and **12**, respectively,[7] thus further supporting the assignment of configuration to *ent*-kaurene. The transformation of **1** into **12** required inversion of the stereochemistry at C-8, accomplished by intramolecular Claisen condensation and cleavage of the resulting 1,3-dione.

11   **12**

An analogous inversion of the centers at the C/D ring junction has been observed in the conversion of hibaene-β-epoxide **13** to *ent*-14-hydroxyisokaurene **14** with BF₃ etherate.[8] The corresponding transformation of the *ent*-hibaene to the kaurene systems has also been observed.[8]

13   **14**

4) F. Dolder, H. H. Lichti, E. Mossettig, P. Quitt, *J. Am. Chem. Soc.*, **82**, 246 (1960).
5) C. Djerassi, P. Quitt, E. Mossettig, R. C. Cambie, P. S. Rutledge, L. H. Briggs, *J. Am. Chem. Soc.*, **83**, 3720 (1961).
6) J. W. Apsimon, O. E. Edwards, *Can. J. Chem.*, **40**, 896 (1962).
7) B. E. Cross, J. R. Hanson, L. H. Briggs, R. C. Cambie, P. S. Rutledge, *Proc. Chem. Soc.*, **17**, (1963).
8) A. H. Kapadi, S. Dev. *Tetr. Lett.*, 1255 (1965); A. Yoshikoshi, M. Kitadani, K. Tahara, *Tetr.* **23**, 1175 (1967).

Synthesis

Total syntheses of ($\pm$)-kaurene have been reported by Ireland and co-workers[9] and by Masamune.[10] The former group (Route I) assembled the intermediate **16** by acidic cyclization of **15** and added the two-carbon bridge of ring D by Claisen rearrangement followed by aldol condensation to give **18**. The synthesis was then completed in a straightforward manner.

Route I[9]

Route II[10]

Masamune, on the other hand, first assembled rings C, D, and E by basic cyclization of the phenol **19**, derived from the corresponding naphthoic acid, to the intermediate **20**. Hydrogenation gave predominantly the desired *cis* B/C ring junction, and then ring A was assembled by carbomethoxylation, annulation with ethyl vinyl ketone and alkylation with methyl iodide. Reduction of the keto and ester functionalities and elaboration of a methylene at C-16 of **21** completed the total synthesis.

Remarks

Kaurene, a tetracyclic diterpene, was first isolated from the leaf oil of New Zealand Kauri (*Agathis australis*)[2] and has since been obtained from *Podocarpus macrophyllus* as well.[11,12] The compound from these sources has been designated ($-$)-kaurene to distinguish it from its enantiomer, ($+$)-kaurene, first isolated from *P. spicatus* in 1939[12] and later from *P. ferrugineus*.[13] The convention for diterpene nomenclature now in use describes the ($+$) and ($-$) antipodes as kaurene and *ent*-kaurene, respectively.

9) R. A. Bell, R. E. Ireland, R. A. Partyka, *J. Org. Chem.*, **27**, 3741 (1962); R. A. Bell, R. E. Ireland, R. A. Partyka, *ibid.*, **31**, 2530 (1966).
10) S. Masamune, *J. Am. Chem. Soc.*, **86**, 288, 289 (1964).
11) K. Nishida, H. Uota, *Bull. Agr. Chem. Soc. Japan*, **6**, 82 (1930); L. H. Briggs, R. W. Cawley, *J. Chem. Soc.*, 1888 (1948).
12) J. M. Butler, J. T. Holloway, *J. Soc. Chem. Ind.*, **58**, 223 (1939).
13) L. H. Briggs, R. W. Cawley, J. A. Loe, W. I. Taylor, *J. Chem. Soc.*, 955 (1950).

4.42   STEVIOL
[Kaurane group]

steviol **1**

mp: 215° α_D: −94.7°
ir(CCl$_4$): 3610(OH), 3521–3333, 1695(COOH),
3058, 1799, 1664, 1425, 886(C=CH$_2$)[3]

1) Stevioside **2** yields steviol **1**, $C_{20}H_{30}O_3$, upon enzymatic hydrolysis and isosteviol **3**, $C_{20}H_{30}O_3$, upon acidic hydrolysis; **1** is converted to **3** upon acid treatment.[1,2]

2) Steviol contains a carboxyl, hydroxyl and vinylidene group (ir); isosteviol is a tetracarbocyclic keto acid (lack of unsaturation, ir) which is dehydrogenated by Pd to pimanthrene **4**.[3]

stevioside **2**

isosteviol **3**

pimanthrene **4**

3) By analogy with phyllocladene **5**, whose ir and uv spectra are similar in certain respects, and assuming no major skeletal rearrangements in the **1**→**3** change, the structure **6** was proposed for steviol.[3]

4) Structure **6** was revised to **7** on the basis of ord comparison of isosteviol **3** with gibberic acid **9** and of **8**, derived from steviol by ozonolysis and methylation, with ketonorallogibberic acid.[4]

5) Degradation of the known alkaloid garryfoline **10** to β-dihydro-*ent*-kaurene **11**, also derived from steviol by reductive transformations, confirmed the steviol carbon skeleton, including the stereochemistry of the A/B ring junction. An intermediate alcohol **12** of the garryfoline degradation was different from the analogous alcohol **13** derived from steviol, demonstrating the C-4 (rather than C-10) location of the steviol carboxyl.[5] This location was further confirmed by the ms fragmentation of stevic acid A methyl ester **14** to lose an A-ring six carbon fragment including the carbomethoxyl.[6]

1) M. Bridel, R. Lavieille, *Bull. Soc. Chim. Biol*, **13**, 636, 781 (1931); M. Bridel, R. Lavieille, *J. Pharm. Chim.*, **14**, 99, 154, 321, 369 (1931).
2) H. B. Woods Jr., R. Allerton, H. W. Diehl, H. G. Fletcher Jr., *J. Org. Chem.*, **20**, 875 (1955).
3) E. Mosettig, W. R. Nes, *J. Org. Chem.*, **20**, 884 (1955).
4) F. Dolder, H. Lichti, E. Mosettig, P. Quitt, *J. Am. Chem. Soc.*, **82**, 246 (1960).
5) F. Mosettig, P. Quitt, U. Beglinger, J. A. Waters, H. Vorbrueggen, C. Djerassi, *J. Am. Chem. Soc.*, **83**, 3163 (1961).
6) E. Mosettig, V. Beglinger, F. Dolder, H. Lichti, P. Quitt, J. A. Waters, *J. Am. Chem. Soc.*, **85**, 2305 (1963).

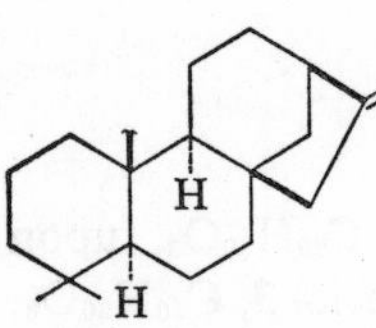

phyllocladene 5

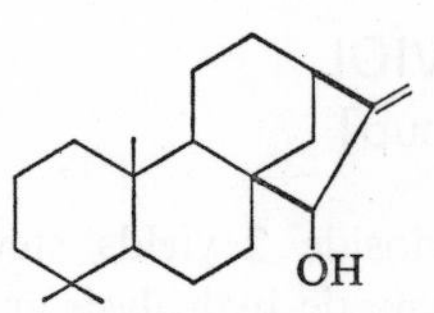

6

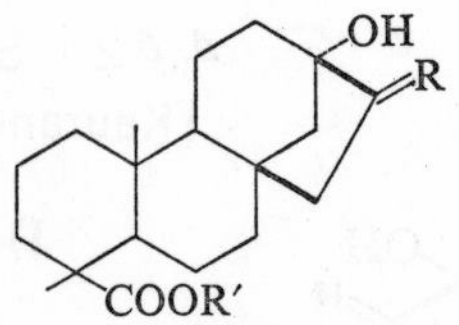

7: R = CH₂, R' = H
8: R = O, R' = Me

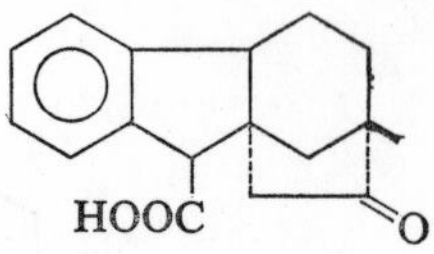

gibberic acid 9

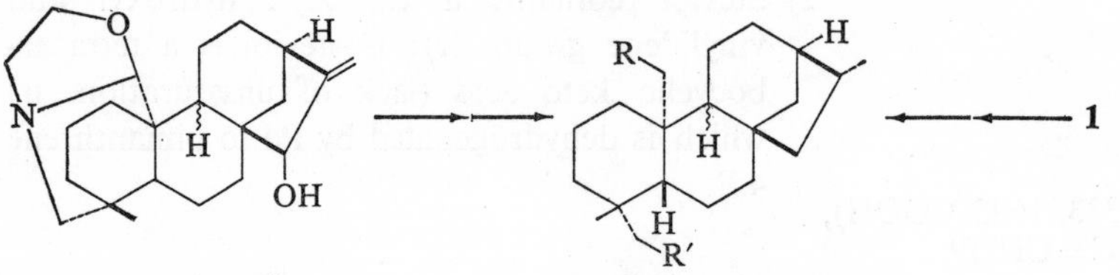

garryfoline 10

11: R, R' = H
12: R = OH, R' = H
13: R = H, R' = OH

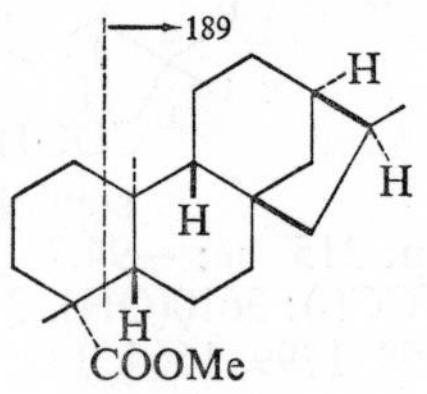

14

6) The pK_{mcs} (measured in methyl cellosolve) of the steviol carboxyl compared with those of known compounds confirmed its α (axial) configuration.[5]

7) The configuration of the 9-H was established as β[7] by application of the octant rule to keto acids 15 and 17 (derived from 8) and keto acid 19 (derived from *ent*-isokaurene 18). Compounds 15 and 19 displayed weak, positive Cotton effects, while 17 showed a negative effect. These observations could also be correlated with ord measurements on certain steroids. In accordance with the octant rule, these results could only have been obtained if the 9-H configuration in 15, 17, and 19 was β. This clarified the 9-H stereochemistry in a number of compounds, among them steviol, *ent*-kaurene and garryfoline.

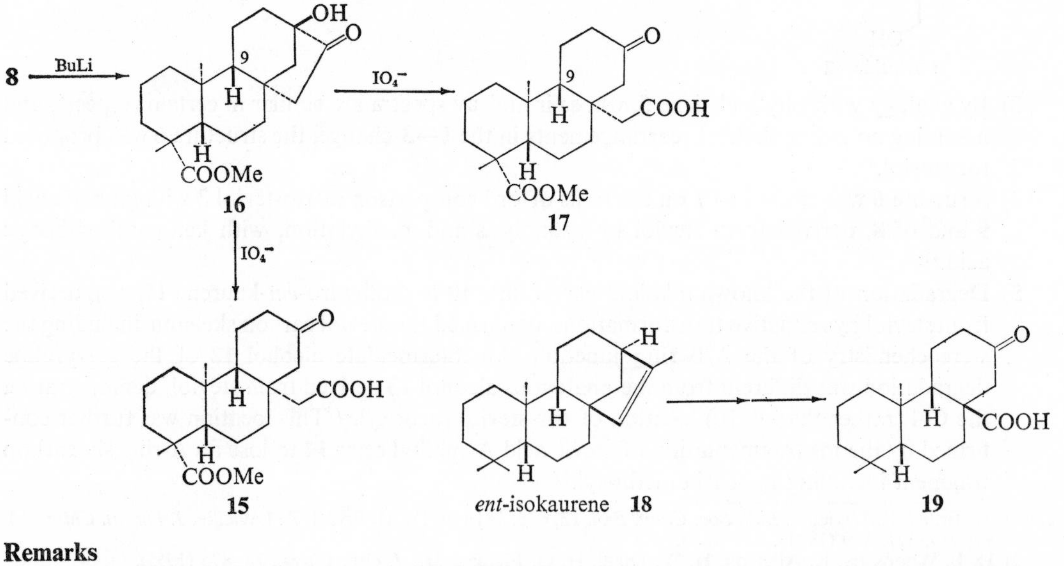

8 →(BuLi)→ 16 →(IO₄⁻)→ 17

16 →(IO₄⁻)→ 15

ent-isokaurene 18 → 19

Remarks

Stevioside 2 was first isolated in pure form from the leaves of *Stevia rebaudiana*. The gluco-

7) C. Djerassi, P. Quitt, E. Mosettig, R. C. Cambie, P. S. Rutledge, L. H. Briggs, *J. Am. Chem. Soc.*, 83, 3720 (1961).

side is about 300 times sweeter than sugar. Steviol has been shown to have gibberellin-like activity.[8]

Steviol shares with many of the gibberellins a bicyclo[1.2.3]octane C/D ring structure characterized by a bridgehead hydroxyl adjacent to a site of unsaturation. This structural feature is susceptible to facile rearrangement, either under acid (**1**→**3**) or base (**8**→**16**) catalysis.

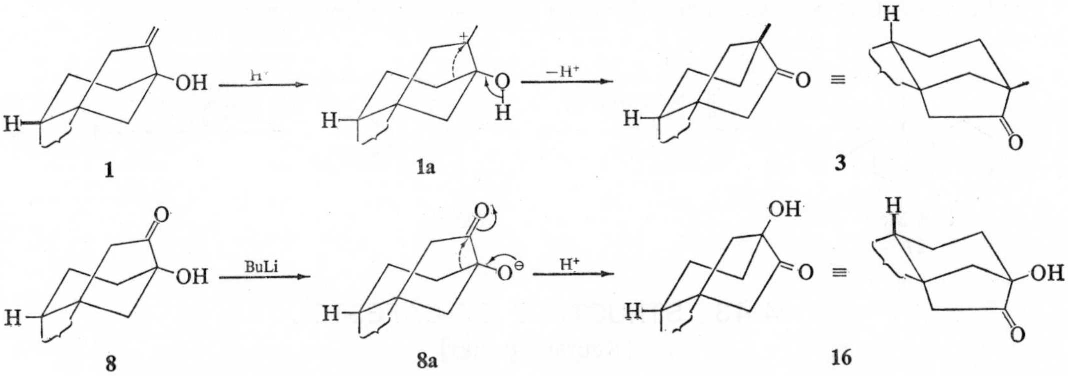

This is of special interest here and in the gibberellin case in that the stereochemistry of the C/D ring junction is inverted in the course of this reaction, thus facilitating the preparation of compounds useful in the mutual correlation of several different families of diterpenoids.

Synthesis

Mori *et al.*[9] reported the total synthesis of steviol starting from (±)-deoxypodocarpic acid **20**, which was converted to **21** and thence to steviol.

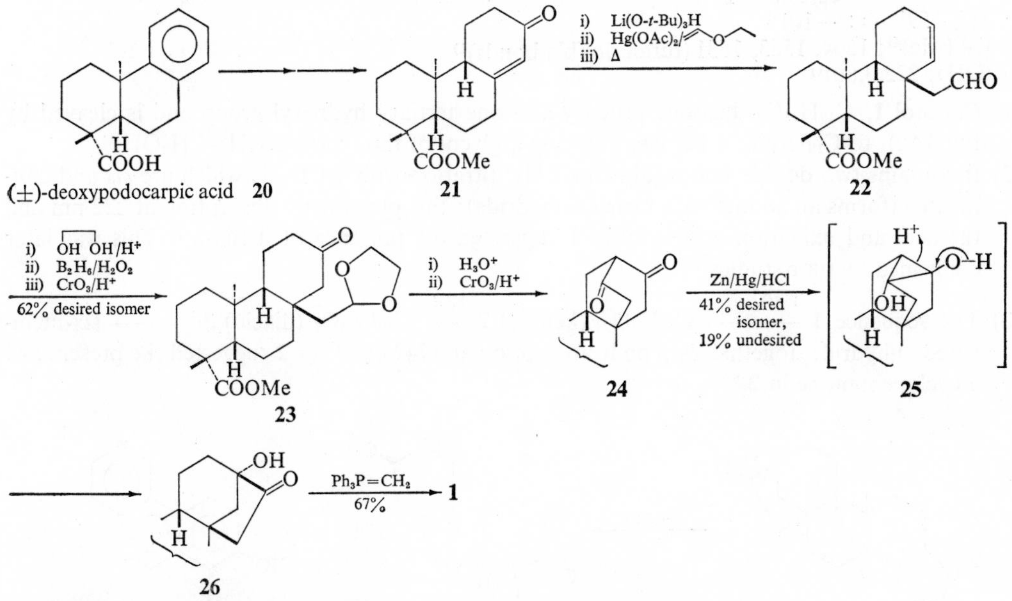

8) M. Rudd, A. Lang, E. Mosettig, *Naturwissenschaften*, **50**, 23 (1963).
9) K. Mori, Y. Nakahara, M. Matsui, *Tetr. Lett.*, 2411 (1970); K. Mori, M. Matsui, *ibid.*, 2347 (1965).
10) I. F. Cook, J. R. Knox, *Tetr. Lett.*, 4091 (1970).

21→22: Reduction of the ketone gave the β alcohol required to fix the correct stereochemistry at C-8 via the Claisen rearrangement.

24→26: The Clemmensen reduction of the dione probably proceeds through the trachylobane-type diol **25**.

A partial synthesis by Cook and Knox[10] started with the naturally derived keto diester **27** and gave the C/D ring system directly via an acyloin-like condensation.

i) Na/liq. NH$_3$/THF
ii) t-BuOH
27%

i)   [O]
ii) Me$_3$SiCl/base
iii) Ph$_3$P=CH$_2$ → **1**

27

4.43 STRUCTURE OF CAFESTOL
[Kaurane group]

cafestol **1**

mp: 158–159° α_D: −101°

ir: 1384 (Me)[4]; 1504, 1563, 1631 (furan double bonds)[1]

uv(EtOH): 222(3.80)[1]

1) Cafestol **1**, $C_{20}H_{28}O_3$, has one primary and one tertiary hydroxyl group and is cleaved by Pb(OAc)$_4$ to $C_{19}H_{24}O_2$, a ketone, implying a glycol of the type $>$C(OH)–CH$_2$OH.[1]

2) It contains two double bonds (absorbs 2 H$_2$; titration with RCO$_3$H) which are *cis* and conjugated (forms an adduct with maleic anhydride); this plus the uv absorption at 222 nm and the acid and oxidation sensitivity of **1** suggested the presence of a furan.[1] This was later confirmed by nmr studies.[2]

3) The sequence **1** $\xrightarrow{\text{Pb(OAc)}_4}$ $C_{19}H_2O_{24}$ (ketone) **2** $\xrightarrow{\text{KOI}}$ $C_{19}H_{28}O_5$ (diacid) **3** $\xrightarrow{\triangle}$ six-membered anhydride, together with an ir absorption at 1742 cm^{-1} for **2** suggested the presence of a cyclopentanone in **2**.[1]

KOI →

2   **3**   **4**

---

1) R. D. Haworth, A. H. Jubb, J. McKenna, *J. Chem. Soc.*, 1983, (1955); and references therein.

4) Degradation experiments supplied information about the furan ring environment; the eventually successful dehydrogenation over Pd of a derivative of **1** gave phenanthrol **4**, indicating the basic A/B/C ring structure and placing the furan ring at C-3/C-4 of ring A.[3,4]

5) The remaining four carbon atoms (C-15,16,17,20) were located as follows: A Me group is present (ir: 1384).[4] This Me (C-20) and the two-carbon fragment (required to form a C-5 ring) were placed by analogy with phyllocladene.[4] Other workers[5] put the C-20 methyl at C-5 (not C-10) on the basis of degradation evidence. The methylol was fixed at C-16 (not C-15) because the ketone **2** could be tribrominated, suggesting 3 (not 2) active hydrogens.[1] This evidence was refuted as violating Bredt's rule and other circumstantial evidence was considered;[6] the degradation of **5**, derived from **1**, to the keto ester **6** proved the point of attachment conclusively.[9]

i) RCO$_3$H
ii) base
iii) CH$_2$N$_2$
iv) CrO$_3$/py

i) Wolff-Kishner
ii) H$_2$/PtO$_2$/HOAc
iii) CrO$_3$

5:R=H
5a:R=Br

6

7

6) Ketone **7**, prepared from **2**, gave alcohols (upon reduction by various methods) and bromination products that were consistent, on the basis of comparison with known cases, only with a methyl at C-10 (not C-5). Ord comparison of **7** with a steroid model suggested that the A/B stereochemistry in cafestol is antipodal to that in the steroids.[6,7]

7) Firm evidence for placement of the cyclopentane ring bridge β at C-8 and C-13 was adduced from the coincidence of the ord curves of ketones **5** and **8**, derived from phyllocladene **9**,[8] in conjunction with application of the octant rule to the keto ester **6**. The 9-H was erroneously fixed as α on the basis of ord comparison with **8**.[9]

8:R=O
9:R=CH$_2$

10

8) An x-ray crystallographic analysis of **5a**, undertaken because the proposed *trans-syn-trans* backbone of cafestol (9α-H) contravened basic concepts in terpenoid biogenesis, proved the structure to be in fact *trans-anti-trans* (9β-H) as shown.[10]

9) The configuration of the C-16 hydroxyl in cafestol **1**, unelucidated by the x-ray analysis, had

2) E. J. Corey, G. Slomp, S. Dev, S. Tobinaga, E. R. Glazier, *J. Am. Chem. Soc.*, **80**, 1204 (1958).
3) C. Djerassi, H. Bendas, P. Sengupta, *J. Org. Chem.*, **20**, 1046 (1955).
4) H. Bendas, C. Djerassi, *Chem. Ind.*, 1481 (1955).
5) R. D. Haworth R. A. W. Johnstone, *J. Chem. Soc.*, 1492 (1957).
6) C. Djerassi, M. Cais, C. A. Mitscher, *J. Am. Chem. Soc.*, **81**, 2386 (1959).
7) C. Djerassi, M. Cais, L. A. Mitscher, *J. Am. Chem. Soc.*, **80**, 247 (1958).
8) The asymmetric centers of phyllocladene in question here were already known.
9) R. A. Finnegan, C. Djerassi, *J. Am. Chem. Soc.*, **82**, 4342 (1960).
10) A. I. Scott, G. A. Sim, G. Ferguson, D. W. Young, F. McCapra, *J. Am. Chem. Soc.*, **84**, 3197 (1962).

been assigned on the basis of the stereospecificity of hydroxylation by OsO_4 of the olefin **10**.[11]

10) Kahweol **11** occurs naturally as a minor impurity of cafestol and posed purification problems in the isolation of **1**. Its structure was initially postulated as **12**,[6] but later proved to have only one double bond in conjugation with the furan system.[12]

11   **12**

Remarks

Cafestol is the main constituent of the non-saponifiable portion of coffee bean oil. It was early reported to have estrogenic activity and a steroid-like structure was postulated. This was abandoned with the disproof of such activity and its investigation as a diterpene then developed.[1]

Valuable information concerning the structure of natural products is often available from studies of their biogenesis. The diterpenoids provide a good illustration of such biogenetic influence on structure determination.

According to the biogenetic isoprene rule as formulated for terpenes,[13] the backbone of diterpenoids should be formed by cyclization of geranyl geraniol **13** in its most stable conformation to give a *trans-anti* backbone.[15]

13   **14**

15   **16**   *ent*-kaurene' **17**

18   **19**

---

11) R. A. Finnegan, *J. Org. Chem.*, **26**, 3057 (1961).
12) H. P. Kaufman, A. K. Sen Gupta, *Chem. Ber.*, **96**, 2489 (1963); *ibid.*, **97**, 2652 (1964).
13) A. Eschenmoser, L. Ruzicka, O. Jeger, D. Arigoni, *Helv. Chem. Acta.*, **38**, 1890 (1959); L. Ruzicka, *Proc. Chem. Soc.*, 341 (1959); L. Ruzicka, *Pure Appl. Chem.*, **6**, 493 (1963).
14) E. Wenkert, *Chem. Ind.*, 282 (1955).
15) A. I. Scott, F. McCapra, F. Comer, S. A. Sutherland, D. W. Young, G. A. Sim, G. Ferguson, *Tetr.*, **20**, 1339 (1964); and references therein.

This further cyclizes to the protonated cyclopropane **16**, suggested by Wenkert,[14] finally giving one of the tetracyclic structures (e.g. *ent*-kaurene **17**). While it is formally possible that biogenetic cyclization could proceed via a B-ring boat **18** to give a *trans-syn* skeleton **19**, the large number of examples of *trans-anti* compounds known in the diterpene series and the intrinsic stability of a chair *vs.* a boat conformation made the *trans-syn* arrangement seem unreasonable.

The known exceptions to the *trans-anti* stereochemistry before 1964 (gibberellic acid, isopimaric acid, stachenone, eperuic acid and cafestol) deviated from the all *trans* cyclization rule only at 9-H. The structures of these were systematically reexamined by Scott *et al.*[15] using cd and x-ray techniques; all the "exceptions" to the biogenetic rule were found in fact to have a 9-H stereochemistry antipodal to that originally proposed.

Many of the former errors in the structures of the above compounds arose because of ambiguities in the interpretation of ord data. Since a specific cd curve gives rotatory information concerning only the immediate carbonyl environment, the use of cd measurements on diterpenoid ketones allowed the resolution of considerable fine structure. From these studies, Scott *et al.*[15] derived a general rule enabling direct determination by cd of the 9-H stereochemistry relative to the two-carbon bridge in compounds incorporating a bicyclo [1.2.3] octane C/D ring system. The validity of this rule, representing the first assignment of stereochemistry by cd, was confirmed by x-ray in several instances.

4.44 ENMEIN
[Kaurane group]

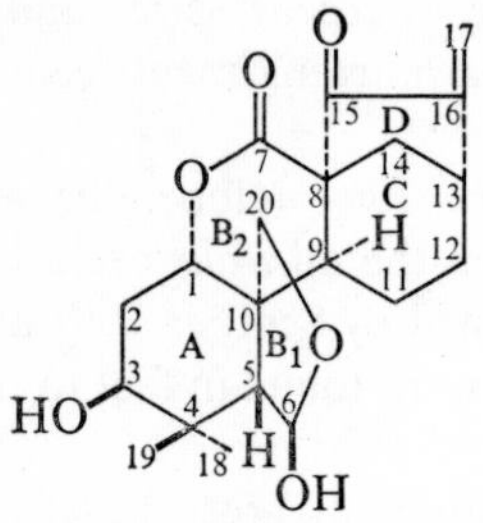

enmein **1**

mp: 274–275° α_D: −156°

1) Enmein was the first of a host of similar compounds to be structurally elucidated; it has a unique 6,7-seco kaurane skeleton and presented a challenging problem. Studies initiated by Yagi in 1910 culminated in identification of the structure as **1** through the joint chemical efforts of several groups,[1] and also by independent x-ray studies.[2] An outline of the structural studies and some interesting transformations encountered during these investigations are summarized below.

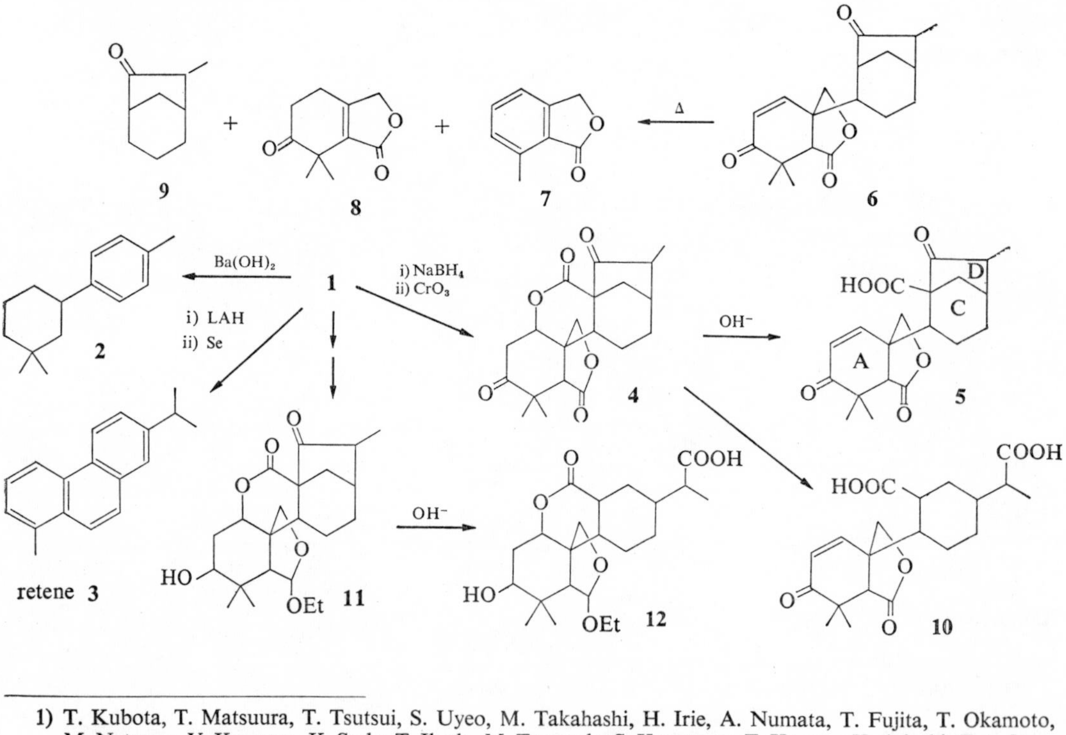

1) T. Kubota, T. Matsuura, T. Tsutsui, S. Uyeo, M. Takahashi, H. Irie, A. Numata, T. Fujita, T. Okamoto, M. Natsume, Y. Kawazoe, K. Sudo, T. Ikeda, M. Tomoeda, S. Kanatome, T. Kosuge, K. Adachi, *Tetr. Lett.*, 1243 (1964); T. Kubota, T. Matsuura, T. Tsutsui, S. Uyeo, H. Irie, A. Numata, T. Fujita, T. Suzuki, *Tetr.*, **22**, 1659 (1966).
2) Y. Iitaka, M. Natsume, *Tetr. Lett.*, 1257 (1964).
3) S. Kanatomo, *Yakugaku Zasshi*, **81**, 1437 (1961).

2) Dry distillation of **1** to afford **2** indicated the nature of the carbon skeleton.[3] However, the production of a minute amount of retene **3** led to the proposal of a phyllocladene ring.[3]

3) Pyrolysis of **6** gave products **7–9**. The keto-lactone **8** and ketone **9** account for 19 of the 20 carbon atoms.

4) Formation of the α,β-unsaturated keto-acid **5**, which decarboxylates upon pyrolysis, suggested the location of the δ-lactone O and C termini, and the relationship of rings C and D to A. These conclusions were supported by formation of the enone dicarboxylic acid **10** and the acid **12**.

5) The presence of the five-membered hemiacetal ring B_1 was indicated by the conversion of dihydroenmein diacetate **13** to dihydrofuran **14** and formation of the dilactone **15**.

6) The relative stereochemistry of **1** was assigned by nmr analysis,[1] and x-ray studies of the monobromoacetate of dihydroenmein-monoacetate.[2]

7) The absolute configuration was deduced by application of Prelog's atrolactic acid method to **4**, the conversion of **1** to (−)-kaurene **20**,[4,5] and by x-ray analysis.[6]

Remarks

Enmein is the main bitter substance of the plant *Isodon trichocarpus*,[7] a home remedy for gastrointestinal disturbances.

4) K. Shudo, M. Natsume, T. Okamoto, *Chem. Pharm. Bull.*, **13**, 1019 (1965).
5) E. Fujita, T. Fujita, K. Fuji, N. Ito, *Chem. Pharm. Bull*, **13**, 1023 (1965); *Tetr.* **22**, 3423 (1966).
6) Y. Iitaka, M. Natsume, *Acta Cryst.*, **20**, 197 (1966).
7) M. Takahashi, T. Fujita, T. Koyama, *Yakugaku Zasshi*, **78**, 699 (1958); T. Ikeda, S. Kanatomo, *ibid.*, **78**, 1123 (1958); K. Naya, *Nippon Kagaku Zasshi*, **79**, 885 (1958).

4.45 TRANSFORMATIONS OF ENMEIN
[Kaurane group]

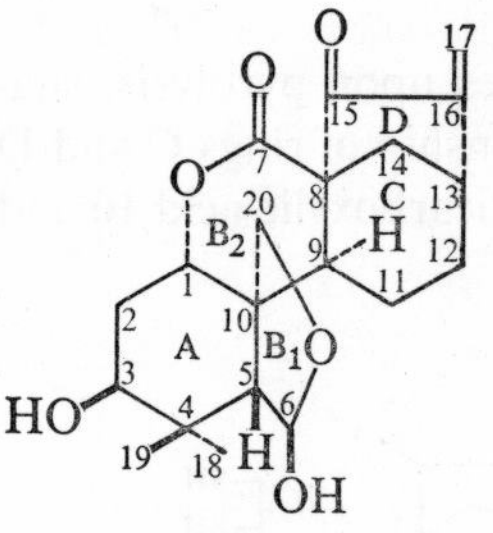

enmein **1**

The polyoxygenated functions of the 6,7-*seco* diterpene enmein **1**, and its kaurane-type congeners, e.g. trichokaurin **22** (presumably a biogenetic precursor of **1**), provide routes to numerous interesting interconversions. Some of the compounds thus obtained have been used in the synthesis of various diterpenes and diterpene alkaloids.

Enmein to ent-kaurane[1, 2]

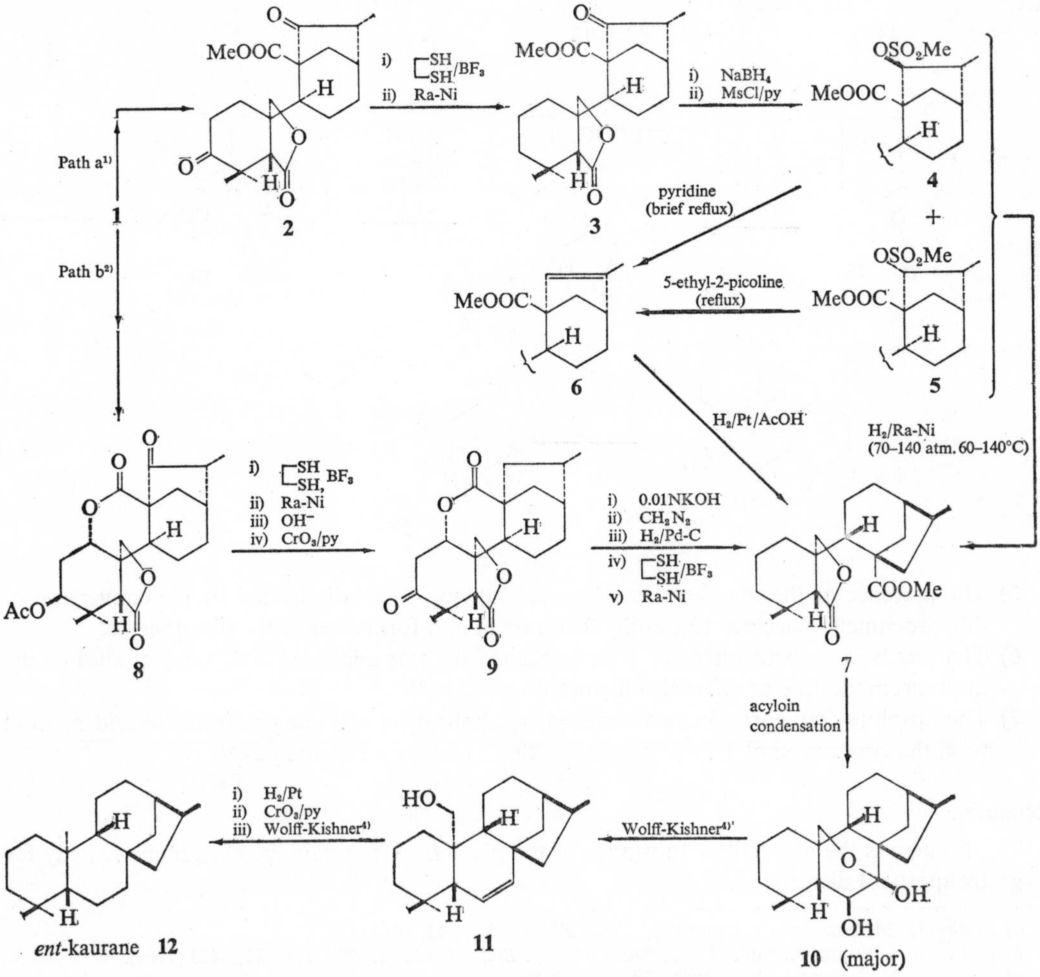

1) K. Shudo, M. Natsume, T. Okamoto, *Chem. Pharm. Bull.*, **13**, 1019 (1965).
2) E. Fujita, T. Fujita, K. Fuji, N. Ito, *Chem. Pharm. Bull.*, **13**, 1023 (1965).

1) Compound **7**, the key intermediate for conversion of enmein into the kaurane skeleton, was prepared by two routes utilizing compounds **2**[3] and **8**[3], both of which played essential roles in structural studies. The major problem involved removal of the 15-oxo function in **1**.

2) The keto-ester function in **3** resisted thioketalization but this was not the case for the keto-lactone **8**; presumably, the lactone in **8** fixes the 15-oxo in an exposed conformation.

3) 4/5→**7** could be carried out directly by employment of vigorous conditions.

4) **7**→**10** was achieved by Na/NH₃[2] or by Na in xylene/toluene/THF.[1]

5) Conversion to **12** furnished the first chemical evidence for the absolute configuration of enmein.

Enmein to *ent*-kaurane, *ent*-15-kaurene and *ent*-16-kaurene[5]

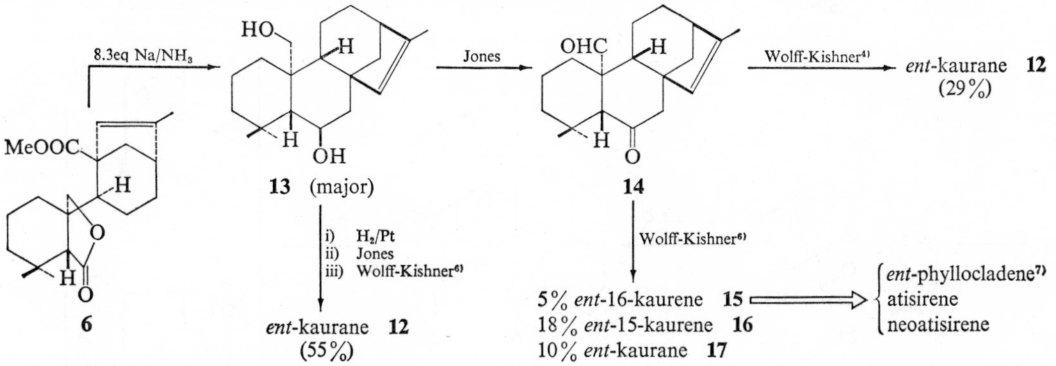

Ent-16-kaurene **15** was used for syntheses of atisirene, etc.[7]

Enmein and trichokaurin to keto acid 25[8,9]

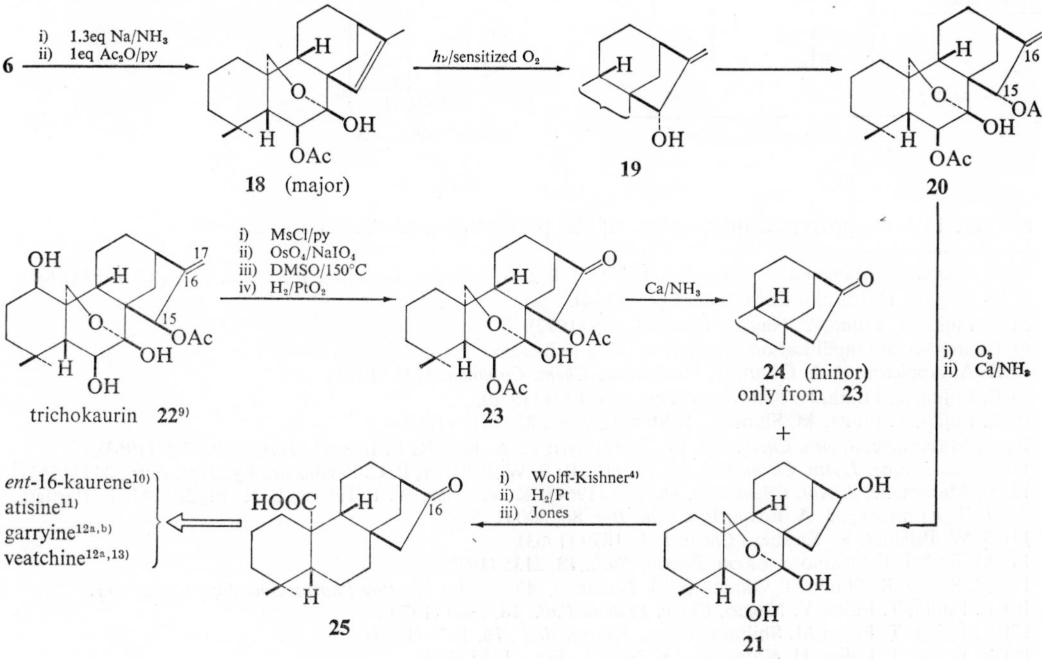

1) Trichokaurin is one of the minor congeners of enmein.
2) Lemieux oxidation of **20** resulted in acyl migration from C-15 to C-17; this migration did not occur in **22**, epimeric at C-15.
3) Keto acid **25** was the key intermediate employed in further conversions to atisine, etc.

Enmein to GA$_{15}$[14)]

See ref. 14 for a proposed mechanism of the photochemical conversion **28→29**.

3) T. Kubota, T. Matsuura, T. Tsutsui, S. Uyeo, H. Irie, A. Numata, T. Fujita, T. Suzuki, *Tetr.*, **22**, 1659 (1966).
4) W. Nagata, H. Itazaki, *Chem. Ind.*, 1194 (1964).
5) E. Fujita, T. Fujita, Y. Nagao, *Tetr.*, **28**, 555 (1972).
6) Huang-Minlon modification of Wolff-Kishner reduction.
7) R. A. Appleton, P. A. Gunn, R. McCrindle, *Chem. Commun.* 1131 (1968).
8) E. Fujita, T. Fujita, H. Katayama, *Tetr.*, **26**, 1009 (1970).
9) E. Fujita, T. Fujita, M. Shibuya, T. Shingu, *Tetr.*, **25**, 2517 (1969).
10) S. Masamune, *J. Am. Chem. Soc.* **86**, 289 (1964); R. A. Bell, R. E. Ireland, *Tetr. Lett.*, 269 (1963).
11) S. Masamune, *J. Am. Chem. Soc.* **86**, 291 (1964); S. W. Pelletier, P. C. Parthasarathy, *Tetr. Lett.*, 205 (1963).
12) S. Masamune, *J. Am. Chem. Soc.* **86**, 290 (1964); K. Wiesner, W. I. Taylor, S. K. Figdor, M. F. Bartlett, J. R. Armstrong, J. A. Edwards, *Chem. Ber.* **86**, 800 (1953).
13) S. W. Pelletier, K. Kawazu, *Chem. Ind.*, 1879 (1963).
14) M. Somei, T. Okamoto, *Chem. Pharm. Bull.*, **18**, 2135 (1970).
15) M. Somei, K. Shudo, T. Okamoto, M. Natsume, *Abstr. 24th Meeting Pharm. Soc. Japan*, 460, 1967.
16) E. Fujita, T. Fujita Y. Nagao, *Chem. Pharm. Bull.*, **18**, 2343 (1970).
17) E. Fujita, T. Fujita M. Shibuya, *Chem. Pharm. Bull.*, **16**, 1573 (1968).
18) E. Fujita, T. Fujita, H. Katayama, Y. Nagao, *Tetr.*, 1335 (1969).

Trichokaurin to isodocarpin[16]

35   **36**   isodocarpin **37**

Isodocarpin[17] is another minor congener.

Enmein to *ent*-abietane[18]

38

ent-abietane **39**

Reaction with NaH opened ring D to give **38**, which was eventually converted to *ent*-abietane; this was the first preparation of a saturated abietane **39**.

4.46 INTRODUCTION TO GIBBERELLINS

The gibberellin group compounds are very widely used plant growth-promoting substances and are manufactured on a large scale by fermentation. The first scientific description of the anomalous growth of rice seedlings, "baka-nae" (meaning "stupidly overgrown seedling") disease, was reported in 1898.[1] In 1926,[2] the disease was correctly attributed to a "toxin" produced by a fungus which was classified later as *Gibberella fujikuroi* (*Fusarium moniliforme*). A key achievement was the isolation of two crystalline substances in 1938 by Yabuta and Sumiki[3] for which the name "gibberellins" was coined. The crystals with mp 245–246° (dec), the active substance, were named gibberellin A (1941),[4] but were subsequently found to be a mixture of three components (1955),[5] gibberellins A_1 **2**, A_2 **3** and A_3 **1** (GA₃, the most representa-tive of the gibberellins, is also called gibberellic acid). The second substance isolated in 1938, mp 194–196°, gibberellin B, was devoid of activity, and was later found to correspond to al-logibberic acid **4**.[6]

1) S. Hori, *Noji Shikenjo Seiseki* (Reports from the Exptl. Agric. Station), **12**, 110 (1898).
2) E. Kurosawa, *Trans. Nat. Hist. Soc. Formosa*, **16**, 213 (1926).
3) T. Yabuta, Y. Sumiki, *J. Agr. Chem. Soc. Japan*, **14**, 1526 (1938).
4) T. Yabuta, Y. Sumiki, K. Aso, T. Tamura, H. Igarashi, K. Tamari, *J. Agr. Chem. Soc. Japan*, **17**, 721, 894, 975 (1941).
5) N. Takahashi, H. Kitamura, A. Kawarada, T. Seta, M. Takai, S. Tamura, Y. Sumiki, *Bull. Agr. Chem. Soc. Japan*, **19**, 267 (1955).
6) B. E. Cross, *J. Chem. Soc.*, 4670 (1954).
7) T. P. Mulholland, G. Ward, *J. Chem. Soc.*, 4676 (1954).
8) *Gibberellins* (in Japanese) (ed. S. Tamura), University of Tokyo Press, 1969.

Early structural studies by the Japanese (University of Tokyo) were hampered by the poor yield from the strain used (50–100 mg crude crystals from 10 l of culture liquid), inconsistencies arising from the fact that gibberellin A was a mixture, and World War II.[7,8] One significant achievement during this period was the isolation of a substituted fluorene (gibberene 5[7]) following Se dehydrogenation of gibberic acid[4] (structure established as 6[9]). Subsequent studies were resumed by Sumiki, Tamura, Takahashi and co-workers, Stodola and co-workers (US Dept. of Agriculture),[10] and the British Imperial Chemical Industries (ICI) group, and culminated in establishment of the structure by ICI. See refs 8, 11, 12 and 13 for reviews of gibberellin chemistry.

9) B. E. Cross, J. F. Grove, J. MacMillan, T. P. C. Mulholland, *J. Chem. Soc.*, 2520 (1958).

10) F. H. Stodola, K. B. Raper, C. I. Fennel, H. D. Conway, V. E. Sohns, C. T. Langford, R. W. Jackson, *Arch. Biochem. Biophys.*, **54**, 240 (1955).

11) F. H. Stodola, *Source Book on Gibberellins, 1828–1957*, Agr. Res. Serv., U. S. Dept. Agr., 1958.

12) P. W. Brian, J. F. Grove, J. MacMillan, *Progress in the Chemistry of Organic Natural Products* (ed. L. Zechmeister), vol. XVIII, p. 350–433, Springer-Verlag, 1969.

13) J. R. Hanson, *The Tetracyclic Diterpenes*, ch. 3, Pergamon Press, 1968.

14) T. P. Mulholland, *J. Chem. Soc.*, 2693 (1958).

15) A. J. Birch, R. W. Rickards, H. Smith, J. Winter, W. B. Turner, *Chem. Ind.*, 3049 (1960).

16) J. F. Grove, T. P. Mulholland, *J. Chem. Soc.*, 3007 (1960).

17) J. F. Grove, J. MacMillan, T. P. C. Mulholland, W. B. Turner, *J. Chem. Soc.*, 3049 (1960).

18) B. E. Cross, J. F. Grove, J. MacMillan, J. S. Moffatt, T. P. C. Mulholland, J. C. Seaton, N. Sheppard, *Proc. Chem. Soc.*, 302 (1959).

19) See structures in ref. 12.

20) F. McCapra, A. I. Scott, G. Sim, D. W. Young, *Proc. Chem. Soc.*, 185 (1962).

21) A. I. Scott, F. McCapra, F. Comer, S. A. Sutherland, D. W. Young, G. A. Sim. A. Ferguson, *Tetr.* **20**, 1339 (1964).

22) K. Schreiber, J. Weiland, G. Sembdner, *Tetr. Lett.*, 4285 (1967).

23) J. MacMillan, N. Takahashi, *Nature*, **217**, 170 (1968).

Determination of the structure of GA_3 1 (gibberellic acid) started with clarification of the structure of gibberene 5 (by synthesis),[7] followed by gibberic acid 6[9] and allo-gibberic acid 4[14]; the conversion of 4 to 6 by acid, a reaction well-known in gibberellin chemistry, involves a Wagner-Meerwein rearrangement.[15] Chemical studies and measurements of optical rotatory dispersion curves correctly indicated the B/C ring fusion and absolute configurations of 4 and 6.[16,17] Extensive studies on the original gibberellins, which centered around GA_1 2 and GA_3 1 finally led to the correct planar representation 1 in 1959.[18] There were controversies in the lactone configuration, and 9-H was considered to have an α-configuration[19] until the present full structure was established by x-ray and circular dichroism studies.[20,21]

About 40 gibberellins have been isolated to date as metabolites of the fungus *G. fujikuroi*, and as growth factors from a wide variety of higher plants. Most of these are C_{19} gibberellins, but some contain a carbon substituent at C-10 and are called C_{20} gibberellins. Major structural variations are the presence or absence of a double bond and hydroxyls at C-1, C-2, C-3, and of hydroxyls at C-13 and C-16; gibberellin glycosides are also found, e.g., 3-O-β-D-glucosyl gibberellin A_3.[22] The gibberellins are named with a subscript after A, as in gibberellin A_{12} (or abbreviated to GA_{12}).[23] The numbering system adopted here follows those of other diterpenes and not the conventional gibberellin system 8.

4.47 BIOSYNTHESIS OF GIBBERELLINS
[Kaurane group]

Birch *et al.*[1] first showed that gibberellin A_3 17 (GA_3) followed the general pattern of diterpenoid biosynthesis by isolation and degradation of radioactive GA_3 from a culture of *Gibberella fujikuroi* to which [2-^{14}C]mevalonic lactone 1 had been added.

The overall biogenetic scheme of various gibberellins, as clarified by feeding labelled precursors, including labelled gibberellins, to *G. fujikuroi*, is summarized overleaf. Mevalonic lactone 1,[1] geranyl pyrophosphate 2,[2] geranylgeraniol 3,[3] ($-$)-labda-8,13-dien-15-ol 4,[3,4] ($-$)-pimara-8(14),15-diene 5,[3,5] ($-$)-kaurene 6,[6] ($-$)-kaurene-19-ol 7,[7] ($-$)-kaurene-19-oic acid 9,[8] ($-$)-7β-hydroxykauren-19-oic acid 10,[9,10] and GA_{12}[3] have been shown to be intermediates by radioactive feeding experiments and isolation of various gibberellins.

Continued on p. 270.

1) A. J. Birch, R. W. Richards, H. Smith, A. Harris, W. B. Whalley, *Tetr.*, **7**, 241 (1959).
2) R. Evans, J. R. Hanson, A. F. White, *J. Chem. Soc. C*, 2601 (1970).
3) J. R. Hanson, A. F. White, *J. Chem. Soc. C*, 981 (1969).
4) J. R. Hanson, A. F. White, *Chem. Commun.*, 103 (1969).
5) J. R. Hanson, A. F. White, *Chem. Commun.*, 1689 (1968).
6) B. E. Cross, R. H. B. Galt, J. R. Hanson, *J. Chem. Soc.*, 295 (1964).
7) A. J. Verbiscar, G. Cragg, T. A. Geissman, B. O. Phinny, *Phytochem.*, **6**, 807 (1967).
8) T. A. Geissman, A. J. Verbiscar, B. O. Phinny, G. Cragg, *Phytochem.*, **5**, 933 (1966).
9) J. R. Hanson, A. F. White, *Chem. Commun.*, 410 (1969).
10) J. R. Hanson, A. F. White, *Chem. Commun.*, 208 (1971).
11) J. R. Hanson, A. F. White, *Chem. Commun.*, 1071 (1969).

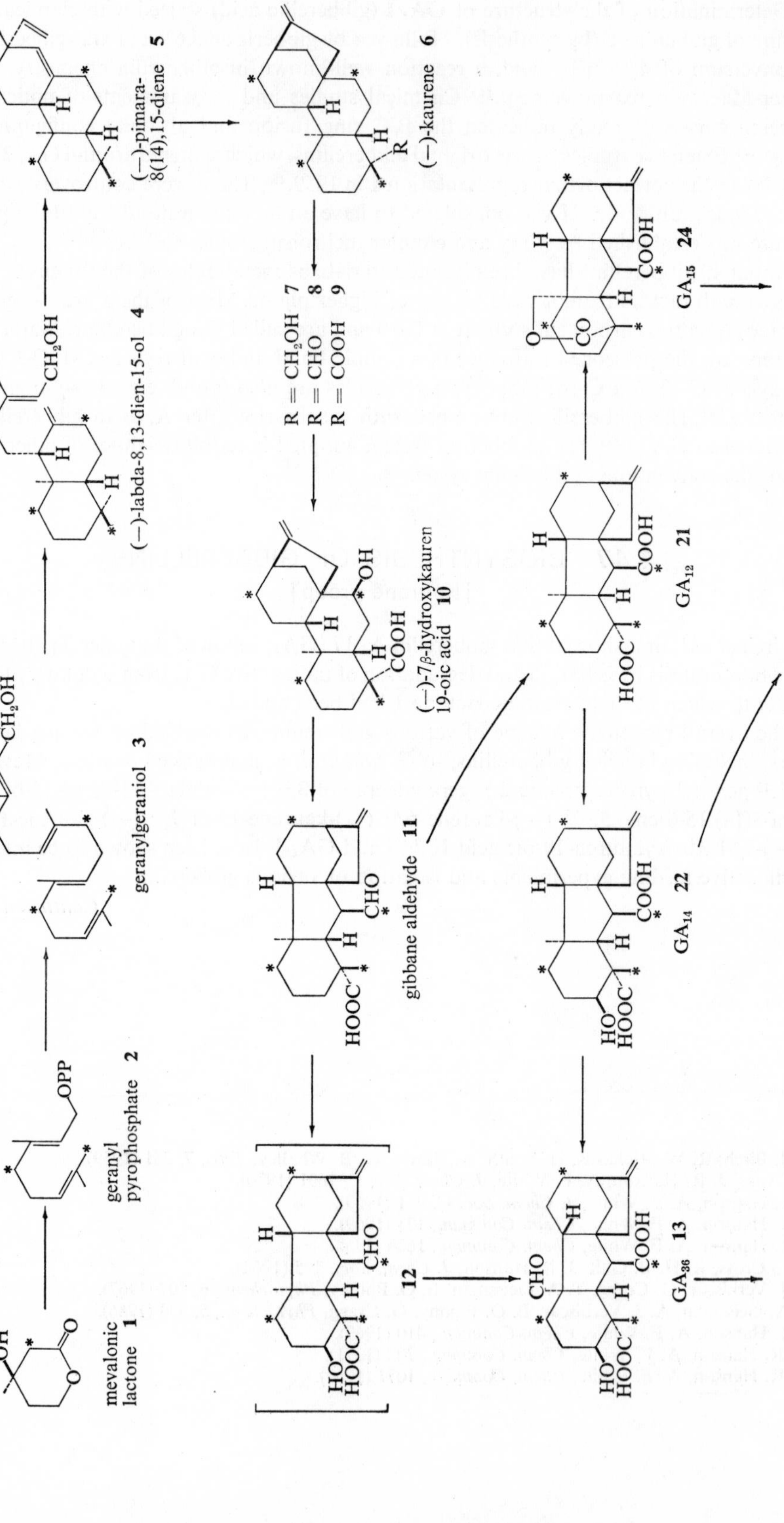

(−)-pimara-8(14),15-diene **5**

(−)-kaurene **6**

(−)-labda-8,13-dien-15-ol **4**

geranylgeraniol **3**

geranyl pyrophosphate **2**

mevalonic lactone **1**

R = CH$_2$OH **7**
R = CHO **8**
R = COOH **9**

(−)-7β-hydroxykauren 19-oic acid **10**

gibbane aldehyde **11**

GA$_{15}$ **24**

GA$_{12}$ **21**

GA$_{14}$ **22**

GA$_{36}$ **13**

12

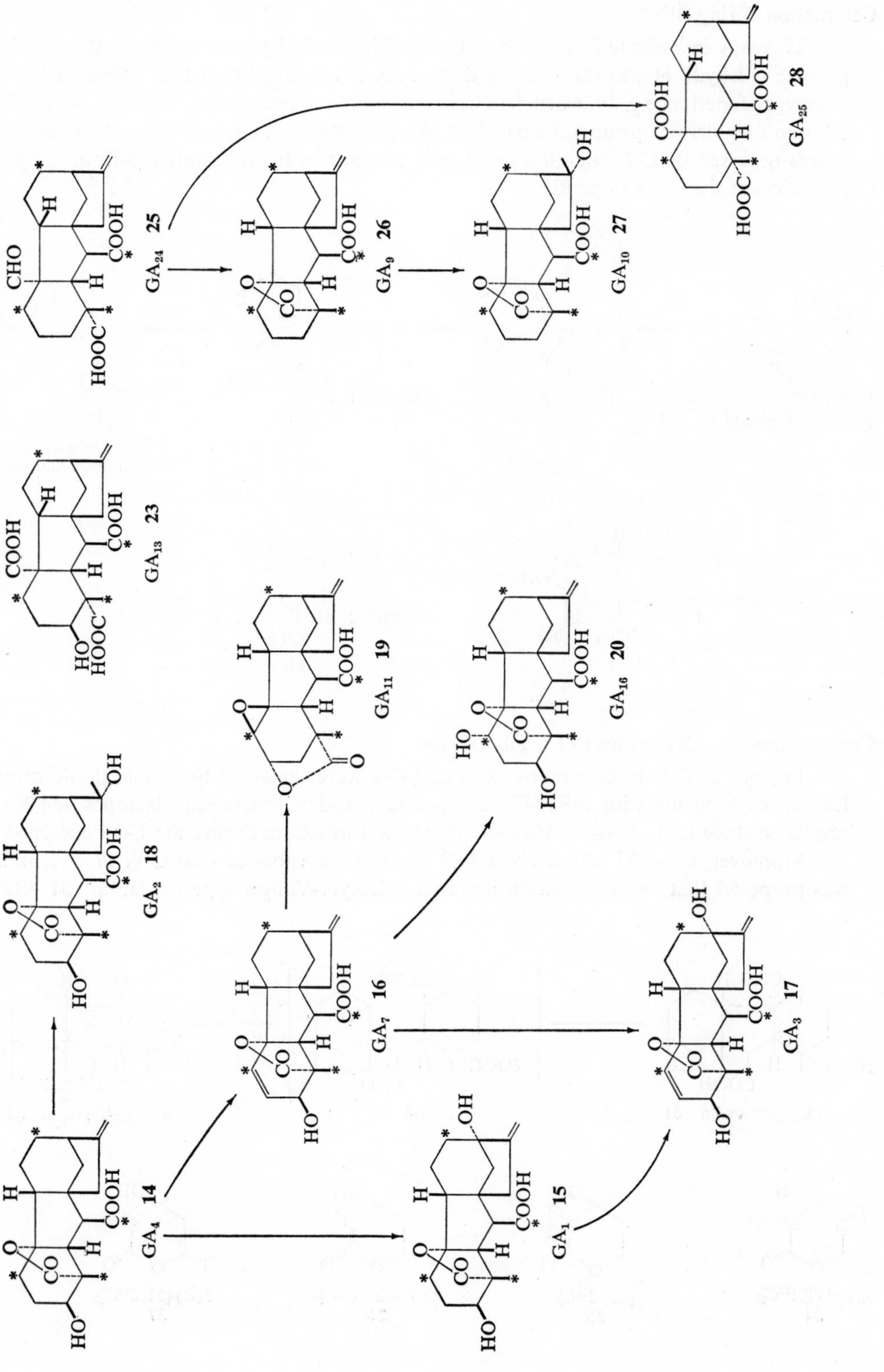
GA4 14
GA1 15
GA7 16
GA3 17
GA2 18
GA11 19
GA16 20
GA13 23
GA24 25
GA9 26
GA10 27
GA25 28

Contraction of ring B[10]

This was investigated by usage of the doubly labelled precursor **30**. As the T/^{14}C ratio in gibbane aldehyde **11** was the same as that of **30**, it was concluded that kaurenolides with the 6α-oxygen function, e.g. **29**, which had been considered as precursors, could be excluded. As one T was in the aldehyde group (shown by loss of one T in **11**→**17**), it was shown that contraction of ring B occurred by C-7 oxidation to **10**, and subsequent hydride shift (3β-T) accompanied by migration of the C-7/C-8 bond.

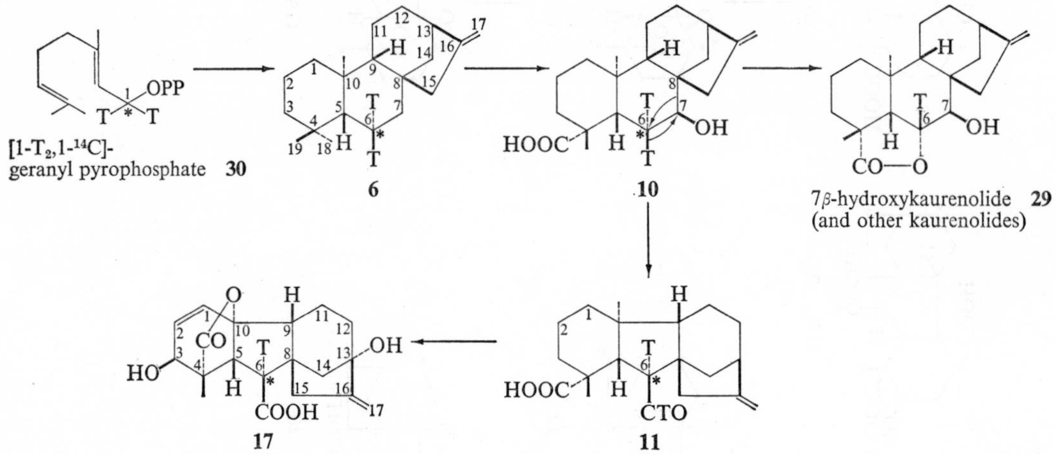

[1-T$_2$,1-^{14}C]-
geranyl pyrophosphate **30**

6

10

7β-hydroxykaurenolide **29**
(and other kaurenolides)

17

11

Conversion of C$_{20}$ gibberellins to C$_{19}$ gibberellins

Protons at C-1 (both protons), C-5 and C-9 were shown to be of mevalonic origin from feeding experiments with (4R)-[4T, 2^{14}C]- and [2T$_2$,2^{14}C]-mevalonic lactones (**34,35**). These results preclude 1,10-, 5,10-, and 9,10-double bond formations during the C-10 decarboxylation.

Moreover, GA$_{13}$ **23** with a 10-COOH was not incorporated into GA$_3$ **17**. Accordingly, it was proposed that decarboxylation involved a Baeyer-Villiger type oxidation (**31**→**32**→**33**).[5]

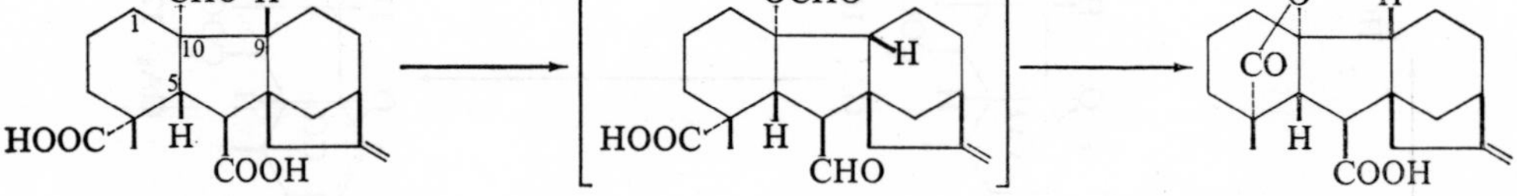

a C$_{20}$-gibberellin **31**

32

a C$_{19}$-gibberellin **23**

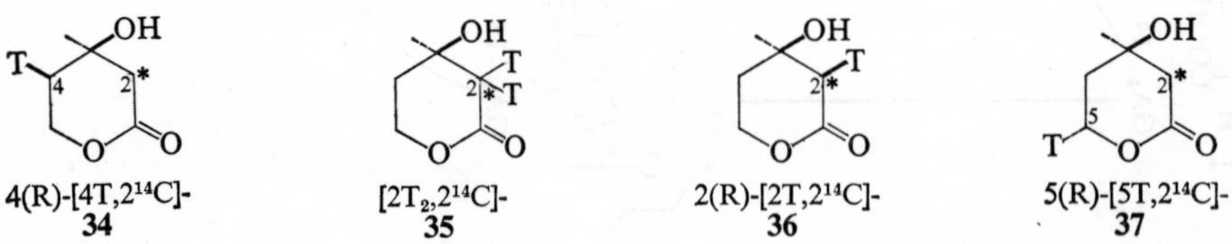

4(R)-[4T,2^{14}C]-
34

[2T$_2$,2^{14}C]-
35

2(R)-[2T,2^{14}C]-
36

5(R)-[5T,2^{14}C]-
37

Feeding with 2(R)-[2T,2^{14}C]-mevalonic lactone **36**, which should lead to aldehyde **11** with a 1β-T, gave GA$_3$ **17** with the T retained at C-1. This showed that the olefinic 1-H in **17** is derived from the 1β-H in **11**. On the other hand, feeding with 5(R)-[5T, 2^{14}C]-mevalonic lactone **37**, which should afford **11** with a 2α-T, gave GA$_3$ **17** with no T at C-2. This showed that the olefinic 2-H in **17** is derived from 2β-H in **11**. Hence the double bond in GA$_3$ **17** results from *cis*-elimination of 1α-H and 2α-H of a saturated gibberellin.[11]

4.48   SYNTHESIS OF GIBBERELLINS
[Kaurane group]

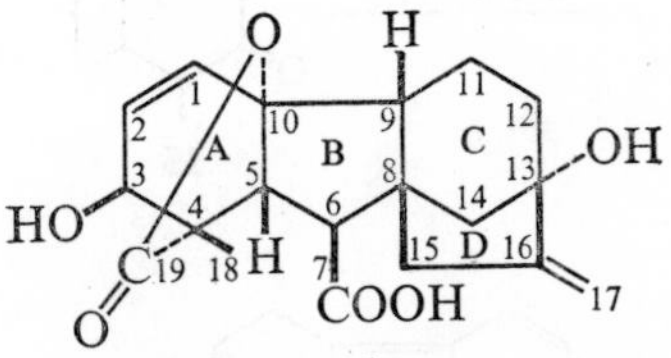

gibberellic acid (GA$_3$)   **1**

Two total syntheses of gibberellins have been carried out. The lengthy synthesis of gibberellin A$_4$ by Mori and co-workers was accomplished in five stages,[1] and is shown in route I. Nagata's synthesis of gibberellin A$_{15}$ is given in route II,[4] and finally some other approaches to gibberellin synthesis are mentioned.

Route I

The first stage, in which epigibberic acid **9** was synthesized from *o*-xylene in twenty-one steps, is summarized below.[1]

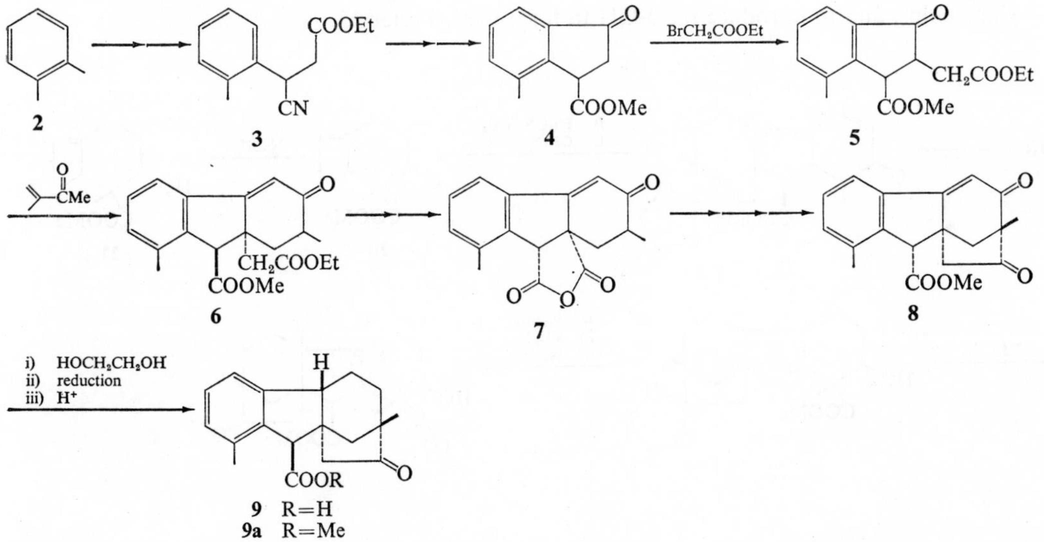

The gibbane skeleton having been formed, the next step was to activate the A ring in order that it could be functionalized. This was accomplished by nitration and eventual conversion into the ketone **12**.

---

1) K. Mori, M. Shiozaki, N. Itaya, Y. Sumiki, *Tetr.*, 1293 (1969).

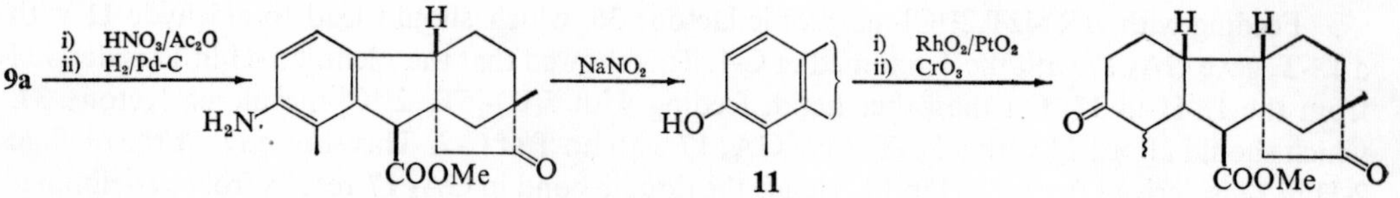

9a → [i) HNO₂/Ac₂O, ii) H₂/Pd-C] → **10** → [NaNO₂] → **11** → [i) RhO₂/PtO₂, ii) CrO₃] → **12** (epimers at C-4)

The third stage involved introduction of unsaturation into the molecule to form the dienone **18** by successive bromination and dehydrobromination reactions.

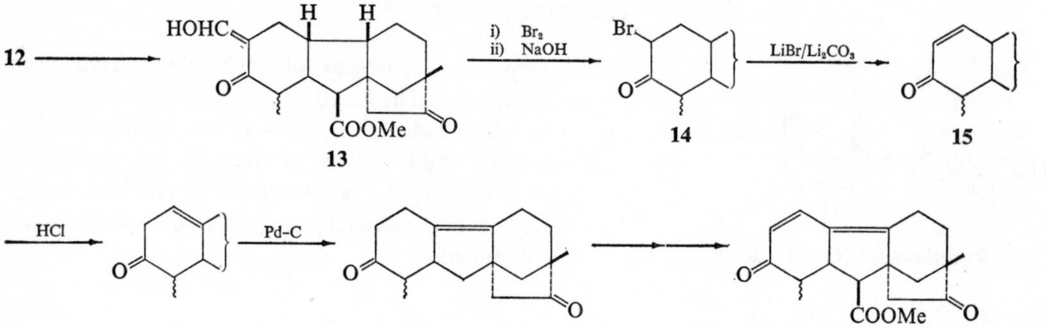

12 → **13** → [i) Br₂, ii) NaOH] → **14** → [LiBr/Li₂CO₃] → **15**

→ [HCl] → **16** → [Pd-C] → **17** → **18** (epimers at C-4)

The next stage was to form the C-4 to C-10 lactone bridge, which is characteristic of all the gibberellins, and to introduce the 3-OH to form methyl ester **23**.

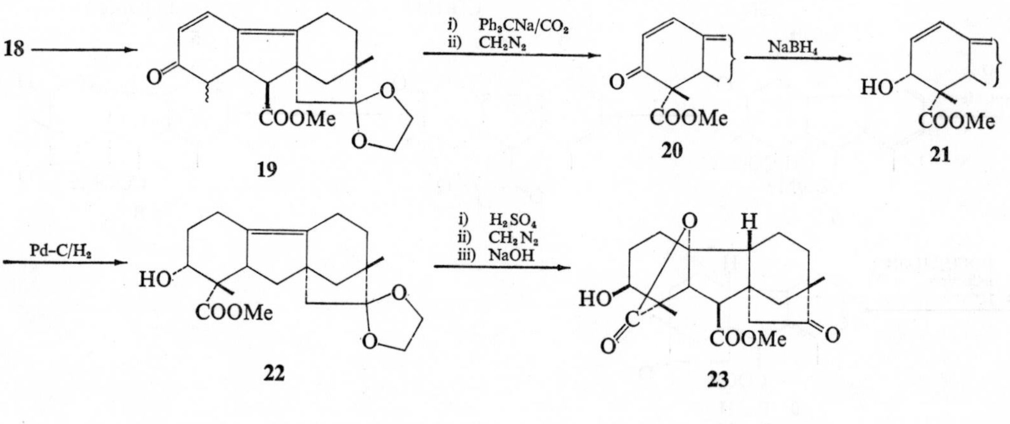

18 → **19** → [i) Ph₃CNa/CO₂, ii) CH₂N₂] → **20** → [NaBH₄] → **21**

→ [Pd-C/H₂] → **22** → [i) H₂SO₄, ii) CH₂N₂, iii) NaOH] → **23**

The final stage of this synthesis utilized a rearrangement of the C/D rings which has received much attention.[2] Cross and co-workers[3] had developed the method for the conversion of **23** to gibberellin A₄ methyl ester **25**, which Mori hydrolyzed to the free acid **26**.

2) J. MacMillan, *Ann. Reports*, **63**, 449 (1966); and references cited therein.
3) B. E. Cross, J. R. Hanson, R. M. Speake, *J. Chem. Soc.*, 3555 (1965).

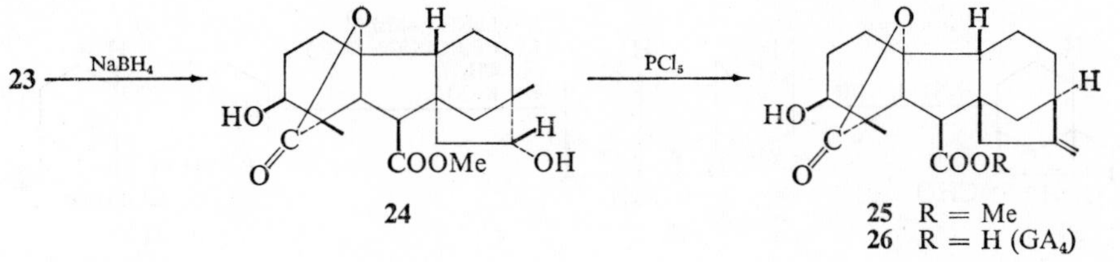

23
NaBH$_4$
PCl$_5$
HO
O
COOMe
OH
H
24
HO
COOR
25 R = Me
26 R = H (GA$_4$)

Route II

Nagata's synthesis of *dl*-gibberellin A_{15} **45**[4] utilized compound **27**, easily formed from the tricyclic ketone **27a**, as the starting material. Compound **27** was a key intermediate in a previous diterpene alkaloid synthesis.[4] Thus the A ring was essentially complete and the B and C rings could be built up without difficulty. Several interesting synthetic methods were developed in the course of this work: the hydrocyanation of an α,β-unsaturated ketone to form **33**,[5] the formylolefination of an aldehyde to form **35**,[6] and the azomethine conversion to the hemiacetals **43** and **44**.[7] (In conversion **43**/**44**→**45**, triphenylphosphine prevented double-bond migration.)

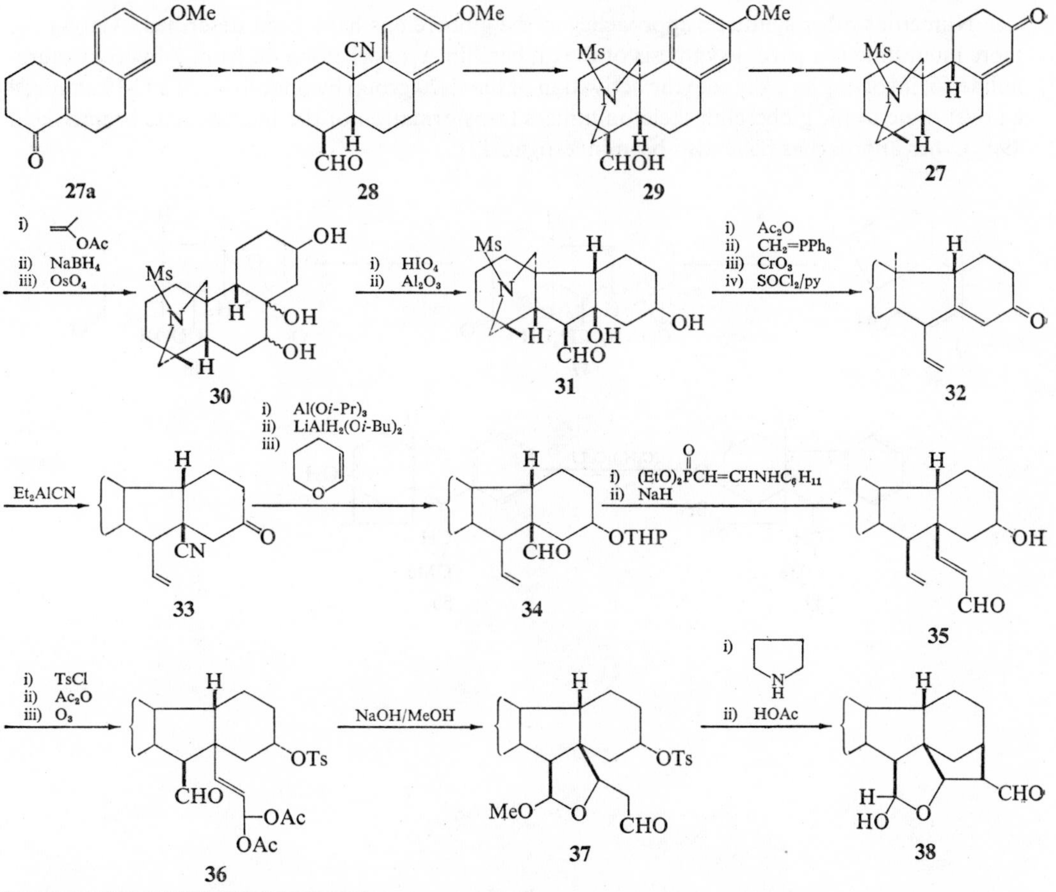

OMe
CN
CHO
Ms
N
CHOH
27a
28
29
27
i) OAc
ii) NaBH$_4$
iii) OsO$_4$
OH
30
i) HIO$_4$
ii) Al$_2$O$_3$
31
i) Ac$_2$O
ii) CH$_2$=PPh$_3$
iii) CrO$_3$
iv) SOCl$_2$/py
32
Et$_2$AlCN
33
i) Al(O*i*-Pr)$_3$
ii) LiAlH$_2$(O*i*-Bu)$_2$
iii)
OTHP
34
i) (EtO)$_2$PCH=CHNHC$_6$H$_{11}$
ii) NaH
35
i) TsCl
ii) Ac$_2$O
iii) O$_3$
OTs
OAc
36
NaOH/MeOH
MeO
37
i) N-H
ii) HOAc
HO
38

4) W. Nagata, T. Wakabayashi, M. Narisada, Y. Hayase, S. Kamata, *J. Am. Chem. Soc.*, **93**, 5740 (1971).
5) W. Nagata, M. Yoshioka, S. Hirai, *Tetr. Lett.*, 1913 (1966).
6) W. Nagata, Y. Hayasi, *J. Chem. Soc.*, 1400 (1969).
7) J. W. ApSimon, O. E. Edwards, R. Howe, *Can. J. Chem.*, **40**, 630 (1962).

39 40 41

42

43 R_1=H, R_2=OH
44 R_1=OH, R_2=H

45 GA_{15}
(separated from 20-oxo isomer)

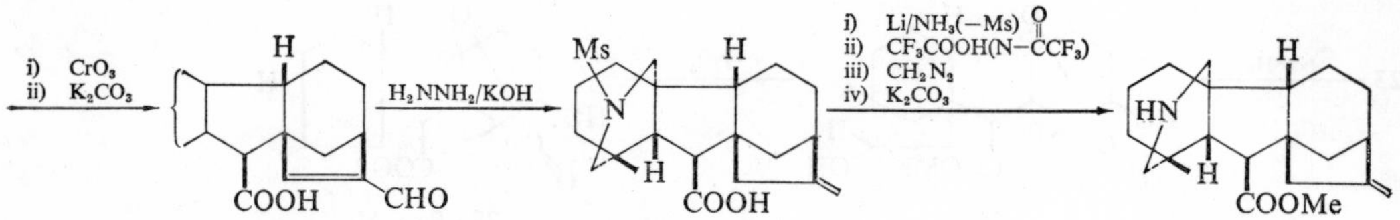

Other routes

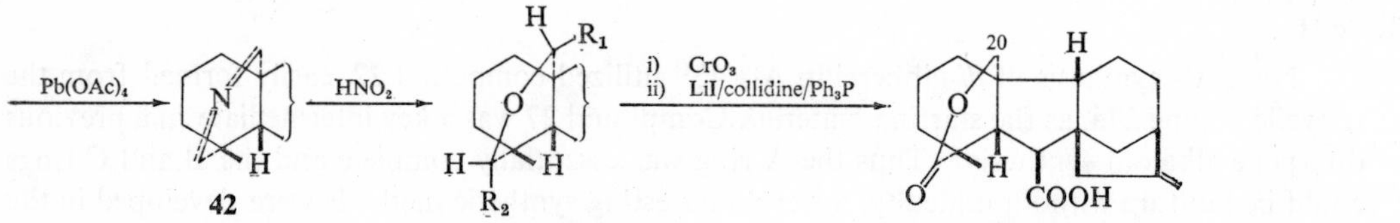

Numerous other synthetic approaches to the gibberellins have been described. Among the more interesting is a partial synthesis of the gibberellin A_{15} norketone **48** from 7-hydroxykaurenolide **46**, involving as a key step the activation of the C-20 group by photolysis of a C-19 amide;[8] a novel route to the gibberellin skeleton utilizes transformation of the intermediate bromoolefin **49**.[9] Other approaches have also been investigated.[10]

46 **47** **48**

49 **50**

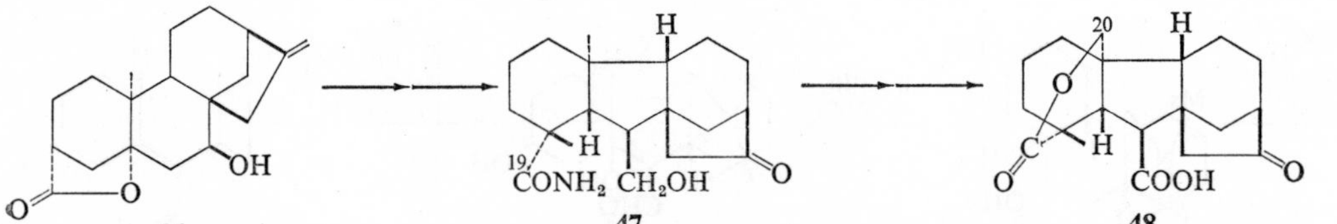

8) B. E. Cross, I. L. Gatfield, *Chem. Commun.* 33 (1970).
9) E. J. Corey, M. Narisada, T. Hiraoka, R. A. Ellison, *J. Am. Chem. Soc.* **92**, 396 (1970); see also *ibid.*, **93**, 7316 (1971).
10) See refs. 1, 4 and references cited therein.

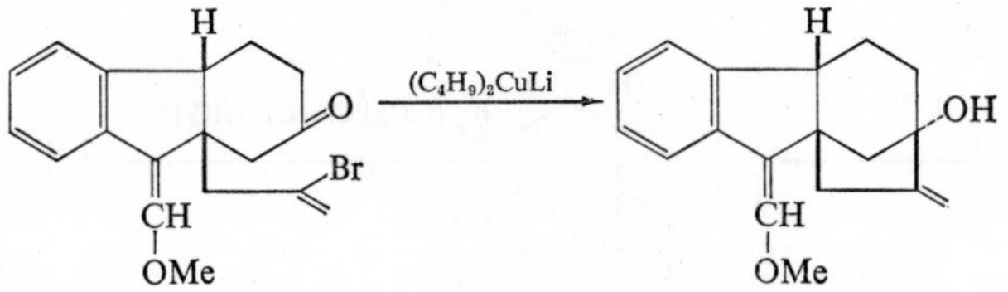

4.49 A SWEET SUBSTANCE

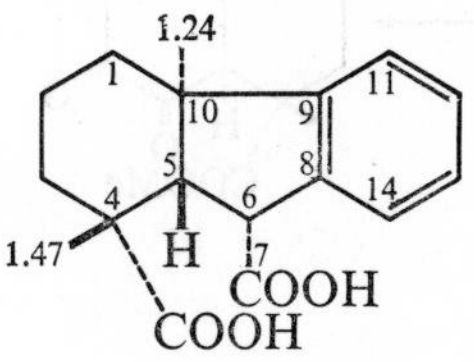

a sweet substance **1**
mp: 222–224°

1) The configurations at C-5 and C-6 in **1** were assigned as follows: *trans* A/B ring junction was deduced in the light of the chemical shift of 10-Me in the dimethyl ester (0.87 ppm) of the more stable 6-epimer of **1**; *cis* 5-H and 6-H was deduced from the formation of the acid anhydride (*trans* 5-H, 6-H would give no anhydride formation).

Remarks

Compound **1** and its sodium salt showed sweetness of 1800 and 2000, respectively, relative to sucrose. Stereoisomers at C-5 and/or at C-6 were reported to be without sweetness.[1] The diacid **1** could also be a versatile synthetic intermediate to C_{20} gibberellins (e.g. GA_{12}).

Synthesis

Route I

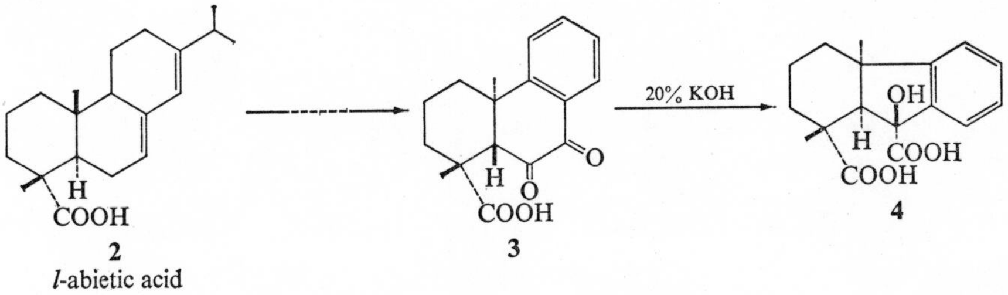

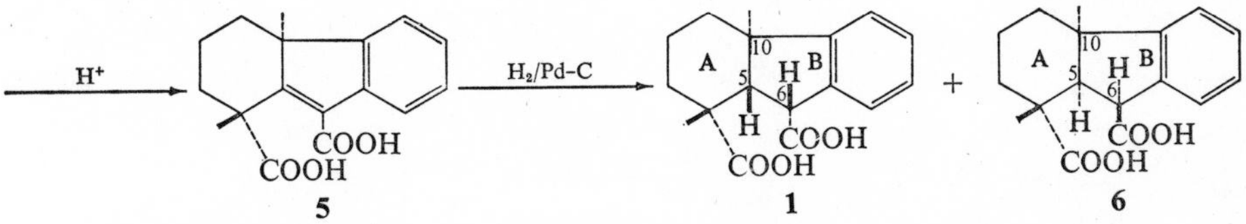

l-Abietic acid **2** was converted into the diketoester **3** following the Wenkert,[4] Ohta[5] method, which involves $AlCl_3$-catalyzed deisopropylation and isomerization of 10-Me. Catalytic dehydrogenation of **5** gave a 1 : 2 mixture of diacids **1** and **6**.

Route II

An improved synthesis in which all the intermediates in parentheses were not isolated afforded 2 kg of the sweet substance **1** starting from 20 kg of dehydroabietic acid **7**.[3]

1) A. Tahara, T. Nakata, Y. Ohtsuka, *Nature*, **233**, 619 (1971).
2) A. Tahara, Y. Ohtsuka, *Chem. Pharm. Bull.*, **18**, 859 (1970).
3) A. Tahara, Y. Ohtsuka, Y. Nakata, S. Takada, 15*th Symp. Chem. Natural Products, Abstracts*, 303, 1971.
4) E. Wenkert, B. G. Jackson, *J. Am. Chem. Soc.*, **80**, 211 (1958).
5) M. Ohta, L. Ohmori, *Chem. Pharm. Bull.*, **5**, 91 (1957).

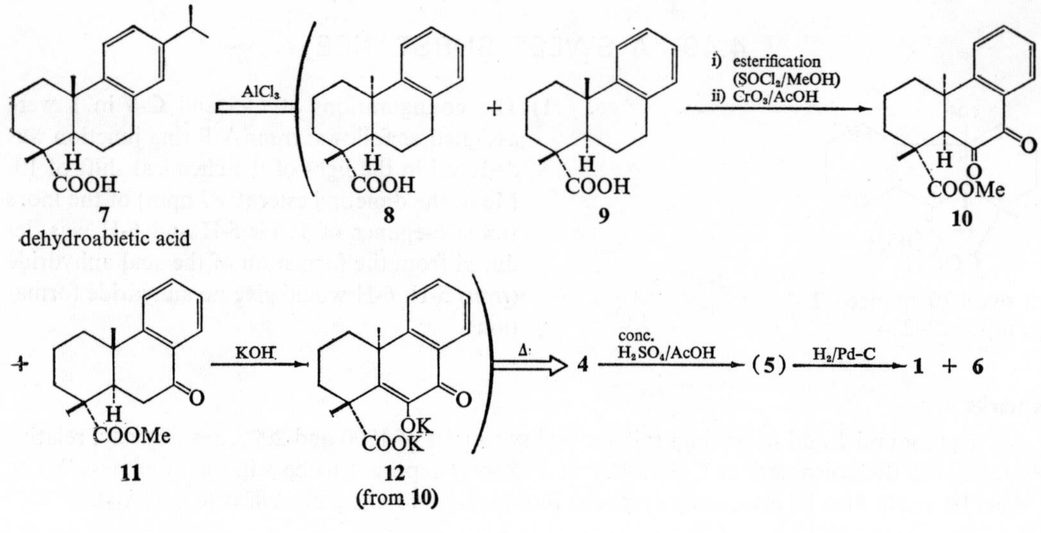

AlCl3
i) esterification
(SOCl2/MeOH)
ii) CrO3/AcOH
COOH
7
dehydroabietic acid
8
9
COOMe
10
KOH
11
OK
COOK
12
(from 10)
Δ
4
conc.
H2SO4/AcOH
(5)
H2/Pd-C
1 + 6

4.50 THE ANTHERIDIOGEN OF ANEMIA PHYLLITIDIS

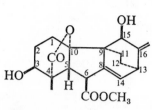

antheridiogen-An (A$_{An}$) **1**[1]
uv (EtOH) : end absorption
ir (KBr): 1758 (γ-lactone), 1723 (hindered acid)
Me ester:
ms: 360.1584 (M$^+$ C$_{20}$H$_{24}$O$_6$)(calcd. 360.1584)
cd (MeOH) : 215 ($\Delta\epsilon + 13.17$)

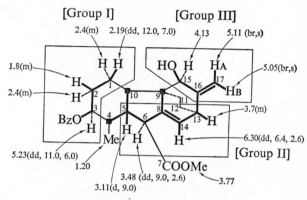

nmr of ester benzoate **3** (CDCl$_3$)

1) The starting material (18 mg, powder) was converted into the Me ester **2**, ester 3-benzoate **3**, ester dibenzoate **4**, and ester 3-benzoate 15-one **5**, and each was studied extensively by 100, 200 and 300 MHz nmr.

2) Groups I–III (shown in **3**) were elucidated by nmr. The 100 and 300 MHz nmr spectra of the ester benzoate **3** are shown below; the high-field region is well-separated in the 300 MHz trace.

3) The Me ester 15-oxo-3-benzoate **5** had a uv (EtOH) peak at 217 nm (ε 4,700), and the exocyclic methylene nmr peaks had shifted to 5.78 (17-H$_A$) and 5.21 ppm (17-H$_B$). These data corroborate group III in **3**.

4) In the ester benzoate **3**, irradiation of 13-H at 3.17 ppm induced a small but crucial noe (+2.7%) at the 5.05 ppm 17-H$_B$ signal. This links groups II and III.

5) Micro Se-dehydrogenation gave 1-methylfluorene **6** (uv, tlc, Fourier nmr) which indicated the 6/5/6 carbocycle.

6) A COOH group and γ-lactone remain to be accounted for. The 9.0–9.6 Hz values for $J_{5,6}$ preclude C-6 from being at a bridgehead and this leads to 6-COOH.

7) Similarity of the 4-Me shifts in gibberellins (1.13–1.22 ppm) and A$_{An}$ derivatives (1.17–1.28 ppm) shows the lactone to be at C-4/C-10 with the CO terminal at C-4.

8) Configuration of Part A in **7**: The OH is at C-3 and not at C-1. If it were at C-1, the proximity of C-1 and C-15 would have produced an exciton-split cd spectrum in the dibenzoate **4**. This was not the case: 205 ($\Delta\varepsilon + 23.0$), 229 ($\Delta\varepsilon + 8.0$). 3-Benzoylation to give **3** induces a 0.26 ppm downfield shift of 5-H and hence 3-OH and 5-H are *cis*. The absolute configuration rests on application of the benzoate sector rule to the ester 3-benzoate **3**.

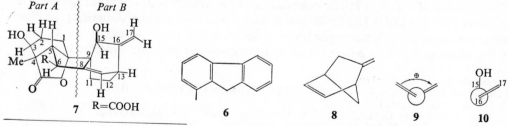

1) K. Nakanishi, M. Endo, U. Naf, L. F. Johnson, *J. Am. Chem. Soc.*, **93**, 5579 (1971).

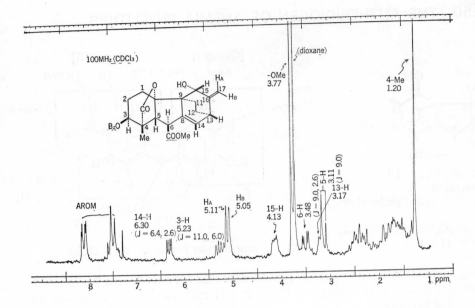

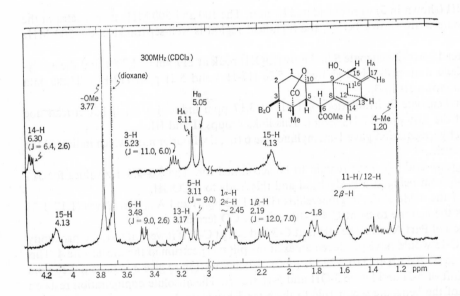

9) Configuration of Part B in **7**: The estimated[2] cd of bicyclo-8[3] is $\Delta\varepsilon + 20$ at ca. 220 nm. The cd of the Me ester **2** ($\Delta\varepsilon + 13.2$ at 220 nm) indicated it to be composite, comprising the diene contribution (positive, **9**) and allylic hydroxyl contribution[4] (negative, **10**). Hence the absolute configuration of Part B is as shown, and this completes proposed structure **1**.

2) Estimated from published ord curve in ref. 3.
3) L. S. Forster, A. Moscowitz, J. G. Berger, K. Mislow, *J. Am. Chem. Soc.*, **84**, 4353 (1962).
4) A. I. Scott, A. D. Wrixon, *Tetr.*, **27**, 4787 (1971).
5) M. Endo, K. Nakanishi, U. Naf, W. McKeon, R. Walker, *Physiol. Plant.*, **26**, 183 (1972).
6) P. R. Zanno, M. Endo, K. Nakanishi, *Naturwiss.*, in press.
7) Dr. G. Kitt, *Kitchawan Botanical Gardens*, unpublished results.

Remarks

Antheridiogen is the name coined for the factor which induces antheridium ("male organ") formation in ferns; it also substitutes for light in spore germination. Antheridiogen-An, the first such factor to be characterized, was isolated[5] from 53-day cultures of *Anemia phyllitidis* spores (18 mg from 57,000 ml of culture). Cultures of *A. hirsuta* (same genus) also give A_{An}[6] and it appears that the structural diversity does not reach the species level. The threshold activity of A_{An} against *A. phyllitidis* is 0.3 g/l.[7] The A_{An} skeleton is new for a diterpene, but it can be derived biogenetically from a gibbane skeleton.

4.51 STRUCTURE AND SYNTHESIS OF TRACHYLOBANES
[Atisane group]

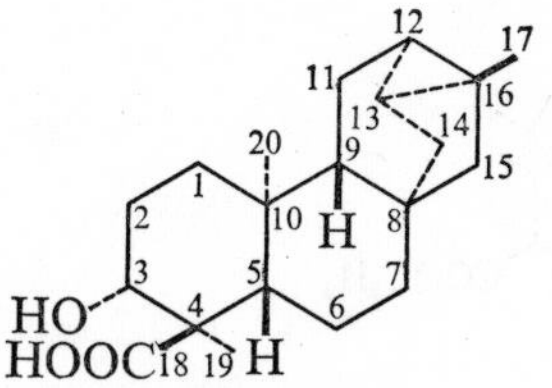

hydroxytrachylobanic acid **1**[1]
  $C_{20}H_{30}O_3$
1 methylester, mp 160°
  α_D (CHCl$_3$) $-61°$

1) The pentacyclic cyclopropane-containing skeleton was revealed by nmr (no olefinic H's, high field cyclopropane H's).
2) The presence of β-hydroxyacid was shown by spontaneous decarboxylation to **2** on oxidation.
3) Correlation of **1** with ($-$)-kaurene **8** gave further information; **4** and **7** (free acid forms) are natural products isolated with **1**, and allow correlation of **1** with ($-$)-kaurene **8**, using the ketone **6**, as shown below.

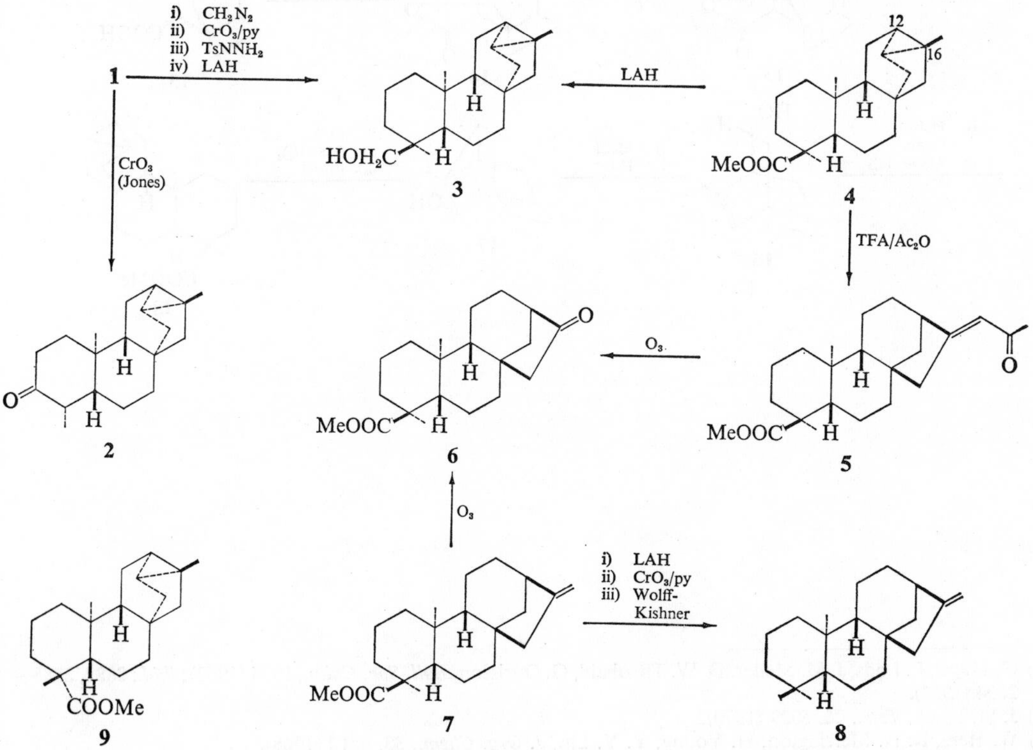

4) Equatorial 3-OH was found from nmr and was supported by determination of the C-3 absolute configuration by Horeau's method.

5) The equatorial carboxyl group was shown by the pK_a of 7.79 and by characteristic ir at 1240 cm^{-1} (CS$_2$).

Remarks

Hydroxytrachylobanic acid and trachyloban-18-oic acid (free acid of **4**) have been isolated from the resin of the Madagascan tree *Trachylobium verrucosum*.[1] Trachyloban-19-oic acid **9** has been isolated from *Helianthus annus*.[2]

Synthesis[3]

The enantiomer **18** of methyl trachylobanate **9** has been synthesized[1] from methyl levo-pimarate **10**.

---

1) G. Hugel, L. Lods, J. M. Meller, D. W. Theobald, G. Ourisson, *Bull. Soc. Chim.*, 1974 (1963); *ibid.*, 2882, 2888, 2894 (1964).
2) J. St. Pyrek, *Tetr.*, **26**, 5029 (1970).
3) W. Herz, R. N. Merrington, H. Young, Y. Y. Lin, *J. Org. Chem.*, **33**, 4210 (1968).

4.52 TAXINES
[Taxane group]

O-Cinnamoyltaxicin-I[1,2]

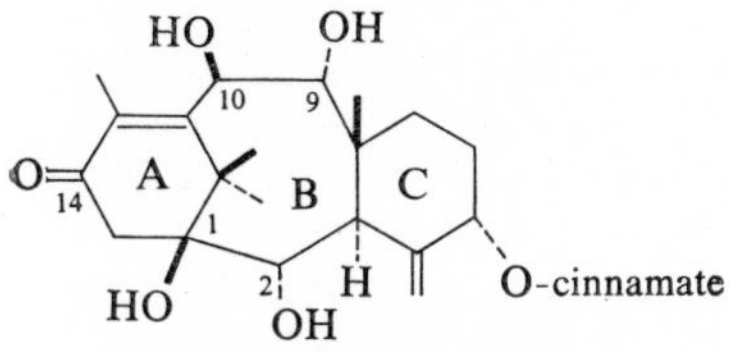

O-cinnamoyltaxicin-I **1**

1) Periodate oxidation afforded two fragments **2** and **3**, both of which were correlated chemically with the respective authentic specimens (including absolute configurations at C-3, C-5, and C-8).

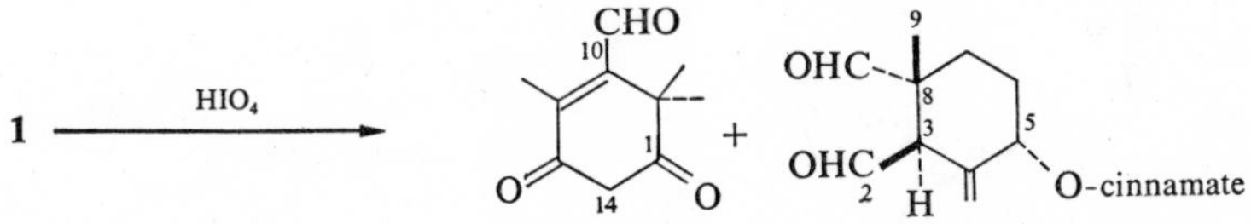

2) The configuration at C-1 was based on the Cotton effect of the enone; those at C-2, C-9 and C-10 were based on $J_{2,3} = 11$ Hz and $J_{9,10} = 10$ Hz.

Taxinine 4 (or O-cinnamoyltaxicin-II triacetate)[1,3,4]

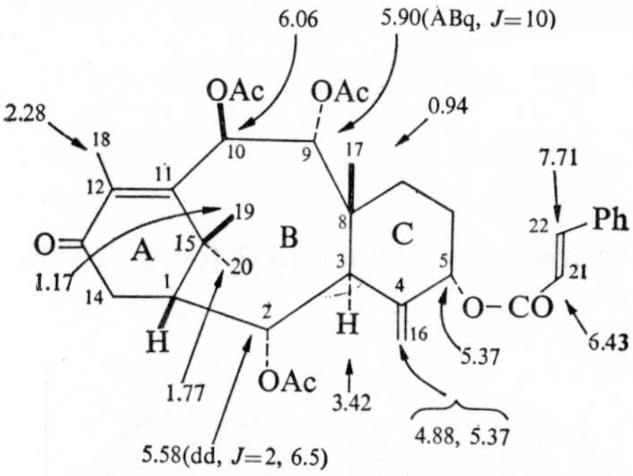

taxinine **4**

mp: 265–267° α_D (CHCl$_3$): +137°
ir (KBr): 1745 (OAc), 1720, 1644 (cinnamate), 1674 ($\Delta^{\alpha\beta}-CO$), 911 (exo-methylene)
uv (MeOH): 218 (4.28), 223 (4.22) 280 (4.45)
nmr in CDCl$_3$

1) The planar structure was based on the changes in functionalities accompanying the formation of degradation products: C-9 to C-13, reduction product **7**; C-13, C-14, C-1 to C-3, the conversion of **9** to **11**; C-5 cinnamate, hydrogenolysis product **6**; C-4 exo-methylene, the long-range coupling between 3-H and 6-H.

1) B. Lythgoe, *The Alkaloids* (ed. R. H. E. Manske), vol. X, p. 597, Academic Press, 1968.
2) B. W. Langley, B. Lythgoe, B. Scales, R. M. Scrowston, S. Trippett, D. Wray, *J. Chem. Soc.*, 2972 (1962); D. H. Eyre, J. W. Harrison, R. M. Scrowston, B. Lythgoe, *Proc. Chem. Soc.*, 271 (1963); D. H. Eyre, J. W. Harrison, B. Lythgoe, *J. Chem. Soc. C*, 452 (1967).
3) M. Kurono, Y. Nakadaira, S. Onuma, K. Sasaki, K. Nakanishi, *Tetr. Lett.*, 2153 (1963); K. Nakanishi, M. Kurono, N. S. Bhacca, *ibid.*, 2161.
4) K. Ueda, S. Uyeo, Y. Yamamoto, Y. Maki, *Tetr. Lett.*, 2167 (1963).

2) Studies of anhydrotaxinol **15** confirmed the planar structure.

3) The relative configuration was determined by x-ray analysis.[5] A B/C *cis* ring junction was incorrectly assigned at first.[6]

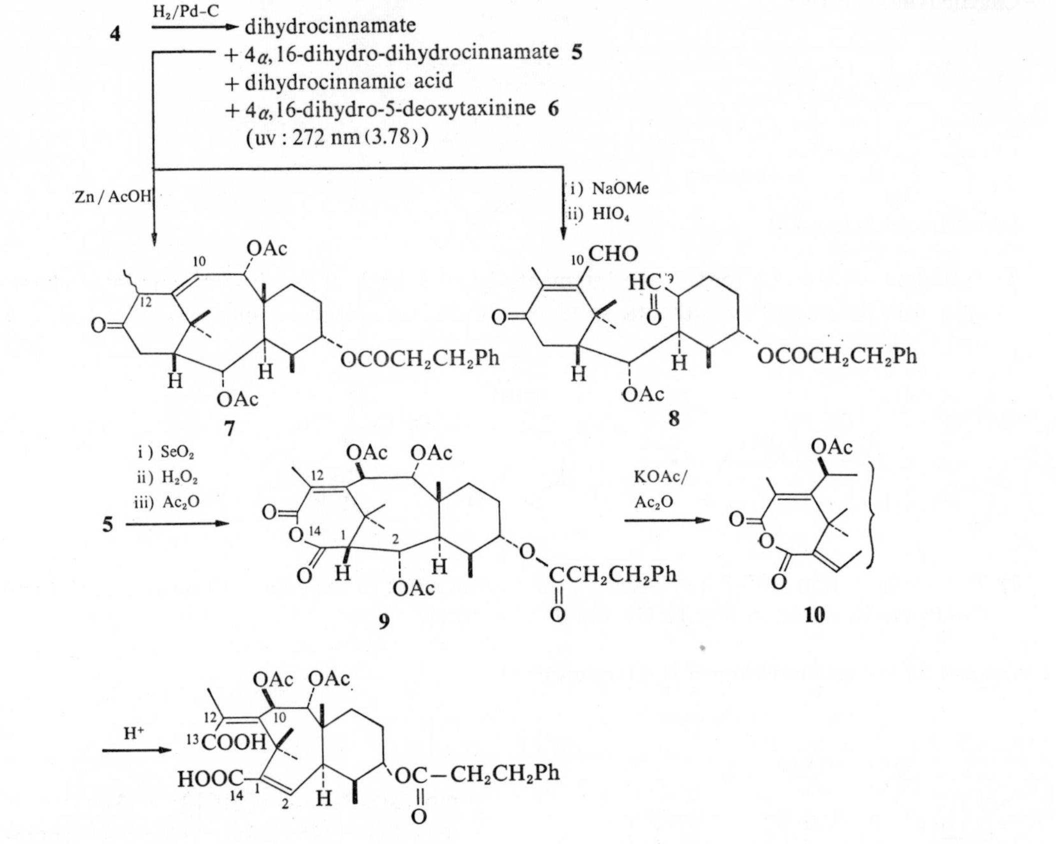
4
H2/Pd-C
dihydrocinnamate
+4α,16-dihydro-dihydrocinnamate 5
+dihydrocinnamic acid
+4α,16-dihydro-5-deoxytaxinine 6
(uv : 272 nm (3.78))
Zn / AcOH
i) NaOMe
ii) HIO4
OAc
OCOCH2CH2Ph
OAc
7
CHO
OCOCH2CH2Ph
OAc
8
i) SeO2
ii) H2O2
iii) Ac2O
5
OAc OAc
CCH2CH2Ph
OAc
9
KOAc/
Ac2O
OAc
10
H+
OAc OAc
COOH
HOOC
OC−CH2CH2Ph
11

4) Alkaline treatment of bis-deacetyl taxinine **12** gave anhydrotaxininol **15**[7], for which the following mechanism has been proposed.[8]

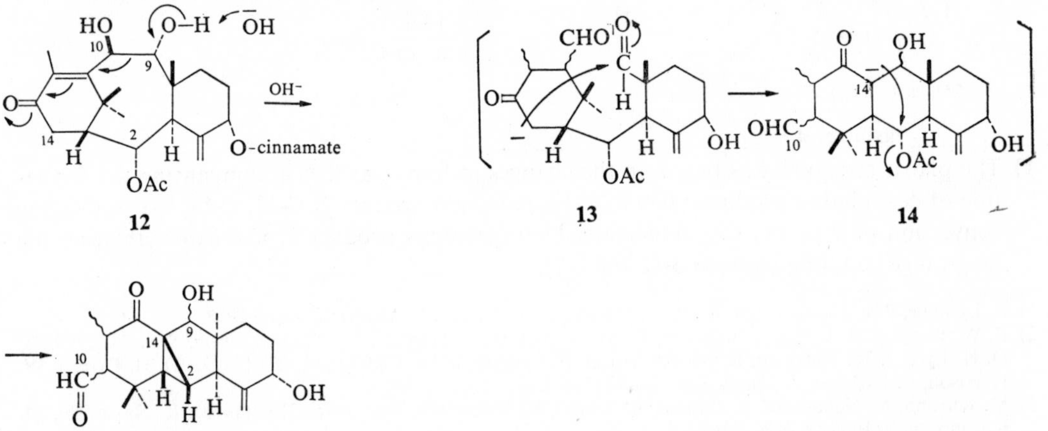
HO
OH
OH−
O-cinnamate
OAc
12
CHO
OH
OAc
13
OH
OHC
OAc
OH
14
OH
HC
OH

anhydrotaxininol **15**

Taxusin[9]

The structure of this compound, which was isolated from *Taxus baccata*, was determined by x-ray analysis of the rearranged derivative **17**.

i) hydrolysis
ii) H_2
iii) HIO_4/acetone
iv) *p*-Br-benzoylation

taxusin **16**   **17**

Related compounds

Related compounds include the baccatins, isolated from *T. baccata*,[10] and taxol, isolated from *T. brevifolia*.[11]

The structure of baccatin V was determined by x-ray analysis of hexahydro-**18** acetate. Taxol **19** is the most complex taxine, and shows both antileukemic and antitumor activities. The structure was determined by x-ray analysis.

baccatin V **18**   taxol **19**

Remarks

Taxines[12,13] comprise a group of amorphous "basic" compounds contained in the European *Taxus baccata* (yew tree), the Japanese *T. cuspidata*, and related species (Taxaceae). They are responsible for the poisonous property of the plants (cattle poison). The "basicity" is due to the fact that the two principal constituents, *O*-cinnamoyltaxicin-I triacetate **1** (2,9,10-triacetate) and taxinine **4**[14] are contained in the plant as "alkaloids," i.e., esters of β-dimethyl-amino-β-phenylpropionic acid, $Me_2NCH(Ph)-CH_2COOH$,[15] which are converted to cinnamates by loss of dimethylamine during isolation. Many related compounds, some of which exist without the dimethylamino group, have been isolated in recent years. The diterpenoid skeleton is called taxane and a numbering system has been proposed.[16] Most of the British work was carried out on **1** from *T. baccata*, while the Japanese work was on **4** from *T. cuspidata*.

5) M. Shiro, T. Sato, H. Koyama, Y. Maki, K. Nakanishi, S. Uyeo, *Chem. Commun.*, 98 (1966).
6) M. Kurone, Y. Maki, K. Nakanishi, M. Ohashi, K. Ueda, S. Uyeo, M. C. Woods, Y. Yamamoto, *Tetr. Lett.*, 1917 (1965).
7) J. Taga, *Chem. Pharm. Bull.*, **8**, 934 (1960); *ibid.*, **12**, 389 (1964); *Yakugaku Zasshi*, **84**, 1067 (1964).
8) Y. Yamamoto, S. Uyeo, K. Ueda, *Chem. Pharm. Bull.*, **12**, 386 (1964).
9) W. R. Chan, T. G. Halsall, G. M. Hornby, A. W. Oxford, W. Sabel, K. Bjamer, G. Ferguson, J. M. Robertson, *Chem. Commun.*, 923 (1966).
10) D. P. C. C. de Marcano, T. G. Halsall, *Chem. Commun.*, 1381 (1970).
11) M. C. Wani, H. L. Taylor, M. E. Wall, P. Coggon, A. T. McPhail, *J. Am. Chem. Soc.*, **93**, 2325 (1971).
12) H. Lucas, *Arch. Pharm.*, **95**, 145 (1856).
13) E. Graf, *Angew. Chem.*, **68**, 249 (1956).
14) H. Kondo, T. Takahashi, *J. Pharm. Soc. Japan*, **45**, 861 (1925).
15) E. Winterstein, A. Guyer, *Z. Physiol. Chem.*, **128**, 175 (1923).
16) B. Lythgoe, K. Nakanishi, S. Uyeo, *Proc. Chem. Soc.*, 301 (1964).

Taxinine **4** is the main constituent (in the form of the "alkaloid") of *T. cuspidata* (Japanese yew tree) and is also contained in *T. baccata*.[2] More than ten congeners have been isolated from *T. cuspidata*[14]; of these, taxinine A undergoes an interesting photochemical transannular reaction to yield another minor constituent, taxinine L **21**.[17]

Application of the benzoate chirality method[18] has confirmed the absolute configuration of taxinines; the olefin octant rule has also been applied.[19] Extensive noe measurements indicated the conformation of taxinines in chloroform solution and also permitted deduction of the structures of some minor congenors (see taxinine **B 22**)[20] (noe's are indicated by arrows).

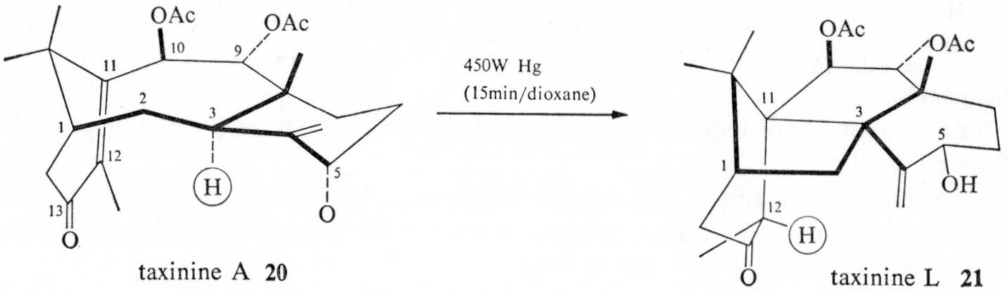

taxinine A **20**

450W Hg
(15min/dioxane)

taxinine L **21**

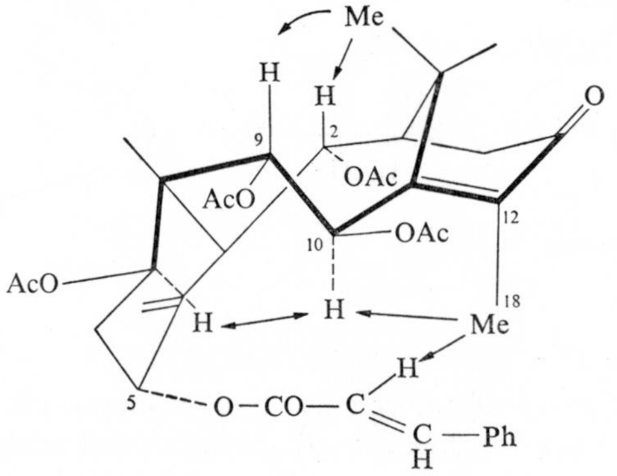

taxinine B **22**

The following biogenesis, similar to that for cembrene, has been postulated starting from geranylgeranyl pyrophosphate.[21]

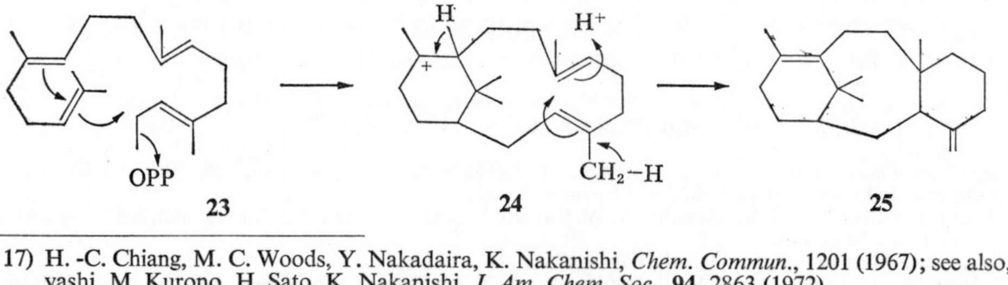

23   24   25

17) H. -C. Chiang, M. C. Woods, Y. Nakadaira, K. Nakanishi, *Chem. Commun.*, 1201 (1967); see also, T. Kobayashi, M. Kurono, H. Sato, K. Nakanishi, *J. Am. Chem. Soc.*, **94**, 2863 (1972).
18) N. Harada, K. Nakanishi, *J. Am. Chem. Soc.*, **91**, 3989 (1969).
19) D. P. C. C. de Marcano, T. G. Halsall, A. I. Scott, A. D. Wrixon, *Chem. Commun.*, 582 (1970).
20) M. C. Woods, H. -C. Chiang, Y. Nakadaira, K. Nakanishi, *J. Am. Chem. Soc.*, **90**, 522 (1968).
21) J. W. Harrison, R. M. Scrowston, B. Lythgoe, *J. Chem. Soc.* C, 1933 (1966); also H. Erdtman, T. Norin, M. Sumimoto, A. Morrison, *Tetr. Lett.*, 3879 (1964).

4.53 GRAYANOTOXINS

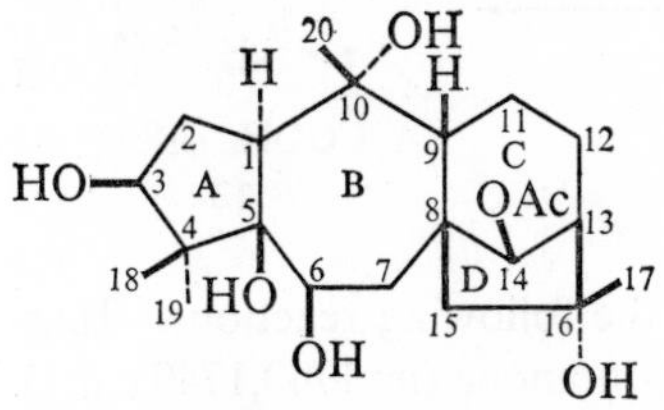

grayanotoxin-I **1**
(G-I) $C_{22}H_{36}O_7$
mp: 272°

Planar structure

Two independent structure determinations of grayanotoxins were reported, one using G-II, the other G-I. Extensive usage was made of uv and ir.

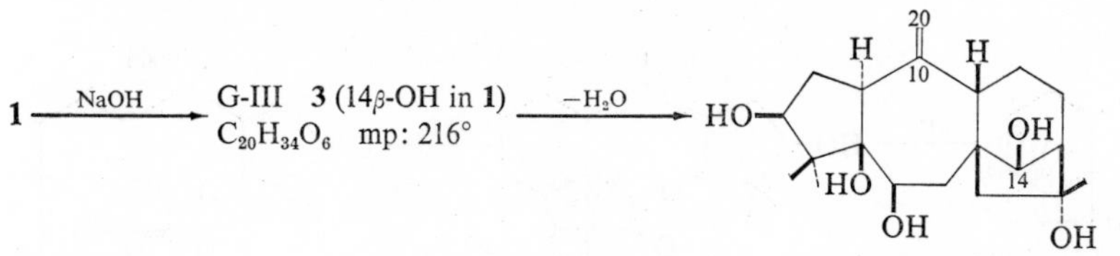
1
NaOH
G-III 3 (14β-OH in 1)
$C_{20}H_{34}O_6$ mp: 216°
$-H_2O$
G-II 2

$C_{20}H_{32}O_5$ mp: 198°

G-II[1]

1) There are five hydroxyls present, including one α-glycol ($NaIO_4$). The 16-OH is readily eliminated on 5,6-acetonide formation. The 3-OH and 14-OH are in a five-membered ring since oxidation gives cyclopentanone (ir).
2) G-II has one exocyclic double bond and gave a saturated dihydro product (uv), suggesting that G-II is a tetracarbocyclic diterpene.
3) Ring A structure is based on the glutamate derivative **6**.
4) Structural analyses of the following reaction products **7–13** led to the entire structure of G-II as **2**.

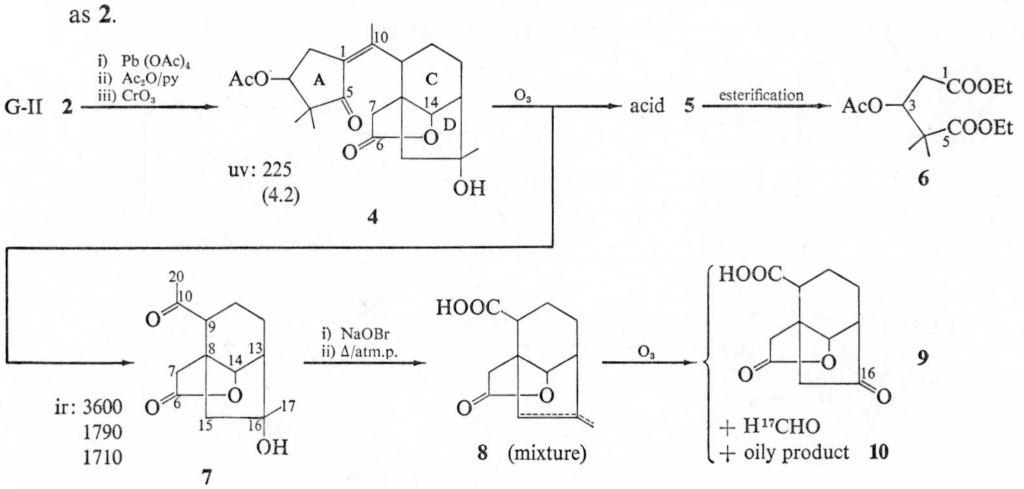
G-II 2
i) Pb(OAc)₄
ii) Ac₂O/py
iii) CrO₃
uv: 225
(4.2)
4
O₃
acid 5
esterification
6
ir: 3600
1790
1710
7
i) NaOBr
ii) Δ/atm.p.
8 (mixture)
O₃
9
+ H¹⁷CHO
+ oily product 10

1) J. Iwasa, Z. Kumazawa, M. Nakajima, *Chem. Ind.*, 511 (1961).

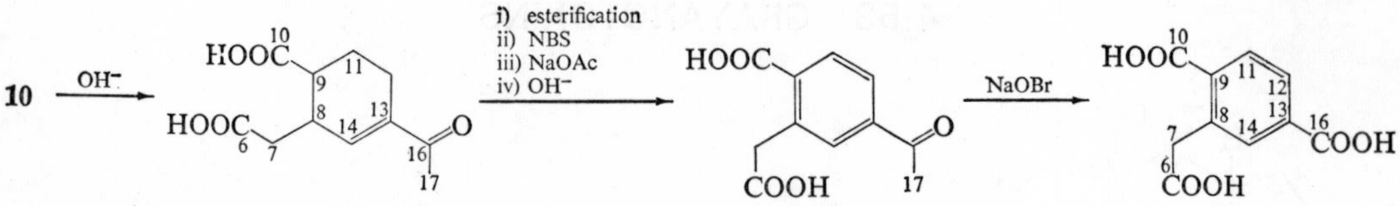

10 $\xrightarrow{OH^-}$

i) esterification
ii) NBS
iii) NaOAc
iv) OH$^-$

11

12

$\xrightarrow{NaOBr}$

13

G-I[2)]

1) Partial structures **14** (in bold lines) and **20** were derived from the following reactions. Rings A and D were five-membered because oxidation gave a dicyclopentanone (ir: 1740,1740 cm^{-1}).

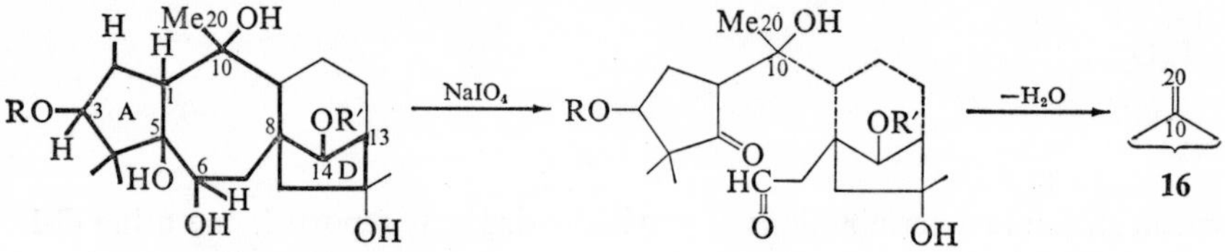

$\xrightarrow{NaIO_4}$

$\xrightarrow{-H_2O}$

16

14 R, R' = H, OAc

15

$\xrightarrow{NaOH}$

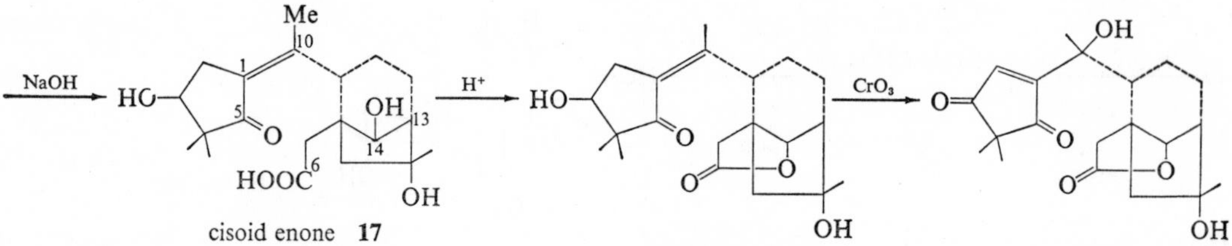

cisoid enone **17**
uv: 258 (ϵ 10,500)
ir: 1700
     1602 (strong)

$\xrightarrow{H^+}$

γ-lactone **18**
ir: 1782

$\xrightarrow{CrO_3}$

19
uv: 235 (ϵ 10,600)

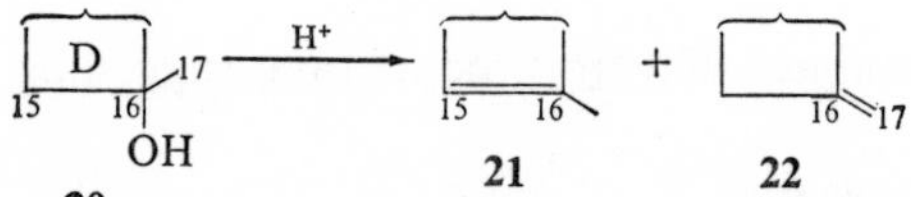

$\xrightarrow{H^+}$

21 + 22

20

2) **14** was extended further to **23**

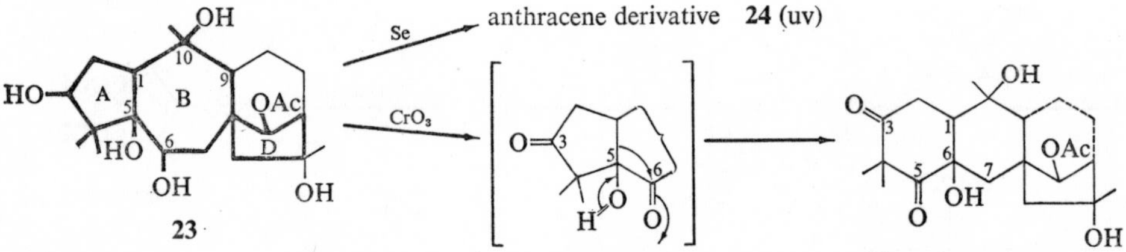

$\xrightarrow{Se}$ anthracene derivative **24** (uv)

$\xrightarrow{CrO_3}$

25

blocked β-diketone **26**

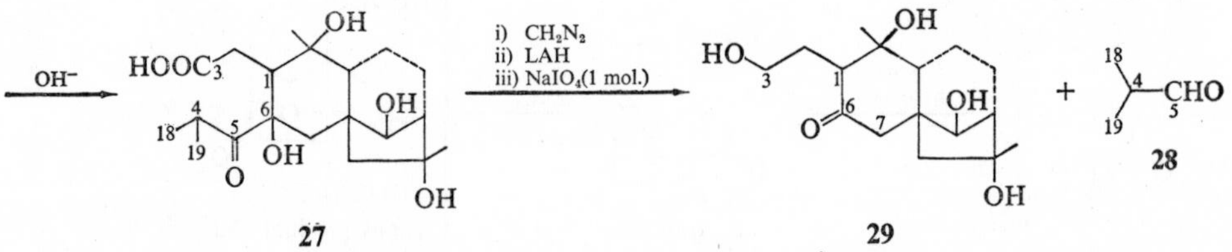

23

$\xrightarrow{OH^-}$

27

i) CH$_2$N$_2$
ii) LAH
iii) NaIO$_4$(1 mol.)

29
ir: 1707

+ 28

2) H. Kakisawa, *J. Chem. Soc. Jap.*, **82**, 1096, 1216 (1961).

3) The remaining structure (rings C/D) is based on the following reaction products. The only possibility in **32** to complete the structure was to link C-11 and C-12.

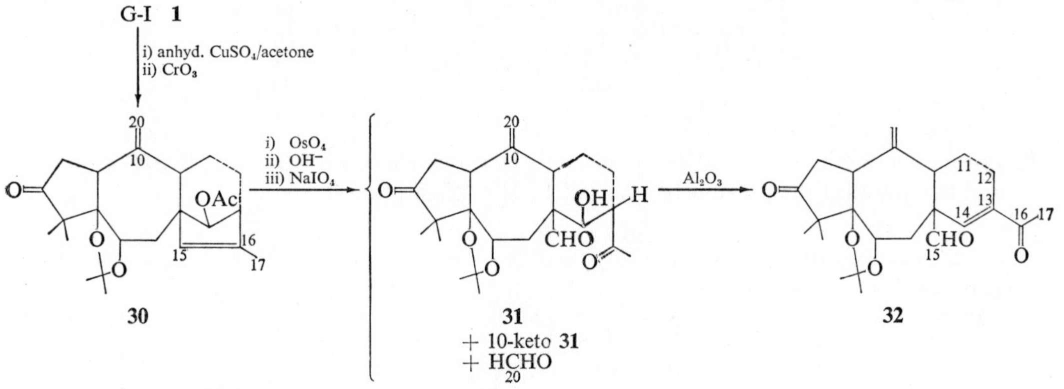

G-I **1**

i) anhyd. CuSO$_4$/acetone
ii) CrO$_3$

i) OsO$_4$
ii) OH$^-$
iii) NaIO$_4$

Al$_2$O$_3$

30   **31**   **32**

+ 10-keto **31**
+ HCHO
  20

Stereochemistry of grayanotoxins[3-6]

1) 3β-OH was determined by Horeau's method.

2) Formation of the diethylacetal **33** from G-II indicated 5β-OH (*cis* to 3-OH).

3) 6β-OH was determined from the benzoate rule (Δ benzoate: +13.5°).

4) 14-OAc or OH and C-7 are *cis*-fused with respect to ring D since the lactone **4** is easily formed.

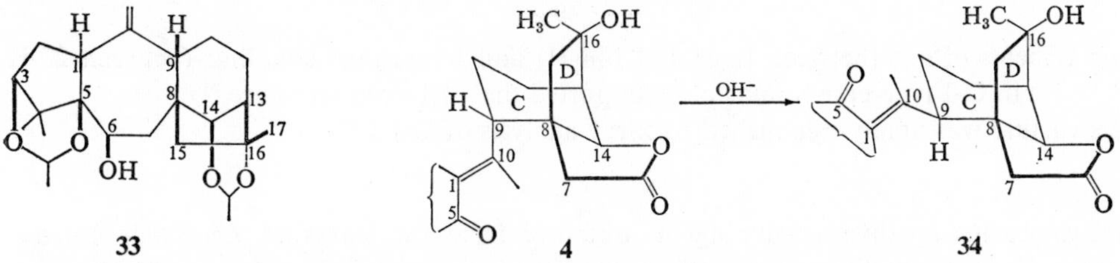

33   **4**   **34**

5) C-10 is axial to ring C since **34** is more stable than **4** upon equilibration.

6) Absolute stereochemistries at C-8, 9 and 13, or β-*cis* D ring were indicated by the weak, negative Cotton effect of the 14-monoketone **35** (derived from G-I).

7) 14β-OR and 16α-OH were indicated by the formation of **33**.

8) 10α-OH in G-I, III was found by the reaction of G-I with TsCl/py to give the α-ether **37**, presumably through intermediate **38**.

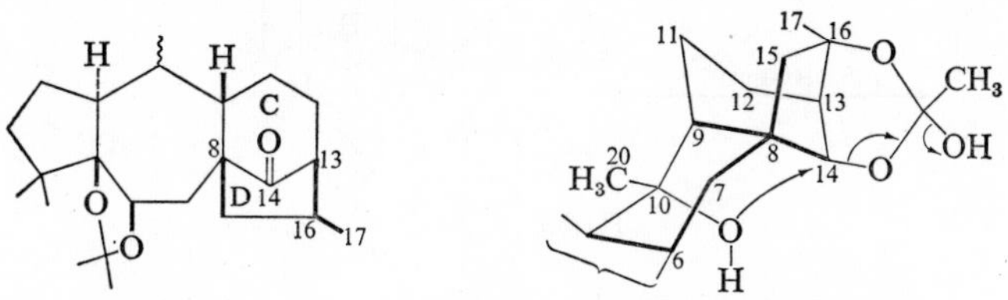

35   **38**

3) H. Kakisawa, M. Yanai, T. Kozima, K. Nakanishi, H. Mishima, *Tetr. Lett.*, 215 (1962).
4) T. Kozima, K. Nakanishi, M. Yanai, H. Kakisawa, *Tetr. Lett.*, 1329 (1964).
5) J. Iwasa, Y. Nakamura, *Tetr. Lett.*, 3973 (1969).
6) Z. Kumazawa, R. Iriye, *Tetr. Lett.*, 927, 931 (1970).

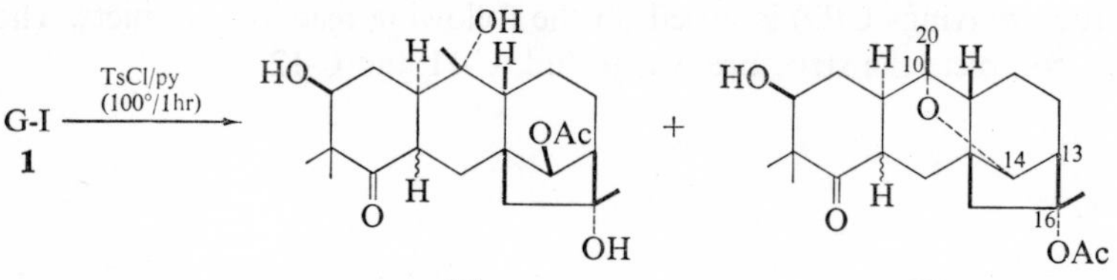

36   37

9) Formation of ethers **39**[5] and **40**[6] could only be explained by an A/B *trans* (1α-H) structure, and this reversed the previously assigned 1β-H configuration.[4]

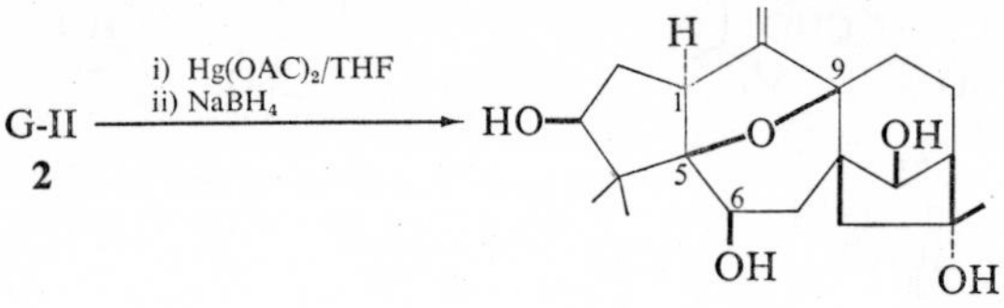

39

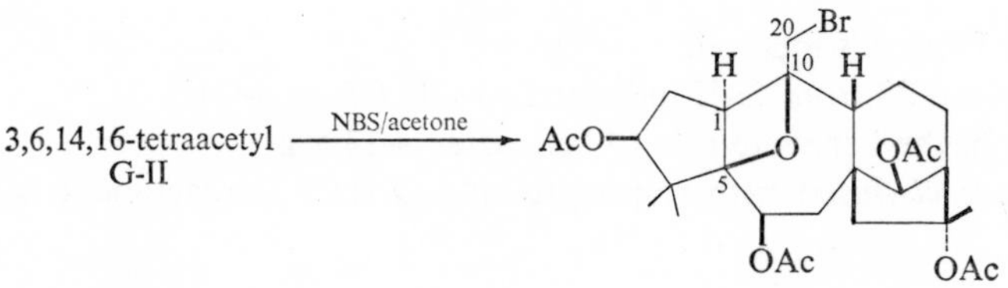

40

10) Measurements of noe (between 1α-H and 14α-H) and long-range coupling (between 20-H and 1α-H) in G-I type compounds also supported the A/B-*trans* structure.[7]

11) The structure was finally established by x-ray analysis of G-I **1**.[8]

Remarks

Grayanotoxins are hypotensive agents obtained from the leaves of *Leucothoe grayana* (Ericaceae). About thirty diterpenes of this group have been isolated from *L. grayana* and other Ericaceaous plants.[9]

Synthetic approaches

1) Solvolysis of mesylate **41** caused rearrangement to yield **42** (A/B *cis*) together with the α-elimination product **43**.[10]

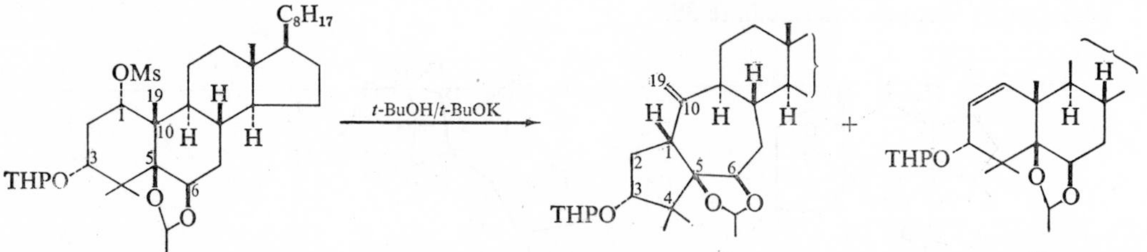

41   42   43

7) H. Hikino, M. Ogura, T. Ohta, T. Takemoto, *Chem. Pharm. Bull.*, **18**, 1071 (1970).
8) P. Narayanan, M. Rohrl, K. Zechmeister, W. Hoppe, *Tetr. Lett.*, 3943 (1970).
9) T. Okuno, N. Hamanaka, H. Miyakoshi, T. Matsumoto, *Tetr.*, **26**, 4765 (1970); J. Sakakibara, K. Ikai, M. Yasue, *Chem. Pharm. Bull.*, **20**, 861 (1972); H. Hikino, T. Ohta, Y. Hikino, T. Takemoto, *ibid.*, **20**, 1090 (1972).
10) T. Okuno, T. Matsumoto, *Tetr. Lett.*, 4077 (1969).

2) The mixture of **44** and **45** (3:1), upon irradiation followed by base treatment, afforded a mixture of **46** and **47** (12%, 3:1).[11]

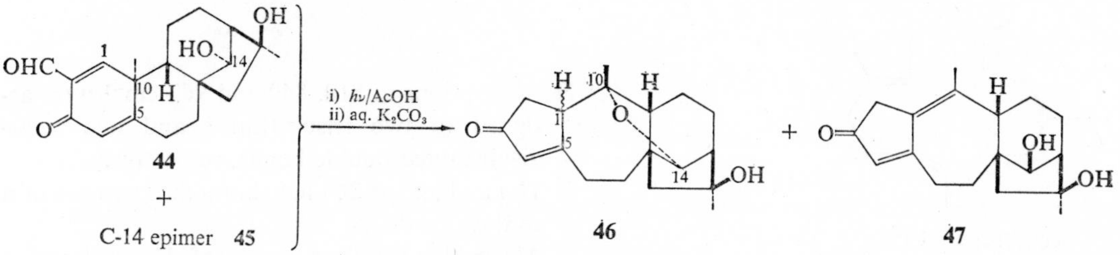

44
+
C-14 epimer **45**

i) *hν*/AcOH
ii) aq. K$_2$CO$_3$

46

47

3) G-II **2** was partially synthesized starting from the tricyclic relay compound **48**, which was obtained by a six-step degradation of G-I.[12]

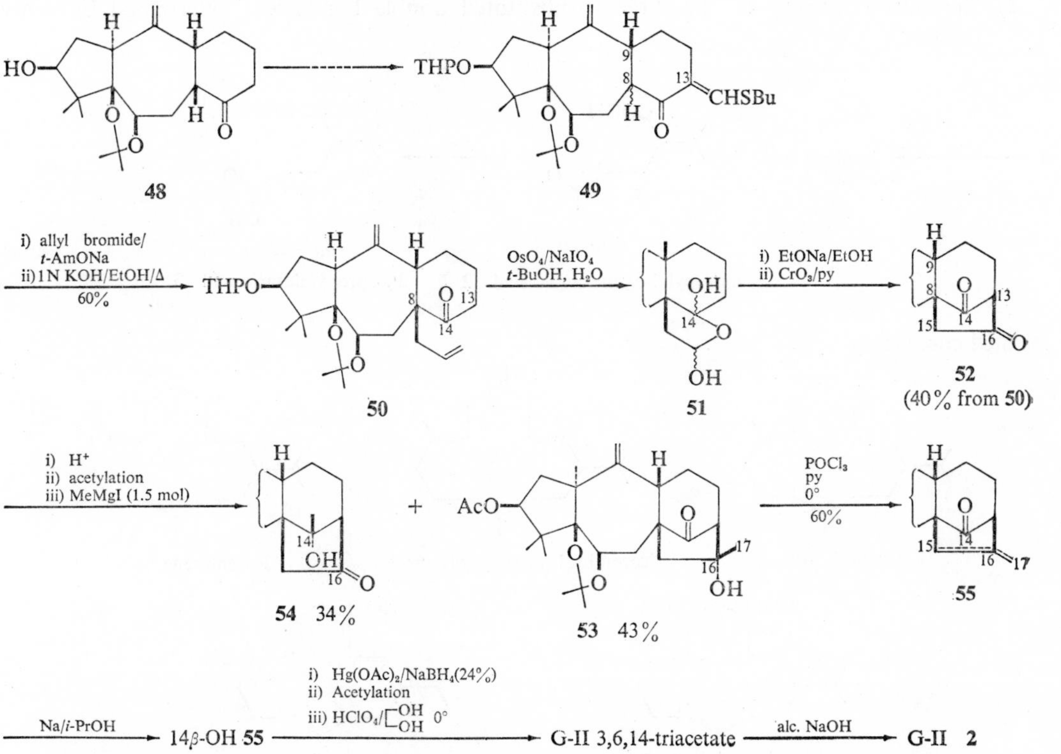

48

49

i) allyl bromide/
t-AmONa
ii) 1 N KOH/EtOH/Δ
60%

50

OsO$_4$/NaIO$_4$
t-BuOH, H$_2$O

51

i) EtONa/EtOH
ii) CrO$_3$/py

52
(40% from **50**)

i) H$^+$
ii) acetylation
iii) MeMgI (1.5 mol)

54 34%

+

53 43%

POCl$_3$
py
0°
60%

55

i) Hg(OAc)$_2$/NaBH$_4$(24%)
ii) Acetylation
iii) HClO$_4$/ $\begin{smallmatrix}OH\\OH\end{smallmatrix}$ 0°

Na/*i*-PrOH ⟶ 14β-OH **55** ⟶ G-II 3,6,14-triacetate ⟶ alc. NaOH ⟶ G-II **2**

11) M. Shiozaki, K. Mori, M. Matsui, T. Hiraoka, *Tetr. Lett.*, 657 (1972).
12) N. Hamanaka, T. Matsumoto, *Tetr. Lett.*, 3087 (1972).

4.54   CEMBRENE (THUNBERGENE) AND RELATED COMPOUNDS

cembrene **1**[1,2]

mp: 60° α_D: +233°
uv: 246   (ε17,100)
ir: 963, 840, 810

1) Absorptions at 810, 840 and 963 cm⁻¹ were assigned to two trisubstituted and one *trans*-disubstituted double bonds, respectively.
2) The uv band at 246 nm shows the presence of a conjugated diene.
3) Absorption of four moles of H_2 indicates a monocyclic structure.
4) The stereochemistry at C-1 was based on the formation of (−)-2-isopropylglutaric acid **3**.
5) The stereochemistry of the three trisubstituted double bonds was determined by x-ray analysis.[3]

$$\mathbf{1} \xrightarrow{\;O_3\;} \text{levulinic acid} \; + \; \text{(compound 2)} \xrightarrow{\;OBr^-\;} \text{(compound 3)}$$

2-isopropyl-5-oxohexanoic acid **2** α_D −30°

2-isopropylglutaric acid **3** α_D −20°

Related compounds

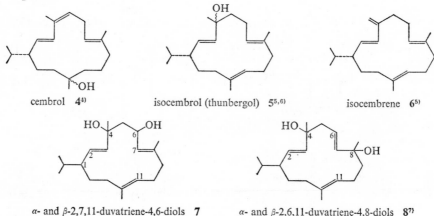

cembrol **4**[4]

isocembrol (thunbergol) **5**[5,6]

isocembrene **6**[5]

α- and β-2,7,11-duvatriene-4,6-diols **7**

α- and β-2,6,11-duvatriene-4,8-diols **8**[7]

1) H. Kobayashi, S. Akiyoshi, *Bull. Chem. Soc. Japan*, **35**, 1044 (1962); *ibid.*, **36**, 823 (1963).
2) W. G. Dauben, W. E. Thiessen, P. R. Resnick, *J. Org. Chem.*, **30**, 1693 (1965).
3) M. G. B. Drew, D. H. Templeton, A. Zalkin, *Acta Cryst.*, **25B**, 261 (1969).
4) A. I. Lisina, A. I. Rezvukhin, V. A. Pentegova, *Khim. Prir. Soedin., Akad. Nauk Uz. SSR.*, 250 (1965).
5) N. K. Kashtonova, A. I. Lisina, V. A. Pentegova, *Khim. Prir. Soedin.*, **4(1)**, 52 (1968).
6) B. Kimland, T. Norin, *Acta Chem. Scand.*, **22**, 943 (1968).
7) D. L. Roberts, R. L. Rowland, *J. Org. Chem.*, **27**, 3989 (1962); *ibid.*, **28**, 1165 (1963).

α-8,11-oxido-
2,6,12-duvatrien-4-ol **9**[8]

α- and β-8,11-oxido-
2,6,12(17)-duvatrien-4-ols **10**[8]

incensole **11**[9]

incensole oxide **12**[10]

ovatodiolide **13**[11]

eunicin **14**[12]

crassin acetate **15**[13]

eupalmerin acetate **16**[14]

jeunicin **17**[14]

Remarks

Cembrene **1** has been isolated from various sources, mostly from the genus *Pinus*.[1,2] Cembrene was the first naturally occuring C_{14}-monocyclic diterpene to be characterized.

The related compounds **4**, **5** and **6** have been isolated from *P. sibirica*[5] and **5** has also been obtained from *Pseudotsuga menziesii*.[6] The diol derivatives **7** and **8** were present in tobacco (*Nicotinana tabacum*). Several hydroxy-ether derivatives are known: **9** and **10** were also found in *N. tabacum*, and **11** and **12** in *Boswellia carteri*. Other related compounds include the lactone derivatives **13** to **17**. Compound **13** was obtained from *Anisomeles ovata*, **14** from *Eunicea mammosa* (marine origin), **15** from *Pseudoplexaura porasa* (marine origin), and **16** and **17** from *E. plameri* (Caribbean gorgonian).

The structures of **14** and **15** were determined by x-ray methods.[15] **15** has antibacterial activity.

8) R. L. Rowland, A. Rodgman, J. N. Schumacher, D. L. Roberts, L. C. Cook, W. E. Walker Jr., *J. Org. Chem.*, **29**, 16 (1964).
9) S. Corsano, R. Nicoletti, *Tetr.*, **23**, 1977 (1967).
10) R. Nicoletti, M. L. Forcellese, *Tetr.*, **24**, 6519 (1968).
11) H. Immer, J. Polonsky, R. Toubiana, H. DocAnn, *Tetr.*, **21**, 2117 (1965).
12) A. J. Weinheimer, R. E. Middlebrook, J. O. Bledsoe Jr., W. E. Marsico, T. K. B. Kaus, *Chem. Commun.*, 384 (1968).
13) M. B. Hossain, D. van der Helm, *Rec. Trav. Chim.*, **88**, 1413 (1969).
14) A. J. Weinheimer, S. J. Rehm, L. S. Ciereszko, *4th Natural Products Symp.*, Jamaica, 1972.
15) M. B. Hossain, A. F. Nicholas, D. van der Helm, *Chem. Commun.*, 385 (1968).

4.55   STRUCTURE AND BIOSYNTHESIS OF PLEUROMUTILIN

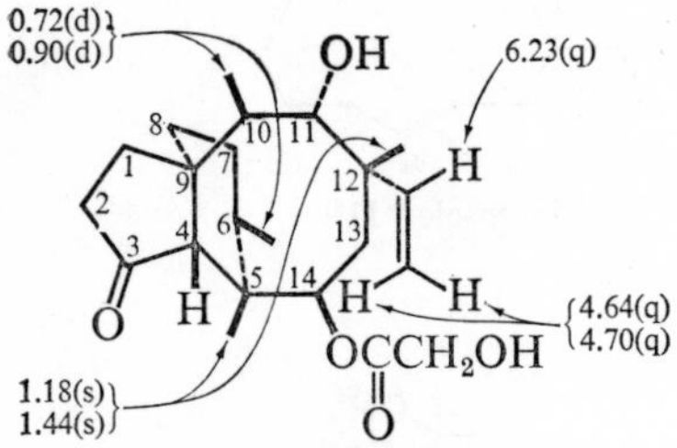

0.72(d)
0.90(d)

OH   6.23(q)

8   10   11
7
2   9   12   H
6   13
3   4   14
O   H   H   H   4.64(q)
OCCH₂OH   4.70(q)
1.18(s)   ‖
1.44(s)   O

pleuromutilin   **1**
mp: 170–171°
ir(CS₂): 3550(OH), 1736(C=O)
3080, 1642, 914(C=CH₂)[1]

1) Pleuromutilin **1** $(C_{22}H_{34}O)_5$ is an ester of glycolic acid and mutilin **2**, a tricyclic diol containing one carbonyl and one vinyl group.[1,2]

2) In an unusual application of tracer experiments to structure determination, [2-¹⁴C]-mevalonic lactone, a specific terpene precursor, was incorporated *in vivo* into pleuromutilin to the extent of 5.5%. [1-¹⁴C]-acetic acid was also effectively utilized (1.7% incorporation) in a fashion (assayed by degradation) that strongly suggested the diterpene nature of mutilin.[1]

3) Dehydrogenation of **2** over Se gave dimethyl indanone **4**, defining the 1-indanone system and the location of two Me groups.[2]

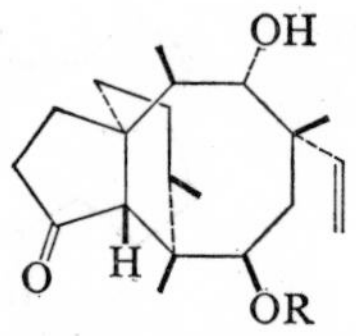

mutilin   **2**   R = H
3   R = OCCH₂Br
‖
O

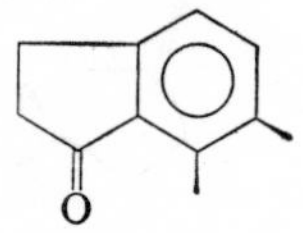

4

4) An elegant series of degradations revealed the structural framework of **1**.[2] One example of these reactions is the sequence **2→8**, which twice involves a transannular Cannizzaro-type reaction and liberates glycollic aldehyde.[2]

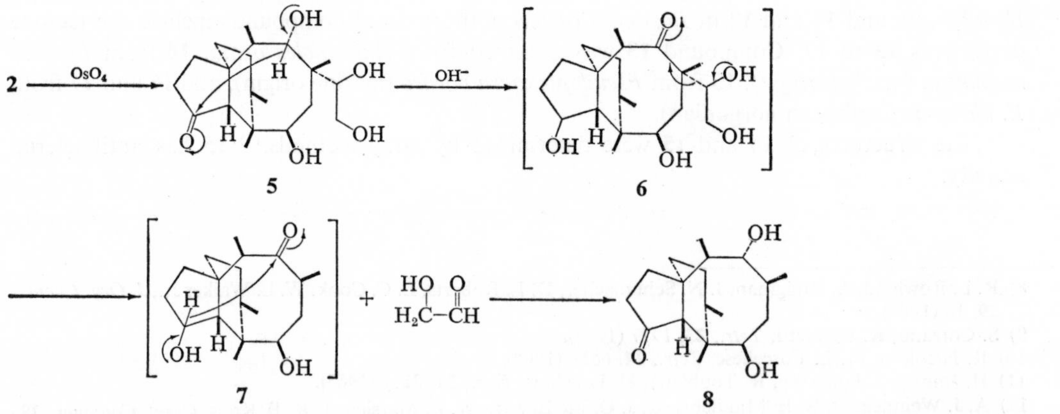

2 $\xrightarrow{\text{OsO}_4}$ **5** $\xrightarrow{\text{OH}^-}$ [**6**   ]

[**7**   ]   +   HO   O
H₂C–CH   →   **8**

1) A. J. Birch, *Proc. Chem. Soc.*, 3 (1962); A. J. Birch, O. W. Cameron, C. W. Holzapfel R. W. Rickards, *Chem. Ind.*, 374 (1963); A. J. Birch, C. W. Holzapfel R. W. Rickards, *Tetr.*, **22**, suppl. 8, part II, 359 (1966).
2) D. Arigoni, *Gazz. Chim. Ital.*, **92**, 884 (1962).

5) This and other transannular reactions, characteristic of medium-ring compounds (C_8–C_{11} carbocycles), played a major role in fixing the stereochemistry of several centers in **1**.[2]

6) Birch and co-workers arrived at the same conclusion as Arigoni as to the structural framework of **1**, on the basis of a different set of degradative experiments.[1]

7) The entire stereochemistry of pleuromutilin was eventually established chemically and then confirmed by x-ray analysis of **3**, although the details of this investigation remain unpublished.[3]

Remarks

Pleuromutilin, an antibiotic inhibitory to Gram-positive bacteria, was isolated from *Pleurotus mutilus* and other Basidiomycetes.[4]

Compound **1** is a most unusual diterpene in that its structure and biogenesis clearly do not fall into established patterns. Investigation of the biosynthetic pathways leading to pleuromutilin was a particularly challenging problem. While some progress in this field was recorded by Birch and his group,[1] Arigoni and co-workers[3] elucidated the entire biosynthetic route to **1** in detail, as outlined below:

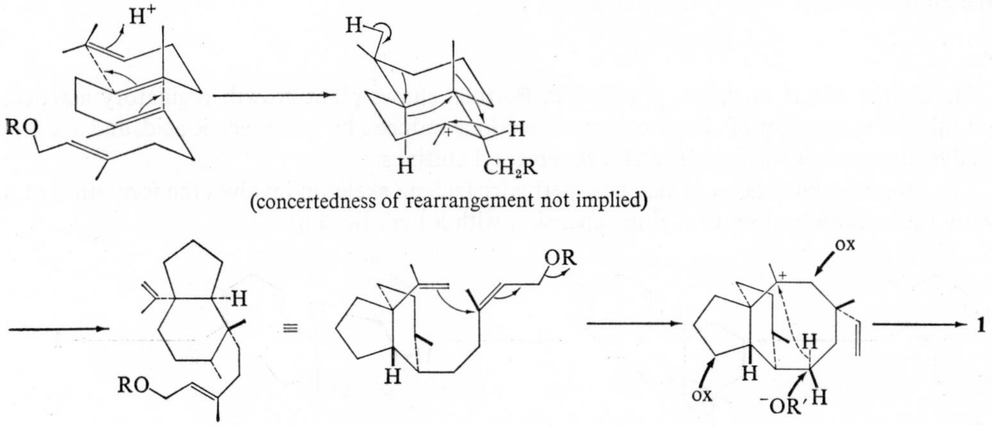

(concertedness of rearrangement not implied)

3) D. Arigoni, *Pure Appl. Chem.*, **17**, 331 (1968), and references therein.
4) F. Kavanagh, A. Hervey, W. J. Robbins, *Proc. Nat. Acad. Sci. U.S.*, **37**, 570 (1951); *ibid.*, **38**, 555 (1952).

4.56   PORTULAL

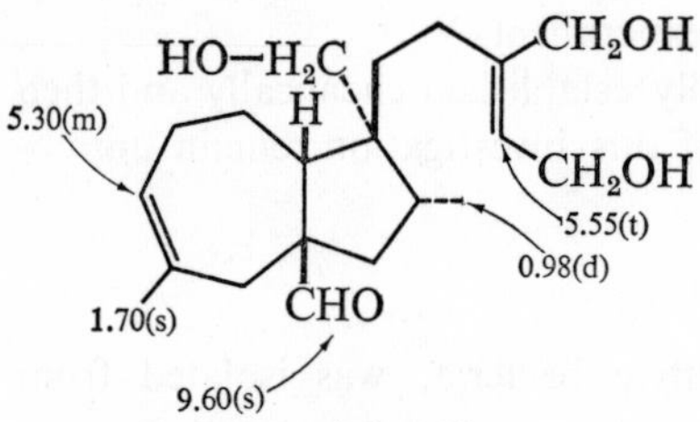

portulal   **1**¹⁾

mp: 113° $\alpha^{23}{}_D$ (EtOH): −50°
ir (nujol): 3260(OH), 2720, 1720(CHO), 1030
uv (EtOH): 285 (ϵ 27)
nmr in acetone-d_6: 1.1–2.8 (14H's)

1) The molecular structure including absolute configuration was established by x-ray studies on the *p*-bromophenyl sulfonylhydrazone of **1**.

Remarks

The source of **1** is *Portulaca grandiflora*. Portulal shows plant growth regulatory activity; i.e., it inhibits elongation of *Avena* coleoptile sections induced by indole acetic acid, and accelerates adventitious root formation of *Azukia* epicotyl cuttings.

The proposed biogenesis of this new perhydroazulene skeleton involves the formation of a *primary* cation **3** and subsequent ring expansion with a hydride shift.

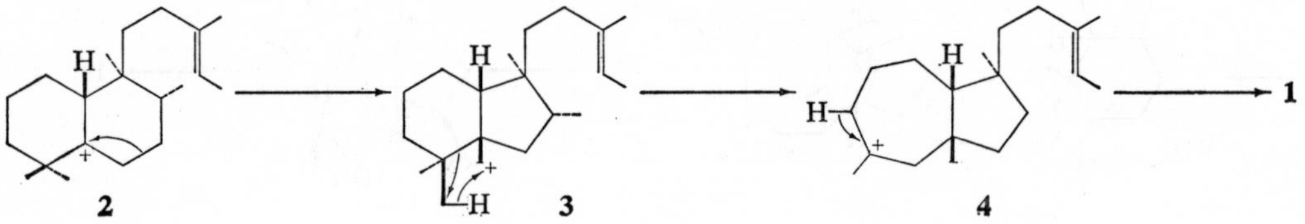

1) S. Yamazaki, S. Tamura, F. Marumo, Y. Saito, *Tetr. Lett.*, 359 (1969).

4.57   GINKGOLIDES

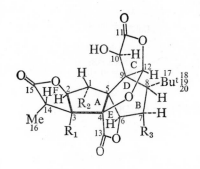

	R_1	R_2	R_3	sec-OH	tert-OH
GA 1	OH	H	H	one	one
GB 2	OH	OH	H	two	one
GC 3	OH	OH	OH	three	one
GM 4	H	OH	OH	three	—

ginkgolides 1–4[1,2]
mp: dec. >280°
most nmr measured in TFA (100 MHz) for
solubility reasons

1) A number of ginkgolides were isolated from 100 kg of root bark peeled from five ginkgo trees of 30 cm diameter: GA (10 g), GB (10 g), GC (20 g) and GM (0.2 g). GA and GB were separated after a 10–15 step fractional recrystallization.

2) The molecular formulae were based on the relatively volatile GA 10-dimethyl ether, $C_{22}H_{28}O_9$, ms: 436.168 (calcd. 436.173). They are all C_{20} compounds.

3) The presence of the *t*-Bu group was deduced as follows. The nmr of all 50 derivatives except for photodehydro-GA 13 showed a 9H singlet at 1.2–1.3 ppm; the base ms peak was at 57.074 (calcd. for *t*-Bu cation, 57.070); Kuhn-Roth oxidation of GC gave pivalic acid (*t*-BuCOOH).

4) The single carbonyl ir peak at 1780 cm⁻¹ (KBr) is replaced by carboxylate bands at 1540 and 1400 cm⁻¹ in Na salts, which were prepared by evaporating aqueous NaOH solutions. Thus all CO functions are lactones. The first indication for three lactones came from the nmr of GA ether 5 (spectrum A). This was corroborated by a modified lactone titration which forced all lactones to open prior to titration.

5) The hydroxyl functions were clarified from comparisons of nmr spectra taken in DMSO-d_6 before and after D_2O addition, and by identification of nmr peaks which underwent downfield shifts upon acetylation. The remaining oxygen was assigned to an ether.

6) The GA ether 5 was a key compound in many respects. It was obtained accidentally when the syrup (an octaol) resulting from LAH reduction of GA was heated in a malfunctioning oven (pyrolysis). The peaks designated cc′, ee′, ff′ in the nmr spectrum A are all absent in spectrum B, showing that they were introduced during the conversion C=O $\xrightarrow{\text{LAH/pyrolysis}}$ CH₂ (or CD₂). The appearance of three new methylenes suggests that GA has three lactones. The close similarity between the nmr of GA and GA ether-d_6 (spectra C and B) shows that the original cage structure is retained in the ether 5.

   The newly introduced CH₂ groups in 5 complicated the nmr spectra, but simultaneously clarified the linking pattern of the otherwise isolated protons, i.e. H_I, or proton groups. The thorough nmr studies required to fully analyze spectrum A also led to the discovery of noe (overleaf).

1) K. Nakanishi, *Pure Appl. Chem.*, **14**, 89 (1967); M. Maruyama, A. Terahara, Y. Itagaki, K. Nakanishi, *Tetr. Lett.*, 299 (1967); *ibid.*, 303 (1967); M. Maruyama, A. Terahara, Y. Nakadaira, M. C. Woods, K. Nakanishi, *ibid.*, 309 (1967); M. Maruyama, A. Terahara, Y. Nakadaira, M. C. Woods, Y. Takagi, K. Nakanishi, *ibid.*, 315 (1967); M. C. Woods, I. Miura, Y. Nakadaira, A. Terahara, M. Maruyama, K. Nakanishi, *ibid.*, 321 (1967);
2) Y. Nakadaira, Y. Hirota, K. Nakanishi, *Chem. Commun.*, 1467 (1969).

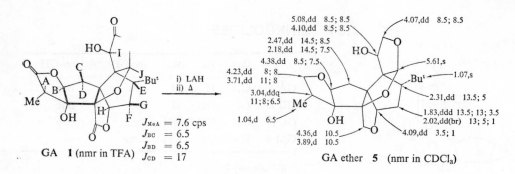

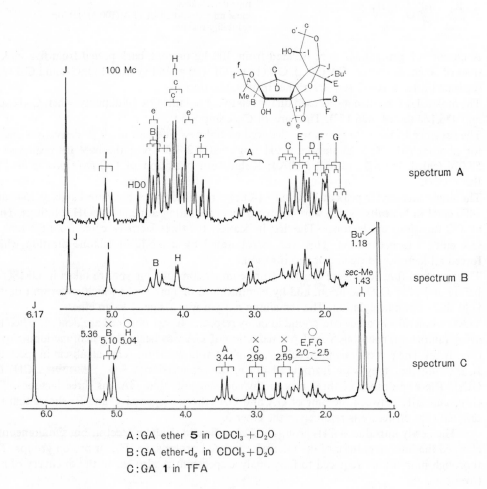

A: GA ether **5** in CDCl$_3$ + D$_2$O
B: GA ether-d$_6$ in CDCl$_3$ + D$_2$O
C: GA **1** in TFA

7) Rings F and A and protons A–D were determined as follows. In **7**, the Hc signal was W-coupled to the 14-Me. Ring F is lost on hydrogenolysis to give **8**, the ozonolysis of which generates a 5-membered ketone (Cotton effect at 290 nm). This sequence fully characterises rings F and A.

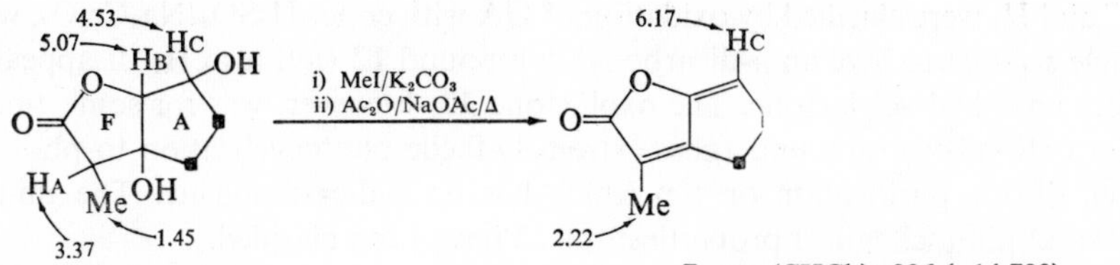

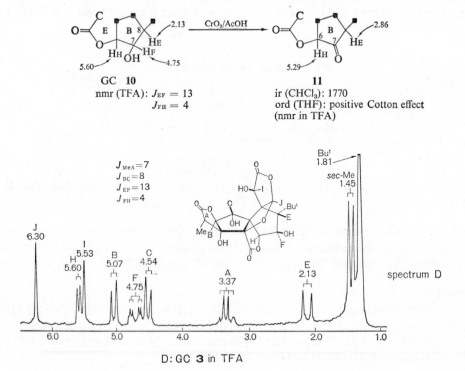

GC 6 (nmr in TFA)

i) MeI/K₂CO₃
ii) Ac₂O/NaOAc/Δ

7 uv (CHCl₃): 286 (ε 14,700)
ir (CHCl₃): 1686, 1630 (C=C),
nmr (TFA): $J_{CMe} = 0.7$

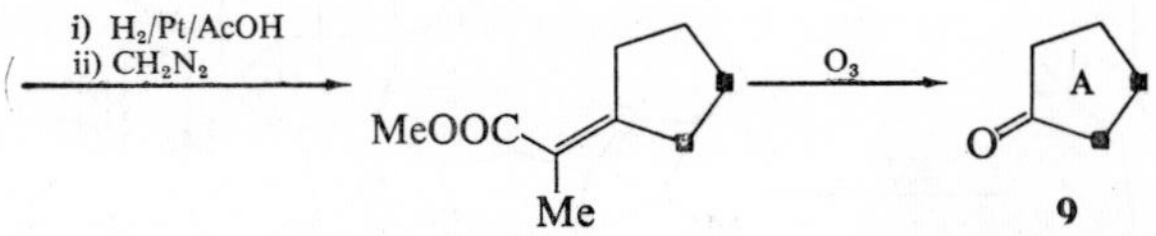

i) H₂/Pt/AcOH
ii) CH₂N₂

8

O₃

9
ir (CHCl₃): 1745
ord (THF): positive Cotton effect

8) Ring B and protons E, F and H were based on nmr (spectrum D). The substituent pattern on ring B is clear from this GC nmr. The ir of dehydro-GC **11** suggests a 4- or 5-membered ketone; later, the high frequency for the 5-membered ketone was attributed to the *t*-Bu (noe below). H_H is linked to a lactone *O*-terminal, because no additional coupling appears in going from GA **1** to ether **5** (spectra C and A).

GC 10
nmr (TFA): $J_{EF} = 13$
$J_{FH} = 4$

CrO₃/AcOH

11
ir (CHCl₃): 1770
ord (THF): positive Cotton effect
(nmr in TFA)

$J_{MeA} = 7$
$J_{BC} = 8$
$J_{EF} = 13$
$J_{FH} = 4$

spectrum D

D: GC **3** in TFA

9) Ring C and H_I were clarified by oxidation of GA with conc. $H_2SO_4/Na_2Cr_2O_7$, which oxidizes the single *sec*-OH to give an α-dicarbonyl compound **12** (uv; also H_I disappears). Ring C is therefore an α-hydroxylactone. The oxolactone **12**, however, was for some time a source of misinformation because it undergoes extremely facile photocyclization to photodehydro-GA **13** (even during purification on tlc) which has no α-dicarbonyl uv. The unusual spectroscopic[2] and photochemical properties[3] of **12** have been clarified.

dehydro-GA **12**
uv (THF): 391 (ϵ 42)
411 (ϵ 34)

photodehydro-GA **13**
(nmr in TFA)

10) Alkali fusion of GA resulted in the loss of 2 carbons (captured as oxalic acid) to give bisnor GA **14**, which is a dilactone hemiacetal (oxidizable to trilactone **15**). Nmr showed that ring C had been destroyed, but other rings were intact; moreover, it was clear from decoupling that the newly formed H_N was strongly coupled to H_E. Ring B is thus linked to hydroxylactone C by sharing C-9.

Regarding the remaining hydrogen, H_J, its low chemical shift (spectra A–D) and the fact that the singlet character is retained in the ether **5** (spectra A and B) requires it to be flanked by two oxygens. The only way to do this is to place it on C-12 and link C-12 to the remaining ethereal oxygen.

GA **1**

bisnor-GA **14**
(nmr in TFA)

dehydrobisnor-GA **15**
(nmr in TFA)

11) The linking of *t*-Bu, structure **6**, rings B/C and ethereal oxygen in a systematic manner leads to eight structures; seven of these can be readily eliminated on the basis of nmr peak shifts in various derivatives and noe, thus leading to "planar" structures **1–4**.

Stereochemistry[1]

The photocyclization of **12** to **13** indicates that *t*-Bu is *cis* to the 9,10 bond. From the ease of ginkgolide regeneration upon acidification of a conc. NaOH solution, all the lactones are *cis*-fused. In addition, rings F and D are *cis* fused onto ring A since 14-H in **14** undergoes a 0.27 ppm shift upon oxidation of the 12-OH to 12-carbonyl (**15**).

The 1-OH was deduced to be α by the following reasoning. H_H is at 5.04 ppm in GA, which lacks the 1-OH; in contrast it is at 5.73 (GB), 5.60 (GC) and 5.49 (GM) when the 1-OH is present (nmr in TFA). Molecular models show that 1α-OH and H_H are in a cyclohexane 1,3-diaxial relationship, and therefore H_H is shifted downfield by the 1α-OH.

With regard to the 7-OH, acetylation of GC in boiling Ac_2O/NaOAc gives the triacetate **16** isomeric with the normal triacetate. GC is regenerated on hydrolysis. Studies of noe for **16** (see also **18**) showed that a translactonization from C-6 to C-7 had occurred: 6-OH and 7-OH are therefore *cis*, i.e. 7β-OH.

GC **3**   iso-GC triacetate **16**

The absolute configuration was derived by application of the octant rule to the positive Cotton effects of GC F-norketone **9** and 7-oxo-GC monomethyl ether (see **11**).

The structure **1** was deduced independently by x-ray studies on GA 3-bromobenzoate.[4]

Intramolecular nuclear Overhauser effect (noe)[1,5]

Thorough studies of complex nmr spectra such as that shown in spectrum A led to the discovery and first application of noe in natural products. It is exemplified by **17** and **18** (nmr in TFA). The 22% noe on H_J in **17** shows that, as might be expected, the *t*-Bu group adopts an e'-position (otherwise, the distance is too great for an noe). This strain on ring B also explains the 1770 cm^{-1} ir band in dehydro-GC **11** (see **8** above). In the isomeric acetate **18**, the *t*-Bu takes the sterically favored a'-conformation, and consequently it exerts an noe on H_H but not on H_J.

GC tetraacetate **17**

irradiated proton	noe in observed H	
t-Bu	I	33%
t-Bu	J	22%
t-Bu	F	16%
t-Bu	E	6%
t-Bu	H	zero

iso-GC tetraacetate **18**

irradiated proton	noe in observed H	
t-Bu	I	27%
t-Bu	J	zero
t-Bu	F	19%
t-Bu	E	13%
t-Bu	H	6%
J	E	10%
E	J	23%

Remarks

Ginkgolides are bitter substances contained in leaves and root barks of *Ginkgo biloba* ("fossil tree"), which has remained unchanged for several million years. They are extremely inert and crystals can be recovered after boiling off a nitric acid solution. They are the first natural products to have a *t*-Bu. The 6 five-membered rings make a rigid cage, with C-1 to C-9 forming the spirocarbocyclic skelton. A related sesquiterpene, bilobalide, also having 3 γ-lactones and a *t*-Bu has been isolated from the same source.[6]

Biosynthetic studies carried out with [2-^{14}C]-acetate (*), *dl*-[2-^{14}C]-mevalonic acid (▲), and [Me-^{14}C]-methionine (●) suggest the following scheme.[7]

Migration of Me from C-3 to C-14 (ginkgolide numbering) in the entpimaradienone cation **19** is consistent with the even label distribution in acetic acid **24**.

The *t*-Bu group is surprisingly formed by cleavage of the usual diterpenoid ring A, loss of one C (e.g. COOH in **21**), and re-introduction of one C from methionine (**21**). The intermediacy of such structures is necessary to explain the inversion at C-8 in going from precursor **19** to GB **2**.

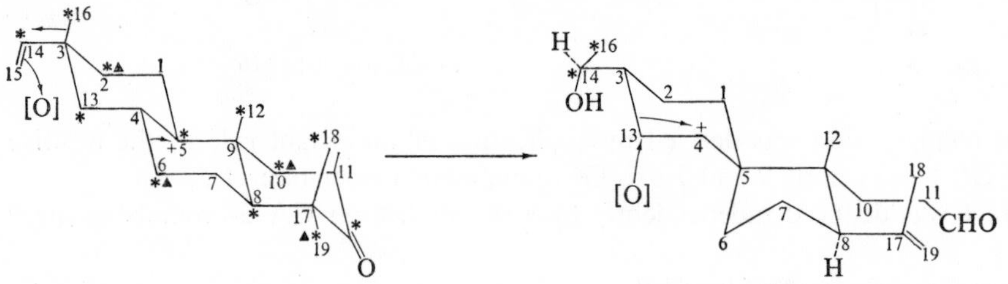

19

20

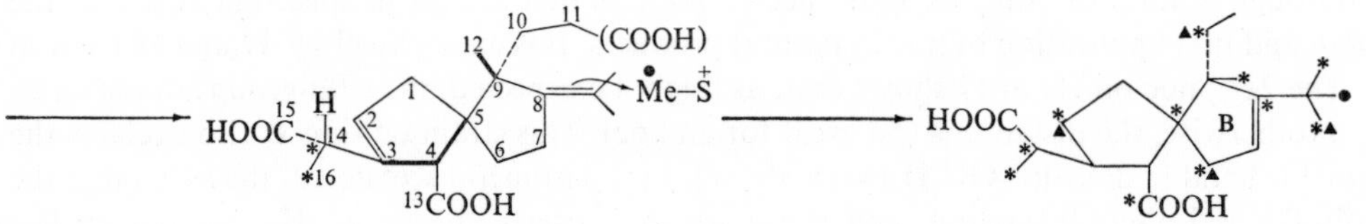

21

22

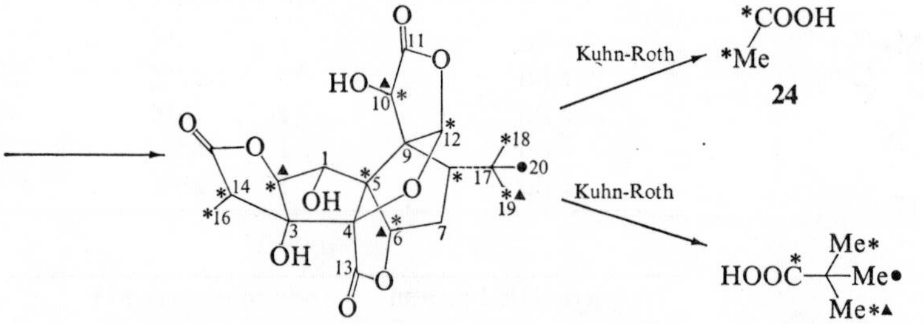

24

2

25

3) Y. Nakadaira, Y. Hirota, K. Nakanishi, *Chem. Commun.*, 1469 (1969).
4) N. Sakabe, S. Takada, K. Okabe, *Chem. Commun.*, 259 (1967); K. Okabe, K. Yamada, S. Yamamura, S. Takada, *J. Chem. Soc.*, 220 (1967).
5) J. H. Noggle, R. E. Schirmer, *The Nuclear Overhauser Effect*, Academic Press, 1971.
6) K. Nakanishi, K. Habaguchi, Y. Nakadaira, M. C. Woods, M. Maruyama, R. T. Major, M. Alauddin, A. R. Patel, K. Weinges, W. Bahr, *J. Am. Chem. Soc.*, **93**, 3544 (1971).
7) K. Nakanishi, K. Habaguchi, *J. Am. Chem. Soc.*, **93**, 3546 (1971).

4.58 STRUCTURE OF CYATHIN A₃

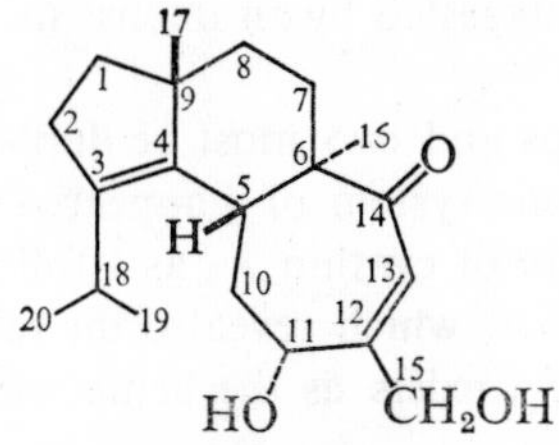

cyathin A₃ **1**[1]
mp: 148–150°   α$_D$(MeOH): −160°
ms (high resoln): M⁺ C$_{20}$H$_{30}$O$_3$
ir (CHCl$_3$, KBr): 1650

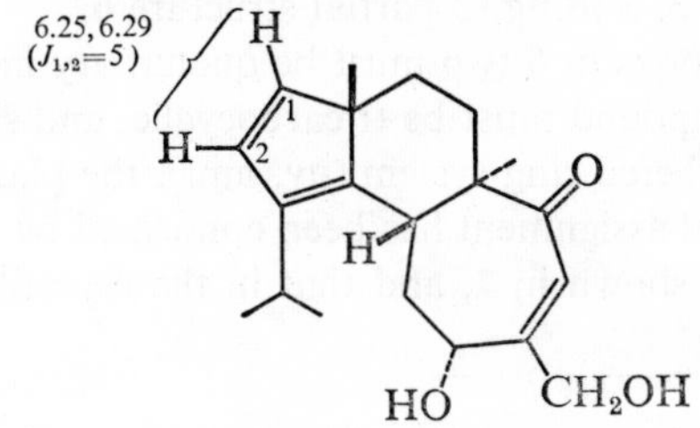

6.25, 6.29
($J_{1,2}$=5)

1,2-dehydrocyathin A₃ **4**
(congener of **1**)

1) Formation of the diacetate **2** indicated the presence of two hydroxyl groups. Its uv was characteristic of an α,β-unsaturated ketone.

2) Nmr studies of **2** revealed an isolated –CH–CH$_2$–CH(OAc)–C(CH$_2$OAc)=CH– system, the allylic protons 5-H and 18-H, and two quaternary methyls.

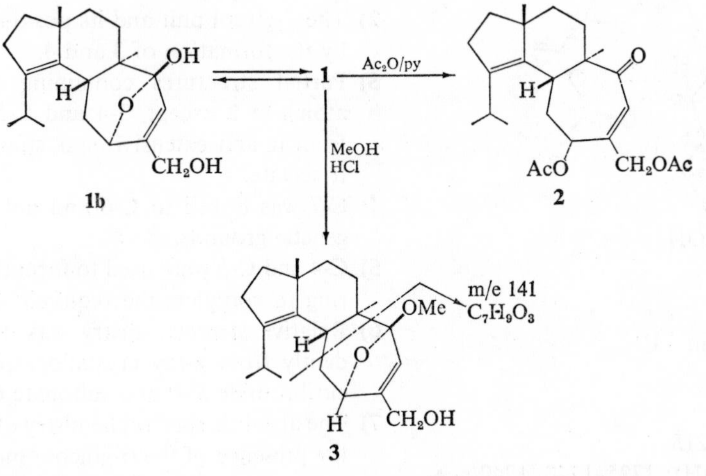

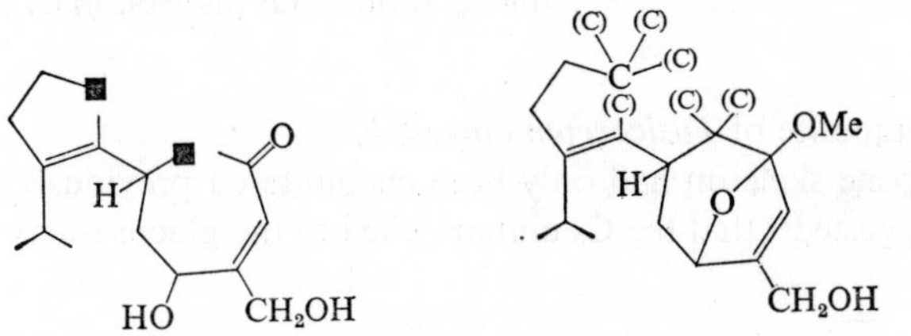

1b

Ac$_2$O/py

2

MeOH
HCl

m/e 141
C$_7$H$_9$O$_3$

3

5

6

1) W. A. Ayer, H. Taube, *Tetr. Lett.*, 1917 (1972).
2) A. D. Allbutt, W. A. Ayer, H. J. Brodie, B. N. Johri, H. Taube, *Can. J. Microbiol.*, **17**, 1401 (1971).

3) Concurrent nmr analysis of 1,2-dehydrocyathin A₃ **4** (a congener of **1**) suggested the presence of a fully substituted double bond in **1**. By inference, if 5-H and 18-H are allylic to this double bond, one can write the partial structure **5**.

4) Location of the enone in a seven-membered ring system was suggested by an intense ms peak at m/e 141 in **3**, leading to partial structure **6**.

5) Of the 22 carbons in **6** two must be quaternary methyl groups and two must be duplicates. Since the compound must be tricarbocyclic, and since the diene system of **4** appeared to be in a five-membered ring system (uv, nmr), the planar structure of cyathin A₃ as **1** follows.[1]

6) This structural assignment has been confirmed by x-ray analysis, which revealed the relative configuration shown in **1**, and that in the crystalline form, A₃ exists as the hemiacetal **1b**.

Remarks

Cyathin A₃ and its congener **4** are antibiotic substances isolated from the bird's-nest fungus *Cyathus helenae*.[2]

4.59   STRUCTURE OF FUSICOCCIN

fusiococcin   **1**[1]

$C_{36}H_{56}O_{12}$
mp: 155–156°
uv: none above 215
ir: 3400–3450 (OH); 1725–1740, 1240(acetate); 1635(olefinic); 920(vinyl)

1) The presence of the sugar moiety was derived from ms and nmr of **1** and its simple glycoside derivatives (e.g. acetates) and from hydrolysis of **1** to the aglycone **2**.

2) The α-glycol unit and its position were revealed by the formation of **3** and **4**.

3) Partial structures containing all the carbon atoms in **2** except C-4 and C-5 were assigned from ir and extensive nmr studies of **2** and its triacetate.

4) C-7 was linked to C-6 and not to C-3 on biogenetic grounds.

5) C-4 and C-5 were used to form the cyclopentane ring to complete the required carbon skeleton.

6) Relative stereochemistry was derived independently from x-ray crystallography of the mercuribromide **5**[2,3] and sulfonate **6**.[1]

7) The absolute spectrochemistry of **1** was based on the presence of the D-glucose moiety,[1] and also on x-ray studies of **6** (from Bijvoet anomalies due to anomalous dispersion by Hg and Br[3]).

Remarks

Fusiococcin **1** is a metabolite of *Fusicoccum amygdali*.

The carbocyclic diterpene skeleton had only been encountered previously in the sesterterpenes (C_{25}). It has been suggested[1] that the C_5 unit attached to the glucose may have originated from a sesterterpene.

1) A. Ballio, M. Brufani, C. G. Casinovi, S. Cerrini, W. Fedeli, R. Pellicciari, B. Santurbano, A. Vaciago, *Experientia*, **24**, 635 (1968).
2) K. D. Barrow, D. H. R. Barton, E. B. Chain, U. F. W. Ohnsorge, R. Thomas, *Chem. Commun.*, 1198 (1968).
3) E. Hough, M. B. Hursthouse, S. Neidle, D. Rogers, *Chem. Commun.*, 1197 (1968).

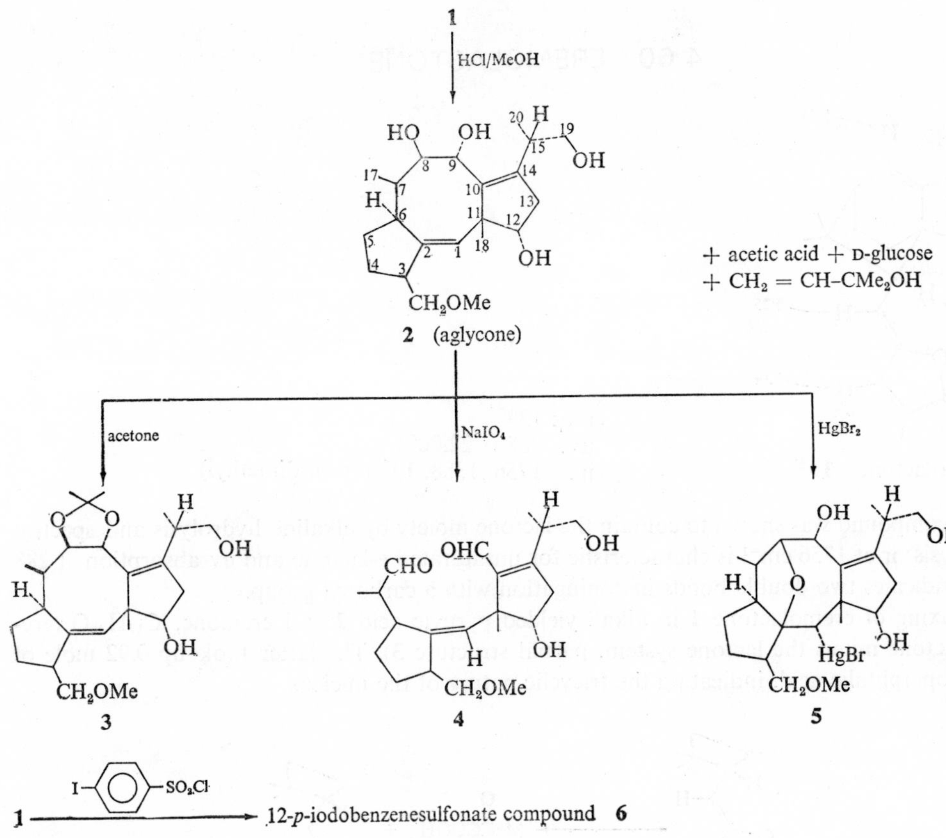

1

HCl/MeOH

2 (aglycone)

+ acetic acid + D-glucose
+ CH₂ = CH–CMe₂OH

acetone

NaIO₄

HgBr₂

3

4

5

1 ──── I–⟨⟩–SO₂Cl ────→ 12-p-iodobenzenesulfonate compound 6

4.60   EREMOLACTONE

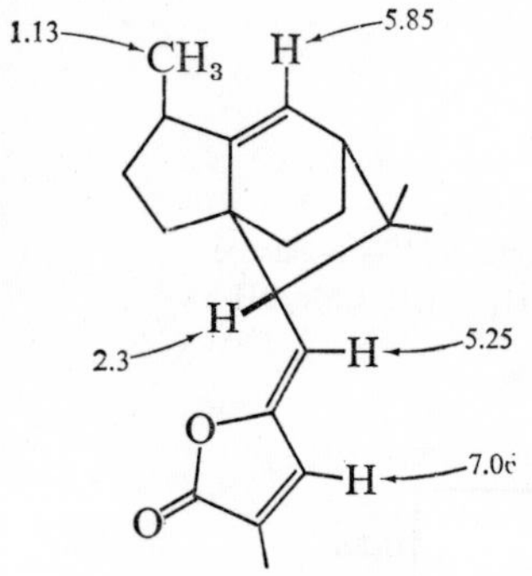

eremolactone   **1**[1,2]

mp: 139°
uv: 288 (ϵ 22,000)
ir: 1756, 1388, 1370 (*gem*-dimethyl)

1) The compound was shown to contain the lactone moiety by alkaline hydrolysis and spectral analysis: ir at 1756 cm⁻¹ is characteristic for unsaturated γ-lactone and uv absorption at 288 nm indicates two double bonds in conjugation with a carbonyl group.

2) Refluxing of eremolactone **1** in alkali yielded pyruvic acid **2** and eremone, $C_{17}H_{26}O$ (eremolactone minus the lactone system, partial structure **3**). The latter took up 0.92 mole of monoperphtalic acid, indicating the tricyclic nature of the nucleus.

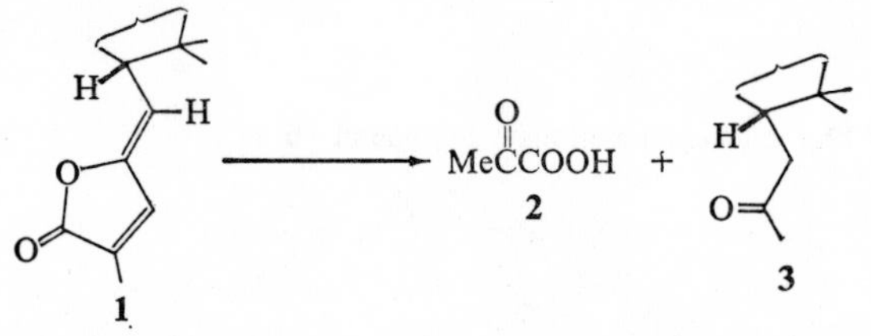

3) Acidic bond isomerization converted eremolactone **1** to isoeremolactone, which was shown to contain the partial structure **4** by the following reactions:

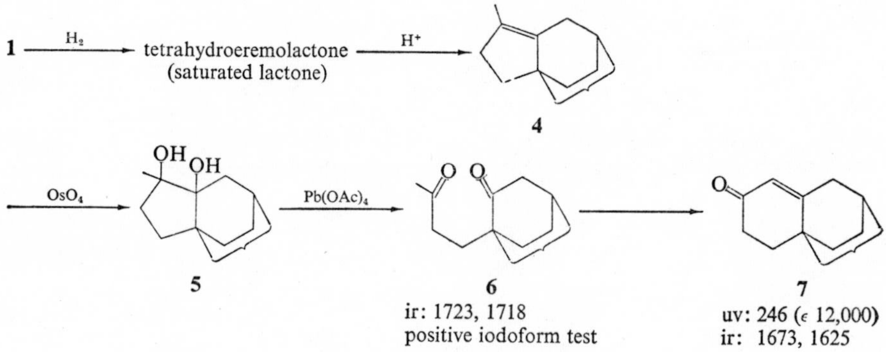

1) A. J. Birch, J. Grimshaw, J. P. Turnbull, *J. Chem. Soc.*, 2412 (1963).
2) A. J. Birch, G. S. R. Subba Rao, J. P. Turnbull, *Tetr. Lett.*, 39, 4749 (1966).

4) The nmr olefinic quartet at 5.85 ($J_2 = 7.2$) disappeared upon isomerization, and the nuclear methyl signal shifted downfield. Such changes were rationalized as follows:

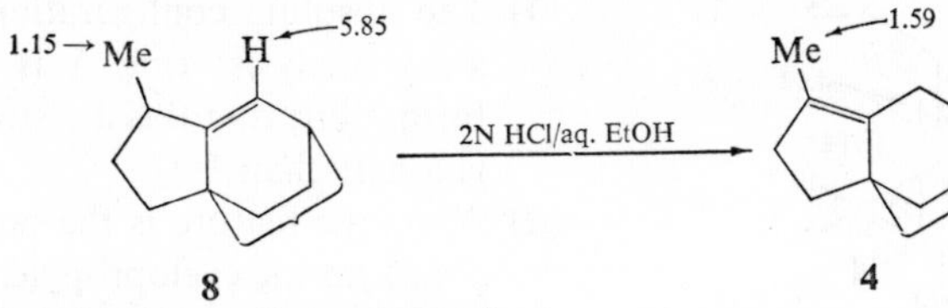

8
(partial structure of eremolactone)

4

5) The structure and stereochemistry of isoeremolactone **9** was determined by x-ray analysis. The three six-membered rings bridged together are in the boat configuration.[3]

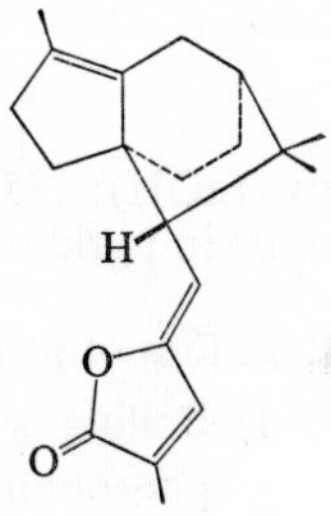

9

6) Reaction of eremolactone **1** with neutral permanganate afforded a ketodicarboxylic acid (m/e 282) **10**. Deuteration with NaOD introduced only one D (m/e 283), **11**. This result established that the bridge termination is at a, and not at b, as was first proposed.

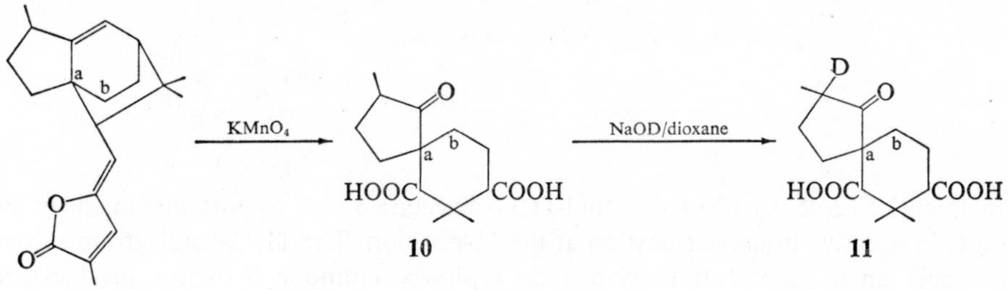

1   **10**   **11**

Remarks

Eremolactone was isolated from the leaves of *Eremophila freelingii*.[4]

3) Yow-lam Oh, E. N. Maslen, *Tetr. Lett.*, **28**, 3291 (1966).
4) P. R. Jefferies, J. R. Knox, E. J. Middleton, *Aust. J. Chem.*. **15**, 532 (1962).

4.61 PHORBOL AND RELATED COMPOUNDS

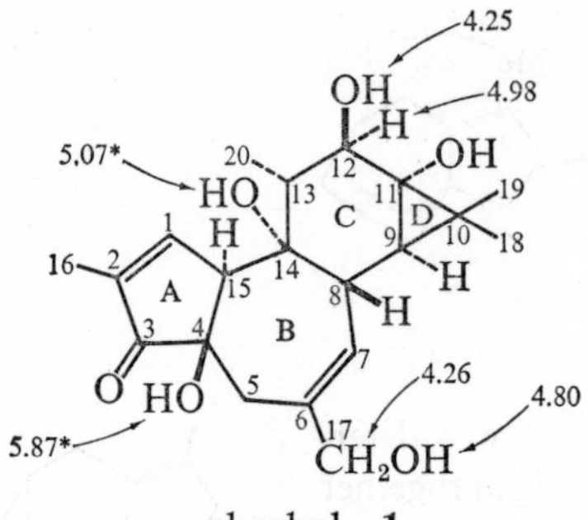

phorbol **1**
uv (EtOH): 235 (5200), 334 (70)[2]
nmr: in pyridine-d_5; data with * in DMSO-d_6

1) The absolute configuration was determined by x-ray analysis: ring A is in an envelope conformation; ring B is a distorted boat, and ring C is a half-chair.[1,2]

2) A unique feature is the presence of a cyclopropanol and a cyclopropylcarbinol system at the C/D ring moiety.

Reactions of phorbol compounds

1) Boiling with 0.02 N sulfuric acid affords crotophorbolone **2**, and phorbobutanone (crotophorbolone K) **3**.[2, 4]

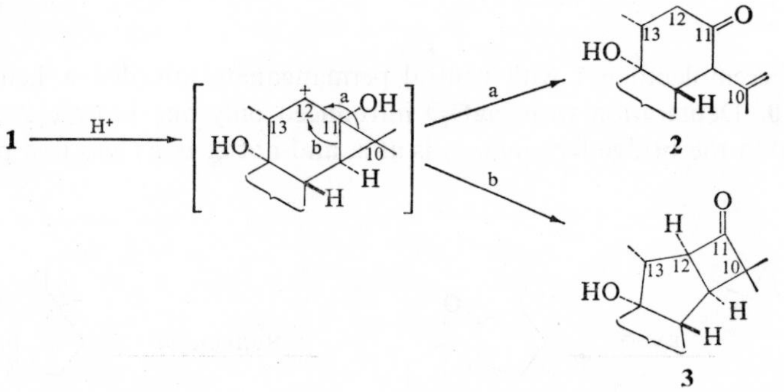

2) Treatment of 12-deoxy-12-oxophorbol-11,17-diacetate **4** with sodium methoxide in methanol results in selective transesterification at the 11-position. The 11,12-ketol group subsequently undergoes an acyloin shift to give hydroxyphorbobutanone **5** (major product), and the hemi-acetal **6**.[5]

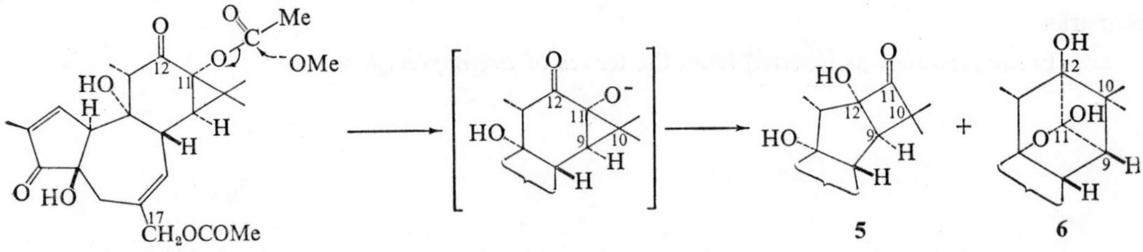

1) R. C. Pettersen, G. Ferguson, L. Crombie, M. L. Games, D. J. Pointer, *Chem. Commun.*, 716 (1967); W. Hoppe, F. Brandl, I. Strell, M. Rohrl, I. Gassman, E. Hecker, H. Bartsch, G. Kreibich, Ch. von Szczepanski, *Angew. Chem. Intern. Ed.*, **6**, 809 (1967); E. Hecker, H. Bartsch, G. Kreibich, Ch. von Szczepanski, *Annalen*, **725**, 130 (1969); E. Hecker, R. Schmidt, *Fortschritte J. Chem. Org. Naturst.*, **31**, p. 378, Springer-Verlag, New York (1974).

3) Reaction of phorbol with lead tetraacetate gives four products: tiglophorbol **7**, and tiglophorbol A, B and T (**8**, **9**, and **10**). The relative amount of each depends on the reaction conditions.[2]

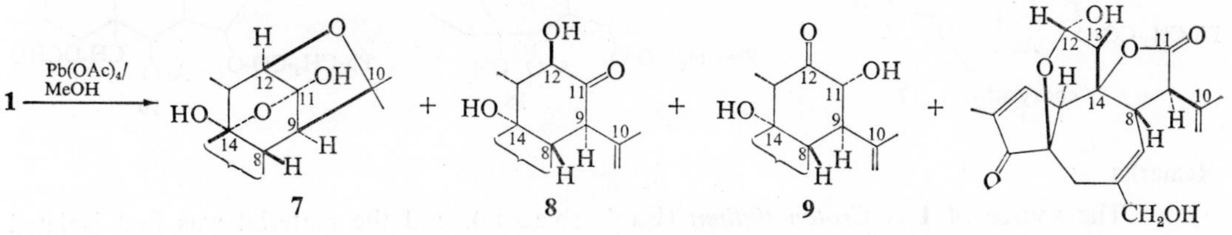

4) Treatment of phorbol pentaacetate with osmium tetroxide gave the expected $6\beta,7\beta$ diol **11** and $7\beta,11,12,14,17$ pentaacetate **12** as the main products, in which a C-4 to C-7 acyl migration occurred.[6]

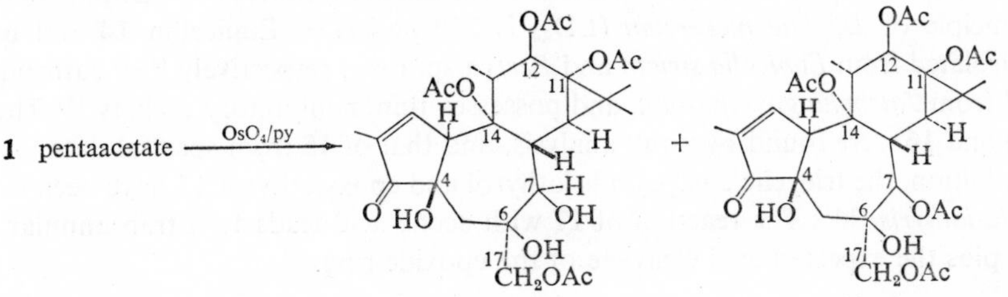

Related compounds[7-11]

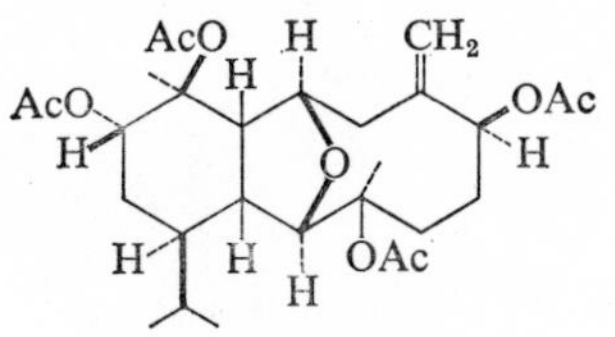

eunicellin **14**[13]

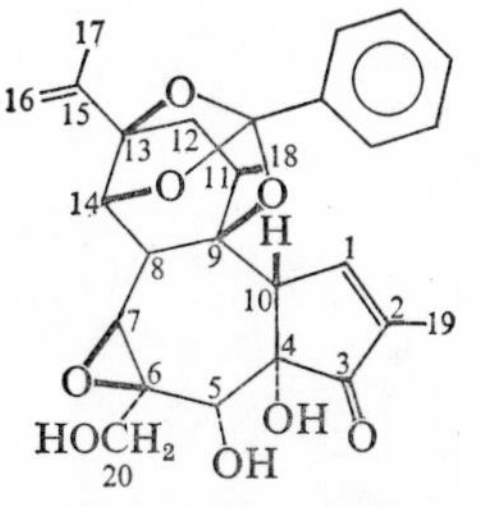

daphnetoxin **13**[12]

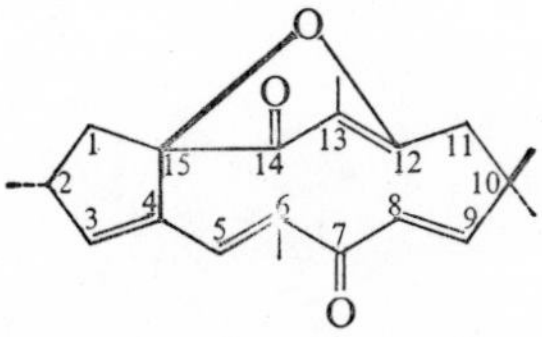

jatrophone **16**[15]

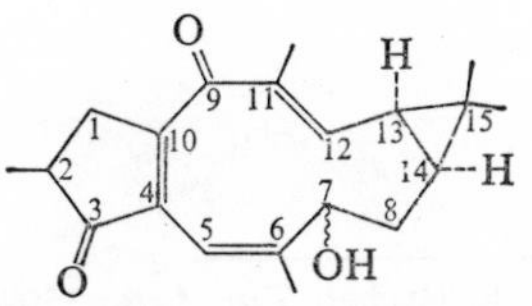

bertyadionol **15**[14]

2) L. Crombie, M. L. Games, D. J. Pointer, *J. Chem. Soc. C*, 1347 (1968).
3) B. Flaschentrager, G. Wigner, *Helv. Chim. Acta*, **25**, 569 (1942); B. Flaschentrager, *German Patent*, 638,004 (1936); *cf.* T. Kauffmann, H. Neumann, *Chem. Ber.*, **92**, 1715 (1959).
4) H. W. Thielmann, E. Hecker, *Annalen*, **728**, 158 (1969).
5) H. Bartsch, E. Hecker, *Annalen*, **725**, 142 (1969).
6) H. W. Thielmann, E. Hecker, *Annalen*, **735**, 113 (1970).
7) K. L. Stuart, L. M. Barrett, *Tetr. Lett.*, 2399 (1969).
8) M. Gschwendt, E. Hecker, *Tetr. Lett.*, 567 (1970).

epoxylathyrol **17**   **18**   **19**

Remarks

The source of **1** is *Croton tiglium* (Euphorbiaceae), and the material was first isolated from the oil of *Croton tiglium* seeds.[3] Croton oil is interesting because of its co-carcinogenic activity. The active components of croton oil have been proved to be various fatty acid esters of phorbol. Esterification occurs at C-11 and C-12 with a pair of long chain and short chain fatty acids.

Related compounds have been isolated from different sources.[7-9] Daphnetoxin **13** is the toxic principle of *Daphne mezereum* (LD_{50} is 250 $\mu g/kg$).[12] Eunicellin **14** and bertyadionol **15** were isolated from *Eunicella stricta* and *Bertya* sp. nov., respectively.[13,14] Jatrophone **16** was obtained from *Jatropha gossypiifolia*, and possesses tumor inhibitory activity.[15] The structures of **13**, **14** and **16** were found by x-ray analysis, and that of **15** from spectral data.

In addition, the tricyclic compounds lathyrol and epoxylathyrol **17** have been isolated from *Euphorbia lathyris*.[10,11] The reaction of **17** with acetic acid leads to a transannular cyclization reaction plus the expected acid cleavage of the epoxide ring.

9) M. Gschwendt, E. Hecker, *Tetr. Lett.*, 3509 (1969).
10) W. Adolf, E. Hecker, A. Balmain, M. F. Lhomme, Y. Nakatani, G. Ourisson, G. Ponsinet, R. J. Pryce, T. S. Sonthakrishnan, L. G. Matyukhina, I. A. Saltikova, *Tetr. Lett.*, 2241 (1970).
11) K. Zechmeister, M. Rohal, F. Brandl, S. Hechfischer, W. Hoppe, E. Hecker, W. Adolf, H. Kubinyi, *Tetr. Lett.*, 3071 (1970).
12) G. H. Sout, W. G. Balkenhol, M. Poling, G. L. Hickernell, *J. Am. Chem. Soc.*, **92**, 1070 (1970).
13) O. Kennard, D. G. Watson, L. Riva di Sanseverino, B. Tursch, R. Bosmans, C. Djerassi, *Tetr. Lett.*, 2879 (1968).
14) E. L. Ghisalberti, P. R. Jefferies, T. J. Payne, G. K. Worth, *Tetr. Lett.*, 4599 (1970).
15) S. M. Kupchan, C. W. Sigel, M. J. Matz, J. A. Saenz Renauld, R. C. Haltiwanger, R. F. Bryan, *J. Am. Chem. Soc.*, **92**, 4476 (1970).

4.62   RYANODINE

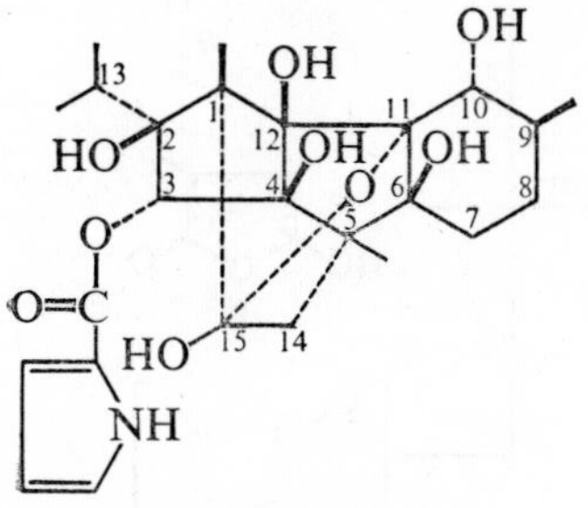

ryanodine   **1**
mp: 219–220 (dec.)
α_D: +26°
uv (EtOH): 286.5 (4.18)

Planar structure of anhydroryanodol 4[1-6]

1) The relationship between ryanodine **1**, anhydroryanodine **3**, ryanodol **2**, and anhydroryanodol **4** is shown in the sequence below.

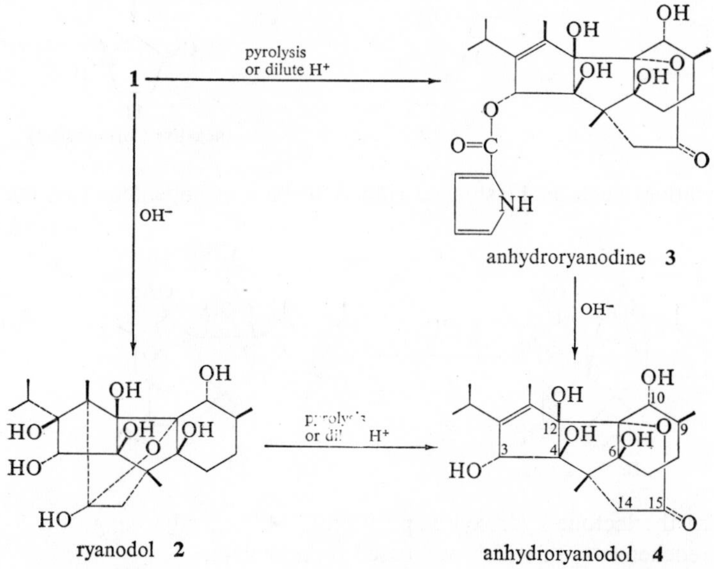

anhydroryanodine   **3**

ryanodol   **2**   anhydroryanodol   **4**

2) The structure of the more stable **4** was determined first according to the following scheme.

3) Compounds **14** and **15** were determined by spectral data and verified by further degradations.
Combined with the formation of compounds **5** through **13** they provided the first clue for the
skeletal structure of **4** and some of its methyl positions.

1) E. F. Rogers, F. R. Koniuszy, J. Shovel, K. Folkers, *J. Am. Chem. Soc.*, **70**, 3086 (1948).
2) R. B. Kelly, D. J. Whittingham, K. Wiesner, *Can. J. Chem.*, **29**, 905 (1951).
3) D. R. Babin, J. A. Findlat, T. P. Forrest, F. Fried, M. Gotz, Z. Valenta, K. Wiensner, *Tetr. Lett.*, **15**, 31 (1960).
4) D. R. Babin, T. Bogri, J. A. Findlay, H. Reinshagen, Z. Valenta, K. Wiesner, *Experientia*, **21**, 425 (1965).
5) K. Wiesner, *Pure Appl. Chem.*, **7**, 285 (1964).
6) K. Wiesner, *Collect. Czech. Chem. Commun.*, **33**, 2656 (1968).

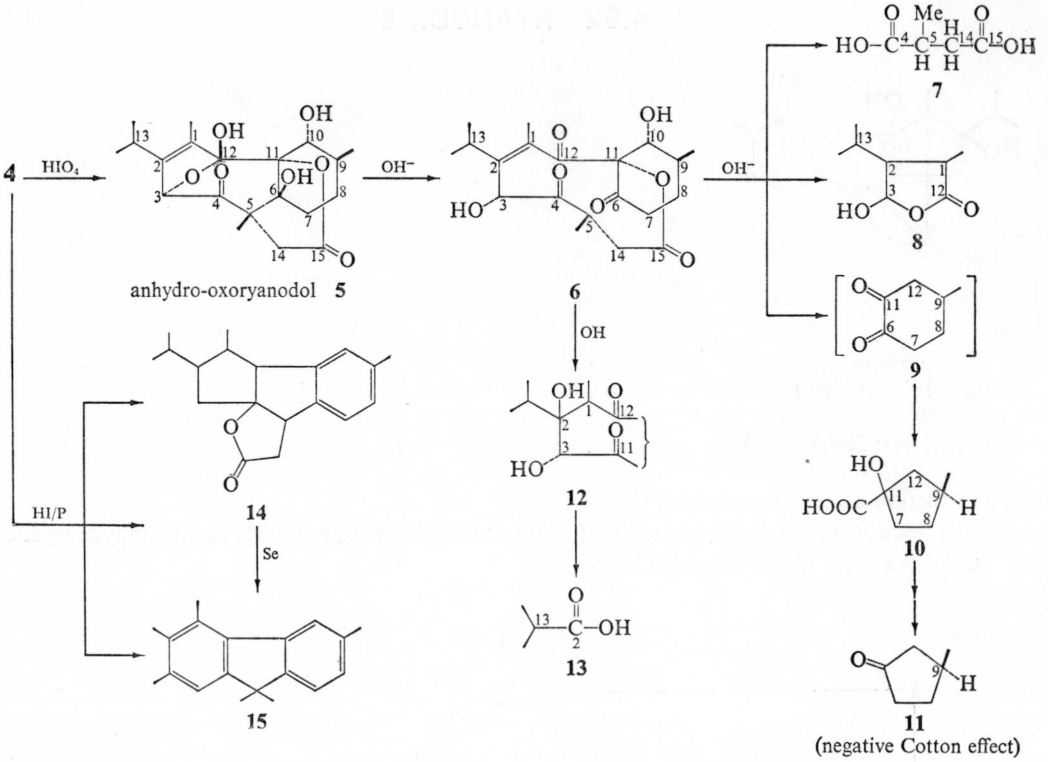

anhydro-oxoryanodol **5**

4) Formation of several derivatives such as **17** showed ring A to be a cyclopentene (uv, ir).

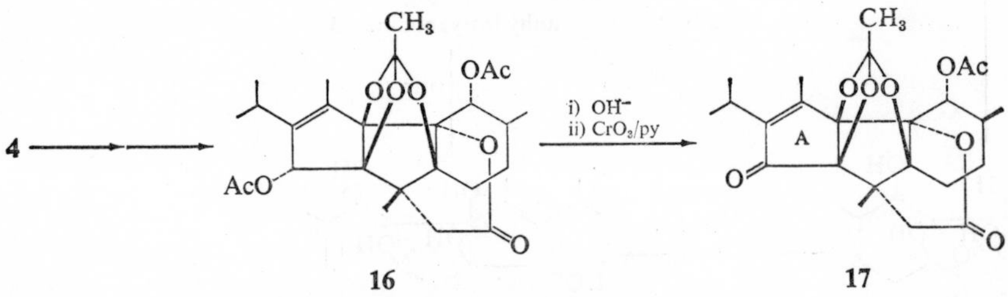

16   **17**

5) The sequence below defined the lactone hydroxyl at position C-11, i.e., adjacent to the 9-Me/10-OH moiety, the stereochemistry of which was based on nmr data.

6) The position of the hydroxyls was determined by nmr measurements and the formation of derivatives such as **18**.

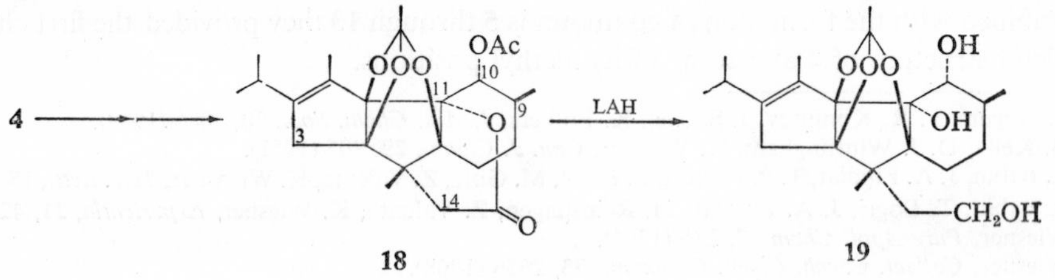

18   **19**

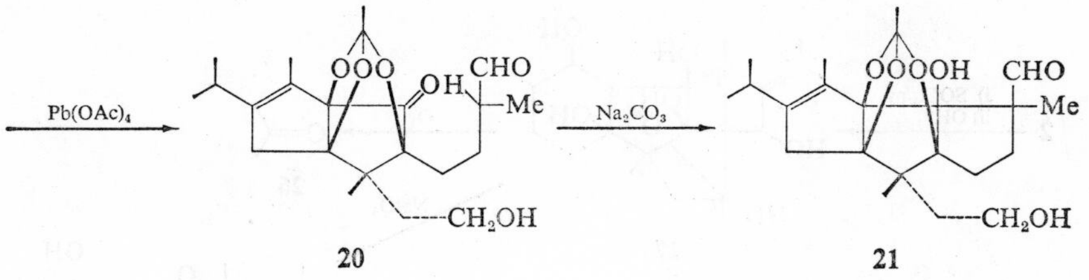

Stereochemistry of 4

1) The asymmetric center at C-9 is determined from **11**.

2) $J_{9,10}$ is ca. 9.5 Hz. Thus 10-OH is α. This value also accounts more readily for an α-lactone (and 4,6,12β-hydroxyls) rather than a β-lactone (and 4,6,12α-hydroxyls).

3) The 3-OH was deduced to be *trans* to the 4,12-diol for the following reasons. Firstly, only 1 mole of periodate reacts, with concurrent formation of **5**. If it was 3β-OH, hemiketal formation would be sterically impossible. Secondly, the nmr of 14-H is (d,d). However, it is a singlet in **18** and other compounds where the 3-OH is removed. Hence 3-OH and 14-CH$_2$ are *cis*.

Ryanodol[1-6]

1) Both **23** and **24** were isolated. The spectroscopic data were in full agreement with the structures shown. The formation of **25** by hydrogenolysis was particularly important in that it showed the hydroxyl lost in going from ryanodol **2** to anhydroryanodol **4** to be attached to C-2.

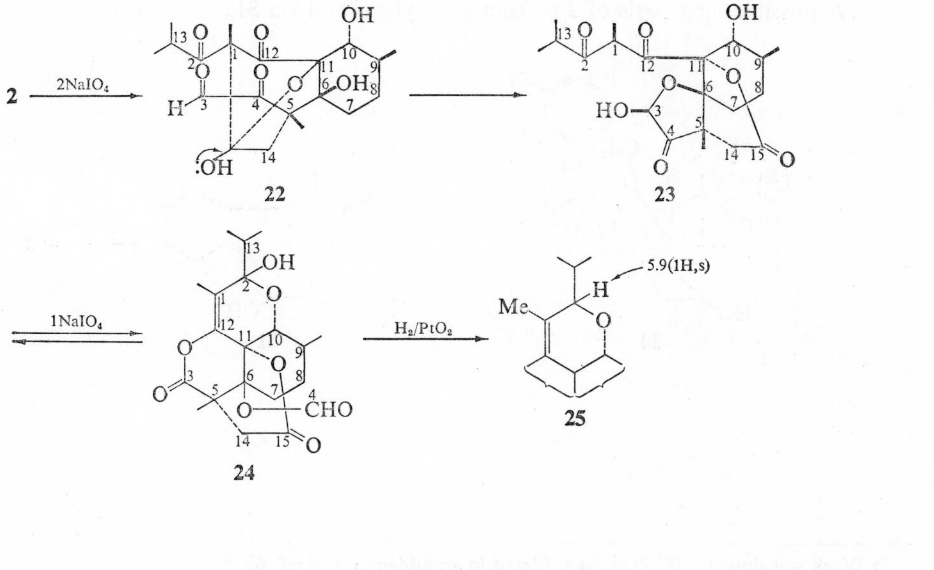

2) The sequence below was important in that it showed ryanodol to contain one C-C bond more than anhydroryanodol (C-1/C-15 bond in **27**), which is cleaved by a retro-aldol reaction.

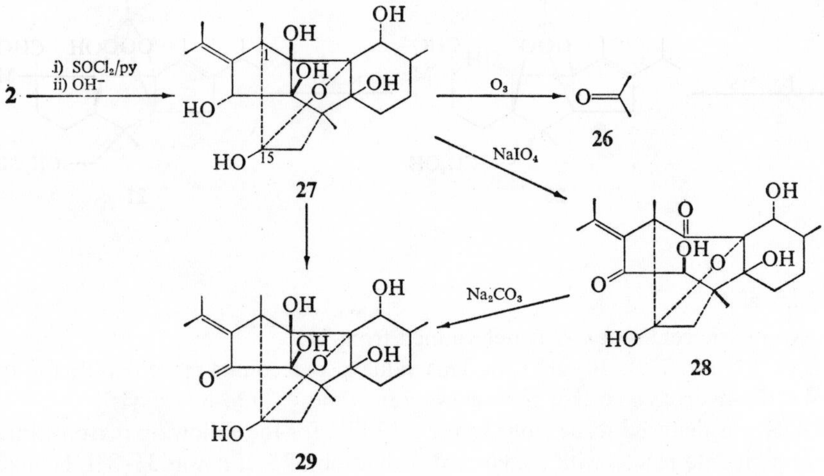

3) Structure **1** explained all these observations. The stereochemistry followed from **2**. The only question was the configuration of the 2-OH. It was originally thought that the hydroxyl group was α because the formation of **3** also proceeded during high vacuum sublimation, and that consequently, a cyclic mechanism with internal protonation should be preferable. However, an x-ray analysis of **30**,[7] besides verifying structure **2**, showed that 2-OH was β.

Remarks

Ryanodine is derived from *Ryania speciosa*, and is used as an insecticide for the European corn borer. Lately it has been shown to inhibit the binding of calcium to muscle protein[8] and to retard circulation by vascular constriction. It is believed that ryanodine affects the area of the brain which participates in the control of the peripheral circulation.[9]

A possible biogenesis of **2** is from geranylgeraniol via **31**.

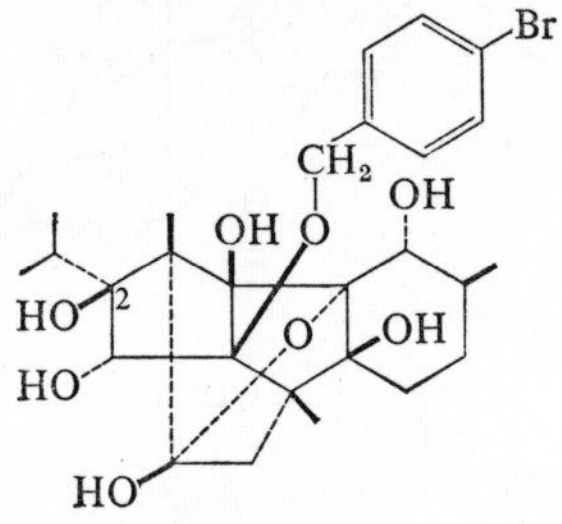

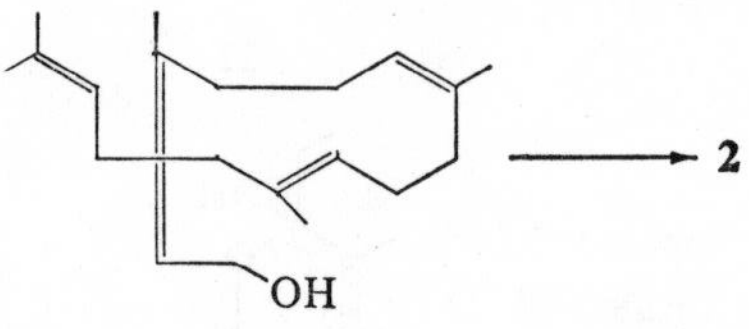

7) Work was done by M. Przbylska. Stated in an addendum to ref. **6**.
8) A. S. Fairhurst, D. A. Ralus, D. J. Jenden, *Circ. Res.*, **21**, 433 (1967).
9) L. Procite, B. J. Stibler, R. L. Marois, P. Lindgren, *J. Pharmacol. Exp. Ther.*, **159**, 335 (1968).

CHAPTER 5

Sester-, Tri- and Higher Terpenoids

5.1   INTRODUCTION

This chapter deals with sester- and triterpenoids and carotenoids. Sesterterpenoids are the newest and smallest class in the terpenoid family: since ophiobolin was recognized as the first sesterterpenoid in 1965, a number of members of this class have been isolated as fungal metabolites and in fern, marine sponge, lichen and insect secretion. A total of five different carbon skeletons are known in this class, which are constructed from five isoprene units joined in a head-to-tail manner, and biogenetically derived from geranylfarnesol.

The triterpenoids form the largest group among the terpenoid classes, and are widely distributed in the plant kingdom, either in the free state or as esters or glycosides, although a few important members have been found in the animal kingdom. These include squalene, first isolated from shark liver oil, and a number of tetracyclic compounds, including lanosterol, obtained from wool fat.

Although studies on the triterpenoids started in the late nineteenth century, it is only 50 years since active structural work was initiated by Ruzicka. All the triterpenoids originate biogenetically from squalene, a tail-to-tail condensate of farnesol, which is a sesquiterpene alcohol. However, great structural variety has been found in nature. With the accumulation of structural data, Ruzicka was able to rationalize the biogenesis of this group of compounds and develop the basic concept of terpenoid biosynthesis[1] ("biogenetic isoprene rule"). The following correlation charts are based essentially on his biogenetic views. Various modifications (C-C bond cleavage) in these carbon skeletons have been observed. Omitted in these charts are the seco- and nortriterpenes, which appear frequently in certain groups of plants. A number of reference works on triterpenoids have been published.[2]

Carotenoids are a family of natural coloring materials which are widely distributed in both the animal and plant kingdoms. They are polyisoprenoids with a long sequence of conjugated double bonds, biogenetically derived by the tail-to-tail condensation of geranylgeraniol. Structural modification occurs mainly at both termini of the carbon chain. Acetylenic and allenic bonds, both of which are rather rare in natural products, are sometimes incorporated in the conjugation. Nor-, seco- and homocarotenoids have also been reported.

---

1) L. Ruzicka, *Experientia*, **9**, 357 (1953), *Proc. Chem. Soc.*, 341 (1959). A. Eschenmoser, L. Ruzicka, O. Jeger, D. Arigoni, *Helv. Chim. Acta*, **38**, 1890 (1955).
2) J. L. Simonsen, W. C. J. Ross, *The Terpenes*, Vol. IV, V, Cambridge University Press, 1957; T. K. Devon, A. I. Scott, *Handbook of Naturally Occurring Compounds*, Vol. II, Academic Press, 1972; *Rodd's Chemistry of Carbon Compounds* (ed. S. Coffey), 2nd Ed., 11c, p. 406 (1969), 11e, 93 (1971); K. H. Overton, *Terpenoids and Steroids*, Vol. 1, 2, *A Specialist Periodical Report*, Chemical Society (1971, 1972).

Schematic Correlation of Main Triterpene Skeletone

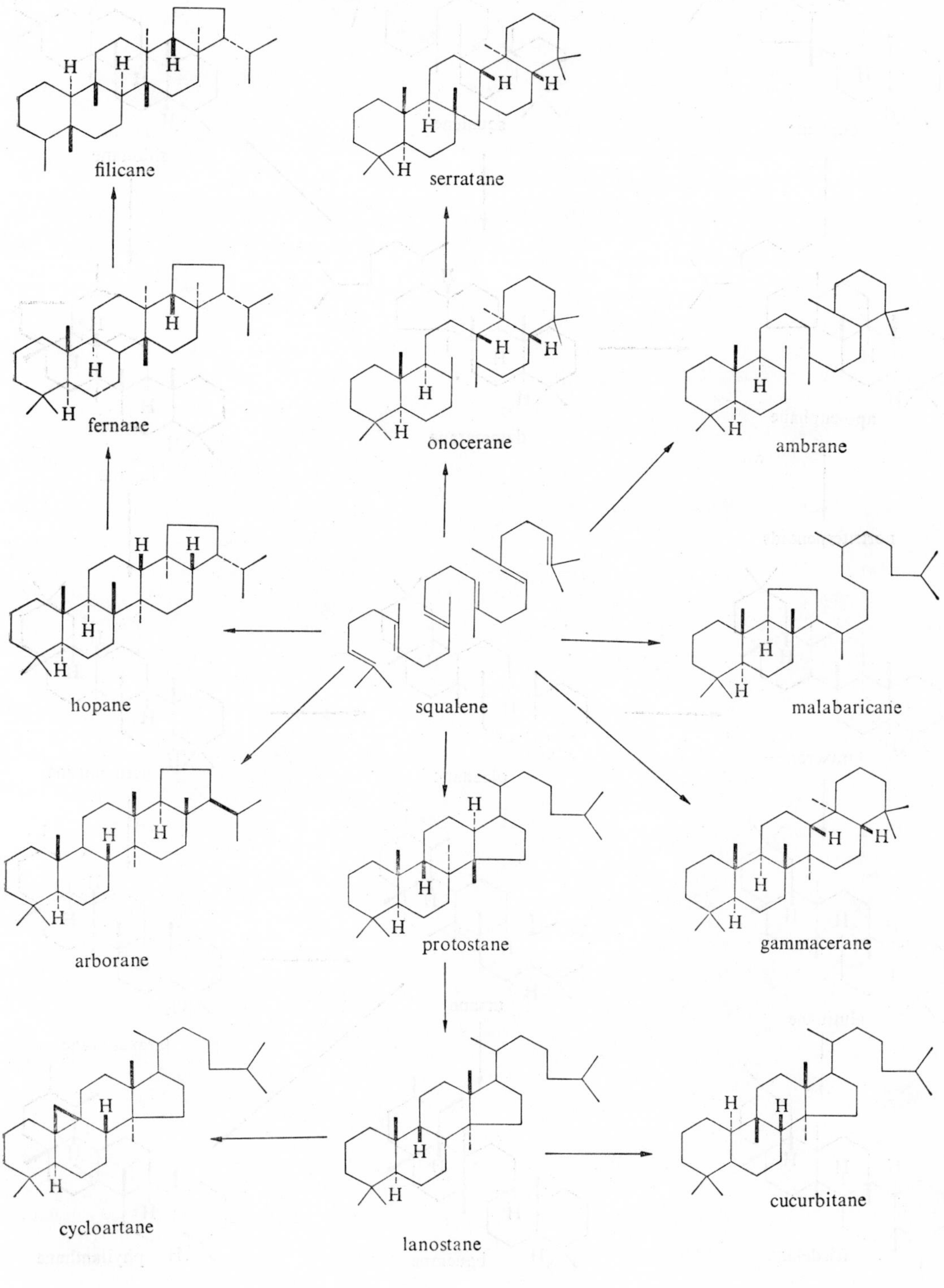

filicane

serratane

fernane

onocerane

ambrane

hopane

squalene

malabaricane

arborane

protostane

gammacerane

cycloartane

lanostane

cucurbitane

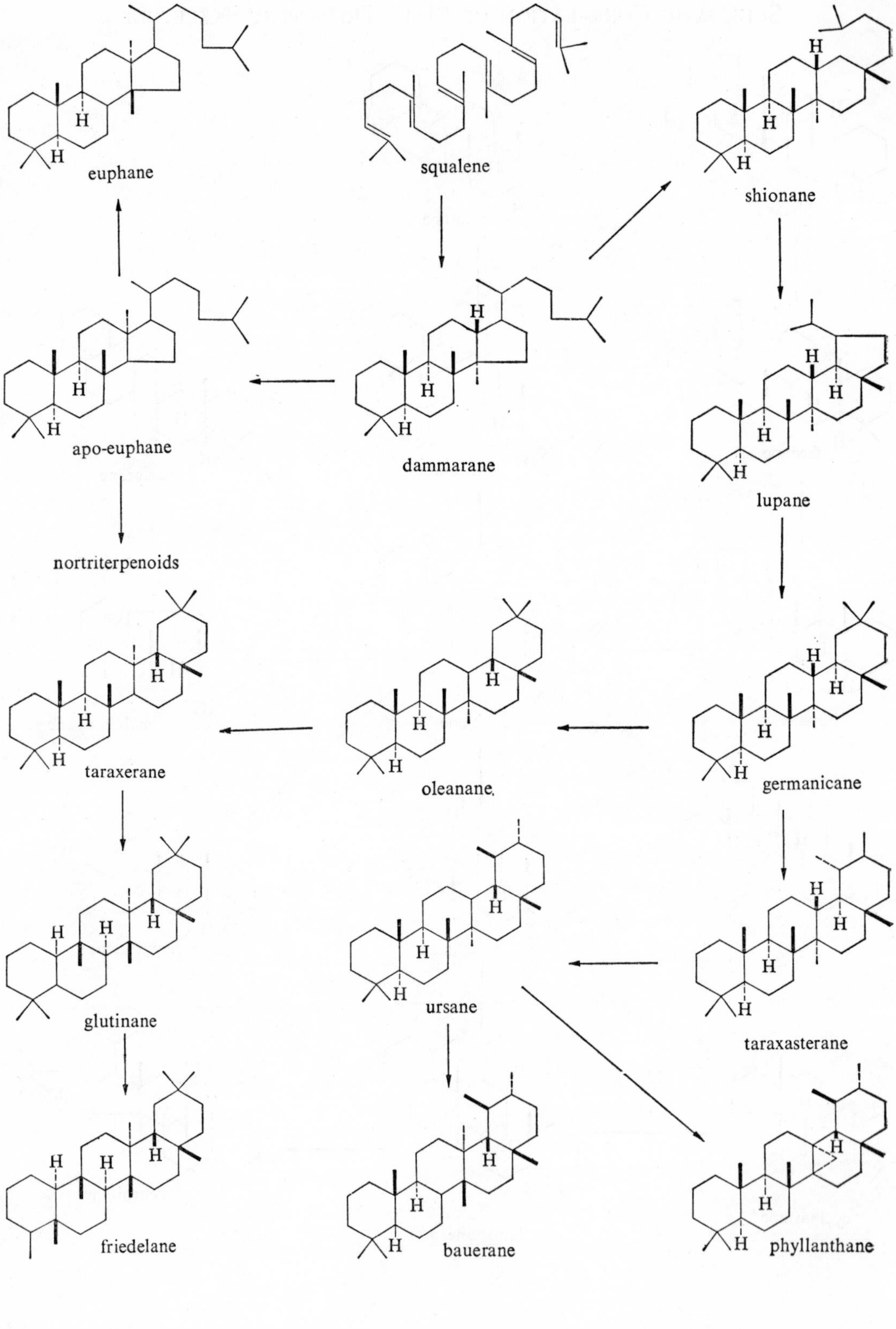

euphane

squalene

shionane

apo-euphane

dammarane

lupane

nortriterpenoids

taraxerane

oleanane.

germanicane

glutinane

ursane

taraxasterane

friedelane

bauerane

phyllanthane

5.2 STRUCTURE OF IRCININS

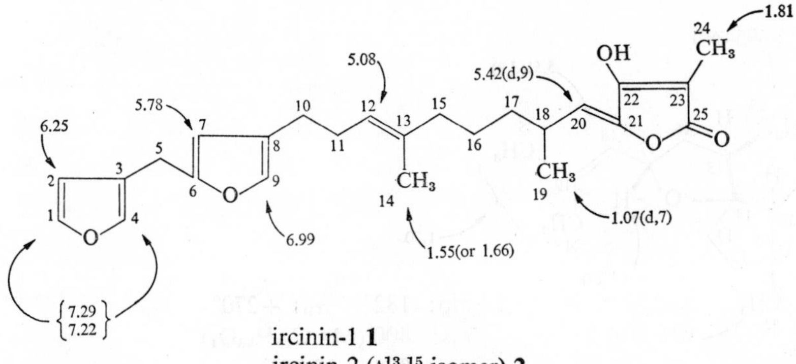

ircinin-1 **1**
ircinin-2 ($\Delta^{13,15}$-isomer) **2**

ms: 410 (M⁺ $C_{25}H_{30}O_5$)
uv (MeOH): 260 (ε 12,500)
uv (MeOH/OH⁻): 248 (ε 10,500), 310
  (ε 7,500)
ir (CHCl₃): 3150 (OH), 1735 ($\alpha\beta$-unsat.
  lactone), 1635 (C=C)

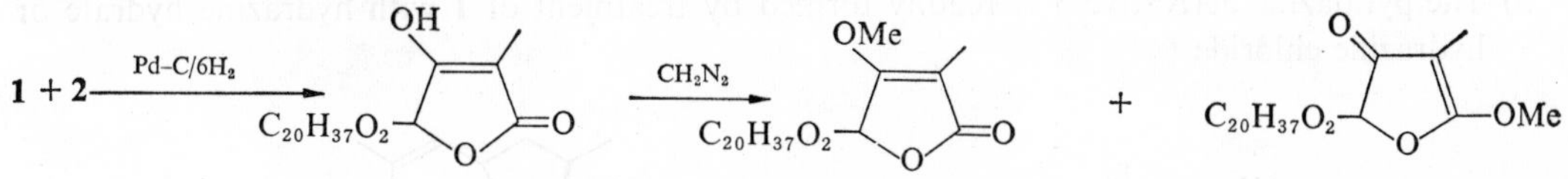

1 + 2 $\xrightarrow{\text{Pd-C/6H}_2}$ **3** uv: 230 (ε7,320), $C_{20}H_{37}O_2$
                                                   258 (ε6,490)

$\xrightarrow{\text{CH}_2\text{N}_2}$ **4** uv: 230 (ε10,200) $C_{20}H_{37}O_2$

+ **5** uv: 263 (ε 9,300) $C_{20}H_{37}O_2$

1 + 2 $\xrightarrow[\text{ii) CH}_2\text{N}_2]{\text{i) OH}^-}$ **6**

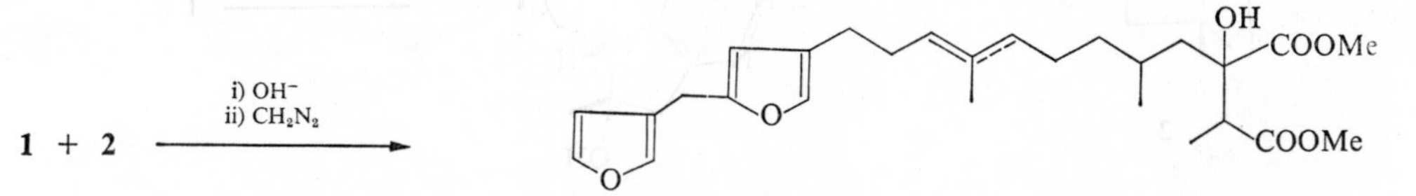

1) Ircinin was isolated as a mixture of two isomeric compounds, **1** and **2**.[1,2]
2) Hydrogenation of the mixture gave a single dodecahydro derivative **3**, whose uv spectrum indicates the presence of a tetronic acid residue in the molecule.
3) Methylation of **3** yielded two isomeric methyl ethers, **4** and **5**.
4) Alkaline hydrolysis followed by esterification afforded **6**, formation of which can be explained by benzilic rearrangement of the α-diketones formed by hydrolysis.

Remarks

    Ircinins were isolated from the marine sponge *Ircinia oros*. C_{21}-furanoterpenes obtained from *Spongia nitens*, *S. officinalis* and *Hippospongia communis* are considered to be derived from sesterterpenes by degradation *in vivo*.

1) G. Cimino, S. De Stefano, L. Minale, E. Fattorusso, *Tetr.*, **28**, 333 (1972).
2) E. Fattorusso, L. Minale, G. Sodano, E. Trivellone, *Tetr.*, **27**, 3909 (1971); *ibid.*, **27**, 4673 (1971).

5.3   STRUCTURE OF OPHIOBOLIN-A AND RELATED COMPOUNDS

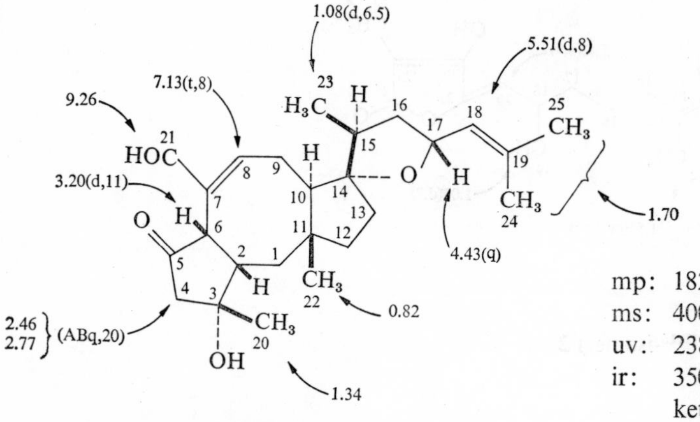

ophiobolin-A 1[1]

mp: 182° α_D: +270°
ms: 400 (M$^+$ $C_{25}H_{36}O_4$)
uv: 238 (ε 13,800)
ir: 3500 (OH), 1743 (5-membered ring ketone), 1674, 1633 ($\alpha\beta$-unsat. carbonyl)

1) Bromination (Br$_2$/MeOH) in the presence of NaOAc affords the methoxy-bromide **2**.
2) The structure and absolute configuration of **1** was established by x-ray crystallographic analysis of **2**.
3) The pyridazine derivative **3** is readily formed by treatment of **1** with hydrazine hydrate or hydrazine chloride.

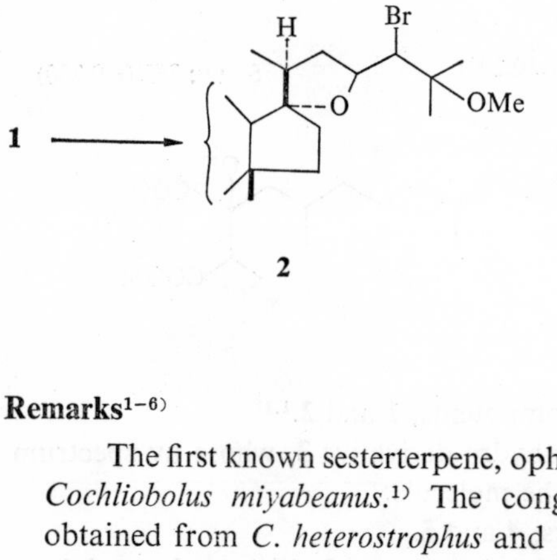

Remarks[1-6]

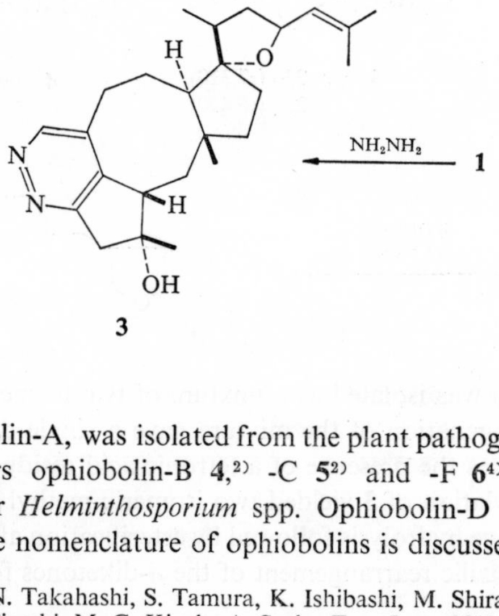

   The first known sesterterpene, ophiobolin-A, was isolated from the plant pathogenic fungus *Cochliobolus miyabeanus*.[1] The congeners ophiobolin-B **4**,[2] -C **5**[2] and -F **6**[4] were also obtained from *C. heterostrophus* and other *Helminthosporium* spp. Ophiobolin-D **7** was isolated from *Cephalosporium caerulens*.[3] The nomenclature of ophiobolins is discussed in ref. 5.

1) S. Nozoe, M. Morisaki, K. Tsuda, Y. Iitaka, N. Takahashi, S. Tamura, K. Ishibashi, K. Shirasaka, *J. Am. Chem. Soc.*, **87**, 4968 (1965); L. Canonica, A. Fiecchi, M. G. Kienle, A. Scala, *Tetr. Lett.*, 1211, 1329 (1966).
2) S. Nozoe, K. Hirai, K. Tsuda, *Tetr. Lett.*, 2211 (1966).
3) A. Itai, S. Nozoe, K. Tsuda, S. Okuda, Y. Iitaka, Y. Nakayama, *Tetr. Lett.*, 4111 (1967).
4) S. Nozoe, M. Morisaki, K. Fukushima, S. Okuda, *Tetr. Lett.*, 4457 (1968).
5) K. Tsuda, S. Nozoe, M. Morisaki, K. Hirai, A. Itai, S. Okuda, L. Canonica, A. Fiecchi, M. G. Kienle, A. Scala, *Tetr. Lett.*, 3369 (1967).
6) S. Nozoe, M. Morisaki, K. Tsuda, S. Okuda, *Tetr. Lett.*, 3365 (1967).

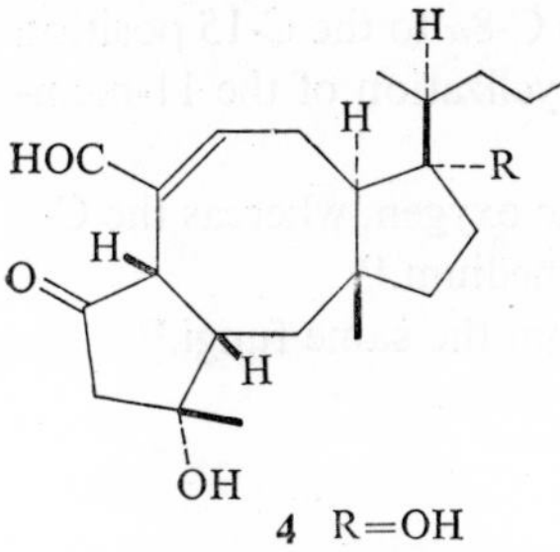

4 R=OH
5 R=H   6 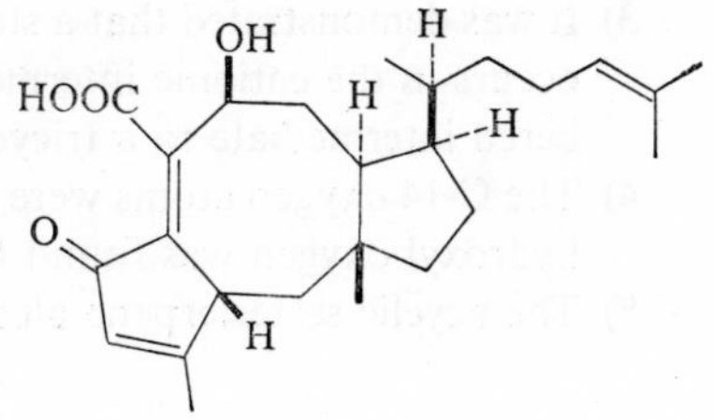   7

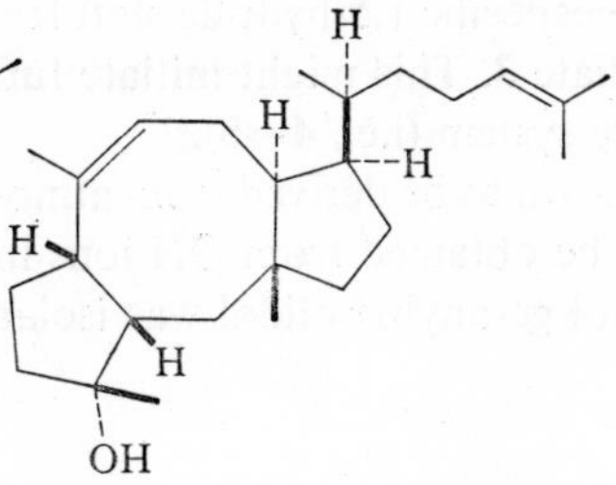

5.4 BIOSYNTHESIS OF OPHIOBOLINS

The biosynthesis of the tricyclic sesterterpene ophiobolin-A from mevalonic acid lactone
has been investigated. The results were interpreted in terms of the biosynthetic route outlined
below.

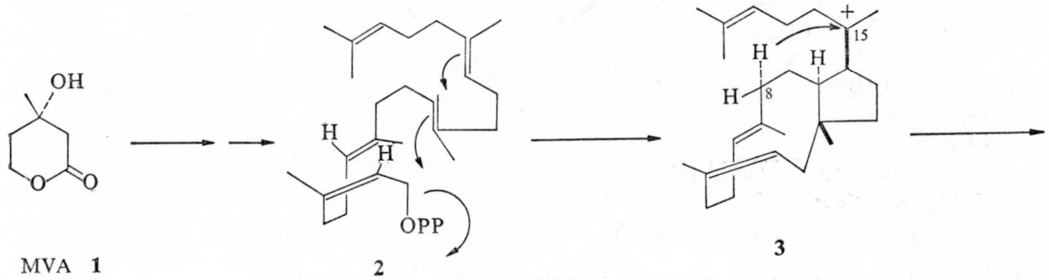

MVA 1   2   3

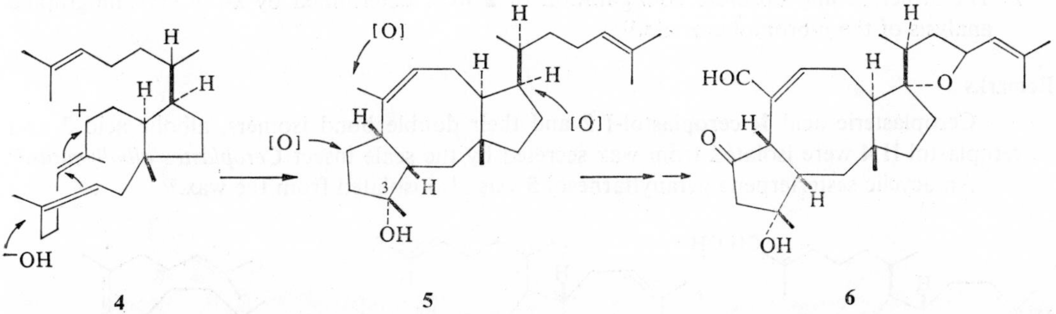

4   5   6

1) The ^{3}H/^{14}C ratio in ophiobolins biosynthesized from doubly labeled mevalonate was consistent with the above cyclization mechanism.[1]
2) The tricyclic alcohol ophiobolin-F 5 was shown to be formed from labeled mevalonate and
   geranylfarnesyl pyrophosphate 2 by a cell-free system prepared by the disruption of
   Cochiliobolus heterostrophus cells.[2]

1) S. Nozoe, M. Morisaki, S. Okuda, K. Tsuda, *Tetr. Lett.*, 2347 (1968).
2) S. Nozoe, M. Morisaki, *Chem. Commun.*, 1319 (1969).
3) L. Canonica, A. Fiecchi, M. G. Kienle, B. M. Ranzi, A. Scala, *Tetr. Lett.*, 4657 (1967); *ibid.*, 3371 (1967).
4) S. Nozoe, M. Morisaki, K. Tsuda, S. Okuda, *Tetr. Lett.*, 3365 (1967).
5) S. Nozoe, M. Morisaki, K. Fukushima, S. Okuda, *Tetr. Lett.*, 4457 (1968).

3) It was demonstrated that a stereospecific 1,5-hydride shift from the C-8α to the C-15 position occurs in the cationic intermediate **3**. This might initiate further cyclization of the 11-membered intermediate to a tricyclic system (i.e., **4→5**).[3]

4) The C-14 oxygen atoms were shown to be derived from atmospheric oxygen, whereas the C-3 hydroxyl oxygen was found to be obtained from OH ions in the medium.[4]

5) The acyclic sesterterpene alcohol geranylnerolidol was isolated from the same fungi.[5]

5.5 STRUCTURE OF CEROPLASTERIC ACID

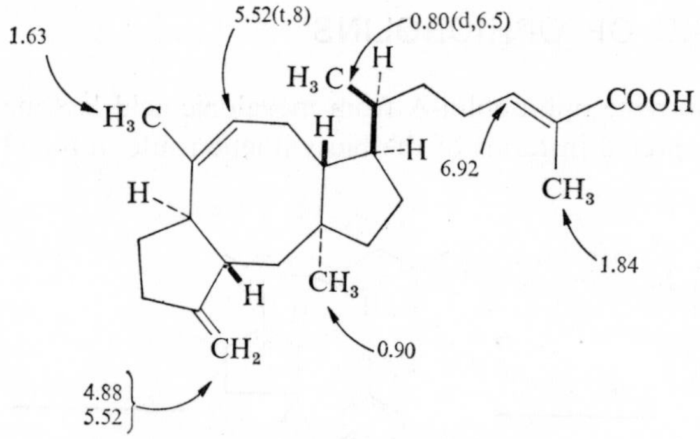

ceroplasteric acid **1**[1]

$C_{25}H_{38}O_2$ $\alpha_D: +87°$

ir (film): 874 (exocyclic methylene)

1) Reduction of **1** affords a naturally occuring alcohol, ceroplastol-I, **2**.

2) The structure and absolute configuration of **2** were determined by x-ray crystallographic analysis of the *p*-bromobenzoate.[1]

Remarks

Ceroplasteric acid **1**, ceroplastol-I **2** and their double bond isomers, albolic acid **3** and ceroplastol-II **4** were isolated from wax secreted by the scale insect *Ceroplastes albolineatus*.[2] An acyclic sesterterpene geranylfarnesol **5** was also isolated from the wax.[3]

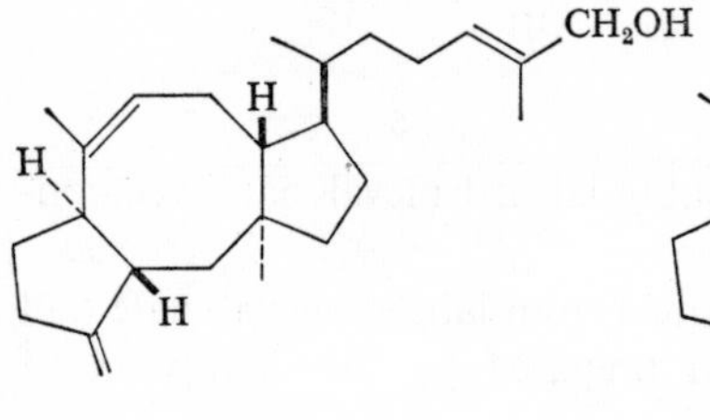

2

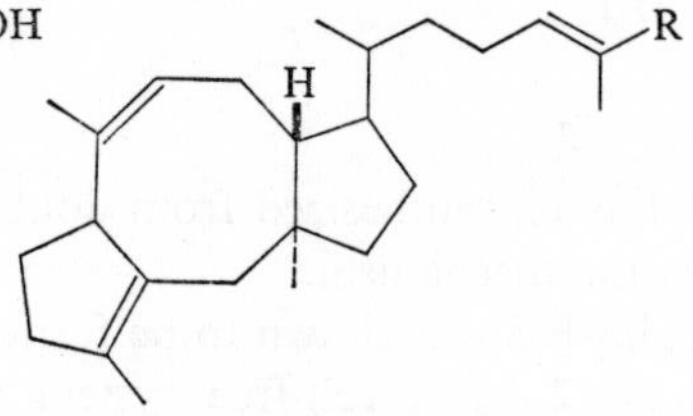

3 R=COOH
4 R=CH₂OH

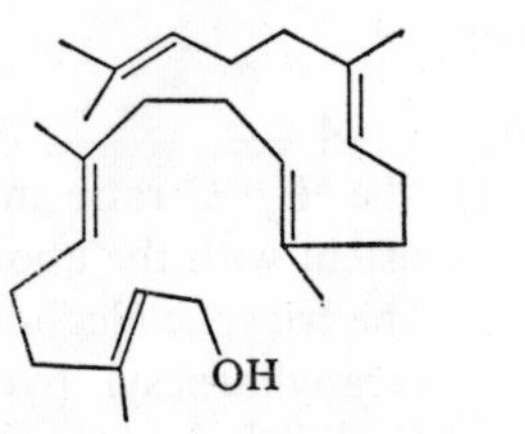

5

1) Y. Iitaka, I. Watanabe, I. T. Harrison, S. Harrison, *J. Am. Chem. Soc.*, **90**, 1092 (1968).
2) T. Ríos, F. Columga, *Chem. Ind.*, 1184 (1965); T. Ríos, L. Quijano, *Tetr. Lett.*, 1317 (1969).
3) T. Ríos, S. Pérez, *Chem. Commun.*, 214 (1969).

5.6 STRUCTURE OF RETIGERANIC ACID

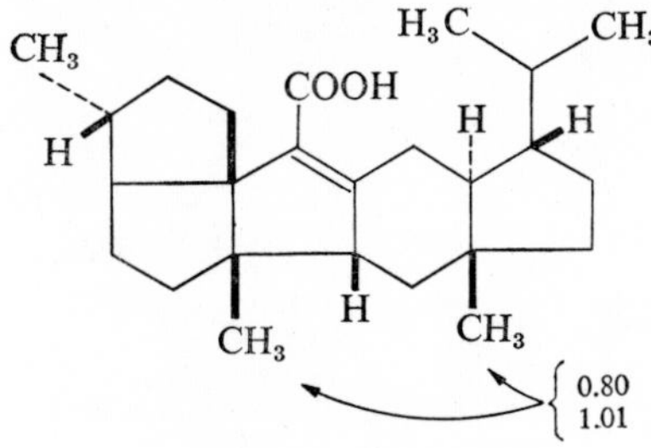

retigeranic acid **1**

mp: 221–222° α_D (CHCl$_3$): −99.2°
ms: 370.287 (M$^+$ C$_{25}$H$_{38}$O$_2$)
uv: 242 (3.91)
ir (KBr): 1662, 1608
nmr in d$_5$-pyridine

1) Nmr of **1** showed two *tert*-Me and three *sec*-Me groups.
2) Uv, ir and mass spectra (ms: 325 (M−COOH)) indicated the presence of an α,β-unsaturated carboxylic acid.
3) X-ray analysis of the *p*-bromoanilide of **1** established the structure and absolute configuration.

Remarks

Retigeranic acid **1** was isolated from the lichen, *Lobaria isidiosa* var. *subisidiosa*.[1] It was found to be identical with the compound isolated from *L. retigera* by Seshadri *et al.*[2] The biogenesis of this unique pentacyclic sesterterpene from geranylfarnesyl pyrophosphate was proposed as to be follows.[1]

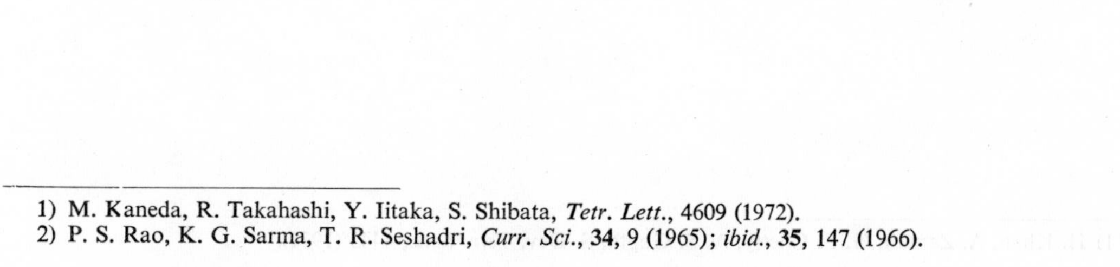

1) M. Kaneda, R. Takahashi, Y. Iitaka, S. Shibata, *Tetr. Lett.*, 4609 (1972).
2) P. S. Rao, K. G. Sarma, T. R. Seshadri, *Curr. Sci.*, **34**, 9 (1965); *ibid.*, **35**, 147 (1966).

5.7 STRUCTURE OF CHEILANTHATRIOL

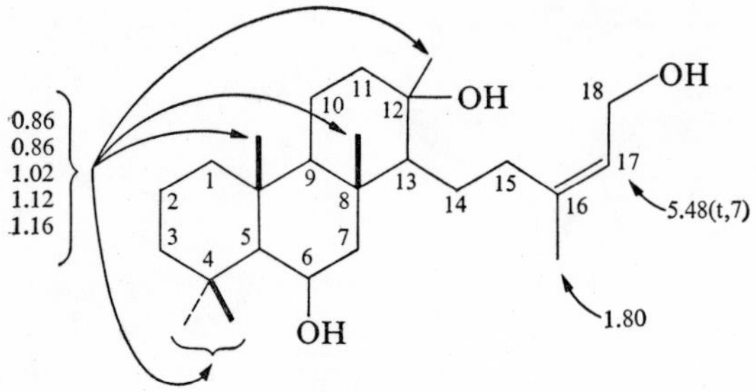

cheilanthatriol **1**

mp: 182–183° α_D (CHCl$_3$): +30.4°
ms: 374 (M − H$_2$O for C$_{25}$H$_{44}$O$_3$)
ir: 3450, 3360, 1125, 1075, 1055, 1003,
  990
nmr in CDCl$_3$

1) The monoacetate of **1** shows nmr signals at 3.97 (t, 11) due to 6-H and at 4.65 (2H, bd, 7) due to 18-H.

2) MnO$_2$ oxidation of **1** afforded an α,β-unsaturated aldehyde, nmr: 9.9 (d, 8) and 5.86 (dq, 8 and 1).

3) Se-dehydrogenation of **1** yielded 1,7-dimethylphenanthrene and 1,7,8-trimethyl-phenanthrene.

4) The biogenetic considerations illustrated below also support the structure.

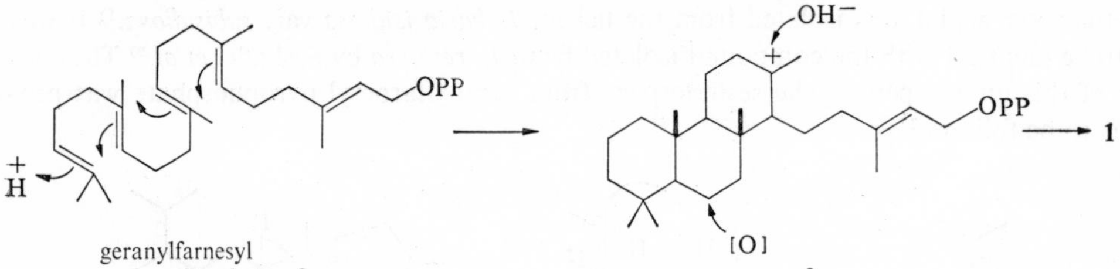

geranylfarnesyl
pyrophosphate **2**   **3**

Remarks

Cheilanthatriol **1** was isolated from the fern, *Cheilanthes farinosa.*[1]

1) H. Khan, A. Zaman, G. L. Chetty, A. S. Gupta, S. Dev, *Tetr. Lett.*, 4443 (1971).

5.8 SYNTHESIS OF SQUALENE

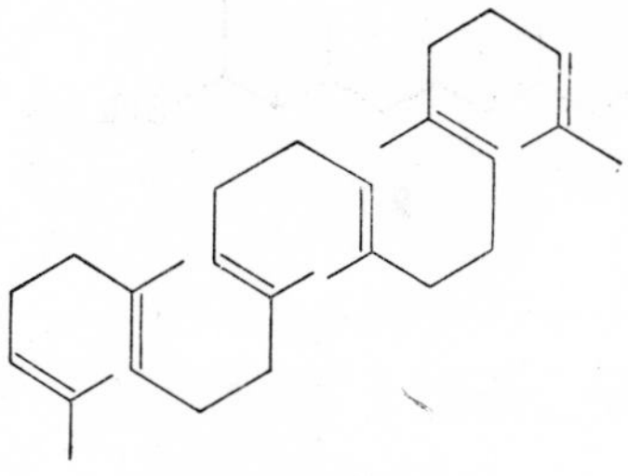

squalene **1**

In addition to the classical syntheses[1] of this biogenetically important hydrocarbon, which is widely distributed in nature, syntheses using newly developed reactions have been achieved recently.

Route I[2]

Squalene **1** has been synthesized via the intramolecular rearrangement of a sulfonium ylide, as shown below.

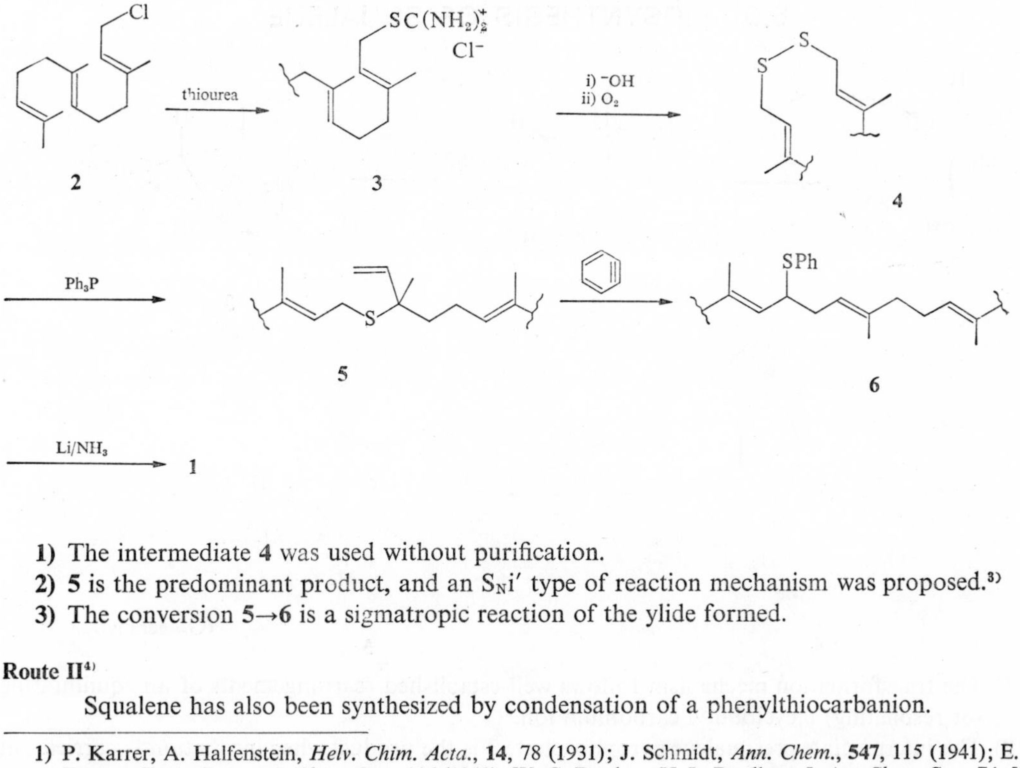

1) The intermediate **4** was used without purification.
2) **5** is the predominant product, and an S_Ni' type of reaction mechanism was proposed.[3]
3) The conversion **5→6** is a sigmatropic reaction of the ylide formed.

Route II[4]

Squalene has also been synthesized by condensation of a phenylthiocarbanion.

1) P. Karrer, A. Halfenstein, *Helv. Chim. Acta.*, **14**, 78 (1931); J. Schmidt, *Ann. Chem.*, **547**, 115 (1941); E. H. Farmer, D. A. Sutton, *J. Chem. Soc.*, 116 (1942); W. G. Dauben, H. L. Bradlow, *J. Am. Chem. Soc.*, **74**, 5204 (1952).
2) G. M. Blackburn, W. D. Ollis, C. Smith, I. O. Sutherland, *Chem. Commun.*, 99 (1969).
3) M. B. Ivans, G. M. C. Higgins, G. Moore, M. Porter, B. Saville, J. F. Smith, B. Trego, A. A. Watson, *Chem Ind.*, 897 (1960).
4) J. F. Biellmann, J. B. Ducep, *Tetr. Lett.*, 3707 (1969); *ibid.*, 5629 (1968).

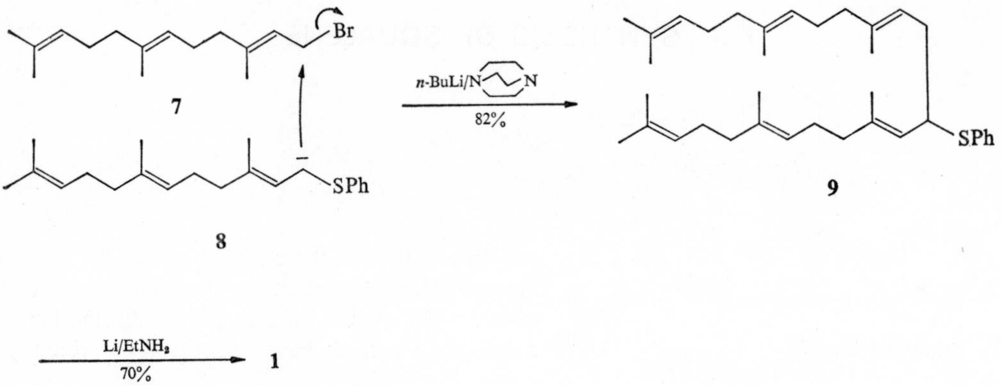

1) **8** was prepared by the action of thiophenol on the corresponding 2,4-dinitrophenyl ether.
2) No double bond isomerization occurs during the reaction sequence.

5.9   BIOSYNTHESIS OF SQUALENE

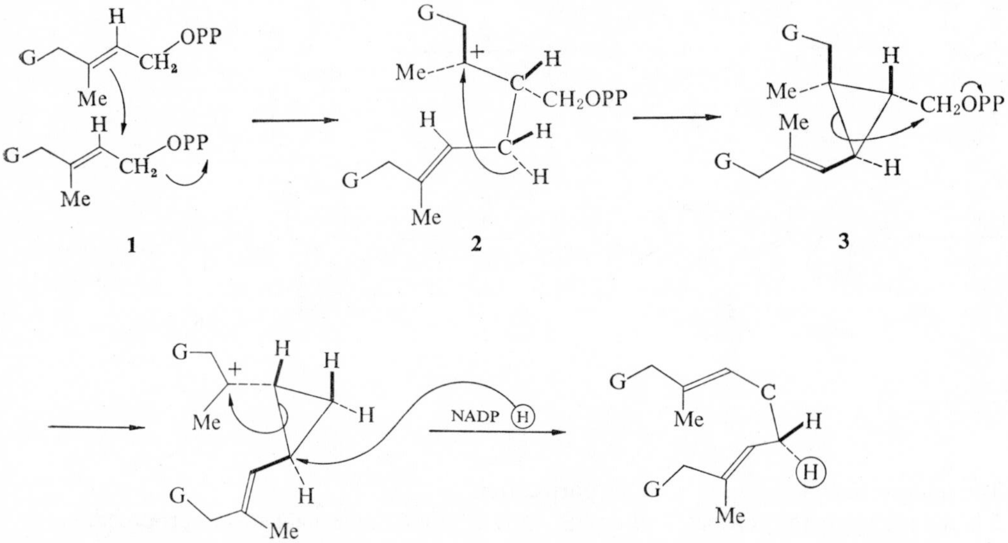

(G = geranyl)

1) The transformation mechanism follows well-established rearrangements of an equilibrating (or resonating) bicyclobutyl carbonium ion.[2]
2) The intermediate presqualene pyrophosphate **3** is the product when incubation is carried out in the absence of NADPH.[3]

1) E. E. van Tamelen, M. A. Schwartz, *J. Am. Chem. Soc.*, **93**, 1780 (1971); L. J. Altman, R. C. Kowerskii, H. C Rilling, *ibid.*, **93**, 1782 (1971).
2) *inter alia* C. D. Poulter, E. G. Friedrich, S. Winstein, *J. Am. Chem. Soc.*, **92**, 4274 (1970); and references therein.
3) H. C. Rilling, *J. Biol. Chem.*, **241**, 3233 (1966).

5.10   SYNTHESIS OF PRESQUALENE ALCOHOL

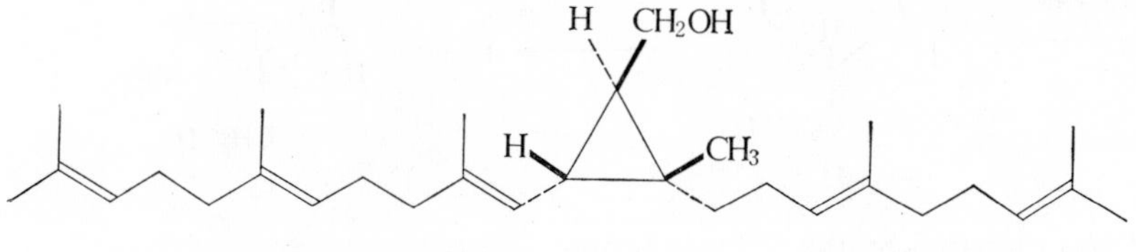

presqualene alcohol **1**

The absolute configuration of **1** has been proved[1] to be enantiometric to the one proposed earlier.[2]

The pyrophosphate of **4** links farnesyl pyrophosphate and squalene in the biosynthesis of sterols.[3]

Route I[2,4]

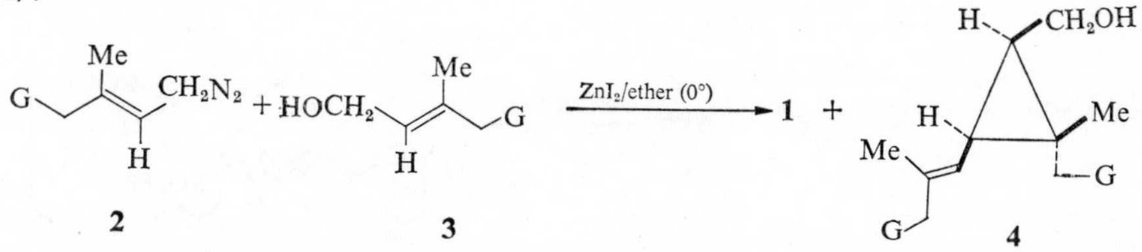

1) **1** and **4** were obtained in 70:30 ratio (G=geranyl).

2) Pyrophosphate ester of synthetic radioactive **1** was enzymatically transformed into squalene in 68% yield (34% as *dl*-mixture).

Route II[5]

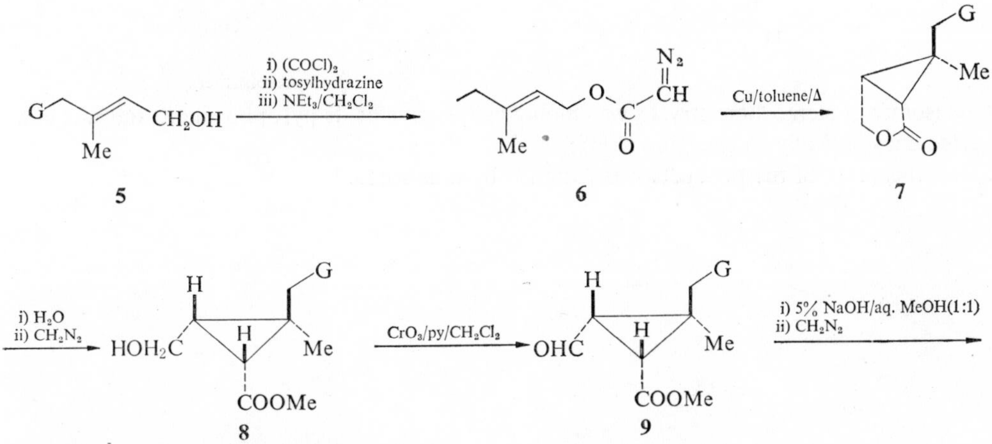

1) G. Popják, J. Edmond, S.-M. Wong, *J. Am. Chem. Soc.*, **95**, 2713 (1973).
2) H. C. Rilling, C. D. Poulter, W. W. Epstein, B. Larsen, *J. Am. Chem. Soc.*, **93**, 1783 (1971).
3) H. C. Rilling, *J. Biol. Chem.*, **241**, 3233 (1966); H. C. Rilling, W. W. Epstein, *J. Am. Chem. Soc.*, **91**, 1041 (1969); W. W. Epstein, H. C. Rilling, *J. Biol. Chem.*, **245**, 4597 (1970); G. Popjak, J. Edmond, K. Clifford, V. Williams, *ibid.*, **244**, 1897 (1969).
4) L. J. Altman, R. C. Kowerskii, H. C. Rilling, *J. Am. Chem. Soc.*, **93**, 1782 (1971).
5) R. M. Coates, W. H. Robinson, *J. Am. Chem. Soc.*, **93**, 1785 (1971).

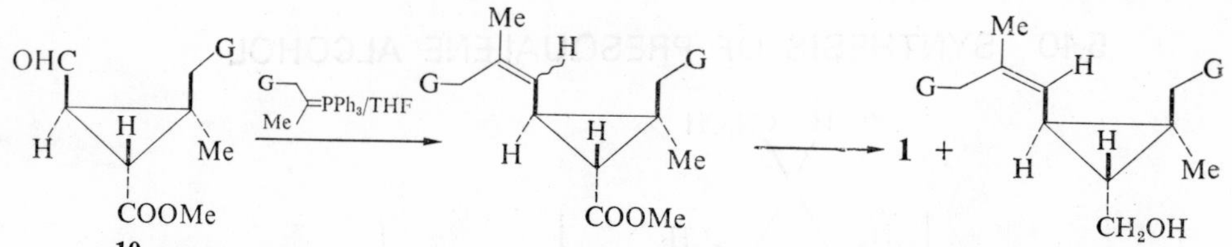

10 → **11** → **1** + **12**

1) The Wittig reagent was prepared by the following route.

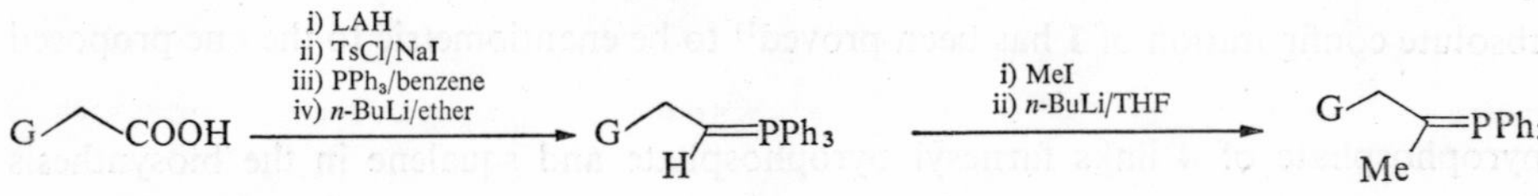

i) LAH
ii) TsCl/NaI
iii) PPh₃/benzene
iv) n-BuLi/ether

i) MeI
ii) n-BuLi/THF

2) Synthetic **1** was incorporated into squalene in 66% yield.

Route III [6]

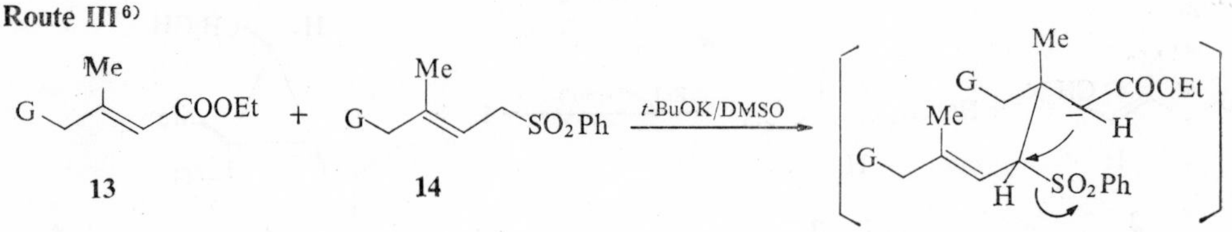

13 + **14** 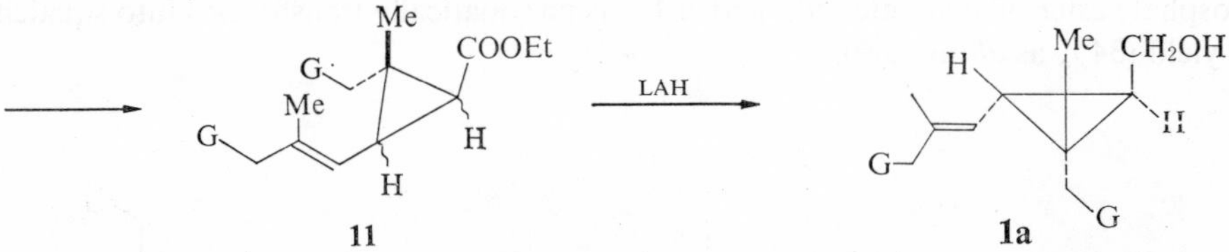 → **11** → **1a**

(four stereoisomers)

1) One isomer of stereochemistry **1a** was biologically active as its pyrophosphate and was converted enzymatically to squalene (63%).
2) Stereochemistry of the products was assigned by nmr studies.

6) R. V. M. Campbell, L. Crombie, G. Pattenden, *Chem. Commun.*, 218 (1971).

5.11   SQUALENE-2,3-EPOXIDE

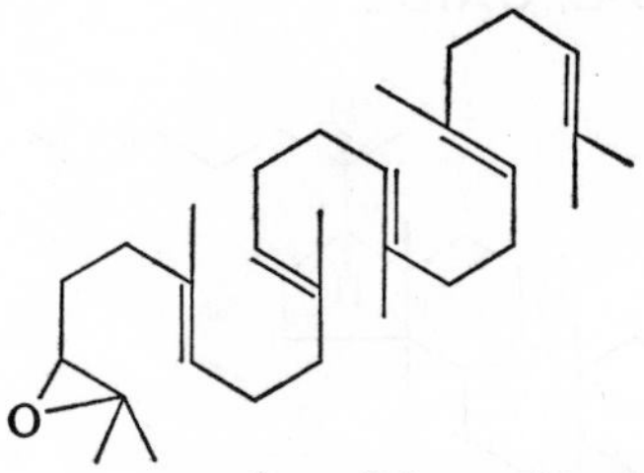

squalene-2,3-epoxide **1**

Biologically, squalene-2,3-epoxide is produced by an epoxidase in the microsomes of liver which requires molecular oxygen and NADPH.[1]

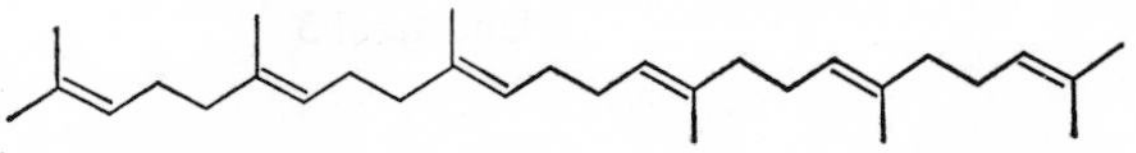

2a
i) NBS/glyme/H$_2$O
ii) OH$^-$
⟶ **1**

(95% or more terminal oxidation)

The following table indicates the regiospecificity of the above oxidation in a highly polar medium.[2]

olefin	in glyme/water		in petr. ether/AcOH	
	terminal	internal	terminal	internal
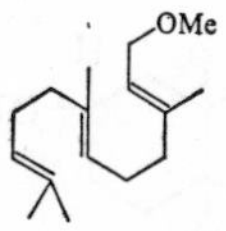 OMe	96	4	81	19
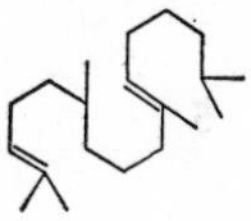	98.5	1.5	62	38
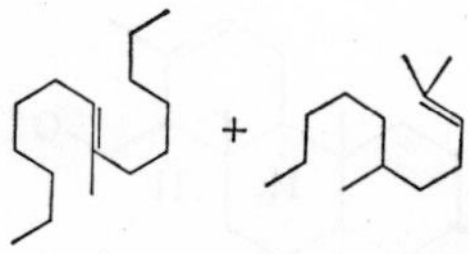 +	100	0		

This large solvent effect in the model compounds suggested that **2** exists in a coiled state (**2b**) in a very polar medium rather than the stretched form (**2a**) in order to account for the terminal oxidation.[2]

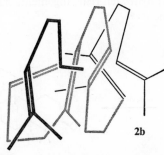

2b

1) S. Yamamoto, K. Bloch, *Natural Substances Formed from Mevalonic Acid* (ed. T. W. Goodwin), p. 35, Academic Press, 1969.
2) E. E. van Tamelen, T. J. Curphey, *Tetr. Lett.*, 121 (1962); E. E. van Tamelen, K. B. Sharples, *ibid.*, 2655 (1967).

5.12   CYCLIZATION OF SQUALENE-2,3-EPOXIDE

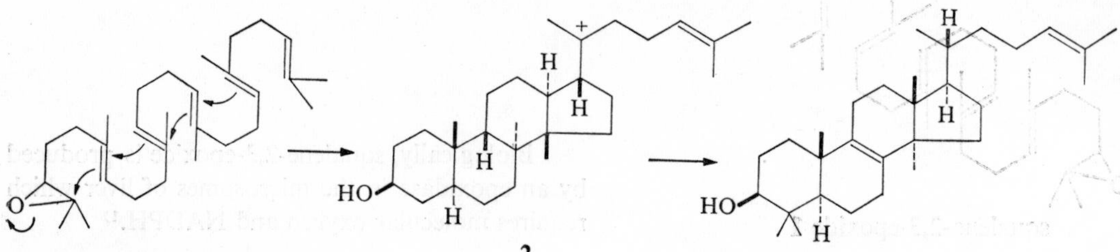

squalene-2,3-epoxide **1**   **2**   lanosterol **3**

Enzymatic cyclization

1) The microsomal enzyme catalyzing this reaction has been isolated.[1]
2) The enzyme requires no cofactors for the cyclization and is inhibited by 2,3-imino squalene.[2]
3) The mechanism for the transformation was proposed by Stork[3] and Eschenmoser[4] and has received some experimental support.[5]

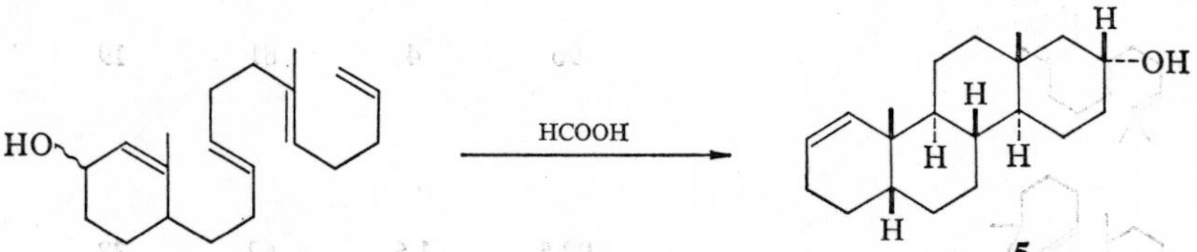

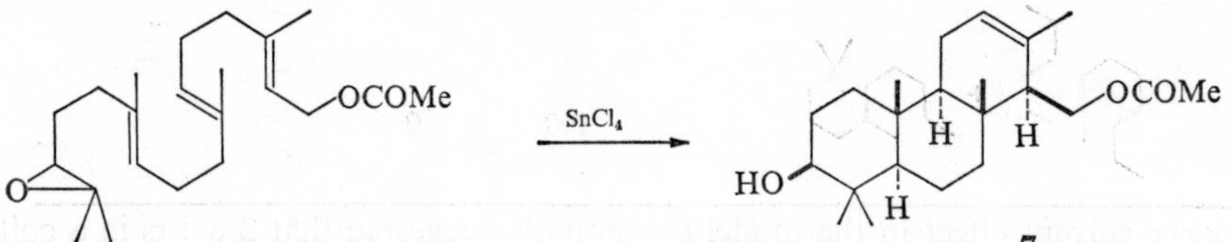

4) The factors affecting methyl and hydrogen rearrangements have been discussed with regard to the following *in vitro* experiments.[6]

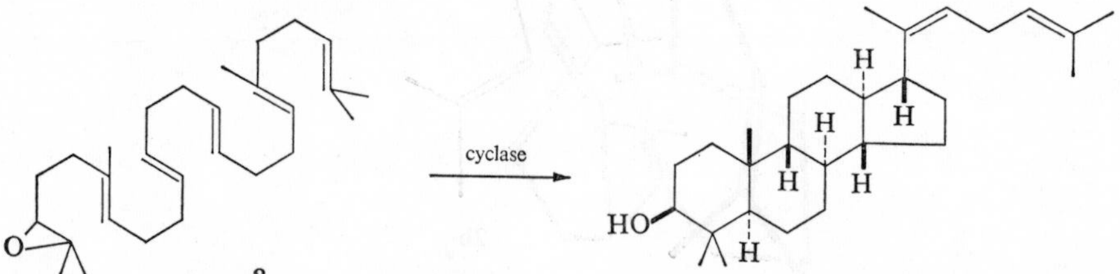

1) P. D. G. Dean, P. R. Ortiz de Montellano, K. Bloch, E. J. Corey, *J. Biol. Chem.*, **242**, 3014 (1967); J. D. Willett, K. B. Sharpless, K. E. Lord, E. E. van Tamelen, R. B. Clayton, *ibid.*, **242**, 4182 (1967).
2) E. J. Corey, P. R. Ortiz de Montellano, K. Lin, P. D. G. Dean, *J. Am. Chem. Soc.*, **89**, 2797 (1967).

10 11

12 13

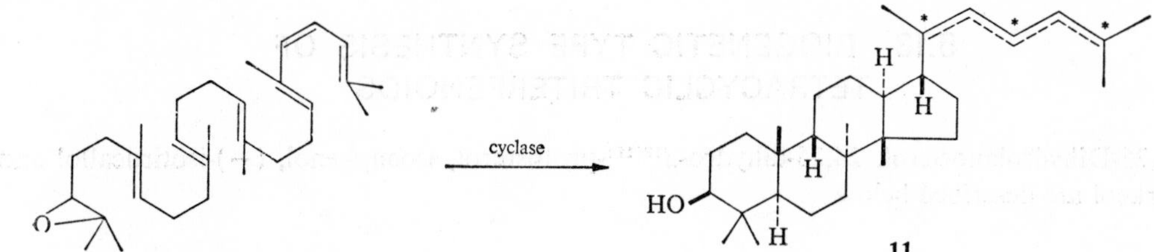

Non-enzymatic cyclization[7,8]

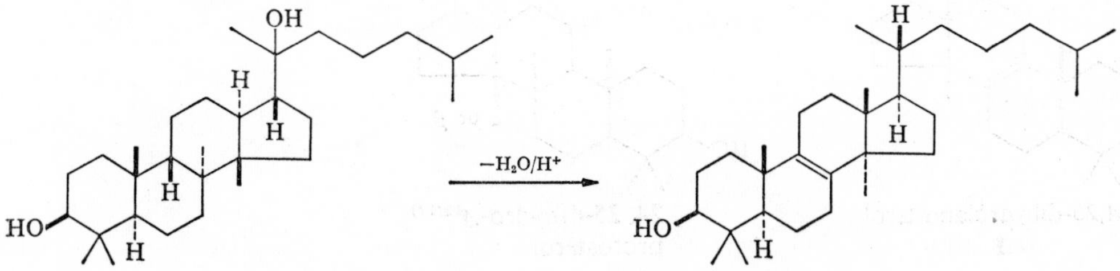

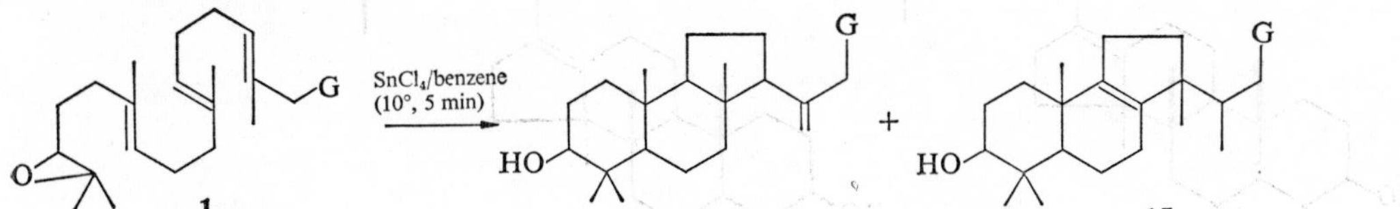

1

G = geranyl

14 15

16 17

18 19 20

3) G. Stork, A. W. Burgstahler, *J. Am. Chem. Soc.*, **77**, 5068 (1955).
4) A. Eschenmoser, L. Ruzicka, O. Jeger, D. Arigoni, *Helv. Chim. Acta*, **38**, 1890 (1955).
5) J. W. Cornforth, R. H. Cornforth, C. Donninger, G. Popjak, Y. Shimizu, S. Ichii, E. Forchielli, W. S. Johnson, *Accounts Chem. Res.*, **1**, 1 (1968); E. E. van Tamelen, *ibid.*, **1**, 111 (1968).
6) E. J. Corey, P. R. Ortiz de Montellano, H. Yamamoto, *J. Am. Chem. Soc.*, **90**, 6254 (1968); E. J. Corey, H. Yamamoto, *Tetr. Lett.*, 2385 (1970); E. J. Corey, K. Lin, H. Yamamoto, *J. Am. Chem. Soc.*, **91**, 2132 (1969).
7) E. E. van Tamelen, J. Willet, M. Schwartz, R. Nodeau, *J. Am. Chem. Soc.*, **88**, 5937 (1966).
8) M. Kishi, T. Kato, Y. Kitahara, *Chem. Pharm. Bull.*, **15**, 1072 (1967).

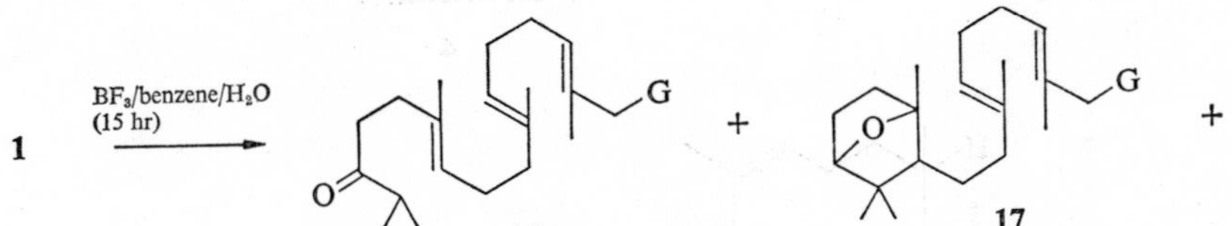

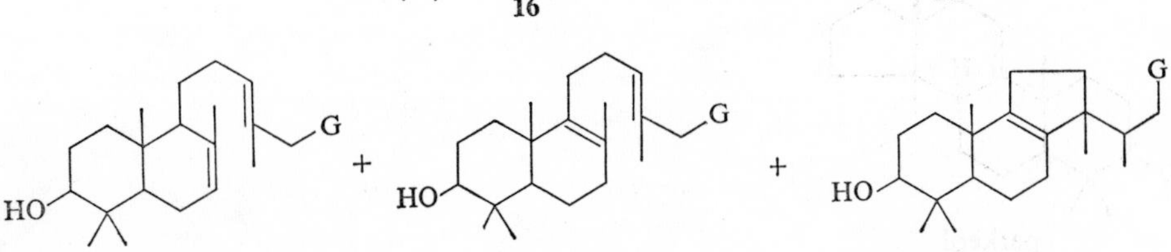

5.13   BIOGENETIC TYPE SYNTHESIS OF
TETRACYCLIC TRITERPENOIDS

24,25-Dihydrolanosterol, 24,25-dihydro-$\Delta^{13(17)}$-protosterol, isoeuphenol, ($-$)-isotirucallol and parkeol are described below.

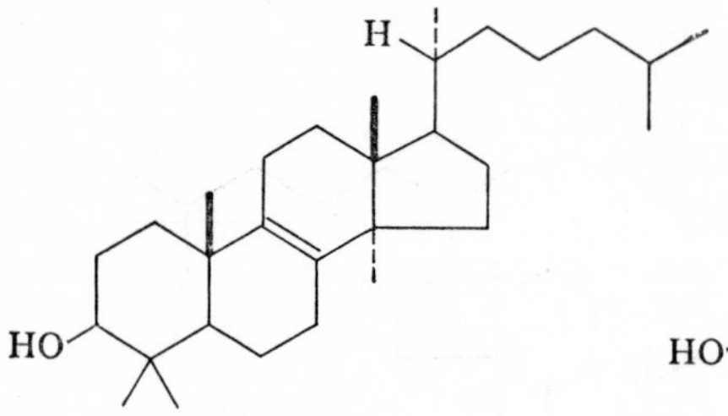

24,25-dihydrolanosterol

1

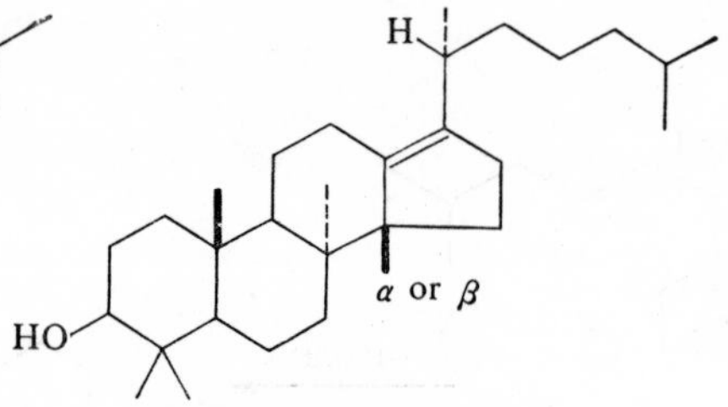

24, 25-dihydro-$\Delta^{13(17)}$
protosterol

2

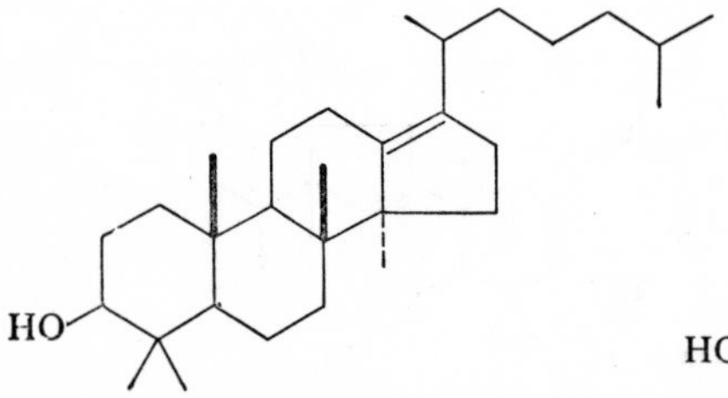

α or β

isoeuphenol

3

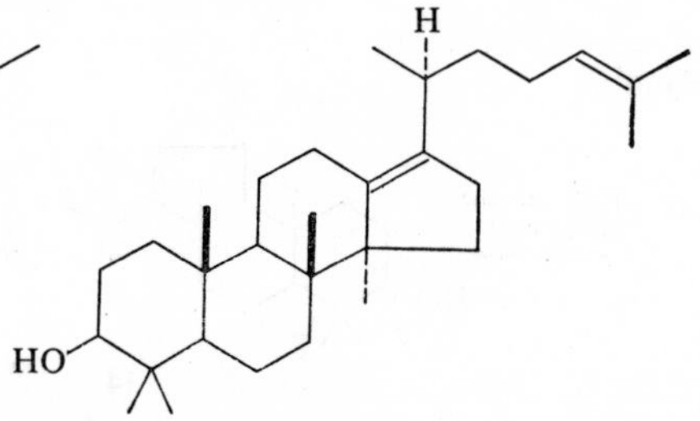

isotirucallol

4

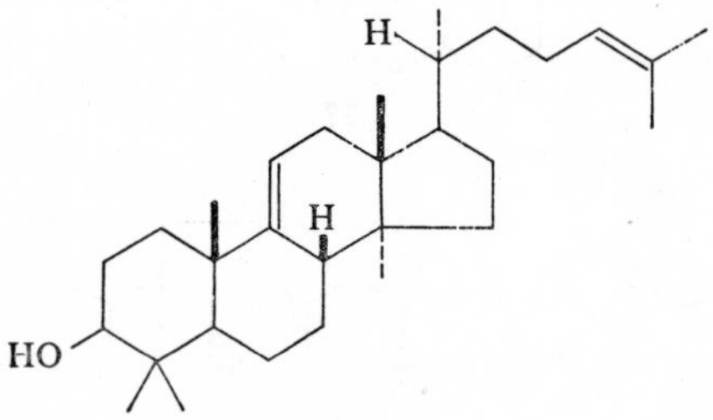

parkeol

5

Total syntheses of compounds **1–5** have been achieved by the following routes.[1–3]
The starting material **6** was previously obtained from (S)-($-$)-limonene.[2]

1) E. E. van Tamelen, R. J. Anderson, *J. Am. Chem. Soc.*, **94**, 8225 (1972).
2) B. A. Pawson, H-C. Cheung, S. Gurbaxani, G. Saucy, *J. Am. Chem. Soc.*, **92**, 336 (1970).
3) R. P. Hanzlik, E. E. van Tamelen, *J. Am. Chem. Soc.*, **90**, 209 (1968).

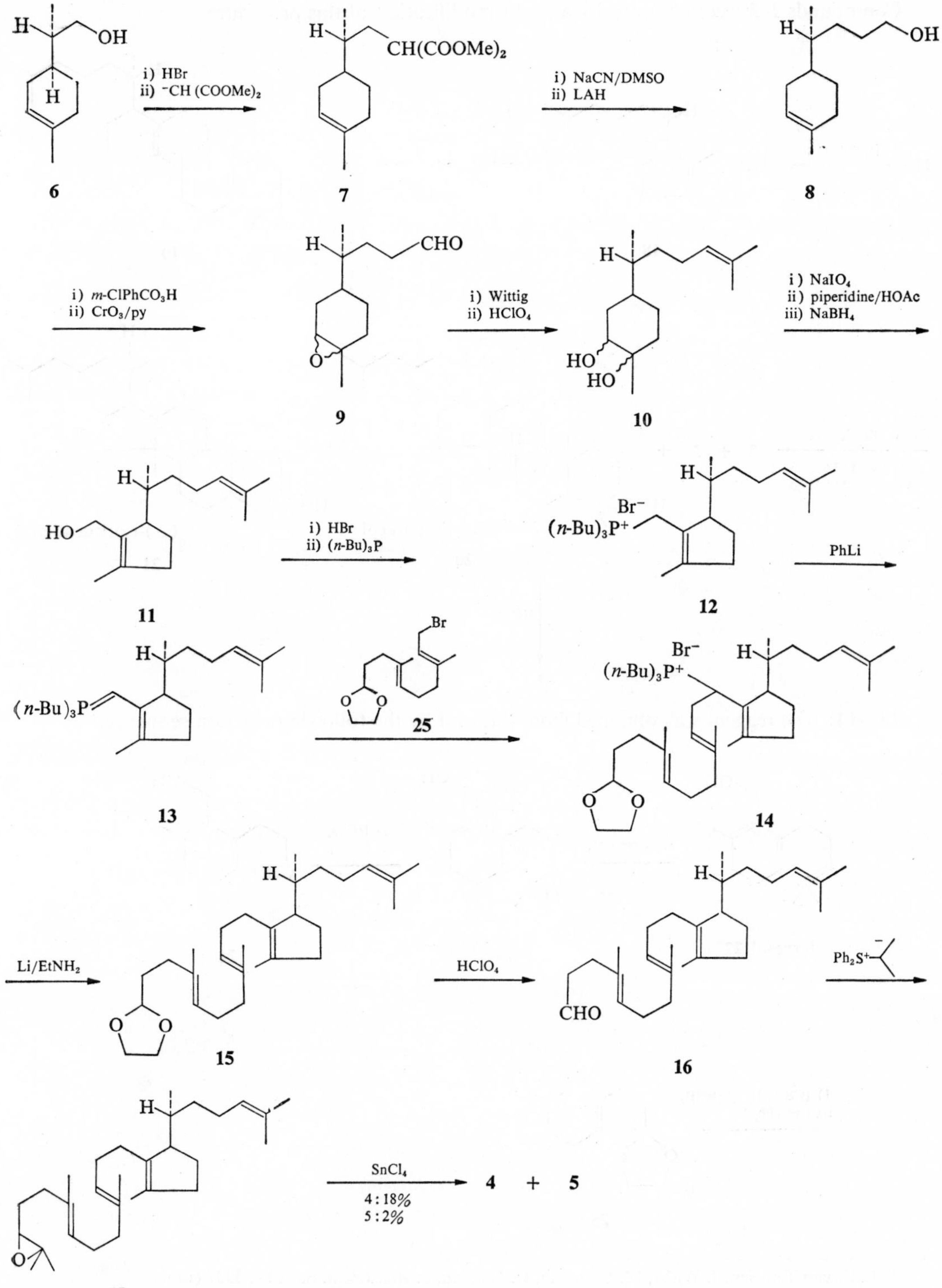

6
i) HBr
ii) ⁻CH (COOMe)₂
7
CH(COOMe)₂
i) NaCN/DMSO
ii) LAH
8
i) m-ClPhCO₃H
ii) CrO₃/py
9
CHO
i) Wittig
ii) HClO₄
10
i) NaIO₄
ii) piperidine/HOAc
iii) NaBH₄
11
i) HBr
ii) (n-Bu)₃P
12
Br⁻
(n-Bu)₃P⁺
PhLi
13
(n-Bu)₃P
25
Br
14
Li/EtNH₂
15
HClO₄
16
CHO
Ph₂S⁺
17
SnCl₄
4 : 18%
5 : 2%
4 + 5

Compounds **1–3** were obtained by a slight modification of this procedure.

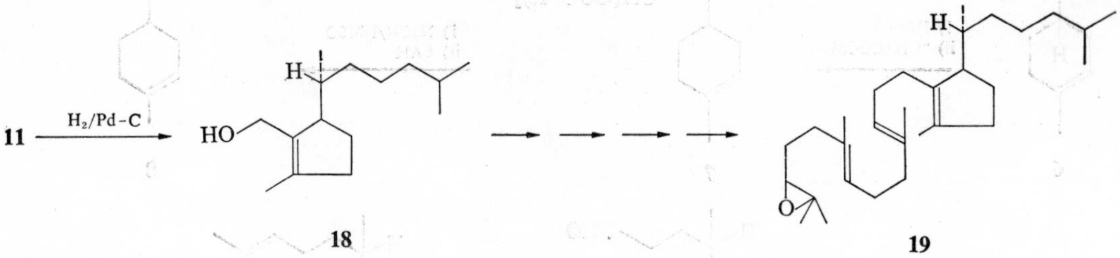

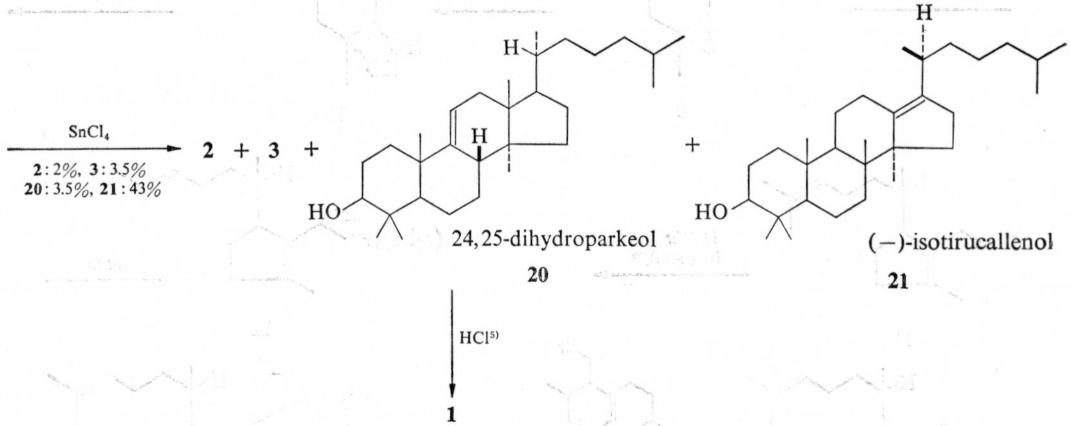

11 $\xrightarrow{\text{H}_2/\text{Pd}-\text{C}}$ ... **18**

→ → → → **19**

$\xrightarrow[\substack{2:2\%,\ 3:3.5\%\\20:3.5\%,\ 21:43\%}]{\text{SnCl}_4}$ **2 + 3 +** 24,25-dihydroparkeol **20** + (−)-isotirucallenol **21**

$\downarrow$ HCl[5]

1

13→14: The reagent was obtained from farnesol by the following reaction sequence.[3]

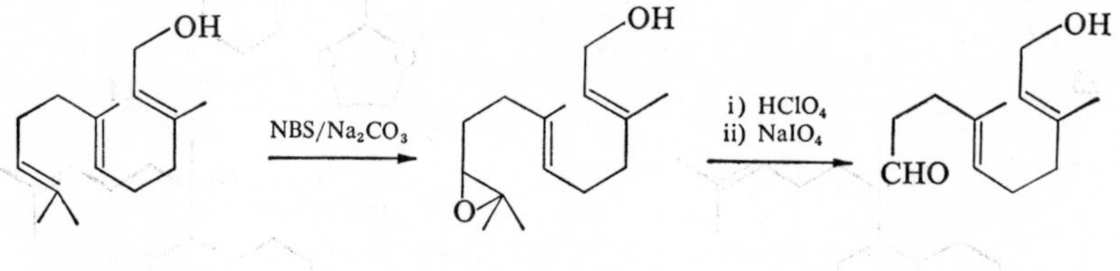

farnesol **22** $\xrightarrow{\text{NBS}/\text{Na}_2\text{CO}_3}$ **23** $\xrightarrow[\text{ii) NaIO}_4]{\text{i) HClO}_4}$ **24**

$\xrightarrow[\text{ii) CBr}_4/\text{Ph}_3\text{P}]{\text{i) (CH}_2\text{OH)}_2/\text{(COOH)}_2}$ **25**

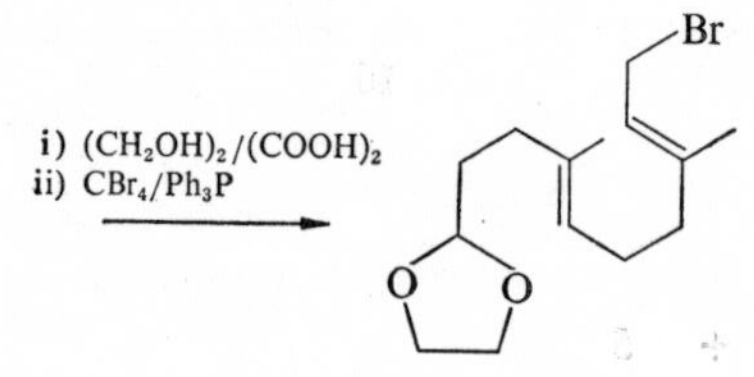

4) E. E. van Tamelen, J. Willet, M. Schwartz, R. Nadeau, *J. Am. Chem. Soc.*, **88**, 5937 (1966).
5) E. E. van Tamelen, J. W. Murphy, *J. Am. Chem. Soc.*, **92**, 7204 (1970).

17→4, 5: The non-enzymatic cyclization has been thoroughly examined previously.[4] Although the reaction with squalene 2,3-epoxide followed the Markownikov rule and furnished tricyclic compounds with the five-membered C ring, the present reaction with a tetrasubstituted double bond yielded compounds with a six-membered C ring.

4 was formed from the "all-chair" conformation of one diastereoisomer of **17**, while **5** resulted from the "chair-boat-chair" conformation of the other diastereoisomer.

"all-chair" conformation **26**   "chair-boat-chair" conformation **27**

18→19: This reaction sequence is exactly the same as **11→17**.

19→2, 3, 20, 21: **3** was formed from the all-chair conformation of one diastereoisomer of **19**, while **21** was formed from the same conformation of the other diastereoisomer. **2** and **20** were obtained from the chair-boat-chair conformation of the former diastereoisomer.

5.14 STRUCTURE OF MALABARICOL AND SYNTHESIS OF MALABARICANEDIOL

6 Me groups { 0.78
0.83
0.98
1.13
1.13
1.27

malabaricol **1**[1]

mp: 68–69.5° α_D: +36.1°
ir: 1700 (carbonyl), 3550 (OH)
ms: 458 (M$^+$ C$_{30}$H$_{50}$O$_3$)
nmr in CCl$_4$

1) Jones oxidation of **1** afforded the γ-lactone **2** and methyl-heptenone **3**.
2) Chemical correlation with the known compound (+)-ambreinolide **4** established the structure and configuration.[2]

1) A. Chawla, S. Dev, *Tetr. Lett.*, 4837 (1967).
2) R. R. Sobti, S. Dev, *Tetr. Lett.*, 2215 (1968).

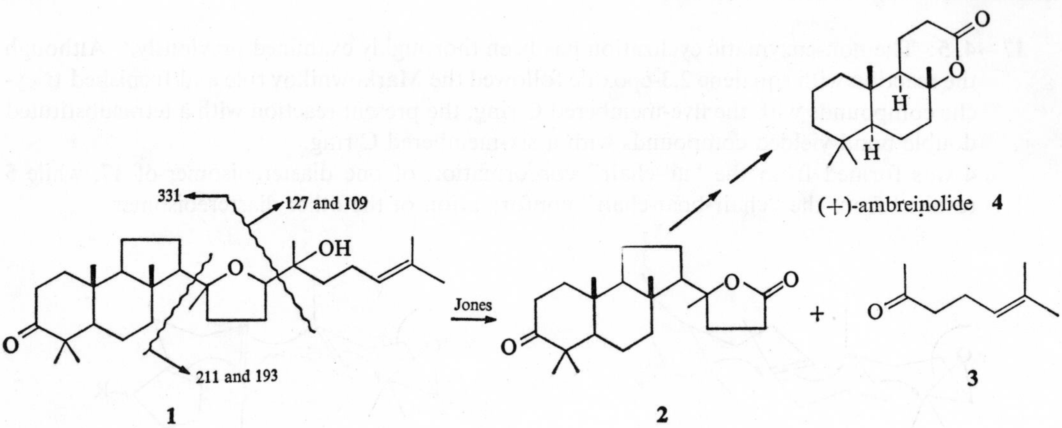

(+)-ambreinolide **4**

1   **2**   +   **3**

Remarks

The tricyclic triterpene malabaricol was isolated from *Ailanthus malabarica*.

Synthesis of malabaricanediol[3]

Biogenetic-type cyclization of the diol **7** yielded malabaricanediol **8**. Non-enzymatic cyclization of 2,3-oxidosqualene[4] and enzymatic cyclization of 18,19-dihydrosqualene 2,3-oxide[5]
afforded the tricarbocyclic malabaricol system (see "Cyclization of squalene-2,3-epoxide").

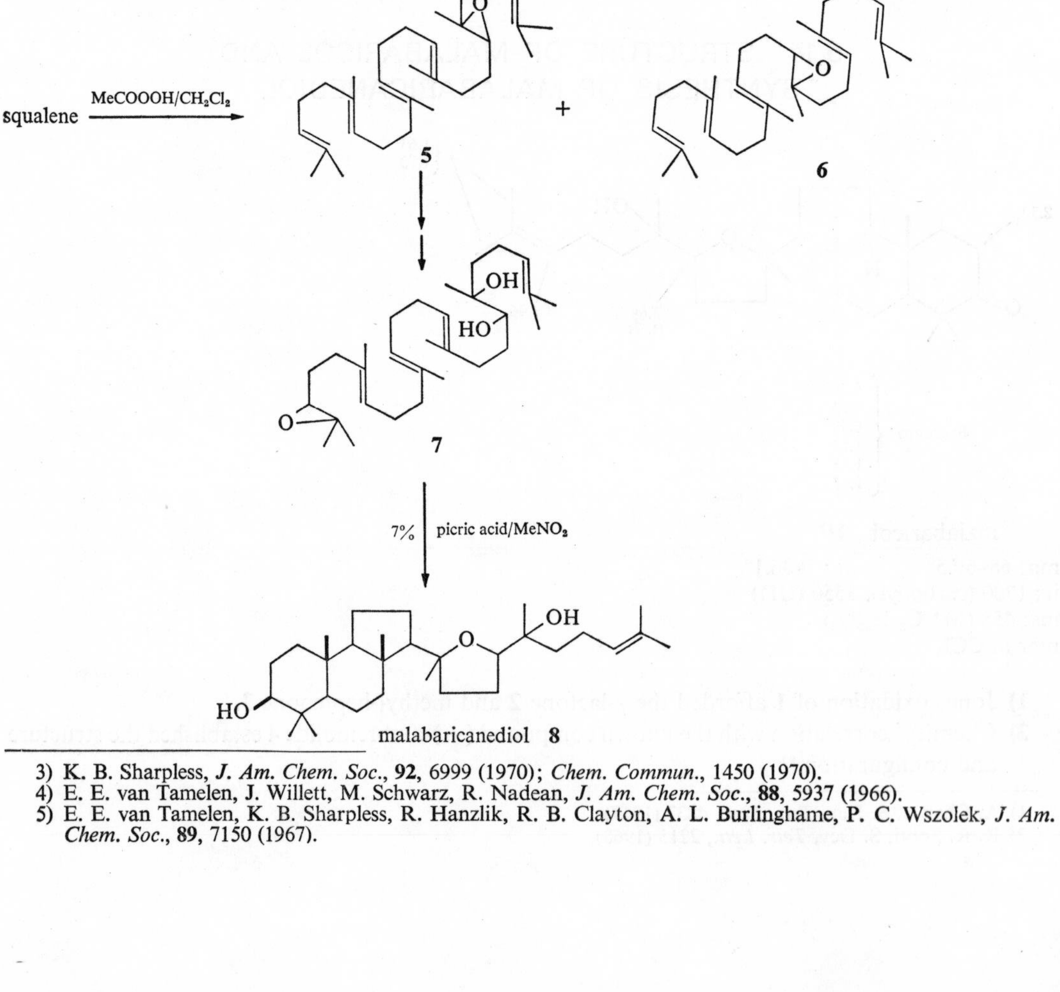

squalene $\xrightarrow{\text{MeCOOOH/CH}_2\text{Cl}_2}$

5   +   **6**

7

7% | picric acid/MeNO$_2$

HO

malabaricanediol **8**

3) K. B. Sharpless, *J. Am. Chem. Soc.*, **92**, 6999 (1970); *Chem. Commun.*, 1450 (1970).
4) E. E. van Tamelen, J. Willett, M. Schwarz, R. Nadean, *J. Am. Chem. Soc.*, **88**, 5937 (1966).
5) E. E. van Tamelen, K. B. Sharpless, R. Hanzlik, R. B. Clayton, A. L. Burlinghame, P. C. Wszolek, *J. Am. Chem. Soc.*, **89**, 7150 (1967).

5.15 SYNTHESIS OF LANOSTEROL

H
12
11
17
16
13
1
9
2
14
15
10
8
3
5
7
HO
4
H
6

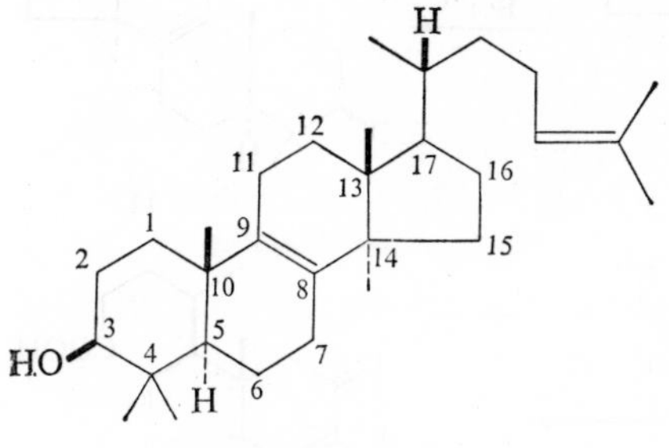

lanosterol **1**

This is the classical transformation of cholesterol **2** to lanosterol **1**.[1] Since the former compound has been synthesized totally,[2] these processes constitute the first total synthesis of a triterpenoid of the lanostane type.

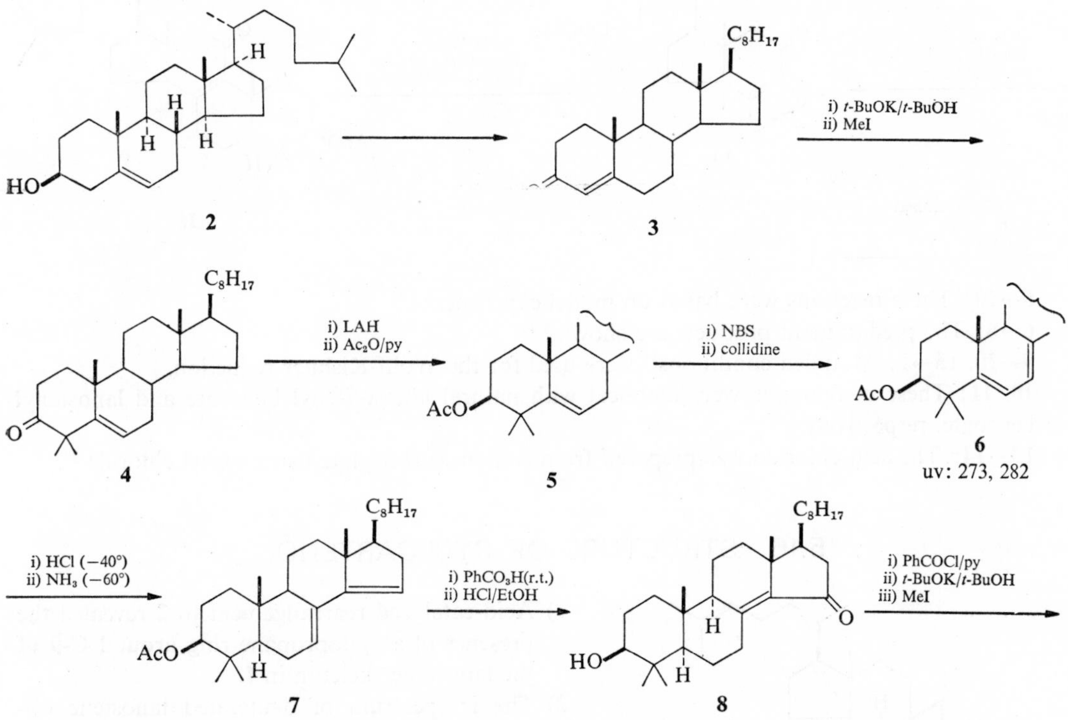

H
H
H
H
HO
2
C_8H_{17}
i) t-BuOK/t-BuOH
ii) MeI
3
C_8H_{17}
O
4
i) LAH
ii) Ac_2O/py
AcO
5
i) NBS
ii) collidine
AcO
6
uv: 273, 282
i) HCl (−40°)
ii) NH_3 (−60°)
C_8H_{17}
AcO
H
7
i) $PhCO_3H$(r.t.)
ii) HCl/EtOH
C_8H_{17}
H
O
HO
H
8
i) PhCOCl/py
ii) t-BuOK/t-BuOH
iii) MeI

1) R. B. Woodward, A. A. Patchett, D. H. R. Barton, R. B. Kelly, *J. Chem. Soc.*, 1131 (1957).
2) R. B. Woodward, F. Sondheimer, D. Taub, *J. Am. Chem. Soc.*, **73**, 3548 (1951); R. B. Woodward, F. Sondheimer, D. Taub, K. Heusler, W. M. McLamore, *ibid.*, **74**, 4223 (1952); H. M. E. Cardwell, J. W. Cornforth, S. R. Duff, H. Holtermann, R. Robinson, *J. Chem. Soc.*, 361 (1953).
3) C. S. Barnes, D. H. R. Barton, G. F. Laws, *Chem. Ind.*, 616 (1953); D. H. R. Barton, G. F. Laws, *J. Chem. Soc.*, 52 (1954).
4) D. H. R. Barton, C. W. J. Brooks, *J. Chem. Soc.*, 257 (1951).
5) D. H. R. Barton, D. A. J. Ives, B. R. Thomas, *J. Chem. Soc.*, 2056 (1955).
6) A. L. Wilds, C. H. Shunk, *J. Am. Chem. Soc.*, **72**, 2388 (1950).

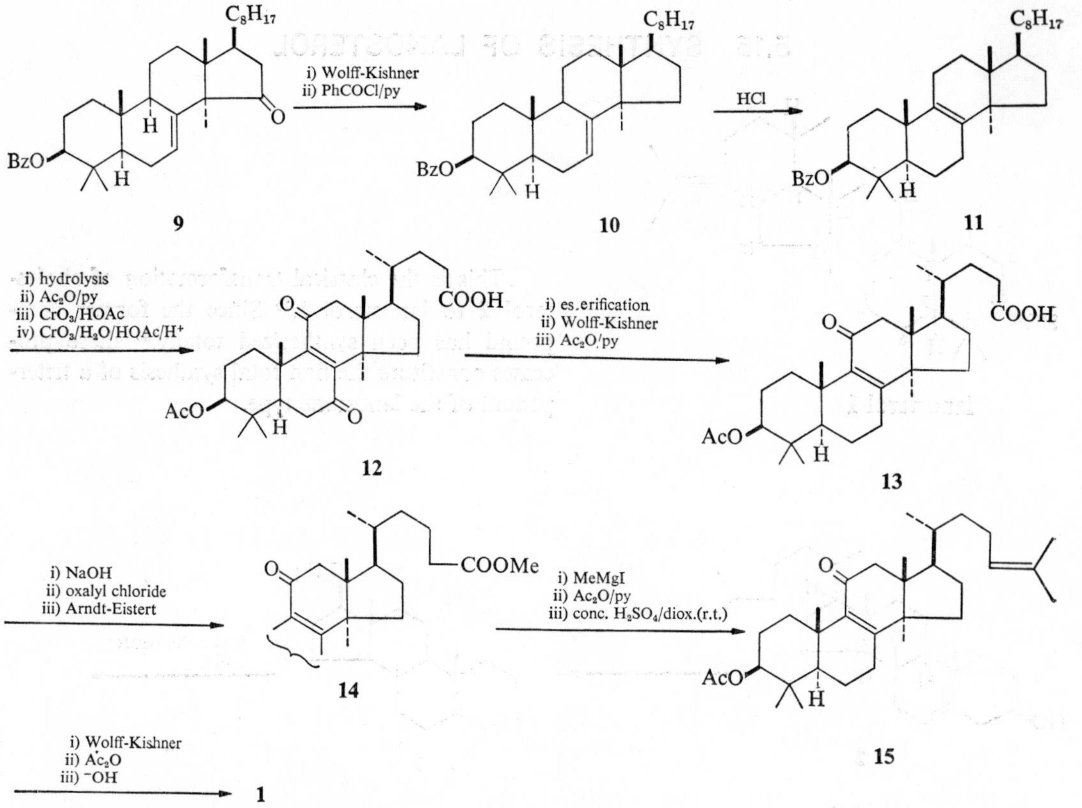

C_8H_{17}

i) Wolff-Kishner
ii) PhCOCl/py

C_8H_{17}

HCl

C_8H_{17}

BzO

BzO

BzO

9   **10**   **11**

i) hydrolysis
ii) Ac$_2$O/py
iii) CrO$_3$/HOAc
iv) CrO$_3$/H$_2$O/HOAc/H$^+$

COOH

i) es.erification
ii) Wolff-Kishner
iii) Ac$_2$O/py

COOH

AcO

AcO

12   **13**

i) NaOH
ii) oxalyl chloride
iii) Arndt-Eistert

COOMe

i) MeMgI
ii) Ac$_2$O/py
iii) conc. H$_2$SO$_4$/diox.(r.t.)

AcO

14   **15**

i) Wolff-Kishner
ii) Ac$_2$O
iii) $^-$OH

1

5→10: These reactions were based on model experiments.[3]

6→8: The predominant products are shown.[3,4]

9→10, 15→1: Modified conditions[5] were used for the Wolff-Kishner reduction.

10, 11: These compounds were identical with natural lanost-7-enyl benzoate and lanostenyl benzoate, respectively.

13→14: The acid chloride was prepared from sodium carboxylate using oxalyl chloride.[6]

5.16 STRUCTURE OF CYCLOARTENOL

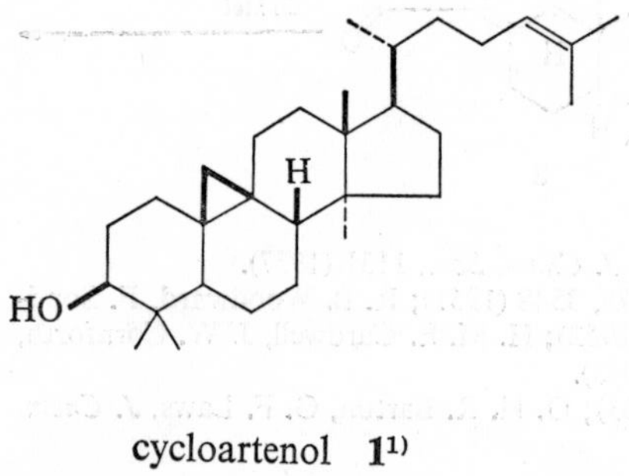

HO

cycloartenol 1[1]

mp: 115° α_D: +54°
ir: 3052–3042 (cyclopropane methylene)

1) Acid-catalyzed rearrangement to **2** revealed the presence of a cyclopropane ring around C-9 of the lanostane skeleton in **1**.

2) The ir spectrum of deuterated lanostene obtained by DCl treatment of cycloartane indicated that the deuterium atom was located at C-19.[2]

3) The carbonyl group in **3** is not conjugated with the cyclopropane ring.

4) The α,β-unsaturated carbonyl system in **4** is conjugated with the cyclopropane ring (uv: 269 nm).

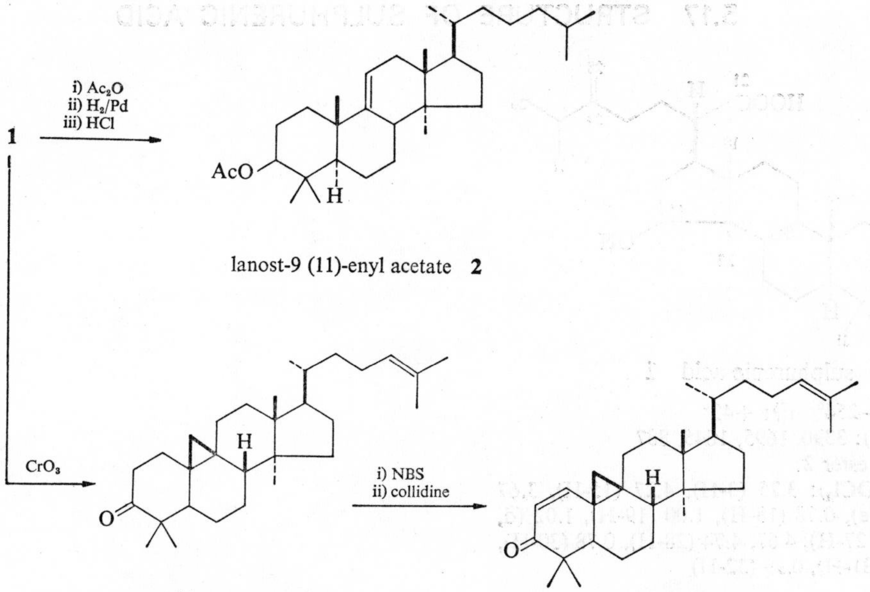
i) Ac₂O
ii) H₂/Pd
iii) HCl
1
AcO
H
CrO₃
H
O
i) NBS
ii) collidine
H
O

lanost-9 (11)-enyl acetate **2**

cycloartenone **3**

4

Remarks

Cycloartenol **1** was isolated from *Artocarpus integrifolia, Strychnos nux-vomica, Euphorbia balsamifera*, etc. It has been shown that cycloartenol is an important biosynthetic precursor of phytosterols.[3] Oxygenated and/or alkylated derivatives of cycloartenol, such as mangiferolic acid, cimigenol, cyclolaudenol, cycloeucalenol, cycloneolitsin, and ambonic acid, are also known. Buxus alkaloids, e.g., **5** and **6**, are presumably derived from cycloartenol.

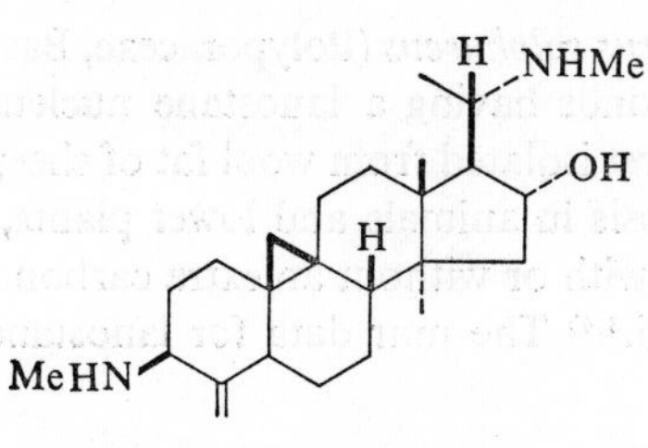
H
NHMe
OH
H
MeHN

cyclobuxine **5**

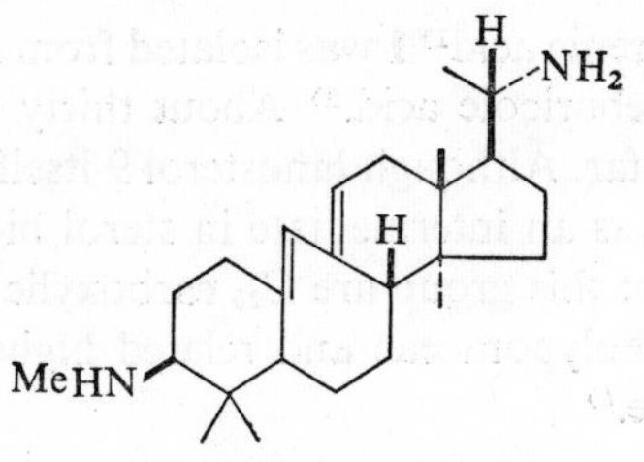
H
NH₂
H
MeHN

buxenine-G **6**

1) H. R. Bentley, J. A. Henry, D. S. Irvine, F. S. Spring, *J. Chem. Soc.*, 3673 (1953); D. S. Irvine, J. A. Henry, F. S. Spring, *ibid.*, 1316 (1955).
2) D. H. R. Barton, J. E. Page, E. W. Warnhoff, *J. Chem. Soc.*, 2715 (1954).
3) See *Terpenoids in Plants* (ed. J. B. Pridham), Academic Press, 1967.

5.17   STRUCTURE OF SULPHURENIC ACID

sulphurenic acid   **1**

mp: 252–254° a_D^{23}: +42°
ir (nujol): 3390, 1695, 1645, 887
For Me ester **2**:
nmr (CDCl$_3$): 3.25 (3-H), 4.27 (15-H), 3.67
   (21-Me), 0.78 (18-H), 1.00 (19-H), 1.02 (d,
   26-H, 27-H), 4.67, 4.74 (28-H), 0.78 (30-H),
   0.96 (31-H), 0.99 (32-H)

1) The methyl ester **2** resists hydrolysis by strong bases. Reaction with Li/liq. NH$_3$ or LiI/collidine regenerates **1**.

2) The presence of two *sec*-OH groups was indicated by the transformation **2→3**.

3) The retropinacol rearrangement **3→4** is characteristic of 3β-hydroxy-4,4-dimethyl steroids.

4) The structure of **5** is known on the basis of a correlation of eburicoic acid with lanosterol.[1,2]

5) The presence of a tetrasubstituted double bond was confirmed by the formation of **6**.

6) The 15α-OH was first suggested by nmr and then confirmed by the formation of **8**.

Remarks

Sulphurenic acid[1] **1** was isolated from *Polyporus sulphureus* (Polyporaceae, Basidiomycetes) along with eburicoic acid.[2] About thirty compounds having a lanostane nucleus have been reported so far. Although lanosterol **9** itself was first isolated from wool fat of sheep and is now well-known as an intermediate in sterol biosynthesis in animals and lower plants, most of the triterpenes of this group are C$_{20}$ carboxylic acids (with or without an extra carbon at C-24) isolated from Polyporaceae and related higher fungi.[3,4] The nmr data for lanostane derivatives are available.[5]

1) J. Fried, P. Grabowich, E. P. Sabo, A. I. Cohen, *Tetr.*, **20**, 2297 (1964).
2) J. L. Simonsen, W. C. J. Ross, *The Terpenes*, vol. 5, p. 1, Cambridge University Press, 1957.
3) G. Ourisson, P. Crabbé, O. R. Rodig, *Tetracyclic Triterpenes*, Herman, 1964.
4) A. Yokoyama, S. Natori, K. Aoshima, *Phytochemistry*, **14**, 487 (1975).
5) A. I. Cohen, D. Rossenthal, G. W. Krakower, J. Fried, *Tetr.*, **21**, 3171 (1965).

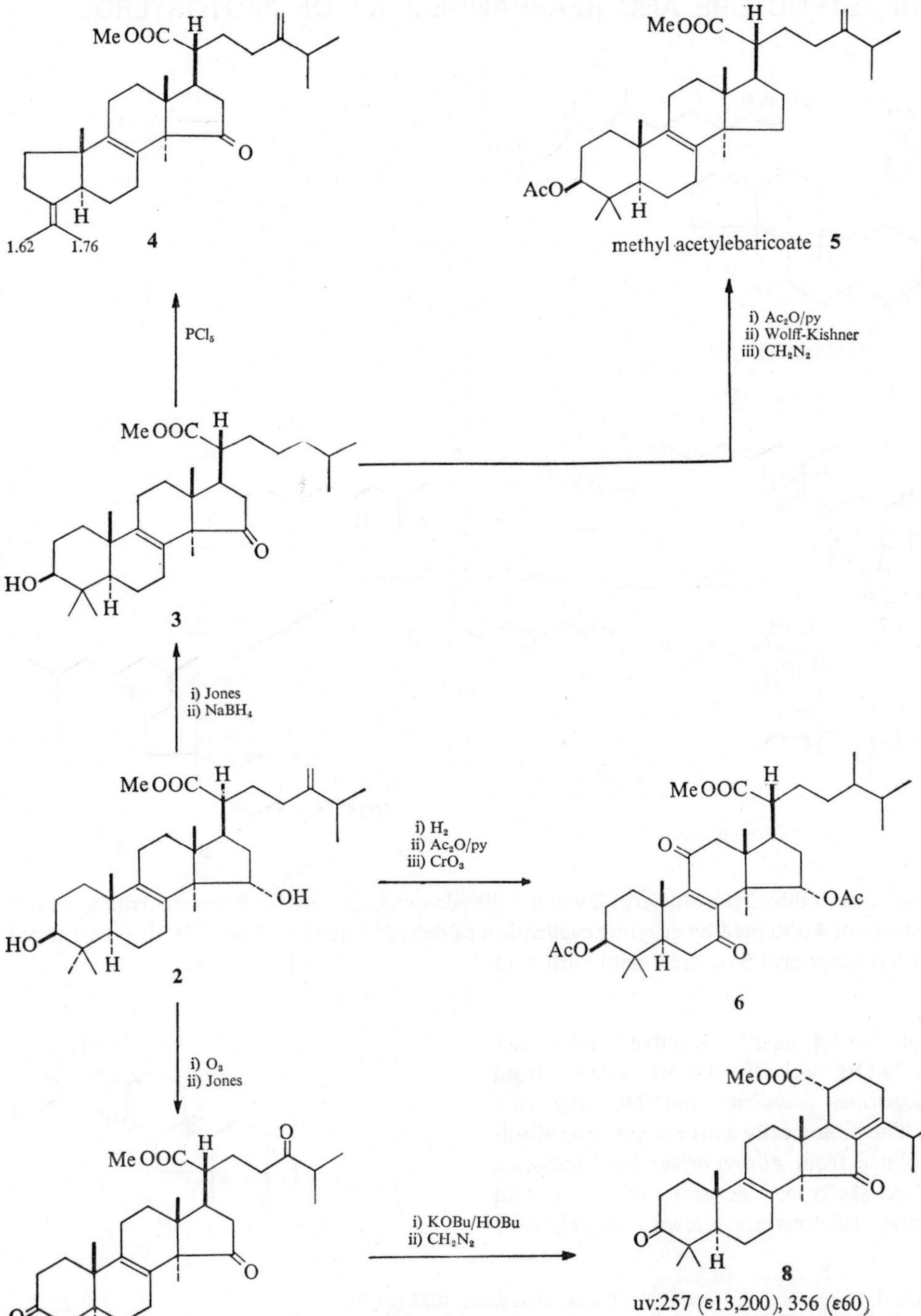
MeOOC
H
1.62
1.76
4
AcO
methyl acetylebaricoate 5
PCl$_5$
i) Ac$_2$O/py
ii) Wolff-Kishner
iii) CH$_2$N$_2$
HO
3
i) Jones
ii) NaBH$_4$
OH
2
i) H$_2$
ii) Ac$_2$O/py
iii) CrO$_3$
OAc
6
i) O$_3$
ii) Jones
7
i) KOBu/HOBu
ii) CH$_2$N$_2$
8
uv:257 (ε13,200), 356 (ε60)
ir:1701, 1664

5.18   STRUCTURE AND REARRANGEMENT OF PROTOSTEROL

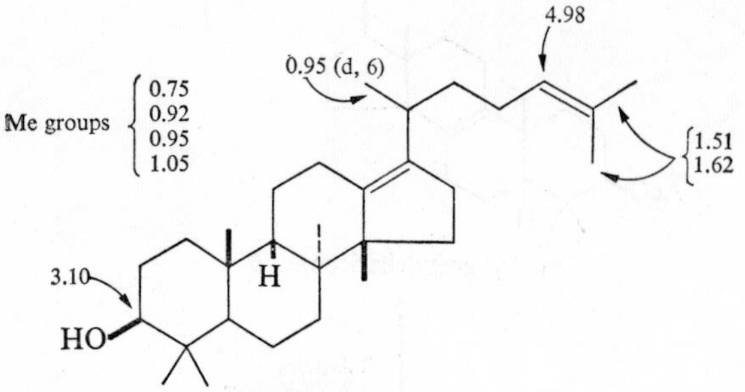

Me groups $\begin{cases} 0.75 \\ 0.92 \\ 0.95 \\ 1.05 \end{cases}$ 0.95 (d, 6)   4.98 $\begin{cases} 1.51 \\ 1.62 \end{cases}$

3.10

HO

protosterol $\mathbf{1}$[1)]

mp: 117–117.5[2)] α_D: $+16°$

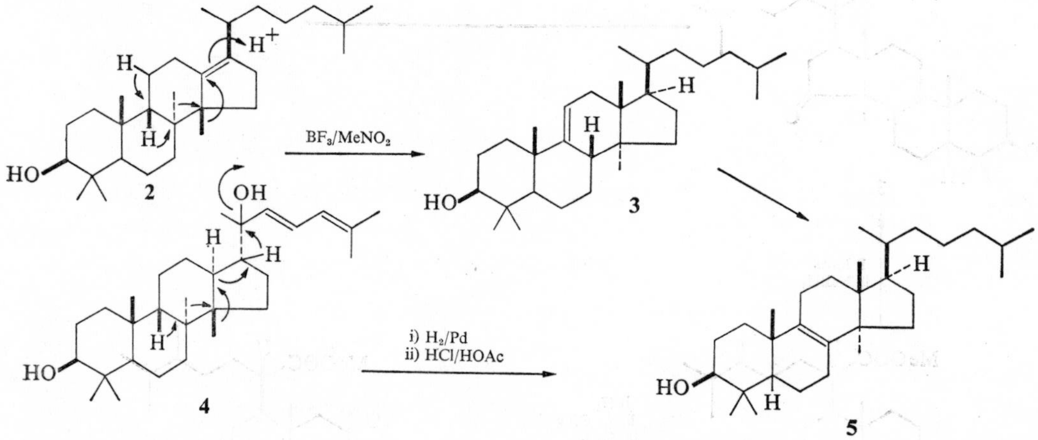

BF$_3$/MeNO$_2$

i) H$_2$/Pd
ii) HCl/HOAc

Acid treatment of dihydroprotosterol **2** yielded dihydroparkeol **3** by backbone rearrangement.[3)] The protosterol **4** obtained by enzymic cyclization of dehydrosqualene 2,3-oxide also rearranged into dihydrolanosterol **5** on treatment with acid.[4)]

Remarks

Protosterol **1**, its $\Delta^{17(20)}$-double bond isomer and its 29-OH derivative[2)] were isolated from *Cephalosporium caerulens* and *Fusidium coccineum*. The biologically active compound alisol-A **6**, isolated from *Alisma orientale,* possesses a protostane skeleton.[5)] Several antibiotics with protostane skeletons are known (see "Helvolic acid").

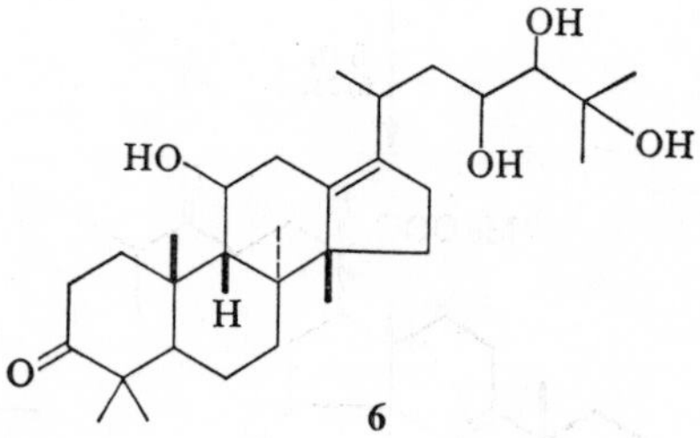

1) T. Hattori, H. Igarashi, S. Iwasaki, S. Okuda, *Tetr. Lett.*, 1023 (1969).
2) S. Okuda, Y. Sato, T. Hattori, H. Igarashi, *Tetr. Lett.*, 4769 (1968).
3) E. E. van Tamelen, R. J. Anderson, *J. Am. Chem. Soc.*, **94**, 8227 (1972).
4) E. J. Corey, H. Yamamoto, *Tetr. Lett.*, 2385 (1970).
5) K. Kamiya, T. Murata, M. Nishikawa, *Chem. Pharm. Bull.*, **18**, 1362 (1970); T. Murata, M. Miyamoto, *ibid.*, **18**, 1354 (1970).

5.19 HELVOLIC ACID

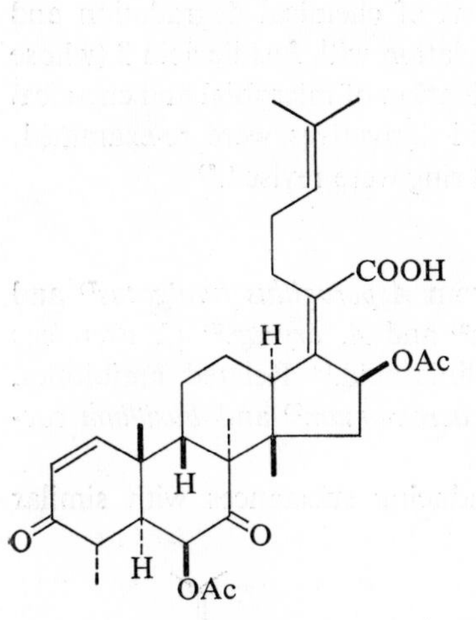

helvolic acid **1**

mp: 211.3–212.1° (dec.)
uv: 231 (ε 17,300), 322 (ε 98)
ir of Me ester (nujol): 1745, 1718, 1675,
   1655, 1255, 1238, 1210
nmr of Me ester (CDCl$_3$): 0.92, 1.18, 1.45,
   1.95, 2.10 (each 3H, s), 1.28 (3H, d, $J=$
   6.5), 1.62, 1.70 (each 3H), 5.22 (1H, s),
   5.83 (1H), 5.83 (1H, d, $J=$11.4), 7.30
   (1H, d, $J=$11.4)

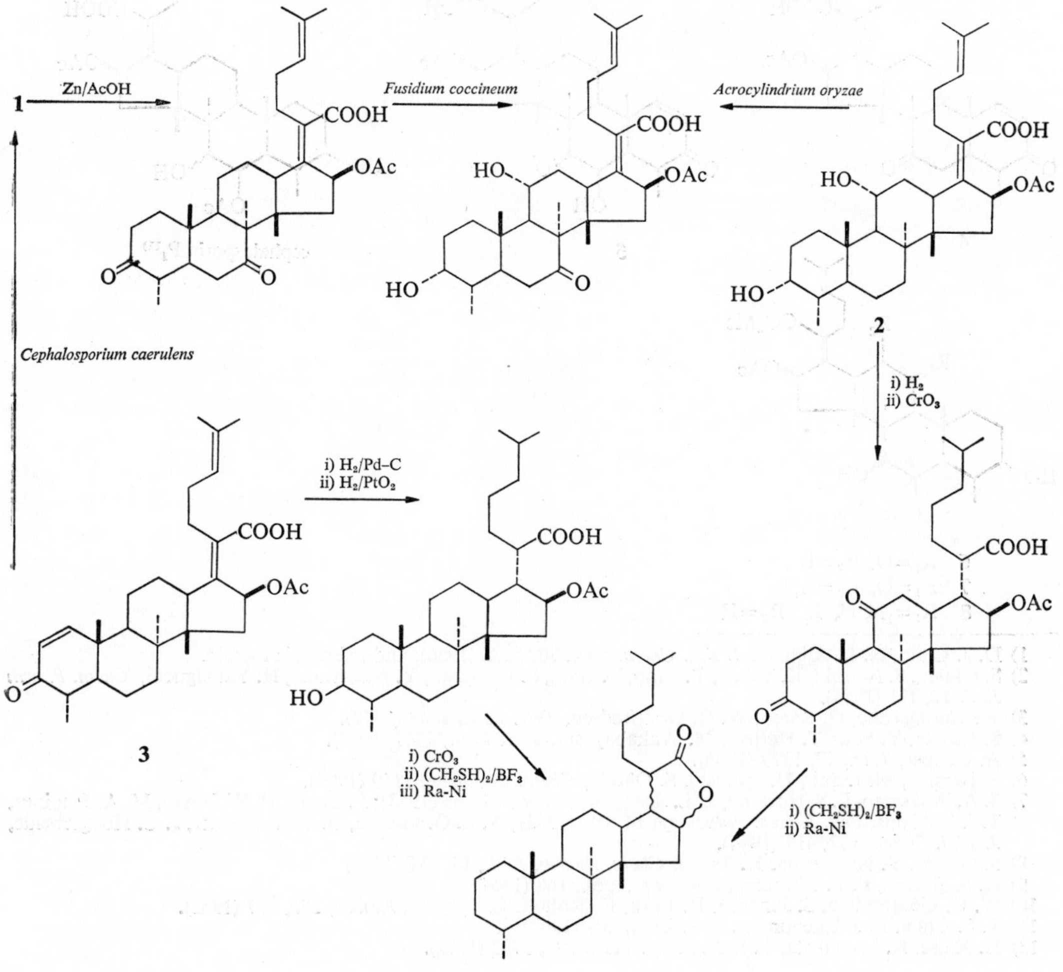

The structure of helvolic acid was proposed on the basis of chemical degradation and spectral data.[1,2] Most of the structure was confirmed by interrelation with fusidic acid **2** (whose structure had been established by x-ray analysis) using a combination of microbial and chemical techniques.[3-5] Subsequently, the nmr spectra of helvolic acid derivatives were re-examined, and the positions of the carbonyl and acetoxy groups on the B ring were revised.[6]

Remarks

Helvolic acid, a steroidal antibiotic, was first isolated from *Asperugillus fumigatus*[7] and later from *Cephalosporium caerulens,*[2] *Emericellopsis terricola*[2] and *A. oryzae.*[3] *C. caerulens* also produces **3**,[4] monodesacetoxy helvolic acid[8] and helvolinic acid.[8] Related antibiotics, cephalosporins and fusidic acid, have been isolated from *C. acremonium*[9] and *Fusidium coccineum,*[10] respectively.

Viridominic acid A, B, and C (**6, 7, 8**)[12], chlorosis-inducing substances with similar structures, were isolated from *Cladosporium* sp.

4   5   cephalosporin P₁[11]

6 R₁=O, R₂=H
7 R₁=O, R₂=OH
8 R₁=β-OH, H   R₂=H

1) D. J. Cram, N. L. Allinger, *J. Am. Chem. Soc.*, **78**, 5275 (1956), and references therein.
2) S. Okuda, S. Iwasaki, K. Tsuda, Y. Sano, T. Hata, S. Udagawa, Y. Nakayama, H. Yamaguchi, *Chem. Pharm Bull.* **12**, 121 (1964).
3) W. von Daehne, H. Lorch, W. O. Godtfredsen, *Tetr. Lett.*, 4843 (1968).
4) S. Okuda, Y. Sato, T. Hattori, M. Wakabayashi, *Tetr. Lett.*, 4847 (1968).
5) A. Cooper, *Tetr.*, **22**, 1379 (1966).
6) S. Iwasaki, M. I. Sair, H. Igarashi, S. Okuda, *Chem. Commun.*, 1119 (1970).
7) S. A. Waksman, E. S. Horning, E. L. Spencer, *J. Bact.*, **45**, 233 (1943); E. Chain, H. W. Forey, M. A. Jennings, T. I. Williams, *Brit. J. Exp. Pathology*, **24**, 108 (1943); A. E. O. Menzel, O. Wintersteiner, J. C. Hoogerheide, *J. Biol. Chem.*, **152**, 419 (1944).
8) S. Okuda, Y. Nakayama, K. Tsuda, *Chem. Pharm. Bull.*, **14**, 436 (1966).
9) H. S. Burton, E. P. Abraham, *Biochem. J.*, **50**, 168 (1951).
10) W. O. Godtfredsen, S. Jahnsen, H. Lork, K. Roholt, L. Tybring, *Nature*, **193**, 987 (1962).
11) T. S. Chou, E. J. Eisenbraun, *Tetr. Lett.*, 409 (1967).
12) H. Kaise, K. Munakata, T. Sassa, *Tetr. Lett.*, 199, 3792 (1972).

5.20 DAMMARENES FROM BETULA, PANAX AND DAMMAR RESIN

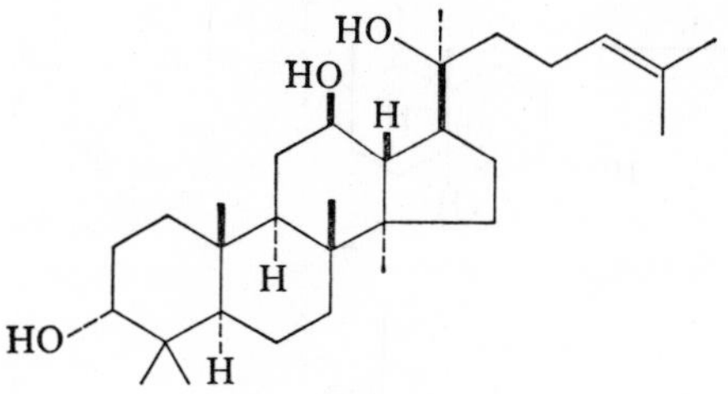

betulafolienetriol 1[1]

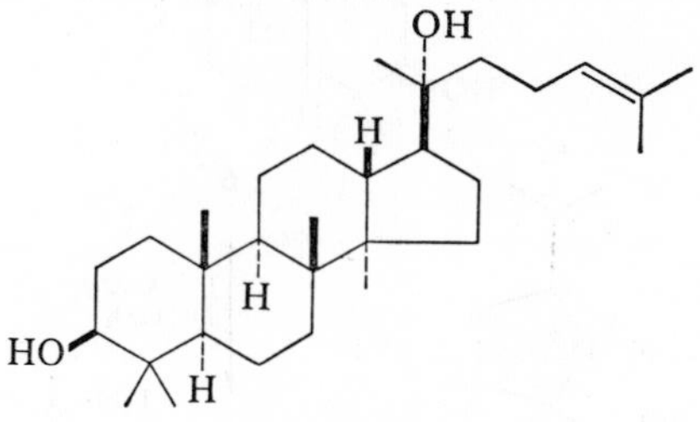

dammarenediol I 2[2]

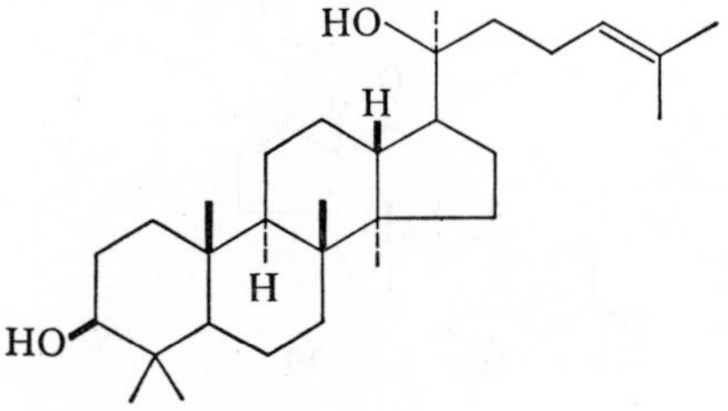

dammarenediol II 3[2]

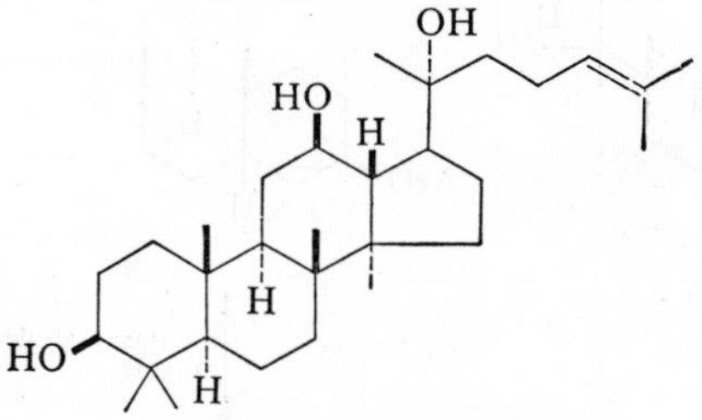

20R-protopanaxadiol 4[4]

1) The structure of 1[1] was finally established by x-ray analysis of 4.[3]
2) The C-20 configuration of 3 was established by chemical correlation with 1.[1]
3) The absolute configuration of 2 was also established by chemical correlation with 1 and 4 through 10.[4]
4) The presence of the C-20 hydroxyl group was suggested to be essential for the facile epimerization of the C-20 configuration of dammarans.

Remarks

2 and 3 are the first dammarans obtained from Dammar resin. 1 was isolated from European white birch and was also found in Japanese white birch. Chemical correlation of these dammarans established their C-20 configurations.

1) F. G. Fischer, N. Seiler, *Ann Chem.*, **626**, 185 (1959); *ibid.*, **644**, 416 (1961).
2) J. S. Mills, A. E. A. Werner, *J. Chem. Soc.*, 3132 (1955); J. S. Mills, *ibid.*, 2196 (1956).
3) O. Tanaka, N. Tanaka, T. Ohsawa, Y. Iitaka, S. Shibata, *Tetr. Lett.*, 4235 (1968).
4) O. Tanaka, M. Nagai, T. Ohsawa, N. Tanaka, K. Kawai, S. Shibata, *Chem. Pharm. Bull.*, **20**, 1204 (1972).

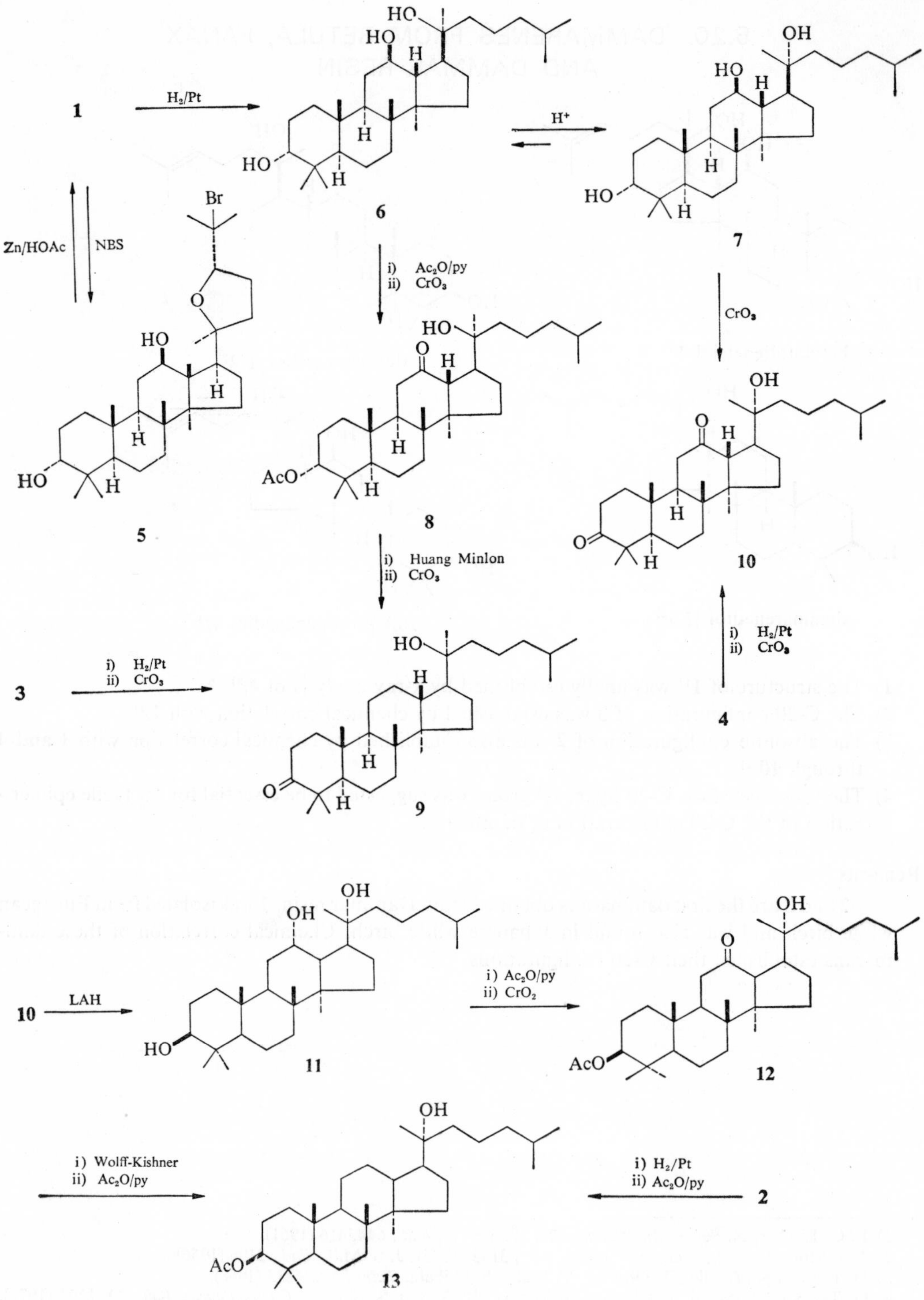
1
H_2/Pt
6
H^+
7
Zn/HOAc
NBS
5
i) Ac_2O/py
ii) CrO_3
8
CrO_3
10
i) Huang Minlon
ii) CrO_3
3
i) H_2/Pt
ii) CrO_3
9
i) H_2/Pt
ii) CrO_3
4
10
LAH
11
i) Ac_2O/py
ii) CrO_2
12
i) Wolff-Kishner
ii) Ac_2O/py
13
i) H_2/Pt
ii) Ac_2O/py
2

5.21 SAPOGENINS FROM GINSENG

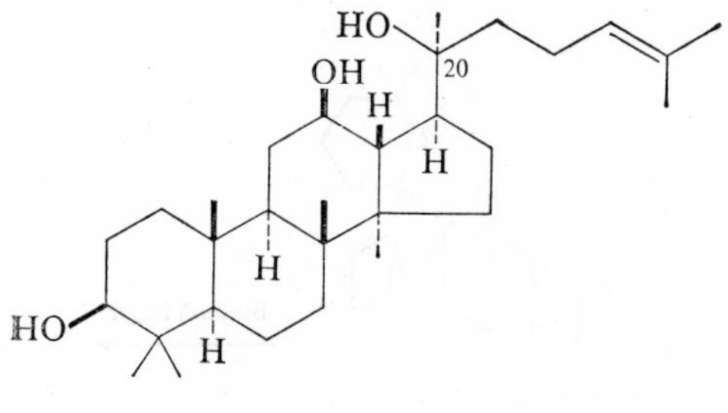

20S-protopanaxadiol **1**[1)]
mp : 197-200°, $\alpha_D + 26.7°$

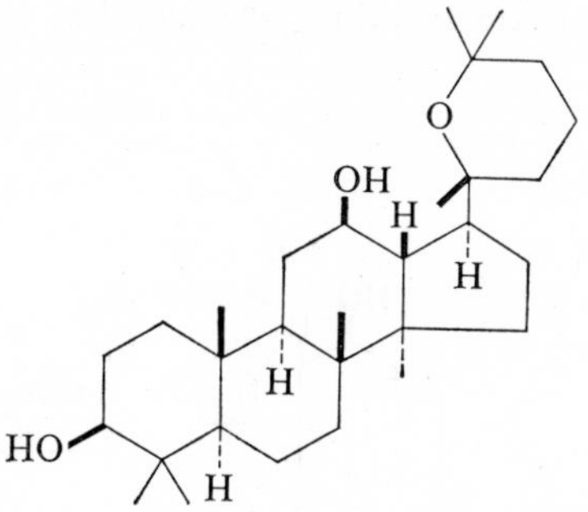

panxadiol **2**[2)]
mp : 250°, $\alpha_D + 1.0°$

1) When 20S-protopanaxadiol **1** was treated with acid, it yielded panxadiol **2**, which had been obtained by the acid hydrolysis of Ginseng saponins. The reaction involves epimerization at C-20 (to **3**) followed by cyclization.

2) Facile epimerization is demonstrated by the dihydro derivative **4**. On acid treatment **4** yields an equillibrium mixture (**4** and **5**), in which **5** is the predominant component.[3)]

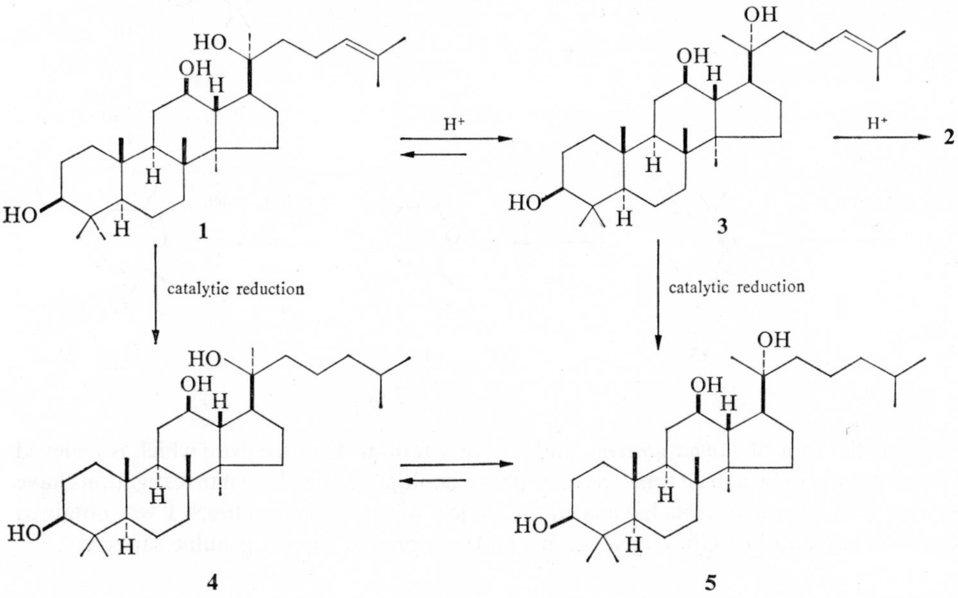

1) M. Nagai, T. Ando, N. Tanaka, O. Tanaka, S. Shibata, *Chem. pharm. Bull.* **20**, 1212 (1972); and references cited therein.
2) S. Shibata, M. Fujita, S. Itokawa, O. Tanaka, T. Ishii, *Tetr. Lett.*, 1239 (1962).
3) O. Tanaka, M. Nagai, T. Ohsawa, N. Tanaka, K. Kawai, S. Shibata, *Chem. Pharm. Bull*, **20**, 1204 (1972).

Absolute C-20 configuration of panaxadiol

The absolute 3-20 configuration of **2** was established chemically by converting **2** to R(−)-methyl cinnenate **11**, which was identical with an authentic sample prepared from R(−)-linalool.[4]

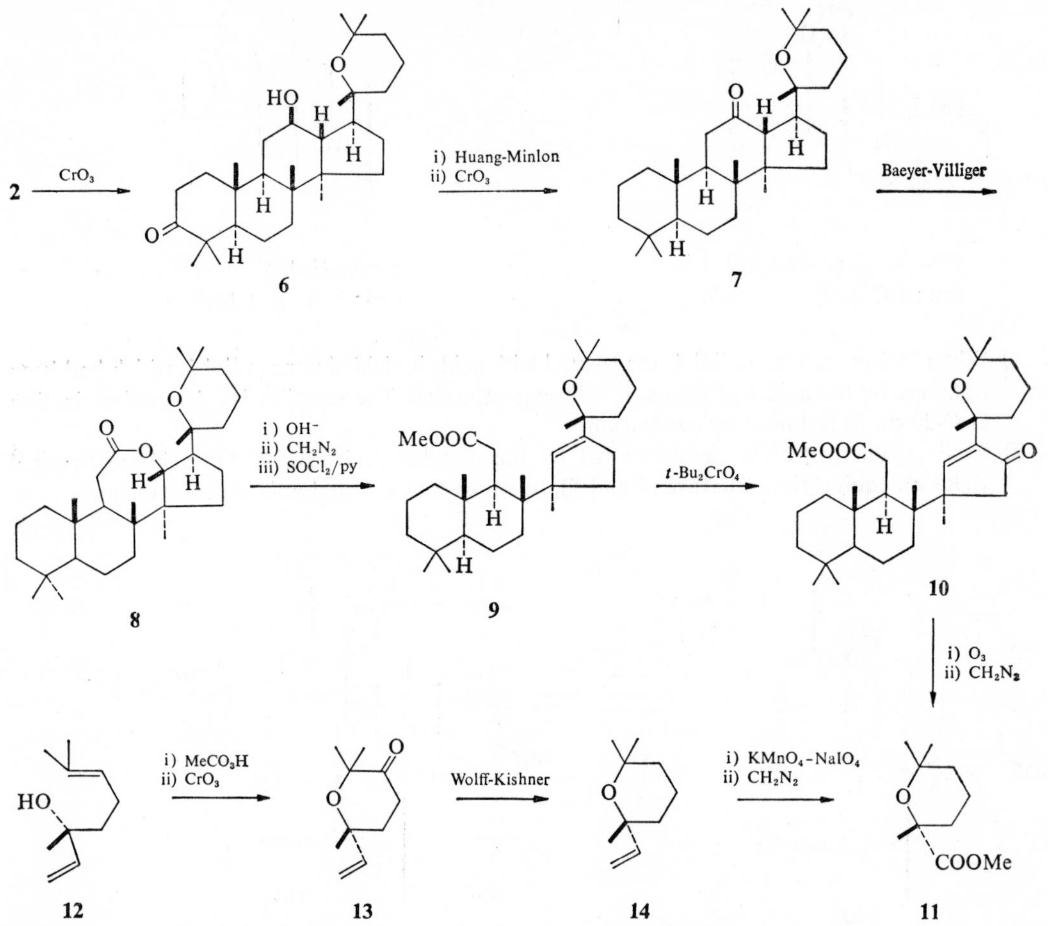

Remarks

Ginseng is the root of *Panax ginseng*, and is a very famous Chinese drug which is believed by East Asian people to be a drug tonic. Several pharmacological and clinical investigations have been reported, and its constituents have been the subject of numerous studies.[5] **1** was obtained by the Smith degradation of Ginseng saponins, and was proved to be a genuine sapogenin.[1]

4) M. Nagai, O. Tanaka, S. Shibata, *Chem. Pharm. Bull.*, **19**, 2349 (1971).
5) S. Shibata, *Some Recent Developments in the Chemistry of Natural Products* (ed. S. Rangaswami, N. V. Subba Row), p. 3, Prentice-Hall of India, 1972.

5.22 BACKBONE REARRANGEMENT OF PENTACYCLIC TRITERPENOIDS

A group of pentacyclic triterpenes, including taraxerol **4**, multiflorenol **6**, glutinol **8** and friedelin **10**, is of considerable interest because compounds of this type undergo a remarkable backbone rearrangement in the presence of acid leading to compounds with oleanane skeletons (e.g., **1–3**). The directions of the stereospecific 1,2-hydride and methyl shifts in these rearrangements are almost exactly the reverse of those postulated for their biogenetic pathway, which involves 1,2-hydride and/or methyl migrations starting from the cation **12** via the ions **13–17**.

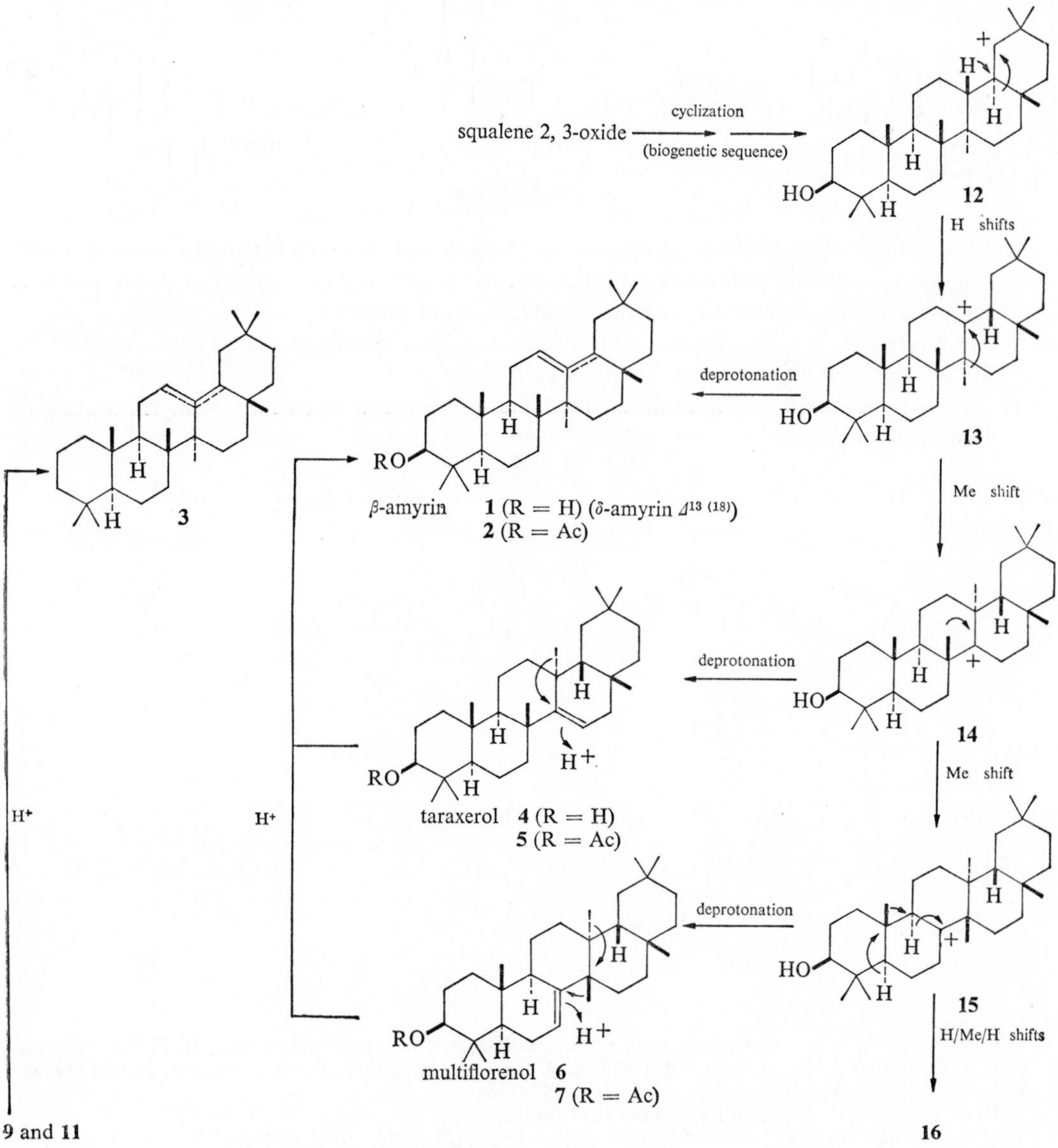

squalene 2, 3-oxide $\xrightarrow[\text{(biogenetic sequence)}]{\text{cyclization}}$ **12**

H shifts

β-amyrin **1** (R = H) (δ-amyrin $\Delta^{13\,(18)}$)
2 (R = Ac)

deprotonation **13**

Me shift

taraxerol **4** (R = H)
5 (R = Ac)

deprotonation **14**

Me shift

multiflorenol **6**
7 (R = Ac)

deprotonation **15**

H/Me/H shifts

3

H⁺

H⁺

9 and 11

16

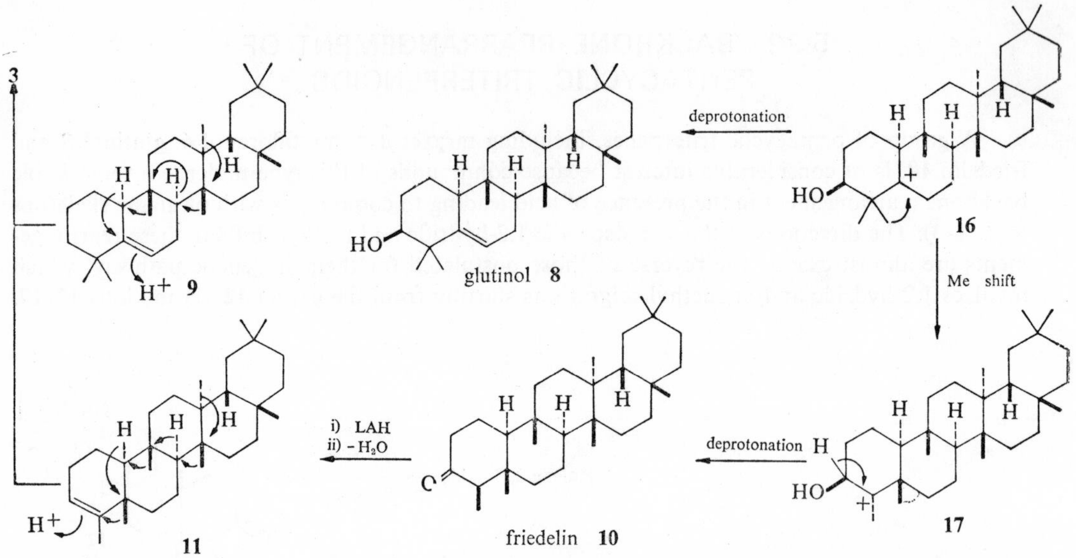

glutinol **8**

16

friedelin **10**

17

11

9

3

1) Friedel-3-ene **11** was converted into olean-12-ene **3** and olean-13(18)-ene by treatment with acid via protonation followed by a series of four methyl and two hydrogen shifts: this may represent the most dramatic backbone rearrangement known.[1]

2) The hydrocarbon **9**, which has a glutinol skeleton, also undergoes three methyl and two hydrogen shifts on acid catalysis, yielding **3**.[2]

3) Taraxeryl acetate **5** and multiflorenyl acetate **7**[3,4] are converted into β-amyrin acetate on treatment with acid.

1) G. Brownlie, F. S. Spring, R. Stevenson, W. S. Strachan, *J. Chem. Soc.*, 2419 (1956); H. Dutler, O. Jeger, L. Ruzicka, *Helv. Chim. Acta*, **38**, 1268 (1955); E. J. Corey, J. J. Ursprung, *J. Am. Chem. Soc.*, **78**, 5014 (1956).
2) F. N. Lahey, M. V. Leeding, *Proc. Chem. Soc.*, 342 (1958).
3) H. N. Khastigir, P. Sengupta, *Chem. Ind.*, 1077 (1961).
4) J. M. Beaton, F. S. Spring, R. Stevenson, J. L. Stewart, *J. Chem. Soc.*, 2131 (1955).

5.23 STRUCTURE AND SYNTHESIS OF TETRAHYMANOL

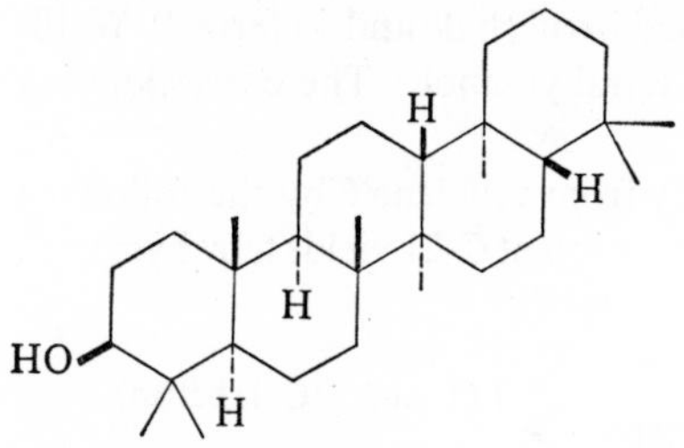

tetrahymanol **1**

mp: 312.5–314.5°
nmr: 0.77(3H), 0.82(9H), 0.85(3H), 0.97(9H)
Data for acetate:
mp: 303–305°
nmr: 0.80(3H), 0.82(3H), 0.84(12H), 0.97(3H)

The structure was established by synthesis from α-onocerin **2**.[1,2] The *dl*-form has been synthesized from farnesol.[3]

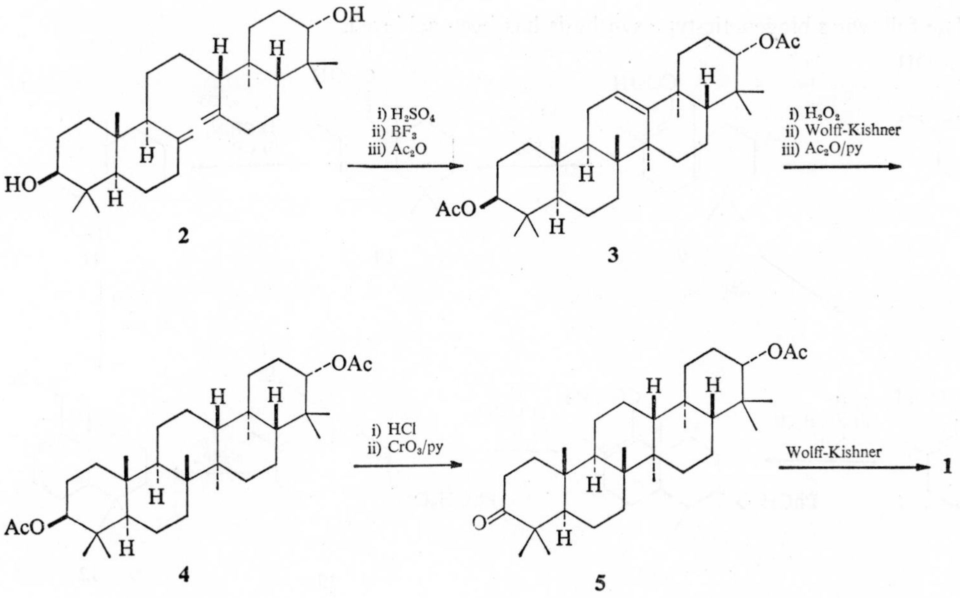

1) Y. Tsuda, A. Morimoto, T. Sano, Y. Inubushi, F. B. Mallory, J. T. Gordon, *Tetr. Lett.*, 1427 (1965).
2) F. B. Mallory, J. T. Gordon, R. L. Conner, *J. Am. Chem. Soc.*, **85**, 1362 (1963).
3) E. E. van Tamelen, R. A. Holton, R. E. Hopla, W. E. Konz, *J. Am. Chem. Soc.*, **94**, 8228 (1972).

Remarks

Tetrahymanol is the only natural triterpenoid known to have a gammacerane skeleton, and the only pentacyclic triterpenoid to be isolated from the animal kingdom (*Tetrahymena periformis*).[2] The presence of tetrahymanol was recently reported in oil shale and in ferns.[4] Wallichiniol, a triterpene from *Oleandra wallichii*, is identical with tetrahymanol.[5] The corresponding hydrocarbon, gammacerane was also found in oil shale.[6]

1 was shown by biosynthetic studies to be formed directly from squalene 6 by the following reaction sequence.[7-9] The attack of the hydroxide ion on the cation at C-21 yields tetrahymanol, while attack at C-22 would lead to 22-hydroxyhopane 7.

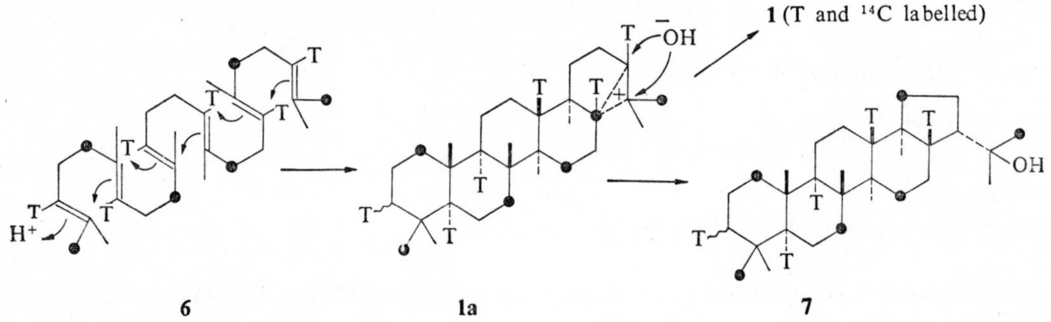

6   1a   7

Synthesis[3]

The following biogenetic-type synthesis has been achieved.

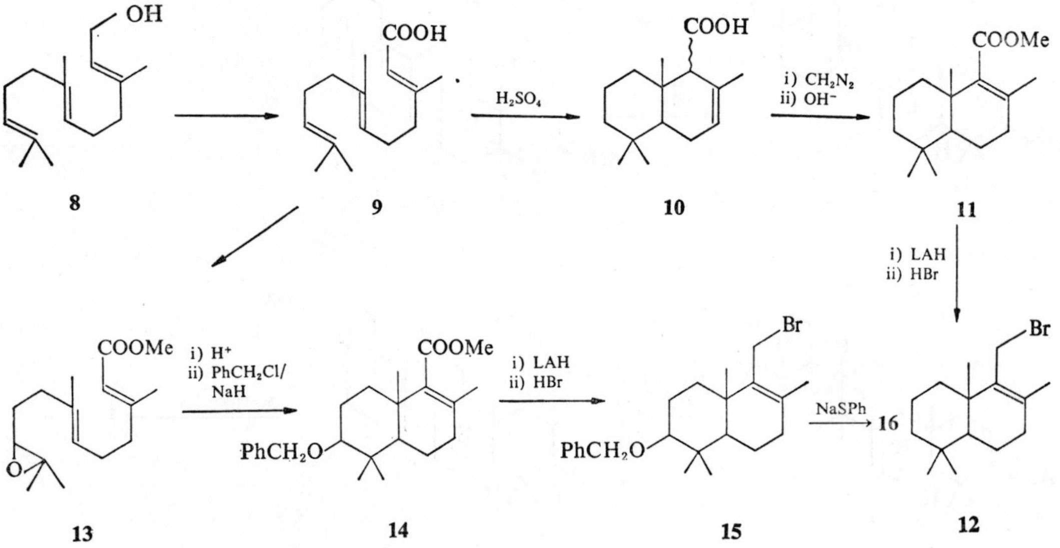

8   9   10   11

13   14   15   12

4) H. Ageta, private communication.
5) J. M. Zander, E. Caspi, G. N. Pandey, C. R. Mitra, *Phytochem.*, **8**, 2267 (1969).
6) E. J. Gallegos, *Anal. Chem.*, **43**, 1151 (1971).
7) E. Caspi, J. M. Zander, J. B. Greig, F. B. Mallory, R. L. Conner, J. R. Landrey, *J. Am. Chem. Soc.*, **90**, 3563 (1968).
8) F. B. Mallory, R. L. Conner, J. R. Landrey, J. M. Zander, J. B. Greig, E. Caspi, *J. Am. Chem. Soc.*, **90**, 3564 (1968).
9) E. Caspi, J. B. Greig, J. M. Zander, A. Mandelbaum, *Chem. Commun.*, 28 (1969).

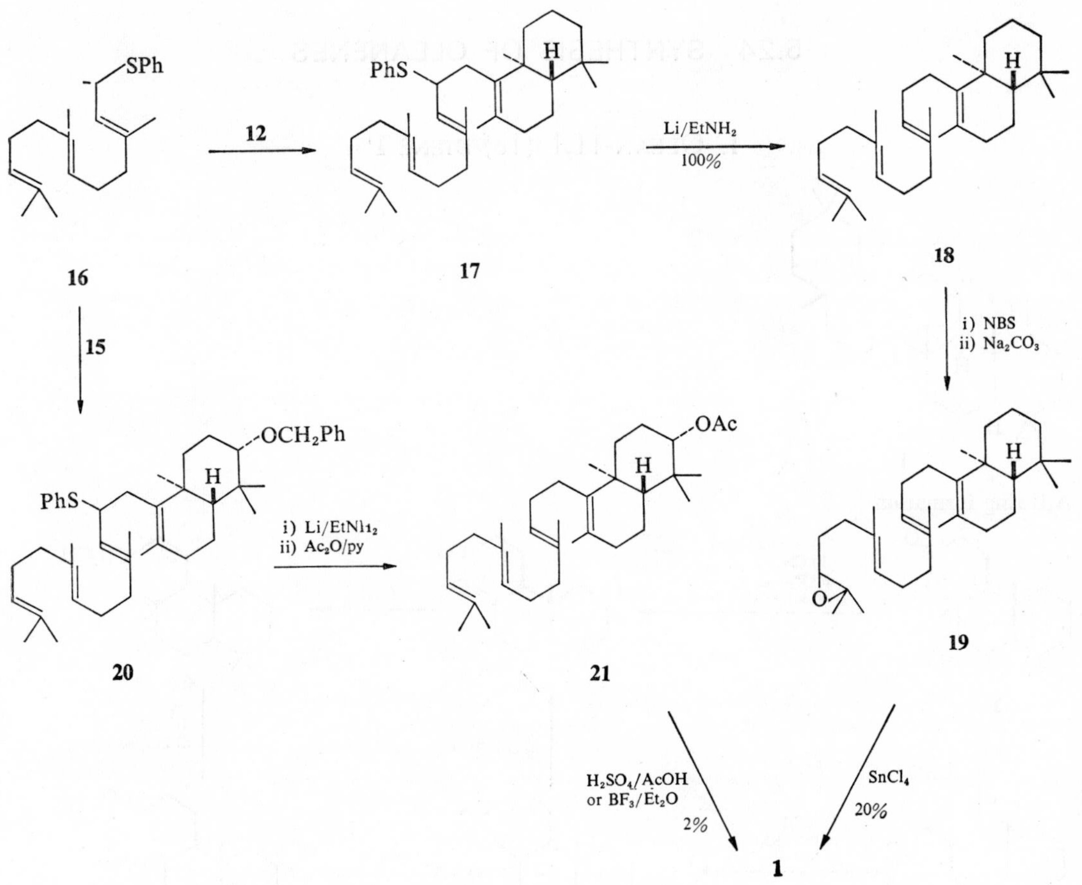
SPh
12
PhS
H
Li/EtNH₂
100%
H
16
17
18
15
i) NBS
ii) Na₂CO₃
OCH₂Ph
OAc
H
H
H
PhS
i) Li/EtNH₂
ii) Ac₂O/py
O
19
20
21
H₂SO₄/AcOH
or BF₃/Et₂O
2%
SnCl₄
20%
1

5.24 SYNTHESIS OF OLEANENES

I. Olean-11,13(18)-diene 1[1)]

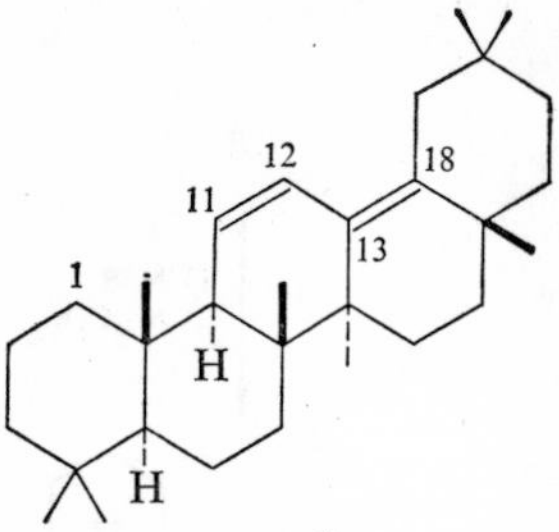
12
18
11
13
1
H
H

1

A,B ring formation

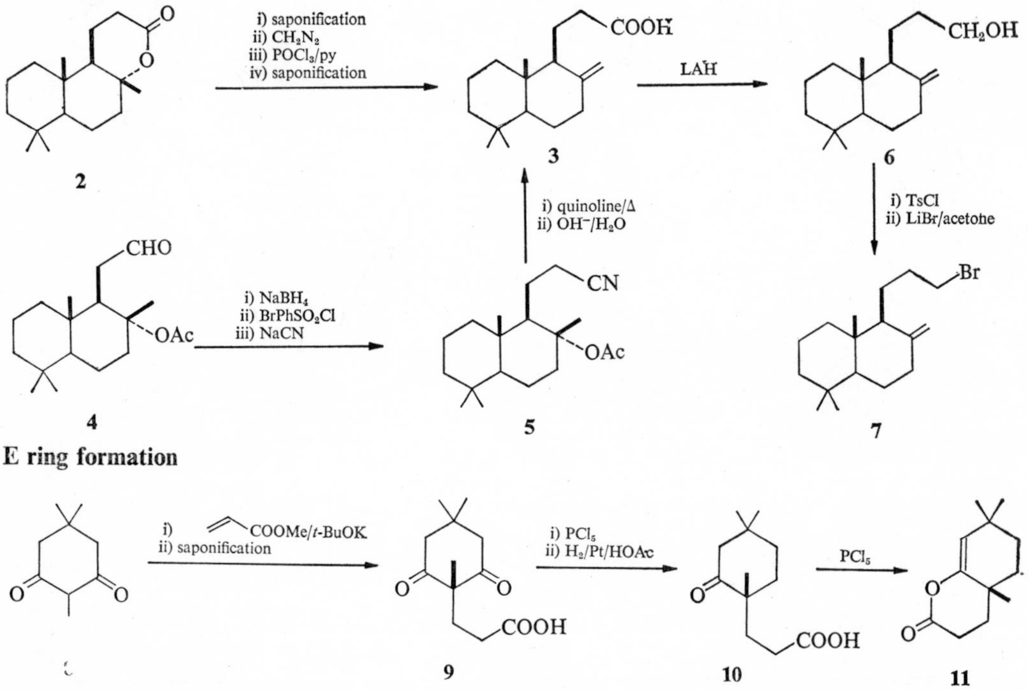
i) saponification
ii) CH_2N_2
iii) $POCl_3$/py
iv) saponification
COOH
LAH
CH_2OH
i) quinoline/Δ
ii) OH^-/H_2O
i) TsCl
ii) LiBr/acetone
CHO
OAc
i) $NaBH_4$
ii) $BrPhSO_2Cl$
iii) NaCN
CN
OAc
Br
2
3
6
4
5
7

E ring formation

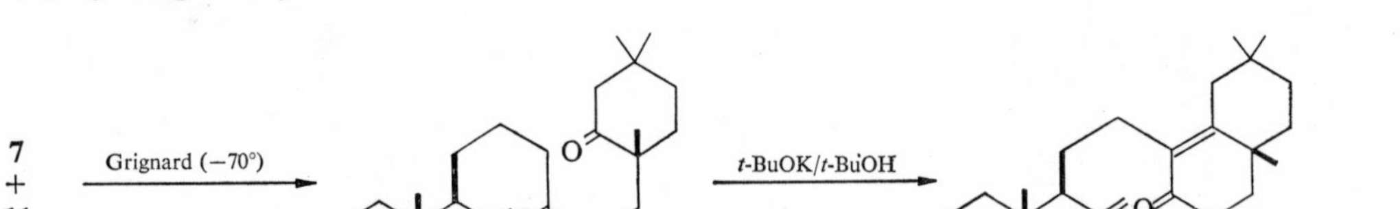
i) COOMe/t-BuOK
ii) saponification
i) PCl_5
ii) H_2/Pt/HOAc
PCl_5
COOH
COOH
9
10
11

A,B/E joining and cyclization

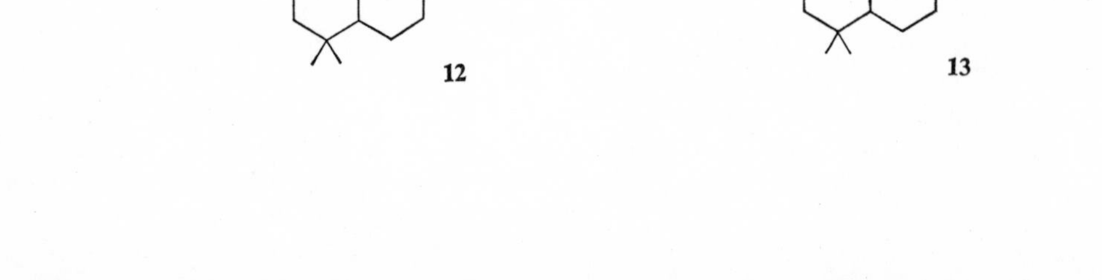
7
+
11
Grignard (−70°)
t-BuOK/t-BuOH
12
13

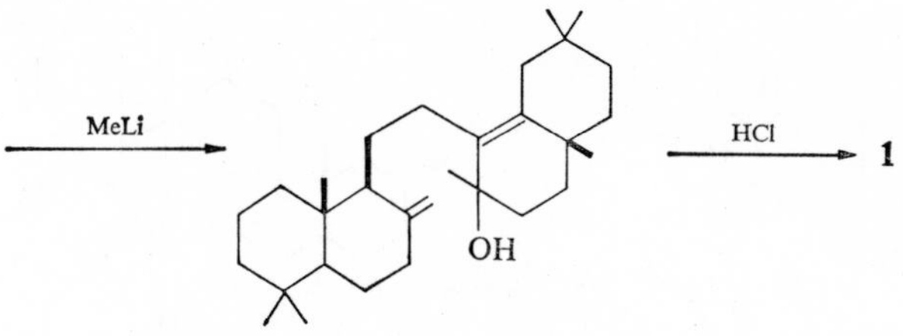

$$\xrightarrow{\text{MeLi}}$$

14

$$\xrightarrow{\text{HCl}} \quad \textbf{1}$$

1) **2** is called ambreinolide, the total synthesis of which has been achieved.[2] In this synthesis the natural product was used.
2) **3** was also obtained from **4** which had previously been synthesized.[3]
3) **8** was previously synthesized.[4]

II. 18α-OLEAN-12-ENE **15** AND OLEAN-13(18)-ENE **16**

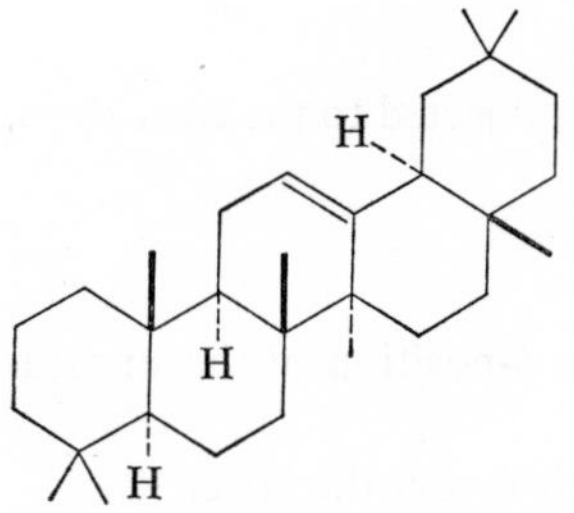

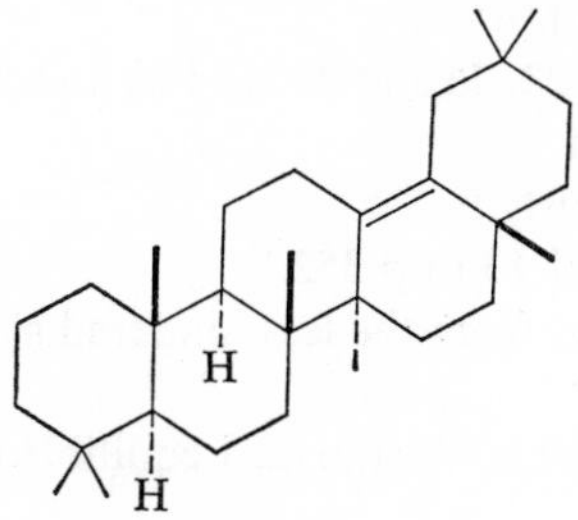

Three synthesis
have been achieved.

18α-olean-12-ene **15**

olean-13(18)-ene **16**

Route I (from natural product)[5]

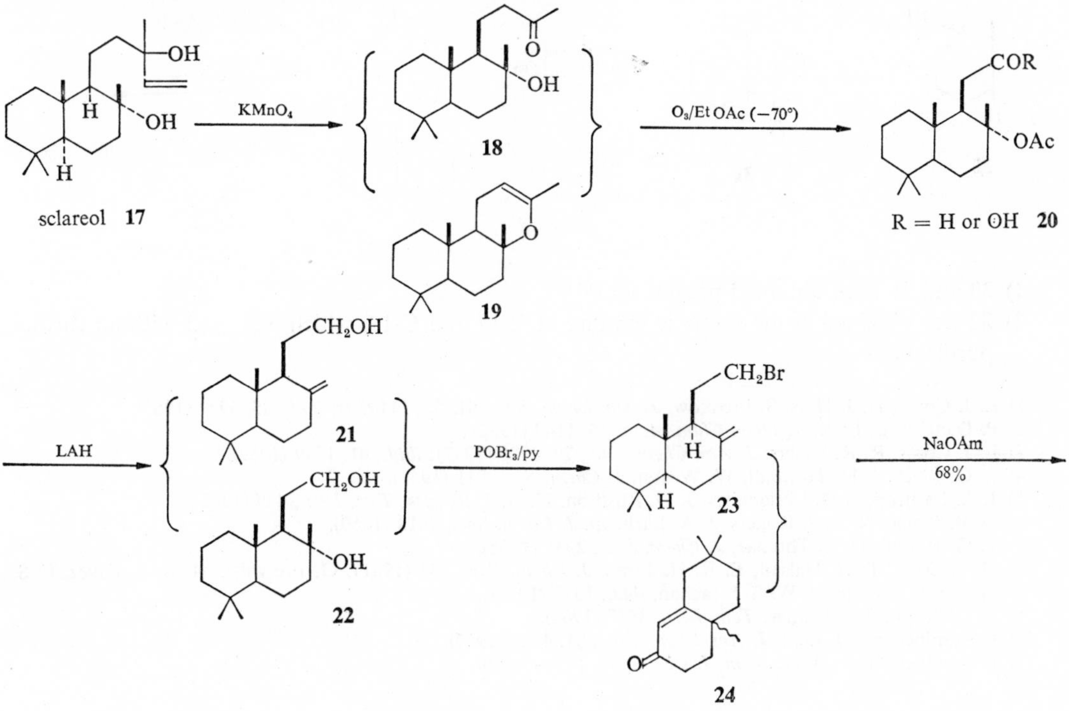

sclareol **17**

$$\xrightarrow{\text{KMnO}_4}$$

18

19

$$\xrightarrow{\text{O}_3/\text{Et OAc} \, (-70°)}$$

R = H or OH **20**

21

22

$$\xrightarrow{\text{LAH}}$$

$$\xrightarrow{\text{POBr}_3/\text{py}}$$

23

$$\xrightarrow[68\%]{\text{NaOAm}}$$

24

$$\xrightarrow[35\%]{\text{AlCl}_3/\text{MeNO}_2/\text{ether (r.t.)}} \quad 15+16 \quad (1:1 \text{ ratio})$$

1) Sclareol **17** has been synthesized.[6,7]

2) The mixtures **18** and **19**, **21** and **22**, **26** and **27**, and **28** and **29** were subjected to the next steps, respectively, without separation.

3) **24** has been synthesized previously.[7]

4) **25** was identical with compound **13** in p 352.[1]

5) **25→26**, **27**: the reduction occurs from the less hindered side at the β-position of the unsaturated ketone.

6) **15** and **16** are known to be in an acid-catalyzed equilibrium which favors the latter.[8]

Route II (total synthesis)[9]

1) **30** and **31** were obtained previously.[10,11]

2) **32** was obtained as an epimeric mixture at C-17 and C-18, which was used without further purification.

1) E. J. Corey, H. J. Hess, S. Proskow, *J. Am. Chem. Soc.*, **81**, 5258 (1959); *ibid.*, **85**, 3979 (1963).
2) P. Dietrich, E. Lederer, *Helv. Chim. Acta*, **35**, 1148 (1952).
3) E. J. Corey, R. R. Sauers, *J. Am. Chem. Soc.*, **79**, 3925 (1957); *ibid.*, **81**, 1749 (1959).
4) E. G. Meck, J. H. Turnbull, W. Wilson, *J. Chem. Soc.*, 811 (1953).
5) J. A. Barltrop, J. D. Littlehailes, J. D. Rushton, N. A. J. Rogers, *Tetr. Lett.*, 429 (1962).
6) D. B. Bigley, N. A. J. Rogers, J. A. Barltrop, *J. Chem. Soc.*, 4613 (1960).
7) T. G. Halsall, D. B. Thomas, *J. Chem. Soc.*, 2431 (1956).
8) G. S. Davy, T. G. Halsall, E. R. H. Jones, *J. Chem. Soc.*, 458 (1951); G. Brownile, M. B. E. Fayez, F. S. Spring, R. Stevenson, W. S. Strachan, *ibid.*, 1377 (1956).
9) E. Ghara, F. Sondheimer, *Tetr. Lett.*, 3887 (1964).
10) F. Sondheimer, D. Elad, *J. Am. Chem. Soc.*, **81**, 4429 (1959).
11) F. Sondheimer, S. Wolfe, *Can. J. Chem.*, **37**, 1870 (1959).

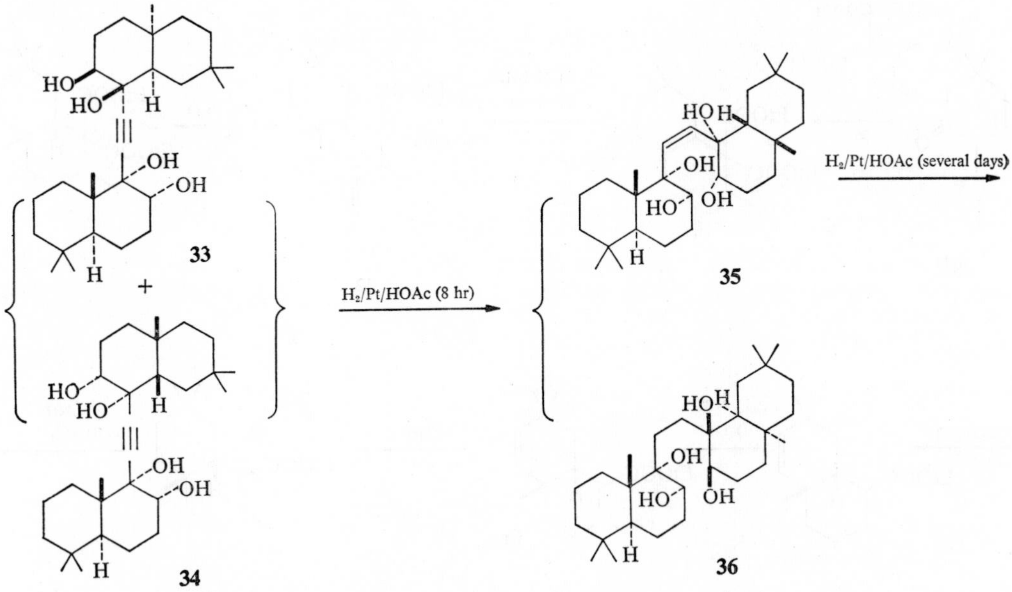

3) **33** and **34** were obtained as an inseparable mixture but **35** could be separated from **36** by chromatography.

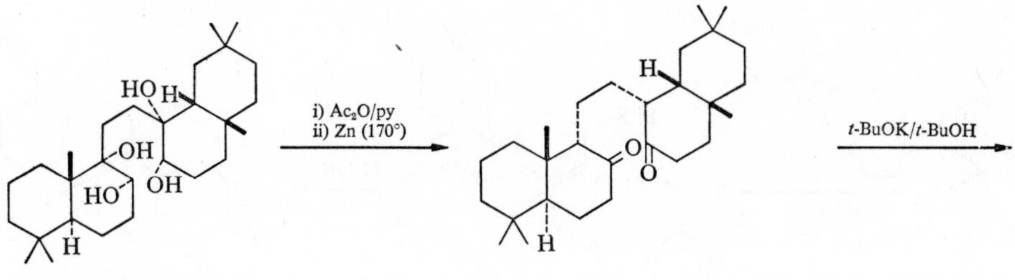

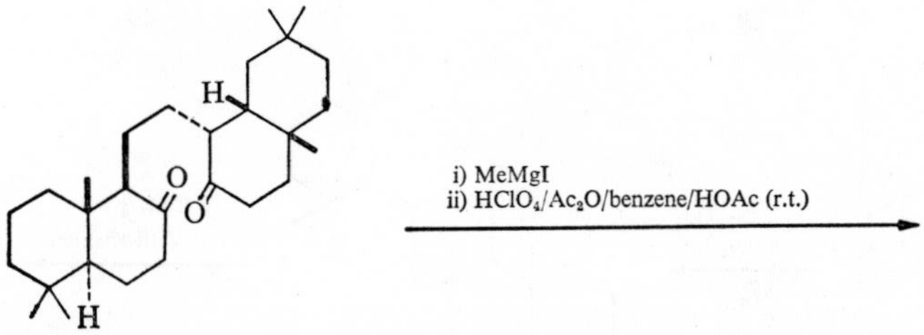

4) **37**→**38**: reaction ii) is the Serini reaction.[12]

5) **38**→**39**: the equilibrium of the base-catalysed reaction is in favor of **39**.

Route III[13]

Route III constitutes a biogenetic-type synthesis of δ-amyrin **51**, a constituent of *Spartium junceum*[14] as well as of the compound **16**.

12) E. Ghara, M. Gibson, F. Sondheimer, *J. Am. Chem. Soc.*, **84**, 2953 (1962).
13) E. E. van Tamelen, M. P. Seiler, W. Wierenga, *J. Am. Chem. Soc.*, **94**, 8229 (1972).
14) O. C. Musgrave, J. Stark, F. S. Spring, *J. Chem. Soc.*, 4393 (1952).

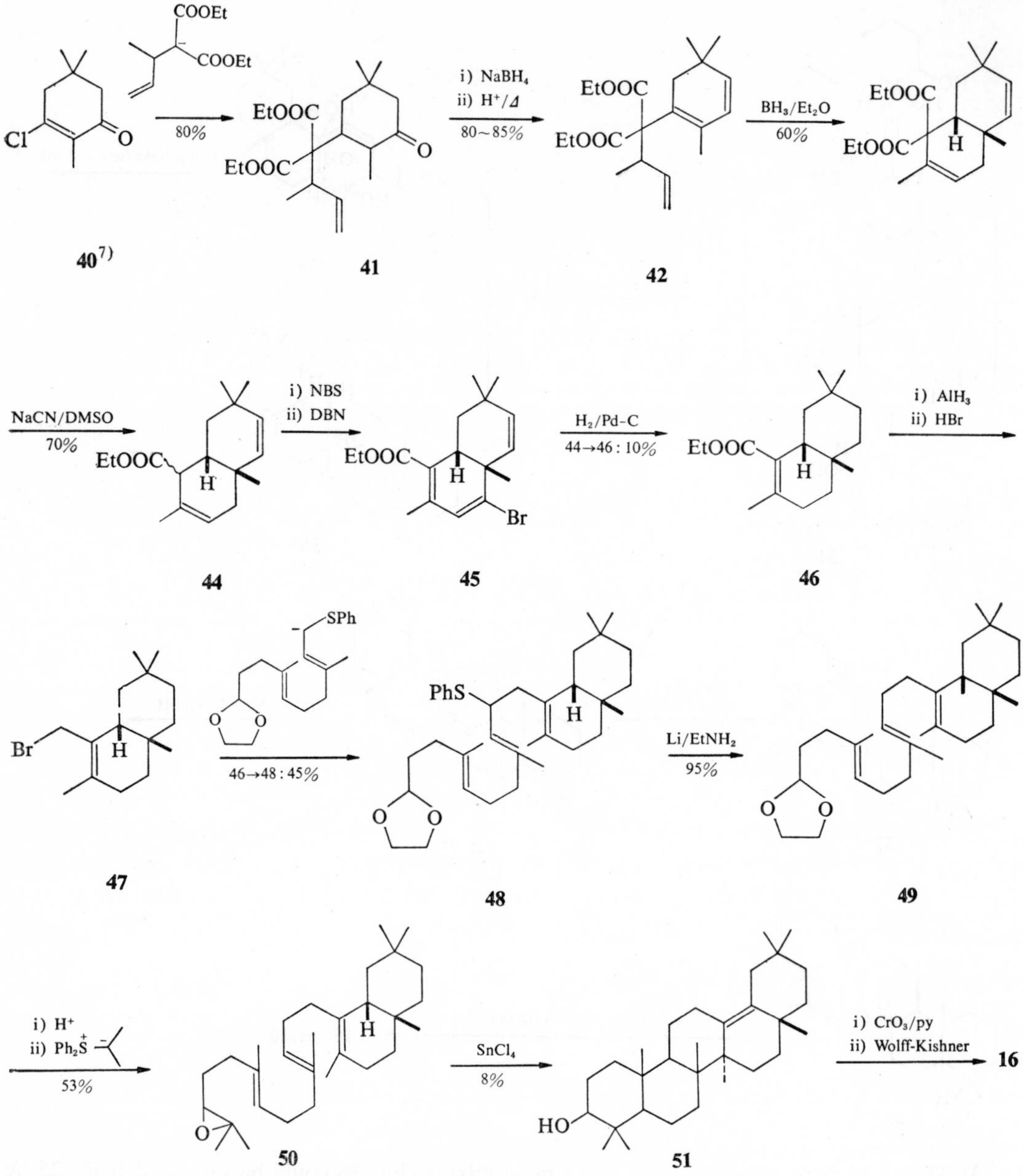

47→48: The reagent was obtained from farnesol.[15]

50→51: The cyclization method was developed with squalene-2,3-epoxide.[16]

51→16: Previously achieved using natural 51.[17] The compound 16 has successfully been converted to β-amyrin[18] and germanicol.[19]

15) E. E. van Tamelen, R. A. Holton, R. E. Hopla, W. E. Konz, *J. Am. Chem. Soc.*, **94**, 8228 (1972).
16) E. E. van Tamelen, J. Willet, M. Schwartz, R. Nodeau, *J. Am. Chem. Soc.*, **88**, 5937 (1966).
17) G. Brownlie, M. B. E. Fayez, F. S. Spring, R. Stevenson, W. S. Strachen, *J. Chem. Soc.*, 1337 (1956).
18) D. H. R. Barton, E. F. Lier, J. F. McGhie, *J. Chem. Soc.* C, 1031 (1968).
19) J. A. Marshall, *Accounts Chem. Res.*, **2**, 33 (1969).

5.25   SYNTHESIS OF β-AMYRIN

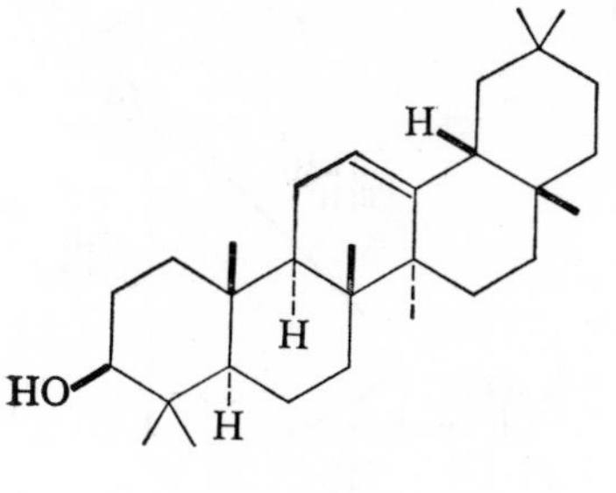

β-amyrin **1**

Formal total synthesis of β-amyrin, one of the most important members of the oleanane series of triterpenoids was accomplished[1] starting from 18α-olean-12-ene.[2]

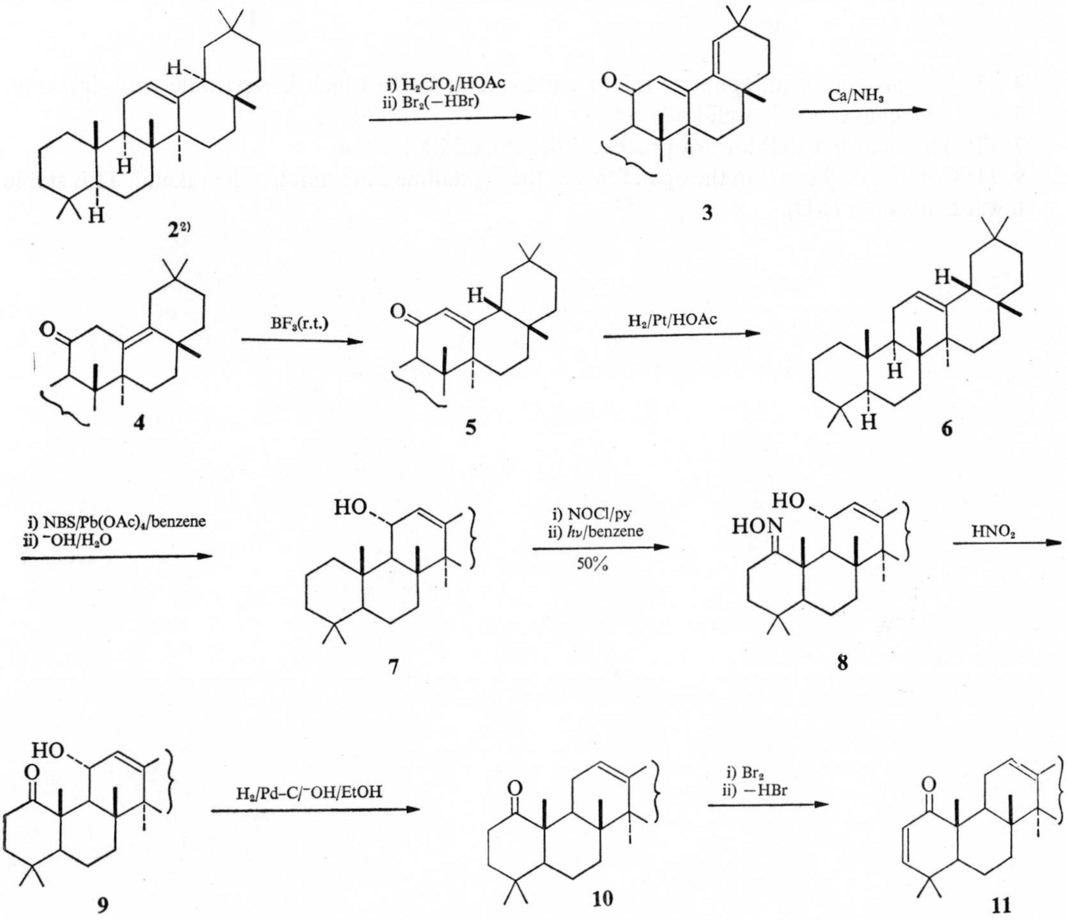

2[2]

i) H$_2$CrO$_4$/HOAc
ii) Br$_2$(−HBr)

3

Ca/NH$_3$

4

BF$_3$(r.t.)

5

H$_2$/Pt/HOAc

6

i) NBS/Pb(OAc)$_4$/benzene
ii) $^-$OH/H$_2$O

7

i) NOCl/py
ii) hν/benzene

50%

8

HNO$_2$

9

H$_2$/Pd–C/$^-$OH/EtOH

10

i) Br$_2$
ii) −HBr

11

1) D. H. R. Barton, E. F. Lier, J. F. McGhie, *J. Chem. Soc.* C, 1031 (1968).
2) J. A. Barltrop, J. D. Littlehailes, J. D. Rushton, N. A. J. Rogers, *Tetr. Lett.*, 429 (1962); E. Ghaha, F. Sondheimer, *ibid.*, 3887 (1964).

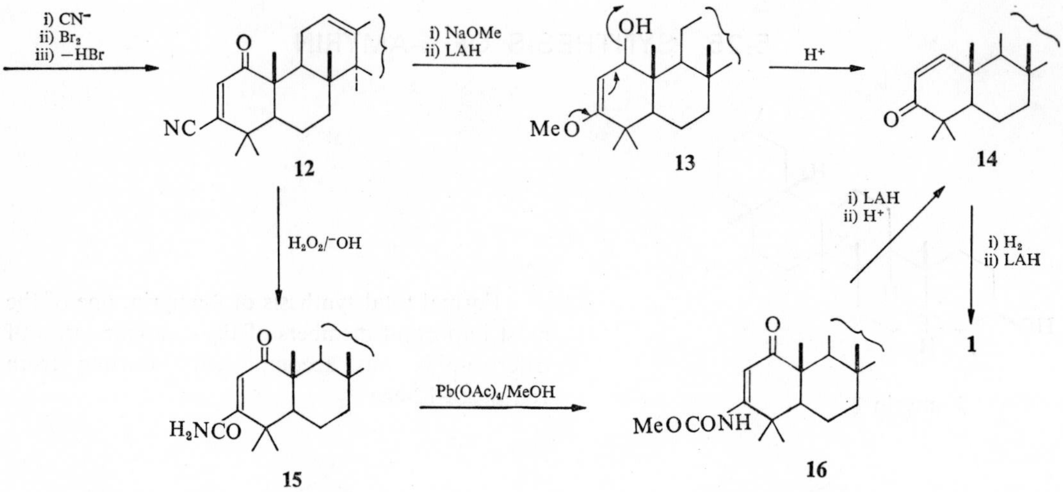
i) CN⁻
ii) Br₂
iii) −HBr
O
NC
12
i) NaOMe
ii) LAH
OH
MeO
13
H⁺
O
14
H₂O₂/⁻OH
i) LAH
ii) H⁺
i) H₂
ii) LAH
1
O
H₂NCO
15
Pb(OAc)₄/MeOH
O
MeOCONH
16

4→5: The product is an epimeric mixture (ca. 1:1 at C-18) which is separable with difficulty.
5→6: This conversion is well-known.[3]
7→8: This is also a well-known process,[4] developed by Barton.
9, 11: Compound **9** exists in the open form in the crystalline state, not as a hemiketal. **11** is stable towards alkaline H_2O_2.

---

3) L. Ruzicka, H. Leuenberger, H. Schellenbeg, *Helv. Chim. Acta*, **20**, 1271 (1937).
4) D. H. R. Barton, J. M. Beaton, L. E. Geller, M. M. Pecket, *J. Am. Chem. Soc.*, **82**, 2640 (1960).

5.26 SYNTHESIS OF OLEANOLIC ACID

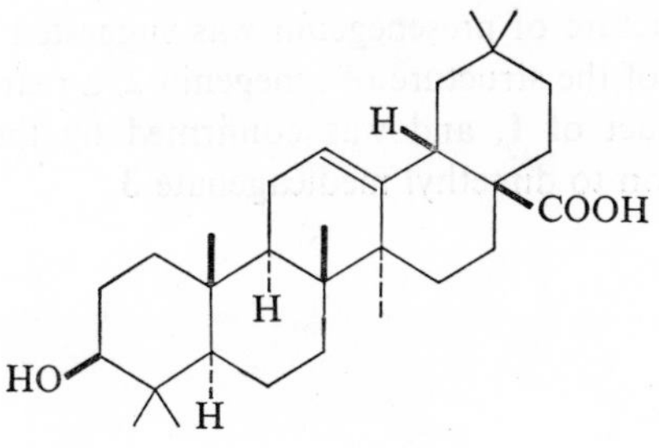

The synthesis of oleanolic acid **1** was recently achieved[1] from β-amyrin **2**. As **2** has been totally synthesized,[2] the process constitutes the formal total synthesis of **1**.

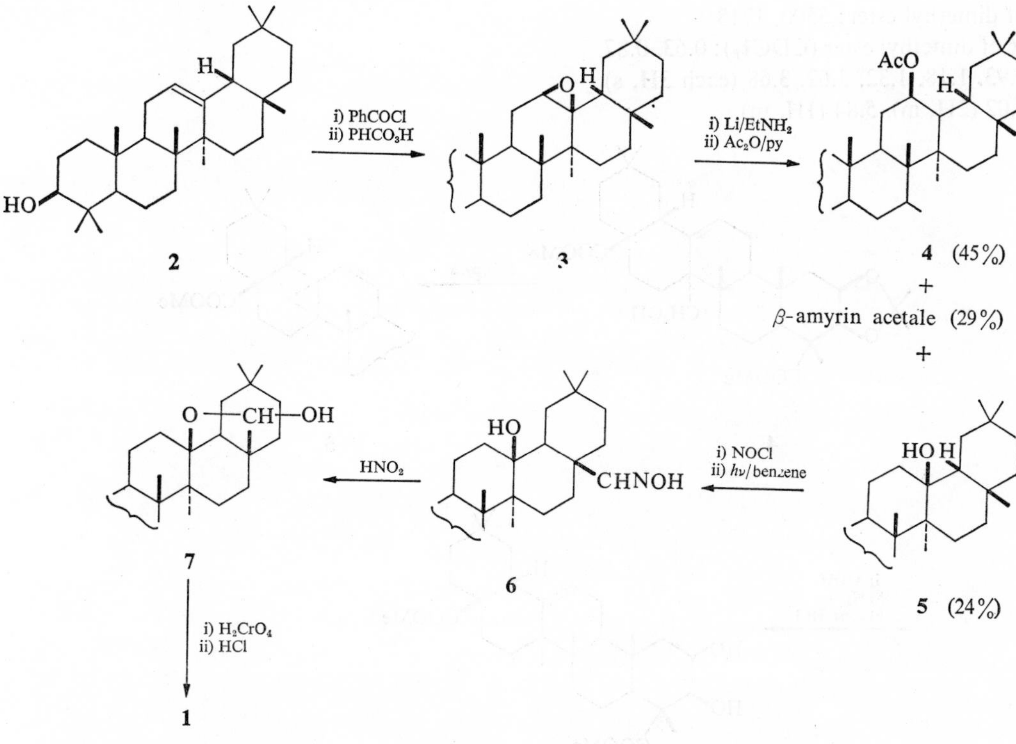

oleanolic acid **1**

5→6: This is a general photo process to functionalize a carbon atom.[3]

1) R. B. Boar, D. C. Knight, J. F. McGhie, D. H. R. Barton, *J. Chem. Soc.* C, 678 (1970).
2) D. H. R. Barton, E. F. Lier, J. F. McGhie, *J. Chem. Soc.* C, 1031 (1968).
3) M. Akhtar, *Adv. Photochem.*, **2**, 263 (1964); D. H. R. Barton, D. Kumari, P. Welzel, L. J. Danks, J. F. McGhie, *J. Chem. Soc.* C, 332 (1969).

5.27   STRUCTURE AND REARRANGEMENT OF PRESENEGENIN

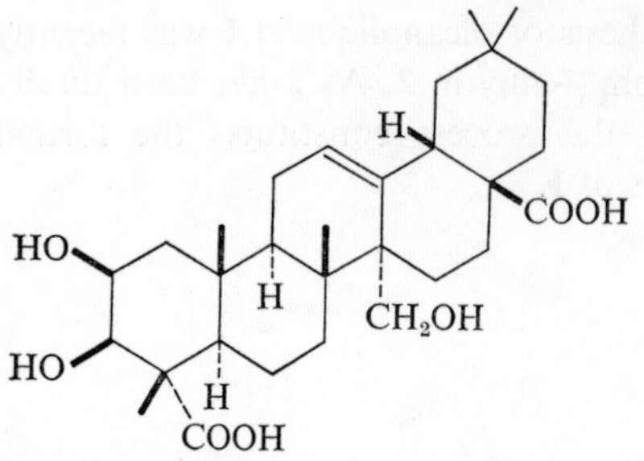

The structure of presenegenin was suggested[1] on the basis of the structure of senegenin 2, a rearranged product of 1, and was confirmed by the transformation to dimethyl medicagenate 3.

presenegenin 1

mp: 310–311°

ir of dimethyl ester: 3500, 1718

nmr of dimethyl ester (CDCl$_3$): 0.63, 0.87, 0.93, 1.18, 1.32, 3.62, 3.68 (each 3H, s), 4.02 (2H, m), 5.84 (1H, m)

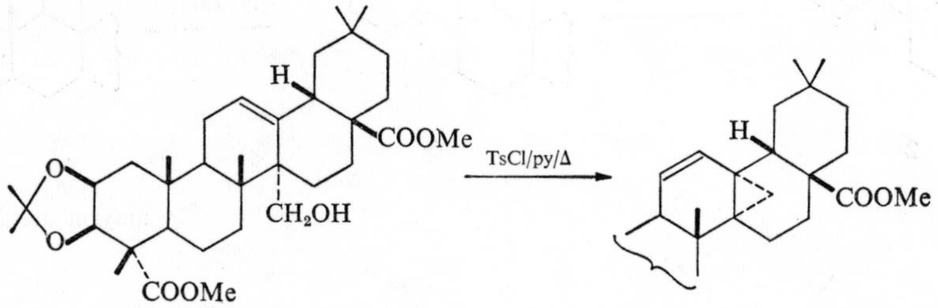

4   5

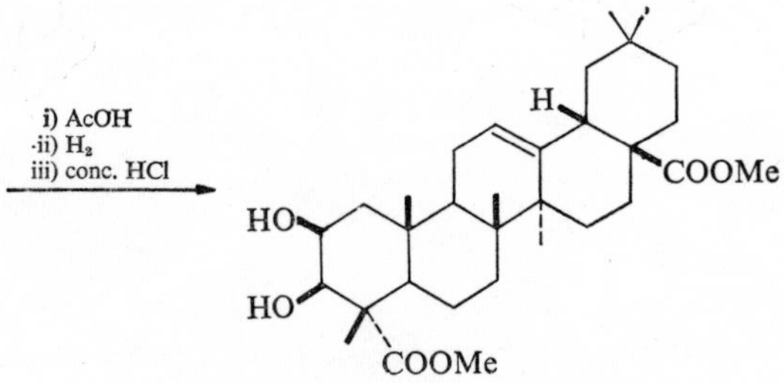

3

1 or its derivatives undergo[1-3] interesting homoallyl-cyclopropyl carbinyl cation-type rearrangements[4] which made the chemistry of *senega* sapogenin more complex. In fact, presenegenin is the only genuine sapogenin obtained from the root of *Polygala senega*.[1] How-

1) J. J. Dugan, P. de Mayo, *Can. J. Chem.*, **43**, 2033 (1965).
2) Y. Shimizu, S. W. Pelletier, *Chem. Ind.*, 2098 (1965).
3) Y. Shimizu, S. W. Pelletier, *J. Am. Chem. Soc.*, **87**, 2065 (1965); S. W. Pelletier, Y. Shimizu, S. Nakamura, *Chem. Commun.*, 727 (1966).
4) R. Breslow, *Molecular Rearrangement* (ed. P. de Mayo), part 1, 259, Interscience, 1963.

ever, depending on the conditions of hydrolysis of the saponin, skeletal rearrangement occurs and various artifacts such as senegenin 2[5], hydroxysenegenin 6[3] or polygalic acid (senegenic acid) 7[1,6] have been isolated. Their structures were determined independently.

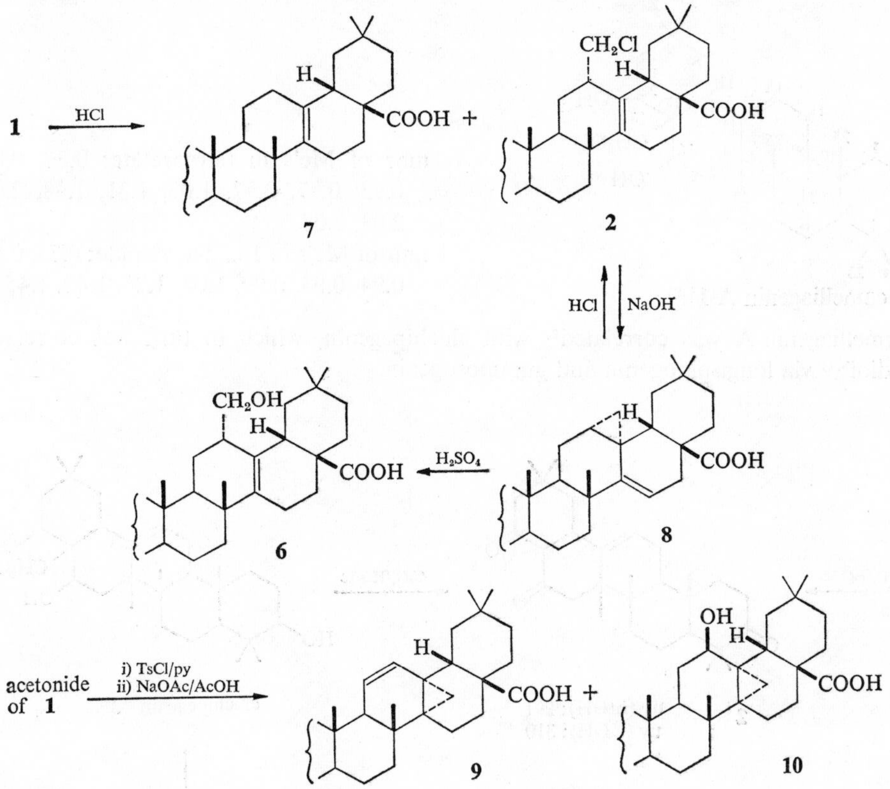

Remarks

Related compounds, polygalacic acid 11 and bredemolic acid 12, were isolated from *Polygala paenea* and *Bredemeyera floribunda*, respectively.

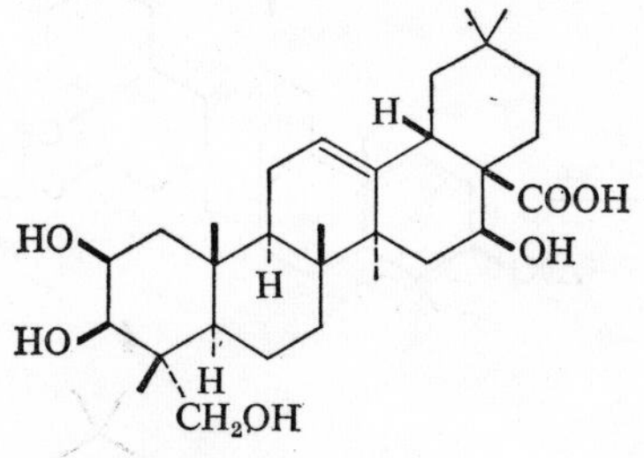

polygalacic acid[7] 11

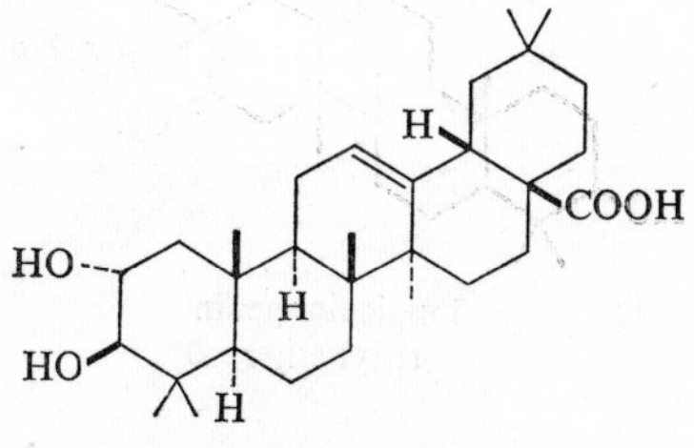

bredemolic acid[8] 12

5) M. Shamma, L. D. Reiff, *Chem. Ind.*, 1272 (1960); J. J. Dugan, P. de Mayo, A. N. Starratt, *Can. J. Chem.*, **42**, 491 (1964); *Proc. Chem. Soc.*, 264 (1964).
6) J. J. Dugan, P. de Mayo, A. N. Starratt, *Tetr. Lett.*, 2564 (1964); S. W. Pelletier, N. Adityachaudhung, M. Tomasz, J. J. Rynold, R. Mechoulan, *Tetr. Lett.*, 3065 (1964); *J. Org. Chem.*, **30**, 4234 (1965).
7) J. Rondest, J. Polonsky, *Bull. Soc. Chim. France*, 1253 (1963).
8) R. Tschesche, E. Henckel, G. Snatzke, *Ann. Chem.*, **676**, 175 (1964).

5.28 CAMELLIAGENIN A

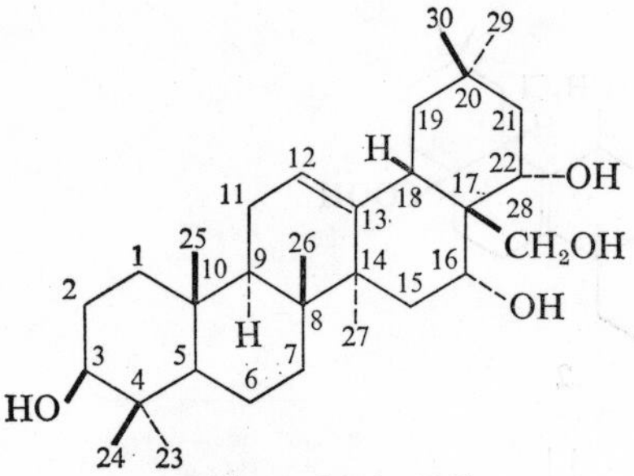

nmr of Me's in tetraacetate: 0.86, 0.86,
0.92, 0.97, 0.97, 1.03, 1.31, 1.98, 2.03,
2.03, 2.08
nmr of Me's in 16,22-acetonide: 0.78, 0.89,
0.94, 0.94, 0.99, 1.07, 1.28, 1.40, 1.45

camelliagenin A 1[1,2]

Camelliagenin A was correlated[2] with chichipegenin, which in turn was correlated with maniladiol[3,4] via longispinogenin and gummosogenin.

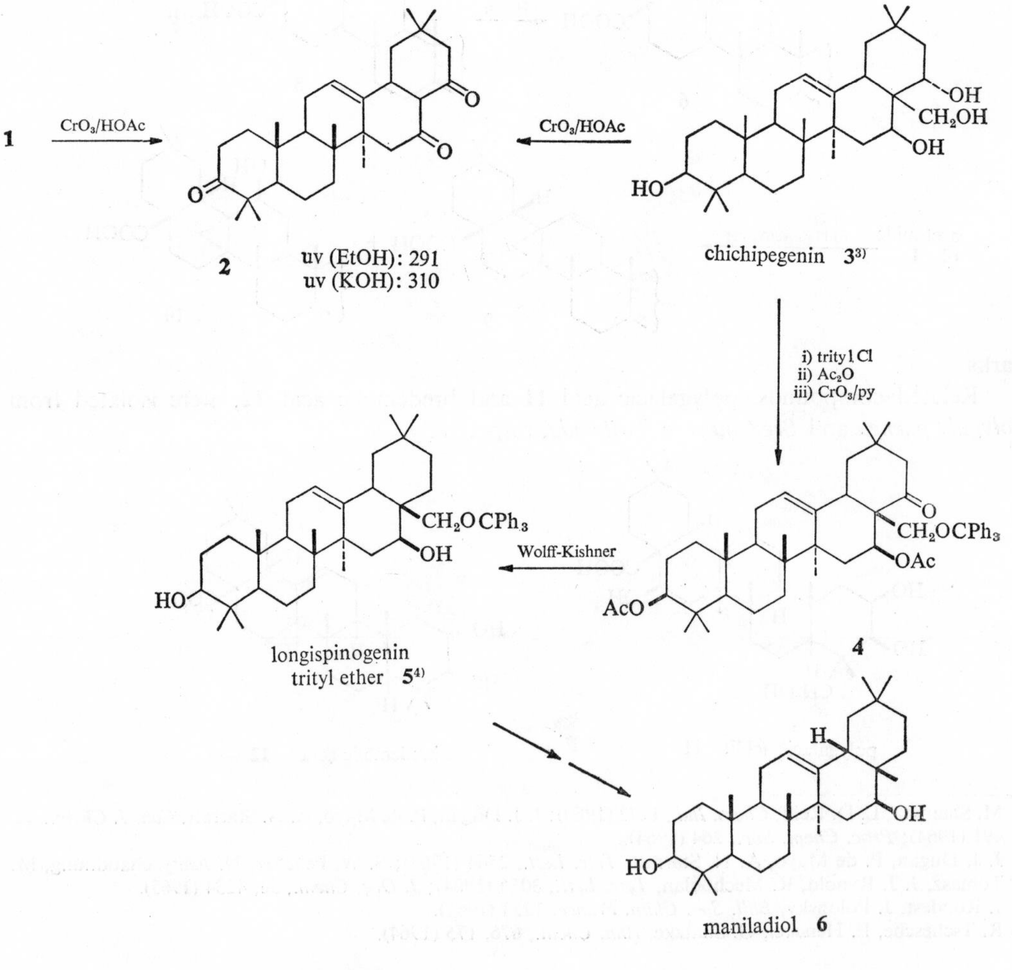

uv (EtOH): 291
uv (KOH): 310

chichipegenin 3[3]

i) trityl Cl
ii) Ac₂O
iii) CrO₃/py

Wolff-Kishner

longispinogenin
trityl ether 5[4]

maniladiol 6

Related compounds

Camellia japonica is the source of two more sapogenols, camelliagenin B and C, which were correlated with **1**.[1,2]

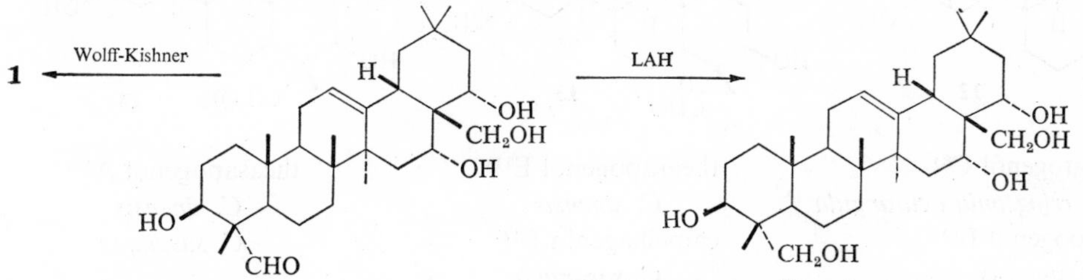

camelliagenin B   **7**

camelliagenin C   **8**

Camelliagenin A was isolated from *Camellia japonica*, *C. sasanqua* and *C. sinensis*. These plants also produce the sapogenols **9** through **18**, which are structurally closely related to each other and have been chemically correlated. Since the same compounds appear in other species under different names, the names and sources are shown below together with the structures. Preference of the first name listed for each compound was proposed.[5]

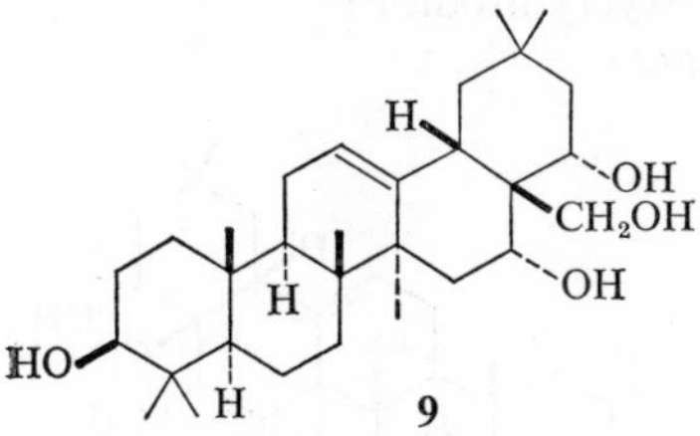

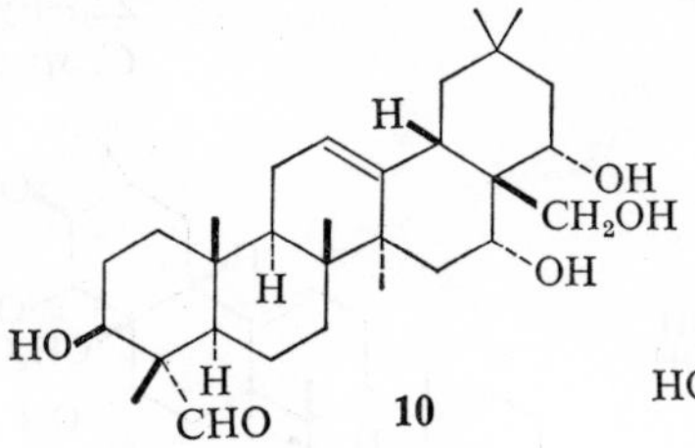

dihydropriverogenin A[5,6]
  Primulaceous plants
camelliagenin A[1,2]
  C. japonica
  C. sasanqua
theasapogenol D[5]
  C. sinensis
A₂-barrigenol[7]
  Barringtonia asiatica

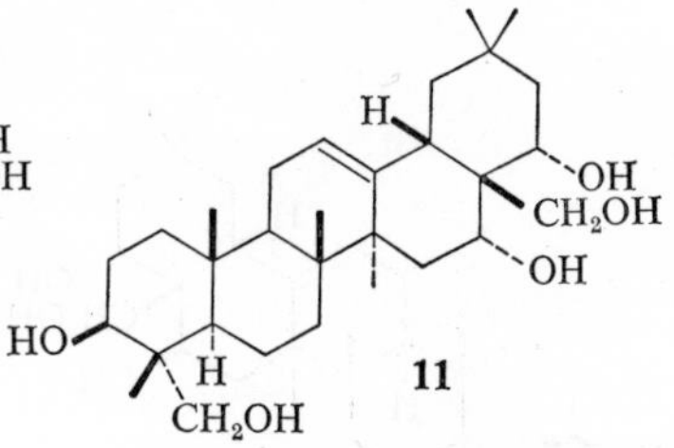

camelliagenin B[1,2]
  C. japonica
  C. sasanqua
  C. sinensis

camelliagenin C[1,2]
  C. japonica
  C. sasanqua
theasapogenol C[5]
  C. sinensis

1) S. Ito, M. Kodama, M. Konoike, *Tetr. Lett.*, 591 (1967).
2) H. Itokawa, N. Sawada, T. Murakami, *Tetr. Lett.*, 597 (1967).
3) A. Sandoval, A. Manjarresm, P. R. Leeming, G. H. Thomas, C. Djerassi, *J. Am. Chem. Soc.*, **79**, 4468 (1957).
4) C. Djerassi, L. E. Geller, A. J. Lemin, *Chem. Ind.*, 161 (1954); *J. Am. Chem. Soc.*, **76**, 4089 (1954).
5) I. Yoshioka, T. Nishimura, N. Watani, I. Kitagawa, *Tetr. Lett.*, 5343 (1967).
6) M. Kodama, S. Ito, *Chem. Ind.*, 1647 (1967).
7) S. Ito, T. Ogino, H. Sugiyama, M. Kodama, *Tetr. Lett.*, 2289 (1967).

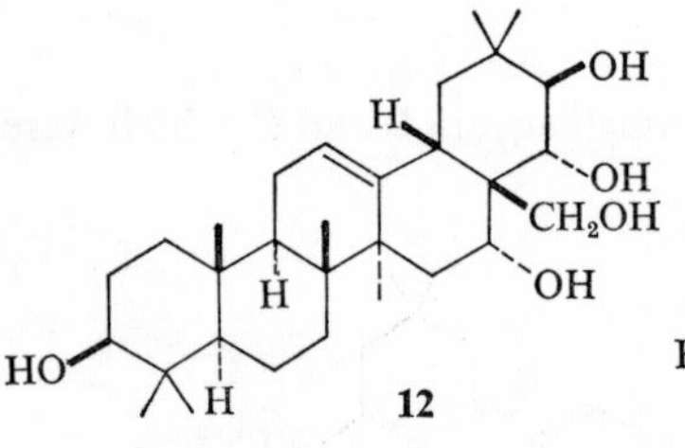

12

barringtogenol C[8)]
  Barringtonia acutangula
theasapogenol B[8)]
  C. sinensis
  C. sasanqua
aescinidin[9)]
  Aesculus hippocastanum
  A. turbinata
jegosapogenol[10)]
  Styrax japonica
R$_2$-barrigenol[7)]
  Barringtonia racemosa

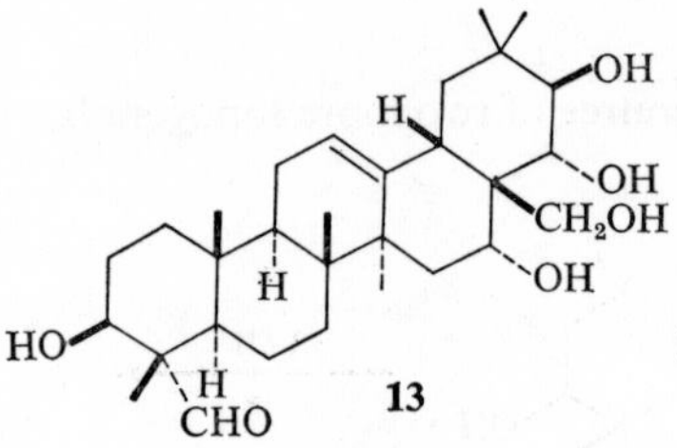

13

theasapogenol E[11)]
  C. sinensis
camelliagenin E[12)]
  C. sasanqua

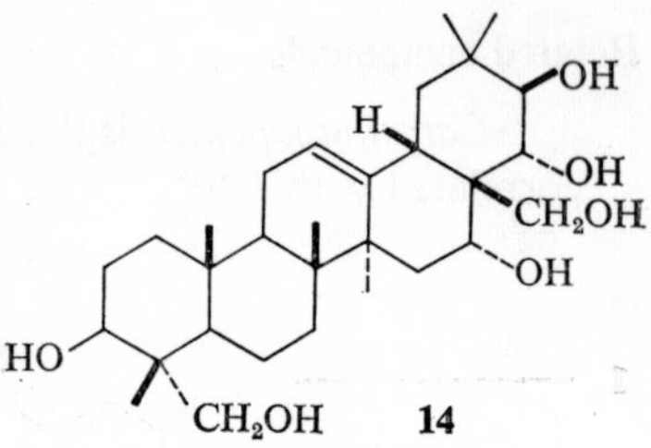

14

theasapogenol A[13)]
  C. sinensis
  C. sasanqua

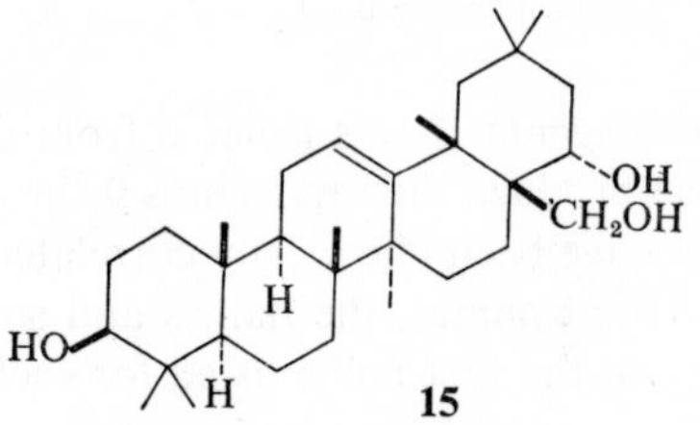

15

22α-hydroxyerythrodiol[14)]
  C. sasanqua

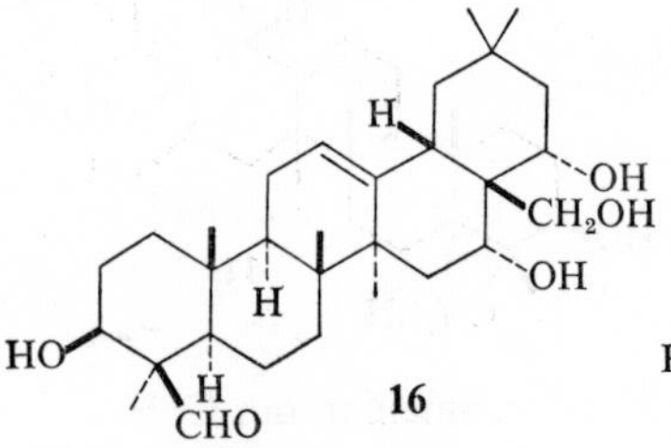

16

camelliagenin D[12)]
  C. sinensis
  C. sasanqua

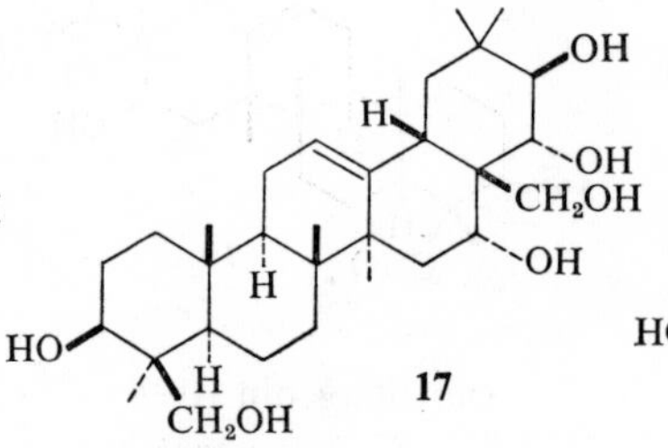

17

protoaescigenin[15)]
  A. turbinate

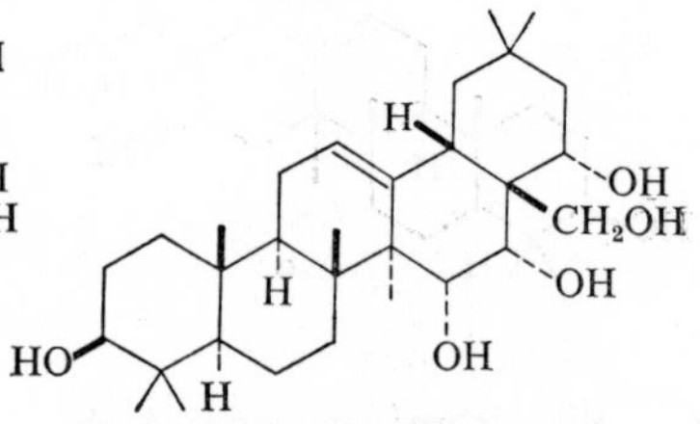

18

A$_1$-barrigenol[7,16)]
  C. sasanqua

8) I. Yoshioka, T. Nishimura, A. Matsuda, I. Kitagawa, *Tetr. Lett.*, 5973 (1966); *Chem. Pharm. Bull.*, **18**, 1610 (1970).
9) R. Tschesche, G. Wulff, *Tetr. Lett.*, 1569 (1965).
10) T. Nakano, M. Hasegawa, T. Fukumaru, S. Tobinaga, C. Djerassi, L. J. Durham, H. Budzikiewicz. *Tetr. Lett.*, 365 (1967).
11) I. Yoshioka, A. Matsuda, T. Nishimura, I. Kitagawa, *Chem. Ind.*, 2202 (1966); *Chem. Pharm. Bull.*, **19**, 1186 (1971).
12) S. Ito, T. Ogino, *Tetr. Lett.*, 1127 (1967).
13) I. Yoshioka, T. Nishimura, A. Matsuda, I. Kitagawa, *Tetr. Lett.*, 5979 (1966); *Chem. Pharm. Bull.*, **18**, 1621 (1970).
14) I. Yoshioka, R. Takeda, A. Matsuda, I. Kitagawa, *Chem. Pharm. Bull.*, **20**, 1237 (1972).
15) I. Yoshioka, T. Nishimura, A. Matsuda, K. Imai, I. Kitagawa, *Tetr. Lett.*, 637 (1967); W. Hoppe, A. Giesen. N. Brodher, R. Tschesche, G. Wulff, *Angew. Chem. Intern. Ed.*, **7**, 547 (1968).
16) S. G. Egrinton, D. E. White, M. W. Fuller, *Tetr. Lett.*, 1289 (1967); T. Ogino, T. Hayasaka, S. Ito, *Chem. Pharm. Bull.*, **16**, 1132 (1968).

5.29 SUBSTITUENT EFFECTS ON THE METHYL SIGNALS IN THE NMR SPECTRUM OF OLEAN-12-ENE

olean-12-ene

Modification of substituents in olean-12-enes is known to be accompanied by systematic changes in the nmr chemical shifts of the Me groups. This substituent effect has been evaluated[1-4] as in the case of steroids.[5] The tables below show selected values which should have potential use in the structural determination of unknown triterpenoids.

Basic chemical shift of the Me groups (ppm from TMS) (Table 1)

Substituent	23-Me	24-Me	25-Me	26-Me	27-Me	28-Me	29-Me	30-Me	ref.
none	0.88	0.84	0.94	0.98	1.15	0.84	0.88	0.88	(1)
3β-OH	0.99	0.80	0.95	1.01	1.15	0.85	0.88	0.88	(1)
	0.99	0.79	0.94	0.97	1.13	0.83	0.87	0.87	(2)
3β-OAc	0.88	0.88	0.96–0.98	0.96–0.98	1.14	0.84	0.88	0.88	(1)
	0.86	0.86	0.96	0.96	1.13	0.83	0.86	0.86	(2)
3β-OBz	0.96	1.01–1.03	1.01–1.03	1.00	1.16	0.85	0.88	0.88	(1)
3-oxo	1.10	1.03	1.07	1.08	1.15	0.85	0.88	0.88	(1)
28-COOH	0.87	0.82	0.91–0.94	0.76	1.15	COOH	0.91–0.94	0.91–0.94	(1)
28-COOMe	0.87	0.82	0.91–0.93	0.73	1.14	COOMe	0.91–0.94	0.91–0.94	(1)

28-Me and 30-Me in olean-12-ene[6] and 23-Me and 24-Me in 3β-hydroxyolean-12-ene[7] have been unambiguously assigned by using selectively deuterated compounds. Other methyl groups are assigned from extensive analysis of many other compounds.

Effects of substituents on the chemical shift of the Me groups (ppm from TMS) (Table 2)

The values shown for each substituent in this table should be added to the figures given in Table 1 for the corresponding "basic structure" to obtain the chemical shifts for the Me groups of disubstituted olean-12-enes.

1) B. Tursch, R. Savoir, R. Ottinger, G. Chiurdoglu, *Tetr. Lett.*, 539 (1967).
2) S. Ito, M. Kodama, M. Sunagawa, T. Oba, H. Hikino, *Tetr. Lett.*, 2905 (1969).
3) H. T. Cheung, D. G. Williamson, *Tetr.* **25**, 119 (1969).
4) G. S. Ricca, G. Russo, *Gazz. Chim. Ital.*, **98**, 602 (1968).
5) R. F. Zucker, *Helv. Chim. Acta*, **46**, 2054 (1963).
6) J. Karliner, C. Djerassi, *J. Org. Chem.*, **31**, 1945 (1966).
7) S. Ito, M. Kodama, M. Sunagawa, *Tetr. Lett.*, 3989 (1967).

"basic structure"	second substituent	23-Me	24-Me	25-Me	26-Me	27-Me	28-Me	29-Me	30-Me
3β-OH	2α-OH	0.03	0.03	0.03	−0.01	−0.01	—	−0.02	−0.02
	2α-OAc	0.03	0.08	0.13	−0.03	−0.03	—	−0.02	−0.02
	2β-OH	?	0.22	0.32	0.02	0	—	0.01	0.01
	6α-OH	0.31	0.19	0.03	0.03	0.01	—	−0.01	−0.01
	7β-OH	0	0.01	0	0.05	0.11	?	?	—
	16α-OH	0	0	−0.03	−0.03	0.17	—	0.05	0.05
	16β-OH	0	0	−0.01	0.05	0.03	—	0.03	0.03
	19α-OH	−0.01	0	−0.01	0	0.11	—	0.04	0.04
	19β-OH	?	0.01	0.02	0.03	0.03	—	0.01	0.01
	21β-OH	−0.01	0	−0.01	−0.01	−0.01	—	0.05	−0.01
	22α-OH	−0.01	0	−0.01	0.01	0.02	0.15	0.06	0.06
	22β-OH	0	0	0.02	0.02	−0.01	0.08 or 0.05	0.01 or 0.04	0.17
	23-OH	—	0.08	0.01	−0.03	−0.01	—	−0.04	−0.03
	28-OH	0.01	0	0.01	−0.02	0.05	—	0	0
	28-COOMe	−0.01	−0.01	−0.02	−0.26	0	—	0.04	0.05
	29-OH	0	−0.01	0	0.01	0	—	—	0.03
3β-OAc	2α-OAc	0.04	0.04	0.13	−0.01	−0.01	—	−0.02	−0.02
	2β-OAc	?	0.18	0.26	0.03	0	—	0	0
	6α-OAc	0.08	0.21	0.12	0.12	0.03	—	0.01	0.01
	7β-OAc	0.01	0.03	0.04	0.13	0.08	(−0.04)	(0.27)	—
	16α-OAc	−0.01	−0.01	0.01	−0.04	0.11	—	0.05	0.05
	16β-OAc	−0.01	−0.01	0	0.05	0.10	—	0.03	0.03
		0.01		0.03	0.01	0.13	—	0.08, 0.02	
	21β-OAc	0	0	0	−0.01	−0.01	—	−0.07	0.05
	22α-OAc	0.01	0.01	0.01	0.01	0.04	0.04	0.06	0.11
	22β-OAc	0.01	0.01	0.02	0.02	0.02	0.15 or −0.01	−0.04 or 0.12	0.01
	23-OAc	—	−0.03	0.03	0	−0.02	—	0.02	0.02
			−0.03	0.03	0.01	−0.03	—	−0.02	−0.02
	24-OAc	0.16	—	−0.01	−0.01	−0.01	—	−0.01	0
		0.12	—	0.05	−0.01	−0.01	—	0	0
	28-OAc	0.01	0.01	0	0	0.04	—	0.01	0.01
	29-OAc	0	0	0.02	−0.01	−0.01	?	—	0.07
	30-OAc	0	0	0	0.01	0	—	0.04	—
	19-oxo	0	0	0.03	0.02	−0.14	—	0.16, 0.26	
3-oxo	28-COOMe	0	0.01–0.03	−0.03–0	−0.28	0.01	—	0.03	0.03
	28-COOH	−0.01	0–0.03	−0.03–−0.01	−0.27	0	—	0.03–0.06	0.03–0.06
	28-CH₂OH	−0.01	−0.03	−0.01	−0.02	0.03	—	0	0
	29-COOMe	0	0	0	−0.02	0	0.02	—	0.32
28-COOH	3β-OAc	0	0.05	−0.03–+0.04	0	0	—	−0.03–+0.04	−0.03–+0.04
	3-oxo	0.06	0.22–0.24	0.09–0.15	0.06	0	—	−0.03–+0.03	−0.03–+0.03
28-COOMe	3β-OH	0.13	−0.03	−0.02–+0.04	0.01	0.01	—	−0.03–+0.04	−0.03–+0.04
	3β-OAc	0	0.05	−0.02–+0.05	0.01	0	—	−0.03–+0.05	−0.03–+0.05
	3-oxo	0.23	0.23–0.25	0.11–0.16	0.04	0.01	—	−0.03–+0.01	−0.03–+0.01

Assignment of methyl signals has also been carried out for other series of triterpenoids, such as lupane,[3,8] dammarane,[9] lanostane,[10] hopane,[11] euphane,[12] isoeuphane,[12] cucurbitane[13] and ursane.[3,14]

8) J. M. Lehn, G. Ourisson, *Bull. Soc. Chim. France*, 1137 (1962); J. M. Lehn, A. Vystrcil, *Tetr.* **19**, 1733 (1963).
9) L. M. Lehn, *Bull. Soc. Chim. France*, 1832 (1962).
10) A. I. Cohen, D. Rosenthal, G. W. Krakower, J. Fried, *Tetr.* **21**, 3171 (1965).
11) S. Huneck, J. M. Lehn, *Bull. Chim. Soc. France*, 1702 (1963).
12) D. Lavie, Y. Shvo, E. Glotler, *Tetr.* **19**, 2255 (1963).
13) D. Lavie, B. S. Benjaminov, Y. Shvo, *Tetr.* **20**, 2585 (1964).
14) R. Savoir, R. Ottinger, B. Tursch, G. Chiurdoglu, *Bull. Soc. Chim. Belges*, **76**, 371 (1967).

5.30 SYNTHESIS OF GERMANICOL

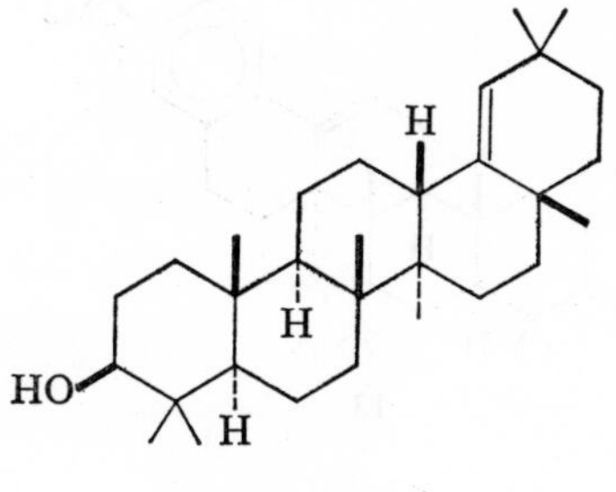

germanicol **1**

Direct synthesis[1] of germanicol[2] **1** was recently achieved, although the synthesis[3] of β-amyrin constitutes a formal synthesis of **1**, since **1** has been prepared from β-amyrin.[4]

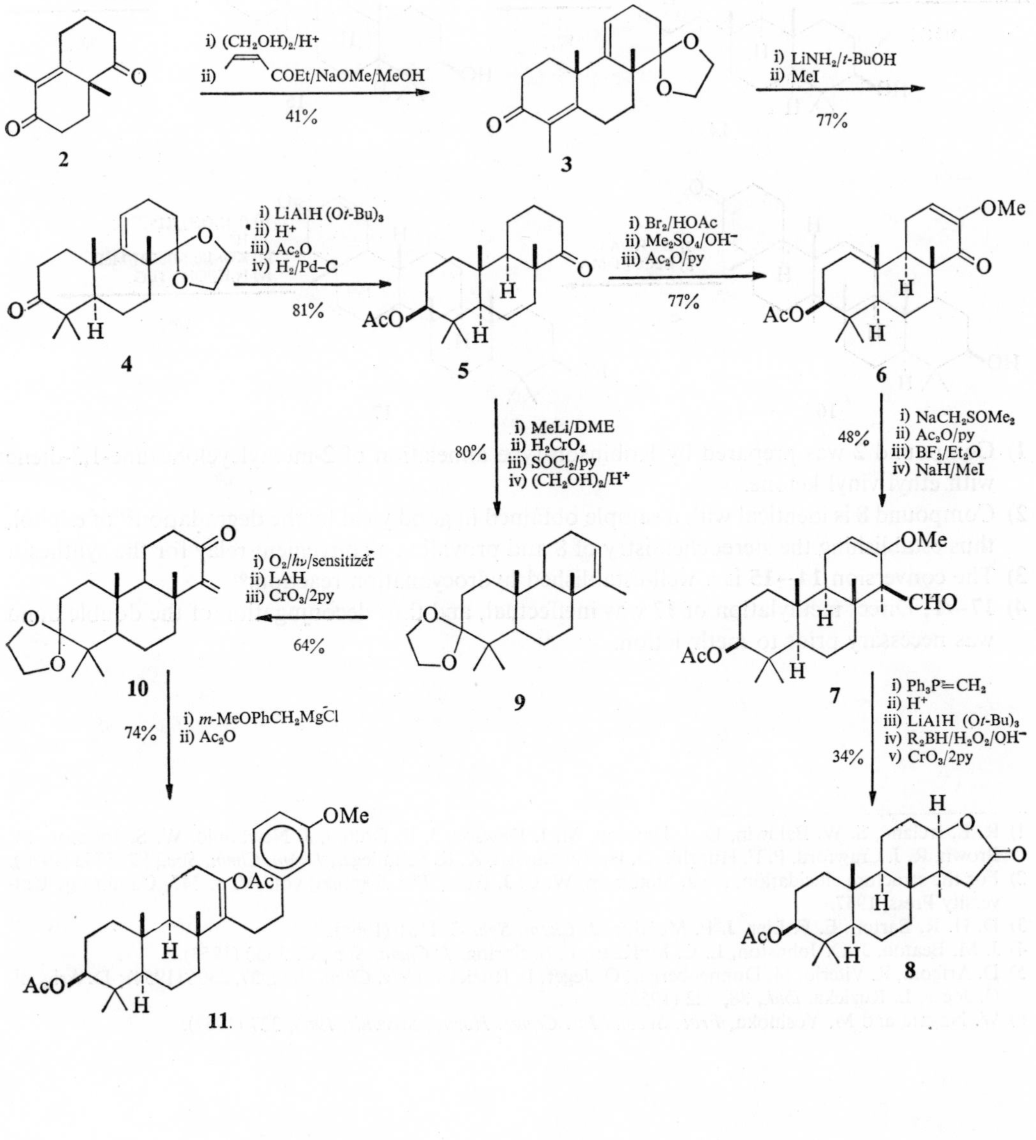

i) OH⁻
ii) CrO₃/2py
iii) (CH₃OH)₂/H⁺
iv) R₂BH/H₂O₂/OH⁻
v) m-MeOPhMgBr/H₂/Pd–C/H⁺
vi) CrO₃/2py

8 ────────────→
57%

i) MeLi/DME
ii) MeI
iii) H⁺
iv) CrO₃

11 ────────────→
61%

PAA (30 min, 25°)
90%

12

13

i) Li/NH₃/HOR
ii) H⁺
31%

AlEt₃/HCN
90%

i) (CH₃OH)₂/H⁺
ii) i-Bu₂AlH
iii) N₂H₄/OH⁻
iv) H⁺
70%

14

15

i) Br₂/HOAc
ii) CaCO₃/DMA
40%

i) (CH₃OH)₂/H⁺
ii) H⁺
iii) KOt-Bu/t-BuOH/MeI
iv) N₂H₄/OH⁻/TEG
20%

16

17

1

1) Compound **2** was prepared by Robinson-type annelation of 2-methylcyclohexane-1,3-diene with ethyl vinyl ketone.

2) Compound **8** is identical with a sample obtained in good yield by the degradation[5] of euphol, thus establishing the stereochemistry of **8** and providing a convenient relay for the synthesis.

3) The conversion **14**→**15** is a well-established hydrocyanation reaction.[6]

4) **17**→**1**: Direct methylation of **17** was ineffectual, and thus deconjugation of the double bond was necessary prior to methylation.

1) R. E. Ireland, S. W. Baldwin, D. J. Dawson, M. I. Dawson, J. E. Dolfini, J. Newbould, W. S. Johnson, M. Brown, R. J. Crawford, P. F. Hudrlik, G. H. Rasmussen, K. K. Schmiegel, *J. Am. Chem. Soc.*, **92**, 5743 (1970).
2) For the structure elucidation, see J. Simonsen, W. C. J. Ross, *The Terpenes*, vol. IV, p. 247, Cambridge University Press, 1957.
3) D. H. R. Barton, E. F. Lier, J. F. McGhie, *J. Chem. Soc. C*, 1131 (1968).
4) J. M. Beaton, J. D. Johnston, L. C. McKean, F. S. Spring, *J. Chem. Soc.*, C, 3660 (1953).
5) D. Arigoni, R. Viterlo, M. Dunnenberger, O. Jeger, L. Ruzicka, *Helv. Chim. Acta*, **37**, 2306 (1954); D. Arigoni, O. Jeger, L. Ruzicka, *ibid.*, **38**, 222 (1955).
6) W. Nagata and M. Yoshioka, *Proc. Second Int. Congr. Horm., Steroids, 1966*, 327 (1967).

5.31 SYNTHESIS OF LUPEOL

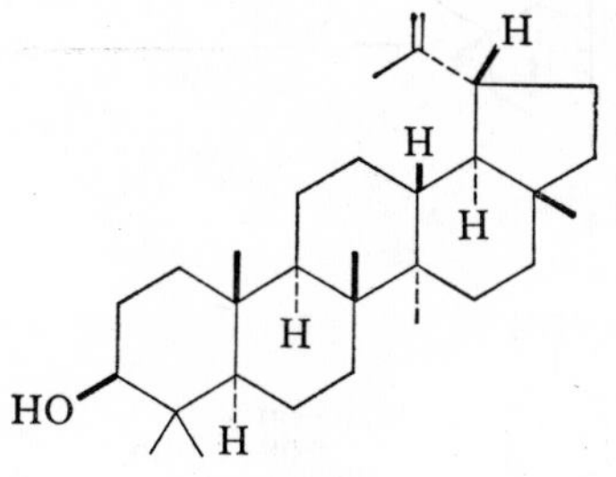

A total synthesis of lupeol, a representative member of lupane-type triterpenoids and the most abundant of the pentacyclic triterpenoids, has recently been achieved[1] by applying the enolate trapping method.[2]

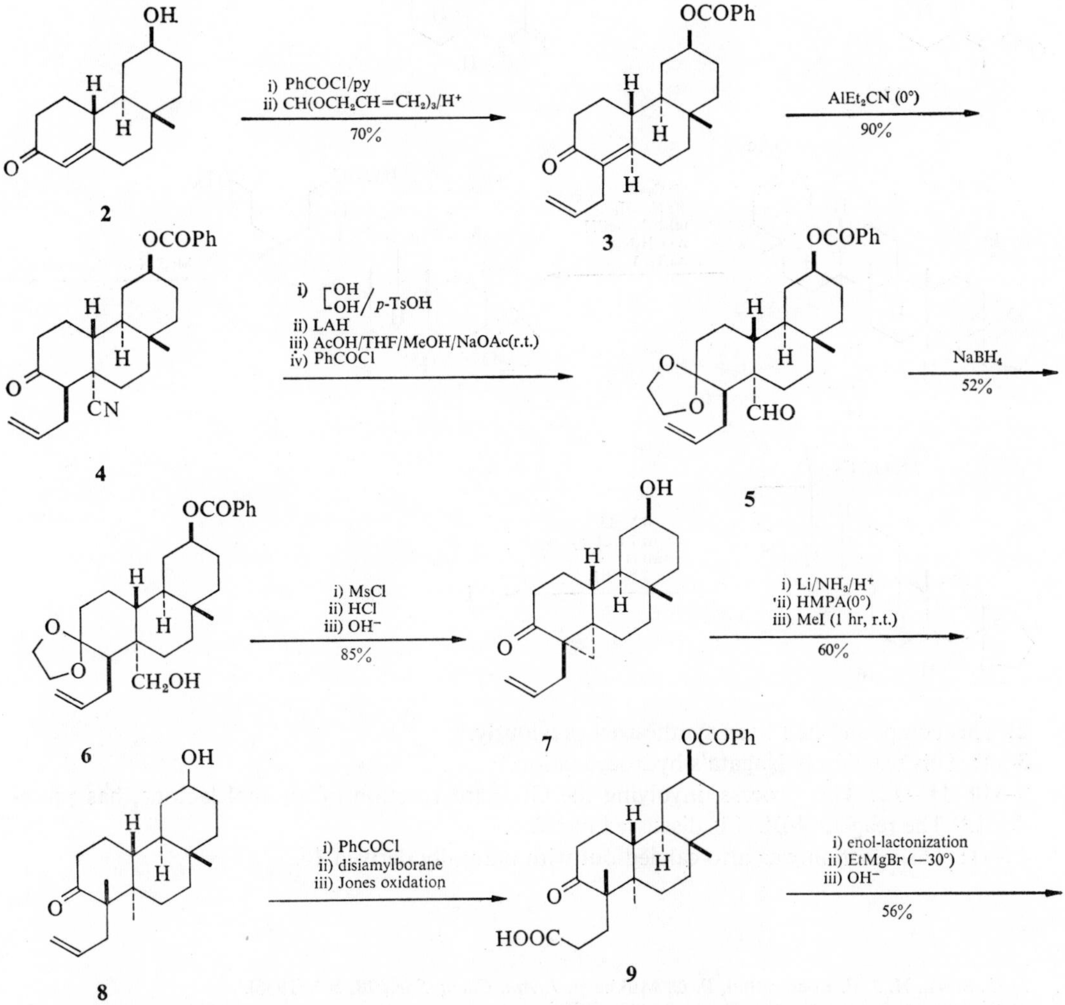

lupeol **1**

i) PhCOCl/py
ii) CH(OCH₂CH=CH₂)₃/H⁺
70%

2 → **3**

AlEt₂CN (0°)
90%

i) ⌈OH⌉/p-TsOH
   ⌊OH⌋
ii) LAH
iii) AcOH/THF/MeOH/NaOAc(r.t.)
iv) PhCOCl

4 → **5**

NaBH₄
52%

i) MsCl
ii) HCl
iii) OH⁻
85%

6 → **7**

i) Li/NH₃/H⁺
'ii) HMPA(0°)
iii) MeI (1 hr, r.t.)
60%

i) PhCOCl
ii) disiamylborane
iii) Jones oxidation

8 → **9**

i) enol-lactonization
ii) EtMgBr (−30°)
iii) OH⁻
56%

1) G. Stork, S. Uyeo, T. Wakamatsu, P. Grieco, J. Lapowitz, *J. Am. Chem. Soc.*, **93**, 4945 (1971).
2) G. Stork, P. Rosen, N. Goldman, R. V. Coombs, J. Tsuji, *J. Am. Chem. Soc.*, **87**, 275 (1965).

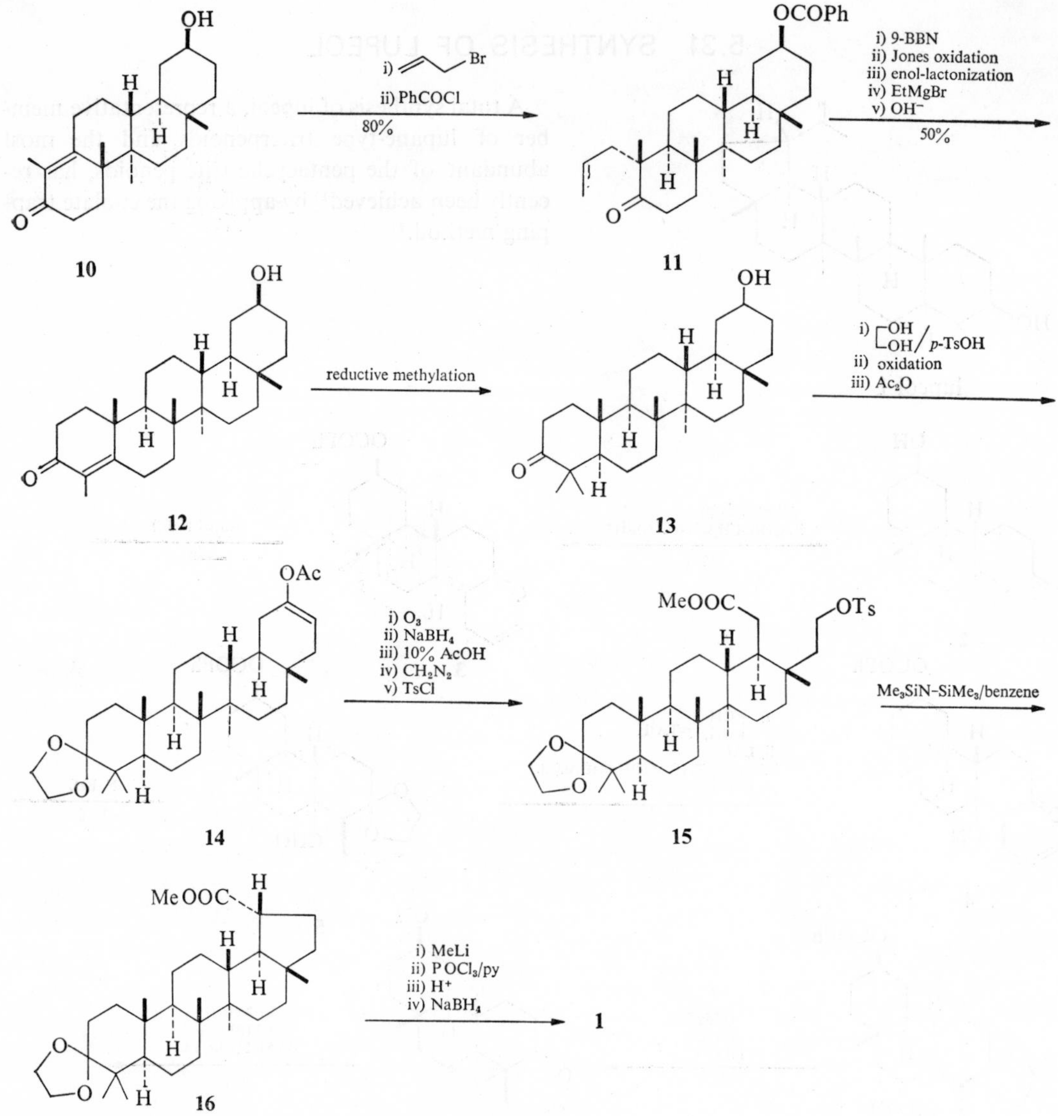

10 → **11**

i) allyl Br
ii) PhCOCl
80%

i) 9-BBN
ii) Jones oxidation
iii) enol-lactonization
iv) EtMgBr
v) OH⁻
50%

12 → **13**

reductive methylation

i) ⌐OH ⌐OH / p-TsOH
ii) oxidation
iii) Ac₂O

14 → **15**

i) O₃
ii) NaBH₄
iii) 10% AcOH
iv) CH₂N₂
v) TsCl

Me₃SiN–SiMe₃/benzene

16 → **1**

i) MeLi
ii) POCl₃/py
iii) H⁺
iv) NaBH₄

2: This compound had been synthesized previously.[3]

3→4: This reaction is Nagata's hydrocyanation.[4]

9→10, 11→12: This process, involving the Grignard reaction of an enol lactone, has precedents.[5] The reagent 9-BBN is discussed in ref. 6.

16→1: This reaction was also carried out with naturally derived **16**.

3) G. Stork, H. J. E. Loewenthal, P. C. Mukharji, *J. Am. Chem. Soc.*, **78**, 501 (1956).
4) W. Nagata, *Nippon Kagaku Zasshi*, **90**, 837 (1969).
5) R. B. Turner, *J. Am. Chem. Soc.*, **72**, 579 (1950); G. I. Fujimoto, *ibid.*, **73**, 1856 (1951).
6) E. H. Knights, H. C. Brown, *J. Am. Chem. Soc.*, **90**, 5280 (1968).

12→13: The two reductive methylation proceedures shown below were employed. The longer route was preferred.

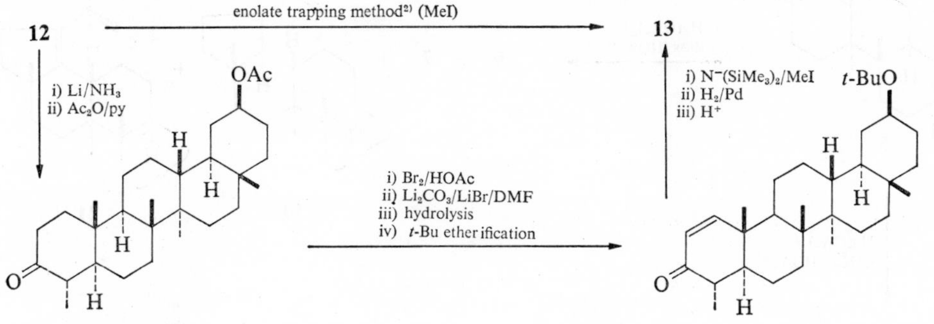

enolate trapping method[2] (MeI)

12

13

i) Li/NH₃
ii) Ac₂O/py

OAc

i) N⁻(SiMe₃)₂/MeI   *t*-BuO
ii) H₂/Pd
iii) H⁺

i) Br₂/HOAc
ii) Li₂CO₃/LiBr/DMF
iii) hydrolysis
iv) *t*-Bu etherification

17

18

5.32   MERCURIC ACETATE OXIDATION IN LUPANES

Mercuric acetate oxidation has been systematically investigated recently on lupane-type triterpenoids.[1-4] Examples showing the types of products are listed below.

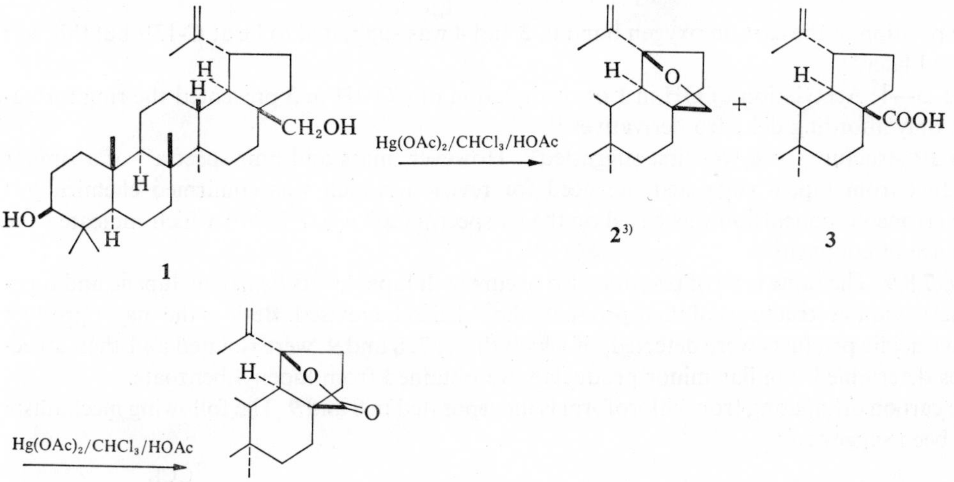

CH₂OH

Hg(OAc)₂/CHCl₃/HOAc

HO

1

2[3]

+

COOH

3

Hg(OAc)₂/CHCl₃/HOAc

4[3]

1) J. M. Allison, W. Lawrie, J. McLean, G. R. Taylor, *J. Chem. Soc.* C, 3353 (1961).
2) J. M. Allison, W. Lawrie, J. McLean, J. M. Beaton, *J. Chem. Soc.* C, 5224 (1961).
3) S. P. Adhikary, W. Lawrie, J. McLean, *J. Chem. Soc.* C, 1030 (1970).
4) S. P. Adhikary, W. Lawrie, J. McLean, M. S. Malik, *J. Chem. Soc.* C, 32 (1971).

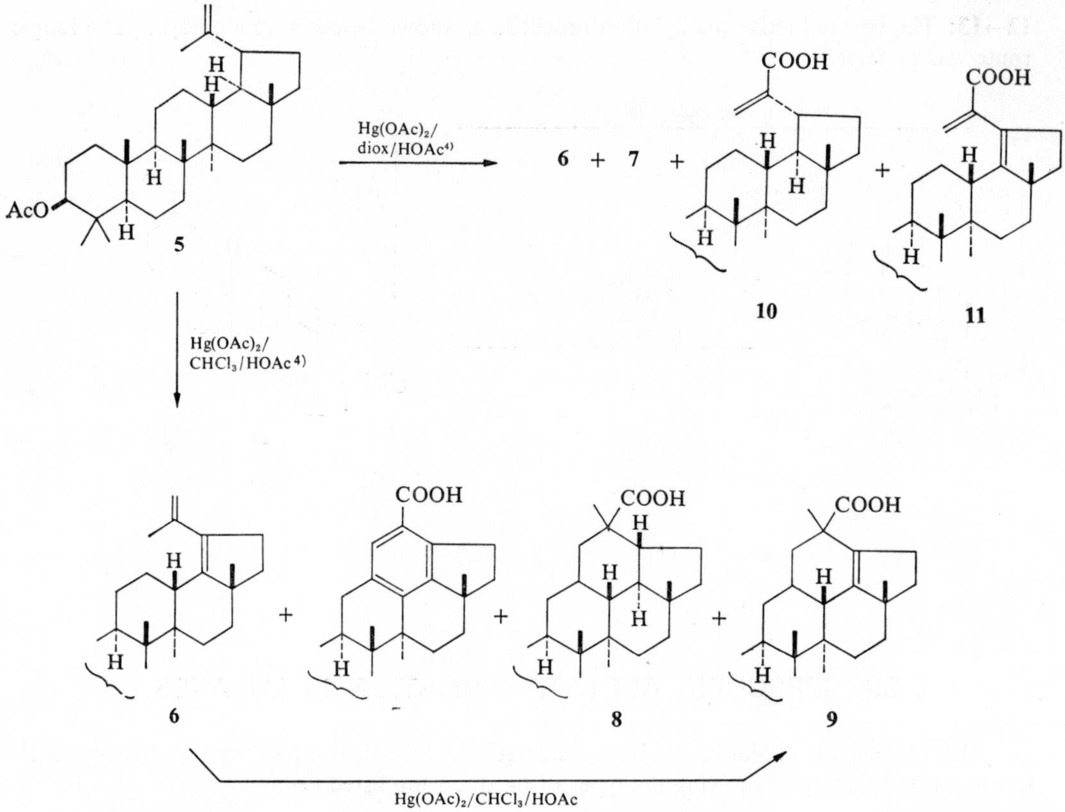

1) The position of the carbon oxygen bond in **2** and **4** was suggested to be at C-13[1] but this was revised later.[3]

2) **1→2, 3→4**: Acetylation of OH in **1** or methylation of COOH in **3** prevented the ring formation only affording dehydro derivatives.[1]

3) The Δ^{12} structure for **6** was first suggested[2]. However, mass and nmr spectra[5] of a similar product from lupeol suggested the need for revision, which was confirmed chemically.[3] The erroneous conclusion was based on the uv spectra (207 nm (ϵ 7000)) which indicated the absence of conjugation.

4) **5→6,7,8,9**: The same type of reaction also occurs with lupeol,[3] its benzoate, lupene and lupenone[2], and the structures of their products should also be revised. Besides the major product **6**, five acidic products were detected, of which three, **7**, **8** and **9**, were isolated and their structures determined. Similar minor products were obtained from lupenyl benzoate.

5) One carbon originating from chloroform is incorporated in **8** and **9**. The following mechanism has been suggested.[4]

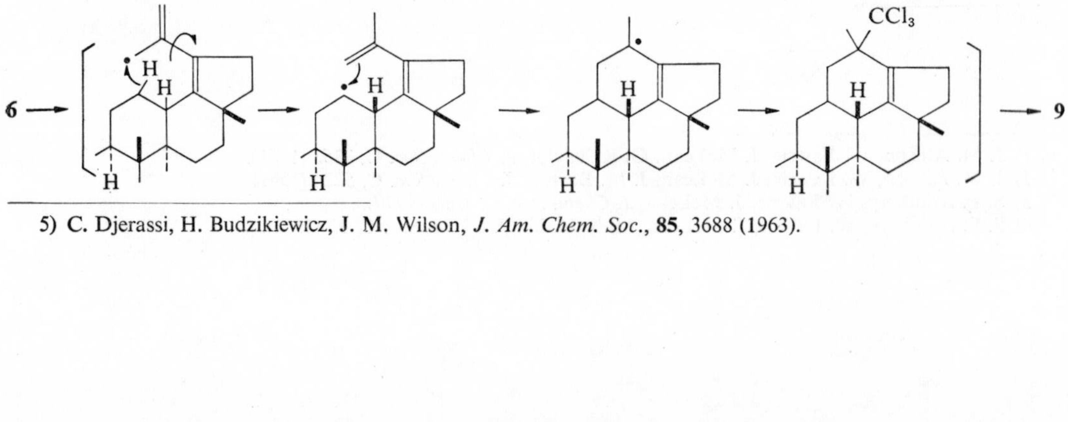

5) C. Djerassi, H. Budzikiewicz, J. M. Wilson, *J. Am. Chem. Soc.*, **85**, 3688 (1963).

5.33 STRUCTURE OF SERRATENEDIOL

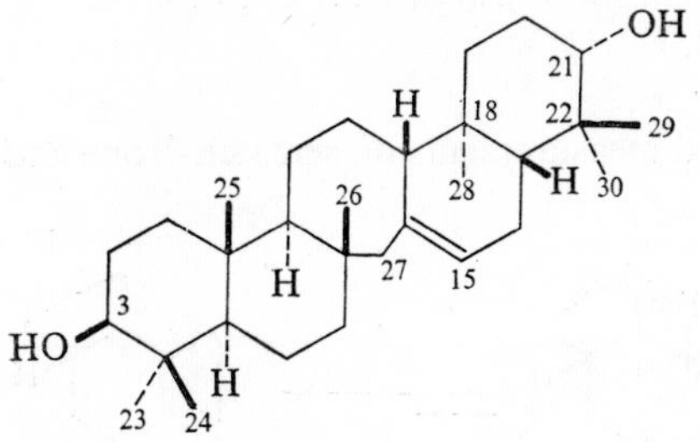

serratenediol **1**[1-3]

Serratenediol was first isolated from *Lycopodium serratum*[4] and was later found in various Lycopodium plants.[5] Pinusenediol, isolated from pine bark, is identical with **1**. [6]

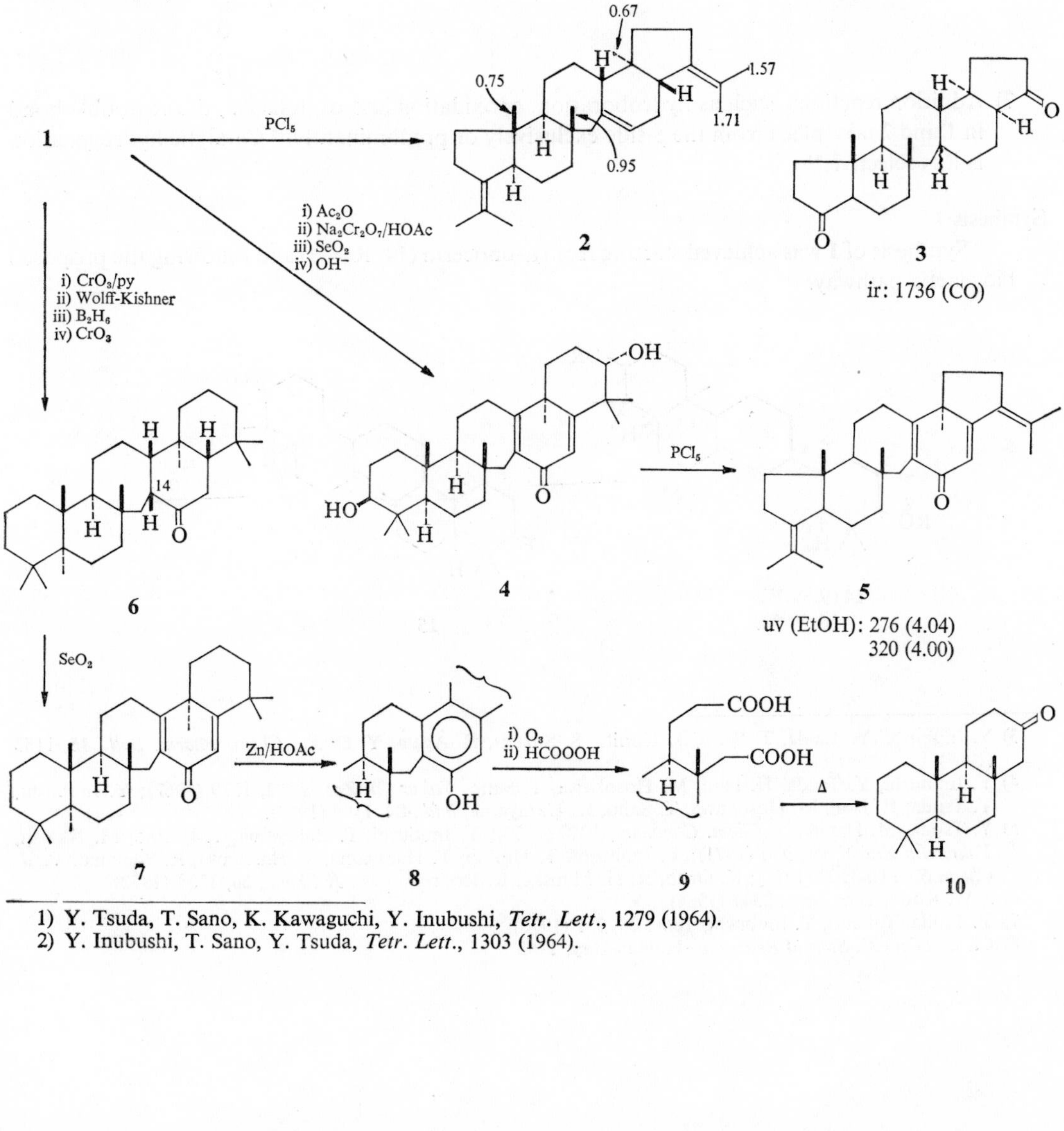

nmr for diacetate:
Me: 0.69, 0.84, 0.84, 0.84, 0.84, 0.84, 0.89, 2.03, 2.03
=CH: 5.31 (1H)

ir: 1736 (CO)

uv (EtOH): 276 (4.04)
320 (4.00)

1) Y. Tsuda, T. Sano, K. Kawaguchi, Y. Inubushi, *Tetr. Lett.*, 1279 (1964).
2) Y. Inubushi, T. Sano, Y. Tsuda, *Tetr. Lett.*, 1303 (1964).

1) The PCl$_5$ reaction of dihydro-**1** (a mixture of C-14 stereoisomers) yielded **3** after oxidation. Inversion at C-5 and C-17 occurs on alumina chromatography.

2) The 14α isomer (*trans*, chair D ring) is more stable than the 14β (*cis*, boat D ring).[7]

3) The reaction **7**→**8** is a dienone-phenol rearrangement.[8]

4) The structure of **10** was established by synthesis.[2,3]

5) The absolute configuration of **1** was deduced from ord measurements on serraten-3-one and 3-acetoxy serraten-21-one.[1,3]

6) Serratene derivatives (general formula **11**) undergo acid-catalyzed isomerization to iso-serratene (Δ^{13}) derivatives (general formula **12**): for example, serratene and isoserratene gave the same 1:3-4 mixture of types **11** and **12** on reaction with 15% HCl/AcOH and CHCl$_3$. Both **11** and **12** afforded unsaturated ketones **13** on sodium dichromate oxidation.[3]

7) Addition reactions, such as hydroboration, epoxidation and osmolation, of the double bond in **1** and **2** take place from the β-side exclusively or predominantly.[7] Catalytic hydrogenation is less selective.[3]

Synthesis[1]

Synthesis of **1** was achieved starting from α-onocerin (**14**, R = H) and following the proposed biogenetic pathway.

3) Y. Inubushi, Y. Tsuda, T. Sano, T. Konita, S. Suzuki, H. Ageta, Y. Otake, *Chem. Pharm. Bull.*, **15**, 1153 (1967).
4) Y. Inubushi, Y. Tsuda, H. Ishii, M. Hosokawa, T. Sano, *Yakugaku Zasshi*, **82**, 1339 (1962); Y. Inubushi, Y. Tsuda, H. Ishii, M. Hosokawa, T. Sano, H. Harayama, *ibid.*, **84**, 1108 (1964).
5) Y. Tsuda, M. Hatanaka, *Chem. Commun.*, 1040 (1969); Y. Inubushi, T. Harayama, T. Hibino, M. Akatsu, *Yakugaku Zasshi*, **91**, 980 (1971); Y. Inubushi, T. Hibino, T. Harayama, T. Hasegawa, R. Somanathan, *J. Chem. Soc.* C, 3109 (1971); K. Orito, R. H. Manske, R. Rodrigo, *Can. J. Chem.*, **50**, 3280 (1972).
6) J. W. Rowe, *Tetr. Lett.*, 2347 (1964).
7) Y. Tsuda, T. Sano, Y. Inubushi, *Tetr.*, **26**, 751 (1970).
8) Cf. C. Djerassi, *Steroid Reactions*, Holden-Day, 1963.

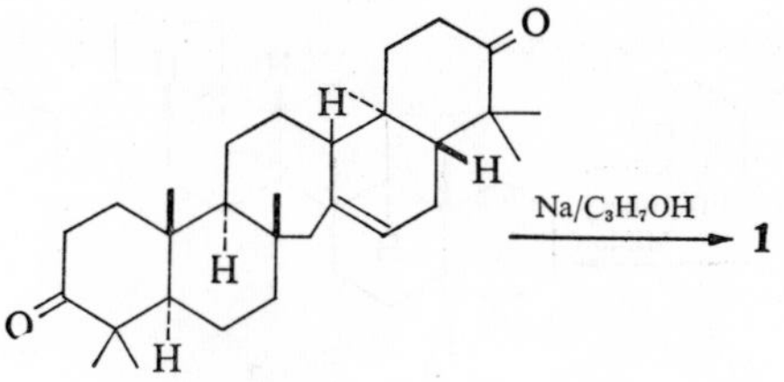

15→16: The reaction produced the following hydrocarbons after removal of oxygen functions. α-, β- and γ-onoceradienes in 7%, 17%, 32% yield, respectively, serratene (33%) and iso-serratene (11%). α-Onocerin diacetate (**14**, R = Ac) yielded, after removal of oxygen functions, serratene and γ-onocerene in a 1 : 2 ratio. Protonic acids failed to cyclize **14** (R = Ac) to serratene derivatives, β-onocerin being formed.

5.34 ZEORIN AND LEUCOTYLIN

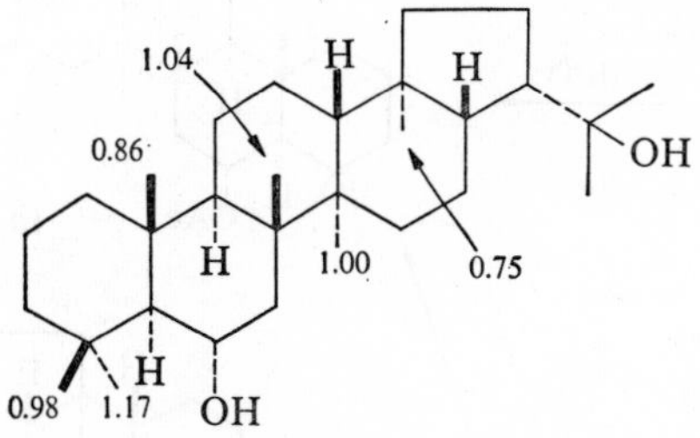

zeorin **1** 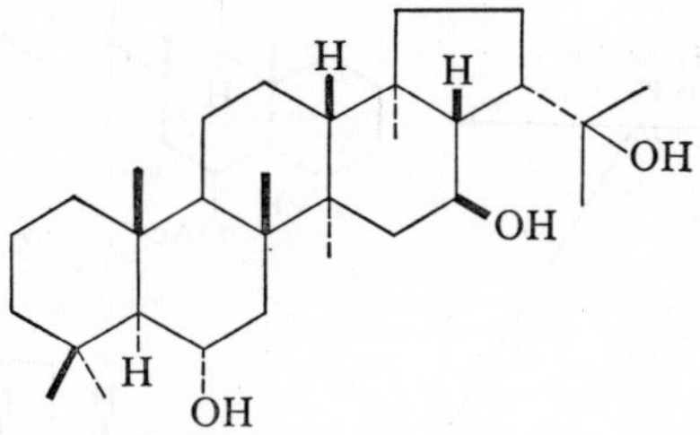   leucotylin **2**

nmr of 6-*O*-acetate:
4.82 (6-H, dt, J=11.6,7.5)

Following the original proposal of the structure of zeorin[1] (without stereochemical detail at C-17 and C-21), spectroscopic and chemical data were accumulated with regard to the stereochemistry.[2-5] Chemical correlation of **1**, leucotylin **2**[6] and hopane **3**[4] lead to some confusion until x-ray analysis of suitable derivatives established the stereochemistry of these compounds.[7-9] Reexamination of the chemical correlation[10] revealed the isomerization depicted below during the dehydration of the tertiary hydroxyl group.

1) D. H. R. Barton, P. de Mayo, J. C. Orr, *J. Chem. Soc.*, 2239 (1958).
2) S. Hunec, *Chem. Ber.*, **94**, 614 (1961).
3) S. Huneck, J. -M. Lehn, *Bull. Soc. Chim. France*, 1702 (1963).
4) I. Yoshioka, T. Nakanishi, I. Kitagawa, *Chem. Pharm. Bull*, **15**, 353 (1967); Y. Tsuda, K. Isobe, S. Fukushima, H. Ageta, K. Iwata, *Tetr. Lett.*, 23 (1967).
5) I. Yoshioka, T. Nakanishi, I. Kitagawa, *Chem. Pharm. Bull.*, **17**, 279 (1969).
6) I. Yoshioka, T. Nakanishi, I. Kitagawa, *Tetr. Lett.*, 1485 (1968); *Chem. Pharm. Bull.*, **17**, 291 (1969).
7) T. Nakanishi, T. Fujiwara, K. Tomita, *Tetr. Lett.*, 1491 (1968).
8) H. Koyama, H. Nakai, *J. Chem. Soc* (B), 546 (1970).
9) T. Nakanishi, H. Yamauchi, T. Fujiwara, K. Tomita, *Tetr. Lett.*, 1157 (1971).
10) I. Yoshioka, T. Nakanishi, H. Yamauchi, I. Kitagawa, *Tetr. Lett.*, 1161 (1971); *Chem. Pharm. Bull.*, **20**, 147 (1972).

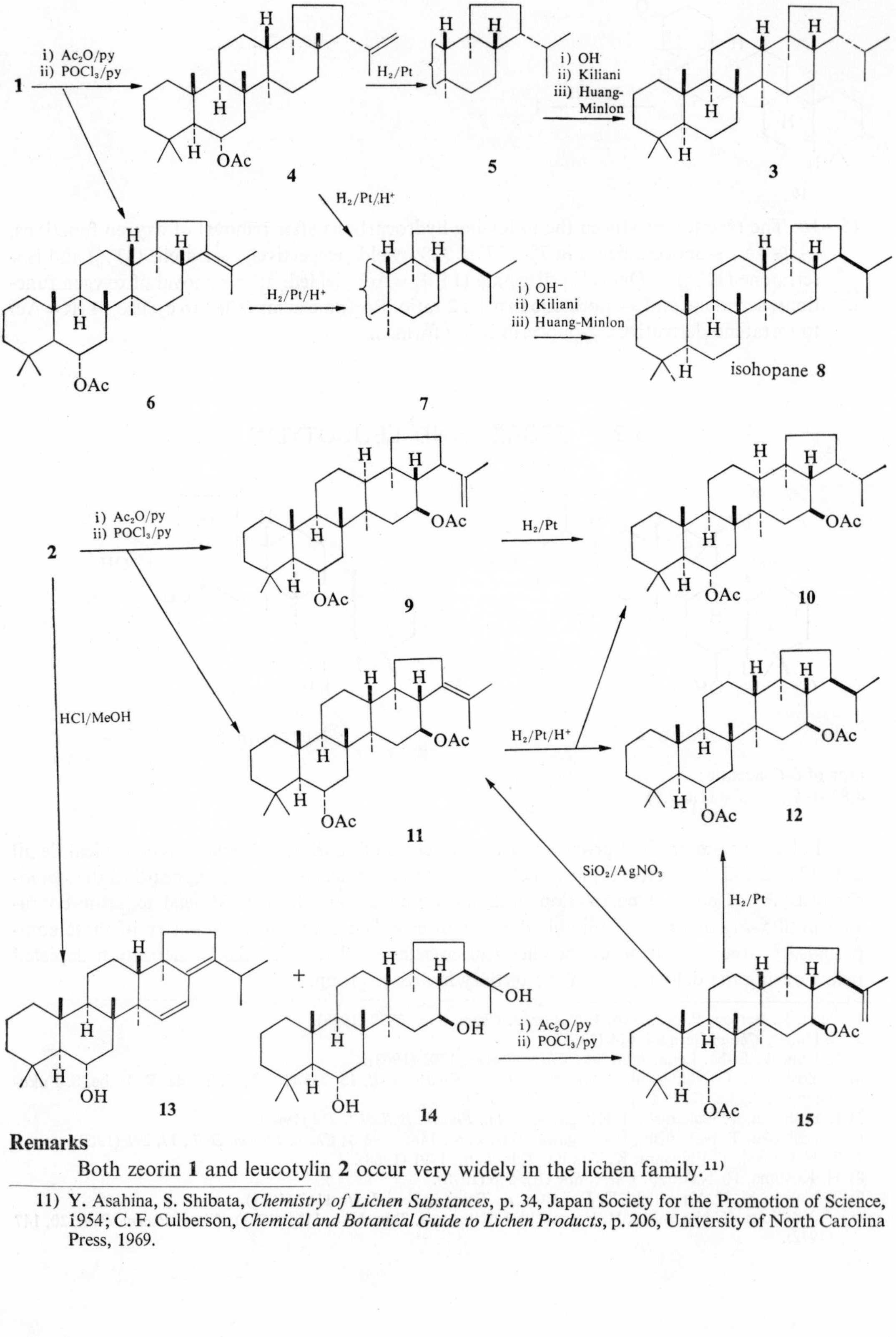

isohopane 8

Remarks

Both zeorin **1** and leucotylin **2** occur very widely in the lichen family.[11]

11) Y. Asahina, S. Shibata, *Chemistry of Lichen Substances*, p. 34, Japan Society for the Promotion of Science, 1954; C. F. Culberson, *Chemical and Botanical Guide to Lichen Products*, p. 206, University of North Carolina Press, 1969.

5.35 HOPENES AND MIGRATED HOPENES IN FERN PLANTS

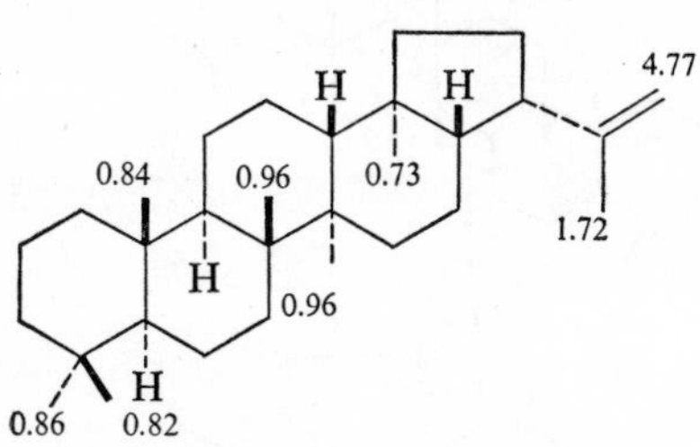

hop-22(29)-ene **1**[1]
(diploptene)
mp : 211° α_D : +61°

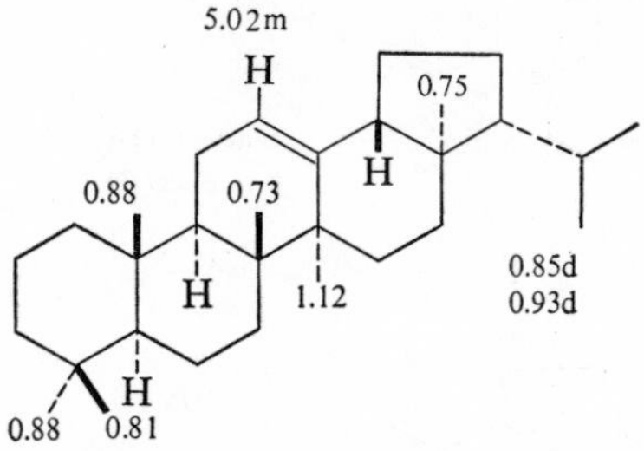

neohop-12-ene **2**[2]
mp : 211° α_D : +41.6°

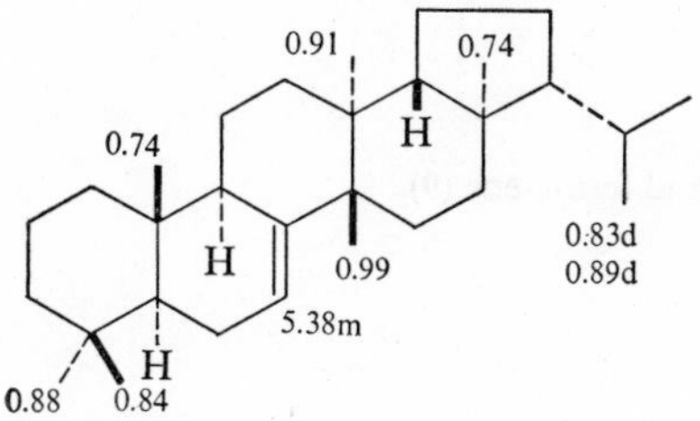

fern-7-ene **3**[3]
mp : 209.5° α_D : −27°

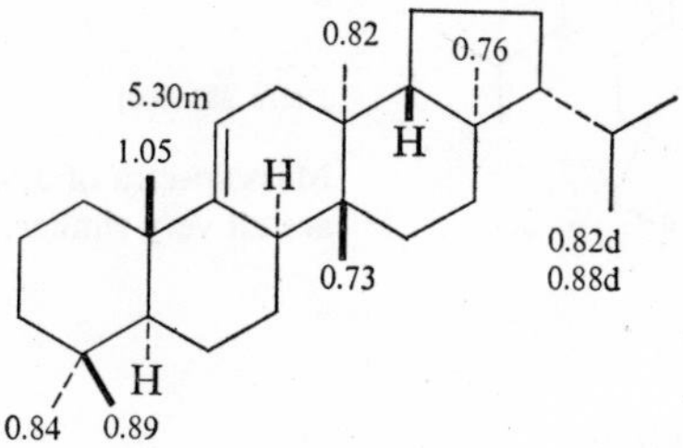

fern-9(11)-ene **4**[1]
mp : 172° α_D : −16.9°

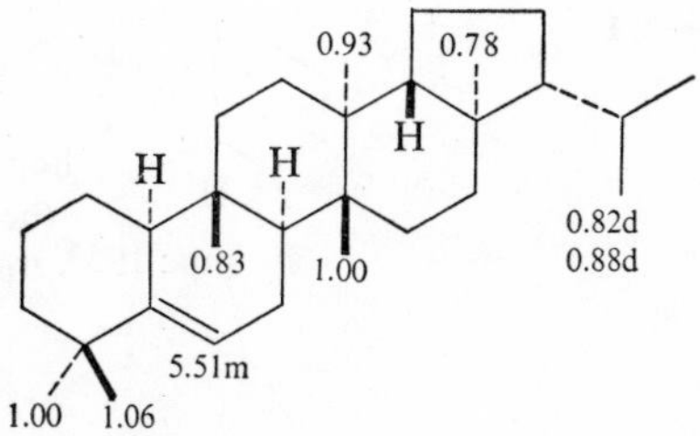

adian-5-ene **5**[3]
mp : 195° α_D + 51.0°

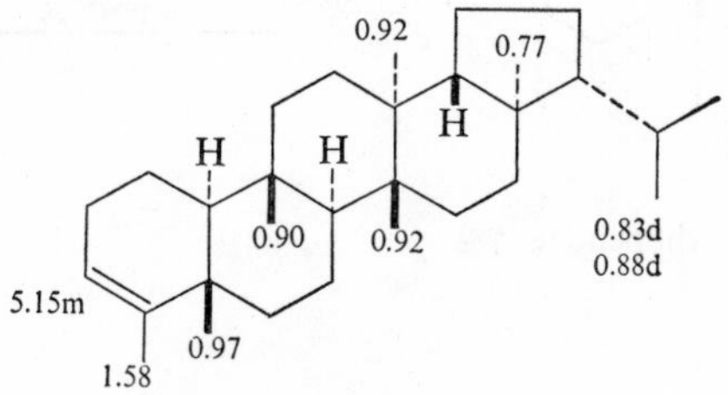

filic-3-ene **6**[3]
mp : 234° α_D + 50.0°

1) H. Ageta, K. Iwata, K. Yonezawa, *Chem. Pharm. Bull*, **11**, 408 (1963); H. Ageta, K. Iwata, S. Natori, *Tetr. Lett.*, 1447 (1963).
2) H. Ageta, K. Shiojima, Y. Arai, *Chem. Commun.*, 1105 (1968).
3) H. Ageta, K. Iwata, S. Natori, *Tetr. Lett.*, 3413 (1964).

A number of triterpenic hydrocarbons have been found in ferns. Most of them lack the ubiquitous oxygen functions at C-3. Some representative members with different carbon skeletons are shown above. These compounds can easily be discriminated by means of their fragmentation patterns in mass spectra.[2,3] Typical fragmentations are shown below.

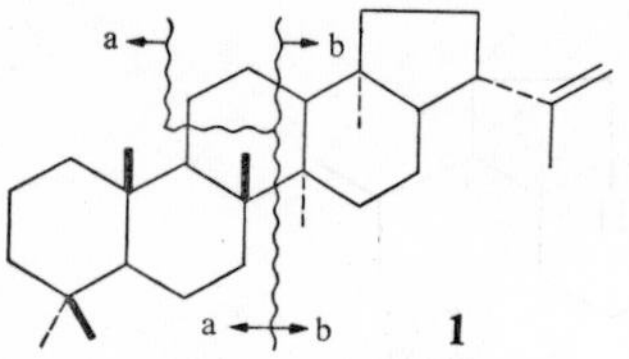
a
b
a
b

1

a m/e 191 (100)
b m/e 189 (94)

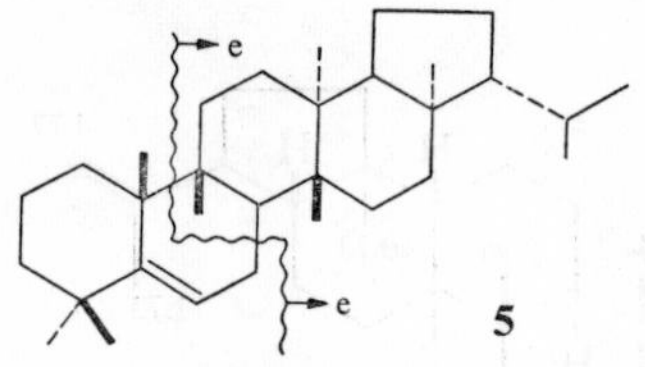
e
e

5

e m/e 274 (100)
e $-CH_3$ 259 (67)

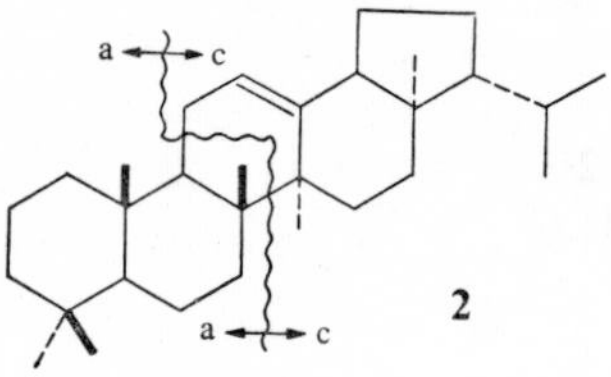
a
c
a
c

2

a m/e 191 (28)
c m/e 218 (100)

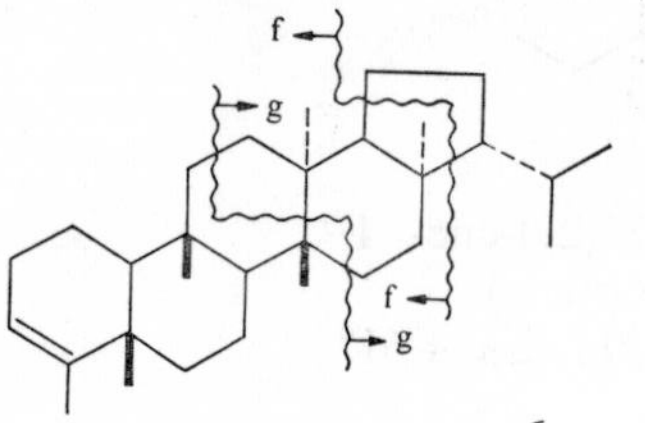
f
g
f
g

6

f m/e 327 (42)
g m/e 191 (100)

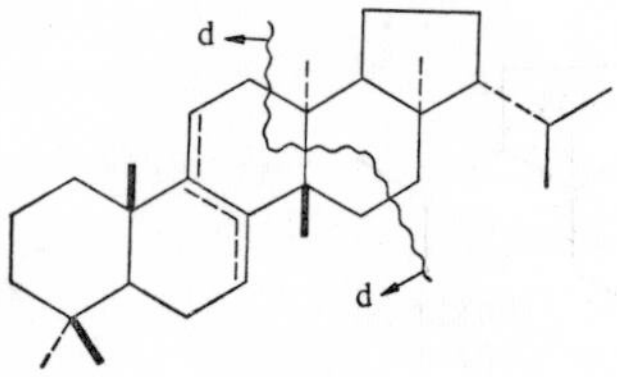
d
d

3 , 4

d m/e 243 (100)

Mass spectra of **3**, **4** and fern-8-ene (**9**) are all very similar.

Chemical correlation

The chemical correlation of these compounds was achieved as follows.[1–4]

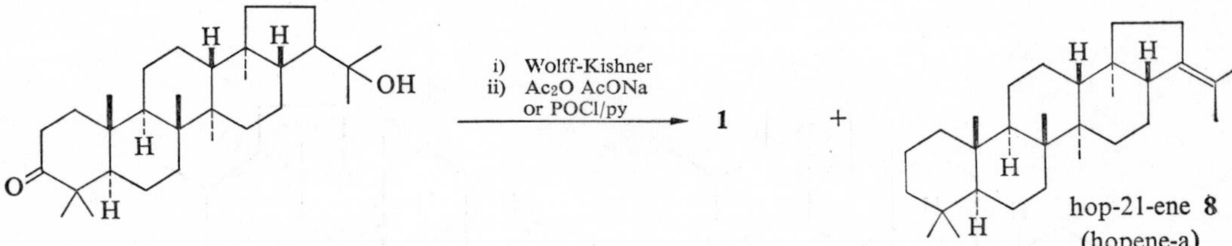
H
H
OH
H
O
H
i) Wolff-Kishner
ii) Ac2O AcONa
or POCl/py
1
+
H
H
H
H

hydroxyhopenone 7[4]

hop-21-ene **8**
(hopene-a)
mp: 195°, α_D: +25°

4) G. V. Baddeley, T. G. Halsall, E. R. H. Jones, *J. Chem. Soc.*, 1715 (1960); *ibid.*, 3891 (1961).
5) H. Ageta, K. Iwata, *Tetr. Lett.*, 6069 (1966).
6) G. Berti, F. Bottari, A. Marsili, I. Morelli, *Tetr. Lett.*, 979 (1966).
7) G. N. Pandey, C. R. Mitra, *Tetr. Lett.* 4683 (1967).
8) F. Bottari, A. Marsili, I. Morelli, M. Pacchiani, *Phytochemistry*, **11**, 2519 (1972).
9) E. L. Ghisalberti, N. J. de Souza, H. H. Rees, T. W. Goodwin, *Phytochemistry*, **9**, 1817 (1970).
10) D. H. R. Barton, G. Mellows, D. A. Widdowson, *J. Chem. Soc.* C, 110 (1971).
11) H. Kakisawa, K. Iguchi, *J. Chem. Soc.* D, 1486 (1970).
12) H. Ageta, K. Iwata, Y. Otake, *Chem. Pharm. Bull.* **11**, 407 (1963).

2   1   8

3, 4, 5, 6

fern-8-ene **9**
(isofernene)
mp: 192°, α_D: +18°

neohop-13(18)-ene **10**
(hopene-II)
mp: 201°, α_D: ±0°

hop-17(21)-ene **11**
(hopene-I)
mp: 189°, α_D: +49°

Remarks

These hydrocarbons have been isolated from the following ferns. **1** and **4** from *Dryopteris crassirhizoma*,[1] **1, 2, 3, 4, 5** and **9** from *Adiantum monochlamys*,[2,3] **2, 3, 4, 5** and **9** from *Adiantum pedatum*,[2,5] **1** and **4** from *Polypodium vulgare*,[6] **10** and neohopa-11,13(18)-diene from *Oleandra wallichii*.[7] Chemotaxonomical observations[8] and biosynthetic studies[9,10] have been reported. A synthesis of **9** and ferna-7,9(11)-diene was achieved starting from α-onocerin.[11]

The following derivatives of the above hydrocarbons have also been isolated from fern plants.

22-hydroxyhopane (diplopterol), mp: 256°, α_D: +44.5° from *Diplopterygium glaucum*.[12]

29-hydroxyhopane (neriifoliol), mp: 244°, α_D: +35°, from *Oleandra neriifolia*.[13]

17β,21β-epoxyhopane, mp: 270° α_D: +47°, from *Polypodium vulgare*.[6]

3β-hydroxyhop-22(29)-en-23-oic acid (Woodwardic acid), mp: 275°, α_D: +126°, from *Woodwardia orientalis*.[14]

30-norhopan-22-one (adiantone), mp: 224°, α_D: +81.1°, from *Adiantum capillis-veneris*,[15] *A. monochlamys*,[16] and *A. pedatum*.[5]

21β-hydroxy-30-norhopan-22-one (hydroxyadiantone), mp: 284°, from *Adiantum venestum, A. capillis-veneris*,[17] and *A. monochlamys*.[16]

22-hydroxy-30-norhopan-22,28-oxide (adipedatol), mp: 188°, α_D: +88°, from *Adiantum pedatum*.[5.18]

fern-9(11)-en-24-oic acid (davallic acid), mp: 283°, from *Davallia divaricata*.[19]

filican-3α,4α-oxide (adiantoxide), mp: 231°, α_D: +46.8°, from *Adiantum capillis-veneris*.[20]

23-oxofilic-3-ene (filicenal), mp: 272° α_D: +74°, from *Adiantum pedatum*.[5]

13) G. N. Pandey, C. R. Mitra, *Tetr. Lett.*, 1353 (1967).
14) T. Murakami, C. -M. Chen, *Chem. Pharm. Bull.* **19**, 25 (1971).
15) G. Berti, F. Bottari, A. Marsili, J. -M. Lehn, P. Witz, G. Ourisson, *Tetr. Lett.*, 1283 (1963).
16) H. Ageta, K. Iwata, Y. Arai, Y. Tsuda, K. Isobe, S. Fukushima, *Tetr. Lett.*, 5679 (1966).
17) A. Zaman, A. Prahash, G. Berti, F. Bottari, B. Macchia, A. Marsili, I. Morelli, *Tetr. Lett.*, 3943 (1966).
18) H. Ageta, K. Shiojima, *Chem. Commun.*, 1372 (1968).
19) K. Nakanishi, Y. -Y. Lin, H. Kakisawa, *Tetr. Lett.*, 1451 (1963); Y. -Y. Lin, H. Kakisawa, Y. Shiobara, K. Nakanishi, *Chem. Pharm. Bull.* **13**, 986 (1965).
20) G. Berti, F. Bottari, A. Marsilli, *Tetr. Lett.*, 1 (1964).

5.36 SOIL BACTERIAL HYDROLYSIS LEADING
TO GENUINE SAPOGENOLS

Acid hydrolysis of saponins is known sometimes to yield sapogenols of modified structure as a result of concomitant secondary reactions. To avoid such secondary reactions, various chemical,[1-3] enzymatic[4] and microbiological[5] methods have been proposed. The soil bacterial hydrolysis method has recently been developed[6] and provides a most reliable way to obtain genuine sapogenols.

The method is based on the finding that a soil bacterium, selected by repeated cultivation on a synthetic medium containing saponin as the only carbon source, hydrolyzes the saponin to liberate the genuine sapogenol. In general, the soil samples are incubated in test tubes containing 3 ml of a synthetic medium containing $(NH_4)_2HPO_4$ (4g), KH_2PO_4 (1g), NaCl (1g), $MgSO_4 \cdot 7H_2O$ (0.7g), $FeSO_4 \cdot 7H_2O$ (0.03g), saponin or glycoside (3g), distilled water (1 liter) and adjusted to pH 6.0 with dil. HCl, and a soil bacterium is selected by repeated cultivation. The selected bacterium is grown in stationary culture at 31° on a large scale for an appropriate period, monitoring by tlc. The culture is then extracted with a suitable organic solvent and the extract is subjected to purification.

By this method, the following genuine triterpenoid sapogenols were elucidated: presenegenin 1 from saponin of *Polygala senega* root,[6] oleanolic acid 2 from saponin of *Panax japonicum* rhizome,[6] protoaescigenin 3, barringtogenol C 4, and 16-desoxy-barringtogenol C 5 from saponin of *Aesculus turbinata* seeds,[7] pomolic acid 6 from ziyu-glycoside I of *Sanguisorba officinalis* root,[8] four acylated derivatives of barringtogenol C (7, 8, 9, 10) from saponin of *Styrax japonica* pericarps,[9] protobassic acid 11 from saponin of *Madhuca longifolia* seeds,[10] and S-protopanaxadiol 12 from saponin of *Panax ginseng* root.[11] Shorter periods of cultivation, in some cases, resulted in the isolation of prosapogenols such as oleanolic acid 28-*O*-β-D-glucopyranoside 13,[6] pomolic acid 3-*O*-α-L-arabopyranoside 14,[8] and S-protopanaxadiol 20-*O*-β-D-glucopyranoside 15.[11]

1) Periodate oxidation: F. Smith, A. M. Unran, *Chem. Ind.*, 881 (1959); G. W. Hay, B. A. Lewis, F. Smith, *Methods in Carbohydrate Chemistry*, **5**, 357 (1965); J. J. Dugan, P. de Mayo., *Can. J. Chem.*, **43**, 2033 (1965); N. Aimi, H. Fujimoto, S. Shibata, *Chem. Pharm. Bull.*, **16**, 641 (1968); T. Kubota, H. Hinoh, *Tetr.*, **24**, 675 (1968); M. Nagai, T. Ando, N. Tanaka, O. Tanaka, S. Shibata, *Chem. Pharm. Bull.*, **20**, 1212 (1972); I. Kitagawa, A. Matsuda, I. Yoshioka, *Chem. Pharm. Bull.*, **20**, 2226 (1972).
2) Mild acid hydrolysis: R. Kuhn, I. Low, *Ann. Chem.*, **669**, 183 (1963); R. Tschesche, F. Inchaurrondo, G. Wulff, *Ann. Chem.*, **680**, 107 (1964).
3) Chemical modification of sapogenol before acid hydrolysis: S. Shibata, O. Tanaka, T. Ando, M. Sado, S. Tsushima, T. Ohsawa, *Chem. Pharm. Bull.*, **14**, 595 (1966); R. Tschesche, G. Lüdke, G. Wulff, *Chem. Ber.*, **102** 1253 (1969).
4) R. Tschesche, G. Wulff, *Tetr.* **19**, 621 (1963); T. Kawasaki, I. Nishioka, T. Yamauchi, K. Miyahara, M. Enbutsu, *Chem. Pharm. Bull.*, **13**, 435 (1965); P. Tunmann, W. Gerner, G. Stapel, *Ann. Chem.*, **694**, 162 (1966).
5) C. H. Hassall, B. S. W. Smith, *Chem. Ind.*, 1570 (1957); Ch. Tamm, A. Gubler, *Helv. Chim. Acta*, **42**, 239 (1959); W. A. Lourens, M. B. O'Donovan, *S. African J. Agr. Sci.*, **4**, 293 (1961) (C. A., **56**, 10196 (1962)).
6) I. Yoshioka, M. Fujio, M. Osamura, I. Kitagawa, *Tetr. Lett.*, 6303 (1966).
7) I. Yoshioka, K. Imai, I. Kitagawa, *Tetr. Lett.*, 2577 (1967).
8) I. Yoshioka, T. Sugawara, A. Ohsuka, I. Kitagawa, *Chem. Pharm. Bull.* **19**, 1700 (1971).
9) I. Yoshioka, S. Saijoh, I. Kitagawa, *Chem. Pharm. Bull.* **20**, 564 (1972).
10) I. Kitagawa, A. Inada, I. Yoshioka, R. Somanathan, M. U. S. Sultanbawa, *Chem. Pharm. Bull.*, **20**, 630 (1972).
11) I. Yosioka, T. Sugawara, K. Imai, I. Kitagawa, *Chem. Pharm. Bull.* **20**, 2418 (1972).

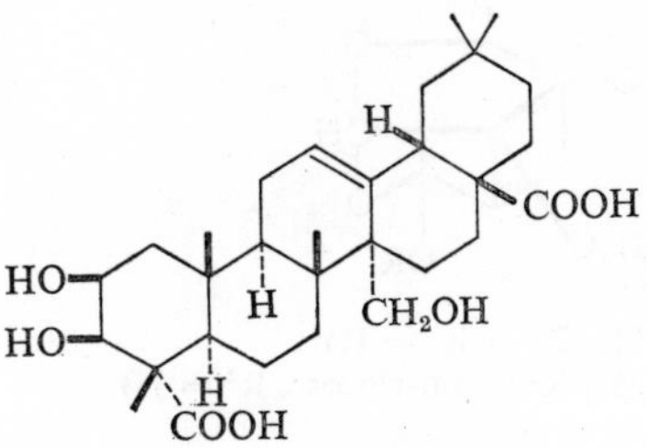

1

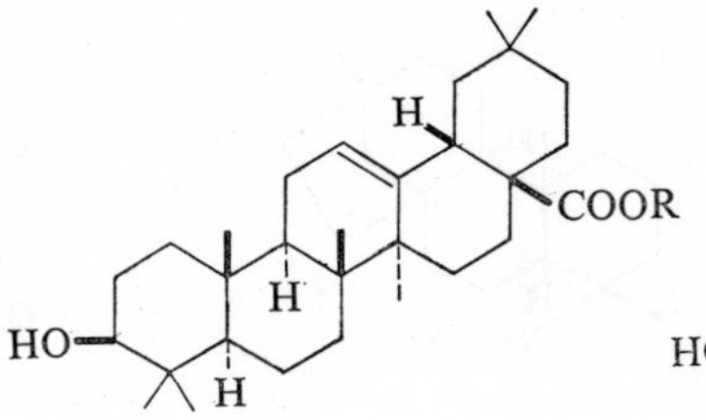

2 (R = H)
13 (R = β-D-glucopyranose)

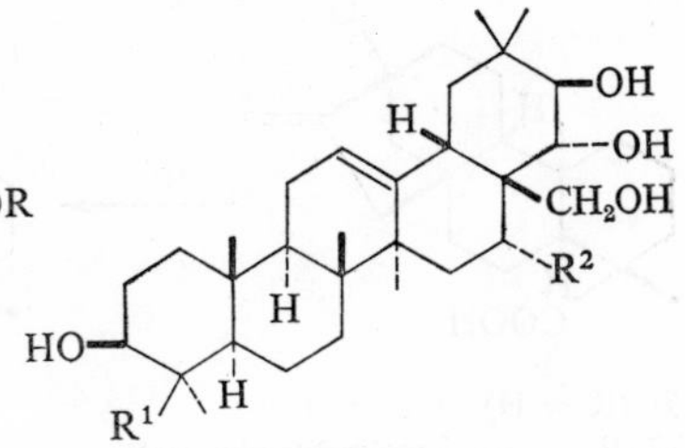

3 (R^1 = CH_2OH, R^2 = OH)
4 (R^1 = Me, R^2 = OH)
5 (R^1 = Me, R^2 = H)

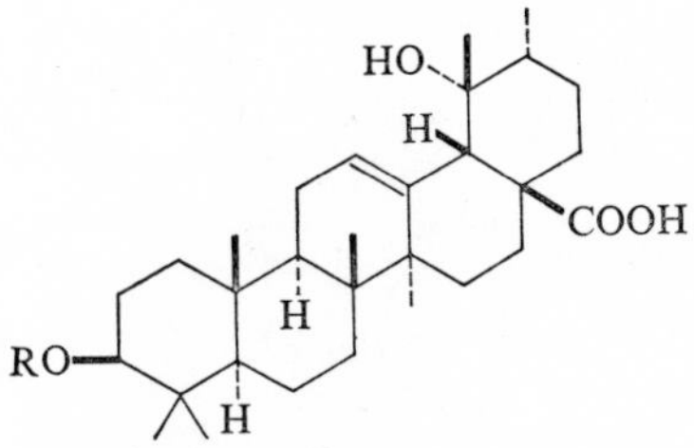

6 (R = H)
14 (R = α-L-arabopyranose)

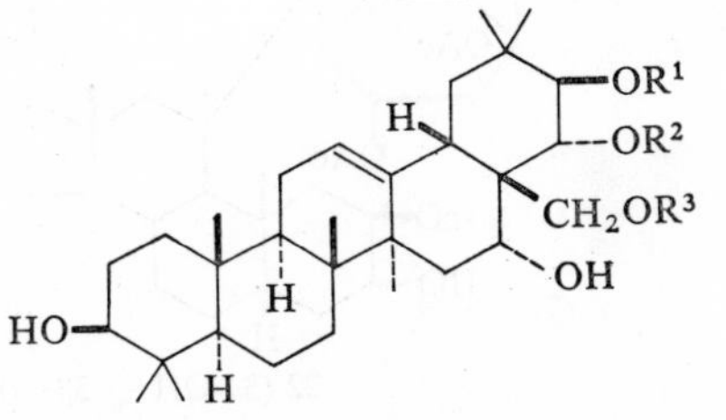

	R^1	R^2	R^3
7	tigloyl	H	acetyl
8	tigloyl	acetyl	H
9	H	H	acetyl
10	2'-cis-hexenoyl	acetyl	H

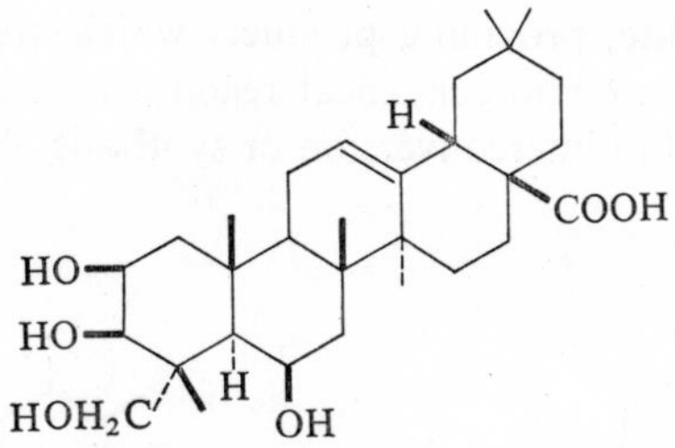

11

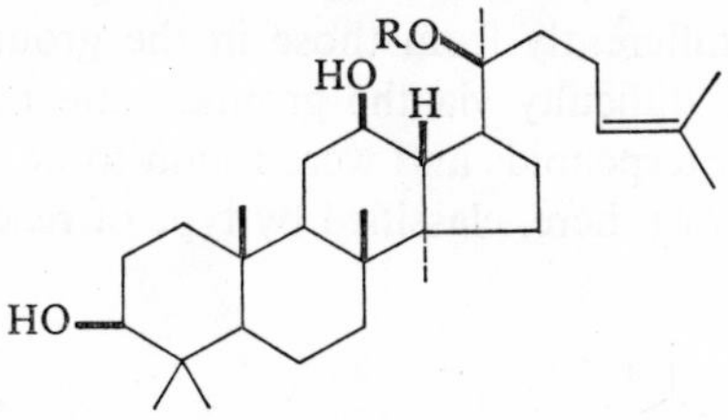

12 (R = H)
15 (R = β-D-glucopyranose)

The usefulness of the method was demonstrated for the cleavage of glycoside linkages of other types, such as diterpenoid glycosides (steviol **16**, steviolbioside **17** from stevioside),[12] monoterpenoid glucosides (a desglucosyl derivative **18** from a desbenzoyl methyl ether **19** of paeoniflorin[13] and dihydroharpagenin **20** from dihydroharpagide **21**[13]), and steroidal glycosides (**22** from glycosides of *Metanarthecium luteo-viride* aerial part[14,15]). Isolation of the prosapogenols **22**, with a fully acetylated arabinoside residue is noteworthy because **22** is inaccesible by the conventional hydrolysis method.

12) I. Yosioka, S. Saijoh, J. A. Waters, I. Kitagawa, *Chem. Pharm. Bull.* **20**, 2500 (1972).
13) I. Yosioka, T. Sugawara, K. Yoshikawa, I. Kitagawa, *Chem. Pharm. Bull.* **20**, 2450 (1972).
14) I. Yosioka, K. Imai, I. Kitagawa, *Tetr. Lett.*, 1177 (1971).
15) Although a soil bacterium selected on a culture medium containing saponin of *Polygala senega* root was identified as *Pseudomonas sp.*, all the other bacteria are so far unidentified.

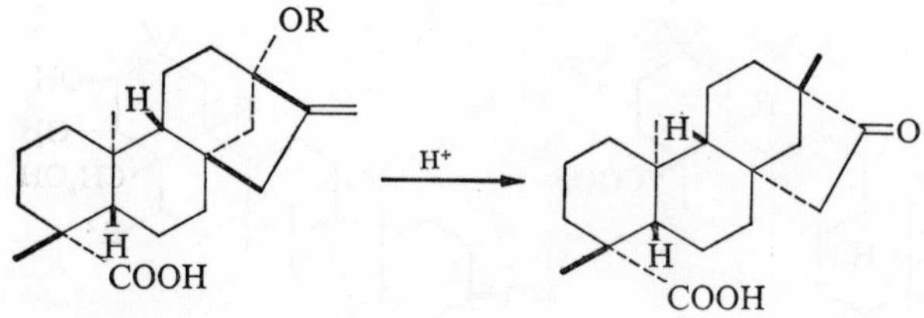

16 (R = H)
17 (R = β-D-glucose-β-D-glucose)

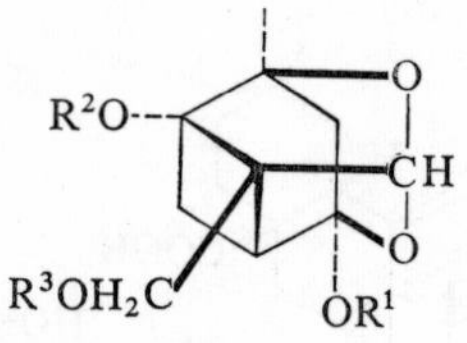

18 (R¹ = Me, R² = R³ = H)
19 (R¹ = Me, R² = β-D-glucose, R³ = H)

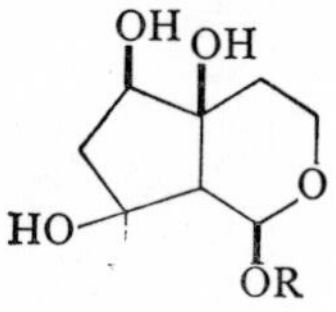

20 (R = H)
21 (R = β-D-glucose)

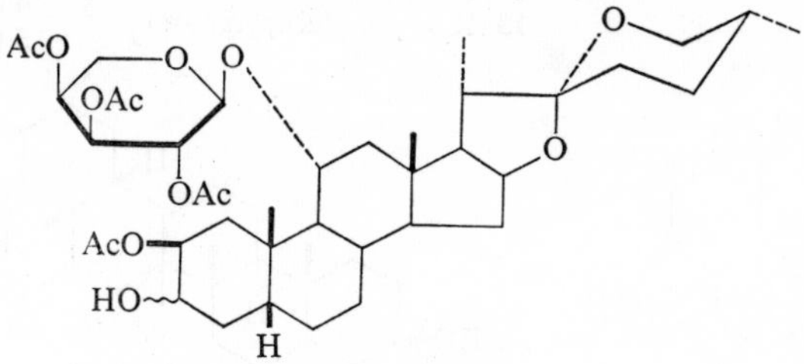

22 (3α-OH + 3β-OH)
(major)

5.37 PHOTOCHEMICAL REACTIONS IN THE TRITERPENE FIELD

Photochemical reactions have been extensively studied in the last decade. Excited molecules behave completely differently from those in the ground state, providing products which are accessible only with difficulty via the ground state reaction.[1] Photochemical reactions were investigated in the triterpenoids and were found to be useful in interconversion or synthesis. A few examples are shown here, classified by type of reaction.

Cyclic dienes

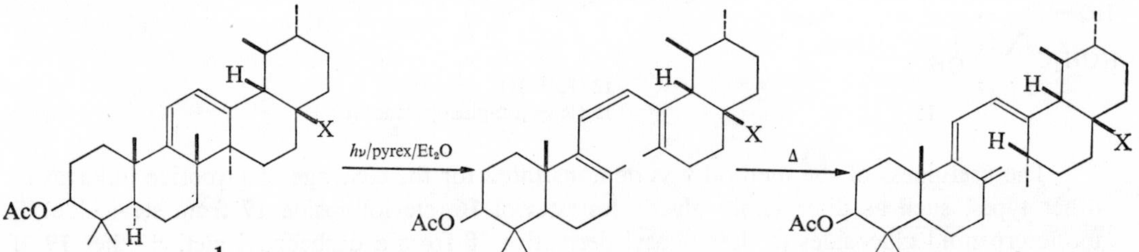

(X = Me or COOMe)

1→2: This is a typical photochemical conrotatory cyclization reminiscent of those in the vitamin D field.[2]

2→3: This is a thermal [1,7]-sigmatropic reaction[3] which partly occurs during irradiation.

1) N. J. Turro, *Molecular Photochemistry*, W. A. Benjamin, 1963; R. O. Kan, *Organic Photochemistry*, McGraw-Hill, 1966; D. O. Neckers, *Mechanistic Organic Photochemistry*, Reinhold, 1967.
2) Cf. L. Fieser, M. Fieser, *Steroids*, Reinhold, 1959; M. P. Rappold, E. Havinga, *Rec. Trav. Chim.*, **79**, 369 (1960).
3) R. L. Autrey, D. H. R. Barton, A. K. Canguly, W. H. Reusch, *J. Chem. Soc.*, 3313 (1961).

Ketone cleavage

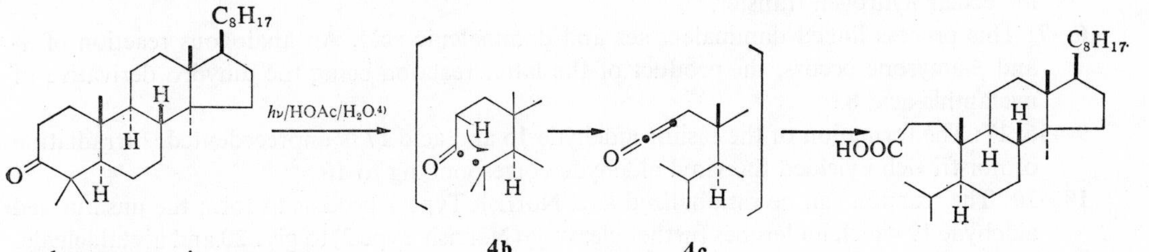

lanostanone **4a** $h\nu/HOAc/H_2O$[4]   **4b**   **4c**   **5**

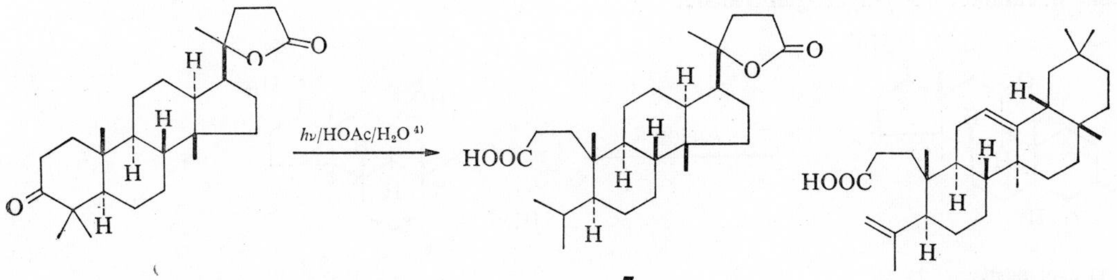

$h\nu/HOAc/H_2O$[4]   **7**   nyctanthic acid **8**[5]

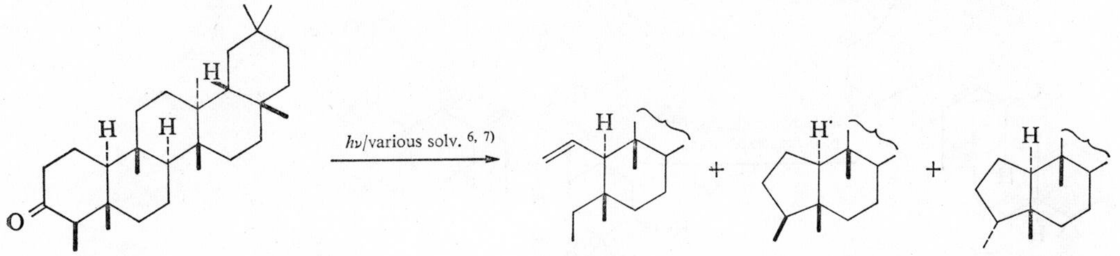

friedelin **9** $h\nu/various$ solv. [6, 7]   **10**   **11**   **12**

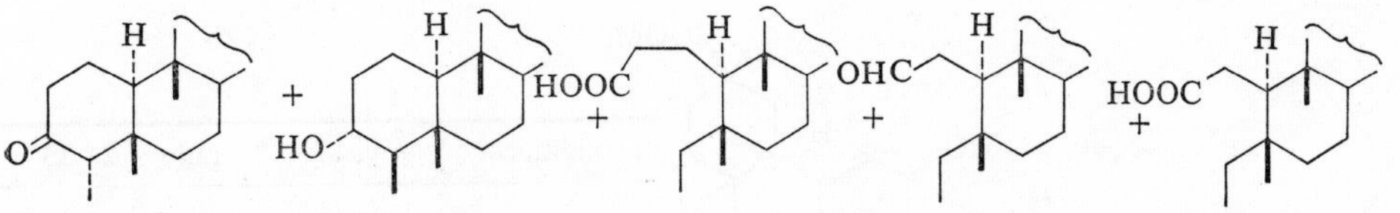

13   **14**   **15**   **16**   **17**

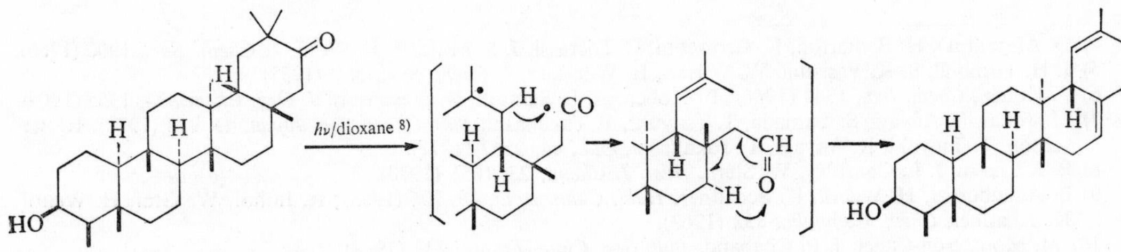

18 $h\nu/dioxane$ [8]   **19**   **20**

4→5: This involves a Norrish Type 1 process (typical of saturated ketones) followed by intra-molecular hydrogen transfer.

6→7: This process linked dammalenones and dammalenic acid. An analogous reaction of α- and β-amyrone occurs, the product of the latter reaction being the dihydro derivative of nyctanthic acid **8**.[5]

9→16, 17: The formation of the bisnor aldehyde **16** and acid **17** is unprecedented.[7] Irradiation of nor-friedelin yielded the vinyl aldehyde corresponding to **16**.

18→20: The reaction can be rationalized as a Norrish Type 1 process to form the unsaturated aldehyde **19** which undergoes further cleavage (Norrish Type 2) to give **20** and acetaldehyde. The latter was actually identified.

Ketone intramolecular γ-hydrogen transfer

11-oxolanostanol **21**   **22**

23   a) $h\nu$/pyrex/dioxane [10]
                b) $h\nu$/quartz/dioxane   **24**

+

conditions	yield	ratio	24:25
a	45%		66:35
b	50%		30:70

25

4) D. Arigoni, D. H. R. Barton, R. Bernsconi, C. Djerassi, J. S. Mills, R. E. Wolff, *J. Chem. Soc.*, 1900 (1960).
5) J. H. Turnbull, S. K. Vasistha, W. Wilson, R. Woodger, *J. Chem. Soc.*, 569 (1957).
6) F. Kohen, *Chem. Ind.*, 1844 (1966); F. Kohen, A. S. Samson, R. Stevenson, *J. Org. Chem.*, **34**, 1355 (1969).
7) M. Takai, R. Aoyagi, S. Yamada, T. Tsuyuki, T. Takahashi, *Bull. Chem. Soc. Japan*, **43**, 972 (1970); R. Stevenson, T. Tsuyuki, R. Aoyagi, T. Takahashi, *ibid.*, **44**, 2567 (1971).
8) B. J. Clarke, J. L. Courtney, W. Stern, *Aust. J. Chem.*, **23**, 1651 (1970).
9) E. Altenburger, H. Wehrli, K. Schaffner, *Helv. Chim. Acta*, **48**, 704 (1965); R. Imhof, W. Graf, H. Wehrli, K. Schaffner, *Chem. Commun.*, 852 (1969).
10) M. Mousseron-Canet, J. P. Chaband, *Bull. Soc. Chim. France*, 239 (1969).
11) B. W. Finucane, J. B. Thomson, *Chem. Commun.*, 380 (1969).

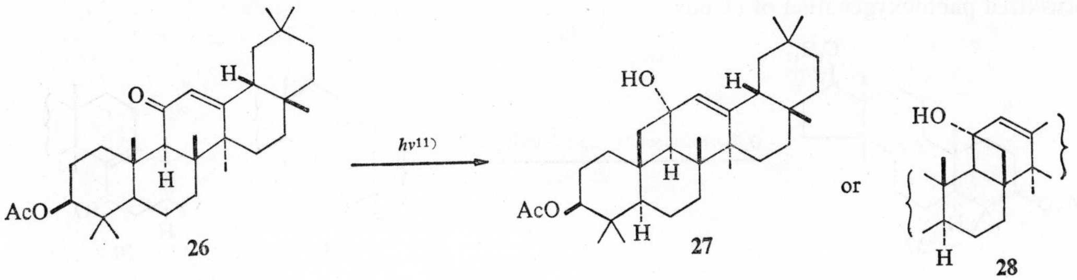

26 → 27 or 28

29 → 30

31 → 32

(R = H or Ac)

21→22: This reaction process was applied in a synthesis of cycloartenol.[9]

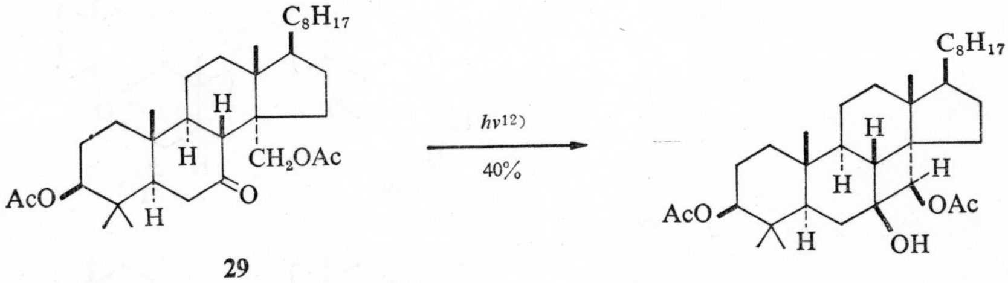

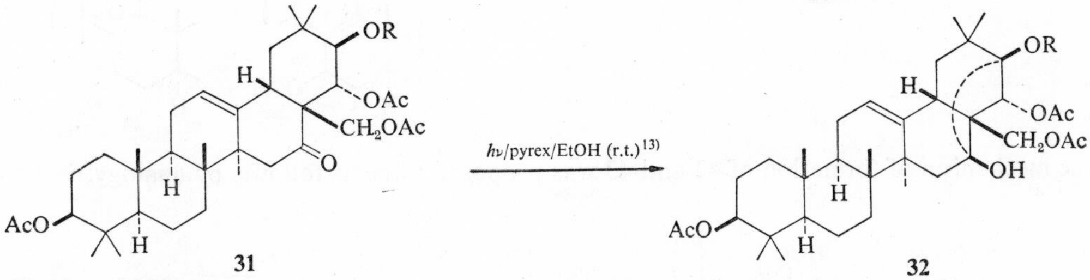

Ketone photoisomerization

This is analogous to the santonin-lumisantonin rearrangement.[15]

33 → 34

35 → 36

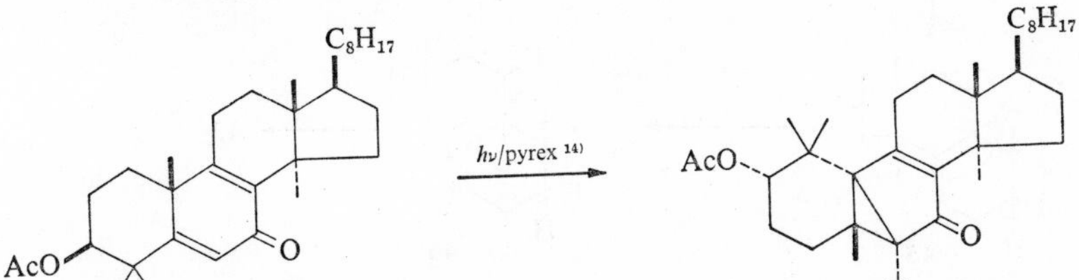

12) J. Fried, J. W. Brown, *Tetr. Lett.*, 925 (1967); *Helv. Chim. Acta*, **42**, 292 (1966).
13) N. Sugiyama, K. Yamada, H. Aoyama, *J. Chem. Soc.* C, 830 (1971).

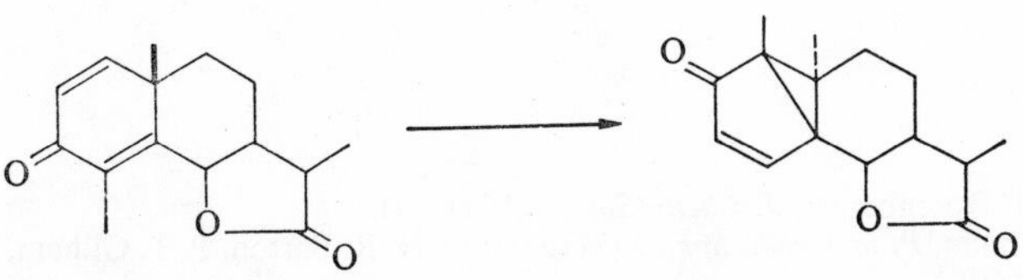

Sensitized photooxygenation of olefins

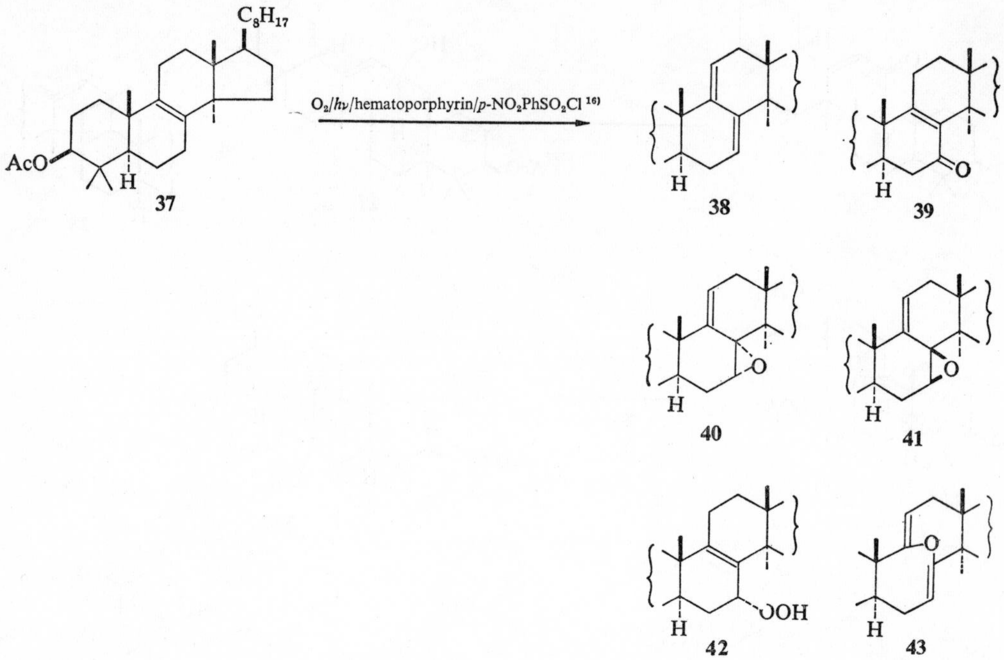

The mechanism of formation of **42** and **43** was proposed to be as follows, by analogy.[17,18]

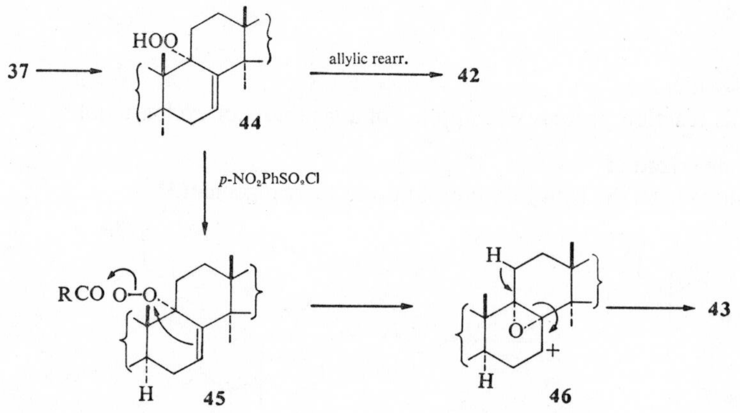

14) D. H. R. Barton, J. F. McGhie, M. Rosenberger, *J. Chem. Soc.*, 1215 (1961).
15) D. H. R. Barton, P. de Mayo, M. Shafiq, *Proc. Chem. Soc.*, 205 (1957); D. H. R. Barton, P. T. Gilham, *ibid.*, 391 (1959); *J. Chem. Soc.*, 4596 (1960).
16) J. E. Fox, A. I. Scott, D. W. Young, *Chem. Commun.*, 1105 (1967); *J. Chem. Soc.*, Parkin 1, 799 (1972).
17) R. Criegee *Chem. Ber.*, **77**, 722 (1944); G. O. Schenck, K. H. Schulte-Elte, *Ann. Chem.*, **618**, 185 (1958).
18) G. O. Schenck, O. A. Menmuller, W. Eisfeld, *Ann. Chem.*, **618**, 202 (1958).

5.38 STRUCTURE OF CUCURBITACIN A

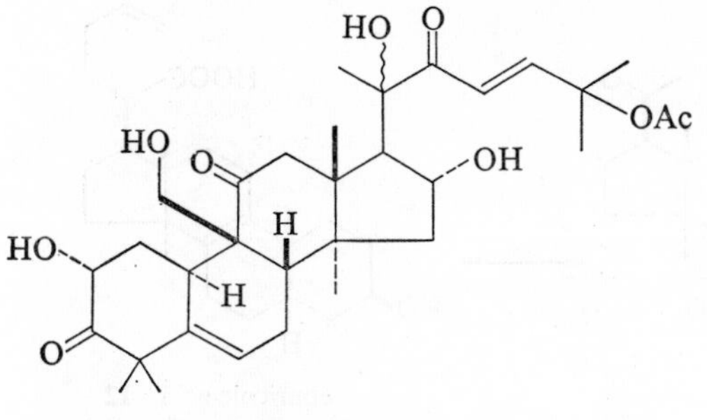

cucurbitacin A **1**

mp: 207–208° α_D (EtOH): +97°
uv: 229 (ϵ 12,200), 290 (ϵ 206)
ir: 1730 (sh), 1715, 1692, 1631, 1258
nmr: 1.06 (3H, s), 1.32 (6H, s), 1.44 (6H, s),
 1.57 (6H, s), 2.05 (3H, s), 5.80 (1H, br), 6.54
 (1H, d, J=14.0), 7.02 (1H, d, J=14.0)

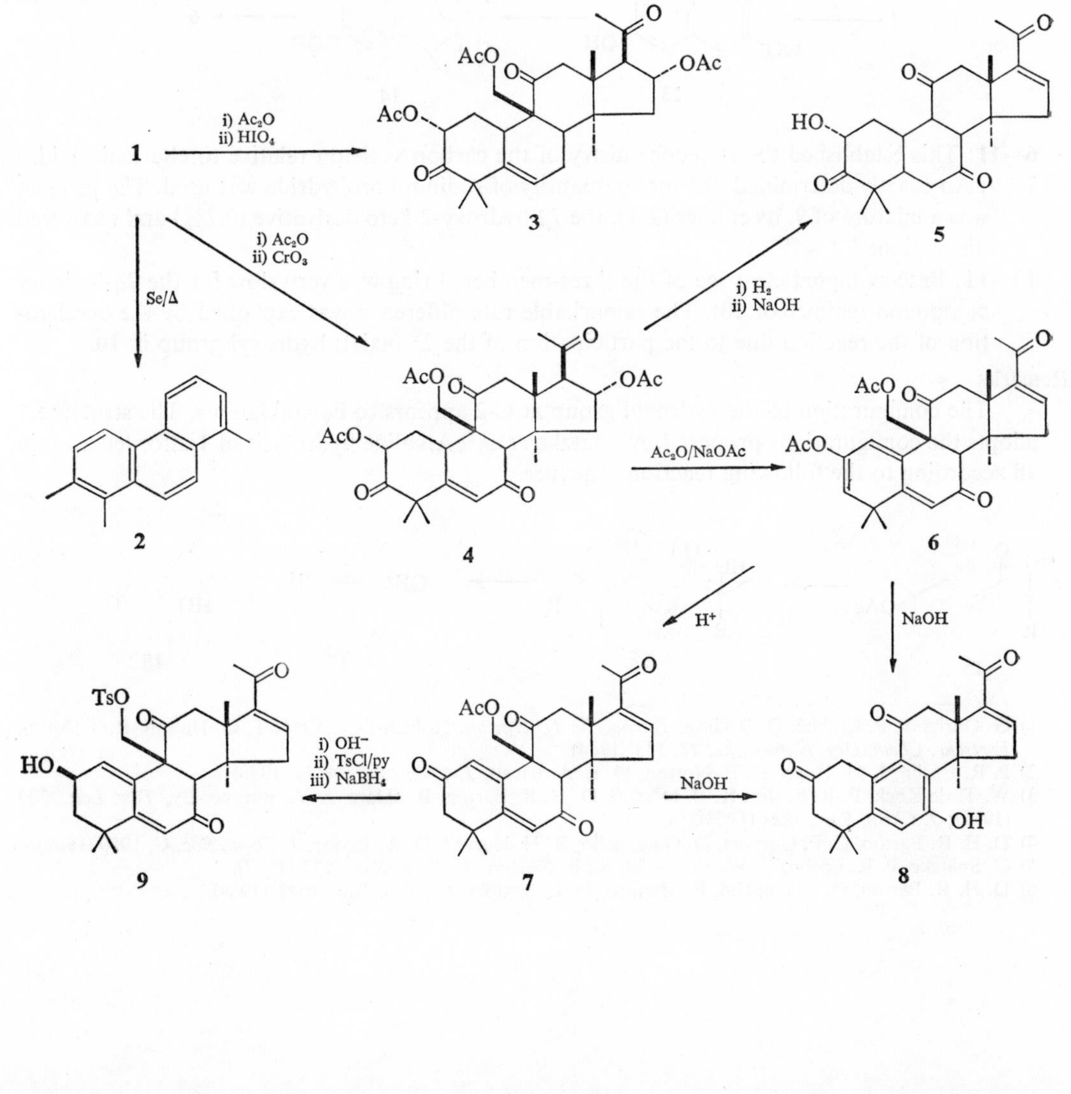

Cucurbitacin A is a major bitter principle found in *Cucurbitaceae*.[1] An oxygenated tetracyclic triterpenoid structure **1** was deduced[2-4] on the basis of extensive degradation reactions as well as chemical correlation with other cucurbitacins.

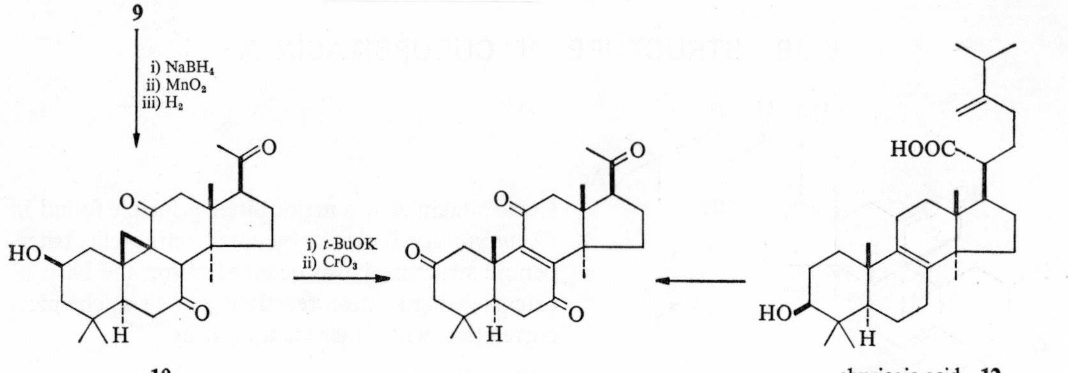

eburicoic acid **12**

1→3: *Trans*-4-acetoxy-4-methylpent-2-enoic acid was also obtained.

4→6: This involves enolization of the 3-keto group and allylic rearrangement as shown.

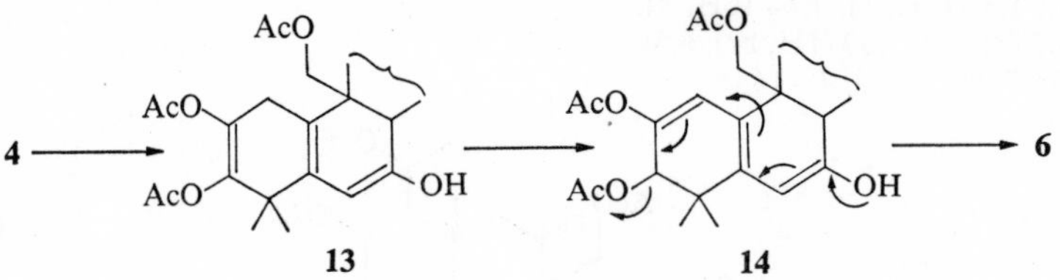

6→11: This established the stereochemistry of the carbon skeleton relative to eburicoic acid.[6]

7→9: An exactly determined (0.5 mole) quantity of sodium borohydride was used. The product was a mixture of **9**, its epimer (2:1), the 7β-hydroxy-2-keto derivative (0.7%) and recovered diene dione **7** ($\sim$45%).

10→11: Base-catalysed cleavage of the three-membered ring was very slow for the 2α-hydroxy compound (epimer of **10**). The remarkable rate difference was explained by the acceleration of the reaction due to the participation of the 2β (axial) hydroxyl group in **10**.

Remarks

The configuration of the hydroxyl group at C-2 appears to be ambiguous. The structure **1** adopts the configurations proposed by Snatzke *et al.*[5] Alkaline hydrolysis of **1** afforded acetoin **18** according to the following reaction sequence.

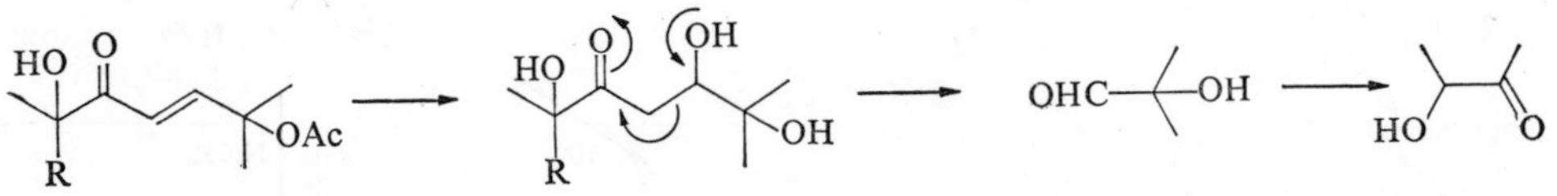

1) G. Ourisson, P. Crabbé, O. Roding, *Tetracyclic Triterpenes*, Holden-Day, 1964; T. G. Halsall, R. T. Aplin, *Fortschr. Chem. Org. Naturstoffe*, **22**, 153 (1964).
2) P. R. Enslin, J. M. Hugo, K. B. Norton, D. E. A. Rivett, *J. Chem. Soc.*, 4779 (1960).
3) W. T. de Kock, P. R. Enslin, K. B. Norton, D. H. R. Barton, B. Sklarz, A. A. Bothner-By, *Tetr. Lett.*, 309 (1962); *J. Chem. Soc.*, 3828 (1963).
4) D. H. R. Barton, C. F. Garbers, D. Giacopello, R. H. Harvey, D. R. Taylor, *J. Chem. Soc. C*, 1050 (1969).
5) G. Snatzke, P. R. Enslin, C. W. Holzapfel, K. B. Norton, *J. Chem. Soc.*, 972 (1967).
6) D. H. R. Barton, D. Giacopello, P. Manitto, D. L. Struble, *J. Chem. Soc.*, 1047 (1969).

5.39 CHEMICAL CORRELATION IN CUCURBITACINS

Some twenty cucurbitacins have been isolated from cucurbitaceaeous plants and their structures determined by chemical correlation[1-5] and from other evidence.

Correlation between cucurbitacins A, B and C

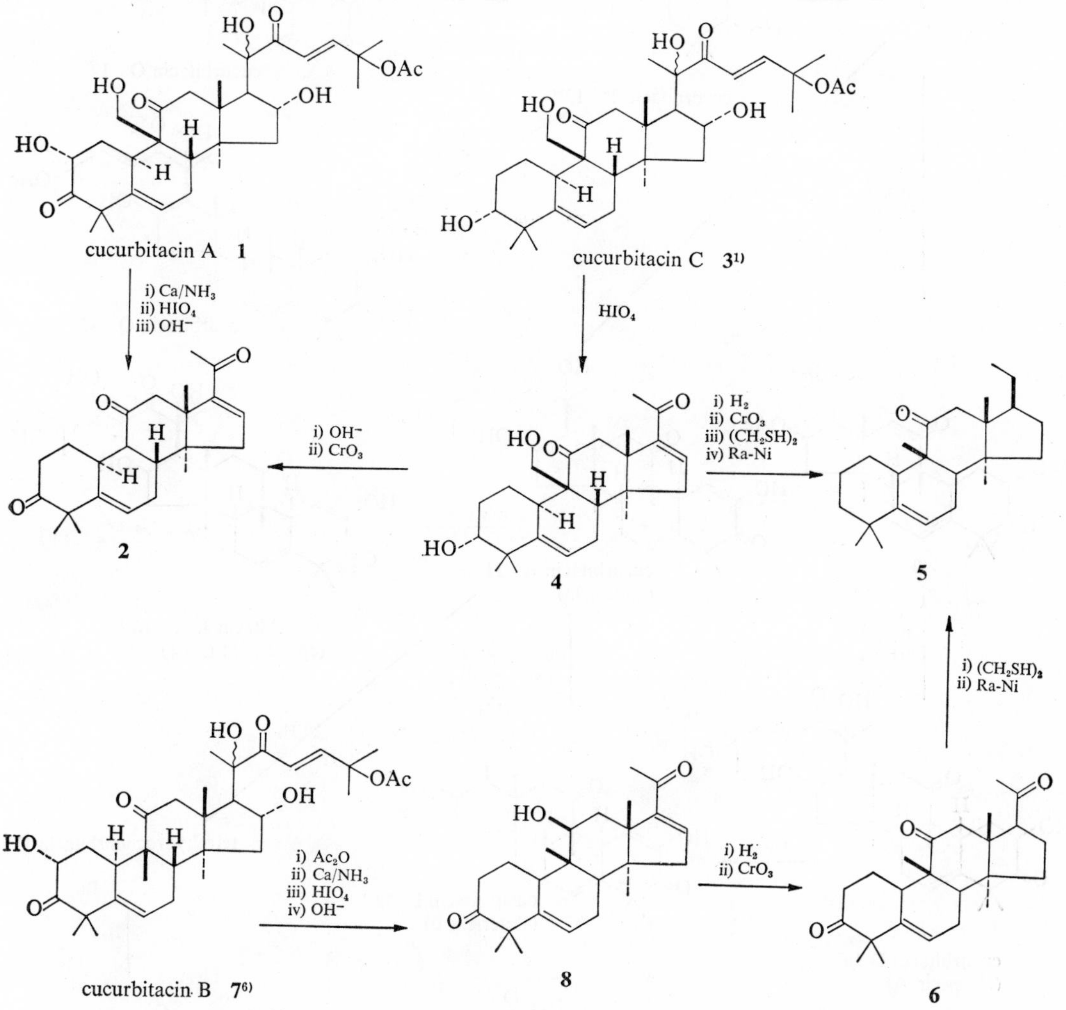

cucurbitacin A **1**

i) Ca/NH₃
ii) HIO₄
iii) OH⁻

cucurbitacin C **3**[1]

HIO₄

i) OH⁻
ii) CrO₃

2

4

i) H₂
ii) CrO₃
iii) (CH₂SH)₂
iv) Ra-Ni

5

i) (CH₂SH)₂
ii) Ra-Ni

cucurbitacin B **7**[6]

i) Ac₂O
ii) Ca/NH₃
iii) HIO₄
iv) OH⁻

8

i) H₂
ii) CrO₃

6

The keto group at C-11 is less reactive than those at the other positions.

1) W. T. de Kock, P. R. Enslin, K. B. Norton, D. H. R. Barton, B. Sklarz, A. A. Bothner-By, *Tetr. Lett.*, 309 (1962); *J. Chem. Soc.*, 3828 (1963).
2) D. Lavie, Y. Shvo, D. Willner, P. R. Enslin, J. M. Hugo, K. B. Norton, *Chem. Ind.*, 951 (1959).
3) D. Lavie, Y. Shvo, O. R. Gottlieb, R. B. Desai, M. L. Khorana, *J. Chem. Soc.*, 3259 (1962).
4) K. J. can der Merwe, P. R. Enslin, K. Pachler, *J. Chem. Soc.*, 4275 (1963).
5) P. R. Enslin, K. B. Norton, *J. Chem. Soc.*, 529 (1964).
6) D. Lavie, Y. Shvo, O. R. Gottlieb, E. Glotter, *Tetr. Lett.*, 615 (1961); *J. Org. Chem.*, **28**, 1790 (1963).

Correlation between cucurbitacins B, D, E, F, I, J, K, O, P and Q and 2-epicucurbitacin B

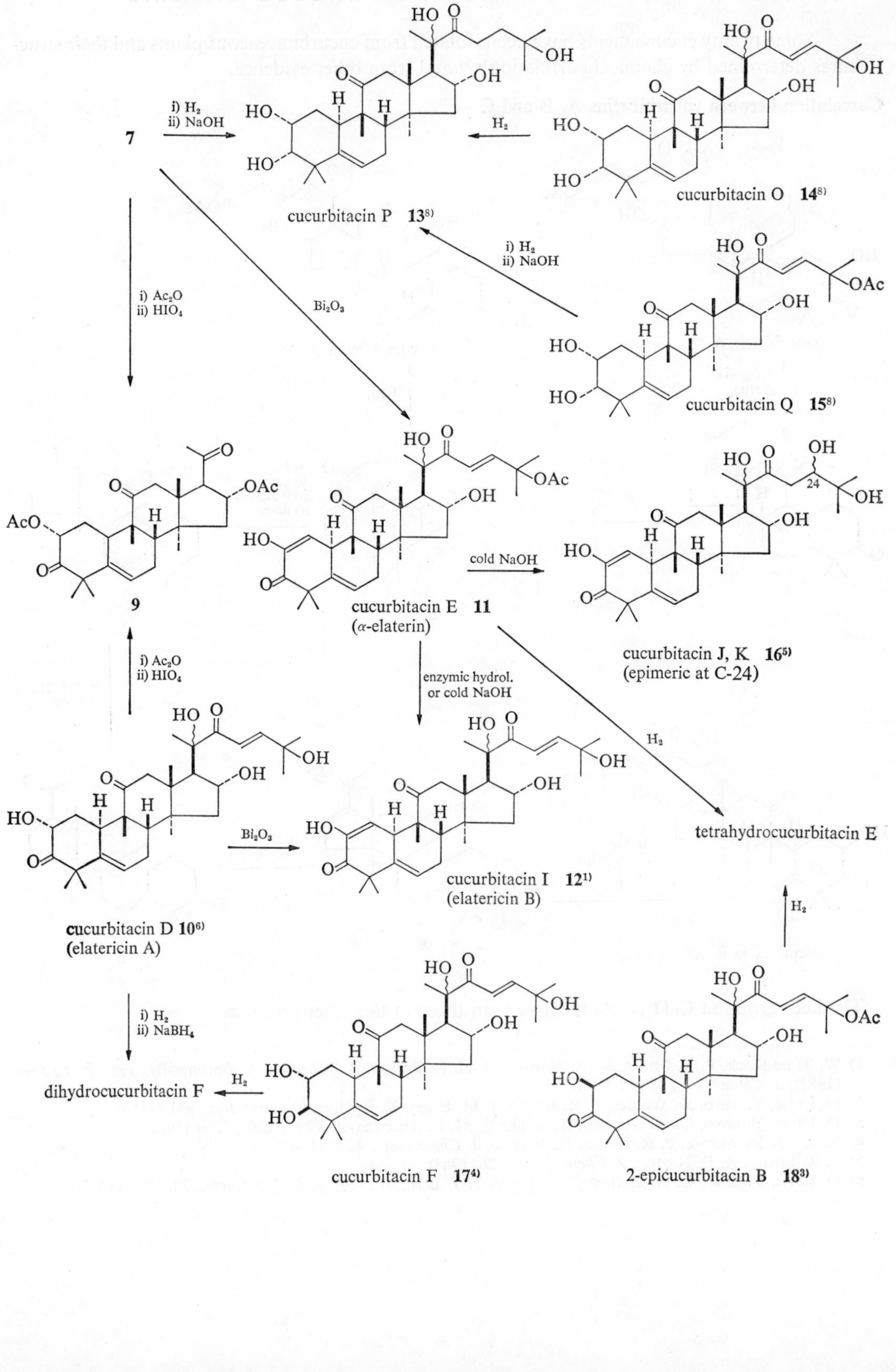

cucurbitacin P 13[8]

cucurbitacin O 14[8]

cucurbitacin Q 15[8]

9

cucurbitacin E 11
(α-elaterin)

cucurbitacin J, K 16[5]
(epimeric at C-24)

tetrahydrocucurbitacin E

cucurbitacin D 10[6]
(elatericin A)

cucurbitacin I 12[1]
(elatericin B)

dihydrocucurbitacin F

cucurbitacin F 17[4]

2-epicucurbitacin B 18[3]

11→12: Fresh juice of Golden Hubbard squash was used for hydrolysis of the acetate. Compound **16** was also isolated from *Iberis amara* (*Cruciferae*).[5] Treatment of **16** with hot alkali afforded **19** and acetoin.[5]

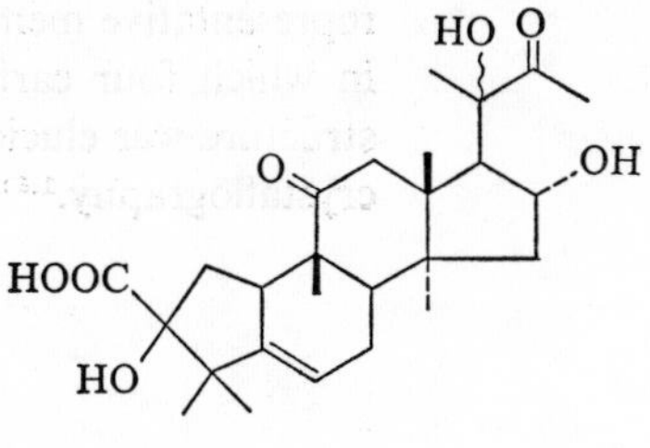

19

Other cucurbitacins

Other cucurbitacins appearing in the literature are as follows:

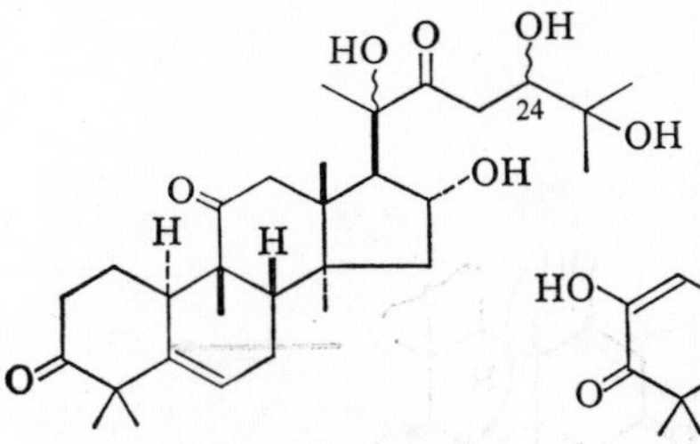

cucurbitacin G, H **20**[7]

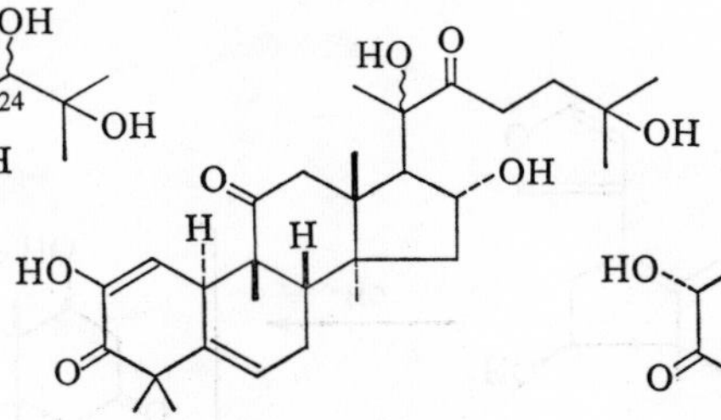

cucurbitacin L **21**[5]

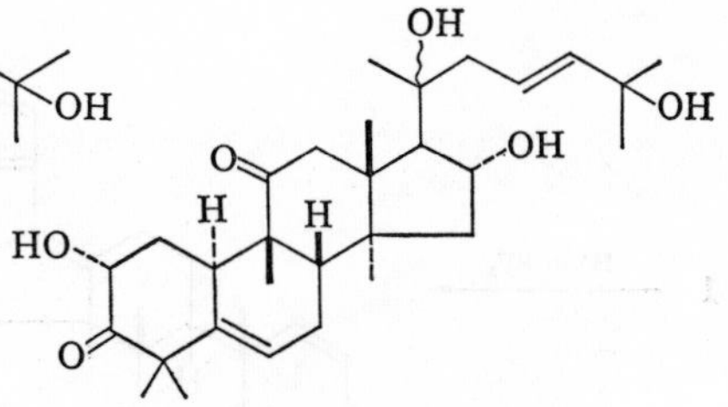

22-deoxocucurbitacin D **22**[9,10]

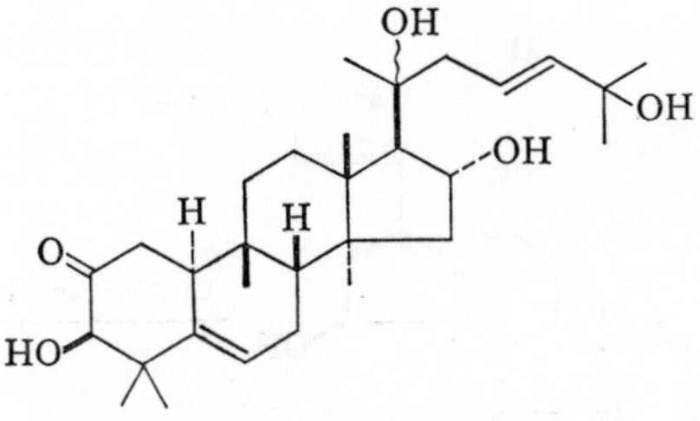

22-deoxoisocucurbitacin D **23**[9,10]

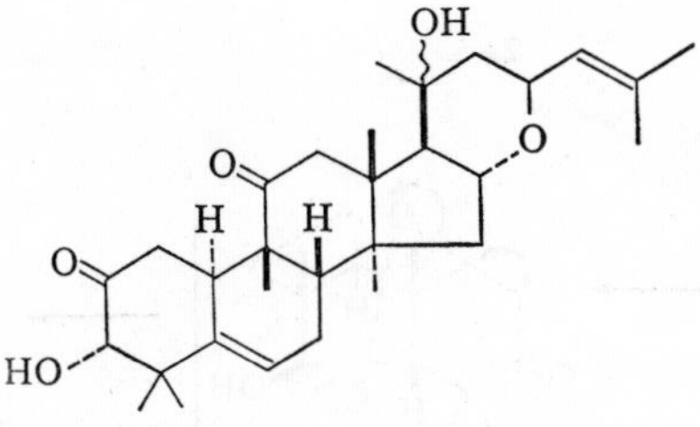

anhydro-3-epi-22-deoxoisocucurbitacin D **24**[9,10]

Systematic nmr studies on some cucurbitacins have been reported.[11]

7) T. K. Devon, A. I. Scott, *Handbook of Naturally Occuring Compounds,* vol. II., *Terpenes,* p. 337, Academic Press, 1972.
8) S. M. Kupchan, R. M. Smith, Y. Aynehchi, M. Maruyama, *J. Org. Chem.,* **35**, 2891 (1970).
9) P. R. Enslin, C. W. Holzapfel, K. B. Norton, S. Rehm, *J. Chem. Soc. C,* 964 (1967).
10) G. Snatzke, P. R. Enslin, C. W. Holzapfel, K. B. Norton, *J. Chem. Soc.,* 972 (1967).
11) D. Lavie, B. S. Benjaminov, Y. Shvo, *Tetr.,* **20**, 2585 (1964).

5.40 CEDRELONE

cedrelone **1**

Cedrelone, isolated from *Cedrela toona*, is a representative member of the limonoids (meliacins) in which four carbocyclic rings remain intact. Its structure was elucidated chemically[1,2] and by x-ray crystallography.[1,3]

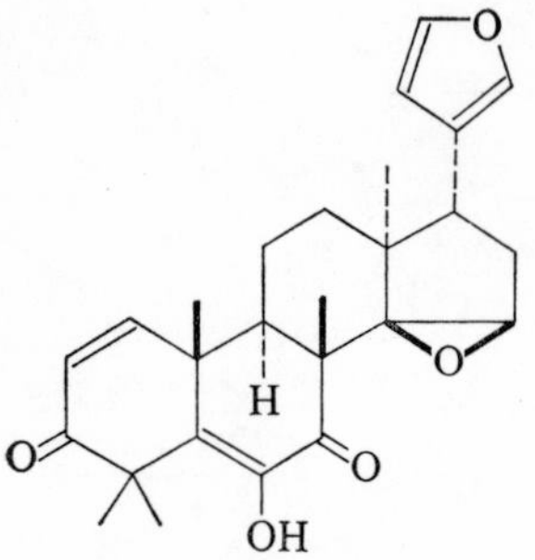

uv (MeOH): 217 (ϵ 11,800), 279 (ϵ 9,100)
uv (NaOH): 327 (ϵ 5,530)
ir: 3400, 3100, 1674, 1505, 878

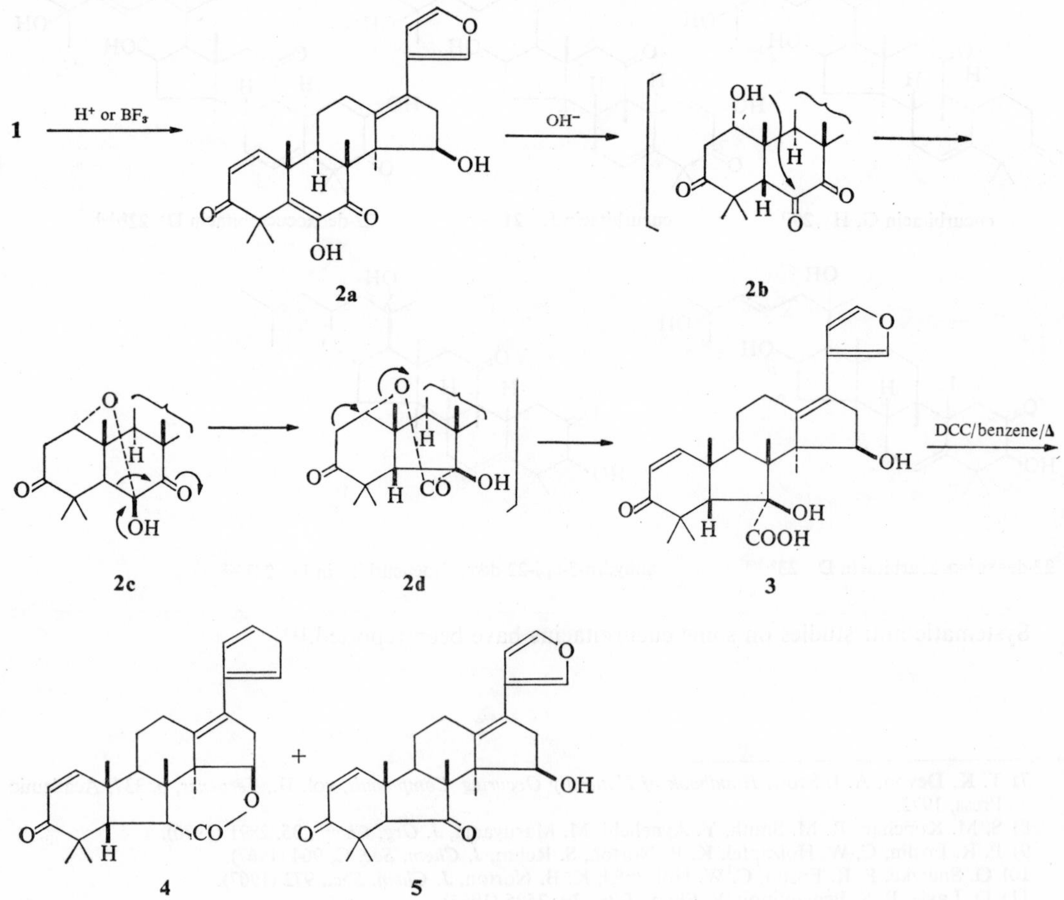

1→2: Facile epoxide ring opening is due to the boat conformation of ring C in **1**.

2→3: This process was suggested because the product has *cis* A/B ring junction, and dihydro-**2** failed to undergo a benzilic acid rearrangement.[1,2]

3→5: **5** is probably derived via a lactide formed from two moles of **3**.[4]

Remarks

Over ten meliacins, a few of which are listed below, have been isolated.

anthothecol **6**[5]   azadirone **7**[6]   nimbolin A **8**[7]

Meliacins appear to be situated biogenetically between tirucallol-type tetracyclic triterpenoids and gedunine-type limonoids. Triterpenoids with oxygen functions in the side chain have been isolated from the sources where limonoids are found (a few are listed below).

flindissol **9**[8]   odoratone **10**[9]   melianone **11**[10]
(*Flindersia dissperma*)   (*Cedrera odorata*)   (*Melia azedarach*)

The above type of triterpenoid has been converted to meliacins in two ways.[11]

1) I. G. Grant, J. A. Hamilton, T. A. Hamor, R. Hodges, S. G. McGeachin, R. A. Raphael, J. M. Robertson, G. A. Sim, *Proc. Chem. Soc.*, 444 (1961).
2) R. Hodges, S. G. McGeachin, R. A. Raphael, *J. Chem. Soc.*, 2515 (1963).
3) I. G. Grant, J. A. Hamilton, T. A. Hamor, J. M. Robertson, G. A. Sim, *J. Chem. Soc.*, 2506 (1963).
4) A. Golomb, P. D. Ritchie, *J. Chem. Soc.*, 838 (1962).
5) C. W. L. Bevan, A. H. Rees, D. A. H. Taylor, *J. Chem. Soc.*, 983 (1963).
6) D. Lavie, M. K. Jain, *Chem. Commun.*, 278 (1967).
7) D. E. U. Ekong, C. O. Fakunle, A. K. Fasina, J. I. Okogun, *Chem. Commun.*, 1166 (1969).
8) A. J. Birch, D. J. Collins, S. Muhammad, J. P. Turnbull, *J. Chem. Soc.*, 2762 (1963).
9) W. R. Chan, D. R. Taylor, G. Snatzke, H. W. Fehlhaber, *Chem. Commun.*, 548 (1967). J. D. Connolly, K. L. Handa, R. McCrindle, K. H. Overton, *Tetr. Lett.*, 3449 (1967).
10) D. Lavie, M. K. Jain, I. Kirson, *J. Chem. Soc. C*, 1347 (1967).
11) J. G. St. C. Buchanan, T. G. Halsall, *Chem. Commun.*, 48 (1969); *J. Chem. Soc. C*, 2280 (1970).

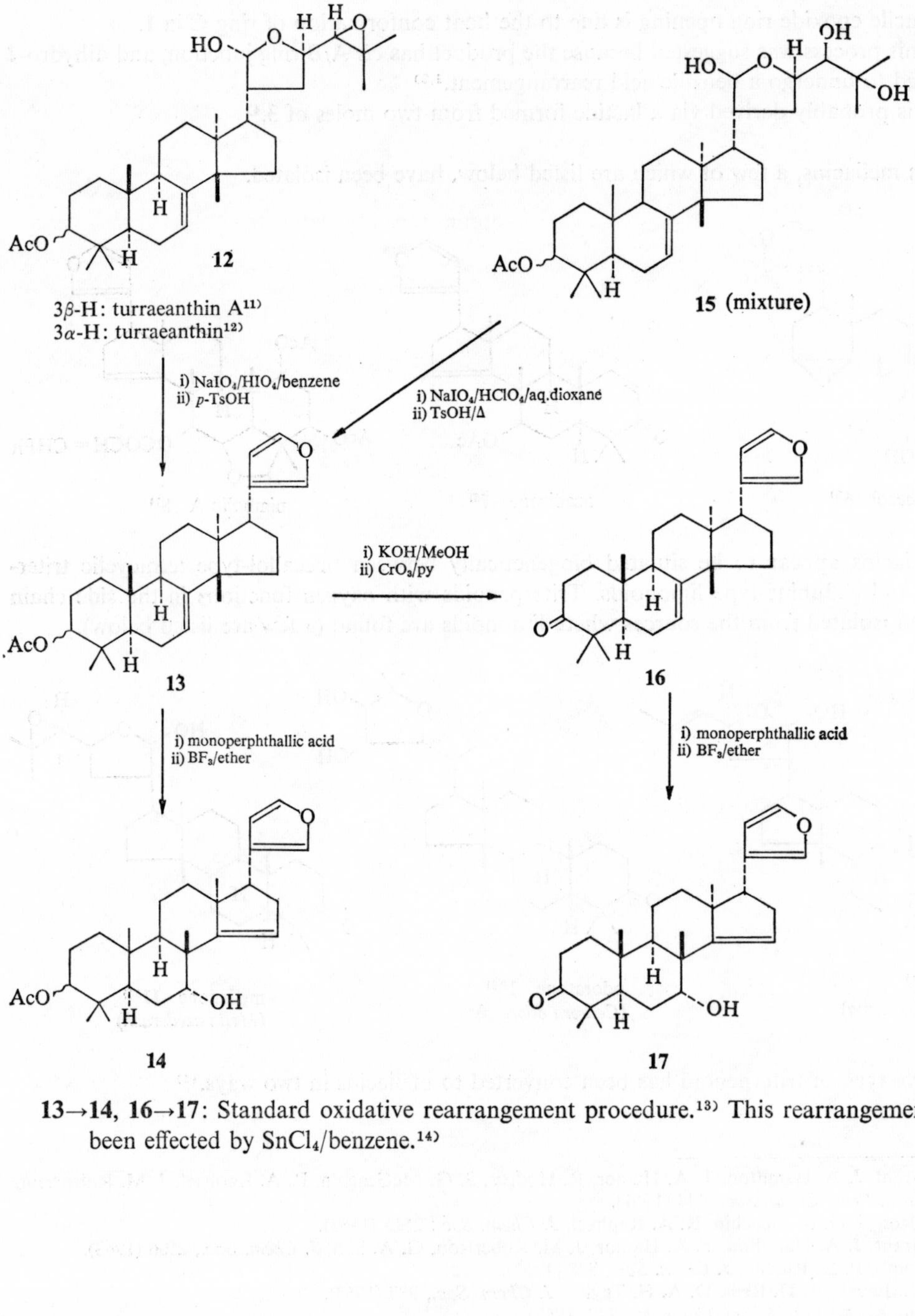

12

3β-H: turraeanthin A[11]
3α-H: turraeanthin[12]

15 (mixture)

i) NaIO$_4$/HIO$_4$/benzene
ii) p-TsOH

i) NaIO$_4$/HClO$_4$/aq.dioxane
ii) TsOH/Δ

13

i) KOH/MeOH
ii) CrO$_3$/py

16

i) monoperphthallic acid
ii) BF$_3$/ether

i) monoperphthallic acid
ii) BF$_3$/ether

14

17

13→14, 16→17: Standard oxidative rearrangement procedure.[13] This rearrangement has **also** been effected by SnCl$_4$/benzene.[14]

12) C. W. L. Bevan, D. E. U. Ekong, T. G. Halsall, P. Toft, *J. Chem. Soc.* C, 820 (1970).
13) G. P. Cotterell, T. G. Halsall, M. J. Wriglesworth, *J. Chem. Soc.* C, 1503 (1970).
14) D. Lavie, E. C. Levy, *Tetr.* **27**, 3941 (1971).

5.41 GEDUNIN

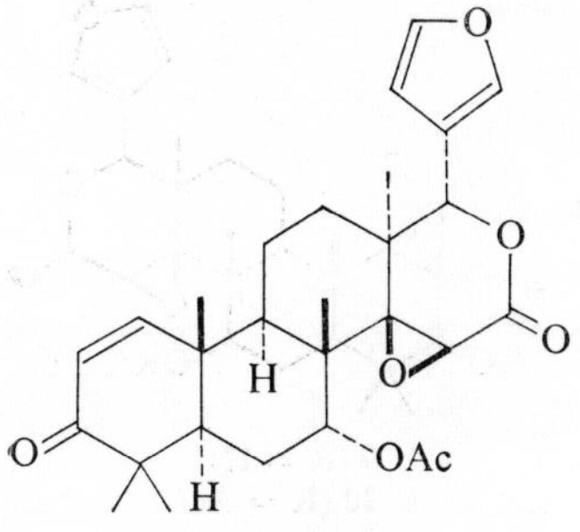

gedunin **1**

Gedunin **1** was isolated from Meliaceae plants.[1] Chemical degradation[2] indicated the presence of the nortriterpenoid structure **1**, which has been confirmed by x-ray analysis.[3]

mp: 218° α_D: −44°
uv: 215 (4.12), 335 (1.8)
ir: 3500, 1740, 1668, 1500, 875

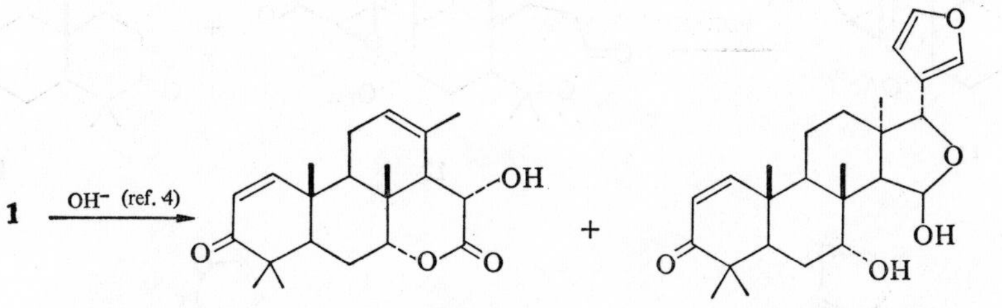

Remarks

The similarity of the skeleton of **2** and that of the qaussins suggests a biogenetic relationship.

Chemical conversion of meliacin-type compounds to gedunin-type compounds in a way reminiscent of the biogenetic pathway of the latter[5] has been achieved.[6,7] The presence of compounds in various oxidation states such as azadirone **4**, azadiradione **5**, epoxyazadiradione **6** and **1** in the same plant supported the stepwise oxidation mechanism of the D ring in the plant.[8]

1) A. Akisanya, C. W. L. Bevan, J. Hirst, T. G. Halsall, D. A. H. Taylor, *J. Chem. Soc.*, 3827 (1960).
2) A. Akisanya, C. W. L. Bevan, T. G. Halsall, J. W. Powell, D. A. H. Taylor, *J. Chem. Soc.*, 3705 (1961).
3) S. A. Sutherland, G. A. Sim, J. M. Robertson, *Proc. Chem. Soc.*, 222 (1962).
4) A. Akisanya, E. O. Arene, C. W. L. Bevan, D. E. U. Ekong, M. N. Nwaji, J. I. Okogun, J. W. Powell, D. A. H. Taylor, *J. Chem. Soc. C*, 506 (1966).
5) D. Arigoni, D. H. R. Barton, E. J. Corey, O. Jeger, L. Caglioti, S. Dev, P. G. Ferrini, E. R. Glazier, A. Melara, S. K. Pradhan, K. Schaffner, F. Sternhell, J. F. Templeton, F. Tobinga, *Experientia*, **16**, 41 (1960).
6) D. Lavie, E. C. Levy, *Tetr. Lett.*, 1315 (1970).
7) J. G. St. C. Buchanan, T. G. Halsall, *Chem. Commun.*, 1493 (1969).
8) D. Lavie, M. K. Jain, *Chem. Commun.*, 278 (1967).

6 → 1; 7 (R = H), 8 (R = Ac); 9 (R = H), 10 (R = Ac)

11 12 13

khayanthone 14 khivorine 15

4

5

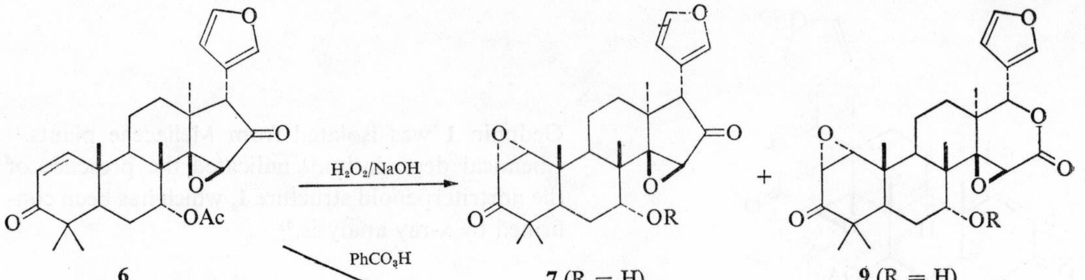

11→**12**: This process implies a possible biogenesis of obacunone.

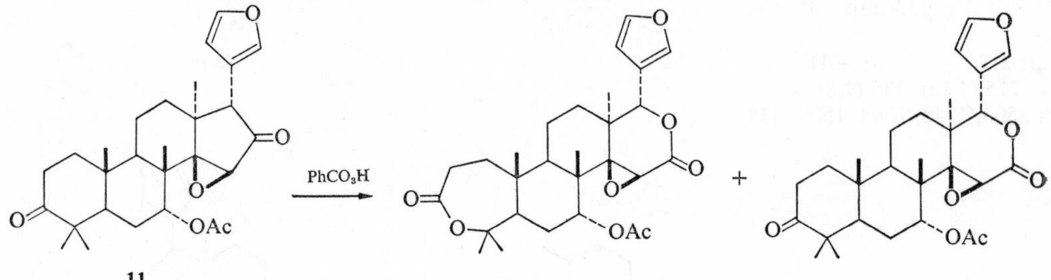

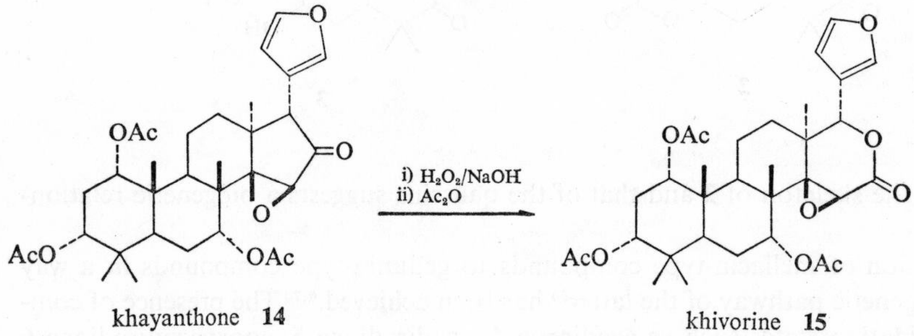

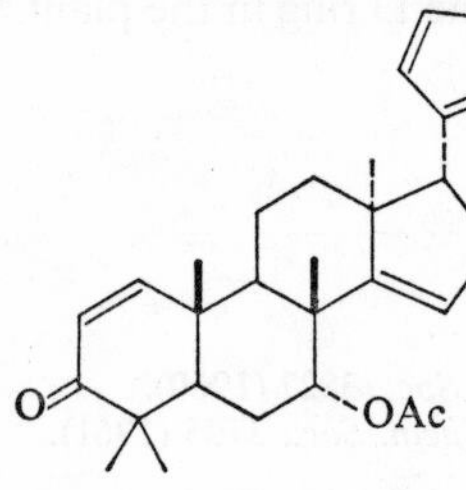

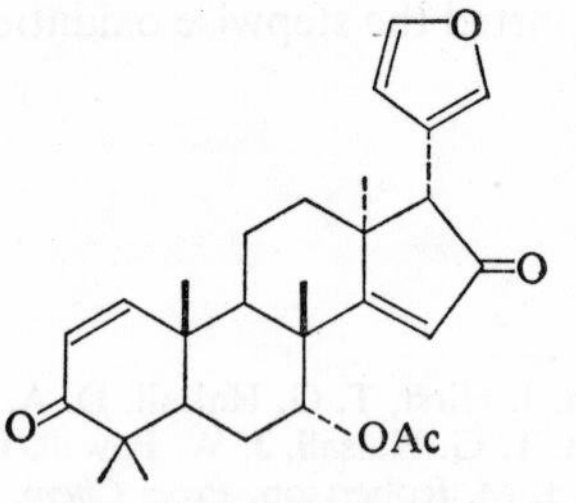

Many related nortriterpenes have been isolated, some of which are illustrated below.

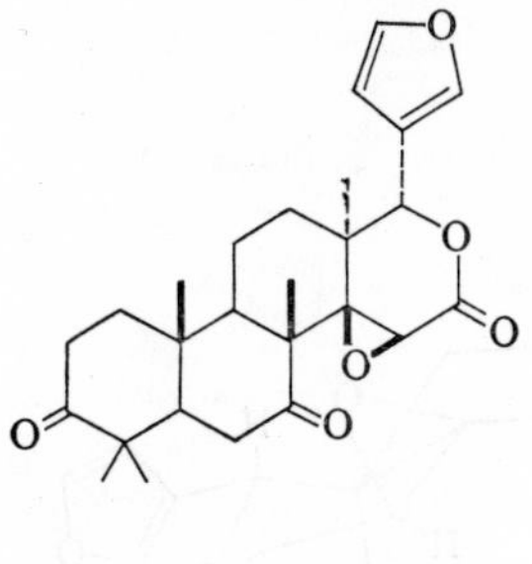

7-oxo-dihydrogedunin **16**[9)]

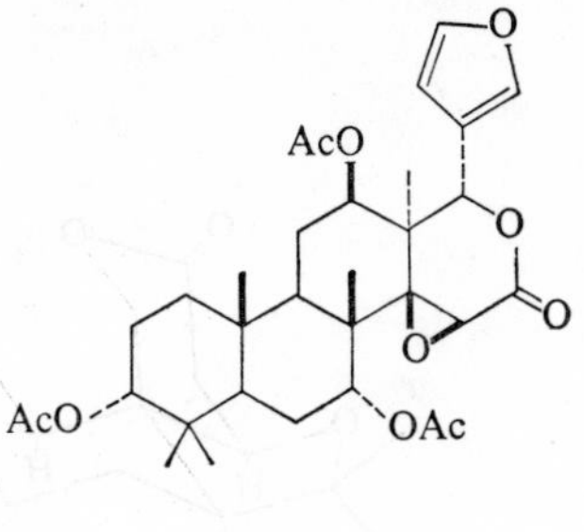

nyasin **17**[10)]

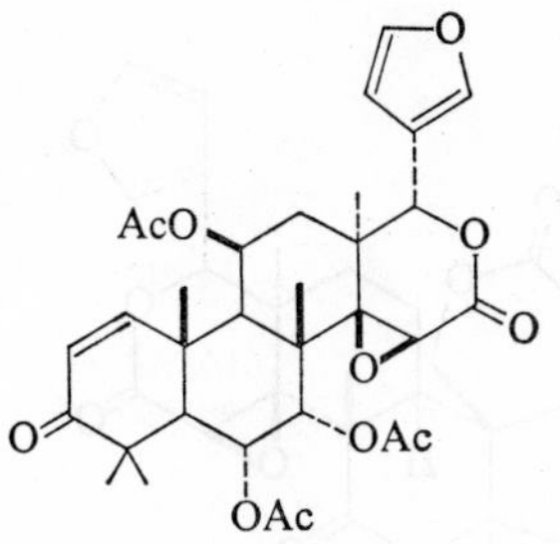

6α, 12β-diacetoxygedunin **18**[11)]

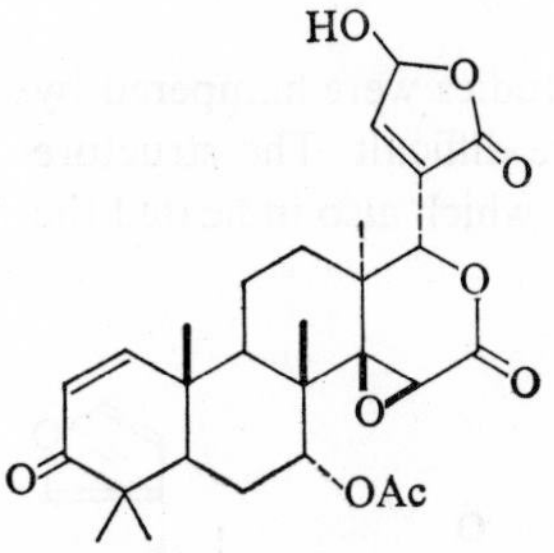

photogedunin **19**[12)]

9) J. R. Housley, F. E. King, T. J. King, P. R. Taylor, *J. Chem. Soc.*, 5095 (1962).
10) D. A. H. Taylor, *Chem. Commun.*, 500 (1967).
11) J. D. Connolly, R. McCrindle, K. H. Overton, J. Feeney, *Tetr.*, **22**, 891 (1966).
12) B. A. Burke, W. R. Chan, K. E. Magnus, D. R. Taylor, *Tetr.*, **25**, 5007 (1969).

5.42   STRUCTURE OF LIMONIN

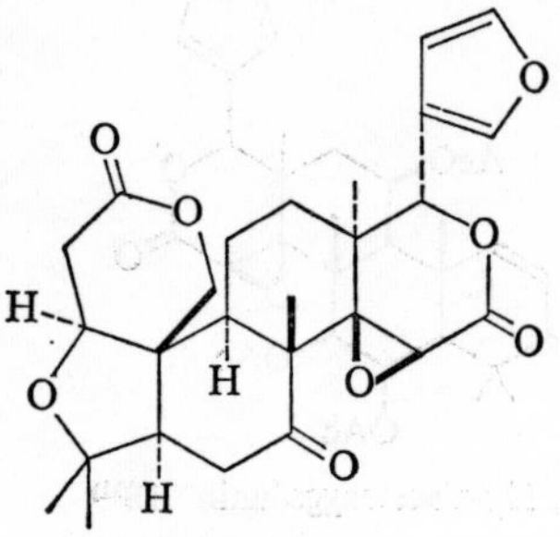

limonin **1**

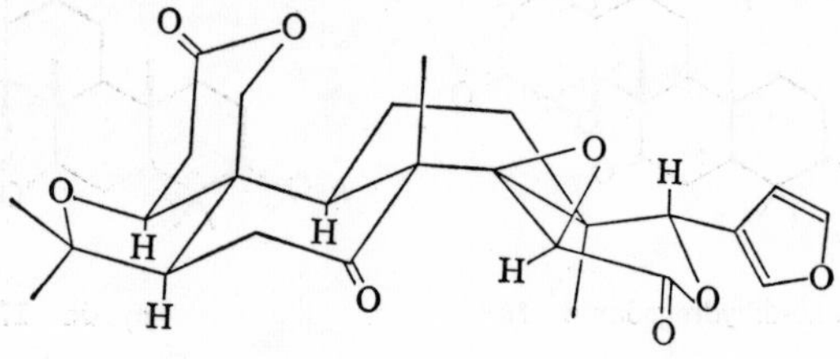

1a

Limonin is the major bitter principle in *Citrus* seeds. Structural studies were hampered by a number of rearrangements[1] which made derivation of the structure difficult. The structure was finally elucidated both chemically[2] and by x-ray crystallography,[3] which also indicated the conformation of the compound **1a**.

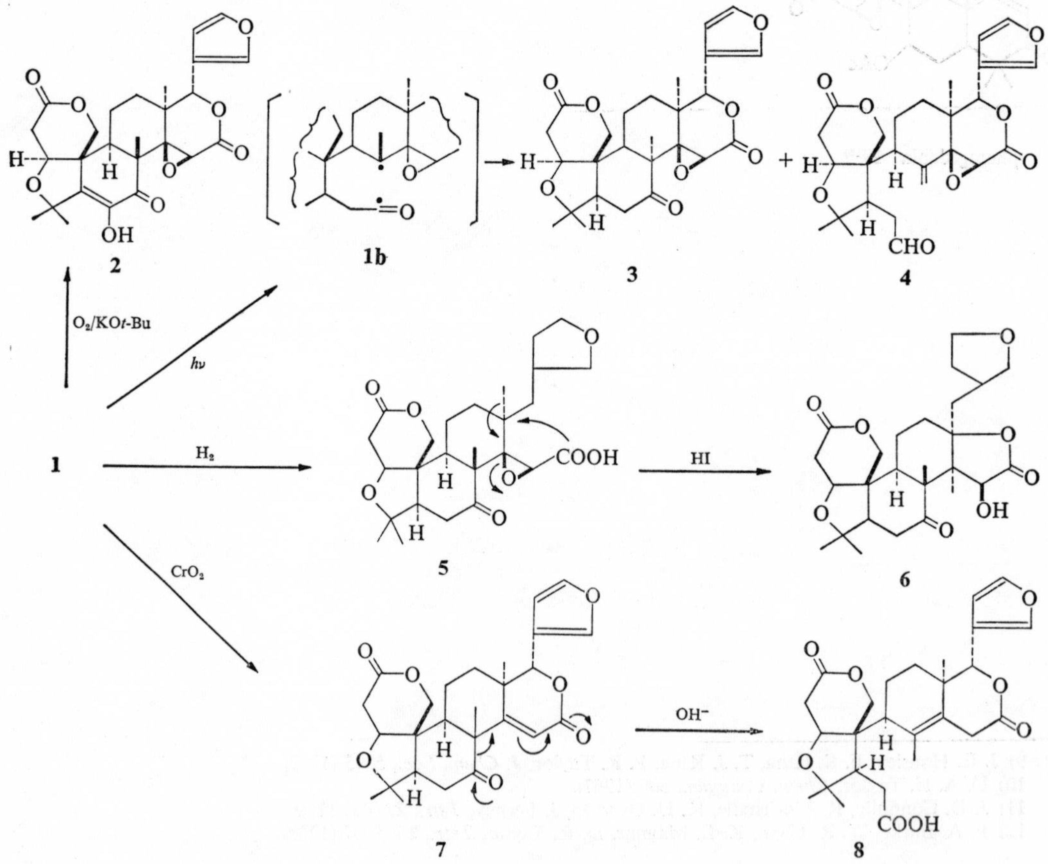

2   **1b**   **3**   **4**

O$_2$/KOt-Bu

$h\nu$

1 $\xrightarrow{\text{H}_2}$

CrO$_3$

5 $\xrightarrow{\text{HI}}$ **6**

7 $\xrightarrow{\text{OH}^-}$ **8**

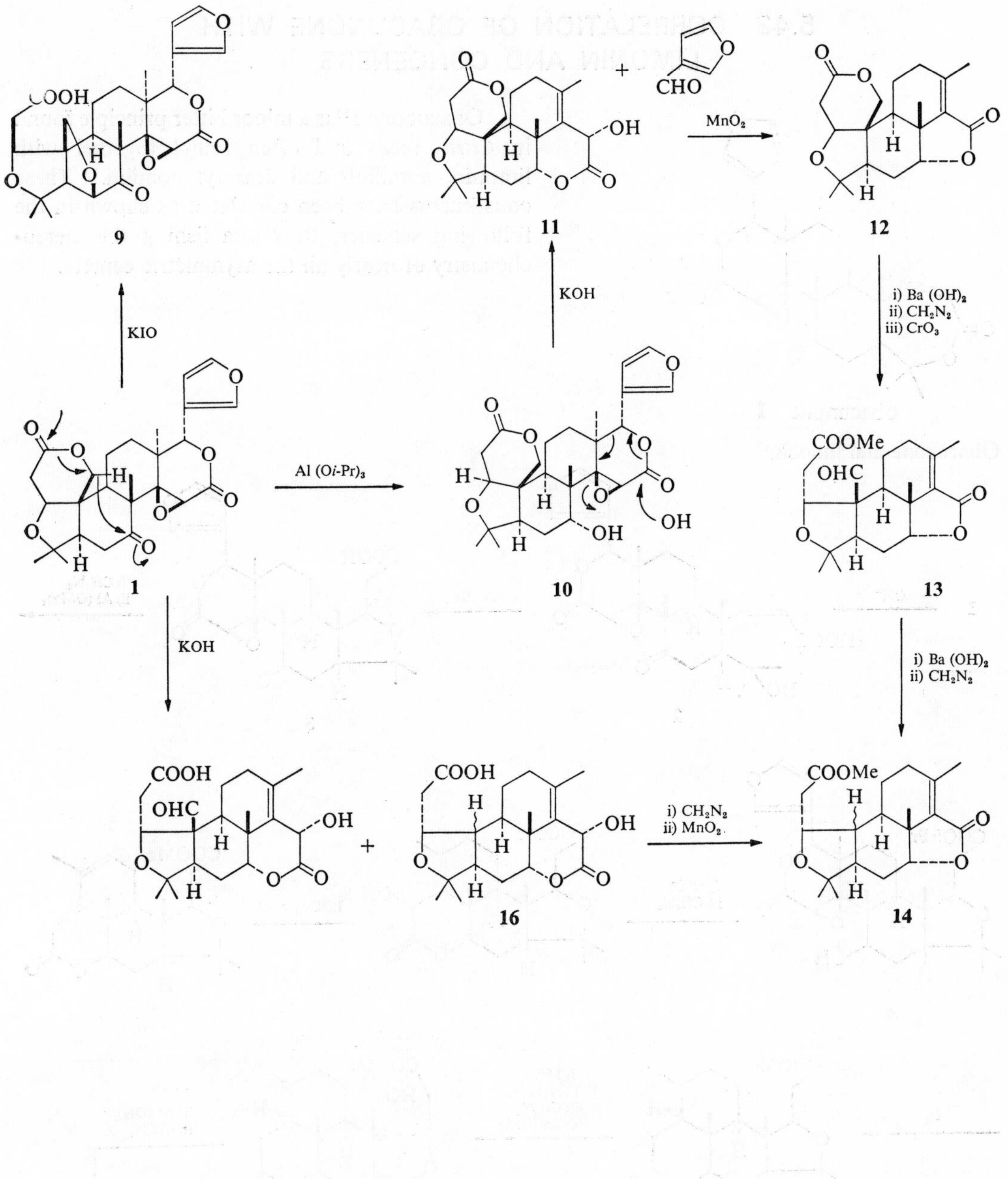

1) J. L. Courtney, *The Terpenoid Bitter Principles, Pure Appl. Chem.*, **11**, 118 (1961); G. Ourisson, P. Crabbé O. Rodig, *Tetracyclic Triterpenes*, Holden-Day, 1964; D. L. Dreyer, *Fortschr. Chem. Org. Naturstoffe*, **26**, 190 (1968).

2) D. Arigoni, D. H. R. Barton, E. J. Corey, O. Jeger, L. Caglioti, S. Dev, P. G. Ferrini, E. R. Glazier, A. Melera, S. K. Pradhan, K. Schaffner, S. Sternhell, J. E. Templeton, S. Tobinaga, *Experientia*, **16**, 41 (1960).

3) S. Arnott, A. W. Davie, J. M. Robertson, G. A. Sim, D. G. Watson, *Experientia.*, **16**, 49 (1960); *J. Chem. Soc.* 4183 (1961).

5.43 CORRELATION OF OBACUNONE WITH LIMONIN AND CONGENERS

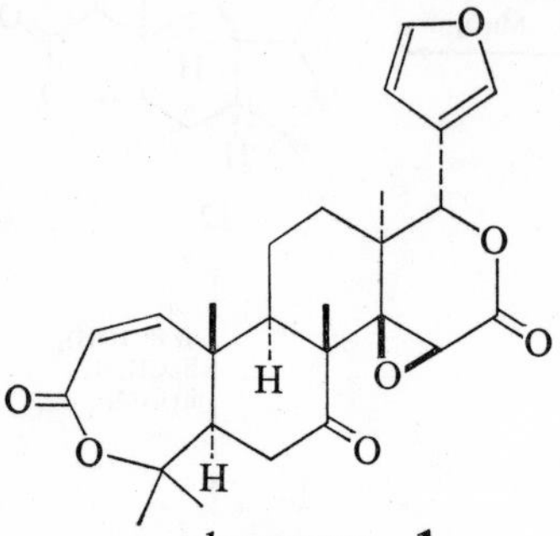

obacunone **1**

Obacunone **1**[1] is a minor bitter principle found in *Citrus* seeds and often found together with limonin, nomilin[1] and deacetyl nomilin.[1] These constituents have been correlated as shown in the following schemes, thus establishing the stereochemistry of nearly all the asymmetric centers.

Obacunone and limonin[2]

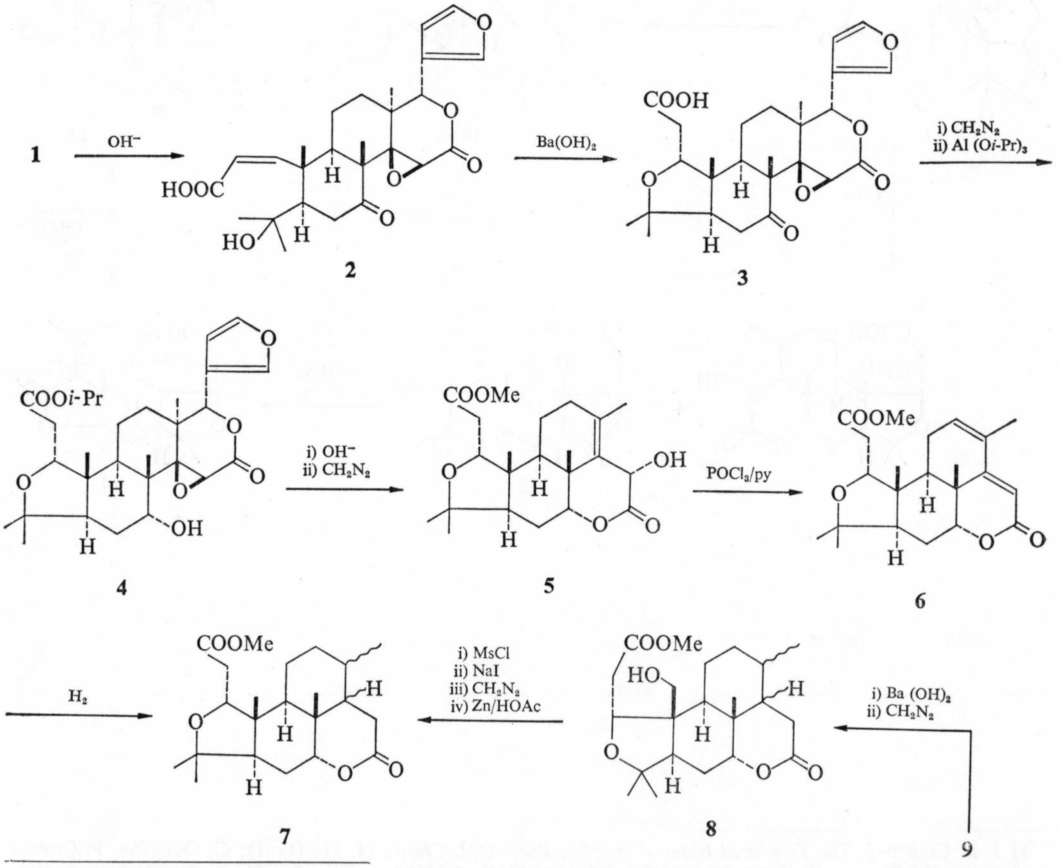

1) O. H. Emerson, *J. Am. Chem. Soc.*, **70**, 545 (1948); *ibid.*, **73**, 2621 (1951); F. Sondheimer, A. Meisels, F. A. Kincl, *J. Org. Chem.*, **24**, 870 (1959); D. Arigoni, D. H. R. Barton, E. J. Corey, O. Jeger, L. Caglioti, S. Dev, P. G. Ferrini, E. R. Glazier, A. Melera, S. K. Pradham, K. Schaffner, S. Sternhell, J. E. Templeton, S. Tobinaga, *Experientia*, **16**, 41 (1960).
2) T. Kubota, T. Matsuura, T. Tokoroyama, T. Komiyama, T. Matsumoto, *Tetr. Lett.*, 325 (1961).
3) O. H. Emerson, *J. Am. Chem. Soc.*, **73**, 2621 (1951); D. L. Drayer, *J. Org. Chem.*, **30**, 749 (1965).

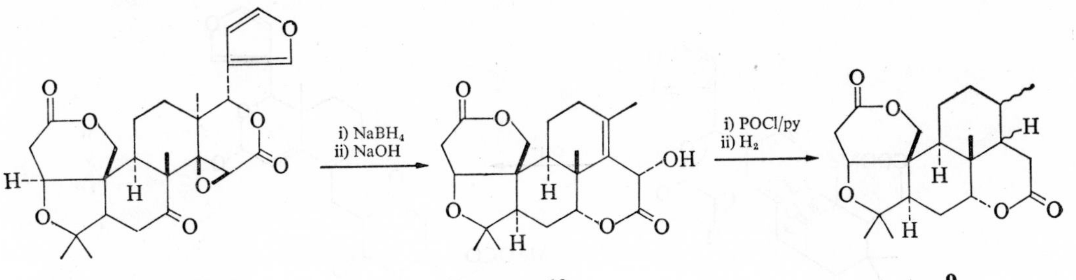

limonin **11**   **10**   **9**

4→5, 11→10: This is a limonol-merolimonol type transformation.

Obacunone, nomilin and deacetyl nomilin[3]

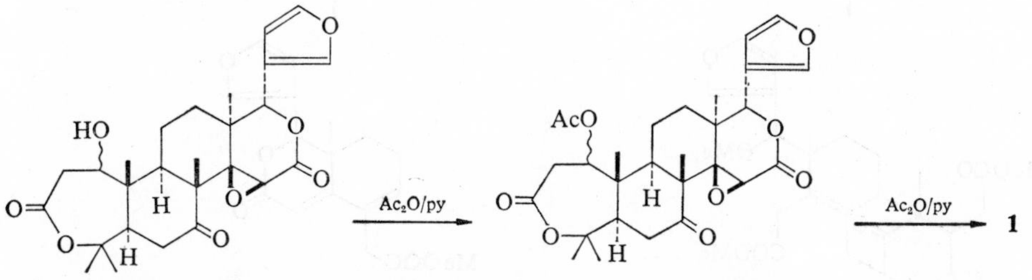

deacetylnomilin **12**   nomilin **13**

12→13: Obacunone was also formed.

Alkaline treatment of **1** and **13**, but not **12**, yielded obacunoic acid **2**, another natural product.

5.44 MEXICANOLIDE

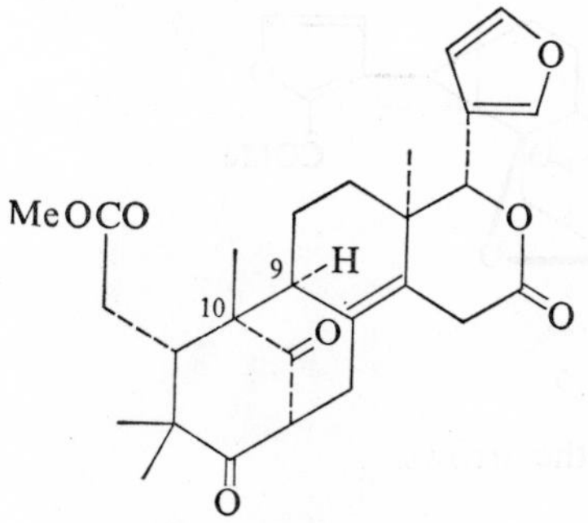

mexicanolide **1**

Mexicanolide **1** was first isolated from *Cedrela adorata*[1] and has been found in plants of the genera *Cedrela* and *Khaya* (Meliaceae). The structure of **1** was deduced from chemical studies[2] and by x-ray studies[3] of a derivative.[4]

1) C. W. L. Bevan, J. W. Powell, D. A. H. Taylor, *J. Chem. Soc.*, 980 (1963).
2) J. D. Connolly, R. McCrindle, K. H. Overton, *Chem. Commun.*, 162 (1965); *Tetr.*, **24**, 1489, 1497 (1968).
3) S. A. Adeoye, D. A. Bekoe, *Chem. Commun.*, 301 (1965).
4) C. W. L. Bevan, J. W. Powell, D. A. H. Taylor, *Chem. Commun.*, 281 (1965).

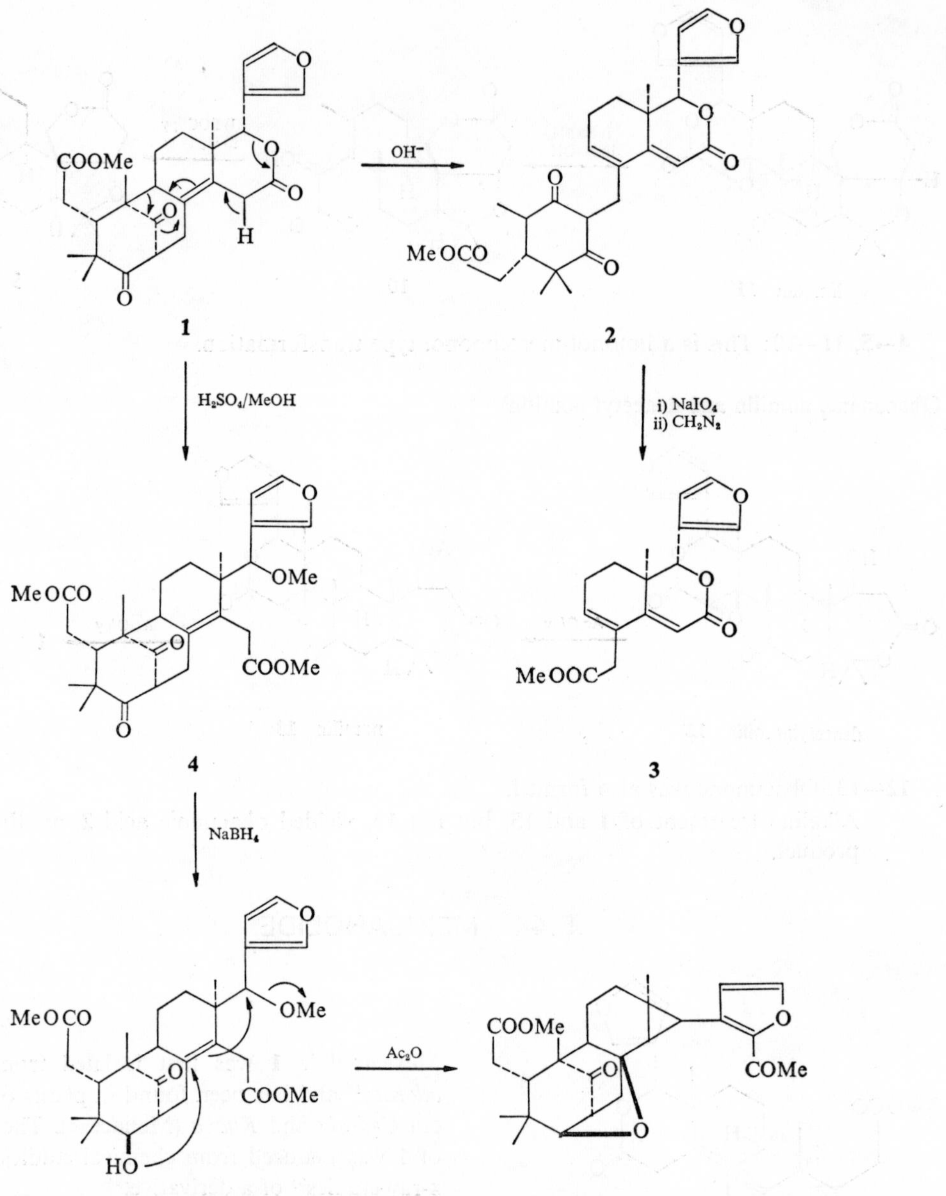

1→2: Alkaline cleavage follows the mechanism indicated by the arrows.

Remarks

A biogenetic route to compounds of this type has been proposed[5] starting from a gedunin-khivorin type limonoid, by cleavage of the C-7 to C-8 bond, followed by reclosure between C-2 and C-8. A partial synthesis of 1 was recently achieved by following this general scheme.[6]

5) J. D. Connolly, R. Henderson, R. McCrindle, K. H. Overton, N. S. Bhacca, *Tetr. Lett.*, 2593 (1964).
6) J. D. Connolly, I. M. S. Thornton, D. A. H. Taylor, *Chem. Commun.*, 17 (1971).

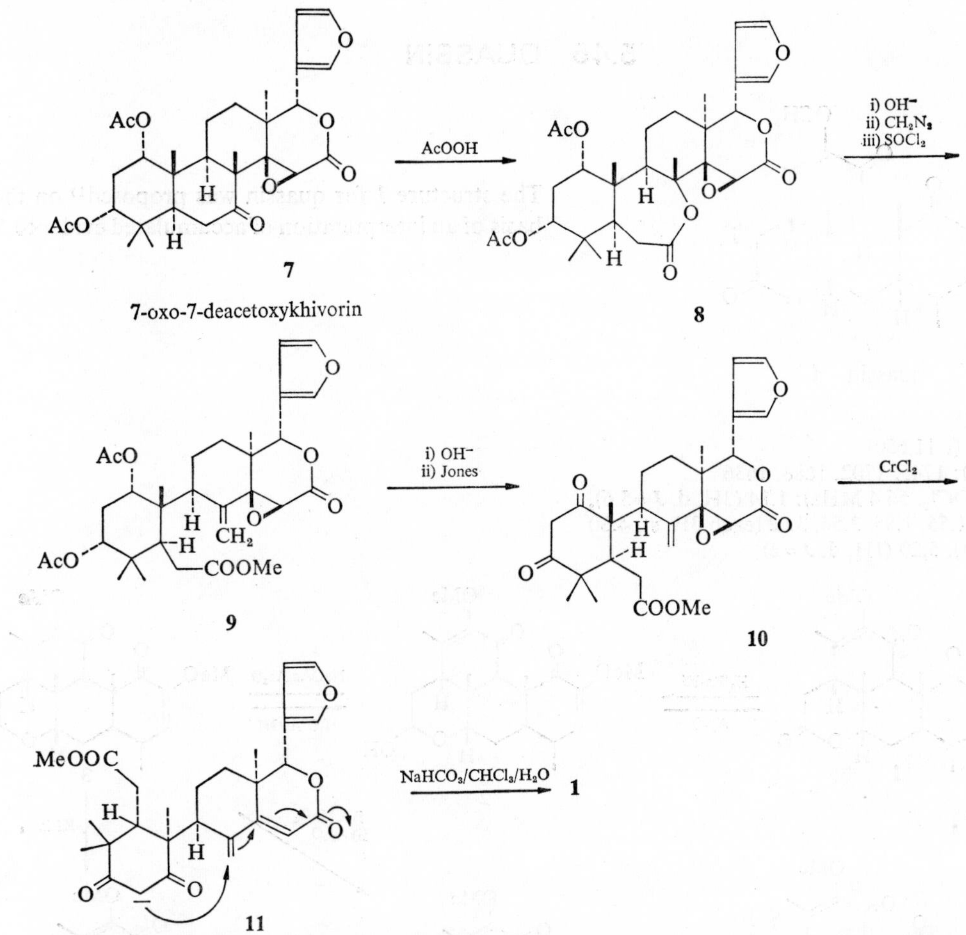

7-oxo-7-deacetoxykhivorin

Over twenty bicyclononanolides have been isolated. Some members of this group of compounds are shown below.

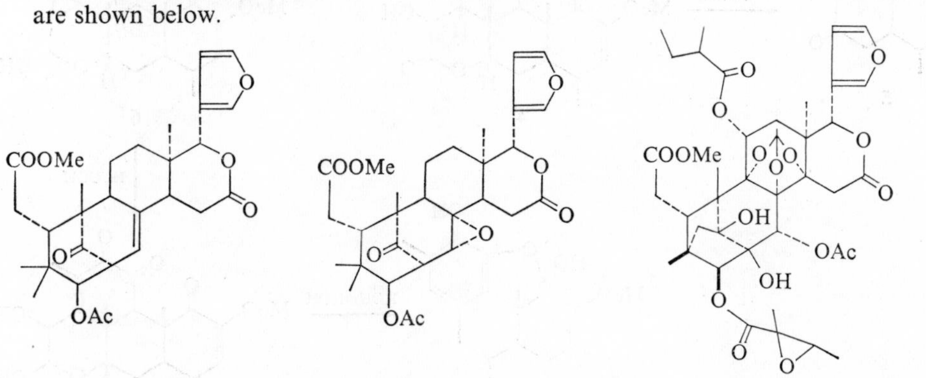

swietenin 12[7]

xylocarpin 13[8]

utilin 14[9]

7) J. D. Connolly, R. Henderson, R. McCrindle, K. H. Overton, N. S. Bhacca, *J. Chem. Soc.*, 6935 (1965).
8) D. A. Okorie, D. A. H. Taylor, *J. Chem. Soc.* C, 211 (1970).
9) H. R. Harrison, O. J. R. Hodder, C. W. L. Bevan, D. A. H. Taylor, T. G. Halsall, *Chem. Commun.*, 1388 (1970).

5.45   QUASSIN

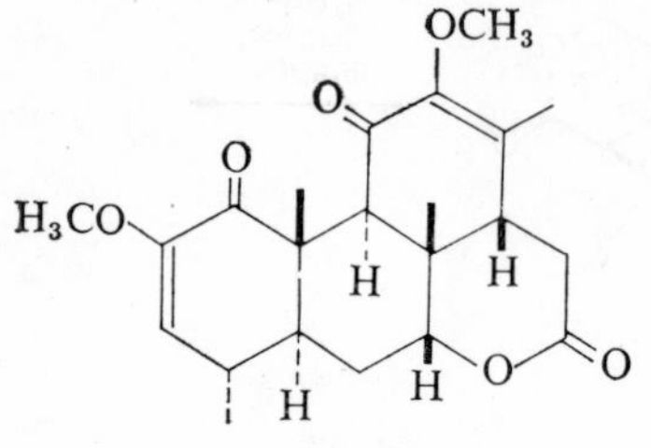

quassin **1**

The structure **1** for quassin was proposed[1] on the basis of an interpretation of accumulated evidence.[2]

mp: 221°
uv: 255 (ε 11,650)
ir (KBr): 1745, 1702, 1688, 1636
nmr (CDCl$_3$, 56.4 MHz): 1.14 (3H, d, $J=5.6$),
   1.21, 1.55, 1.85, 3.56, 3.62 (each 3H, s), 4.30
   (1H, t), 5.29 (1H, d, $J=2$)

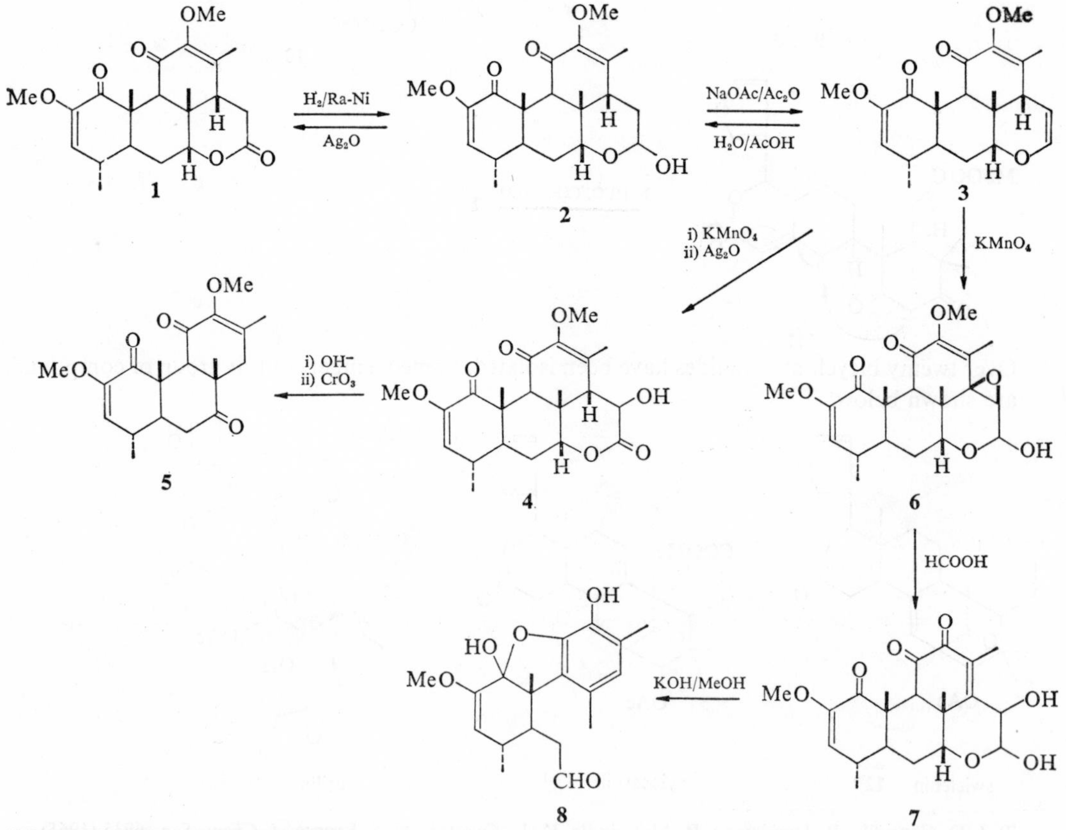

1) Z. Valenta, A. H. Grey, S. Papadopoulos, C. Podesva, *Tetr. Lett.*, 25, (1960); Z. Valenta, S. Papadopoulos, C. Podesva, *Tetr.*, **15**, 100 (1961); Z. Valenta, A. H. Grey, D. E. Orr, S. Papadopoulos, C. Podesva, *ibid.*, **18**, 1433 (1962).

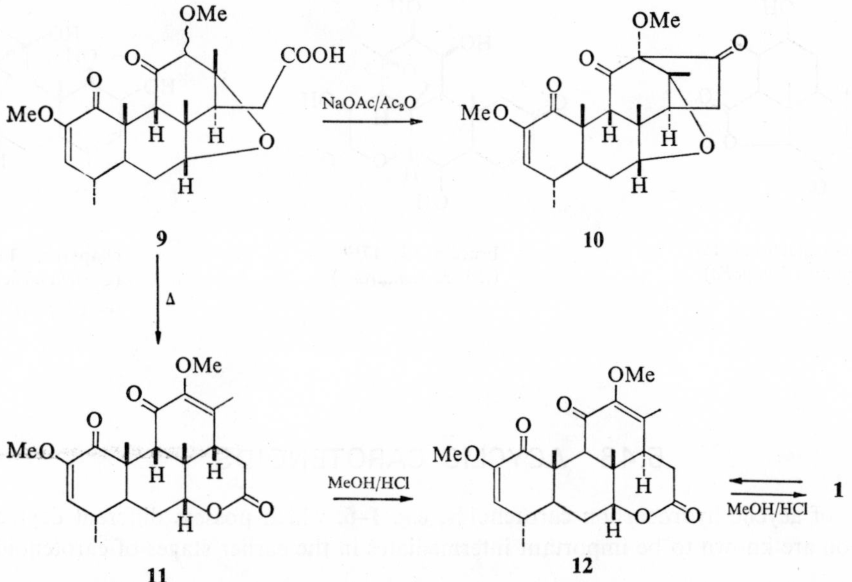

9 → **10**

11 → **12**

1→12: The equilibrium mixture consists of nearly equal amounts of **1** and **12**. **11** isomerizes exclusively to this mixture.

Remarks

Quassin and neoquassin were first isolated from *Quassia amara*.[1,2] Nearly thirty quassinoids have been isolated from Simaroubaceae plants. Some are shown below together with their origin.

Biogenetically, quassinoids are related to limonoids.[3] Similar transformations were achieved with limonin[4] and gedunin.[5]

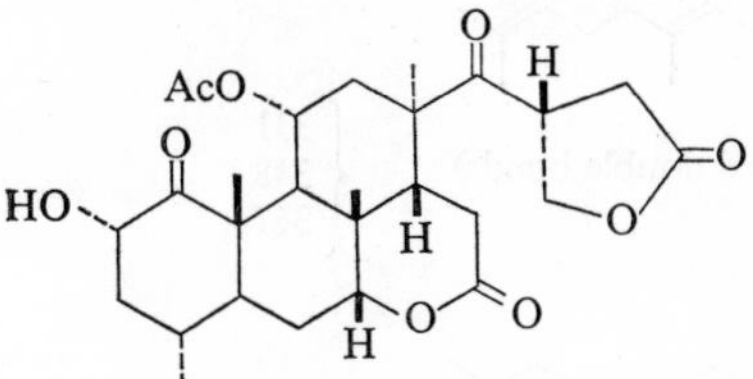

simarolide **13**[6]
(*Simarouba amara*)

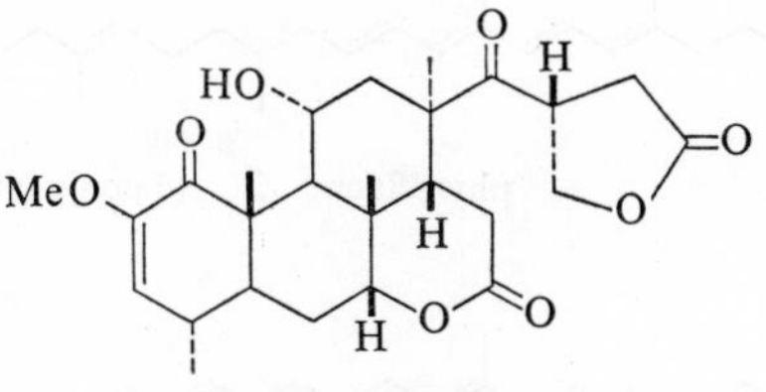

picrasin A **14**[7]
(*Picrasma quassioides*
= *P. ailanthoides*)

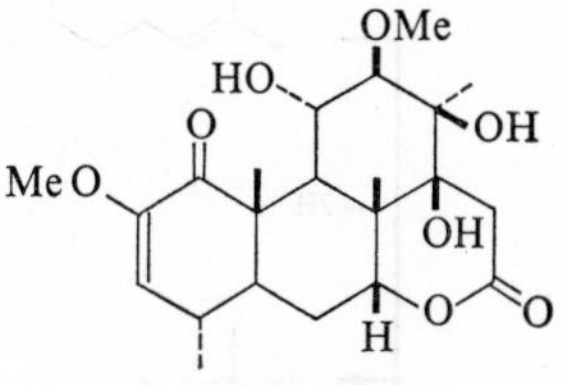

nigakilactone H **15**[8]
(*Picrasma quassioides*
= *P. ailanthoides*)

2) E. Lordon, A. Robertson, H. Worthington, *J. Chem. Soc.*, 3431 (1950); R. J. S. Beer, D. B. G. Jaquiss, A. Robertson, W. E. Savige, *ibid.*, 3672 (1954); K. R. Hanson, D. B. G. Jaquiss, J. A. Lamberton, A. Robertson, W. E. Savige, *ibid.*, 4238 (1954); R. J. S. Beer, K. R. Hanson, A. Robertson, *ibid.*, 3280 (1956); R. J. S. Beer, B. G. Dutton, D. B. G. Jaquiss, A. Robertson, W. E. Savige, *ibid.*, 4850 (1956).
3) T. R. Hollands, P. de Mayo, M. Nisbet, P. Crabbé, *Can. J. Chem.*, 43, 3008 (1965).
4) D. Arigoni, D. H. R. Barton, E. J. Corey, O. Jeger, *Experientia*, 16, 41 (1960).
5) A. Akisanya, E. O. Arene, C. W. L. Bevan, D. E. U. Ekong, M. N. Nwaji, J. I. Okagun, J. W. Powell, D. A. H. Taylor, *J. Chem. Soc.*, 506 (1966).
6) J. Polonsky, *Proc. Chem. Soc.*, 292 (1964); W. A. C. Broun, G. A. Sim, *ibid.*, 293 (1964).
7) H. Hikino, T. Ohta, T. Takemoto, *Chem. Pharm. Bull.* 18, 219, 1082 (1970); *ibid.*, 19, 212, 2212, 2651 (1971); T. Murae, T. Tsuyuki, T. Nishihama, S. Masuda, T. Takahashi, *Tetr. Lett.*, 3013 (1969); T. Murae, T. Tsuyuki, T. Nishihama, T. Takahashi, *Chem. Pharm. Bull.*, 18, 2590 (1970).
8) T. Murae, T. Tsuyuki, T. Nishihama, T. Takahashi, *Bull. Chem. Soc. Japan*, 43, 3021 (1970).
9) Le-Van-Thoi, Nguen-Ngoc-Suong, *J. Org. Chem.*, 35, 1104 (1970).
10) G. R. Duncan, D. B. Henderson, *Experientia*, 24, 768 (1968).

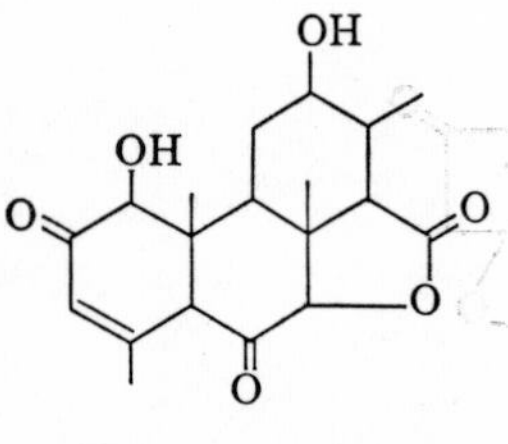

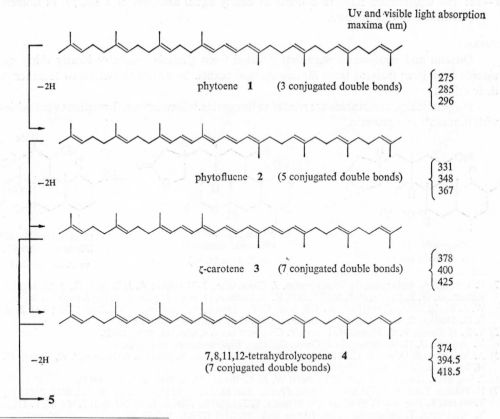

eurycomalactone **16**[9)]
(*Eurycoma longifolia*)

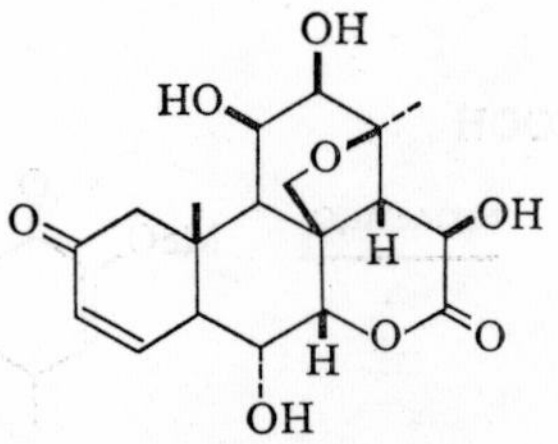

brucein G **17**[10)]
(*Brucea sumatrana*) .

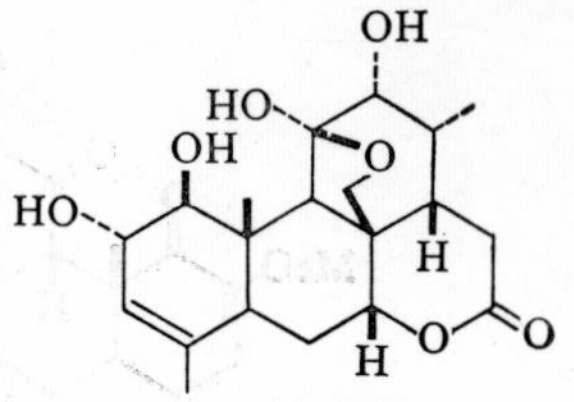

chaparrin **18**[3)]
(*Costela nicholsoni*
= *Chaparro amargoso*)

5.46 ACYCLIC CAROTENOIDS

A series of acyclic hydrocarbon carotenoids, e.g. **1–6**, which possess different degrees of un-saturation are known to be important intermediates in the earlier stages of carotenoid biosynthesis.

Uv and visible light absorption
maxima (nm)

−2H

phytoene **1**   (3 conjugated double bonds)

$$\begin{cases} 275 \\ 285 \\ 296 \end{cases}$$

−2H

phytofluene **2**   (5 conjugated double bonds)

$$\begin{cases} 331 \\ 348 \\ 367 \end{cases}$$

ζ-carotene **3**   (7 conjugated double bonds)

$$\begin{cases} 378 \\ 400 \\ 425 \end{cases}$$

7,8,11,12-tetrahydrolycopene **4**
(7 conjugated double bonds)

$$\begin{cases} 374 \\ 394.5 \\ 418.5 \end{cases}$$

−2H

→ **5**

1) J. B. Davis, L. M. Jackman, P. T. Siddons, B. C. L. Weedon, *Proc. Chem. Soc.*, 261 (1961).
2) J. B. Davis, L. M. Jackman, P. T. Siddons, B. C. L. Weedon, *J. Chem. Soc.*, 2154 (1966).
3) B. H. Davies, E. A. Holmes, D. E. Loeber, T. P. Toupe, B. C. L. Weedon, *J. Chem. Soc.*, 1266 (1969).

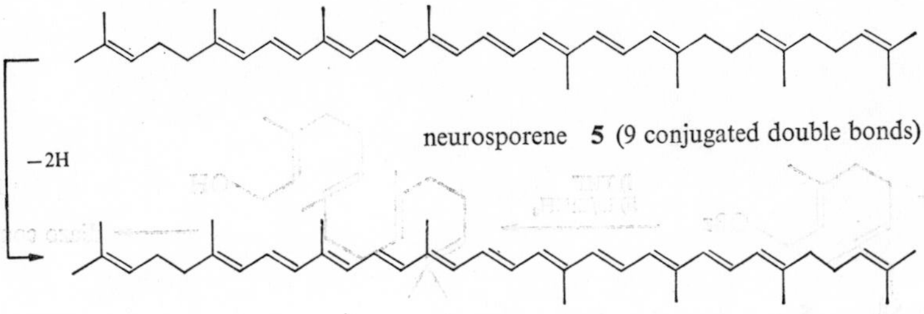

neurosporene **5** (9 conjugated double bonds)

$$\begin{cases} 416 \\ 440 \\ 470 \end{cases}$$

−2H

lycopene **6**   (11 conjugated double bonds)

$$\begin{cases} 446 \\ 472 \\ 505 \end{cases}$$

As the degree of conjugation is extended in polyene systems, the uv and visible light absorption maxima move to longer wavelength regions, providing a useful method for characterizing such carotenoids. The compounds **2–6** are known to be formed *in vivo* by sequential dehydrogenation, as shown above.[1,2] The unsymmetrical tetrahydrolycopene **4**, which has been isolated from *Rhodospirillum rubrum*, might also be involved, at least in some organisms, in such a dehydrogenation sequence.[3]

5.47   STRUCTURE AND SYNTHESIS OF PREPHYTOENE PYROPHOSPHATE

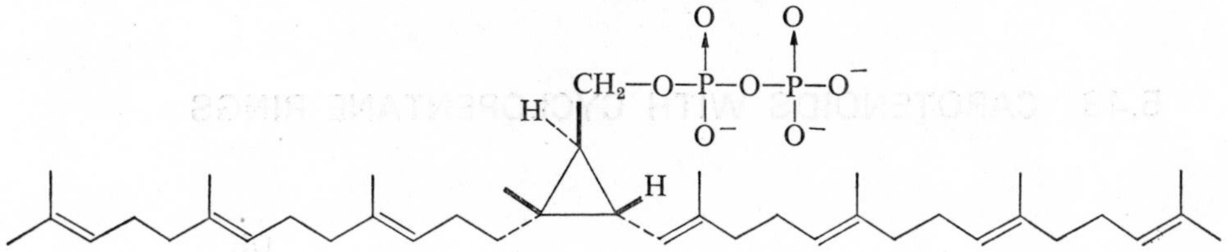

prephytoene pyrophosphate **1**[1]

1 $\xrightarrow{\text{LAH}}$ prephytoene alcohol **2** $\xrightarrow[\text{ii) Ac}_2\text{O/py}]{\text{i) O}_3}$

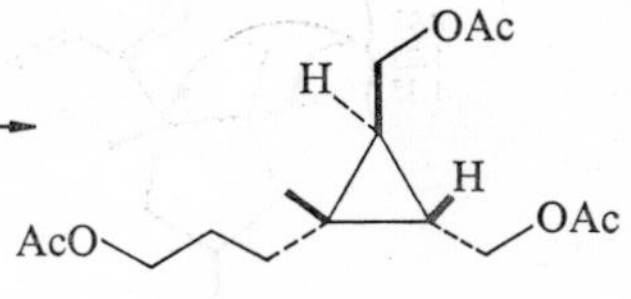

3

1) The identity of naturally occurring **1**, **2** and **3** with the synthetic specimens was confirmed by cochromatography.

2) Both synthetic and natural **1** were converted to phytoene by bacterial extracts.

Remarks

Prephytoene pyrophosphate **1** has been found in butanol extracts of the incubation mixture of geranylgeranyl pyrophosphate with a cell-free preparation from bacterial cells (*Mycobacterium*). It has been demonstrated that this compound is a biosynthetic intermediate between geranylgeranyl pyrophosphate and phytoene. The latter is an important precursor of carotenoids.

1) L. J. Altman, L. Ash, R. C. Koerski, W. W. Epstein, B. R. Larsen, H. C. Rilling, F. Muscio, D. E. Gregonis, *J. Am. Chem. Soc.*, 94, 3257 (1972).
2) L. Crombie, D. A. R. Findley, D. A. Whiting, *Chem. Commun.*, 1045 (1972).

Synthesis

Route I[1)]

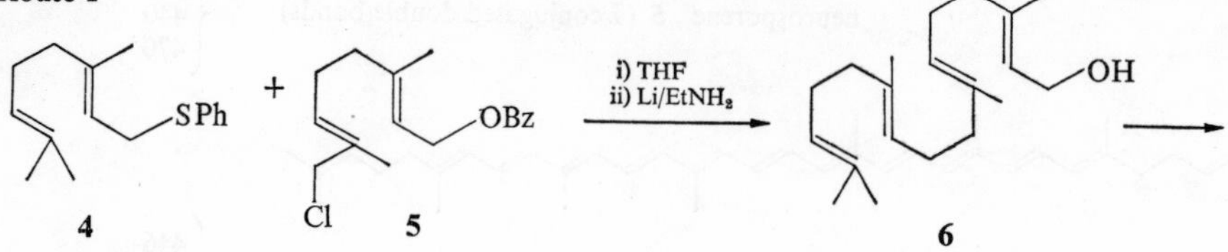

$$7 + 6 \xrightarrow{ZnI_2/Et_2O} 2 \xrightarrow{phosphorylation} 1$$

Route II[2)]

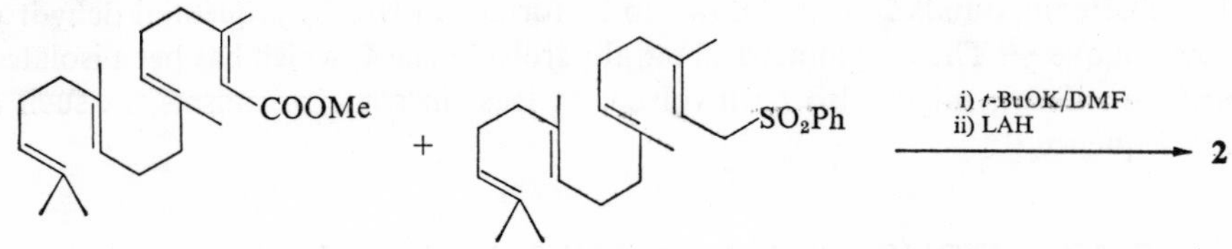

5.48   CAROTENOIDS WITH CYCLOPENTANE RINGS

Type I

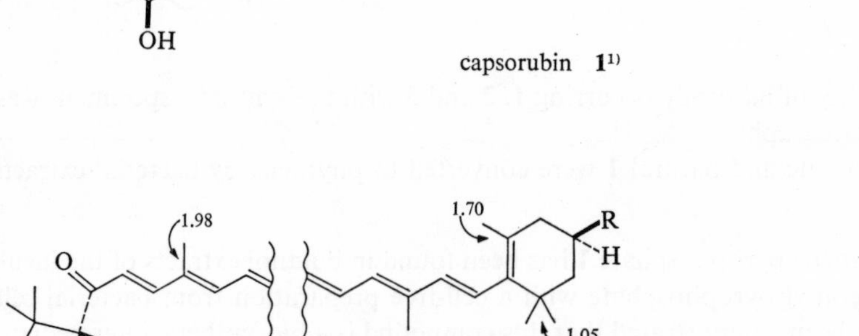

capsorubin **1**[1)]

capsanthin **2** (R = OH)[1)]
kryptocapsin **3** (R = H)[2)]

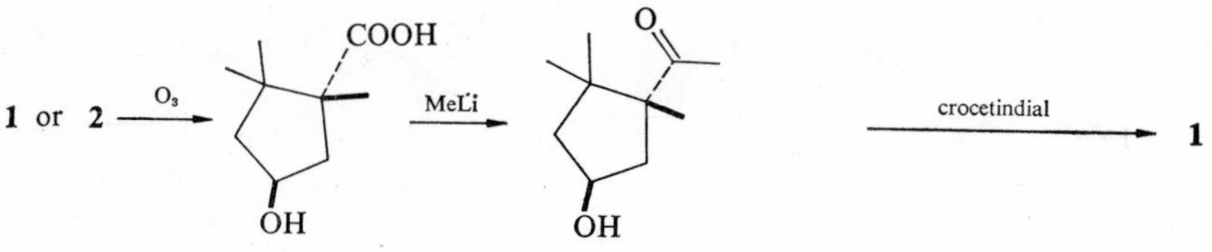

1 or 2 —O₃→ **4** —MeLi→ **5** —crocetindial→ **1**

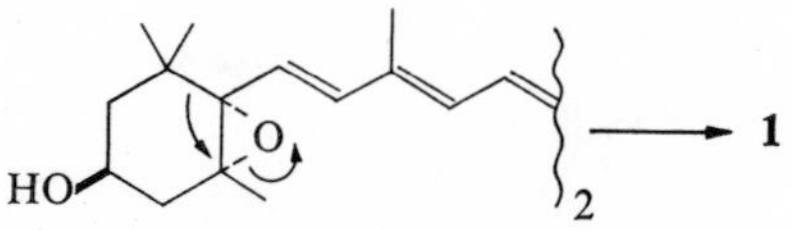

6 → 1

1) The ozonolysis product **4** was chemically correlated with (+)-camphor.[3]

2) (±)-**1** was synthesized from racemic *trans* acid **4** via the methyl ketone **5**.[4]

3) The rearrangement of an epoxide, e.g. violaxanthin **6**, was postulated for the biogenesis of capsorubin **1**.

Type II

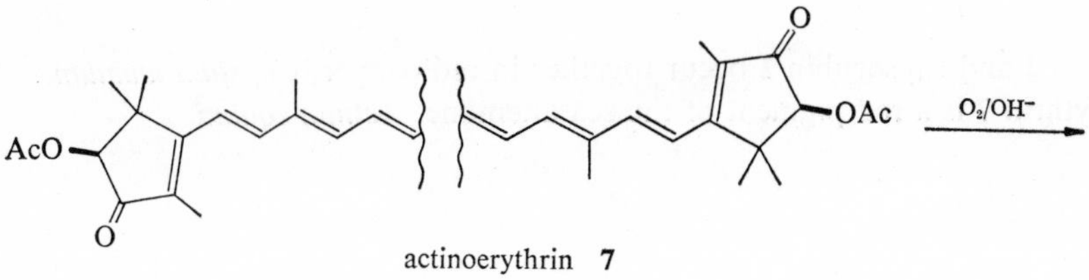

actinoerythrin **7**

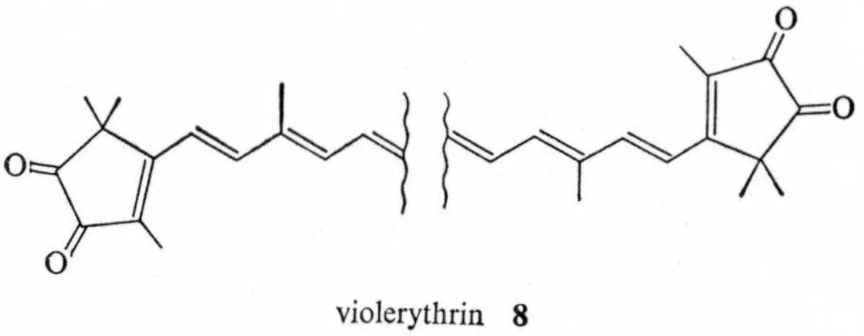

violerythrin **8**

1) M. S. Barber, L. M. Jackman, B. C. L. Weedon, *J. Chem. Soc.*, 4019 (1961); H. Faigle, P. Karrer, *Helv. Chim. Acta*, **44**, 1904 (1961).
2) L. Cholonosky, J. Szabolcs, R. D. G. Cooper, B. C. L. Weedon, *Tetr. Lett.*, 1257 (1963).
3) B. C. L. Weedon, *Chem. Brit.*, **3**, 424 (1967).
4) R. D. G. Cooper, L. M. Jackman, B. C. L. Weedon, *Proc. Chem. Soc.*, 215 (1962).

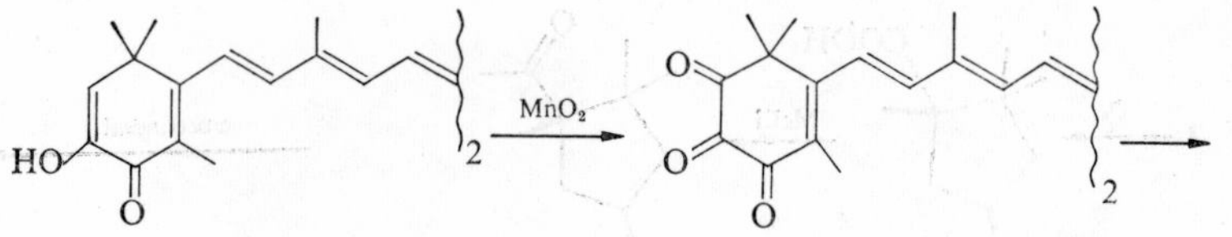

astacene **9**   **10**

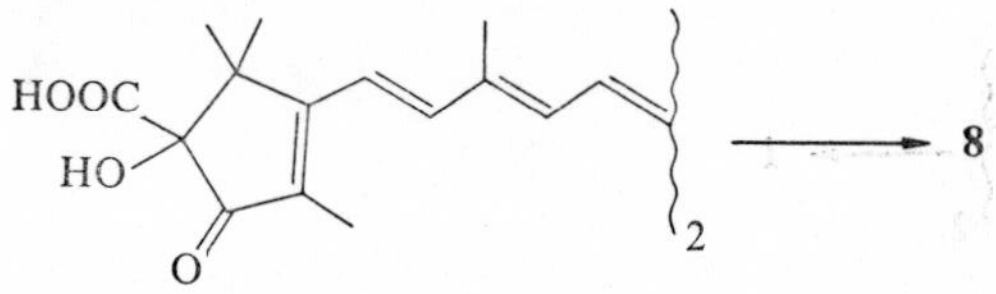

11

1) Actinoerythrin **7**, a red pigment, was transformed to the blue pigment violerythrin **8** by air oxidation.[5]

2) **8** was considered to be formed by benzylic acid rearrangement of the triketone **10** followed by oxidative decarboxylation of the resulting hydroxy-acid **11**. This transformation was achieved chemically by treatment of astacene **9** with MnO_2 in acetone.[6]

Remarks

Capsanthin **2** and capsorubin **1** occur together in red pepper, *Capsium annuum*.
Actinoerythrin **7** is a red pigment of the sea anemone, *Actinia equina*.

5) S. Herzberg, S. Liaaen-Jensen, *Acta. Chem. Scand.*, **22**, 1714 (1968).
6) R. Holzel, A. P. Leftwick, B. C. L. Weedon, *Chem. Commun.*, 128 (1969).

5.49 ALLENIC AND ACETYLENIC CAROTENOIDS

fucoxanthin **1**[1,2]

ms: 658.4201 (M$^+$ $C_{42}H_{58}O_6$)
uv (CS$_2$): 508 (ε 107,000), 478 (ε 134,000)
ir: 3615 (OH), 1927 (allene), 1740 (acetate)
   1660 (unsat. C=O)
nmr in CDCl$_3$

neoxanthin (foliaxanthin) **2**[2,3]

ms: 600.419 (M$^+$ $C_{40}H_{56}O_4$)
uv (EtOH): 467, 439, 416
ir: 3597, 1923, 965
nmr in CDCl$_3$

1) Cleavage of the epoxide ring of **1** by base yields isofucoxanthin **3**.

2) Acid treatment of **4** derived from **1** by LAH reduction afforded a stereoisomer of neochrome **5**. **5** was obtained from neoxanthin **2** by mild acid treatment.

3) On reduction with LAH, the 5,6-epoxy end group of **2**, the allenic end groups of **2**, **5** and **10**, and the dihydrofuran-type end groups in **5** or **9** were all converted into a β-ionone end group of the type shown in **6**. Thus, LAH reduction of **2** and **5** yielded zeaxanthin **6**.

4) The C-6/C-7 bond of **7** (obtained by oxidation of **1**) was cleaved by base, yielding the naturally occurring C$_{31}$ carotenoid, paracentrone **8**.[4]

5) The allenic alcohol function in **2** or **10** could be transformed by treatment with acid into an acetylene function. Thus, acid treatment of **2** and **10** yielded diadinochrome **9** and diaxanthin **11**, respectively.[5]

1) R. Bonnett, A. K. Mallams, A. A. Spark, J. L. Tee, B. C. L. Weedon, A. McCormick, *J. Chem. Soc.*, 429 (1969).
2) For absolute configurations, see T. E. DeVille, M. B. Hursthouse, S. W. Russell, B. C. L. Weedon, *Chem. Commun.*, 1311 (1969).
3) L. Cholnoky, K. Györgyfy, A. Ronai, J. Szabolcs, Cy. Toth, G. Galasko, A. K. Mallams, E. S. Waight, B. C. L. Weedon, *J. Chem. Soc.*, 1256 (1969).
4) J. Hora, T. P. Toube, B. C. L. Weedon, *J. Chem. Soc.*, 241 (1970)
5) H. Nietsche, K. Egger, A. G. Dabbagh, *Tetr. Lett.*, 2999 (1969).

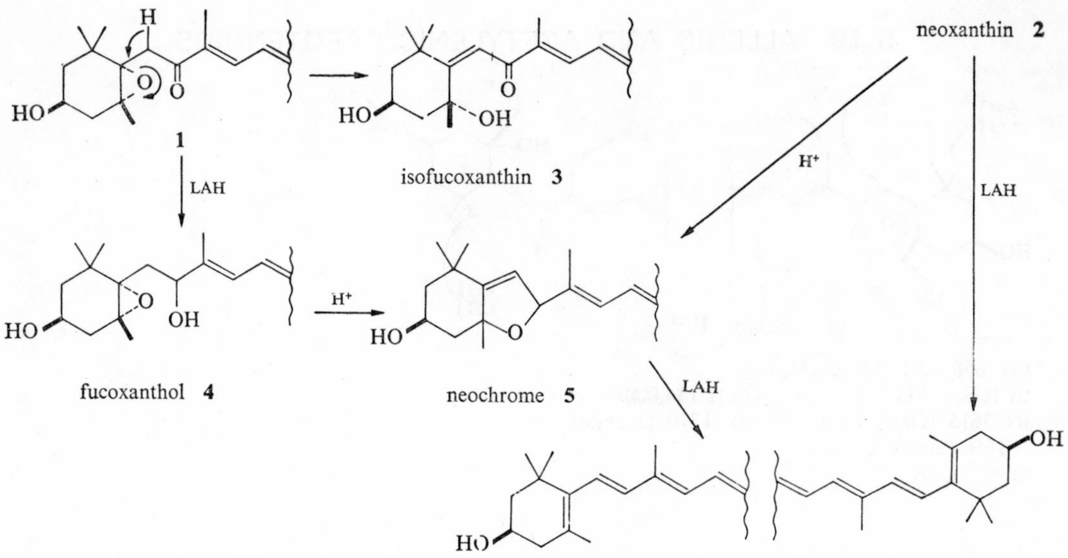

isofucoxanthin 3

neoxanthin 2

fucoxanthol 4

neochrome 5

zeaxanthin 6

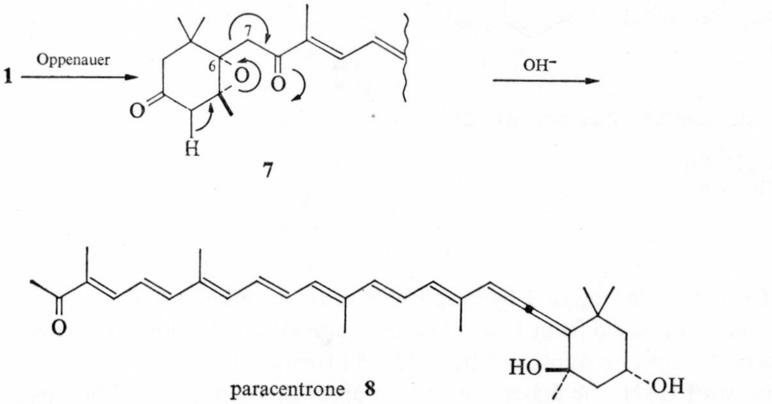

paracentrone 8

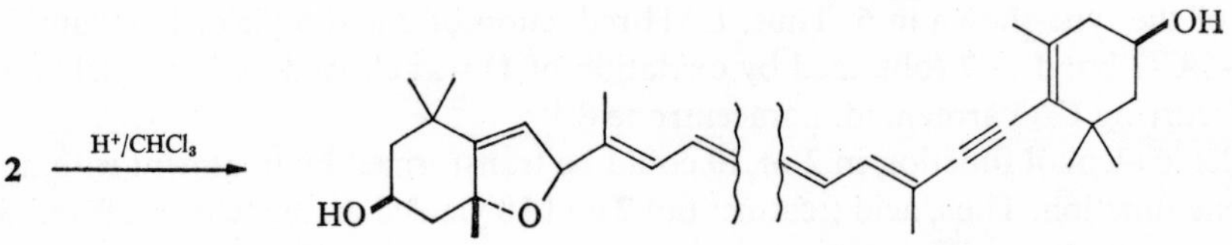

diadinochrome 9

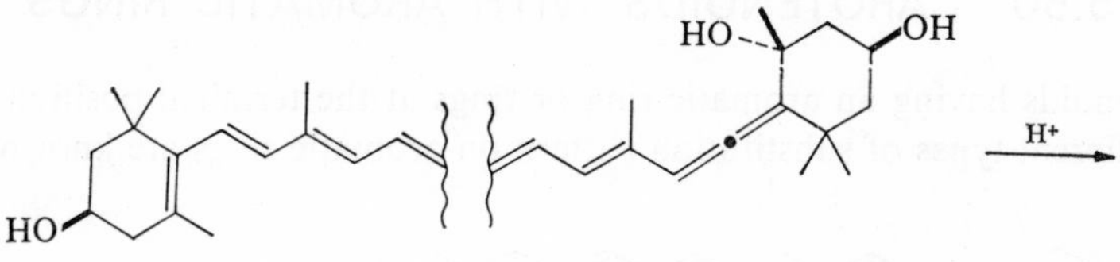

deepoxy-neoxanthin **10**

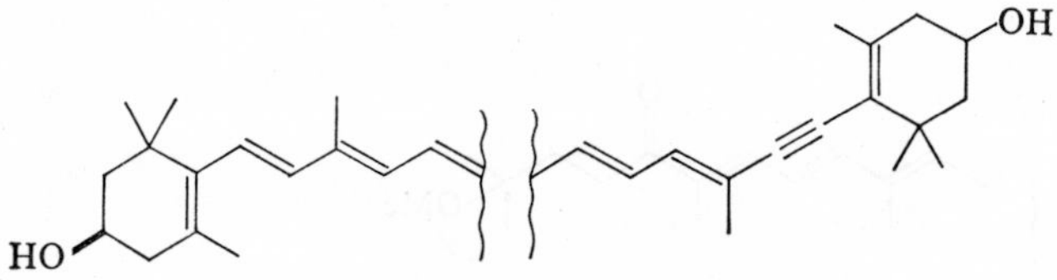

diaxanthin **11**

Remarks

Fucoxanthin **1** occurs in brown algae (Phaeophyceae) and is one of the most abundant carotenoids in nature. Neoxanthin **2** occurs in green leaves of higher plants. Isofucoxanthin **3** was found in the yolks of eggs from chickens fed on seaweed meal. Paracentrone **8** was isolated from sea urchins, *Paracentrotus lividus*. Deepoxy-neoxanthin **10** was found in the flowers of *Mimulus guttatus*. From flagellates of the algal class Cryptophyceae, alloxanthin **12** and monadoxanthin **13** were isolated.

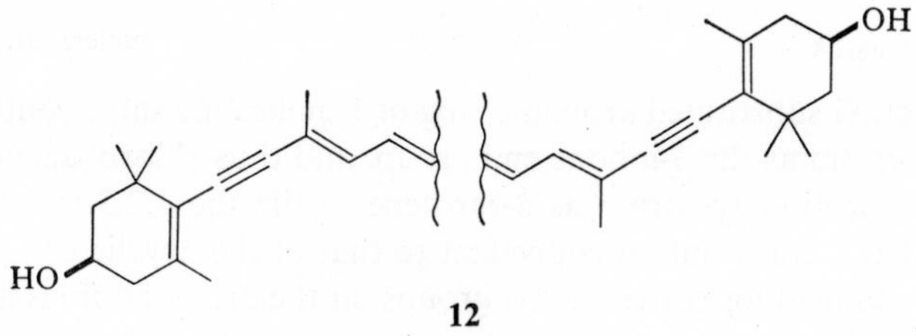

12

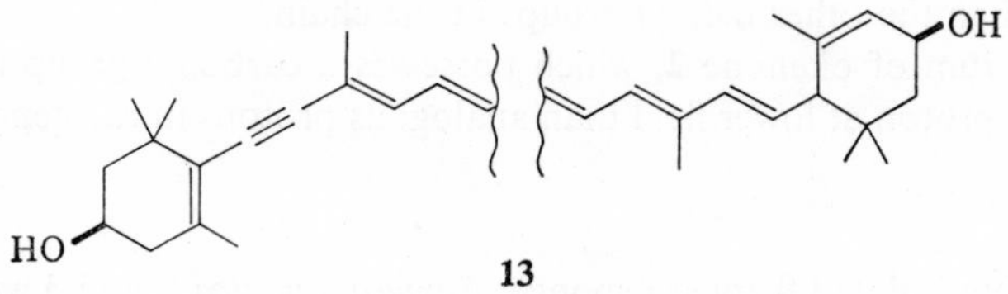

13

5.50 CAROTENOIDS WITH AROMATIC RINGS

Several carotenoids having an aromatic ring or rings at the terminal positions have been isolated. Two different types of substitution pattern on aromatic rings are known.

chlorobactene **1**

okenone **2**

isorenieratene **3**

β-isorenieratene **4**

renieratene **5**

renierapurpurin **6**

1) The 1,2,5-trimethyl substituted aromatic ring of **1** makes the same contribution to the visible absorption spectrum as the β-ionone end group, and thus chlorobactene **1** exhibits the same visible light absorption spectrum as δ-carotene, while the 1,2,3-trimethyl substituted aromatic ring makes a contribution equivalent to that of the acyclic end group.

2) Characteristic deshielding of the methyl groups on the aromatic rings is observed in the nmr spectrum.

3) The signal of the methyl group in the chain which is nearest to the aromatic ring appears at a lower field than the other methyl groups in the chain.

4) The nmr spectrum of okenone **2**, which possesses a carbonyl group in a remote position, shows the 5-H proton at lower field than analogous protons in carotenoids with no carbonyl group.

Remarks

3, **5** and **6** were isolated from sea sponge, *Reniera japonica*,[1] and **3** was also found in *Mycobacterium phlei*. **1**, **2**, **3**, and **4** were isolated from various photosynthetic sulfur bacteria.[2-4]

1) M. Yamaguchi, *Bull. Chem. Soc. Japan*, **33**, 1560 (1960).
2) S. Liaaen-Jensen, E. Hegge, L. M. Jackman, *Acta Chem. Scand.*, **18**, 1703 (1964).
3) S. Liaaen-Jensen, *Acta Chem. Scand.*, **19**, 1025 (1965).
4) S. Liaaen-Jensen, *Acta Chem. Scand.*, **21**, 961 (1967).
5) For synthesis, see R. D. G. Cooper, J. B. Davis, B. C. L. Weedon, *J. Chem. Soc.*, 5637 (1963).

5.51 CAROTENOIDS HAVING ADDITIONAL CARBONS

Carotenoids possessing additional carbon atoms (C_5 or C_{10}) are known. The structures **1–3** were elucidated mainly on the basis of mass spectral fragmentations.

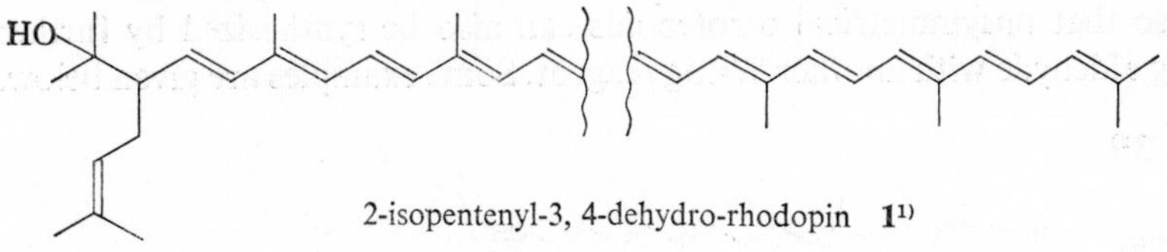

2-isopentenyl-3, 4-dehydro-rhodopin **1**[1]

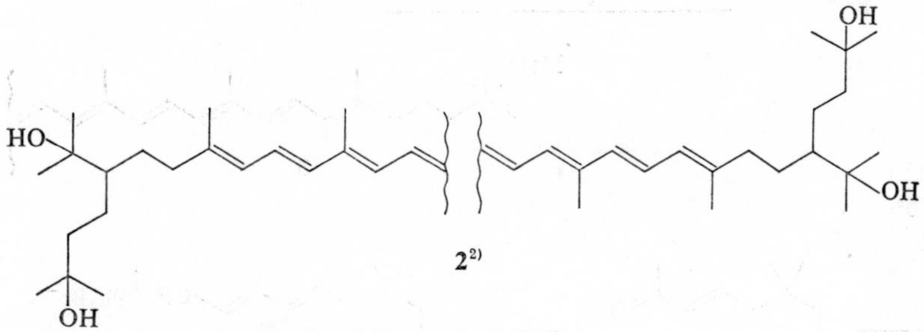

2[2]

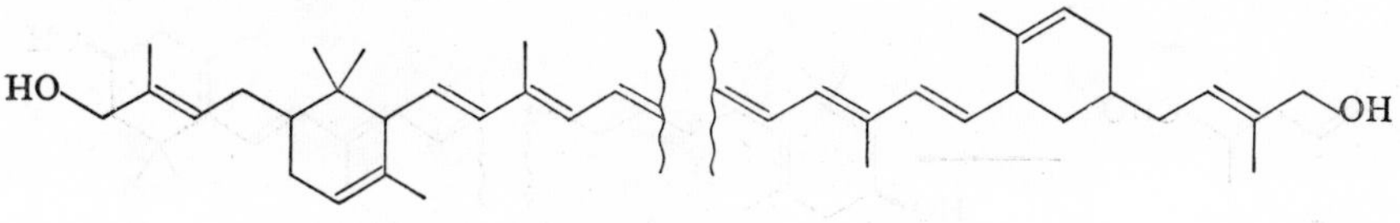

dehydrogenans P439 **3**[3]

Remarks

The C_{45} carotenoid **1** was isolated from *Corynebacterium poinsettiae*.[1] The C_{50} carotenoid **2** was obtained from certain halophilic bacteria,[2] and **3** was isolated from the nonphotosynthetic bacterium *Flavobacterium dehydrogenans*.[3]

1) S. Norgard, S. Liaaen-Jensen, *Acta Chem. Scand.*, **23**, 1463 (1969).
2) M. Kelly, S. Liaaen-Jensen, *Acta Chem. Scand.*, **21**, 2578 (1967).
3) S. Liaaen-Jenden, *Acta. Chem. Scand.*, **21**, 1972 (1967); S. Herzberg, S. Liaaen-Jensen, *ibid.*, **22**, 1714 (1968); U. Schwieter, S. Liaaen-Jensen, *ibid.*, **23**, 1057 (1969).

5.52   SYNTHESIS OF CAROTENOIDS

For the synthesis of symmetrical carotenoids,[1] a Wittig condensation reaction between the appropriate Wittig reagents (e.g., **1**, **6** and **9**) corresponding to the two end groups and a poly-olefinic dial (e.g., **2**, and **7**) is generally employed. Under suitable conditions, the dial reacts selectively at one end, so that unsymmetrical carotenoids can also be synthesized by further reaction of the remaining aldehyde with another Wittig reagent. Some examples are given below.

Synthesis of apirilloxanthin 3[2]

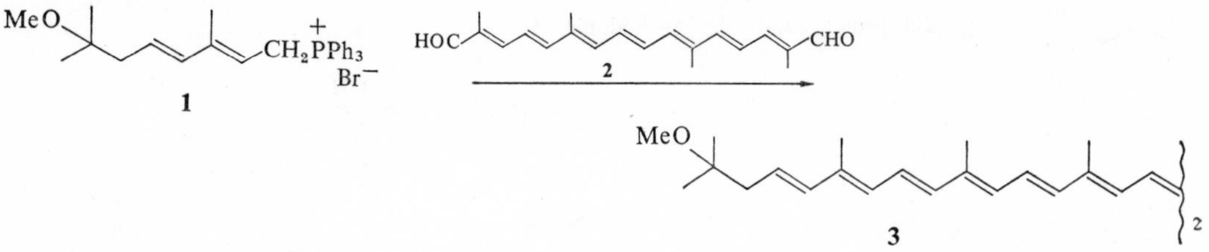

Synthesis of zeaxanthin 8[3]

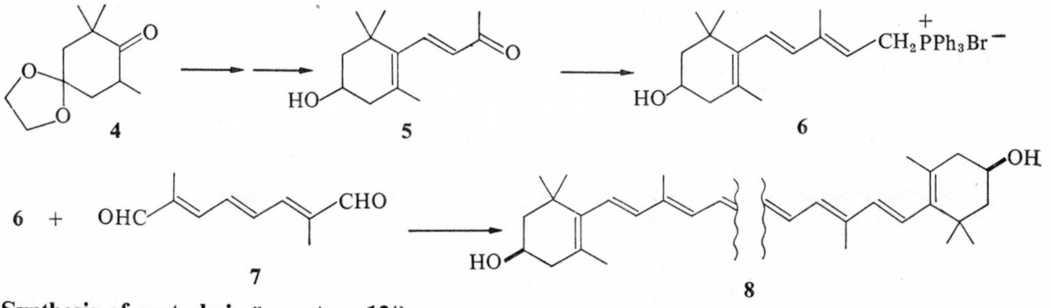

Synthesis of central cis ζ-carotene 12[4]

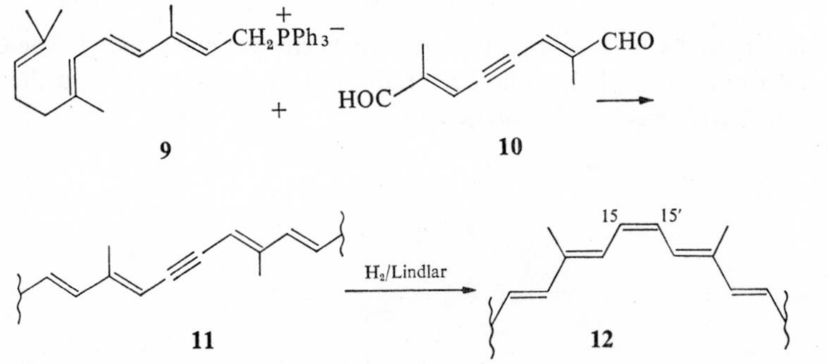

15,15'-*cis* carotenoids (e.g. **12**) can be prepared by condensation of a Wittig reagent such as **9** with an acetylenic dialdehyde such as **10**, followed by catalytic hydrogenation of the acetylenic intermediate **11** using the Lindlar catalyst.

1) For a review, see O. Isler, *Pure Appl. Chem.*, **14**, 245 (1967); B. C. L. Weedon, *ibid.*, **14**, 265 (1967); H. Pommer, *Angew. Chem.*, **72**, 811, 911 (1966); B. C. L. Weedon, *Chem. Brit.*, **3**, 424 (1967).
2) D. F. Schneider, B. C. L. Weedon, *J. Chem. Soc.*, 1686 (1967).
3) D. E. Loeber, S. W. Russell, T. P. Toube, B. C. L. Weedon, J. Diment, *J. Chem. Soc.*, 404 (1971).
4) J. B. Davis, L. M. Jackman, P. T. Siddons, B. C. L. Weedon, *J. Chem. Soc.*, 2154 (1966).

5.53 PHOTOOXYGENATION OF CAROTENOIDS

Photooxygenation of carotenoids, e.g. β-carotene **1**, zeaxanthin **2** and violaxanthin **15**, affords a number of products, some of which are naturally occurring substances.

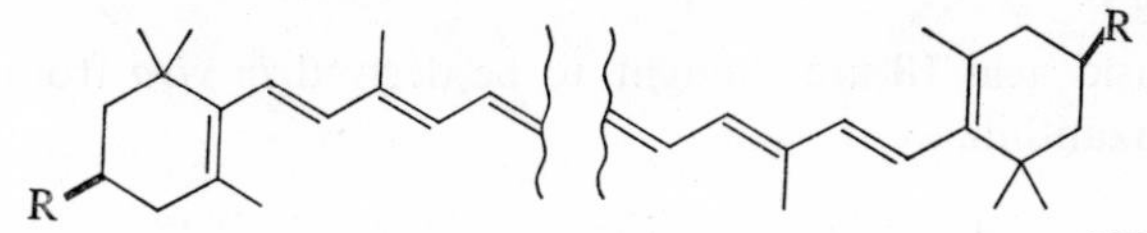

β-carotene **1** (R = H)
zeaxanthin **2** (R = OH)

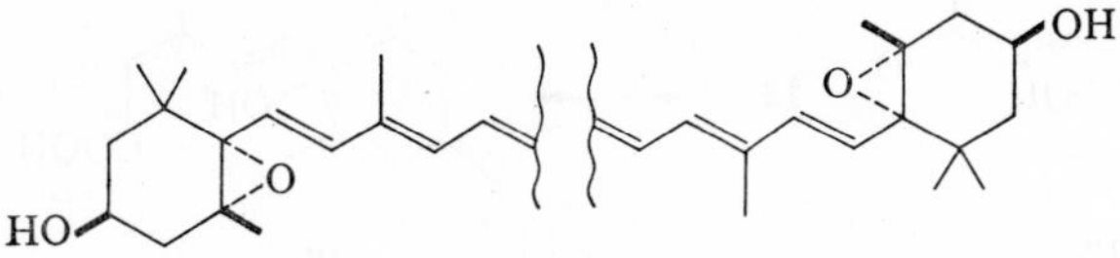

violaxanthin **15**

$1,2 \xrightarrow{h\nu/O_2/\text{benzene}}$

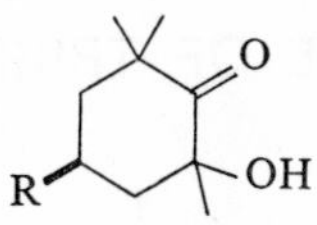

3 (R = H)
4 (R = OH)

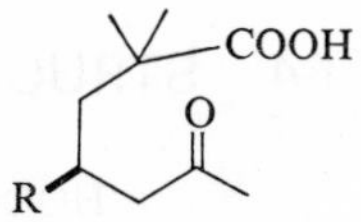

5 (R = H)
6 (R = OH)

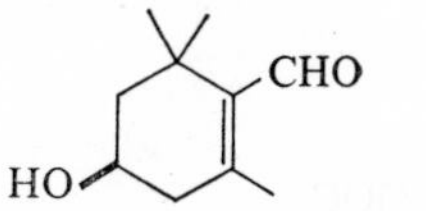

7

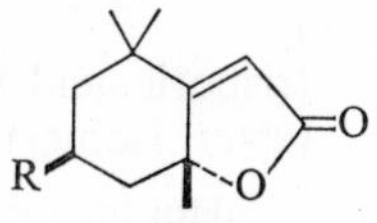

dihydroactini-
diolide **8** (R = H)
loliolide **9** (R = OH)

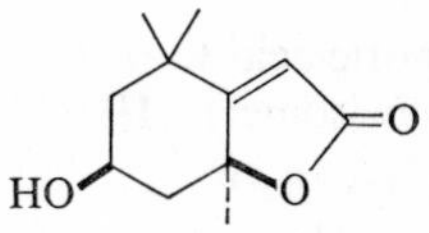

10

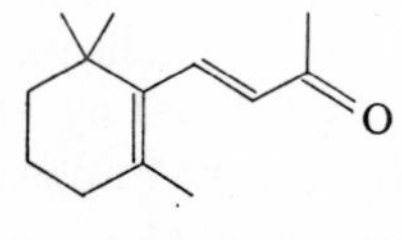

β-ionone **11**

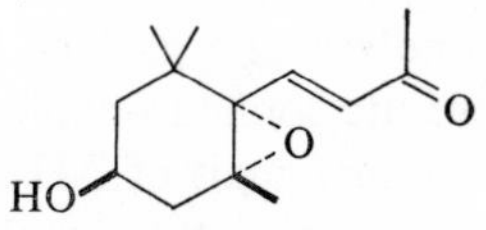

12

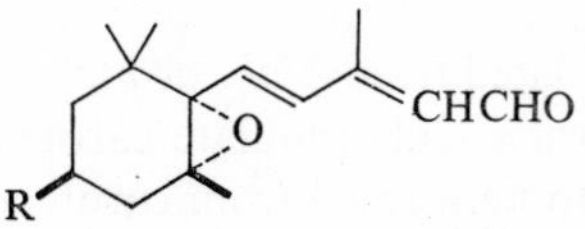

xanthoxin **13** (R = H)
14 (R = OH)

1) S. Isoe, S. B. Hyeon, T. Sakan, *Tetr. Lett.*, 279 (1969).
2) S. Isoe, S. B. Hyeon, S. Katsumura, T. Sakan, *Tetr. Lett.*, 2517 (1972).
3) R. S. Burden, H. F. Taylor, *Tetr. Lett.*, 4071 (1970); H. F. Taylor, R. S. Burden, *Phytochemistry*, **9**, 2217 (1970).

1) Photooxygenation of **1** in benzene in the presence of traces of water yielded **3, 5, 8, 11** and **13**.[1]

2) The compounds **4, 6, 7, 9, 10** and **12** were obtained from zeaxanthin **2** under the same conditions.[2] The α_D value of **9** was identical with that of the natural (−)-loliolide.

3) Violaxanthin **15** afforded **9, 12** and xanthoxin **14** under the same conditions. Xanthoxin **14** was converted into (+)-abscisic acid **18**.[3]

4) Vomifoliol **17**, xanthoxin **14** and abscisic acid **18** are thought to be derived *in vivo* from carotenoids such as zeaxanthin or violaxanthin.

12 ⟶ ... 16   17   14 ⟶ ⟶ ... 18

5.54 STRUCTURE OF TRISPORIC ACIDS

trisporic acid C
(Δ^9-*cis* isomer) **1**[1]

trisporic acid B
(Δ^9-*cis* isomer) **2**[2,4]

anhydrotrisporic acid C
(Δ^9-*cis* isomer) **3**[2]

data for methyl ester[1]
α_D (CHCl$_3$): −11
uv: 320 (conjugated trienone)
ir: 3440 (OH), 1736 (COOMe), 1663, 1601 (C=C–C=O)

data for methyl ester[4]
α_D: 39
uv: 229 (ε 8,600), 322 (ε 17,400)
ir: 2950, 1850, 1730, 1350, 1250

All trisporic acids exist in nature as a mixture of Δ^9-*trans* and Δ^9-*cis* geometrical isomers. The structures were elucidated from their physicochemical properties as well as by extensive degradation.[1,2] The structures were confirmed by synthesis.[3,4]

1) Ozonolysis of **1** gave (−)-valerolactone, indicating that C-13 has the R-configuration.[2]

2) Absolute configuration at C-1 was deduced by cd.[5]

3) The nmr signals of 7-H and 8-H of *trans*-**1** appear at 6.23 and 6.33 ($J = 16$Hz) and the corresponding proton signals of *cis*-**1** appear at 6.41 and 6.84. As 8-H of *cis*-**1** is deshielded by the 11-methylene group, its signal appears downfield with respect to that of the *trans* isomer.[4]

4) The geometry of the Δ^9-double bond was determined by the synthesis of *trans*-**2**.[4]

Remarks

Trisporic acids were isolated from *Blakeslea trispora*. These acids regulate sexual reproduction in heterothallic Mucorales and stimulate carotenoid and steroid synthesis in *B. trispora*. Trisporic acids are known to be formed from carotenoids by way of retinal.[6]

1) L. Cagliotti, G. Cainelli, B. Camerino, R. Mondelli, A. Prieto, A. Quillico, T. Salvadori, A. Selva, *Tetr.* (Suppl.), **7**, 175 (1966).
2) T. Reschke, *Tetr. Lett.*, 3435 (1969).
3) S. Isoe, Y. Hayase, T. Sakan, *Tetr. Lett.*, 3694 (1971).
4) J. A. Edwards, V. Schwarz, J. Fajkos, M. L. Maddox, J. H. Fried, *Chem. Commun.*, 292 (1971).
5) J. D. BuLock, D. J. Austin, G. Snatzke, L. Hruban, *Chem. Commun.*, 255 (1970).
6) D. J. Austin, J. D. BuLock, D. Drake, *Experientia*, **26**, 348 (1970).

5.55 SYNTHESIS OF TRISPORIC ACIDS

Route I[1)]

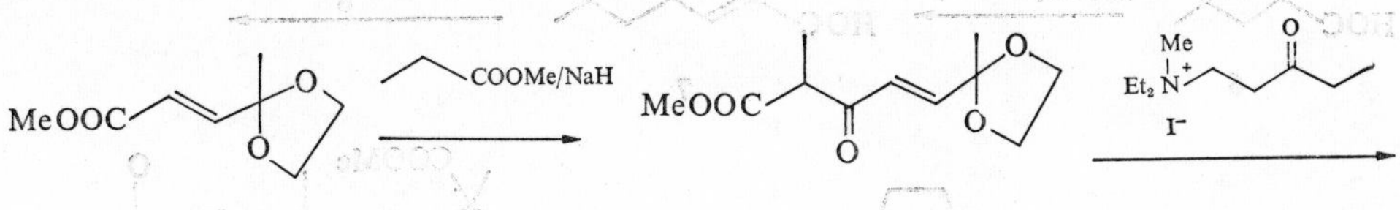
MeOOC
COOMe/NaH
MeOOC
Me
Et_2N^+
I^-

2 3

COOMe
NaOMe
COOMe

4 5

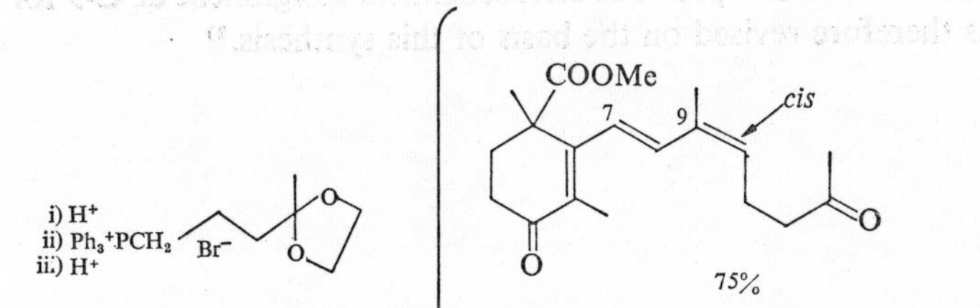
i) H^+
ii) $Ph_3{}^+PCH_2$ Br^-
iii) H^+
COOMe
7
9
cis
75%

Δ^9-*cis* trisporic acid B (*cis*-**1**) Me ester

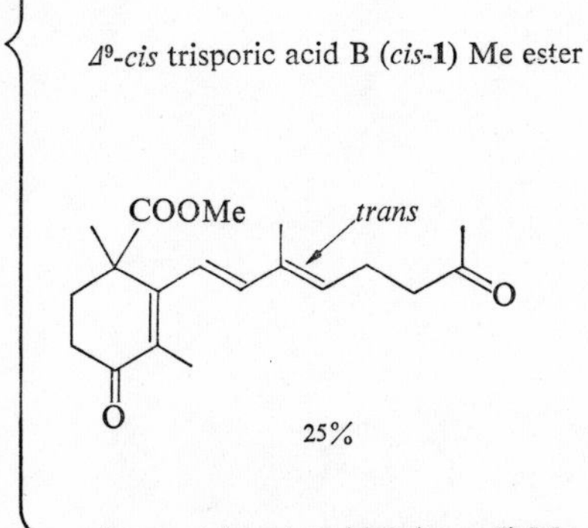
COOMe
trans
25%

Δ^9-*trans* trisporic acid B (*trans*-**1**) Me ester

Nonstereospecific condensation of the diketone derived from **5** with Wittig reagent yielded *cis*-**1** and *trans*-**1** in a ratio of 3:1.

1) S. Isoe, Y. Hayase, T. Sakan, *Tetr. Lett.*, 3694 (1971).

Route II[2)]

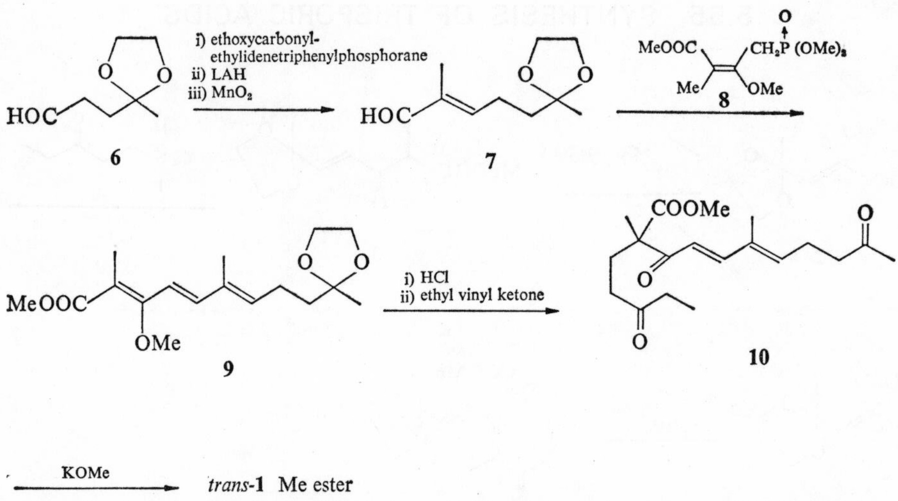

i) ethoxycarbonyl-
   ethylidenetriphenylphosphorane
ii) LAH
iii) MnO₂

6

7

8

MeOOC

OMe

9

i) HCl
ii) ethyl vinyl ketone

COOMe

10

KOMe ⟶ *trans*-**1** Me ester

Condensation of the aldehyde **7** with a new phosphonate reagent **8** which contains a masked
β-keto-ester function afforded the all-*trans* trienyl ester **9**. Synthetic product ($\varDelta^7$-*trans*, $\varDelta^9$-*trans*)
was found to be identical with the Jones oxidation product of natural trisporic acid C which
was believed to have a $\varDelta^9$-*cis* double bond. The previous stereochemical assignment at C-9 for
the isomeric trisporic acid C was therefore revised on the basis of this synthesis.[2)]

2) J. A. Edwards, V. Schwarz, J. Fajkos, M. L. Maddox, J. H. Fried, *Chem. Commun.*, 292 (1971).

Steroids

6.1   INTRODUCTION

Steroids (Gk., *stereos* = solid) are solid alcohols that are widely distributed in the animal and plant kingdoms, the major sterol being cholesterol, a C_{27} compound. The basic skeleton consists of 17 carbon atoms arranged in the form of a perhydrocyclopentenophenanthrene. They include wide variations in structure, and encompass compounds of vital importance to life, such as cholesterol, the bile acids, vitamin D, sex hormones, corticoid hormones, cardiac aglycones, antibiotics, and insect moulting hormones; some of the most potent toxins are steroidal alkaloids.

Because of their academic as well as pharmacological importance, a staggering number of steroids have been prepared in the laboratory in addition to those occurring naturally. Cholesterol (Gk., *chole* = bile) is the characteristic steroid of higher animals and is widely present especially in the brain and nerve tissues, human gall stones, suprarenal glands and egg yolk. It was first isolated from nonsaponifiable animal lipids in 1812 (Chevreul) but it was not until 1903 that active chemical investigations were initiated. The structural studies culminating in the proposal of the correct planar structure in 1932 were mainly due to the immense investigations lead by Windaus (on bile acids) and Wieland (on cholesterol). However, the relative configurations at the chiral centers of key steroids were established only in 1947, while the absolute configuration was not known until that of glyceraldehyde was determined by x-ray analysis in 1952; the complexity of the problem is apparent from the fact that cholic acid, for example, contains eleven centers of asymmetry, and thus 2^{11} (2048) modifications are theoretically possible.

The second stage in the development of steroid chemistry was due to the isolation and structural elucidation of the sex hormones from 1929 to 1935, headed by the work of Butenandt and Ruzicka; this period was followed by the exciting studies on the corticoids in the period 1935–1938 (Kendall, Reichstein, Wintersteiner).

The steroids are ideally suited for chemical and physico-chemical studies and have invariably played vital roles in the development of new reactions, concepts and physical measurements. This arises from: (i) their relatively flat and rigid molecular framework; (ii) their highly crystalline properties; and (iii) the enormous number of derivatives which have been prepared. Thus, they have played indispensable roles in the development of practically every phase of organic chemistry, i.e., isolation (chromatographic methods), organic reactions, total syntheses of complex natural products (Woodward, Robinson), conformational analysis (Barton), ultraviolet spectroscopy (Woodward, Fieser), infrared spectroscopy, optical rotatory dispersion (Djerassi) and circular dichroism, nuclear magnetic resonance spectroscopy, mass spectroscopy, biosynthesis (Bloch, Cornforth, Popjak). The first successful application of x-ray crystallography to natural products was to cholesterol in 1932 (Bernal, Rosenheim, King).

For pedagogic reasons a considerable portion of this chapter is devoted to the classic degradative methods of structure studies.

6.2 NOMENCLATURE OF STEROIDS

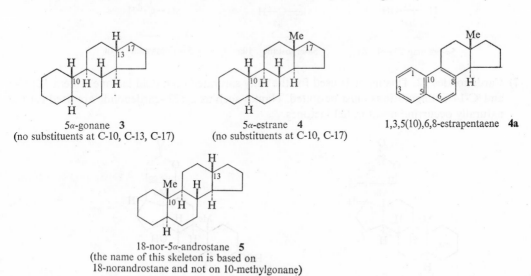

5α-cholestan-3β-ol **1**   5ξ-cholestane **2**

The following is an abbreviated statement of the "IUPAC-IUB 1971 Definitive Rules for Steroid Nomenclature."[1]

1) Numbering of the carbons and lettering of the rings follow **1**.

2) Configurations of atoms or groups attached to the rings are indicated as follows.

α : below the plane, shown by dotted lines (·········).

β : above the plane, shown by solid lines, preferably thickened (——).

ξ : configuration unknown, shown by wavy lines (〜〜) (as in **2**).

3) All hydrogens and methyls attached at ring junctions must be indicated by H and CH₃ (Me may be used if editorial conventions require it). Practices of deleting hydrogens and denoting methyls simply by lines should be abandoned (however, this is not followed in this book).

4) Unless otherwise stated, a steroid name implies configurations at 8, 9, 10, 13, 14 and 17 as shown in **1**. The 5-H is always to be denoted by 5α, 5β, or 5ξ.

5) Names of fundamental carbocycles are as indicated by **3** through **11** (for 5α-series).

5α-gonane **3**
(no substituents at C-10, C-13, C-17)

5α-estrane **4**
(no substituents at C-10, C-17)

1,3,5(10),6,8-estrapentaene **4a**

18-nor-5α-androstane **5**
(the name of this skeleton is based on
18-norandrostane and not on 10-methylgonane)

---

1) *Pure Appl. Chem.*, **31**, 285 (1971).

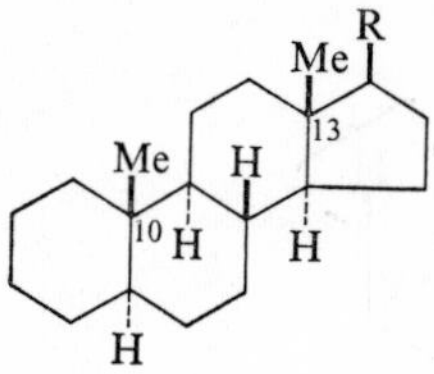

6	R=		Name
7	H		5α-androstane
8	Et		5α-pregnane
9			5α-cholestane[*1] (5β-cholestane should not be called coprostane)
10			5α-ergostane[*2]
11			5α-stigmastane[*3]

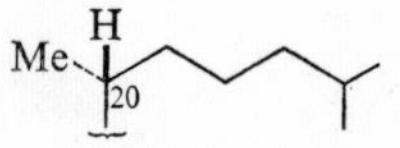

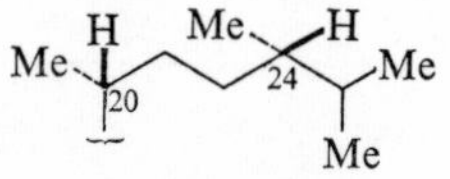

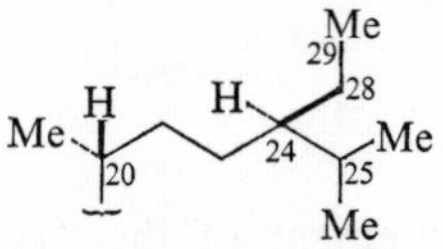

*1 = 20R configuration   *2 = 24S configuration   *3 = 24R configuration

6) C-20 Configurations in pregnanes **8**. When the side chain is drawn according to the Fischer projection with the highest number at the top (**8a**), substituents to the right of C-20 are termed α and those to the left β. When first suggested, this convention also included other side chain carbons, e.g., **10a** and **11a**.[2]

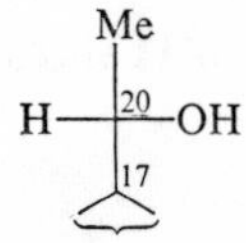

5α-pregnan-20α-ol **8a**

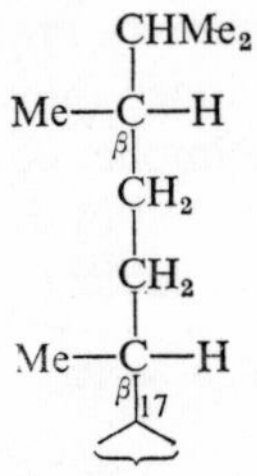

5α-ergostane **10a**

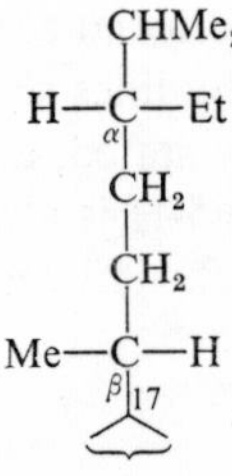

5α-stigmastane **11a**

7) Cardanolides **12**. This name is used for the fully saturated digitaloid lactone system **12**; C-5 and C-14 configurations must be stated. Names such as 20(22)-cardenolide are used for the naturally occurring unsaturated lactones.

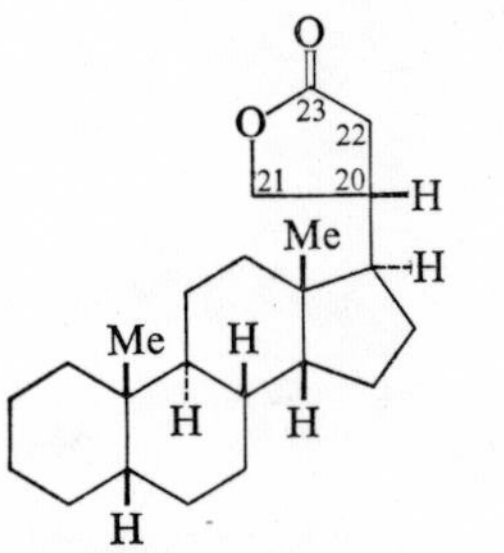

5β,14β-cardanolide **12**

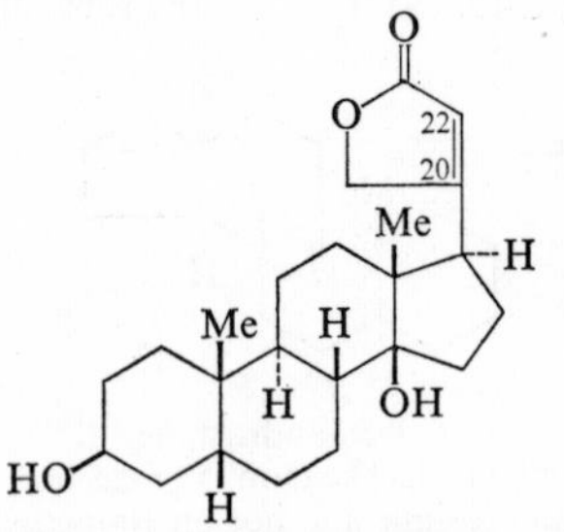

3β,14β-dihydroxy-5β,14β-card-20(22)-enolide **13** (=digitoxigenin)

2) L. F. Fieser, M. Fieser, *Steroids*, p. 337–340, Reinhold, 1959.

8) Bufanolides **14**. This name is used for the fully saturated system of the squill-toad group of lactones; C-5 and C-14 configurations must be stated. Names such as 20(22)-bufadienolide are used for the naturally occurring doubly unsaturated lactones.

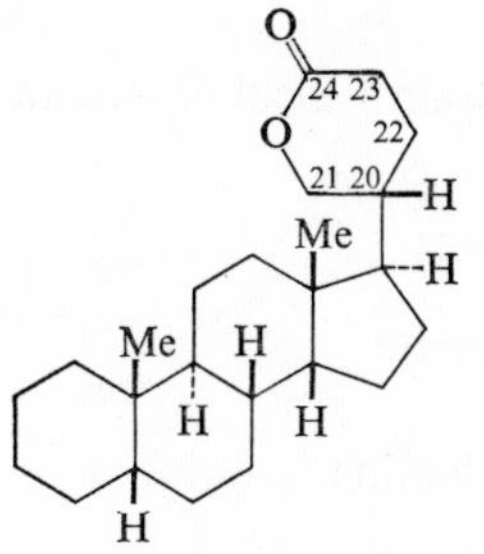

5β,14β-bufanolide   **14**

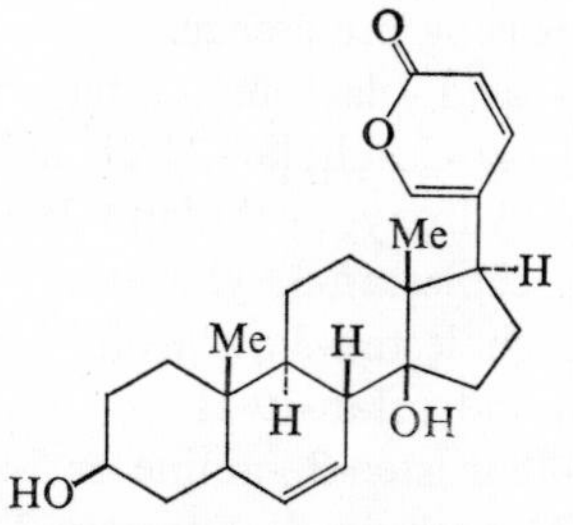

3β,14β-dihydroxy-14β-bufa-
4,20,22-trienolide   **15** (=scillarenin)

9) Spirostan **16**. This specifies the structure shown except for the configurations at C-5 and C-25.

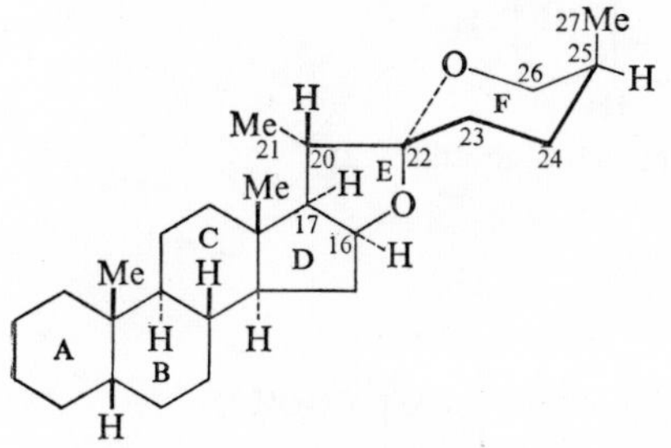

(25S)-5β-spirostan   **16**

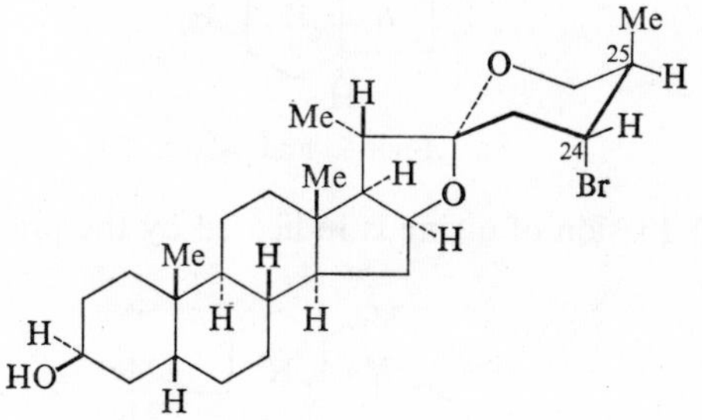

(24R, 25R)-24-bromo-5β-spirostan-3β-ol   **17**

10) Furostan **18**. This specifies the structure shown except for the configurations at C-5, C-22 and C-25.

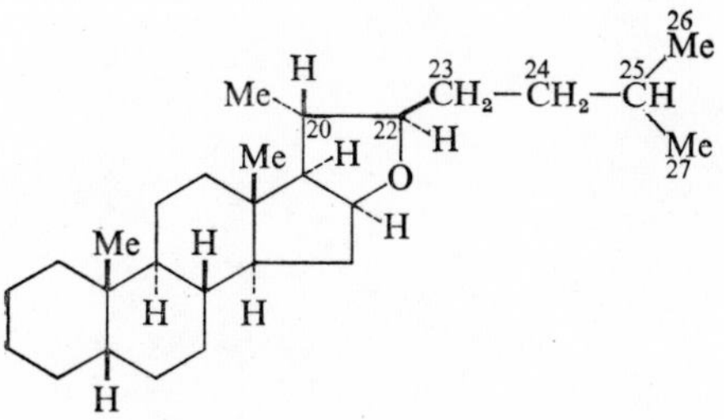

(22R)-5β-furostan   **18**

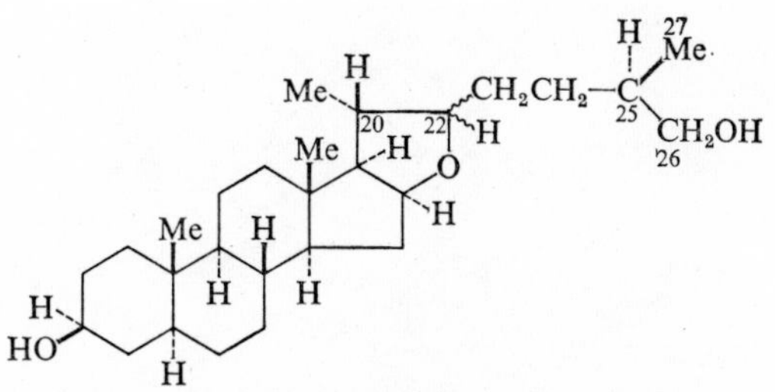

(20S, 22ξ, 25R)-5α-furostan-3β, 26-diol   **19**
(=dihydropseudotigogenin)

11) Unsaturation is shown by changing terminal -an(e) to -en(e), -adien(e), -yn(e), etc. For example 5α-cholest-6-ene, 5β-cholesta-7,9(11)-diene.

12) Substituents can be designated either as suffixes or prefixes; a few, such as halogens, alkyl and nitro groups can only be prefixes. Where possible, one type of substitutent must be designated as a suffix; when more than one type is present, the choice for suffix is as follows, in order of decreasing preference.

        onium salt – acid – lactone – ester – aldehyde – ketone – alcohol – amine – ether.

    Examples: 11-oxo-5α-cholan-24-oic acid

                3β-hydroxy-5α-cholano-24,17α-lactone

                5α-cholestan-3β-yl acetate

                5α-cholestan-3α,12α-diol 3-acetate 12-benzoate

                5α-androstan-19-al

                5-androsten-3β-amine or 5-androsten-3β-ylamine

                17β-methoxy-4-androsten-3-one

13) Ring contraction and ring expansion are indicated by the prefixes nor and homo, respectively, preceded by an italic capital indicating the ring affected.

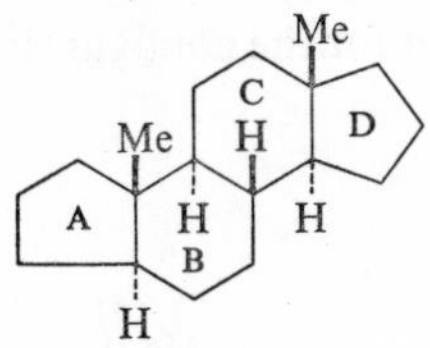

A-nor-5α-androstane **20** 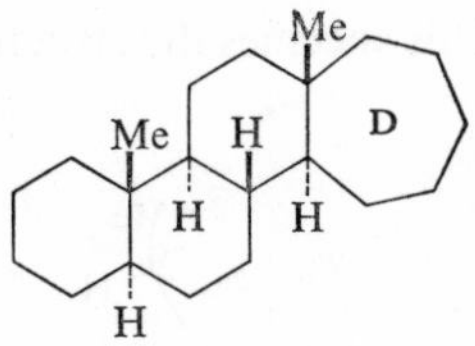   D-homo-5α-androstane **21**

14) Fission of a ring is indicated by the prefix seco.

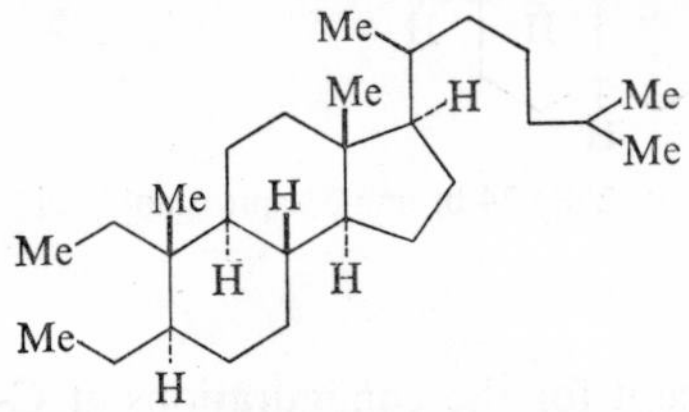

2,3-seco-5α-cholestane **22**

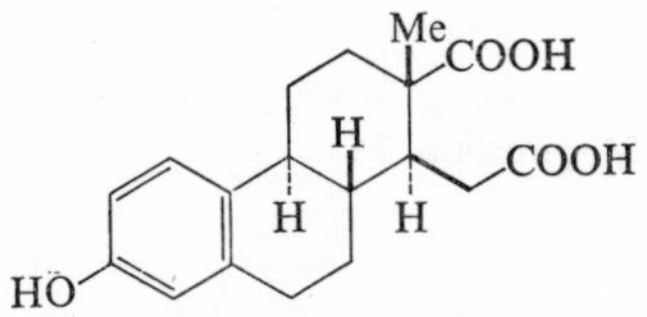

3-hydroxy-16, 17-seco-
1,3,5(10)-estratriene-16, 17-dioic acid **23**

6.3 BILE ACIDS

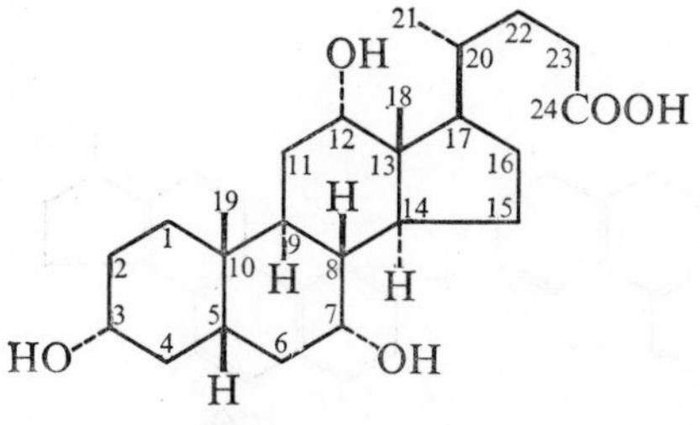

cholic acid **1**
$C_{24}H_{40}O_5$ mp: 198°
α_D (EtOH): +37°

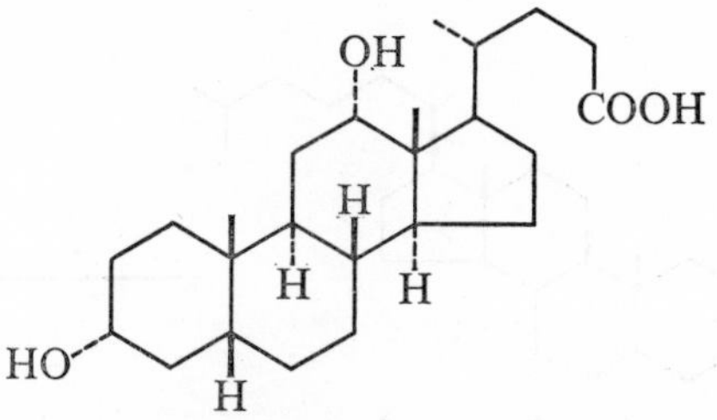

desoxycholic acid **2**
$C_{24}H_{40}O_4$ mp: 177°
α_D (EtOH): +55°

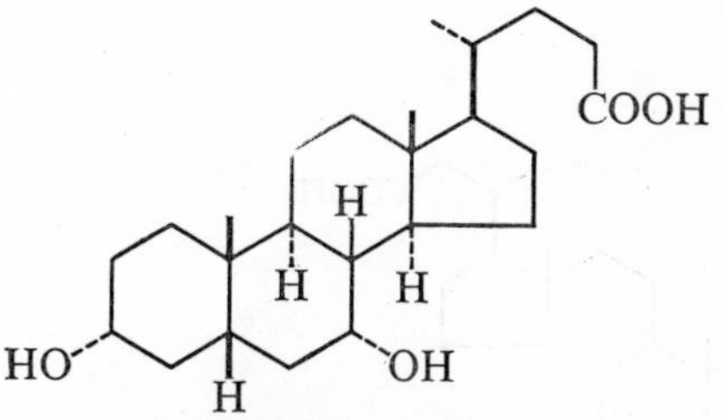

chenodesoxycholic acid **3**
$C_{24}H_{40}O_4$ mp: 143°
α_D (EtOH): +11°

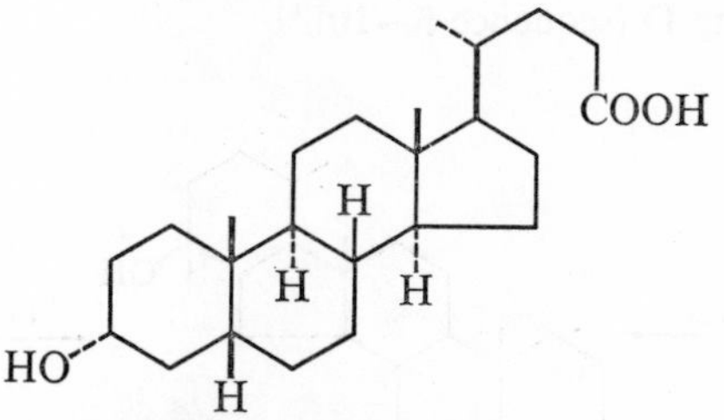

lithocholic acid **4**
$C_{24}H_{40}O_3$ mp: 184–186°
α_D (EtOH): +34°

1) Carbon-hydrogen analysis established the molecular formula of cholic acid as $C_{24}H_{40}O_5$.[1]

2) Structural relationships between cholic **1**, desoxycholic **2**, chenodesoxycholic **3**, and litho-cholic acid **4** were established by their conversion to cholanic acid **5**, e.g. cholic acid **1** to cholanic acid **5**.[2]

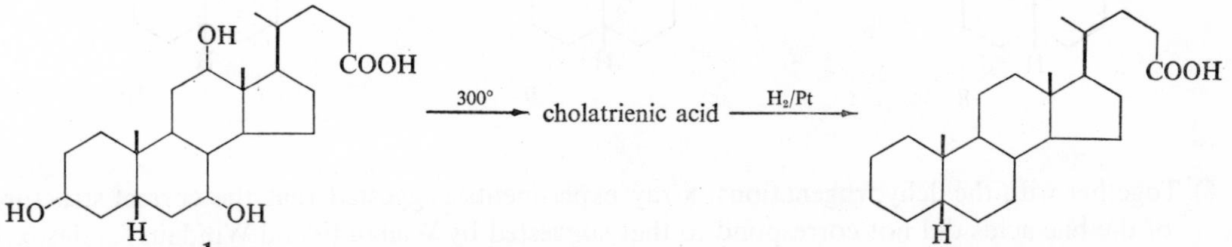

$\xrightarrow{300°}$ cholatrienic acid $\xrightarrow{H_2/Pt}$

3) Cholesterol had been converted to coprostanol, and coprostane was converted to cholanic acid. The interrelation of two series of steroidal compounds enabled structural elements deduced for members of one series to be related to members of the other.[3]

1) H. Wieland, F. J. Weil, *Z. Physiol. Chem.*, **80**, 287 (1912).
2) H. Wieland, P. Weyland, *Z. Physiol. Chem.*, **110**, 136 (1920); H. Wieland, G. Reverey, *ibid.*, **140**, 186 (1924); H. Wieland, K. Kraus, H. Keller, H. Ottawa, *ibid.*, **241**, 47 (1936).
3) see "Cholesterol."

4) Dehydrogenation of cholesterol **6** and cholesteryl chloride gave chrysene **6a** and the Diels hydrocarbon **6b**.[4] The nature of the latter was not recognized until after the elucidation of the structure of the steroid nucleus.[5]

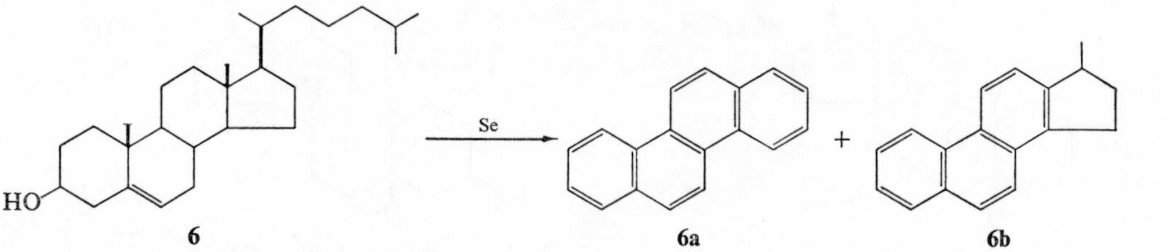

5) Barbier-Wieland degradation of the side chain of cholanic acid **5** to yield etianic acid **10** confirmed the structure of the side chain and provided a point of entry for investigation of ring D (sequence **5→10**).[6]

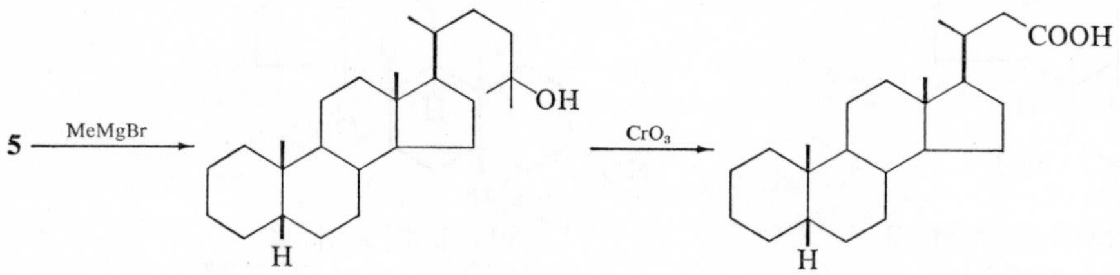

6) Together with the dehydrogenations, x-ray experiments suggested that the correct structure of the bile acids did not correspond to that suggested by Wieland[7] and Windaus for desoxy-cholic acid **11**. Bernal's crystallographic examination of ergosterol **12**[7] suggested a long, thin molecule rather than the more globular shape expected from structure **11**. Based on this evidence, Rosenheim and King[8] suggested the structure **13**, which was amended to **14** following re-examination of degradative experiments.[9]

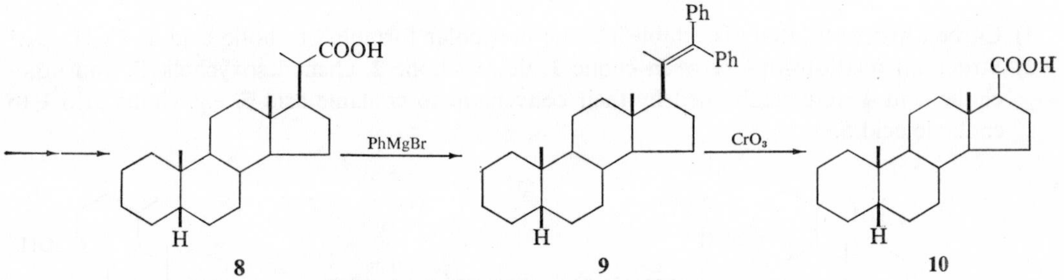

4) O. Diels, W. Gädke, P. Körding, *Ann. Chem.*, **459**, 1 (1927).
5) S. H. Harper, G. A. R. Kon, F. C. J. Ruzicka, *J. Chem. Soc.*, 124 (1934); G. A. R. Kon, E. S. Narracott, *ibid.*, 672 (1938); O. Diels, H. F. Rickert, *Chem. Ber.*, **68**, 267, 325 (1935).
6) H. Wieland, O. Schlichting, R. Jacobi, *Z. Physiol. Chem.*, **161**, 80 (1926).
7) J. D. Bernal, *Nature*, **129**, 277 (1932); H. Wieland, *Angew. Chem.*, **42**, 421 (1929).
8) O. Rosenheim, H. King, *Chem. Ind.*, **51**, 464 (1932).
9) H. Wieland, E. Dane, *Z. Physiol. Chem.*, **210**, 268 (1932); O. Rosenheim, H. King, *Nature*, **130**, 315 (1932).

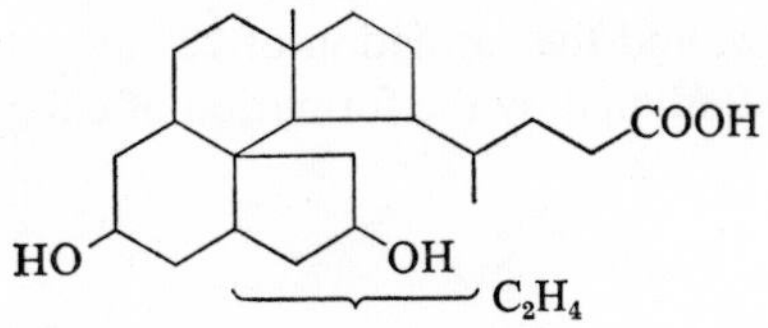

11 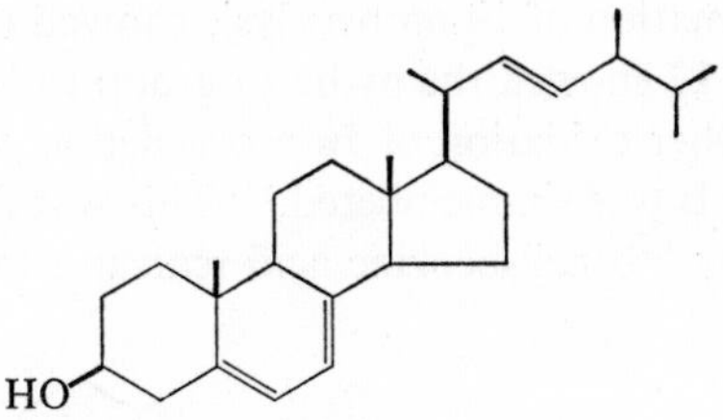   **12**

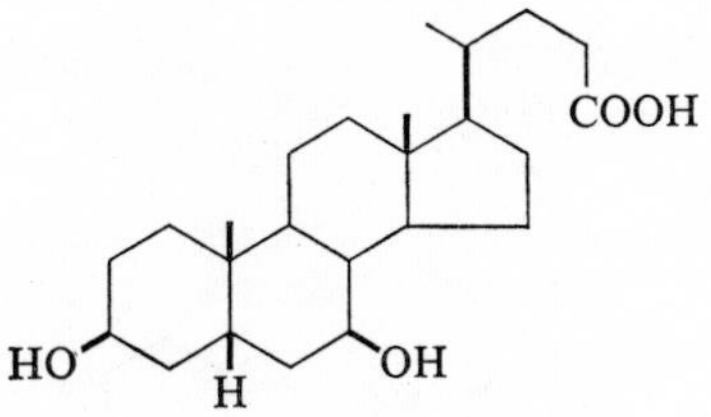

13 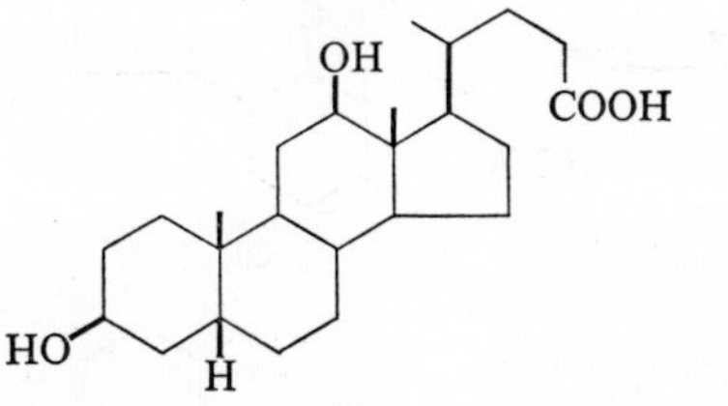   **14**

7) Structure **11** was deduced by Wieland and Windaus on the basis of a long series of degradation experiments, the chief feature of which was oxidative cleavage of each of the rings in turn to a diacid. The behavior of the diacid upon pyrolysis was interpreted in light of the Blanc rule according to which 1,5-diacids yield anhydrides, while 1,6- or higher diacids yield ketones.

8) The nature of ring A was deduced from the transformations shown below.

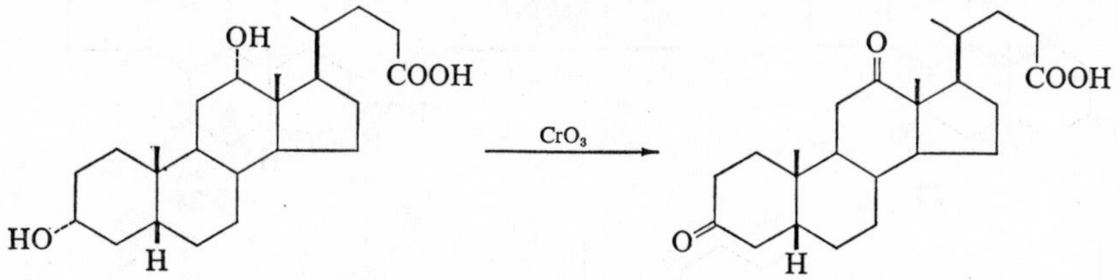

2   **15**

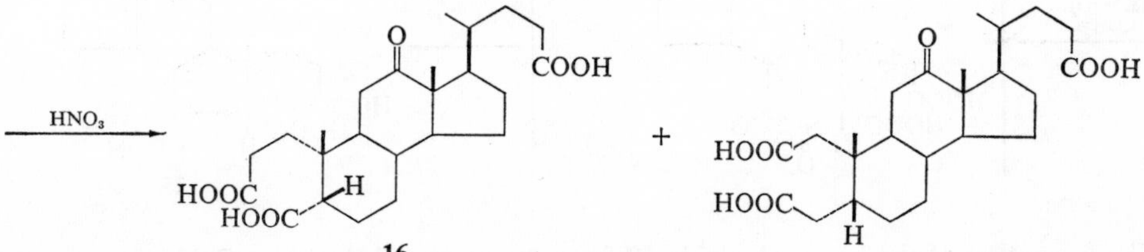

16   +   **17**

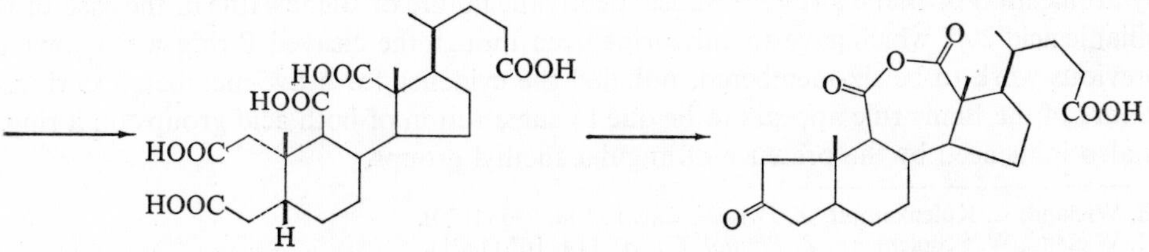

18   **19**

Formation of **19** on pyrolysis showed ring A to be six-membered, while the formation of **16** and **17** showed the hydroxy group to be flanked by two methylene groups.[10]

9) Further oxidation of **16** proceeded as shown below, and the formation of **22** suggested that ring B was six-membered.[11] This was further substantiated by the formation of cilianic acid **25** by a benzil-benzilic acid rearrangement.[12]

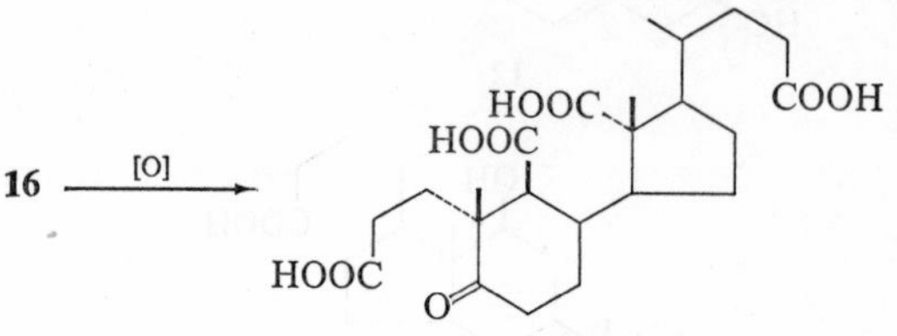

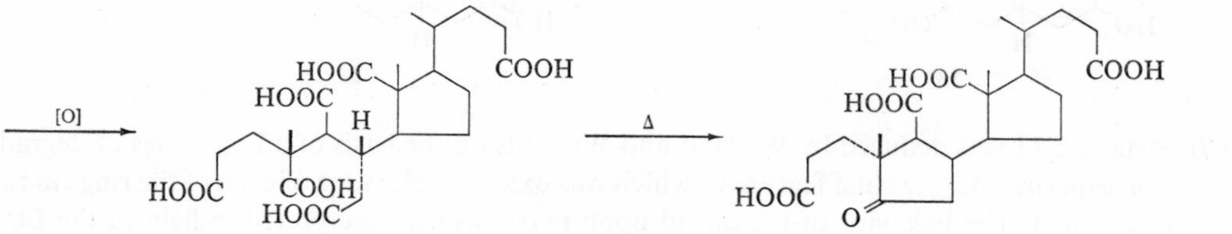

10) The formation of the keto-anhydride **19** from **18** suggested that ring C was five-membered, by application of Blanc's rule.[13] Subsequently the failure of Blanc's rule in the case of thilobilianic acid **26**,[9] which gave an anhydride even though the cleaved B ring was known from previous work to be six-membered, nullified the evidence for a five-membered C ring. The failure of the Blanc rule appears to be due to substitution of both acid groups on a ring, but is also influenced by the presence of angular methyl groups.

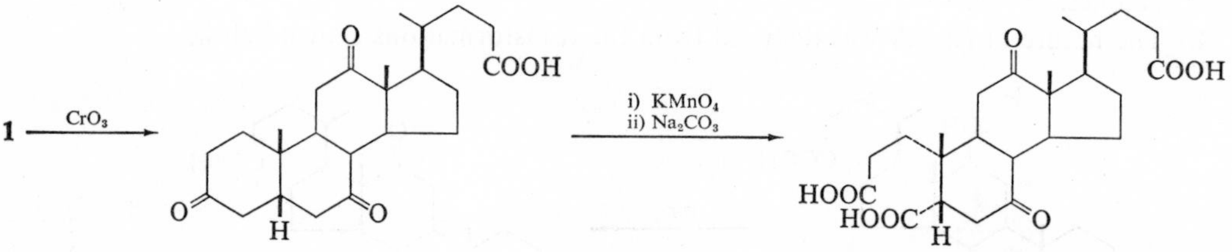

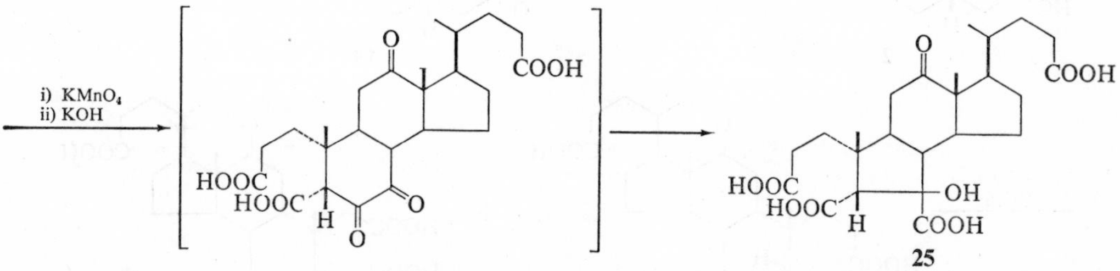

10) H. Wieland, A. Kulenkampff, *Z. Physiol. Chem.*, **108**, 295 (1920).
11) H. Wieland, W. Schulenberg, *Z. Physiol. Chem.*, **114**, 167 (1921).
12) W. Borsche, R. Frank, *Chem. Ber.*, **60**, 723 (1927).
13) H. Wieland, *Z. Physiol. Chem.*, **108**, 306 (1920).

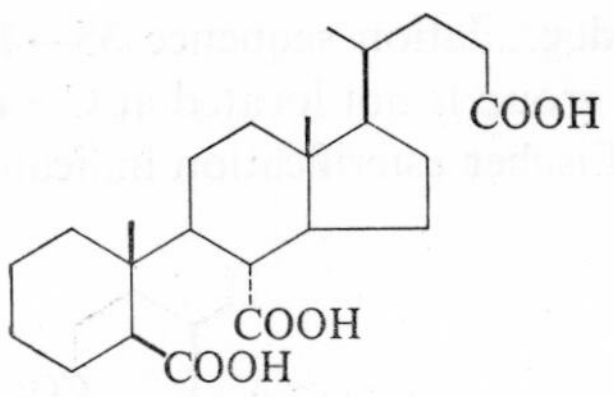

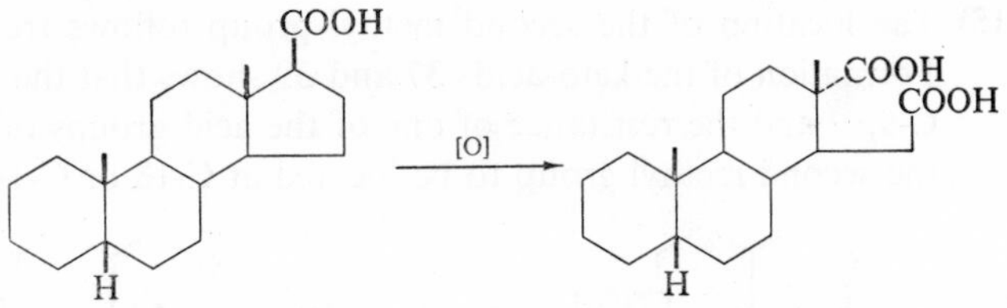

11) Oxidative cleavage of etianic acid **10** yielded etiobilianic acid **27**. The formation of an anhydride on pyrolysis indicated ring D was five-membered.[6]

12) Adoption of a perhydrocyclopentenophenanthrene ring system required the two methyl groups to be at C-1, or at the ring junctions, since positions adjacent to the hydroxyl groups of cholic acid, and positions 15 and 16 had been ruled out by previous oxidation experiments.

13) The location of a methyl group at C-1 was ruled out by the formation of **30** and **31** by successive Barbier-Wieland degradations of the oxidation product **29** from cholestenone **28**. Resistance of **31** to Fischer esterification showed the acid to be tertiary, i.e. a methyl group was situated at C-10.[14] From this sequence of reactions was also deduced the location of the ring A hydroxyl group at C-3.

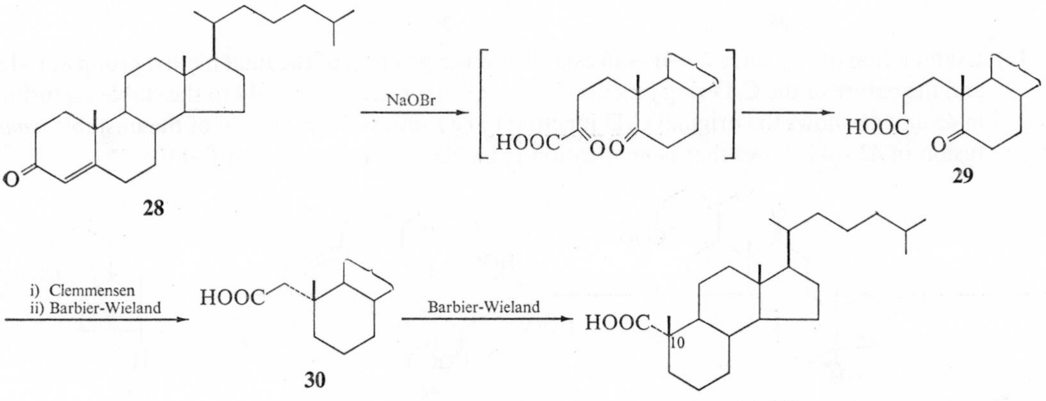

14) The position of the side chain was shown by identification of **34**[15] resulting from dehydrogenation of the Perkin condensation product **33** of the 12-ketocholanic acid **32**.[16]

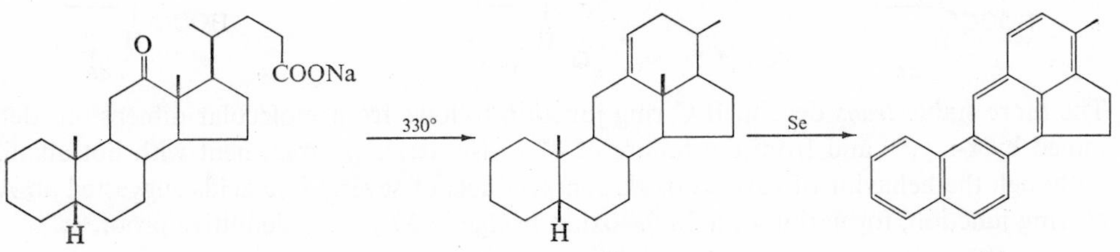

14) R. Tschesche, *Ann, Chem.,* **498,** 185 (1932).
15) J. W. Cook, G. A. D. Haslewood, *J. Chem. Soc.,* 428 (1934); L. F. Fieser, A. M. Seligmann, *J. Am. Chem Soc.,* **57,** 228, 942 (1935).
16) H. Wieland, E. Dane, *Z. Physiol. Chem.,* **219,** 240 (1933).

15) The location of the second methyl group follows from the degradation sequence **35→40**. Formation of the keto-acids **37** and **39** shows that the methyl group is not located at C-8 or C-9,[17] and the resistance of one of the acid groups of **40** to Fischer esterification indicated the second methyl group to be located at C-13 or C-14.[18]

16) Examination of sequence **2→41→46** established the position of the final methyl group at C-13 and the nature of the C/D ring junction as *trans*.[19] Isomerization of **44** to the stable *cis* fusion in **45** and **46** shows the original C/D junction to be *trans*, while retention of the unstable *trans* fusion in **42→43** shows that isomerization is blocked by the methyl at C-13.

17) The more stable *trans* decalin B/C ring junction follows from molecular dimensions determined by x-ray,[20] and from the failure of **47** to isomerize on treatment with hot alkali.[21]

18) Although the behavior of various oxidation products of several bile acids suggested an A/B *cis* ring junction, formation of a 3α,9α-oxido bridge in **52** gave a definitive proof.[22]

17) H. Wieland, T. Posternak, *Z. Physiol. Chem.*, **197**, 17 (1931).
18) H. Wieland, F. Vocke, *Z. Physiol. Chem.*, **177**, 68 (1928).
19) H. Wieland, O. Schlichting, *Z. Physiol. Chem.*, **134**, 276 (1924); H. Weiland, E. Dane, *ibid.*, **216**, 91 (1933).
20) J. D. Bernal, *Chem. Ind.*, **51**, 466 (1932); L. Ruzicka, G. Thomann, *Helv. Chim. Acta*, **16**, 221 (1933).
21) H. Wieland, V. Wiedersheim, *Z. Physiol. Chem.*, **186**, 232 (1930).
22) R. B. Turner, V. R. Mattox, L. L. Engel, B. F. McKenzie, E. C. Kendall, *J. Biol. Chem.*, **166**, 345 (1946).

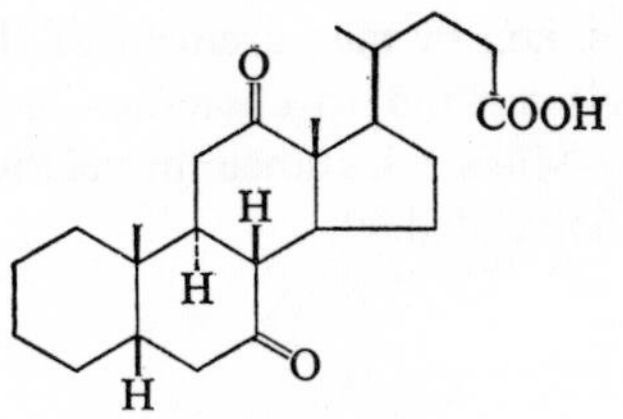

47

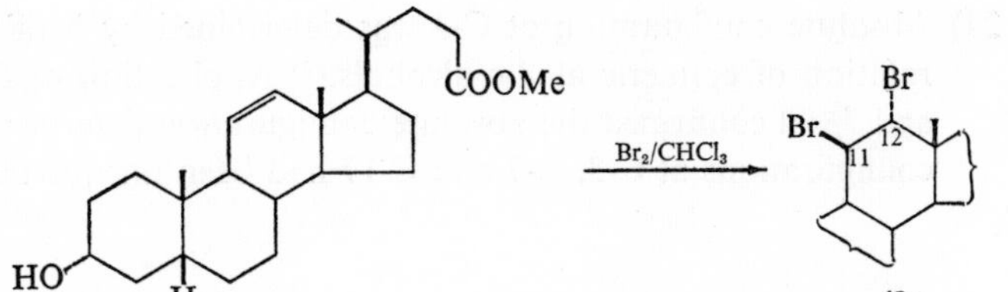

48

Br₂/CHCl₃

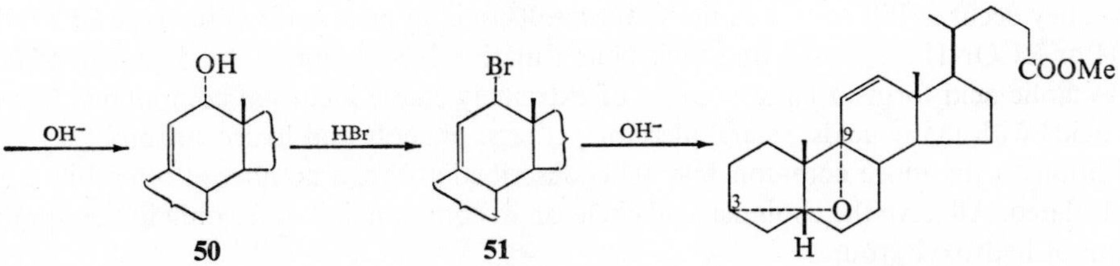

49

OH⁻

50

HBr

51

OH⁻

52

19) Chenodesoxycholic acid **3** could be obtained from cholic acid.[2,23] Isolation of the lactone acid **53** from the oxidation of **3** indicated the *α-cis* relationship of 5-COOH and 7-OH, i.e., the hydroxyl is **7α** in both compounds.[24]

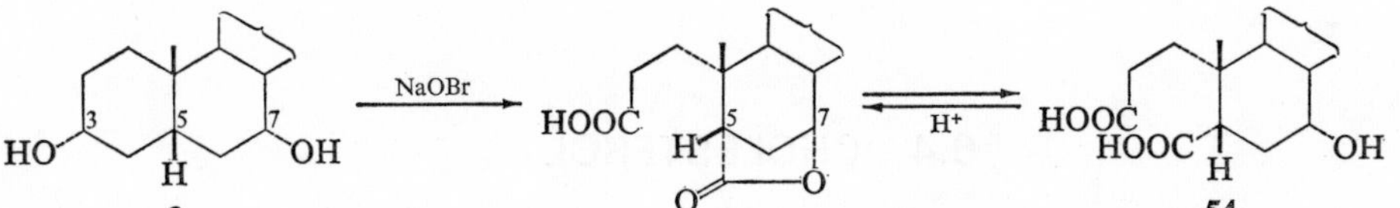

3

NaOBr

53

H⁺

54

20) The absolute configuration at C-20 was determined by degradation of Δ^{14}-cholesteryl benzoate to the aldehyde **57**[25] whose configuration was related to D-(+)-glyceraldehyde through D-(+)-methylsuccinic acid and D-(−)-citronellal.[26]

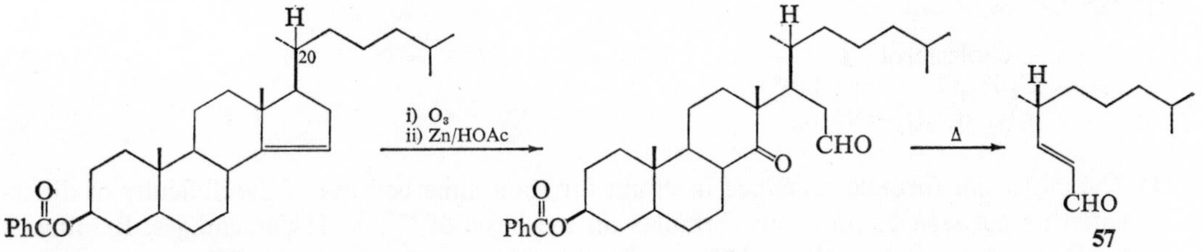

55

i) O₃
ii) Zn/HOAc

56

Δ

57

23) A. Windaus, A. Bohne, E. Schwarzkopf, *Z. Physiol. Chem.*, **140**, 177 (1924).

24) A. Windaus, A. van Schoor, *Z. Physiol. Chem.*, **148**, 225 (1925); H. Lettre, *Chem. Ber.*, **68**, 766 (1933).

25) B. Riniker, D. Arigoni, O. Jeger, *Helv. Chim. Acta*, **37**, 546 (1954); J. W. Cornforth, I. Youhotsky, G. Popjak *Nature*, **173**, 536 (1954).

26) A. Fredga, J. K. Miettinen, *Chem. Scand.*, **1**, 371 (1947).

27) J. A. Mills, *J. Chem. Soc.*, 4976 (1952).

28) V. Prelog, *Helv. Chim. Acta*, **36**, 308 (1953); V. Prelog, H. L. Meier, *ibid.*, **36**, 320 (1953); W. G. Dauben, D. F. Dickel, O. Jeger, V. Prelog, *ibid.*, **36**, 325 (1953).

29) L. F. Fieser, M. Fieser, *Steroids*, Reinhold, 1953.

30) *Chemistry of Carbon Compounds* (ed. E. H. Rodd), vol. II, part B, Elsevier, 1953.

21) Absolute configuration at C-3 was determined by Mill's rules, i.e., by measurement of the rotation of epimeric allylic alcohols.[27] Application of Prelog's method to cholestan-7α-ol and 7β-ol confirmed the absolute configuration determined by Mill's rules since the relative configurations at C-3, C-7 and C-17 had been unequivocally correlated.[28]

Remarks

Bile acids, as their name implies, are isolated from the bile of various animals, as well as humans. They occur as bile salts, i.e., the sodium salts of conjugate acids of the type $RCONH$-CH_2COOH or $RCONHCH_2SO_3H$, and their chief function is to facilitate the digestion of fats.

Desoxycholic acid forms a large number of extremely stable inclusion compounds known as choleic acids with many acids, esters, alcohols, ethers, phenols and hydrocarbons.

In addition to the more common bile acids described above, a number of other bile acids have been isolated. All have the cholanic acid skeleton in common and differ only in the number and position of hydroxyl groups.

Closely related bile alcohols have been isolated from the bile of lower organisms. These compounds include symnol (shark), ranol (frog) and cyprinol (carp).

The references listed pertain to specific items in the above text. For a more comprehensive treatment provided with extensive references, see references 29 and 30.

6.4   CHOLESTEROL

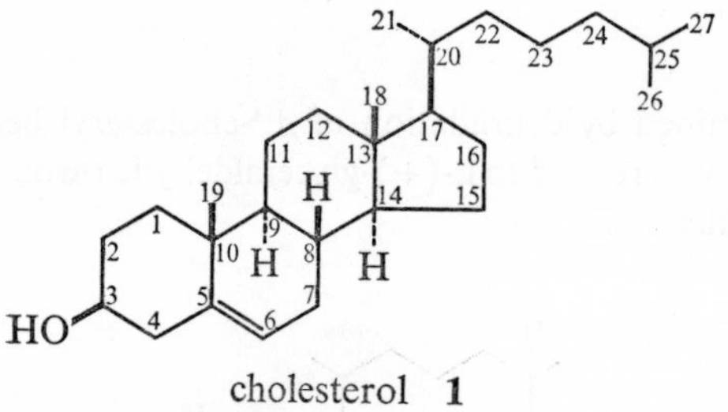

cholesterol **1**

$C_{27}H_{46}O$ mp: 149°

$\alpha_D(CHCl_3)$: −39°

1) The molecular formula remained in doubt for some time because of the difficulty of differentiating between homologous formulas on the basis of C and H percentages. Reinitzer's analyses on the acetate dibromide were finally accepted as definitive.[1]

2) Cholesterol **1** is transformed to coprostanol **7** in the digestive tract.[2] The relationship between these two alcohols and their parent hydrocarbons cholestane **4** and coprostane **8** was established by the interconversions shown below, and **4** and **8** were recognized as C-5 epimers.[3]

1) F. Reinitzer, *Monatsh.*, **9**, 421 (1888).

2) St. von Bondzynski, F. Humnicki, *Z. Physiol. Chem.*, **22**, 396 (1896).

3) J. Mauthner, W. Suida, *Monatsh.*, **15**, 85 (1894); J. Mauther, *ibid.*, **30**, 635 (1909); *ibid.*, **27**, 305 (1906); *ibid.*, **28**, 1113 (1907); A. Windaus, C. Uibrig, *Chem. Ber.*, **48**, 857 (1915).

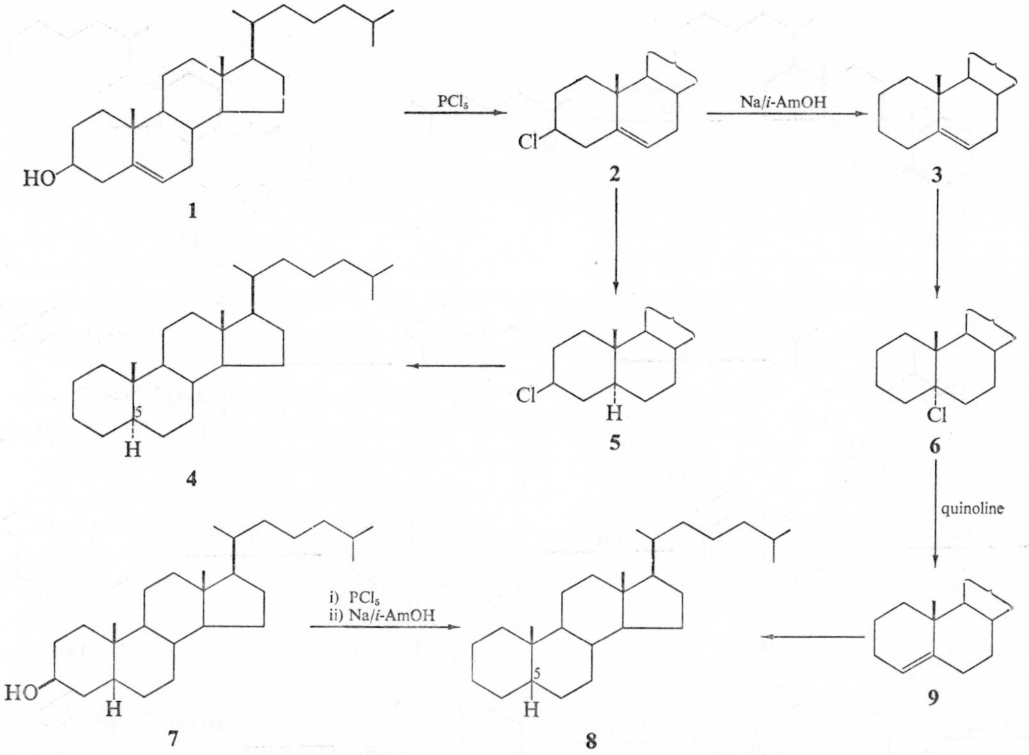

3) Coprostane 8 was converted to cholanic acid 10 by chromic acid oxidation, thus linking the class of sterols with that of the bile acids[4] so that evidence regarding one series could be applied to the other.

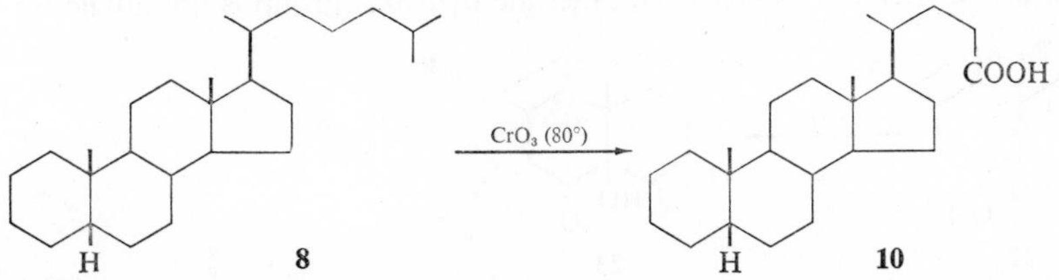

4) Vigorous oxidation of cholesteryl acetate produced methyl isohexyl ketone 12, isolated as the semicarbazone.[5] Ozonolysis yielded the same fragment from cholestanol 13 or coprostanol 7, establishing the nature of the side chain.[6,7]

5) Oxidative cleavage of the chloroketone 18[8] derived from cholesteryl acetate 11,[9] followed by a second oxidative cleavage of the hydroxy-diacid 20 showed the hydroxyl group and double bond to be in different rings.[8]

4) A. Windaus, K. Neukirchen, *Chem. Ber.*, **52**, 1915 (1919).
5) A. Windaus, C. Reseau, *Chem. Ber.*, **46**, 1246 (1913).
6) C. Doree, J. A. Gardner, *J. Chem. Soc.*, **93**, 1328 (1908); C. Doree, *ibid.*, **95**, 638 (1909).
7) O. Diels, *Chem. Ber.*, **41**, 2596 (1908).
8) A. Windaus, G. Stein, *Chem. Ber.*, **37**, 3599 (1904).
9) C. E. Anagnostopoulos, L. F. Fieser, *J. Am. Chem. Soc.*, **76**, 532 (1954); J. Mauthner, W. Suida, *Monatsh.*, **15** 85 (1894); *ibid.*, **24**, 648 (1903); A. Windaus, *Habilitionsschrift*, (1903); *Chem. Ber.*, **36**, 3752 (1903).

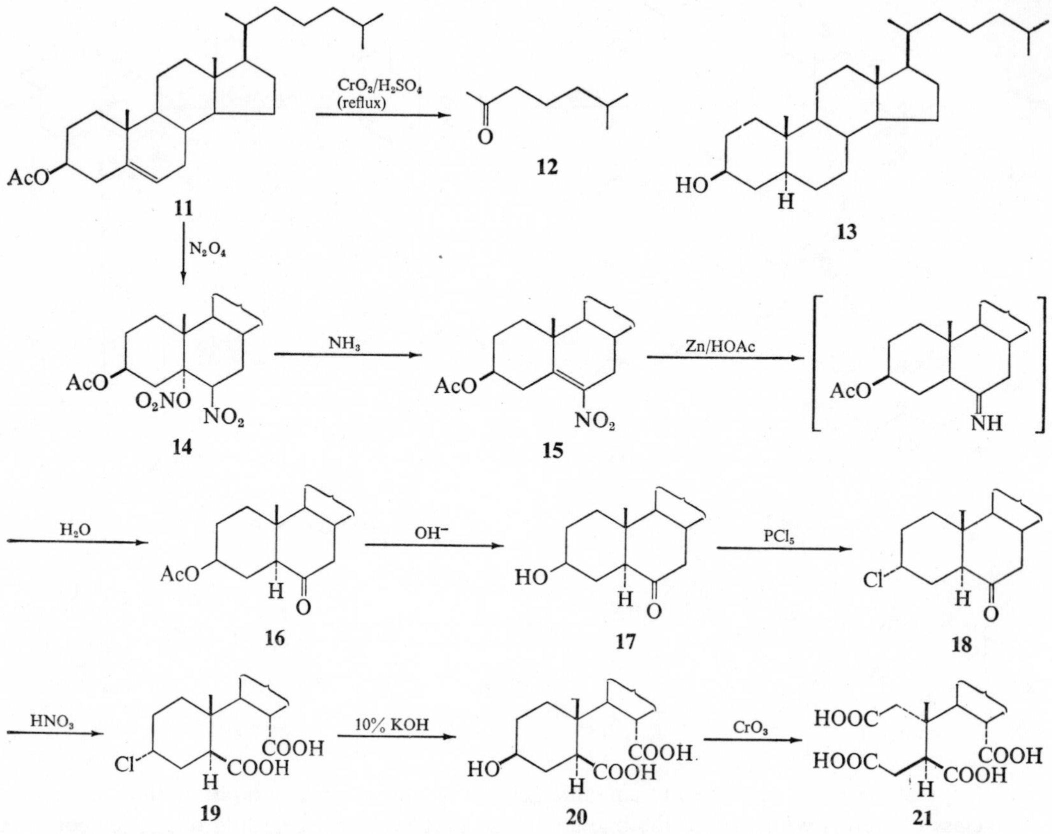

6) The relationship of the hydroxyl group and double bond was shown by the conversion of **1** to **24**, and to **26** which was not a β-keto acid, i.e., the hydroxyl group is not allylic.[10]

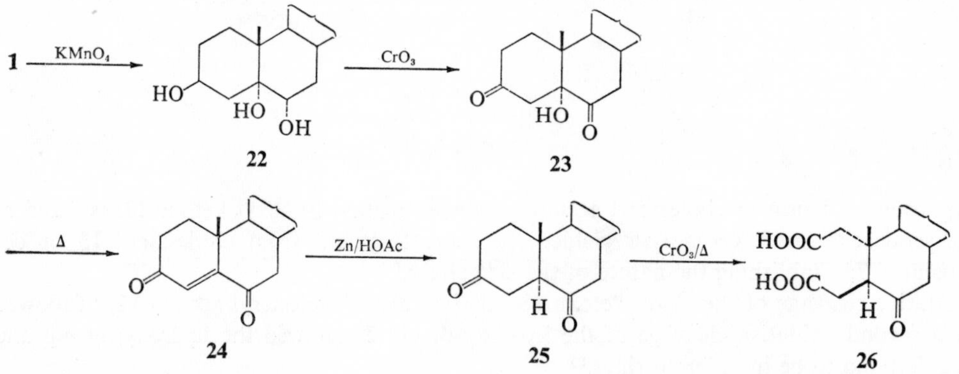

7) Formation of a lactone **27** from **20** showed the C-3 hydroxyl group to be *cis* to the B ring residue, i.e., C-3 hydroxyl is β.[11] The epimeric hydroxy acid **28** yielded the unsaturated anhydride **29**.[8,11]

10) A. Windaus, *Chem. Ber.*, **40**, 257 (1907); V. Prelog, E. Tagman, *Helv. Chim. Acta*, **27**, 1867 (1944); A. Windaus, A. von Staden, *Chem. Ber.*, **54**, 1059 (1921).
11) H. Lettre, *Chem. Ber.*, **68**, 766 (1935); C. W. Shoppee, *J. Chem. Soc.*, 1032 (1948).

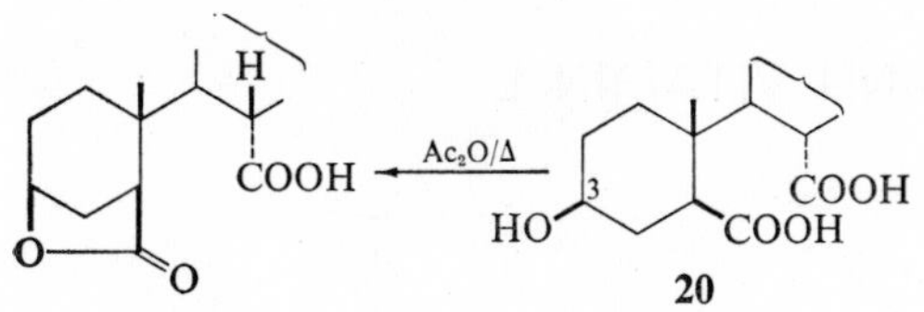

8) Bernal's crucial x-ray experiments provided the key to the correct nuclear structure of cholesterol and the bile acids. In the light of this work the extensive degradations of Wieland and Windaus led to the elucidation of the correct structure of the steroid nucleus (see "Bile Acids").

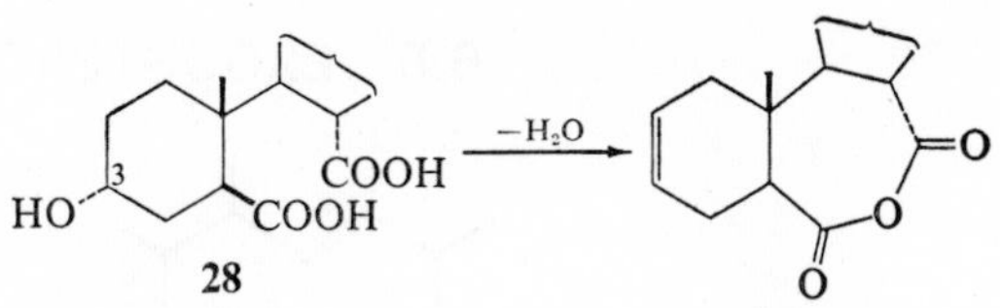

Remarks

Cholesterol is found all body tissues and is the chief constituent of gallstones. Coprostanol is a metabolic product of cholesterol found in human feces.[2]

Both cholestanol **13** and coprostanol **7** are isomerized by heating with alkali metal alcoholates to give mixtures of the C-3 epimers in the ratios 9:1 ($\beta:\alpha$) and 1:9 ($\beta:\alpha$), respectively.[12] The β-epimers both form insoluble complexes with digitonin and this complex formation is remarkably stereospecific for all steroids with a 3β-OH, provided 10-Me has the natural β-configuration.

Derivatives of cholesterol **1** undergo conversion to the interesting i-steroids. Acetolysis of cholesteryl tosylate **30** gives i-cholesteryl acetate **31**; the reversal of the process occurs to give a normal cholesterol derivative (**31→32**). Both reactions are highly stereospecific, and epicholesterol derivatives (3α-OH) do not react in the same manner.

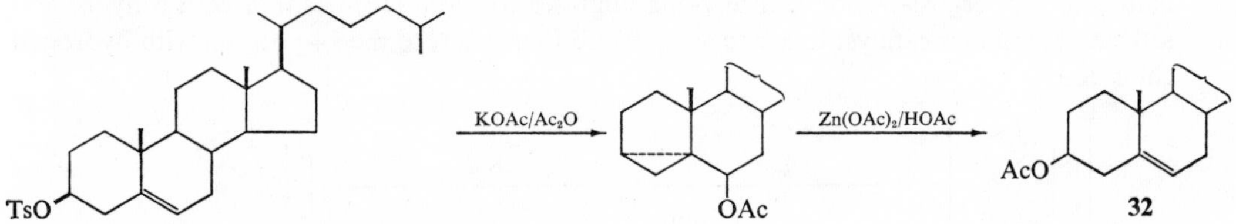

The references cited pertain to specific items in the above text. A more comprehensive treatment with extensive references is provided in refs. 13 and 14.

12) A. Windaus, C. Uibrig, *Chem. Ber.*, **47**, 2384 (1914); *ibid.*, **48**, 857 (1915).
13) L. F. Fieser, M. Fieser, *Steroids*, Reinhold, 1959.
14) *Chemistry of Carbon Compounds* (ed. E. H. Rodd), vol. II, part B, Elsevier, 1953.

6.5   ERGOSTEROL AND VITAMIN D

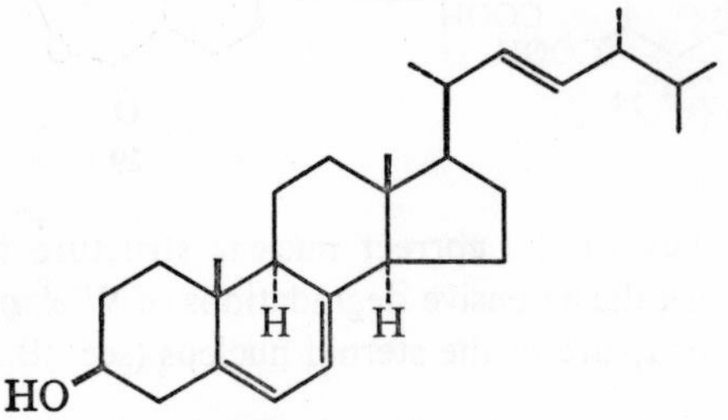

ergosterol   **1**

It had been established around 1920 that rickets responds favorably to treatment either with sunlight or by supplementation of the diet with cod liver oil. At first cholesterol was considered to be the provitamin, which was converted into the vitamin.[1-3] However, it was subsequently shown that ergosterol **1** is the provitamin and that the provitamin nature of cholesterol was due to contamination with small amounts of ergosterol.[4-6]

Reactions of ergosterol

1) Ergosterol **1** is photochemically oxidized to the peroxide **2**,[7] which was later isolated from the mold *Aspergillus fumigatus*.[8]

2) Sensitized irradiation in the absence of oxygen affords bisergosterol **5**,[9,10] dimerization occuring at the less hindered C-7.[11]

3) Upon hydrogenation, the 5-ene is first saturated and this is followed by side-chain hydrogenation to give ergost-7-enol **7**. The 7-ene migrates to the 8 (14)-position with a hydrogen-saturated platinum catalyst in acetic acid,[12] and thereafter, to the 14-position with hydrogen chloride.[13]

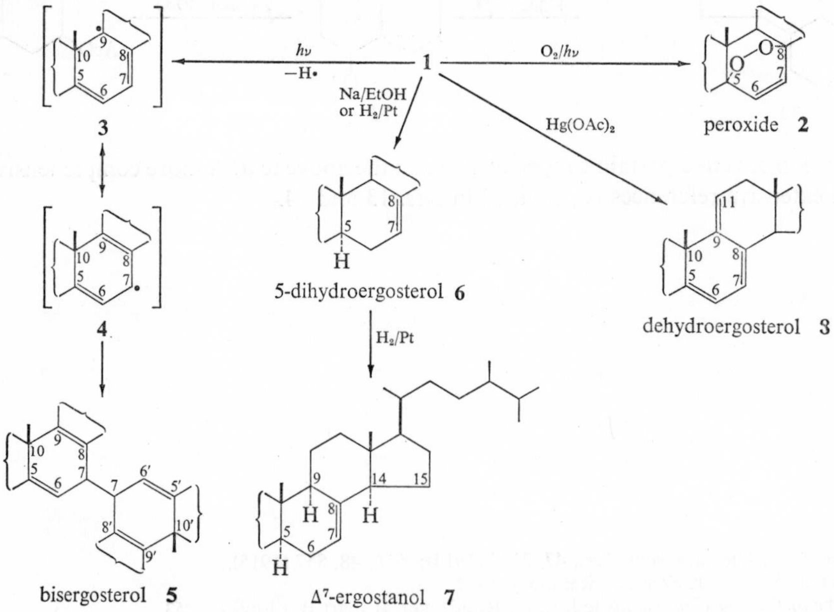

3

dehydroergosterol   **3**

peroxide   **2**

5-dihydroergosterol   **6**

4

bisergosterol   **5**

Δ⁷-ergostanol   **7**

B-Isomers of ergosterol

1) Treatment of ergosterol **1** with hydrogen chloride in chloroform resulted in consecutive rearrangements of double bonds to give a mixture of isomers designated ergosterol-B₁ **10**, -B₂ **8** and -B₃ **9**.[14]

ergosterol **1**   ergosterol-B₂ **8**   ergosterol-B₃ **9**   ergosterol-B₁ **10**

Vitamin D₂

1) Isolation of an antirachitic substance vitamin D_1[14] (or calciferol[15]) from irradiation products of ergosterol was reported in 1931. This was later shown to be a 1:1 mixture of vitamin D_2 **11** and an inactive isomer lumisterol **19**; the active substance was soon isolated in a pure state and named vitamin D_2.[16,17]

2) The ketone **15** was obtained from dimethyl ester **13** of the maleic anhydride adduct. In vitamin D_2, therefore, the original C/D rings are still intact.[18]

3) Formation of the α,β-unsaturated aldehyde **12** upon oxidation showed the presence of the 7-ene.[19,20]

4) From this evidence it was concluded that the ergosterol ring B had cleaved between carbons 9 and 10 to give vitamin D_2.

5) The stereochemistry was deduced by x-ray analysis.[21]

1) For general references, see: L. F. Fieser, M. Fieser, *Steroids*, p. 90, Reinhold, 1959; C. W. Shoppee, *Chemistry of the Steroids* (2nd ed.), p. 62, Butterworth, 1968; E. Heftmann, *Steroid Biochemistry*, p. 31, Academic Press, 1970.
2) H. Steenbock, A. Black, *J. Biol. Chem.*, **68**, 263 (1925).
3) O. Rosenheim, T. A. Webster, *Lancet*, 1025 (1925).
4) R. Pohl, *Nachr. Geo. Wiss. Gottingen, Math-phys. Kl.*, **142**, 185 (1926).
5) O. Rosenheim, T. A. Webster, *Biochem. J.*, **21**, 389 (1927).
6) A. Windaus, A. Hess, O. Rosenheim, R. Pohl. T. A. Webster, *Chem.-ztg.*, **51**, 113 (1927).
7) A. Windaus, J. Brunken, *Ann. Chem.*, **460**, 225 (1928).
8) P. Wieland, V. Prelog, *Helv. Chim. Acta*, **30**, 1028 (1947); J. V. Fiore, *Arch. Biochem.*, **16**, 161 (1948).
9) A. Windaus, P. Bargeaud, *Ann. Chem.*, **460**, 235 (1928).
10) H. H. Inhoffen, *Naturwiss.*, **25**, 125 (1937).
11) J. L. Owades, *Experientia*, **6**, 258 (1950).
12) H. Wieland, W. Benend, *Ann. Chem.*, **556**, 1 (1943).
13) F. Reindel, E. Walter, H. Rauch, *Ann. Chem.*, **452**, 34 (1927); F. Reindel, E. Walter, *ibid.*, **460**, 212 (1928).; I. M. Heilbron, D. G. Wilkinson, *J. Chem. Soc.*, 1708 (1932).
14) A. Windaus, A. Lüttringhaus, M. Deppe, *Ann. Chem.*, **489**, 252 (1931).
15) T. C. Angus, F. A. Askew, R. B. Bourdillon, H. M. Bruce, R. K. Callow, C. Fischmann, J. St. L. Philpot, T. A. Webster, *Proc. Roy. Soc.*, **B108**, 340 (1931).
16) A. Windaus, O. Linsert, A. Lüttringhaus, G. Weidlich, *Ann. Chem.*, **492**, 226 (1932).
17) F. A. Askew, R. B. Bourdillon, H. M. Bruce, R. K. Callow, J. St. L. Philpot, T. A. Webster, *Proc. Roy. Soc.* **B109**, 488 (1932).
18) A. Windaus, W. Thiele, *Ann. Chem.*, **521**, 160 (1936).
19) I. M. Heilbron, F. S. Spring, *Chem. Ind.*, **54**, 795 (1935); I. M. Heilbron, K. M. Samant, F. S. Spring, *Nature* **135**, 1072 (1935).
20) I. M. Heilbron, R. N. Jones, K. M. Samant, F. S. Spring, *J. Chem. Soc.*, 905 (1936).
21) D. Crowfoot, J. D. Dunitz, *Nature*, **162**, 608 (1948); D. Crowfoot Hodgkin, M. S. Webster, J. D. Dunitz, *Chem. Ind.*, 1149 (1957).

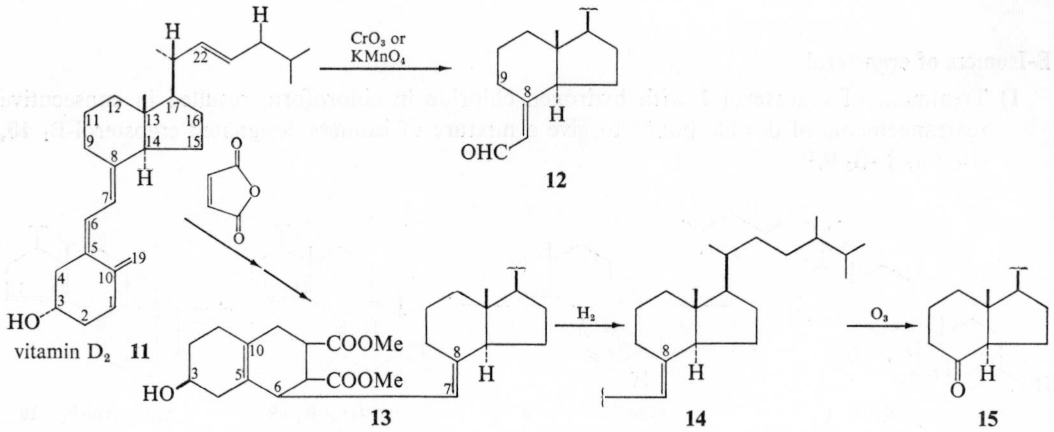

Vitamin D$_3$

Vitamin D$_3$, an antirachitic substance having a saturated side chain, was chemically prepared from cholesterol.[22] The compound was later isolated from tuna liver oil.[23-26]

Tachysterol

1) Tachysterol **16** is one of the by-products isolated from the ergosterol irradiation mixture.[27]
2) It was correlated with vitamin D$_2$ **11** by reduction, which yielded the two dihydro products **17** and **18**.[28]
3) Dihydrotachysterol **17** is devoid of antirachitic activity but shows antitetany activity; dihydrovitamin D$_2$ **18** is inactive.

tachysterol **16** dihydrotachysterol **17** dihydrovitamin D$_2$ **18**

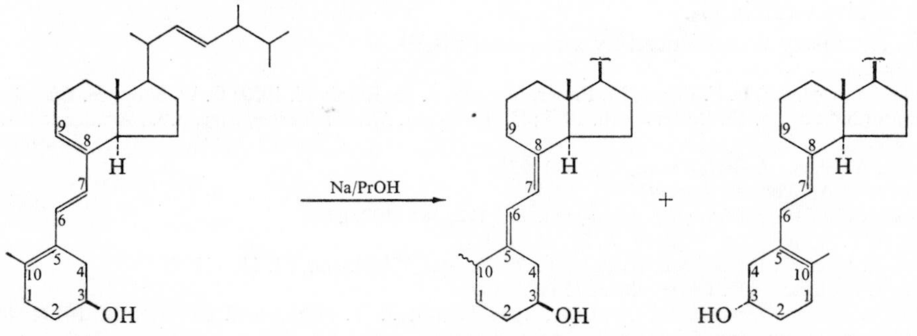

Lumisterol

1) The other irradiation by-products, lumisterol **19**[29] and pyrocalciferol **21**,[30] are dehydrogenated to dehydrolumisterol **20**, while ergosterol **1** and isopyrocalciferol **23** give dehydroergosterol **22**, indicating 10β-methyl in lumisterol **19**.[31]

22) A. Windaus, H. Settre, Fr. Schenk, *Ann. Chem.*, **520**, 98 (1935).
23) H. Brockmann, Y. H. Chen, *Z. Physiol. Chem.*, **241**, 129 (1936).
24) A. Windaus, Fr. Schenk, F. von Werder, *Z. Physiol. Chem.*, **241**, 100 (1936).
25) H. Brockmann, A. Busse, *Naturwiss.*, **26**, 122 (1938).
26) E. J. H. Simons, B. Demarest, *Naturwiss.*, **26**, 11 (1938).
27) A. Windaus, F. von Werder, A. Lüttringhaus, *Ann. Chem.*, **499**, 188 (1932).
28) F. von Werder, *Z. Physiol. Chem.*, **260**, 119 (1939).
29) A. Windaus, K. Dithmar, E. Fernholz, *Ann. Chem.*, **493**, 259 (1932).
30) F. A. Askew, *Proc. Roy. Soc.*, *B***109**, 488 (1932).

2) The tricyclic ketone **27**, derived from both lumisterol **19** and isopyrocalciferol **23** indicates 9β-hydrogen in lumisterol **19**.[32]

lumisterol **19**   dehydrolumisterol **20**   pyrocalciferol **21**

ergosterol **1**   dehydroergosterol **22**   isopyrocalciferol **23**

lumisterol **19** → i) Oppenauer oxidation, ii) acid isomerization → **24** → H$_2$/Pd-C in KOH/EtOH → **25**

i) bromination of side-chain, ii) ozonization, iii) debromination → **26** → pyrolysis in sodium phenylacetate → **27** ← isopyrocalciferol **23**

Intermediates of ergosterol irradiation

1) Previtamin D$_2$ **29** was isolated from the photoequilibrium mixture, and was shown to be convertible into vitamin D$_2$ **11** by heat.

2) The quantitative conversion of **29** into tachysterol **16** by iodine, the standard *cis*→*trans* isomerization catalyst, indicated that previtamin D$_2$ **29** is the *cis* isomer of tachysterol.[33-35]

3) The isolation of tachysterol **16**, lumisterol **19**, and ergosterol **1** from the irradiation mixture of previtamin D$_2$ **29** established the reversibility of the photochemical process.[36,37]

4) Investigation with radioactive ergosterol finally led to the conclusion that lumisterol **19** and tachysterol **16** are formed in side reactions, and are not intermediates leading to vitamin D$_2$.[38,39]

31) A. Windaus, K. Dimroth, *Chem. Ber.*, **70**, 376 (1937).
32) D. A. Shepard, R. A. Donia, J. A. Campbell, B. A. Johnson, R. P. Holyoz, G. Slomp, Jr., J. E. Stafford, R. L. Pederson, A. C. Ott. *J. Am. Chem. Soc.*, **77**, 1212 (1955).
33) L. Velluz, G. Amiard, A. Petit, *Bull. Soc. Chim. Fr.*, **16**, 501 (1949).
34) H. H. Inhoffen, *Naturwiss.*, **93**, 396 (1956).
35) A. L. Koevoet, A. Verloop, E. Havinga, *Rec. Trav. Chim.*, **74**, 788 (1955).
36) L. Velluz, G. Amiard, *Bull. Soc. Chim. Fr.*, **22**, 205 (1955).
37) L. Velluz, G. Amiard, B. Goffinet, *Bull. Soc. Chim. Fr.*, **22**, 134 (1955).
38) E. Havinga, A. L. Koevoet, A. Verloop, *Rec. Trav. Chim.*, **74**, 1230 (1955).
39) A. Verloop, A. L. Koevoet, E. Havinga, *Rec. Trav. Chim.*, **76**, 689 (1957).

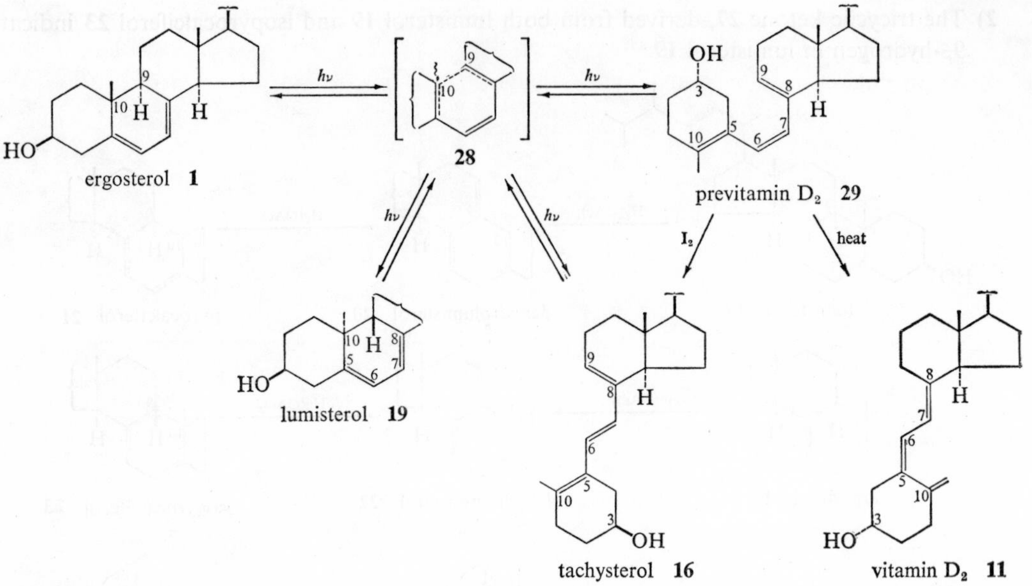
ergosterol 1
28
previtamin D₂ 29
lumisterol 19
tachysterol 16
vitamin D₂ 11
hν
I₂
heat
HO
OH
H

Active analogs

Active analogs with different side chains (vitamin-D_4, -D_5, -D_6, -D_7 and 25-hydroxycalciferol) have been prepared.[1)]

6.6 ESTROGENS

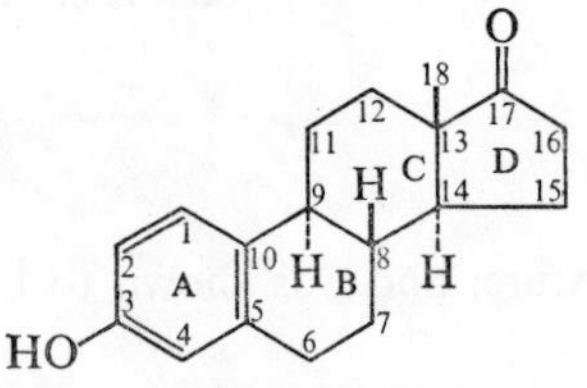

estrone 1[1)]
mp: 251–254°
α_D: +152°
uv: 283–285

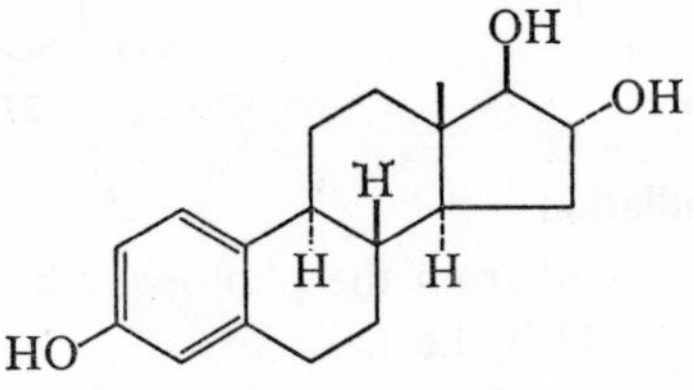

estriol 2

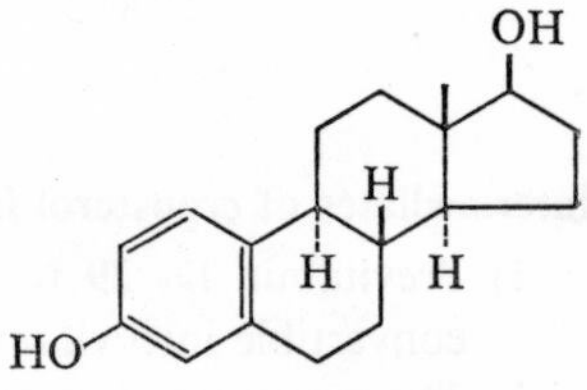

estradiol 3

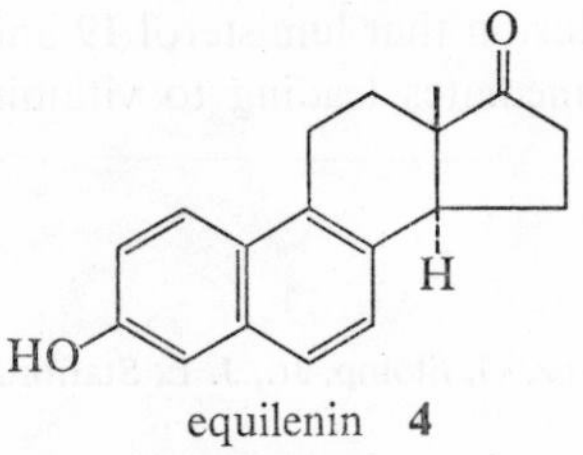

equilenin 4

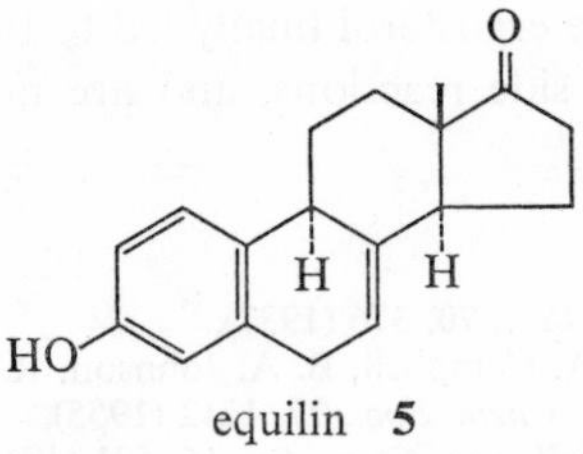

equilin 5

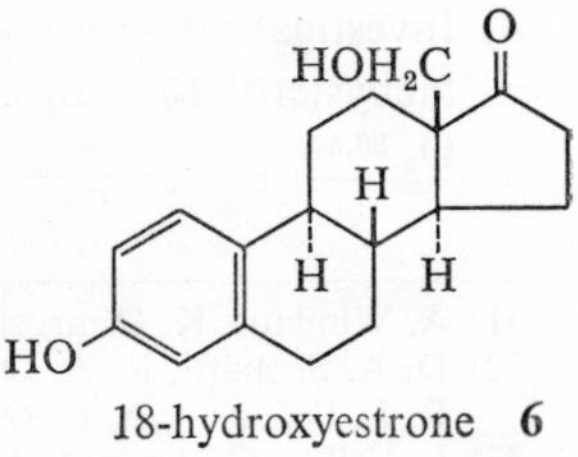

18-hydroxyestrone 6

Structure of estrone 1

1) Elemental analysis gave $C_{18}H_{22}O_2$.[1-3)]
2) The formation of an oxime showed the presence of a ketone; a phenol accounted for the other oxygen.[1-4)]

3) Quantitative hydrogenation and determination of molecular refraction showed the presence of only 3 benzenoid double bonds. Hence estrone must have 4 rings.[5,6]

4) X-ray and surface film measurements suggested a steroid ring structure with the functional groups at opposite ends of the molecule.[7,8]

5) Pregnanediol 7 was isolated with estrone, and its structure was determined by degradations and conversion to 8 which was a known compound. Their co-isolation suggested that 1 may have a similar substituent pattern.[8]

7

pregnane 8
(α_D:+20)

6) Compound 2, which can be converted to 1 by reaction with $KHSO_4$, was degraded to 10 which was then synthesized. This led to the postulated structure 1.[8,9]

7) The structure was finally verified when subsequent reactions were carried out on estriol 2.

Structure of estriol 2

1) Estriol was degraded to 11, the structure of which was proven by synthesis.[9]

2) The formation of an anhydride by pyrolysis of 9-acetate showed ring D to be 5-membered (Blanc's rule).[9]

2 —KOH fusion→

marrianolic acid 9   10   11

3) The position of the phenolic OH was proven by formation of 14, the structure of which was verified by synthesis.[11,12]

1 —MeOH/H⁺→ 12 —Wolff-Kishner→ 13 —Se→ 14

1) A. Butenandt, *Z. Physiol. Chem.*, **191**, 140 (1930).
2) A. Butenandt, F. Hildebrandt, *Z. Physiol. Chem.*, **199**, 243 (1931).
3) S. A. Thayer, L. Levin, E. A. Doisy, *J. Biol. Chem.*, **91**, 791 (1929).
4) G. F. Marrian, G. A. D. Haslewood, *Chem. Ind.*, **10**, 277 (1932).
5) A. Butenandt, *Nature*, **130**, 238 (1932).
6) A. Butenandt, U. Westphal, *Z. Physiol. Chem.*, **223**, 147 (1934).
7) J. D. Bernal, *Chem. Ind.*, **51**, 259 (1932).
8) N. K. Adam, J. F. Danielli, G. A. D. Haslewood, G. F. Marrian, *Biochem. J.*, **26**, 1233 (1932).
9) A. Butenandt, H. A. Weidlich, H. Thompson, *Chem. Ber.*, **66**, 601 (1933).
10) D. W. MacCorquodale, S. A. Thayer, E. A. Doisy, *J. Biol. Chem.*, **99**, 327 (1933).
11) A. Cohen, J. W. Cook, C. L. Hawett, A. Girard, *J. Chem. Soc.*, 653 (1934).
12) J. W. Cook, A. Girard, *Nature*, **133**, 377 (1934).

4) The carbonyl position was shown by formation of **17**, whose structure was proven by synthesis.[13]

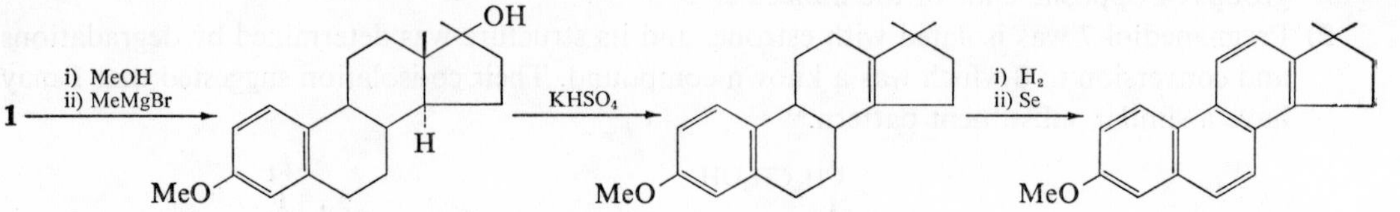

5) The conversion **15→16** showed that rings C and D are *trans*; in model compounds, only when the rings are *trans* does the methyl group shift during dehydration, probably to relieve steric strain.

6) The remaining stereochemistry followed from the preparation of **3** from **18**, the stereochemistry of which was known through its derivation via cholesterol.[14]

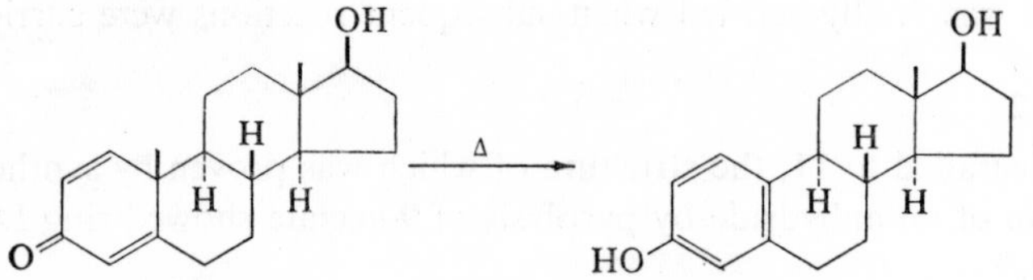

7) The presence of a *trans* α-glycol was shown by cleavage with $Pb(OAc)_4$[15] to an aldehyde acid, by dehydration to estrone **1**,[9] and by its failure to form an acetonide.[16] The 16α,17β-configuration was deduced by partial synthesis.[17]

Other estrogens

1) Estradiol, **3**, the most potent of the estrogens, contains a 17β-OH as shown by formation of **3** from **1** by LAH (α attack directed by steric hindrance), and from rotation data of acetoxyl derivatives.[18]

2) Equilenin **4** was determined by its conversion into **16**[13] (as in **1→15→16**), and by the uv spectrum of its picrate which showed the presence of a naphthalene nucleus.

3) Equilin **5** was determined as follows. Firstly, its uv spectrum was identical to that of estrone; hence there is no α,β-unsaturated ketone.[19] Secondly, equilin was converted to **4** by Pd.[20] Thirdly, the glycol **19** only gave a monoacetate. Hence the double bond must be secondary-tertiary, and therefore at C-7.[21] Fourthly, equilin was transformed into **1**, as shown below.[22]

13) A. Cohen, J. W. Cook, C. L. Hewett, *J. Chem. Soc.*, 445 (1935).
14) H. H. Inhoffen, *Angew. Chem.*, **53**, 471 (1940); H. H. Inhoffen, G. Zuhlsdorff, *Chem. Ber.*, **74**, 1911 (1941).
15) M. N. Huffman, M. H. Lott, *J. Am. Chem. Soc.*, **71**, 719 (1949).
16) A. Butenandt, E. L. Schäffler, *Z. Naturforsch.*, **1**, 82 (1946).
17) M. N. Huffman, *J. Biol. Chem.*, **167**, 273 (1947).
18) T. F. Gallagher, T. H. Kritchevsky, *J. Am. Chem. Soc.*, **72**, 882 (1950).
19) J. W. Cook, E. Roe, *Chem. Ind.*, **54**, 501 (1935).
20) W. Dirocherl, F. Hanusch, *Z. Physiol. Chem.*, **233**, 13 (1935).
21) A. Serini, W. Logemann, *Chem. Ber.*, **71**, 186 (1938).
22) W. H. Pearlman, O. Wintersteiner, *J. Biol. Chem.*, **130**, 35 (1939).

4-OMe $\xrightarrow{\text{OsO}_4}$ **19** $\xrightarrow[\substack{\text{i) distillation} \\ \text{ii) } \text{>}\!-\!\text{OAc} \\ \text{iii) Pd}}}$ **20**

$\xrightarrow[\substack{\text{i) PCl}_5 \\ \text{ii) NaI/py}}]{}$ **21** $\xrightarrow{\text{Pd/H}_2}$ **1**

4) The structure of **6**, determined by formation of **22**, was verified by synthesis.[23]

6 $\xrightarrow{\text{OH}^-}$ **22**

5) Other minor estrogens include 16-epi-estriol, 16α-hydroxyestrone, 16β-hydroxyestrone, 16-ketoestradiol, 16-ketoestrone and 17α-estradiol.

Remarks

The estrogens can be isolated from urine, placentas and reproductive organs of mammals. For example, in an original extraction, about 700 units of activity (which is now known to correspond to 0.07 mg of estrone) were obtained from 1 liter of urine received from pregnant women.[3] The estrogens are responsible for the mature growth, development and maintenance of the female reproductive tract and the secondary sex organs.

The extraction process of keto steroids was greatly facilitated by the use of Girard's reagent,[24] trimethylaminoacetohydrazide hydrochloride [Me$_3$N$^+$CH$_2$CONHNH$_2$]Cl$^-$, which reacts with carbonyl groups, e.g., 17-keto in estrone, to form water-soluble semicarbazones; after workup the semicarbazones can be readily hydrolyzed back to the ketones.

In recent years, there have been numerous attempts to convert the estrogens into new compounds to increase their hormonal activity. Following are some examples of work done on the estrogens.

23) K. H. Loke, E. J. D. Watson, G. F. Marrian, *Biochem. J.*, **71**, 43 (1959).
24) A. Girard, G. Sandulesco, *Helv. Chim. Acta*, **19**, 1095 (1936).
25) K. Miescher, *Helv. Chim. Acta*, **27**, 1727 (1944); J. Iriarte, P. Crabbé, *Chem. Commun.*, 1110 (1972).
26) W. W. Westerfield, *J. Biol. Chem.*, **143**, 177 (1943).
27) R. P. Jacobsen, G. M. Picha, H. Levy, *J. Biol. Chem.*, **171**, 61, 71, 81, (1947).
28) A. Butenandt, A. Wolff, P. Karlson, *Chem. Ber.*, **74**, 1308 (1941).
29) A. Butenandt, L. Poschmann, *Chem. Ber.*, **77**, 392 (1944).
30) A. Blade Font, *Bull. Soc. Chim. Fr.*, 906 (1964).

Estrogenolic acids[4,10,25]

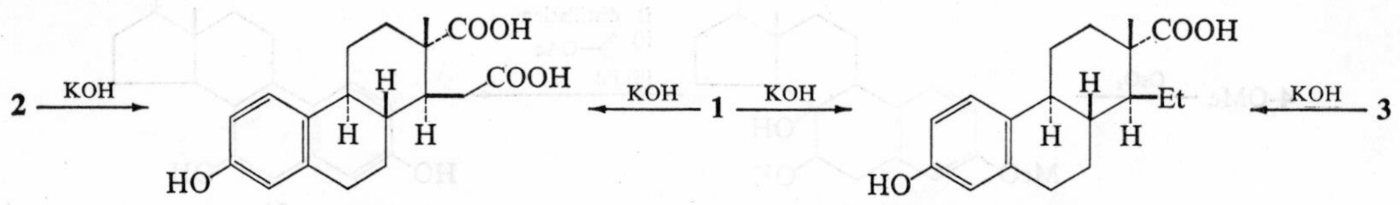

trans-marrianolic acid **23**   trans-doisynolic acid **24**

The stereochemistry of **24** has been established by conversion from estrone.[25]

Estrogenic lactones[26,27]

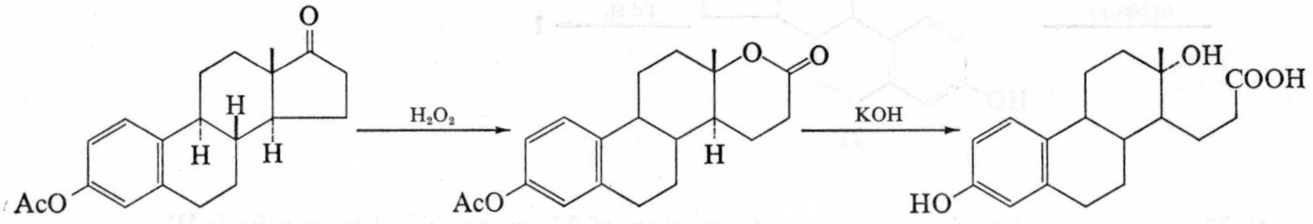

25   **26**   estrolic acid **27**

Epimerization[28,29]

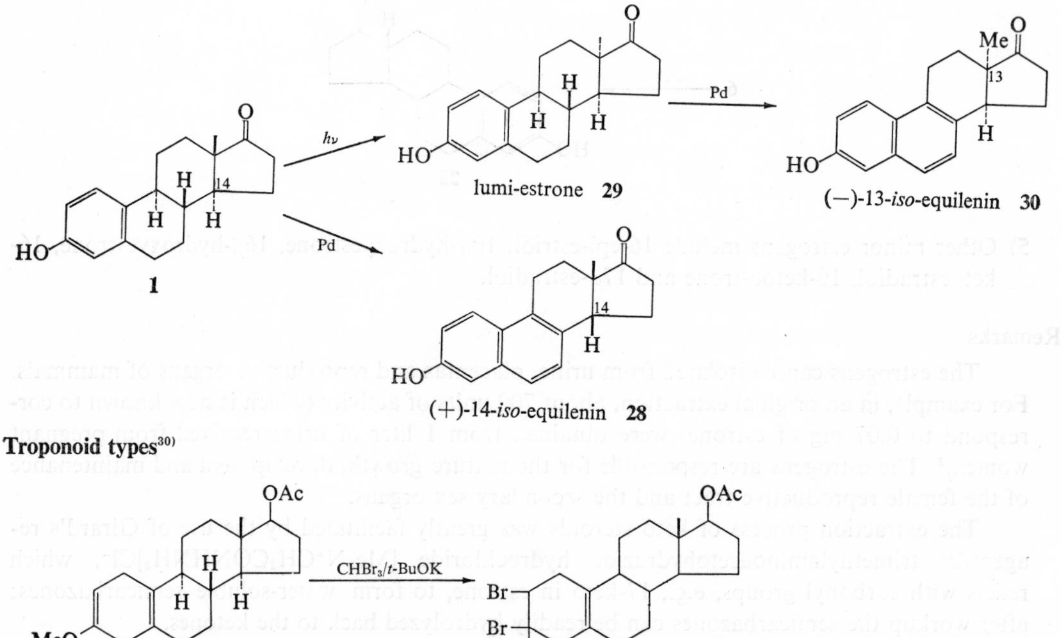

lumi-estrone **29**

(−)-13-iso-equilenin **30**

1

(+)-14-iso-equilenin **28**

Troponoid types[30]

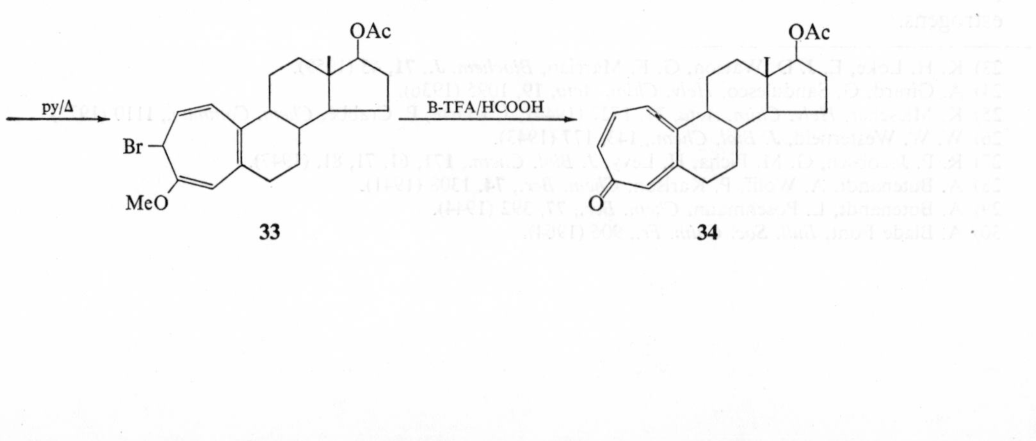

31   **32**

33   **34**

Addition[31]

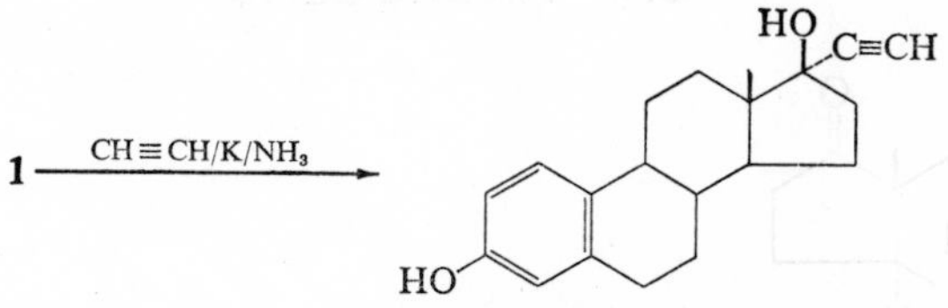

$$1 \xrightarrow{\text{CH}\equiv\text{CH/K/NH}_3}$$

17α-ethynylestradiol **35**

The above compound **35** and its 3-OMe derivative are the only synthetic estrogens now in use. They are at least as strong as estradiol, the most potent natural hormone and have the advantage that they are not destroyed when taken orally, as estradiol is.

Physiological derivatives[32]

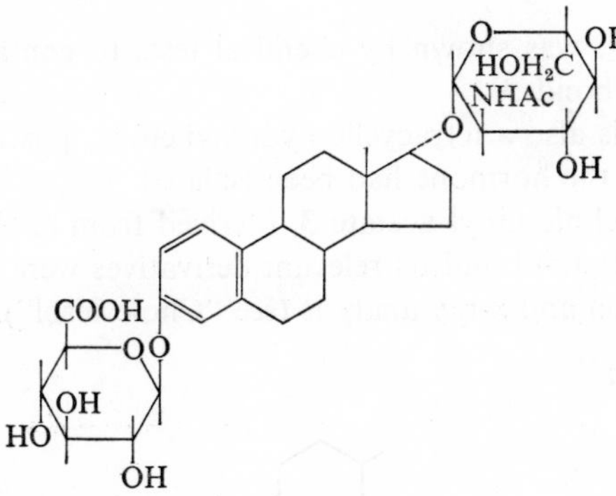

36 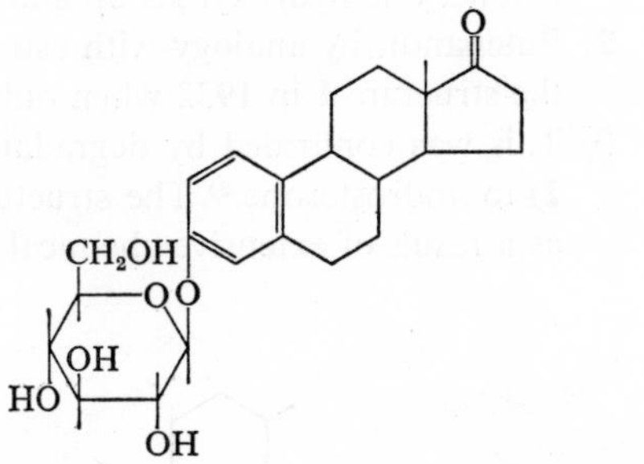   **37**

While **36** is the excretion product of estrone, there is strong evidence both from *in vivo* and *in vitro* experiments that **37** may be an active intermediate in estrogen metabolism. It was found that **37** can penetrate cell membranes while **36** cannot, and it may be possible that steroid glucosides are the important transport forms of steroids.

Non-steroidal estrogens[33]

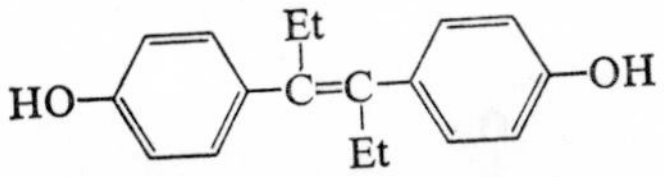

38 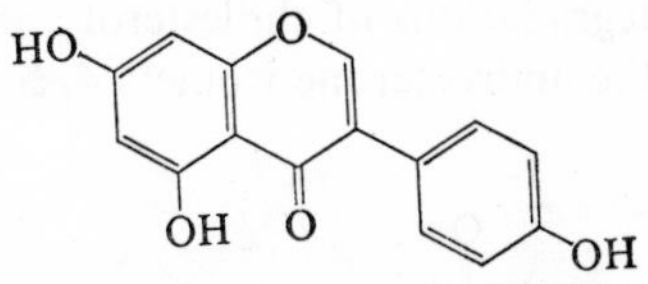   **39**

The above are interesting in that they show estrogen activity even though totally unrelated to the steroid structure. Diethylstilbestrol, **38**, a highly active synthetic estrogen, is an example of those found in the stilbene family, while genistein **39**, which shows weak estrogen activity, is an isoflavone.

For a list of synthetic estrogens, see ref. 34.

31) L. Velluz, G. Amiard, J. Martel, J. Warnant, *Bull. Soc. Chim. Fr.*, 1484 (1957).
32) D. S. Layne, G. A. Quamme, R. S. Labow, J. D. Mellor, A. Polakova, D. G. Williamson, *Hormonal Steroids* (ed. V. H. T. James, L. Martini), 317–322, Excerpta Medica, 1971.
33) J. Grundy, *Chem. Rev.*, 57, 281 (1957).
34) *Handbook of Chemistry and Physics* (52nd ed.), C–734, The Chemical Rubber Company, 1971–1972.

6.7 ANDROGENS

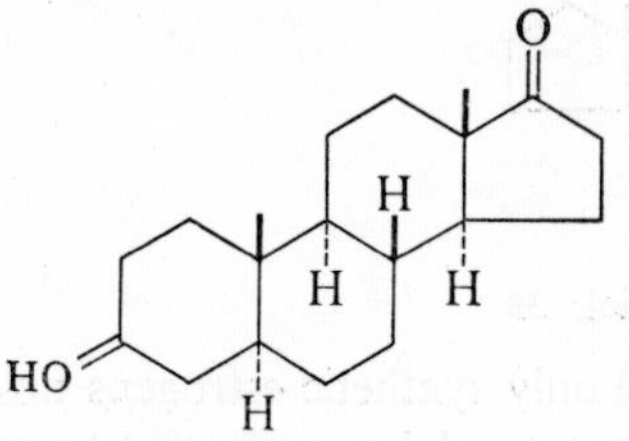

androsterone **1**
mp: 183° αD (EtOH): +95°

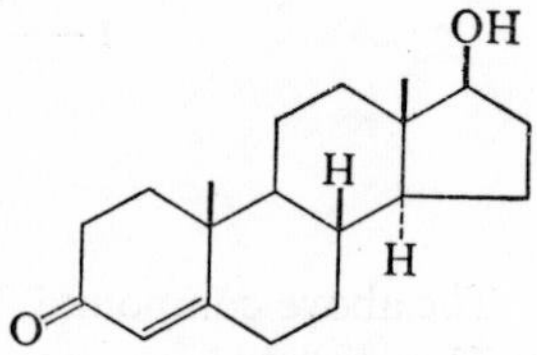

testosterone **13**
mp: 155° αD: +109°

Androsterone

1) Androsterone **1** from male urine, $C_{19}H_{30}O_2$, was shown by chemical tests to contain one ketone, one hydroxyl group and no double bonds.[1]

2) Butenandt, by analogy with estrone which is also a tetracyclic hydroxyketone, postulated[1] the structure **1** in 1932 when only 25 mg of the hormone had been isolated.

3) This was confirmed by degradation of epi-cholestanyl acetate **3**, (derived from cholesterol **2**) to androsterone.[2] The structures of cholesterol and its relevant derivatives were known as a result of extensive chemical investigation and x-ray analysis (see "Cholesterol").

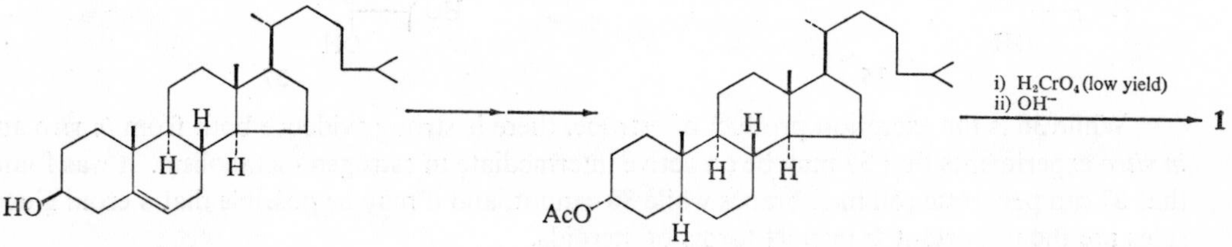

cholesterol **2**

i) H_2CrO_4 (low yield)
ii) OH⁻ → **1**

3

4) Similar degradations of cholesterol and coprostanol derivatives[2] likewise proved the structures of the androsterone isomers **4**, **5** and **6**.

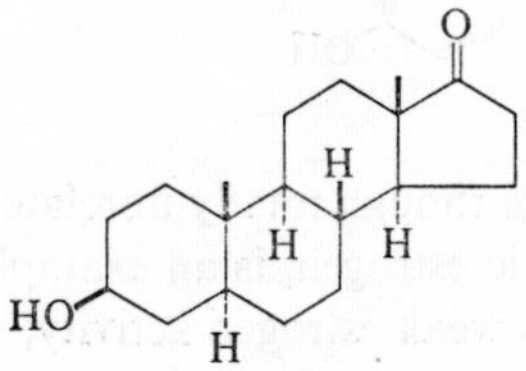

epiandrosterone **4**

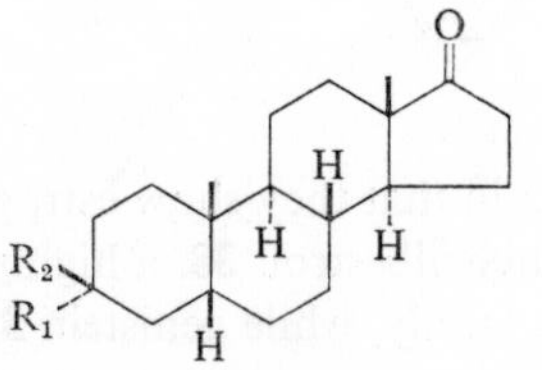

3α-hydroxyetiocholan-17-one **5** (R_1=OH, R_2=H)
3β-hydroxyetiocholan-17-one **6** (R_1=H, R_2=OH)

1) A. Butenandt, *Angew. Chem.*, **44**, 905 (1931); *ibid.*, **45**, 655 (1932); *Naturwiss*, **21**, 49 (1933); A. Butenandt, K. Tscherning, *Z. Physiol. Chem.*, **229**, 167 (1934).

2) L. Ruzicka, M. W. Goldberg, H. Brüngger, *Helv. Chim. Acta*, **17**, 1389 (1934); L. Ruzicka, M. W. Goldberg, J. Meyer, H. Brüngger, E. Eichenberger, *ibid.*, **17**, 1395 (1934); L. Ruzicka, H. Brüngger, E. Eichenberger. J. Meyer, *ibid.*, **18**, 430 (1935).

Dehydro-epiandrosterone

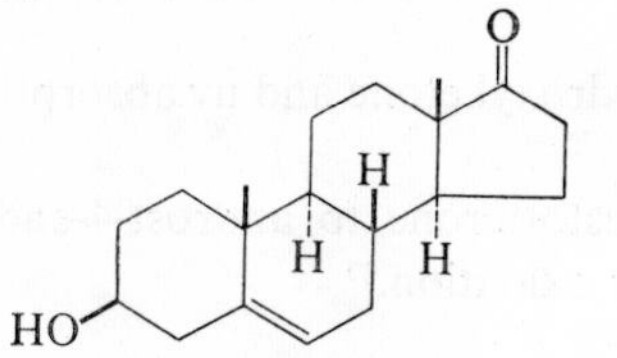

dehydro-epiandrosterone **7**
(mp: 153° α_D: $+11°$)

1) Dehydro-epiandrosterone **7** (also known as androstenolone), $C_{19}H_{28}O_2$, was isolated together with the biologically inactive chloroketone **8**. The latter was shown to have been formed from **7** during the hydrochloric acid hydrolysis of male urine in the isolation procedure and could be prepared directly by PCl_5 treatment of **7**.[1,3]

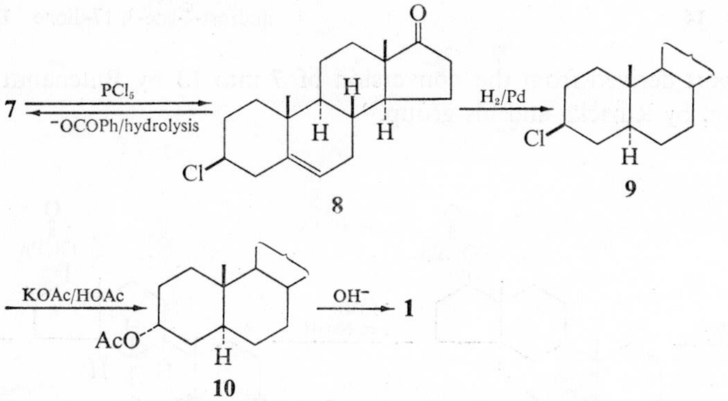

2) The conversion **7→8→9→10→1** proved the structure of dehydroepiandrosterone, although it was only later, when **7** was prepared by degradation of cholesteryl acetate dibromide **11**,[4] that the β configuration of the 3-OH was recognized and the **9→10** conversion was understood to have involved inversion at C-3.

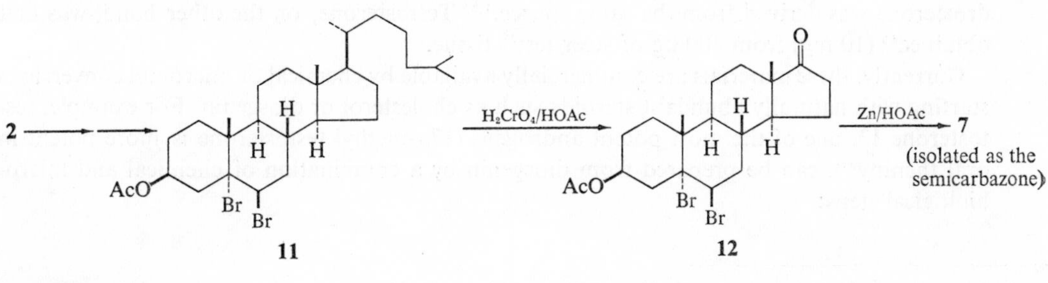

Testosterone

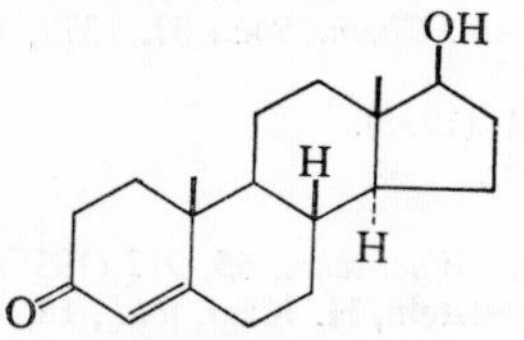

testosterone **13**

1) Testosterone **13**, $C_{19}H_{28}O_2$, loses its biological activity when treated with alkali, a property shared with the known progesterone **14**.[5]

2) The formation of derivatives (acetate and oxime) implied a hydroxyketone and uv absorption at 240 nm suggested an α,β-enone.[5]

3) The structure was conclusively proven by conversion of testosterone to androst-4-ene-3, 17-dione **15**, also obtained from dehydro-epiandrosterone by oxidation.[6]

progesterone **14**   androst-4-ene-3, 17-dione **15**

4) Further support derived from the conversion of **7** into **13** by Butenandt and coworkers[7] and, as shown, by Ruzicka and his group.[8]

Isolation and partial synthesis

Androsterone was first isolated (15 mg) from 15,000 liters of male urine[1]; dehydroepiandrosterone was derived from the same source.[1,3] Testosterone, on the other hand, was first obtained[6] (10 mg) from 100 kg of steer testis tissue.

Currently, these materials are commercially available by chemical or microbial conversions starting with naturally abundant steroids such as cholesterol or diosgenin. For example, testosterone **13**, one of the most potent androgens (17α-methyl testosterone is more potent in oral therapy[9]), can be prepared from diosgenin by a combination of chemical and microbiological steps:

3) A. Butenandt, H. Dannenbaum, *Z. Physiol Chem.*, **229**, 192 (1924).

4) A. Butenandt, H. Dannenbaum, G. Hanisch, H. Kudszus, *Z. Physiol. Chem.*, **237**, 57 (1935); L. Ruzicka, A. Wettstein, *Helv. Chim. Acta*, **18**, 986 (1935); E. S. Wallis, E. Fernholz, *J. Am. Chem. Soc.*, **57**, 1379, 1504 (1935); W. Schoeller, A. Serini, M. Gehrke, *Naturwiss.*, **23**, 337 (1935).

5) K. David, E. Dingemanse, J. Freud, E. Laqueur, *Z. Physiol. Chem.*, **233**, 281, (1935).

6) K. David, *Act. Brevia Neerl. Physiol. Pharmacol. Microbiol.*, **5**, 85, 108 (1935).

7) A. Butenandt, G. Hanisch, *Z. Physiol. Chem.*, **237**, 89 (1935).

8) L. Ruzicka, *J. Am. Chem. Soc.*, **57**, 2011 (1935); A. Wettstein, *Schweiz. Med. Wochschr.*, **65**, 912 (1935); L. Ruzicka, A. Wettstein, *Helv. Chim. Acta*, **18**, 1264 (1935); L. Ruzicka, A. Wettstein, H. Kägi, *ibid.*, **18**, 1478 (1935).

9) L. F. Fieser, M. Fieser, *Steroids*, p. 519, Reinhold, 1959.

diosgenin **18**   **19**   **20**

16-dehydropregnenolone acetate **21**   **22**   **23**

7   **13**

Yields: **18**→**21**: 60%[10])
21→ **7**: 74%[11])
7→**13**: 81%[12])

Biological activity

Androgens (Greek, *andro*, male) are hormones which control the development of the male genital organs and secondary sex characteristics. One of the widely used bioassay techniques applied in research on these substances has been the capon comb-growth test.[13,1] The capon unit (c.u.) is defined as the amount of substance administered to each of 3 capons on two successive days which produces an average 20% increase in comb growth as measured by planimeter on a shadowgraph. Measured thus, the three most important androgens—testosterone, androsterone and dehydro-epiandrosterone—are active to the same degree in the proportion 15γ: 100γ:700γ, respectively.[9]

It was early observed[14] that androgenic hormones also possess protein anabolic (or myotrophic) activity, promoting the storage of protein and the general stimulation of tissues. Since the ratio of anabolic to androgenic activity is not constant from androgen to androgen, research has been devoted to developing synthetic hormones in which this ratio is maximized.

10) R. E. Marker, E. Rohrman, *J. Am. Chem. Soc.*, **61**, 3592 (1939); *ibid.*, **62**, 518 (1940); R. E. Marker, R. B. Wagner, P. R. Olshafer, E. L. Wittbecker, D. P. J. Goldsmith, C. H. Ruof, *ibid.*, **69**, 2167 (1947); A. F. B. Cameron, R. M. Evans, J. C. Hamlet, J. S. Hunt, P. G. Jones, A. G. Long, *J. Chem. Soc.*, 2807 (1955); L. Fieser, M. Fieser, p. 550, *Steroids*, Reinhold, 1959.
11) G. Rosenkranz, O. Mancera, F. Sondheimer, C. Djerassi, *J. Org. Chem.*, **21**, 520 (1956).
12) L. Mamoli, *Chem. Ber.*, **71**, 2278 (1938).
13) A. Pezard, *Compt. Rend.*, **153**, 1027 (1911); T. F. Gallagher, F. C. Koch, *J. Biol. Chem.*, **84**, 495 (1929).
14) C. D. Kochakian, *The Protein Anabolic Effects of Steroid Hormones, Vitamins and Hormones*, **4**, 255 (1946).

These would be suitable for therapy of metabolic and endocrine conditions, particularly in women, where androgenic side effects are undesirable. Two such hormones are **24** (commercially *Halotestin*)[15] and **25**.[16]

24

25   (R=H or Me)

Chemical reactivity

The facile formation of chloroketone **8** from a natural precursor under acidic conditions and its preparation from PCl_5 with retention of configuration at C-3 are examples of a phenomenon generally observed with cholesterol and other 3β-hydroxy-Δ^5-steroids.[17] Furthermore, cholesteryl tosylate solvolyzes in methanol to form *i*-cholesteryl methyl ether **27**.[18]

cholesteryl 3β-tosylate

MeOH/⁻OAc (as a buffer)

26a   **26b**   **27**

The kinetics of these reactions were carefully studied,[19] leading to the tentative conclusion that the homoallylic carbonium ion **26** was involved as the key intermediate responsible for the reaction rate acceleration observed. This finding was one of the earliest demonstrations of possible resonance stabilization by delocalization of σ bonds and was therefore an important contribution to the development of the nonclassical carbonium ion concept.[20]

The *i*-steroids as a class are now referred to as 3,5-cyclosteroids. Recently, the reactivity of this type of compound has been exploited in the synthesis of ecdysone.[21] A 3,5-cyclosteroid intermediate was used to introduce the required functionality at C-3 and C-6 (see "Ecdysones").

15) M. E. Herr, J. A. Hogg, R. H. Levin, *J. Am. Chem. Soc.*, **78**, 500 (1956).

16) H. J. Ringold, E. Batres, O. Halpern, E. Necoechea, *J. Am. Chem. Soc.*, **81**, 427 (1959).

17) C. W. Shoppee, *J. Chem. Soc.*, 1147 (1946).

18) E. S. Wallis, E. Fernholz, F. T. Gephart, *J. Am. Chem. Soc.*, **59**, 137 (1937).

19) S. Winstein, R. Adams, *J. Am. Chem. Soc.*, **70**, 838 (1948); S. Winstein, A. H. Schlesinger, *ibid.*, **70**, 3528 (1948).

20) P. D. Bartlett, *Nonclassical Ions*, p. 16, 22, Benjamin, 1965.

21) D. H. R. Barton, P. G. Feakins, J. P. Poyser, P. G. Sammes, *J. Chem. Soc. C,* 1585 (1970); W. van Bever, F. Kohen, V. V. Ramada, R. E. Counsell, *Chem. Commun.*, 758 (1970).

6.8 GESTOGENS (PROGESTOGENS)

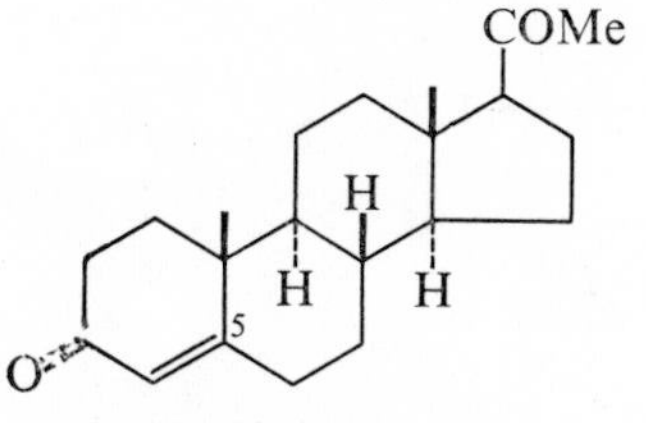

progesterone **1**[1-3]
mp: 121° α_D: 172–182°   uv: 240

1) The uv of **1** showed the presence of an α,β-unsaturated ketone.[1]
2) X-ray measurements demonstrated the presence of a steroid nucleus.[2]
3) Compound **2** was a major contaminant of **1**. Their co-occurrence and close'y related empirical formulae ($C_{21}H_{30}O_2$ and $C_{21}H_{34}O_2$, respectively) suggested a close structural relationship.[1]
4) Compound **2** was converted to **3** (A/B *trans*), which was similar in many ways to pregnanedione **4** (A/B *cis*), a compound whose structure was known because of its correlation with pregnane (A/B *cis*) derived from bisnorcholanic acid. The differences were ascribed to an opposite configuration at C-5.[1]

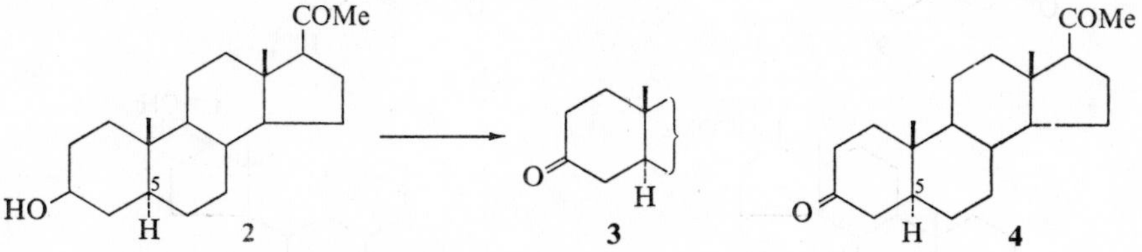

5) Structure **1** was correctly proposed from the above observations.[2]
6) The structure was verified by the following conversions.[3,4]

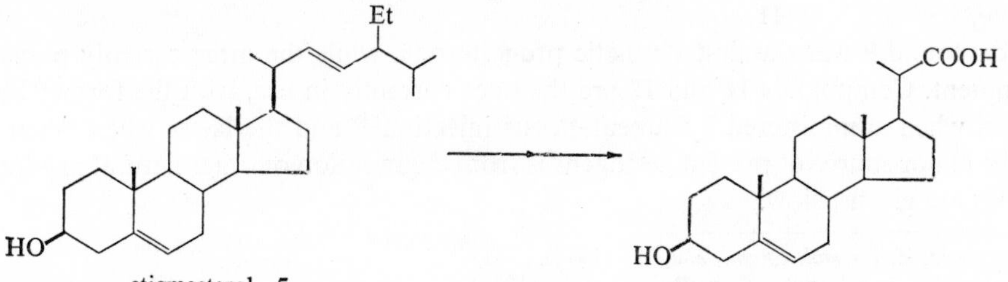

stigmasterol **5**

6

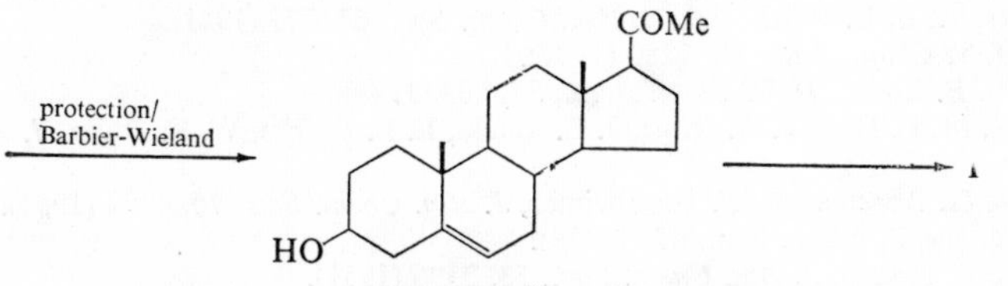

protection/
Barbier-Wieland

pregnenolone **7**

1) K. H. Slotta, H. Ruschig, E. Fels, *Chem. Ber.*, **67**, 1624 (1934).
2) A. Neuhaus, *Chem. Ber.*, **67**, 1627 (1934).
3) A. Butenandt, U. Westphal, *Chem. Ber.*, **67**, 2085 (1934).

pregnanediol **8**

Remarks

Progesterone **1** has been isolated from placentas, the corpora lutea and from pregnancy urine of mammals. In the original isolation, 625 kg of ovaries from 50,000 sows yielded 20 mg of pure progesterone.[5] This hormone is crucial for the maintenance of pregnancy. It is excreted as the pregnanediol glucuronide in urine.[6]

The 3-keto group in ring A is essential for activity, whereas the 20-keto group is not.[7] Analogs of progesterone have been prepared. Some are listed below.[8-11]

9

10

11

12

Compound **9** was the first synthetic progesterone, while the more recently prepared **10** is more potent. Compounds **11** and **12** are the ones currently in use, with the former being most powerful when administered by subcutaneous injection[12] and the latter when taken orally.[13]

The major source of present gestogens is from degradation of other natural products. Some examples are given below.

4) E. Fernholz, *Z. Physiol. Chem.*, **230**, 185 (1934).
5) A. Butenandt, J. Schmidt, *Chem. Ber.*, **67**, 1901 (1934).
6) E. H. Venning, *J. Biol. Chem.*, **119**, 473 (1937).
7) R. E. Marker, H. M. Crooks Jr., E. L. Wittbecker, *J. Am. Chem. Soc.*, **63**, 777 (1941).
8) L. Ruzicka, K. Hofmann, *Helv. Chim. Acta.*, **20**, 1280 (1937).
9) P. A. Plattner, H. Heusser, S. F. Boyce, *Helv. Chim. Acta.*, **31**, 603 (1948).
10) J. C. Babcock, E. S. Gutsell, M. E. Herr, J. A. Hogg, J. E. Stucki, L. E. Barnes, W. E. Dulin, *J. Am. Chem. Soc.*, **80**, 2904 (1958).
11) C. Djerassi, L. Miramontes, G. Rosencranz, F. Sondheimer, *J. Am. Chem. Soc.*, **76**, 4092 (1954).
12) S. P. Barton, B. Ellis, V. Petrow, *J. Chem. Soc.*, 478 (1959).
13) L. F. Fieser, J. E. Herz, W. Y. Huang, *J. Am. Chem. Soc.*, **73**, 2396 (1951).
14) R. E. Marker, R. B. Wagner, P. R. Ulshafer, E. L. Wittbecker, D. P. J. Goldsmith, C. H. Rouf, *J. Am. Chem. Soc.*, **69**, 2167 (1947).
15) K. Miescher, Wettstein, *Helv. Chim. Acta*, **29**, 627 (1946).
16) G. Slomp Jr., J. L. Johnson, *J. Am. Chem. Soc.*, **80**, 915 (1958).
17) V. Regnier, C. R. Seances, *Soc. Biol. Filates*, **127**, 519 (1938).

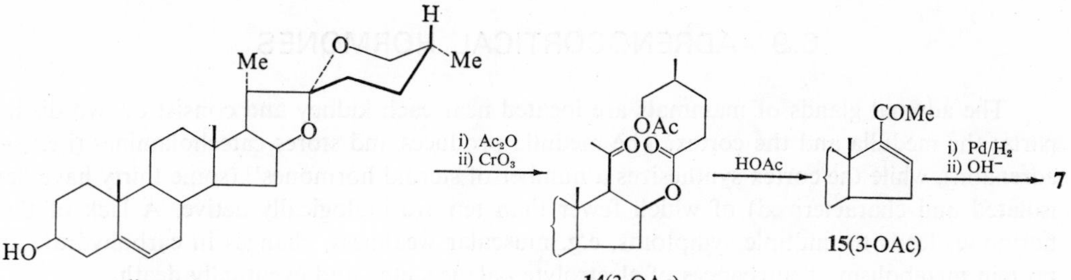

diosgenin 13[14]

14(3-OAc)

15(3-OAc)

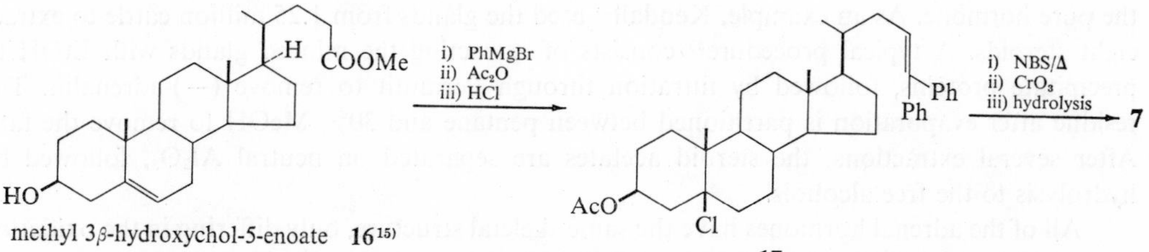

methyl 3β-hydroxychol-5-enoate 16[15]

17

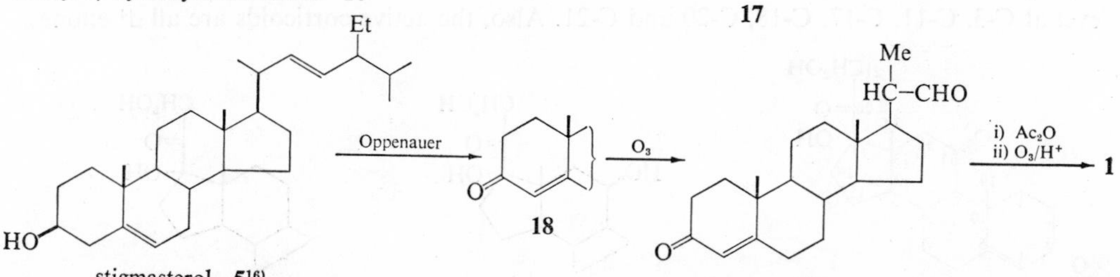

stigmasterol 5[16]

18

19

Diosgenin was the first natural product source used, but recently stigmasterol **5** has been mostly used.

It has been known for some time that several hormones are "antiandrogen" via gonatropin inhibition.[17] The term applies to compounds affecting organ receptors that are dependent on androgens.[18] Antiandrogen compounds include progesterone,[19] 19-norprogesterone,[20] *A*-norprogesterone **20** and 1,2α-methylene-17-acetoxy-4,6-pregnanediene-3,20-dione **21**.[22] These compounds can suppress the development of rudimentary genitalia[23] and cause female characteristics in males.[24] A clinical possibility is the use of antiandrogens to control the appetites of males with disturbed desires.[25]

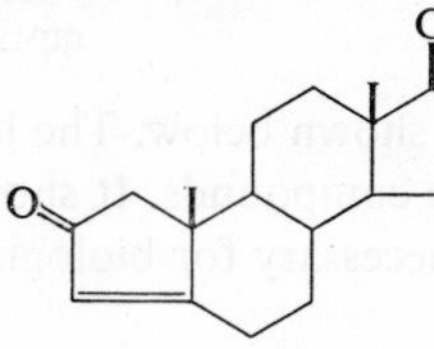

20

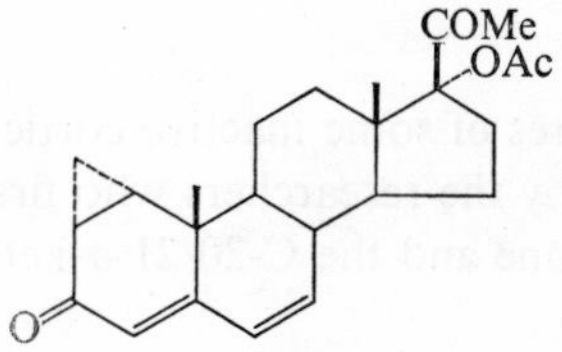

21

18) F. Neumann, *Symposium for Methods on Drug Evaluation*, p. 548, North Holland, 1965.
19) R. J. Dorfmann, *Proc. Soc. Exp. Biol. Med.*, **111**, 441 (1962).
20) R. J. Dorfmann, *Steroids*, **2**, 185 (1963).
21) L. J. Lerner, *Recent Prog. Hormone Res.*, **20**, 435 (1964).
22) R. Wiechert, F. Neumann, *Arzneimittelforsch*, **15**, 244 (1965).
23) F. Neumann, W. Elger, M. Kramer, *Endocrinology*, **78**, 628 (1966).
24) W. Elger, F. Neumann, *Proc. Soc. Exp. Biol. Med.*, **123**, 637 (1966).
25) R. Wiechert, *Angew. Chem. Intern. Ed.*, **9**, 5 (1970).

6.9 ADRENOCORTICAL HORMONES

The adrenal glands of mammals are located near each kidney and consist of two distinct parts: the medulla and the cortex. The medulla produces and stores catecholamines (i.e. (−)-adrenalin), while the cortex synthesizes a number of steroid hormones[1] (some thirty have been isolated and characterized) of which fewer than ten are biologically active. A lack of these hormones leads to multiple symptoms, e.g. muscular weakness, changes in carbohydrate and protein metabolism, disturbances of electrolyte balance, etc., and eventually death.

Since the steroids exist in small amounts, a large number of glands are necessary to isolate the pure hormone. As an example, Kendall[2] used the glands from 1.25 million cattle to extract eight steroids. A typical procedure[3] consists of extracting the minced glands with EtOH to precipitate proteins, followed by filtration through Permutit to remove (−)-adrenalin. The residue after evaporation is partitioned between pentane and 30% MeOH to remove the fats. After several extractions, the steroid acetates are separated on neutral Al_2O_3, followed by hydrolysis to the free alcohols.

All of the adrenal hormones have the same skeletal structure, only differing in the oxidation level at C-3, C-11, C-17, C-18, C-20 and C-21. Also, the active corticoids are all Δ^4 enones.

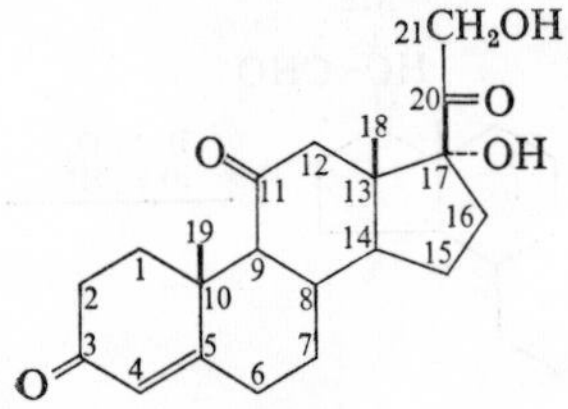

cortisone **1**
mp:215° α_D:+209° uv:240

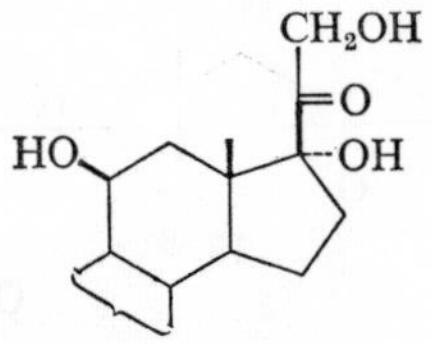

cortisol **2**
mp:220° α_D:+167°

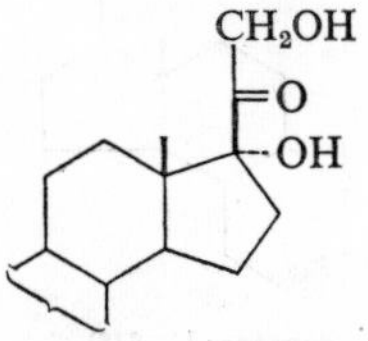

cortexolone **3**
mp:213° α_D:+116°

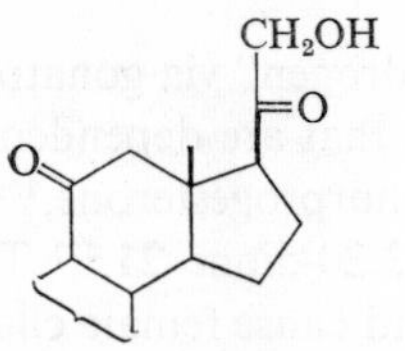

11-dehydrocorticosterone **4**
mp:180° α_D:+258°

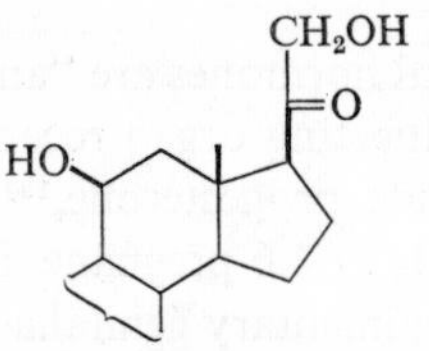

corticosterone **5**
mp:182° α_D:+222°

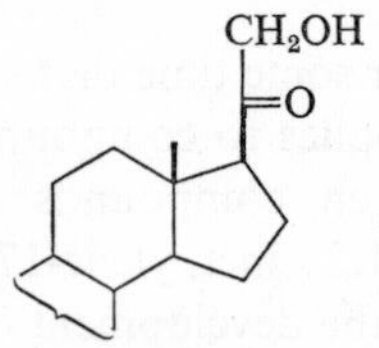

cortexone **6**
mp:142° α_D:+178°

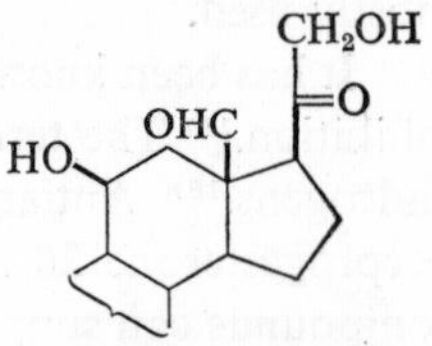

aldosterone **7**
α_D:+152° ir:OH
acetate
mp:99° α_D:+122°

The structures of some inactive corticoids (**8–10**) are shown below. The letter designations are those given by the researchers who first isolated these compounds. It should be noted that both the C-3 enone and the C-20/21 α-ketol seem to be necessary for biological activity.

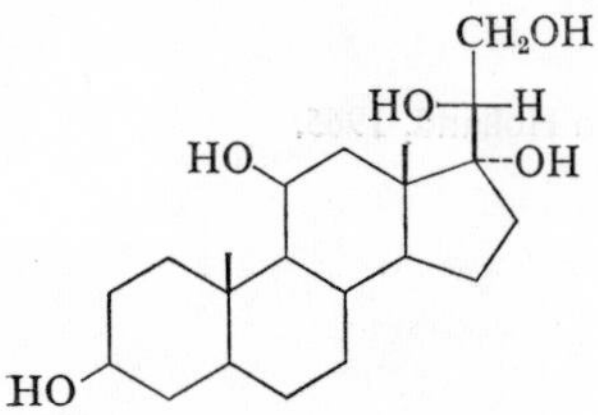

A **8**

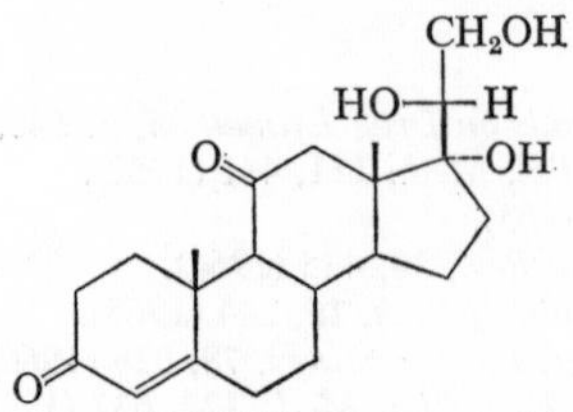

U **9**

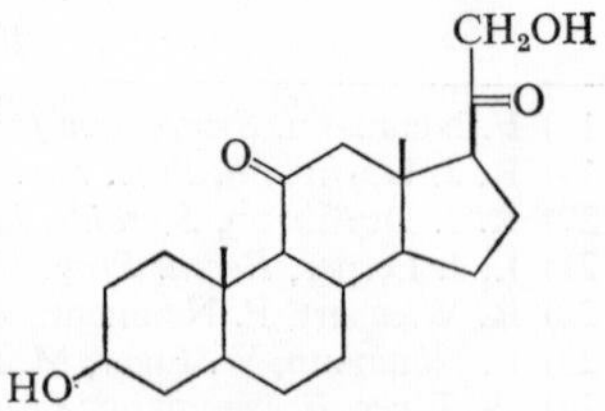

N **10**

Structure of the nucleus

1) The nucleus was determined to be the hydrocarbon androstane **13** from the following sequence of reactions.[4]

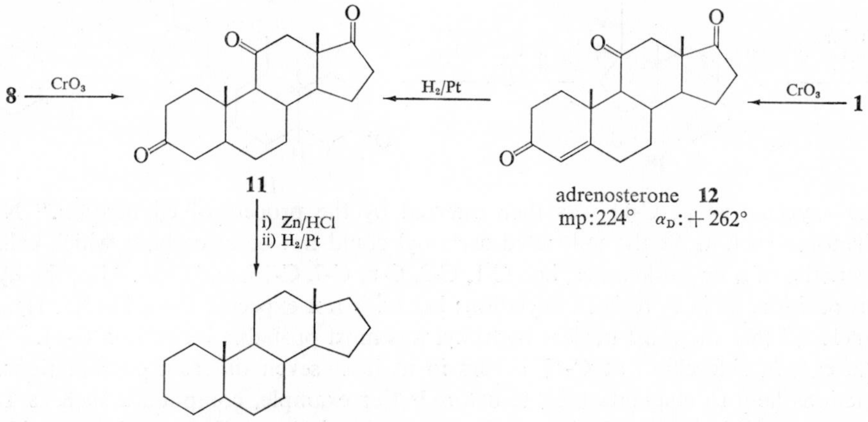

11

i) Zn/HCl
ii) H₂/Pt

adrenosterone **12**
mp:224° α_D: + 262°

androstane **13**

2) The position of the carbon side chain in **5** was found to be at C-17 when 5α-pregnane **15** was isolated.[5]

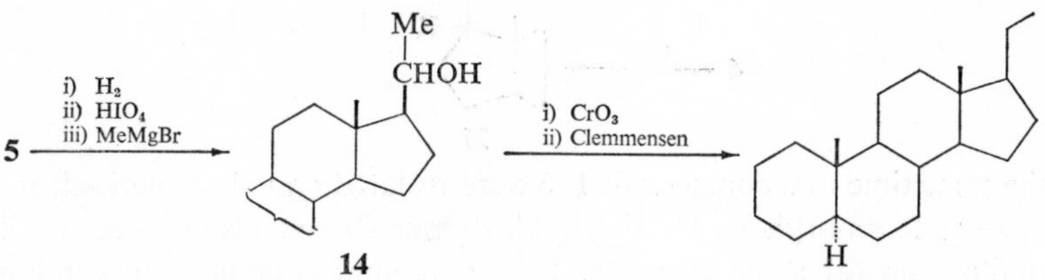

14

5α-pregnane **15**

Oxygen functions

1) The C-3 and C-17 oxygenated positions of the corticoids were determined by conversion to known diols[6] as in **17**.

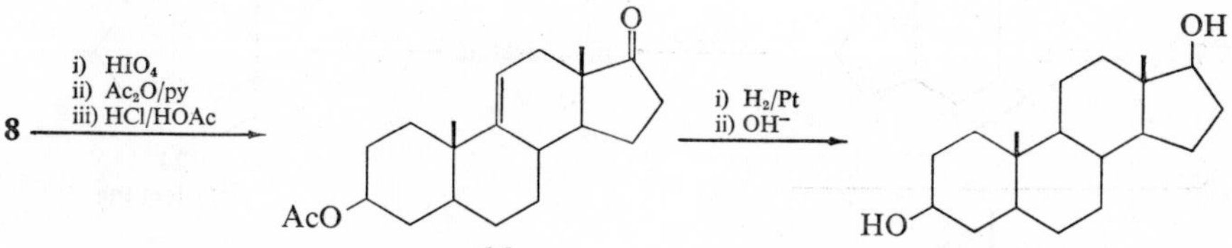

16

17

---

1) R. D. H. Heard, *The Hormones* (ed. G. Pincus, K. U. Thiman), Ch. 8, vol. I, Academic Press, 1948; L. F Fieser, M. Fieser, *Steroids*, Ch. 19, Reinhold, 1959; P. G. Marshall, *Rodd's Chemistry of Carbon Compounds* (ed. D. S. Coffey), p. 286–334, vol. II (2nd ed), part D, Elsevier, 1970.
2) A. C. Kendall, *Cold Spring Harbor Quart. Biol.*, **5**, 299 (1937).
3) J. J. Pfiffner and H. M. Vars, *J. Biol. Chem.*, **106**, 645 (1934); T. Rechstein, *Helv. Chim. Acta*, **21**, 546, 1197 (1938).
4) T. Reichstein, *Helv. Chim. Acta*, **19**, 479 (1936).
5) M. Steiger, T. Reichstein, *Helv. Chim. Acta*, **21**, 161 (1938).
6) C. W. Shoppee, *Helv. Chim. Acta*, **23**, 740 (1940).

2) The position of the double bond in the α,β unsaturated series was shown to be Δ^4 when the reduction of this bond led to the isolation of both epimers of pregnane-3,20-dione, **19** and **20**.[7]

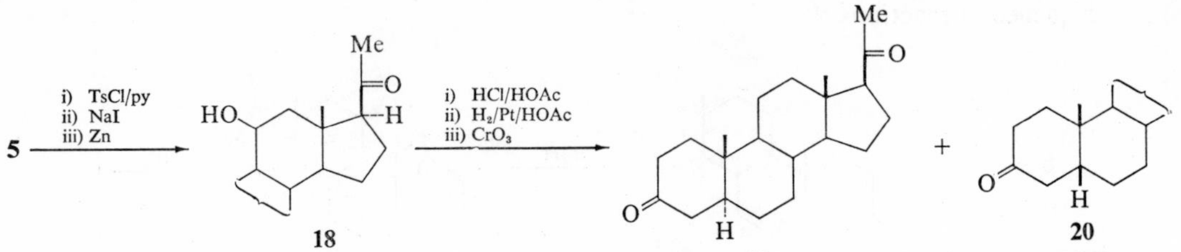

3) The oxygenation at C-11 was then inferred by the process of elimination.[8] Namely, in triketones such as **12** the unlocated carbonyl could not be at carbons which exhibit characteristics of α or β diketones, i.e., C-1, C-2, C-6, C-7, C-15, and C-16. Also, the hydroxyl at this position, as in **5**, resisted acylation; behavior not expected for a 12-OH. Hence, it was concluded that the third nuclear hydroxyl was most probably located on C-11.

4) The carbon side chain at C-17 occurs in at least seven different combinations. Various reactions help to elucidate their structure.[1] For example, compounds such as **1–7**, which have an α-ketal group, exhibit reducing properties similar to fructose. Upon oxidation with CrO_3 the corresponding *etio* acid **21** is obtained.

If the C-17 is hydroxylated this reaction leads to a 17-one as in **11**.

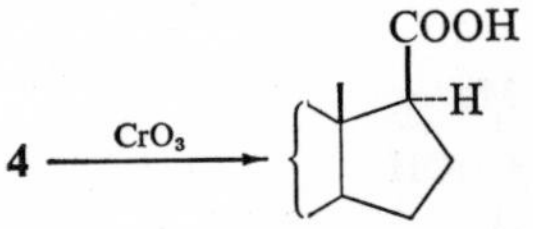

5) At about the same time that compounds **1–6** were isolated and characterized, it was realized that an appreciable amorphous fraction (aldosterone **7**) was also very active. This fraction defied crystallization for some 20 years. The compound is actually in equilibrium with a hemiacetal **22**. Preliminary studies had shown that **7** was a derivative of cortexone **6** with two additional oxygen functions at positions other than C-17. Formation of the γ-lactones **23** and **24**, suggested that the oxygens were at C-11 and C-18.[10]

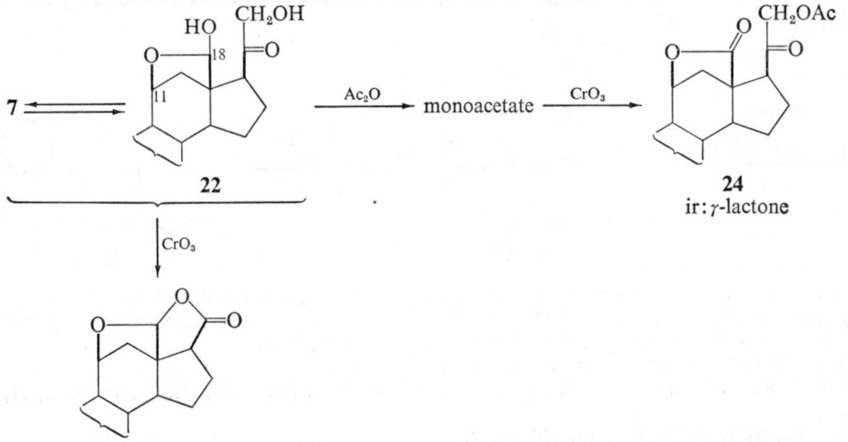

Stereochemistry

1) The assignment of hydroxyl configurations at C-3, C-11 and C-17 resulted from the studies of numerous workers over a long period of time. These conclusions rested on evidence gained from consideration of steric factors involved in reactivities of the hydroxyl groups, reduction of corresponding oxo compounds, etc. The results were confirmed only upon completion of the various syntheses.

2) The configuration of those corticoids that have a 20-ol was determined by Fieser[11] and Sarett.[12] Fieser's method depended on the fact that upon acetylation the α_D values of 20α-ols undergo positive shifts whereas 20β-ols undergo negative shifts. Sarett's method is summarized below.

It involves converting 20,21-diols **25**, 17,20-diols **30** and 17,20,21-triols **29** to 20-ols whose configuration is known. In both sequences cleavage of the epoxide ring involves inversion at the 20-OH position.

In conclusion it should be noted that many of the corticoids can be interconverted by simple reactions. Therefore structural conclusions reached about some of the corticoids are easily extended to others.

Synthesis

One of the major difficulties in the synthesis of corticoids has been the introduction of the 11-OH. This was overcome with the help of very efficient and precise chemical factories: microbes. An example is shown in Route I. Numerous synthetic methods have also been developed to introduce oxygenation at C-11, as shown in Routes II, III and IV.

Route I

It was found that various bacilli can oxidize steroids in a number of different places. For example, a culture of *Rhizopus anhizus* isolated from Kalamazoo air on exposure of an agar plate, converted progesterone to 11α-hydroxyprogesterone in yields of 50%.[13] Later, other

7) T. Reichstein, H. G. Fuchs, *Helv. Chim. Acta*, **23**, 684 (1940).
8) M. Steiger, T. Reichstein, *Helv. Chim. Acta*, **20**, 817 (1937).
9) A. Wettstein, G. Anner, *Experientia*, **10**, 397 (1954).
10) S. A. Simpson, J. F. Tait, A. Wettstein, R. Neher, J. Von Euw, T. Reichstein, *Helv. Chim. Acta*, **37**, 1200, 1163 (1954).
11) L. F. Fieser, M. Fieser, *Experientia*, **4**, 285 (1948).
12) L. H. Sarett, *J. Am. Chem. Soc.*, **71**, 1175 (1949).
13) D. H. Petersen, H. C. Murray, S. H. Eppstein, L. M. Reineke, A. Weintraub, P. D. Meister, H. M. Leigh, *J. Am. Chem. Soc.*, **74**, 5933 (1952).

microorganisms raised this yield to 80–90%. The efficiency of this reaction allowed it to be incorporated into a synthesis of cortisol.[14]

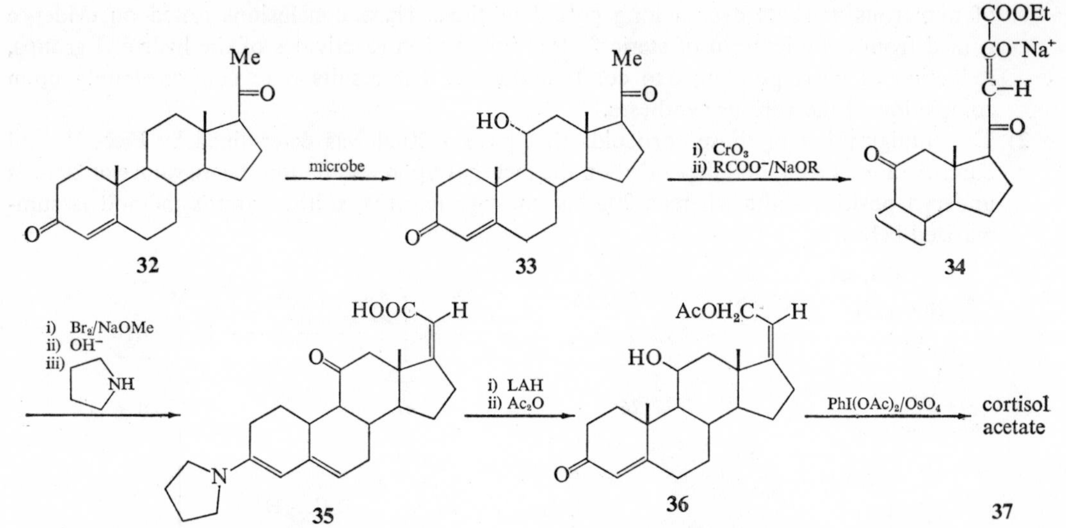

34→35: Sodium methoxide treatment of the bromo derivative effects a Favorsky rearrangement to the methyl cis-Δ^{17}-ene-21-oate.

36→37: A modified Miescher-Schmidlin method was used to effect hydroxylation.

Route II

One of the original methods for the formation of 11-OH involved conversion of C-12 hydroxyl in deoxycholic acid series compounds to an 11-one through a Δ^{11}-ene.[15] This was used to synthesize 11-dehydrocorticosterone **4**.

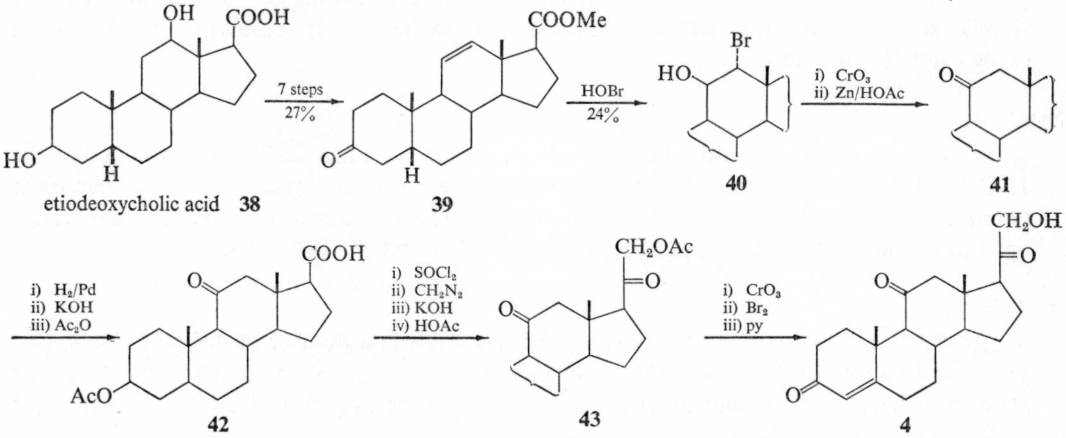

39→40: Originally this reaction also yielded the 11,12-dibromide and the 9-bromo-Δ^{11}-ester as side products. The yield of the desired **40** was improved by addition of H_2SO_4 as a catalyst.

41→42: Reduction of the 3-one proceeded without concomitant reduction of the 11-one.

43→4: The final bromination and dehydrobromination proceeded in low yield (8.5%).

14) O. Mancera, H. J. Ringold, C. Djerassi, G. Rosenkranz, F. Sondheimer, *J. Am. Chem. Soc.*, **75**, 1286 (1953).

15) A. Lardon, T. Reichstein, *Helv. Chim. Acta,* **26**, 747 (1943).

Route III

Another method involved formation of a Δ^{11}-$3\alpha,9\alpha$ epoxide.[16]

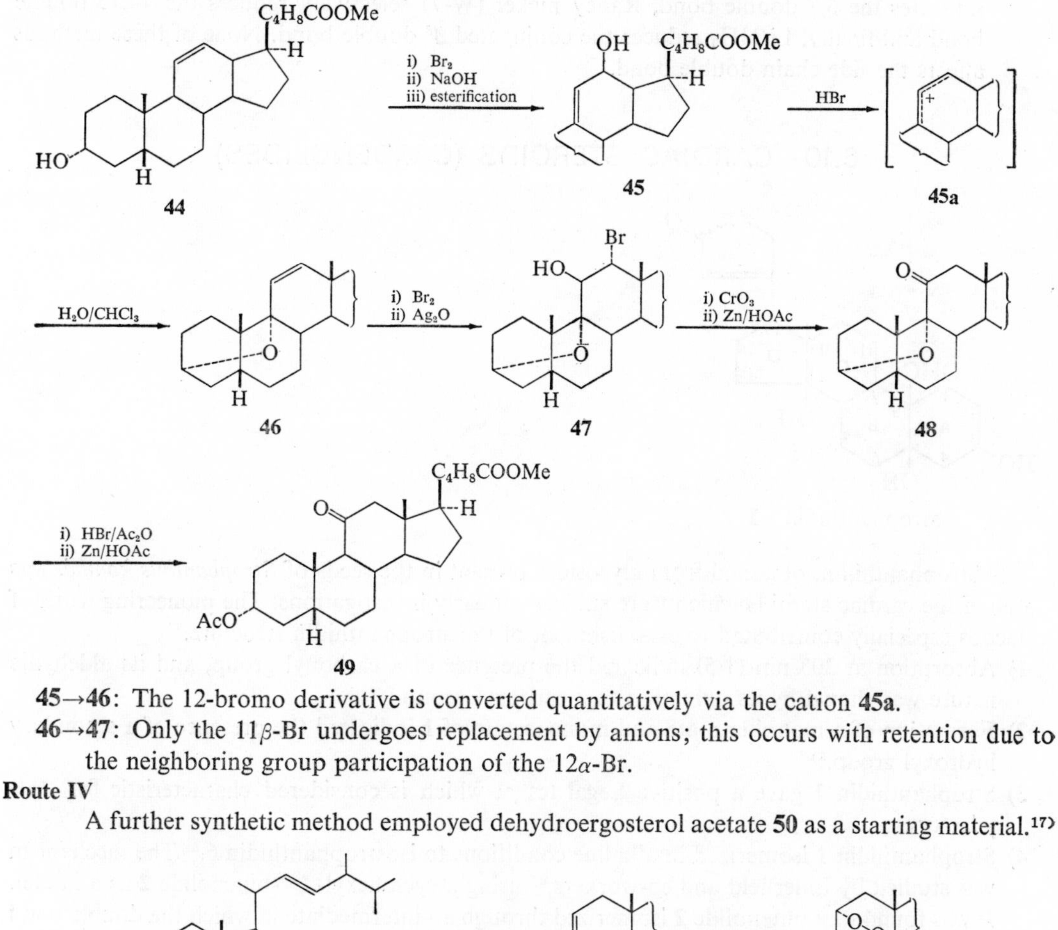

45→46: The 12-bromo derivative is converted quantitatively via the cation 45a.

46→47: Only the 11β-Br undergoes replacement by anions; this occurs with retention due to the neighboring group participation of the 12α-Br.

Route IV

A further synthetic method employed dehydroergosterol acetate 50 as a starting material.[17]

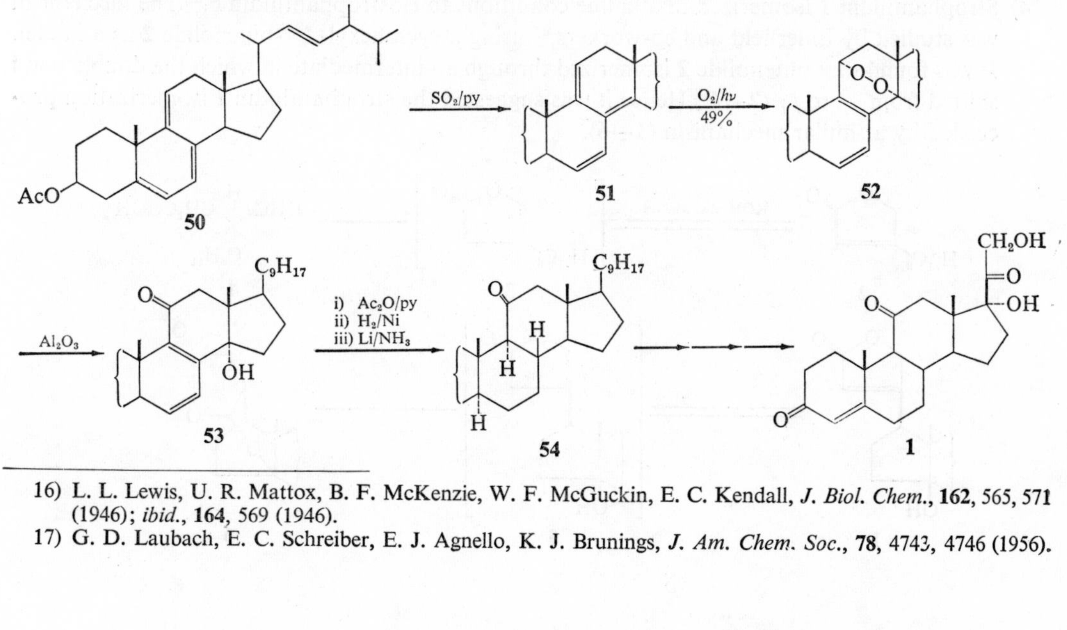

16) L. L. Lewis, U. R. Mattox, B. F. McKenzie, W. F. McGuckin, E. C. Kendall, *J. Biol. Chem.*, **162**, 565, 571 (1946); *ibid.*, **164**, 569 (1946).
17) G. D. Laubach, E. C. Schreiber, E. J. Agnello, K. J. Brunings, *J. Am. Chem. Soc.*, **78**, 4743, 4746 (1956).

50→51: The assignment of structure **51** is based on the formation of transannular adducts with oxygene or malcic anhydride.

53→54: Reduction of the 6-, 8- and 14-enes proceeds in three steps. Raney nickel (W-2) saturates the 6,7 double bond, Raney nickel (W-7) selectively reduces the 14,15 double bond and finally, Li/NH₃ reduces the conjugated Δ^8 double bond. None of these methods affects the side chain double bond.

6.10   CARDIAC STEROIDS (CARDENOLIDES)

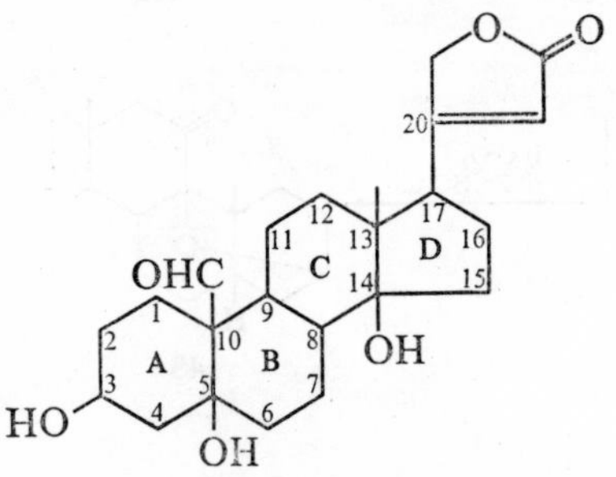

strophanthidin   **1**

Strophanthidin, obtained from glycosides present in the seeds of *Strophanthus kombé*, was one of the cardiac steroids which were subjects of early investigations. The pioneering work of Jacobs especially contributed to establishment of the strophanthidin structure.

1) Absorption at 303 nm (1.5) indicated the presence of a carbonyl group, and its aldehydic nature was shown by its oxidation to a carboxyl group.[1-3]

2) Formation of a monobenzoate and monoacetate of **1** indicated the presence of a secondary hydroxyl group.[3]

3) Strophanthidin **1** gave a positive Legal test,[4] which is considered characteristic for $\Delta^{\alpha,\beta}$-γ lactones.

4) Strophanthidin **1** isomerized in alkaline conditions to isostrophanthidin **6**.[5] The mechanism was studied by Elderfield and co-workers,[6] using β-cyclohexyl-$\Delta^{\alpha,\beta}$-butenolide **2** as a model. It was found that butenolide **2** isomerized through an intermediate in which the double bond shifted from αβ to βγ (**2→4**). Hence it was suggested the strophanthidin **1** isomerization proceeded by a similar mechanism (**1→6**).

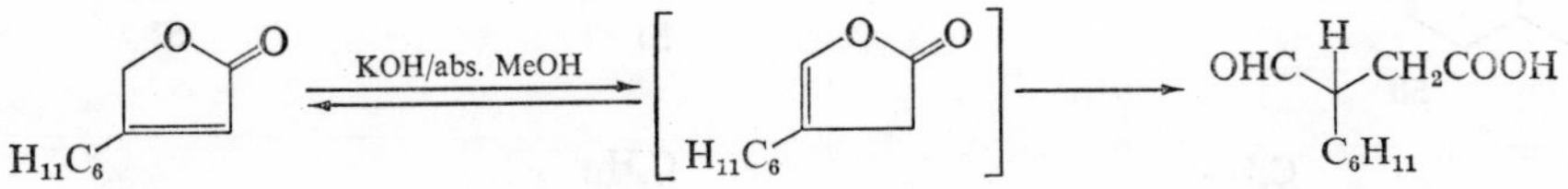

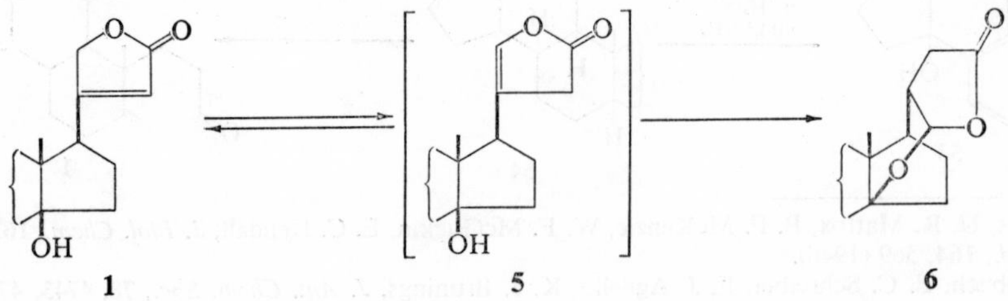

The formation of **6** revealed that one of the tertiary hydroxyl groups is at either C-8 or C-14 and is *cis* to the butenolide moiety (that the 14-OH is β and rings C/D are *cis* fused was proved by degradation of digitoxigenin[7]). Moreover, this hydroxyl group is also the easiest to remove, since monoanhydro-strophanthidin **8** is incapable of forming *iso*-compounds.[3]

7   **8**   **6**

5) Oxidation of the secondary hydroxyl group of **8** activated the remaining *tert*-OH (at C-5), which can be removed easily by gentle heating to give the enone **10**. Hence the *tert*-OH is β with respect to the secondary hydroxyl group.

8   **9**   **10**

6) Reaction of **1** with HCl in cold ethanol resulted in hemiacetal formation between the aldehydic and secondary hydroxyl groups.[8] This is an indication that the two groups are close to each other and have the same steric orientation.

HCl/EtOH

1   **11**

7) Structural information for ring A was obtained through the degradation of **12** in which all functional groups were protected except for the enone moiety.[3]

1) W. A. Jacobs, A. M. Collins, *J. Biol. Chem.*, **65**, 491 (1925).
2) W. A. Jacobs, *J. Biol. Chem.*, **57**, 553 (1923).
3) W. A. Jacobs, E. L. Gustus, *J. Biol. Chem.*, **74**, 795 (1927).
4) E. Legal, *Jarhresber. Fortschr. Chem.*, 1648 (1883).
5) W. A. Jacobs, A. M. Collins, *J. Biol. Chem.*, **61**, 387 (1924).
6) W. D. Paist, E. R. Blout, F. C. Uhle, R. C. Elderfield, *J. Org. Chem.*, **6**, 273 (1941).
7) F. Hunziker, T. Reichstein, *Helv. Chim. Acta*, **28**, 1472 (1945).
8) W. A. Jacobs, A. M. Collins, *J. Biol. Chem.*, **59**, 713 (1924).

12   **13**

14   0.1N NaOH   **15**   [H]   **16**

Oxidation of **12** afforded **14** which decarboxylated easily to give **15**. The keto group in **14** and **15** corresponded to the original position of the *tert*-OH which is therefore β to the aldehydic group in **1**. Hydrogenation of **15** afforded a hydroxyacid which lactonized readily to **16**. Such evidence, plus that established in **5)** and **6)**, can be accounted for by either structure **17** or **18**.

17   **18**

8) Determination of the size of ring A was achieved by using the so-called Blanc rule, as in the cases of cholesterol and bile acids. Windaus *et al.* had shown that the dibasic acid produced by opening ring A of a saturated 3-ketone of gitoxigenin **28c** was converted to a ketone on pyrolysis.[9] According to the Blanc rule, such results would indicate that ring A of cardenolides is six-membered, provided, however, that there is evidence of ring A being terminal. The following degradation reactions furnish such evidence.[10]

15   i) MeCOCl
                          ii) [H]   **19**   **20**

Barbier-Wieland degradation of **19** afforded a dibasic acid **20** with four less carbon atoms. Three of the lost atoms must come from the original lactone ring of **15**, and the remaining one should therefore come from the methylene carbon (C-2) on ring A. The total result is that ring A contains the moiety $-CH_2CH(OH)CH_2-$, and hence must be terminal.

9) Presence of a terminal five-membered ring was established by the following degradation reactions.

9) A. Windaus, K. Westphal, G. Stein, *Chem. Ber.*, **61**, 1847 (1928).
10) W. A. Jacobs, E. L. Gustus, *J. Biol. Chem.*, **84**, 183 (1929); *ibid.*, **92**, 323 (1931); W. A. Jacobs, R. C. Elderfield, *J. Biol. Chem.*, **102**, 237 (1933).

21 → i) KMnO$_4$/OH$^-$ ii) CrO$_3$ → **22** → [H] → **23**

The ready lactonization leading to **23** indicated that one of the hydroxyl groups was γ or δ relative to the carboxyl, and hence the ring being cleaved must be five-membered.[11]

10) Establishment of the steroid nature of strophanthidin was not accomplished, however, until the selenium dehydrogenation method was developed. Thus Elderfield and Jacobs[12] were able to obtain the Diels hydrocarbon **24** by selenium dehydrogenation of **1**.

1 → Se(320°–340°) → **24**

Remarks

Cardenolides are plant steroids which occur as glycosides. Cardiac glycosides possess powerful cardiotonic activity and can be used for treatment of heart disease. Cardiac glycosides occur mostly in plants of the order *Apocynaceae, Scrophulariaceae, Liliaceae, Moraceae,* and *Ranunculaceae.* Certain genera of *Digitalis,* particularly *D. purpurea* (fox glove) are main sources of most of the drugs of therapeutic value.

Enzymatic or chemical hydrolysis of cardiac glycosides affords the aglycones or genins and one or more sugars. For example, *k*-strophanthoside **25**, a trioside present in the seeds of *S. kombé,* is cleaved by the enzyme strophanthobiase (from seeds of *S. courmontii*)[13] to cymarin **26** and two moles of glucose. Acid hydrolysis of **26** affords strophanthidin **1** and the rare sugar cymarose **27**.

β-glucose–β-glucose–cymarose–O **25** → cymarose–O **26**

11) W. A. Jacobs, R. C. Elderfield, *J. Biol. Chem.,* **97**, 727 (1932).
12) R. C. Elderfield, W. A. Jacobs, *Science,* **79**, 279 (1934); *J. Biol. Chem.,* **107**, 143 (1934).
13) W. A. Jacobs, A. Hoffmann, *J. Biol. Chem.,* **69**, 153 (1926).

$$\xrightarrow{\text{H}^+}$$ 27 + 1

Most of the sugars isolated from the hydrolysed products of cardiac glycosides have not been found elsewhere in nature. From molecular rotation analysis, Klyne[14] suggested that all natural cardiac glycosides of D(L)-sugars are $\beta(\alpha)$-glycosides at the aglycone-sugar linkage (Klyne's rule).

The principle feature of cardenolides is the 14β-OH C/D-*cis* 17β-butenolide moiety. Most cardenolides also belong to the 5β series and have a 3β-hydroxyl group. Structural variations in aglycones involve the presence or absence of oxygen function at various positions. For example, digitoxigenin 28a, digoxigenin 28b, and gitoxiginin 28c, three cardenolides from *Digitalis*, differ from each other only in the oxygenation pattern at C-12 and C-16. Cardenolides have now been synthesized successfully.[15-17]

a : R₁ = H,   R₂ = H
b : R₁ = OH, R₂ = H
c : R₁ = H,   R₂ = OH

28

The main biosynthetic route was shown to be cholesterol→pregnenolone 36→progesterone 37→cardenolids. Pregnenolone 36 occurs naturally in plants, and has been isolated from *Xysmalobinur undulatum* and *Trachycalymna fimbriatum*.[18,19] The conversion of radioactive pregnenolone into progesterone and cardenolides has been effected in *D. lanata*,[20,21] etc. That cholesterol is the precursor of pregnenolone was also demonstrated by the conversion of [4-14C]-cholesterol into radioactive pregnenolone by *Haplopappus heterophyllus*[22] and *D. purpurea*.[23] Further evidence of this biosynthetic pathway come from tracer studies with radioactive meva-lonic acid. For example, incorporation of [2-14C]-mevalonic acid 38 into *Digitalis* gave digitoxi-genin 39 labeled at C-1, C-7 and C-15.[24]

14) W. Klyne, *Biochem. J.*, **47**, xli (1950).
15) N. Danieli, Y. Mazur, F. Sondheimer, *J. Am. Chem. Soc.*, **84**, 875 (1962).
16) W. Fritzsch, H. Ruschig, *Ann. Chem.*, **655**, 39 (1962); W. Fritzsch, U. Stache, W. Haede, K. Radsheit, *ibid.*, **721**, 168 (1969).
17) F. Sondheimer, *Chem. Brit.*, **1**, 454 (1965).
18) R. Tschesche, G. Snatzke, *Ann. Chem.*, **636**, 105 (1960).
19) R. Elber, Dissertation, Basel, (1964).
20) E. Caspi, D. O. Lewis, *Science, N. Y.*, **156**, 519 (1967).
21) H. H. Sauer, R. D. Bennett, E. Heftmann, *Phytochemistry*, **6**, 1521 (1967).
22) R. D. Bennett, E. Heftmann, *Phytochemistry*, **5**, 747 (1966).
23) E. Caspi, D. O. Lewis, D. M. Piatak, K. V. Thimann, A. Winter, *Experientia*, **22**, 506 (1966).
24) E. G. Gros, E. Leete, *J. Am. Chem. Soc.*, **87**, 3479 (1965).

The fact that no activity was found in the butenolide moiety supports the hypothesis that C-22 and C-23 originate from a separate acetate unit after 20/22 cleavage of the cholesterol side chain.

The mechanism of introducing oxygen functions at various positions in the biosynthesis of cardenolides from the C_{21} pregnenolone precursor has also been studied extensively. Saver *et al.*[21,25,26] administrated [4-^{14}C]-pregnenolone and [4-^{14}C]-progesterone to *D. lanata* and [4-^{14}C]-progesterone to *S. kombé*. The radioactive compounds **40**, **41** and **42** were isolated as metabolites of *S. kombé*, while only **40** was isolated from *D. lanata*.

They therefore suggested[26] the following pathway for strophanthidin **1** biosynthesis: progesterone **37**→**40**→**41**→**42**→cardenolide.

36   **37**

38

$* = ^{14}C$

39

40   **41**   **42**

Since the 14-OH is β, an inversion at C-14 must occur at some stage, and Caspi *et al.*[27] considered 14α-OH, $\Delta^{8(14)}$- or Δ^{14}-compounds as possible biosynthetic intermediates. However, the following experiments carried out with *D. lanata* have so far been negative. Thus, 14α-hydroxy-progesterone[27] and Δ^{14}-progesterone[28] were not incorporated into cardenolides; on the other hand, both the C-15 protons in progesterone[29] and the tritium atom in [8-^{3}H, 4-^{14}C]-cholesterol[30] were retained in the cardenolides.

25) R. D. Bennett, H. H. Sauer, E. Heftmann, *Phytochemistry*, **7**, 41 (1968).
26) H. H. Sauer, R. D. Bennett, E. Heftmann, *Phytochemistry*, **8**, 69 (1969).
27) E. Caspi, D. O. Lewis, *Phytochemistry*, **7**, 683 (1968).
28) R. Tschesche, R. Hombach, H. Schoten, M. Peters, *Phytochemistry*, **9**, 1505 (1970).
29) L. Canonica, F. Ronchetti, G. Russo, *Chem. Commun.*, 1675 (1970).
30) D. J. Aberhart, J. G. Lloyd-Jones, E. Caspi, private communication.

Synthesis

Early attempts at synthesis were unsuccessful, despite pioneering studies by Ruzicka, Plattner, Reichstein and Elderfield. The difficulties are due to the 14β-OH C/D-*cis* 17β-butenolide system, namely, the ready dehydration of 14β-OH, the thermodynamic instability of the 17β side chain in C/D *cis*-fused steroids, and the interaction between the 17β side chain and the 14β-OH. The first synthesis of digitoxigenin was finally achieved in 1962 as follows.[15]

29 30 31 32

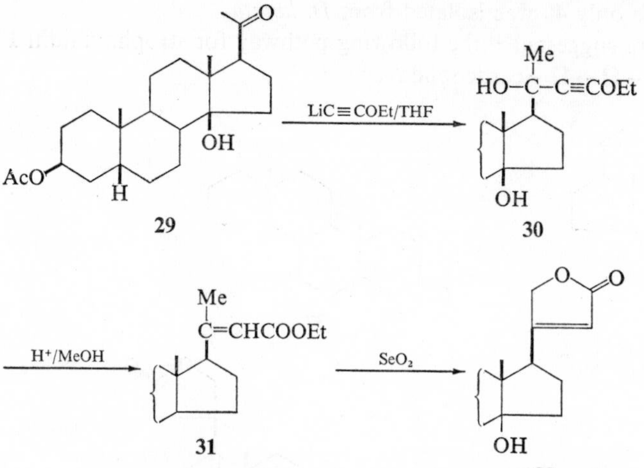

Other synthetic methods have since been developed. A number of these obtained the butenolide ring from corticosterone side chains; for example, intramolecular cyclization of **33** afforded the butenolide **35** via the intermediate **34**.[16] On the other hand, 14 β-OH steroids have become more accessible through microbiological hydroxylation.[17]

33 34 35

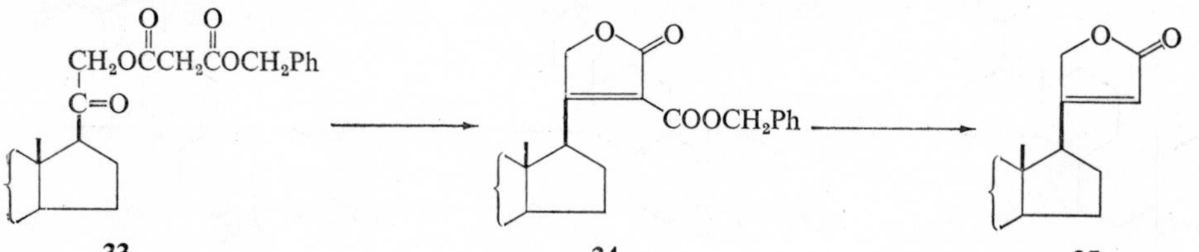

6.11 BUFADIENOLIDES

bufalitoxin **1**

enzyme →

bufalin **2**   mp: 244°

+

$HOOC(CH_2)_6-CONH-\underset{\underset{COOH}{|}}{C}H-(CH_2)_3NH-\underset{\overset{NH}{||}}{C}-NH_2$

3

I. Toad Bufadienolides

Toad venoms secreted from the parotid glands are highly toxic to mammals and frogs. These venoms, which show digitalis-like action towards the heart, have long been used as indispensable folk drugs in China (Ch'an Su) and Japan (Senso) (obtained from the local toad, *Bufo bufo gargarizans*).[1]

The main components of the venom, which could be responsible for the physiological activity are firstly (−)-adrenaline and indole-type bases, e.g. bufotenine, bufothionine, etc., and secondly a group comprising cholesterol, 7α-hydroxycholesterol, ergosterol, β-sitosterol and bufadienolides, both the bufogenins (aglycones) and bufotoxins (suberylarginine conjugates of bufogenins).

Bufotoxins are rapidly hydrolyzed in the body to their aglycones. The aglycones and conjugates are both physiologically active.

Structure

All bufogenins possess an A/B-*cis*-B/C-*trans*-C/D-*cis* steroid skeleton with an α-pyrone at 17β, and either a 14β-hydroxyl or 14β,15β-oxide. The steroid skeleton was established by Se-dehydrogenation (Diels' hydrocarbon or chrysene). About twenty bufogenins from toads are known, of which the following three are typical. Most structure determinations are based on chemical correlations with cardenolides.

Bufalin 2[2-4]

1) The presence of one *sec*-OH and one *tert*-OH was shown by the formations of a mono-acetate, monoketone (CrO_3), monoanhydro compound (HCl) and an acetylanhydro compound (hot Ac_2O).

2) The α-pyrone was indicated by uv (300 nm) and by the formation of a tetrahydro derivative of the acetate.

3) The 14-OH is *cis* to the α-pyrone because of the formation of the ketolactone **3**, indicating 14β-OH.

4) The entire structure was deduced from the formation of **4** upon KMnO$_4$ oxidation, since **4** was also obtained from digitoxigenin acetate, having the known structure **5**.

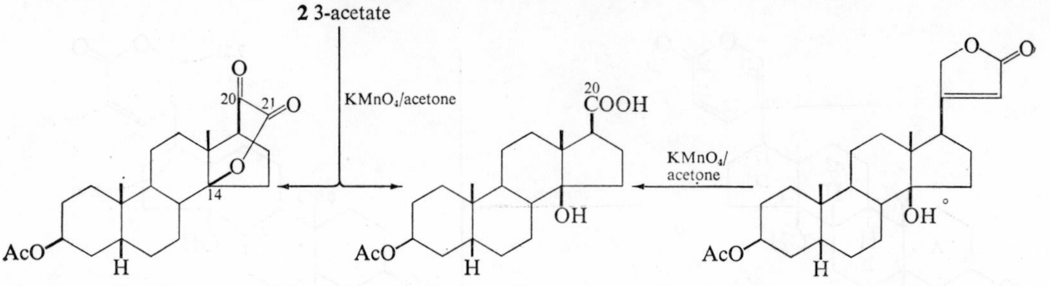

Bufotalin 6[5-7)]

bufotalin **6**
mp: 223–227°
α_D(CHCl$_3$): +5.4°
uv: 300(ϵ 5,500)

uv: 300nm(ϵ 16,600)

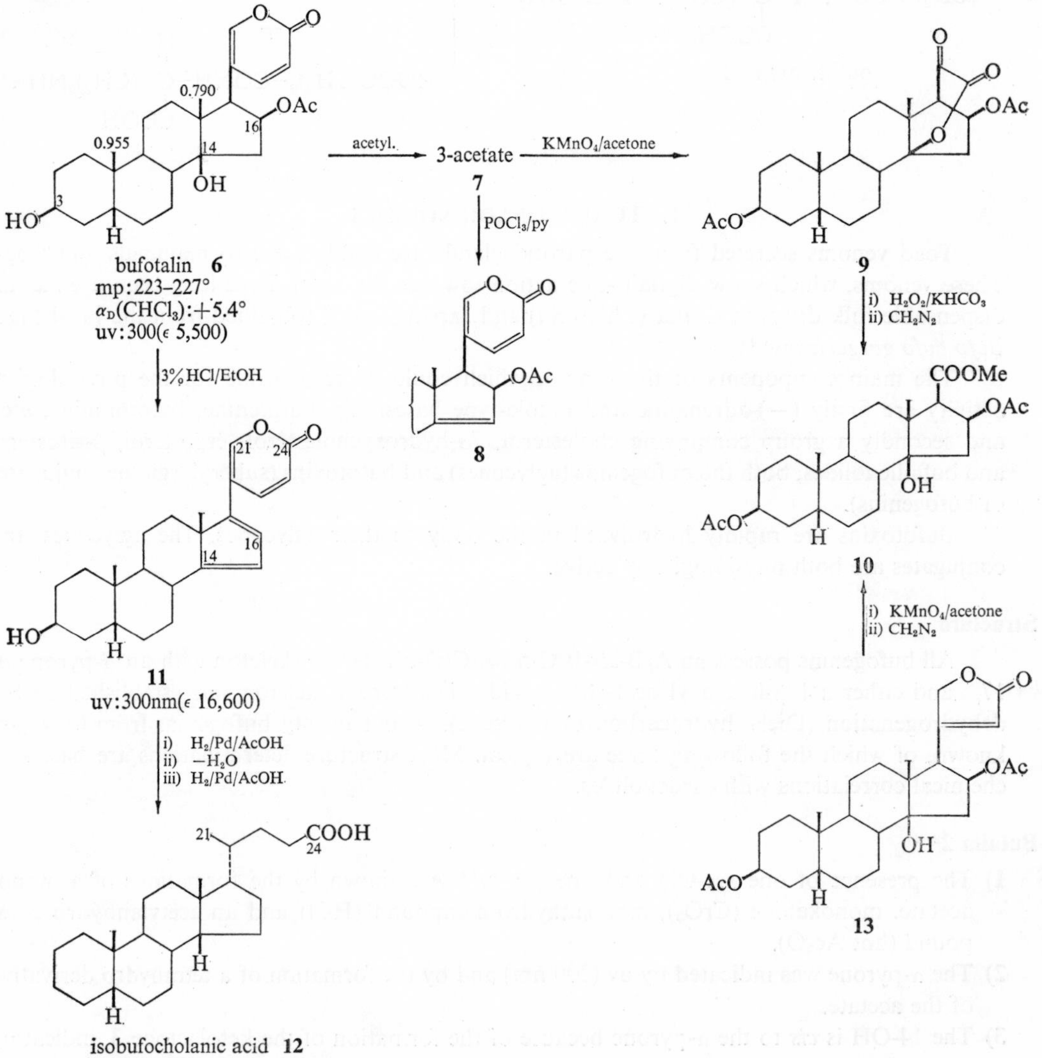

isobufocholanic acid **12**
(=14-iso-17-isocholanic acid)

1) The presence of one *sec*-OH and one *tert*-OH besides an acetyl group was shown by the formation of a monoketone and the monoanhydro compound **8**.

2) The reactions shown lead to the structure **6** for bufotalin, and the structure was finally established by chemical correlation with a known cardenolide, gitoxigenin acetate **13**.

Resibufogenin 14[8–10]

1) $14\beta,15\beta$-oxide: Epoxide (suggested by ir (CH): 3040 cm^{-1}) was placed at C-14/15 since resibufogenin **14** was converted into the known 14β-OH compound **4** after epoxide cleavage at C-15.

2) 15-Ketone **18** (C/D *trans*) epimerizes at C-14 upon base treatment to the more stable epimer **19** (C/D *cis*).

resibufogenin **14**
mp:113–140°, 155–168°
α_D(CHCl$_3$):-8.2°

14α-artebufogenin **18**

14β-artebufogenin **19**

acetate **15**

i) KMnO$_4$
ii) CH$_2$N$_2$

16

LAH

17

i) acetyl.
ii) KHCO$_3$/aq. MeOH
iii) CrO$_3$/AcOH

4

Remarks[11,12]

Spectroscopic methods, especially nmr, have greatly facilitated structural studies of bufogenins (see structures **2**, **6** and **14**).[12]

Bufotoxins, which are suberylarginine conjugates of bufogenins, were partially synthesized (from corresponding bufogenins), thus establishing the position of the conjugate as being at the 3-OH.[11]

1) For reviews including historical descriptions, see: L. F. Fieser, M. Fieser, *Natural Products Related to Phenanthrene* (3rd ed.), p. 552–577, Reinhold, 1949; *Idem., Steroids*, p. 782–799, Reinhold, 1959; K. Sellhorn *Arch. Pharm.*, **285**, 382 (1952); T. Reichstein, *Angew. Chem.*, **63**, 412 (1951).
2) M. Kotake, K. Kuwada, *Sci. Papers Inst. Phys. Chem. Research*, **36**, 106 (1939).
3) K. Meyer, *Experientia*, **4**, 385 (1948).
4) K. Meyer, *Helv. Chim. Acta*, **32**, 1238 (1949).
5) H. Wieland, F. J. Weil, *Chem. Ber.*, **46**, 3315 (1913).
6) H. Wieland, G. Hesse, R. Huttel, *Ann. Chem.*, **524**, 203 (1936).

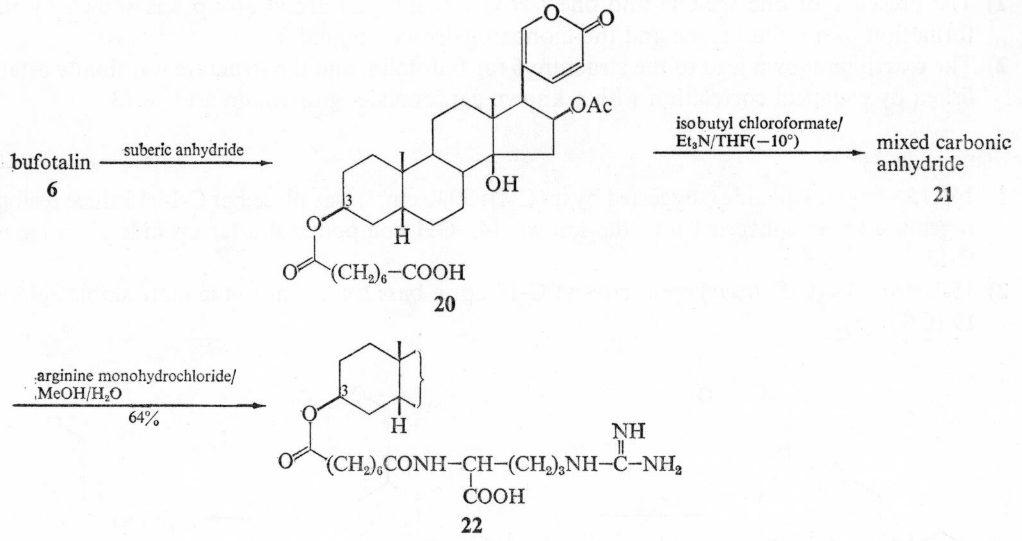
bufotalin
6
suberic anhydride
OAc
OH
3
H
(CH$_2$)$_6$–COOH
20
isobutyl chloroformate/
Et$_3$N/THF(−10°)
mixed carbonic
anhydride
21
arginine monohydrochloride/
MeOH/H$_2$O
64%
3
H
NH
(CH$_2$)$_6$CONH–CH–(CH$_2$)$_3$NH–C–NH$_2$
COOH
22

II. Plant Bufadienolides

Bufadienolides are also contained in plants, mostly in the form of glycosides, but not many are known.

Scillaren A 23[13–15)]

1) A rhamnoside-glycoside from the white squill, *Scilla maritima*, this compound is hydrolyzed enzymatically to glucose and proscillaridin A. However, on acid treatment scillaren A gave the elimination product **26**.
2) The carbon skeleton was derived by conversion into 3β-hydroxyallocholanic acid **25**.
3) Formation of **26** and **27** permitted clarification of the entire structure **23**.

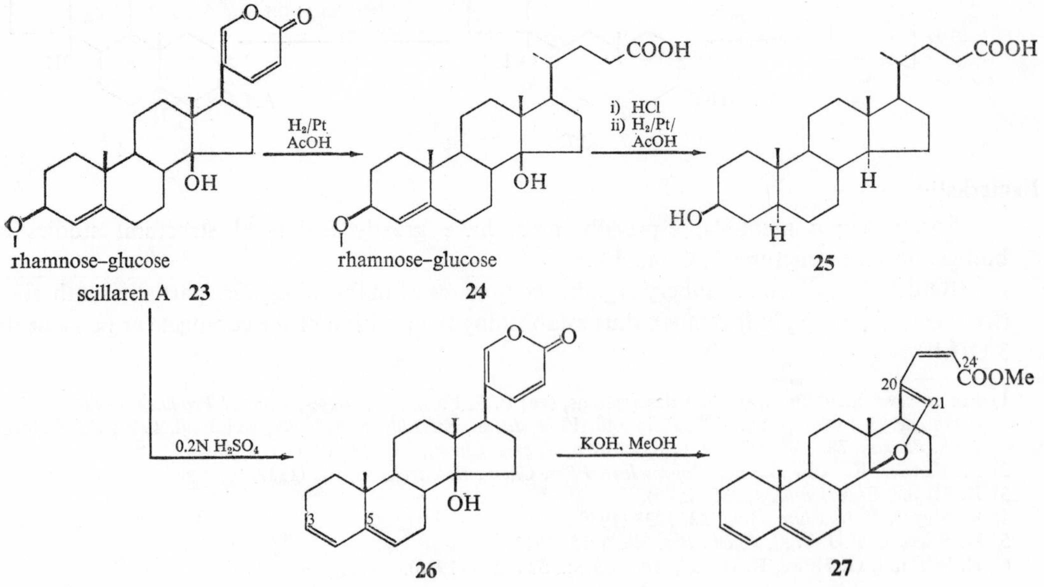
OH
rhamnose–glucose
scillaren A 23
H$_2$/Pt
AcOH
COOH
OH
rhamnose–glucose
24
i) HCl
ii) H$_2$/Pt/
AcOH
COOH
HO
H
H
25
0.2N H$_2$SO$_4$
OH
3
5
26
KOH, MeOH
24
COOMe
20
21
27

Remarks

Hellebrigenin 3-acetate, isolated from *Bersama abyssinica*, and its 3,5-diacetate exhibit tumor-inhibitory activities.[16]

hellebrigenin 3-acetate **28**

Details of the biosynthesis of these compounds remain to be clarified. However, radioactive pregnenolone (21-^{14}C) **29** was incorporated into hellebrin **30**.[17] This suggests a biosynthesis in plants involving incorporation of a C_3 unit (e.g. oxaloacetic acid) to form the α-pyrone, a route which is very similar to cardenolide biosynthesis. In the toad, however, cholanoic acid derivatives are better precursors, suggesting that all the carbons of bufadienolides come from cholesterol.[18,19]

stem of *Helloborus atrorubens*

rhamnose–glucose

30

III. Syntheses and reactions

Synthesis of bufalin 2 and resibufogenin 14

The first synthesis of bufadienolides was accomplished in 1969.[20]

7) K. Meyer, *Helv. Chim. Acta.*, **32**, 1993 (1949).
8) K. Meyer, *Helv. Chim. Acta*, **35**, 2444 (1952).
9) H. Linde, K. Meyer, *Helv. Chim. Acta*, **42**, 807 (1959).
10) H. Linde, K. Meyer, *Experientia*, **14**, 238 (1958).
11) G. R. Pettit, Y. Kamano, *Chem. Commun.*, 45 (1972).
12) nmr summary: L. Gsell, Ch. Tamm, *Helv. Chim. Acta*, **52**, 551 (1969).
13) A. Stoll, E. Suter, W. Kreis, B. B. Bussemaker, A. Hofmann, *Helv. Chim. Acta*, **16**, 703 (1933).
14) A. Stoll, J. Renz, *Helv. Chim. Acta*, **24**, 1380 (1941).
15) A. Stoll, J. Renz, A. Brack, *Helv. Chim. Acta*, **35**, 1934 (1952).
16) S. M. Kupchan, R. J. Hemingway, J. C. Hemingway, *J. Org. Chem.*, **34**, 3894 (1969).
17) R. Tschesche, H. Scholten, M. Peters, *Z. Naturforsch.*, **24b**, 1492 (1969).
18) A. M. Porto, E. G. Gros, *Experientia*, **26**, 11 (1970).
19) C. Chen, M. V. Osuch, *Biochem. Pharmacol.*, **18**, 1797 (1969).
20) F. Sondheimer, W. McCrae, W. G. Salmond, *J. Am. Chem. Soc.*, **91**, 1228 (1969).

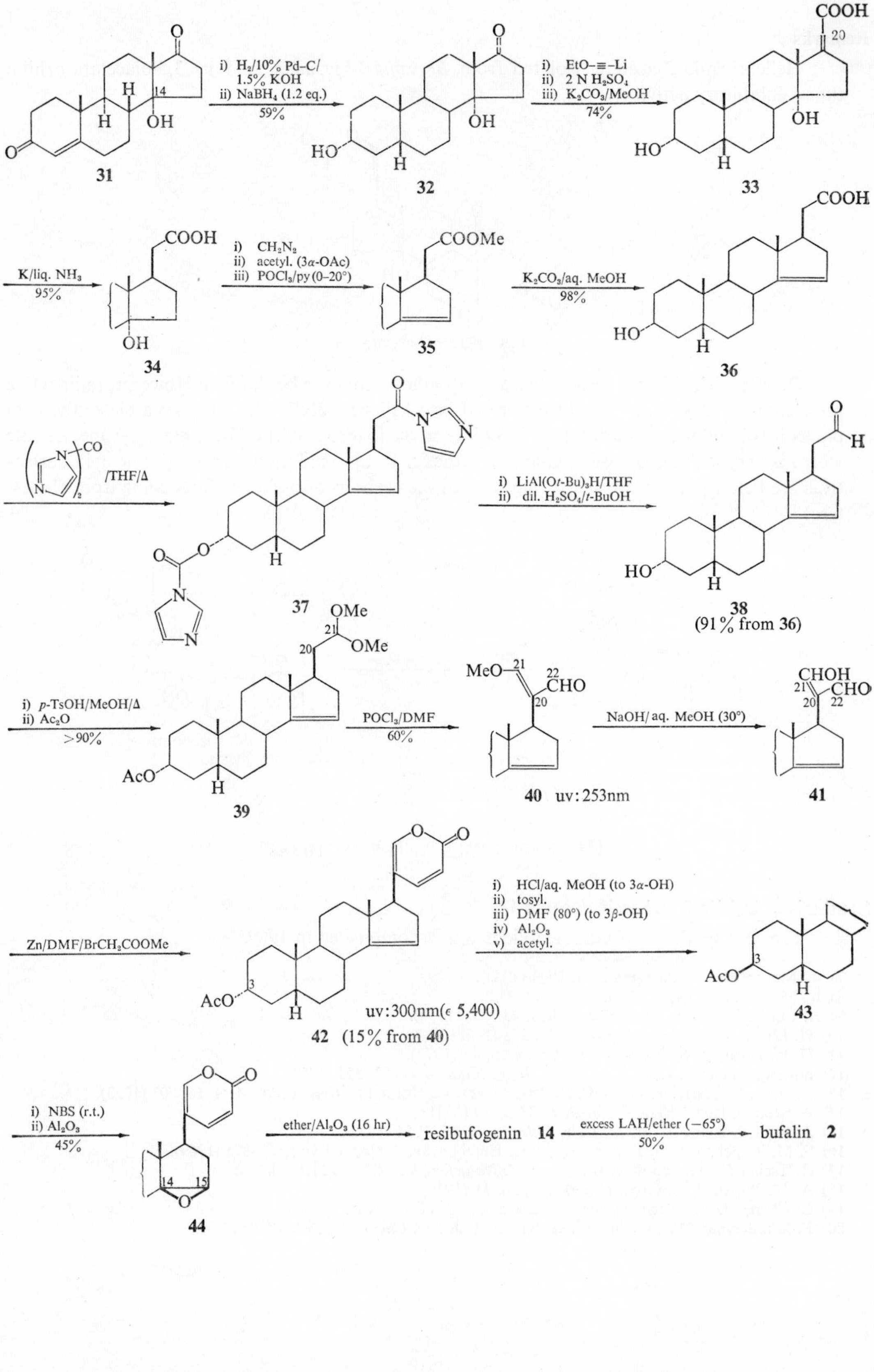

31

i) H$_2$/10% Pd–C/
1.5% KOH
ii) NaBH$_4$ (1.2 eq.)
59%

32

i) EtO–≡–Li
ii) 2 N H$_2$SO$_4$
iii) K$_2$CO$_3$/MeOH
74%

COOH
20

33

K/liq. NH$_3$
95%

COOH

OH

34

i) CH$_2$N$_2$
ii) acetyl. (3α-OAc)
iii) POCl$_3$/py (0–20°)

COOMe

35

K$_2$CO$_3$/aq. MeOH
98%

COOH

36

$\left(\begin{array}{c}N\\N\end{array}\right)_2$CO /THF/Δ

37

i) LiAl(Ot-Bu)$_3$H/THF
ii) dil. H$_2$SO$_4$/t-BuOH

O
H

38
(91% from 36)

i) p-TsOH/MeOH/Δ
ii) Ac$_2$O
>90%

OMe
21
OMe
20

39

POCl$_3$/DMF
60%

MeO 21
22 CHO
20

40 uv: 253nm

NaOH/aq. MeOH (30°)

CHOH
21
20
CHO
22

41

Zn/DMF/BrCH$_2$COOMe

uv: 300nm(ε 5,400)
42 (15% from 40)

i) HCl/aq. MeOH (to 3α-OH)
ii) tosyl.
iii) DMF (80°) (to 3β-OH)
iv) Al$_2$O$_3$
v) acetyl.

AcO 3

43

i) NBS (r.t.)
ii) Al$_2$O$_3$
45%

14 15
O

44

ether/Al$_2$O$_3$ (16 hr) → resibufogenin **14** $\xrightarrow[50\%]{\text{excess LAH/ether (−65°)}}$ bufalin **2**

31: The 14α-hydroxy compound **31** was obtained from cortexolone by microbial oxidation at 14α followed by oxidation with sodium bismuthate, or alternatively, from 3β-acetoxy-5-androsten-17-one by a five-step synthesis.

31→32: This conversion also afforded the 3β-OH isomer in 22% yield.

32→33: The stereochemistry at C-20 in **33** is not defined.

34→35: The 8(14)-ene was obtained as a by-product (30% yield).

39→40: Vilsmeier-Haack reaction gave a Δ²⁰-mixture of *cis* (60%, **40**) and *trans* (19%), in which the *trans* isomer was easily isomerized to **40**.

Conversion of a cardenolide into a bufadienolide

Digitoxigenin **45**, a cardenolide, was transformed into **43** which had been converted into resibufogenin **14** and bufalin **2**.[21]

digitoxigenin **45**

i) acetyl.
ii) −H₂O
iii) NaOMe/MeOH
iv) MeOH/*p*-TsOH
iii, iv: 88%

MeO OMe
20 COOH

46
epimeric mixture at C-20

i) acetyl.
ii) ⌈SH
  ⌊SH /HClO₄
ii: 89%

47

i) (COCl)₂
ii) CH₂N₂
iii) Ag₂O/Na₂S₂O₃
ii: 54%, iii: 81%

21
20 22 23 COOH

S S
21 24 COOH
20 22 23

48

HgCl₂/HgO/AcOH/H₂O
80%

21 CHO COOH
20

49

benzene/*p*-TsOH (warm)
20%

AcO

50

S (210°, 25mm)
28%

43

47→48: Homologation of **47** to **48** involves an Arndt-Eistert sequence.

α-Pyrone syntheses

The routes **51→53** and **54→53** constitute short α-pyrone syntheses[22]; the latter was applied to the synthesis of 14-anhydroscillarenone **59** (60%) from **58**.[23]

21) G. R. Pettit, L. E. Houghton, J. C. Knight, F. Bruschweiler, *Chem. Commun.*, 93 (1970).
22) K. Radscheit, U. Stache, W. Haede, W. Fritsch, H. Ruschig, *Tetr. Lett.*, 3029 (1969).
23) U. Stache, K. Radscheit, W. Fritsch, H. Kohl, W. Haede, H. Ruschig, *Tetr. Lett.*, 3033 (1969).
24) S. Sarel, Y. Shalon, Y. Yanuka, *Chem. Commun.*, 80, 81 (1970).

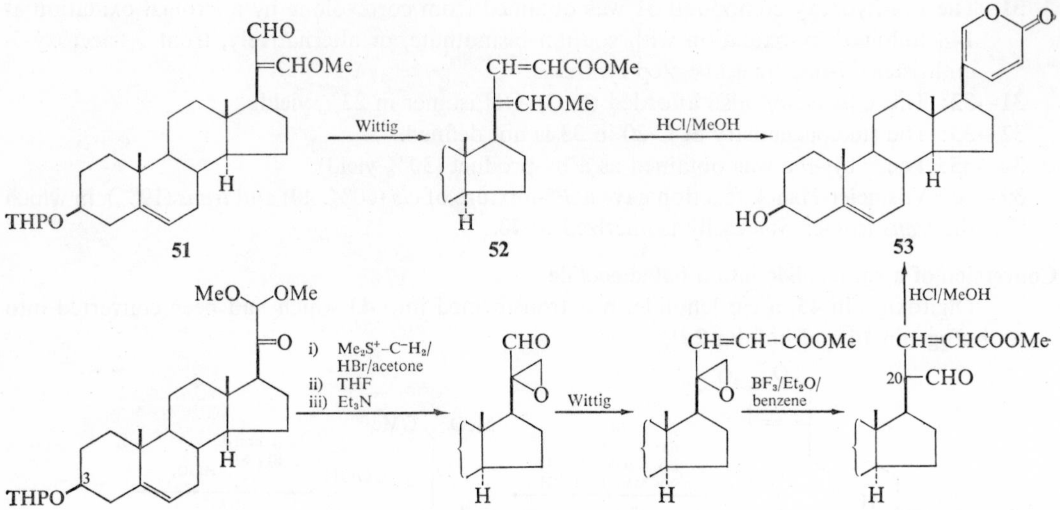

51   52   53

54   55 (3β-OH)   56   57

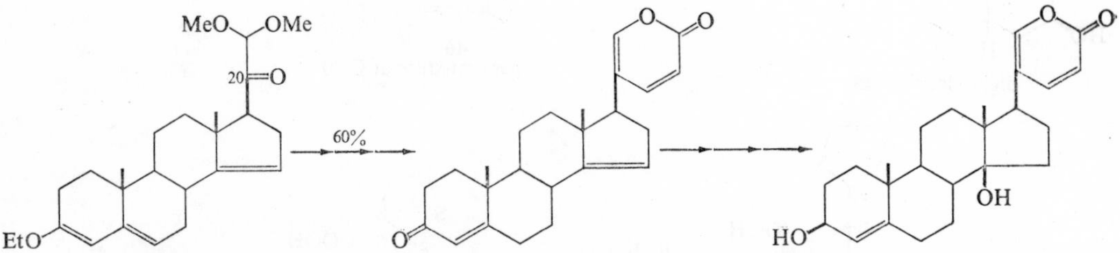

58   14-anhydroscillarenone 59   scillarenin 60

DDQ oxidation of the enolide **62** in the presence of *p*-TsOH also yields an α-pyrone, **63**.[24]

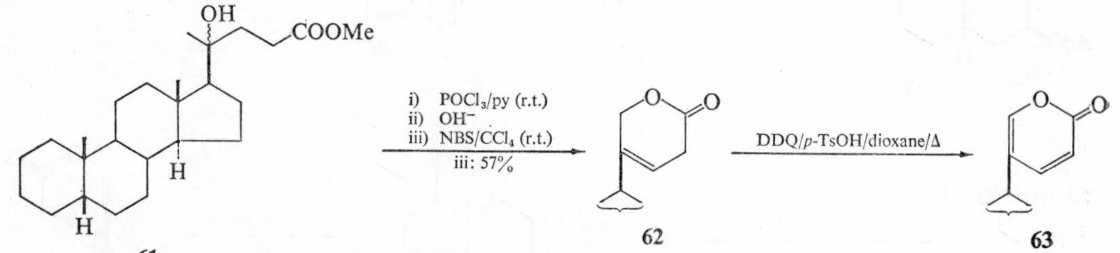

61   62   63

6.12   SAPOGENINS

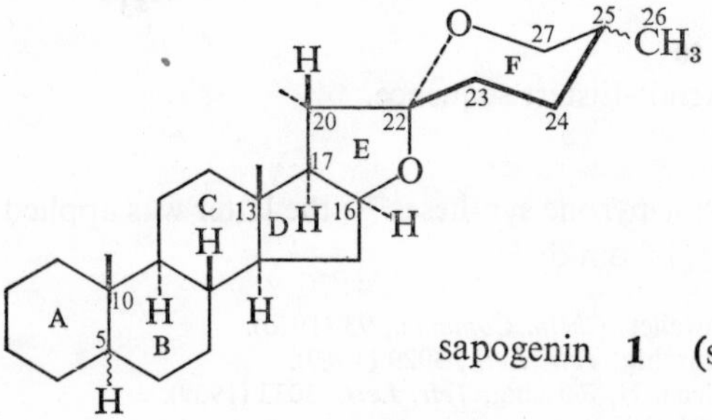

sapogenin **1**   (spirostan)

1) Sapogenins are aglycones of saponins, which have the distinctive property of forming soaping solutions. They were finally shown to be a group of C_{27} compounds after careful purification and repeated analysis.[1]

2) Steroid nucleus: Selenium dehydrogenation[2] of sapogenins gave the Diels hydrocarbon and a ketone with eight carbon atoms. These observations established that sapogenins possess the steroid ring system and a C_8 side chain.

3) Ring system: Two sapogenins, tigogenin[3] and sarsasapogenin,[4] were degraded to etioallobilianic acid **6** and etiobilianic acid **8**, respectively. This established the structure and configuration of the ring system and strongly indicated that the side chain includes a tetrahydrofuran ring fused to ring D.

tigogenin acetate **2** $\xrightarrow{CrO_3}$ **3**

3 steps → **4** $\xrightarrow{PhMgBr}$ **5** $\xrightarrow{CrO_3}$ etioallobilianic acid **6**

sarsasapogenin acetate **7** $\xrightarrow{\text{same degradation as } 2\rightarrow6}$ etiobilianic acid **8**

4) Spiroketal moiety: Marker[5] found that two oxygen atoms in the sapogenin side chain are unusually active in acid media, though they are inert in neutral and alkaline media. Various reactions in acid media, i.e., hydrogenation, bromination and Clemmensen reduction, indicated the presence of a carbonyl group existing as a spiroketal.

1) J. C. E. Simpson, W. A. Jacobs, *J. Biol. Chem.*, **109**, 573 (1935); P. L. Liang, C. R. Noller, *J. Am. Chem. Soc,* **57**, 525 (1935).
2) W. A. Jacobs, J. C. E. Simpson, *J. Biol. Chem.*, **105**, 501 (1934).
3) R. Tschesche, A. Hagedorn, *Chem. Ber.*, **68**, 1412 (1935).
4) S. N. Farmer, G. A. R. Kon, *J. Chem. Soc.*, 414 (1937).

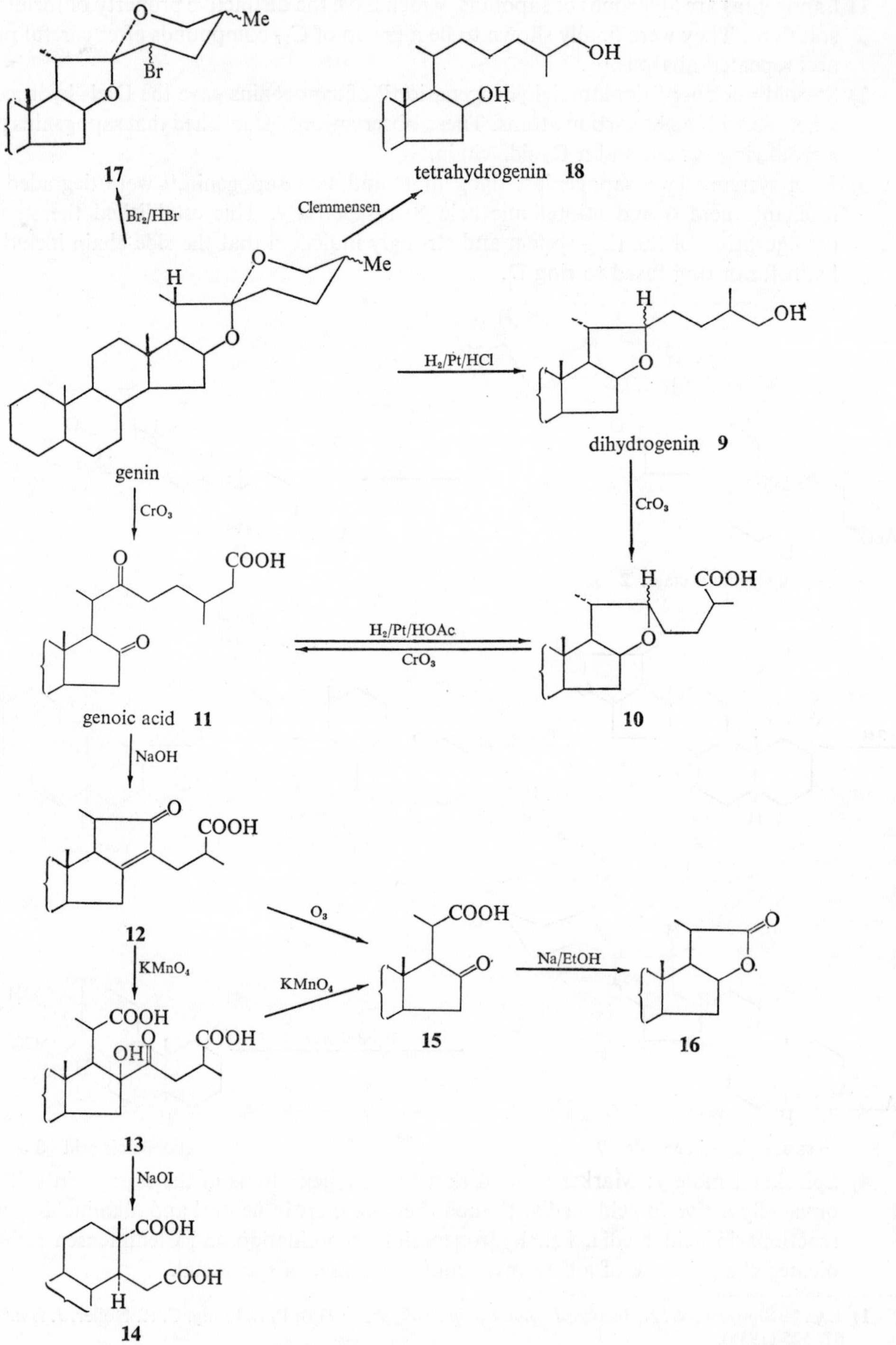

17

Br$_2$/HBr

tetrahydrogenin **18**

Clemmensen

genin

CrO$_3$

H$_2$/Pt/HCl

dihydrogenin **9**

CrO$_3$

genoic acid **11**

NaOH

H$_2$/Pt/HOAc

CrO$_3$

10

12

KMnO$_4$

O$_3$

15

KMnO$_4$

Na/EtOH

16

13

NaOI

14

5) R. E. Marker, E. Rohrmann, *J. Am. Chem. Soc.*, **61**, 846, 2072 (1939); R. E. Marker, A. C. Shabica, *ibid.*, **64**, 147 (1942); R. E. Marker *et al., ibid.*, **64**, 180 (1942).

5) Pseudosapogenins: When sapogenins are heated with acetic anhydride around 200°, a fission of the tetrahydropyran ring takes place with formation of a 20,22-double bond.[6] The resulting products can be degraded further to 20.[7] This degradation of sapogenins provides the key to their use for the large-scale preparation not only of progesterone 23, but also of sex hormones, cortical hormones and analogs.

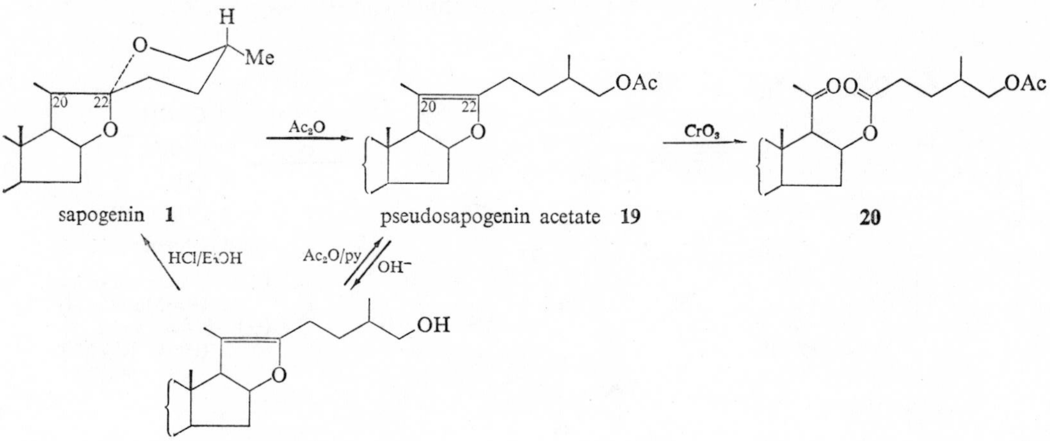

sapogenin 1   pseudosapogenin acetate 19   20

pseudosapogenin 21

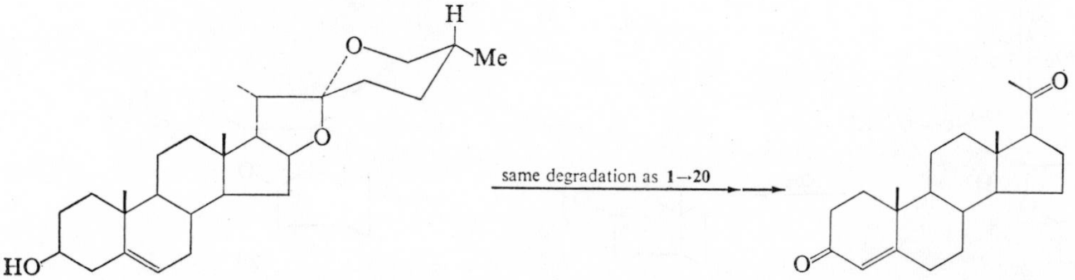

diosgenin 22   same degradation as 1→20   progesterone 23

6) Iso reaction: Several pairs of stereoisomers were found among naturally occurring steroid sapogenins. One isomer of such a pair is converted to the other on being refluxed in HCl/EtOH.[5] The less stable isomers are known as "normal" or "neo" sapogenins, and the more stable as "iso" sapogenins. The iso reaction is reversible, the equilibrium lying well to one side.[8]

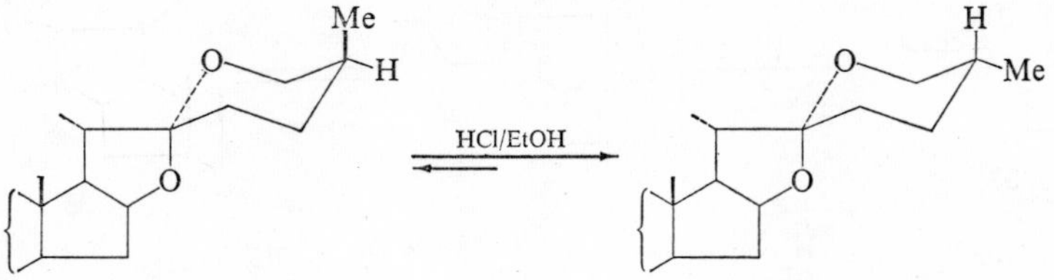

neo sapogenin 24   HCl/EtOH   iso sapogenin 25

6) R. E. Marker *et. al., J. Am. Chem. Soc.*, **61**, 3592 (1939); *ibid.*, **62**, 518, 648, 989, 7525, 2532 (1940); *ibid.*, **63** 774 (1941).

7) D. H. Gould, H. Staendale, E. B. Hershberg, *J. Am. Chem. Soc.*, **74**, 3685 (1952); W. G. Dauben, B. J. Fonken, *ibid.*, **76**, 4618 (1954); M. E. Wall, H. E. Kenney, E. S. Rothman, *ibid.*, **77**, 5655 (1955).

8) M. E. Wall, S. Serota, L. P. Witnauer, *J. Am. Chem. Soc.*, **77**, 3086 (1955).

7) Absolute configuration at C-25: Marker regarded the neo and iso compounds as C-22 epimers. However, both series gave the same **29**, thus showing that they differ only at C-25.[9] Also the same **32** was obtained and therefore the C-22 configuration should be identical.[10] In addition, antipodal acids **28** were obtained from the two series.[9] Finally, the C-25 configuration was established by degradation[11] of enone acid **12**, derived from isosapogenin **26** (R′ = Me, R = H) to the known (+)-methylsuccinic acid.[12]

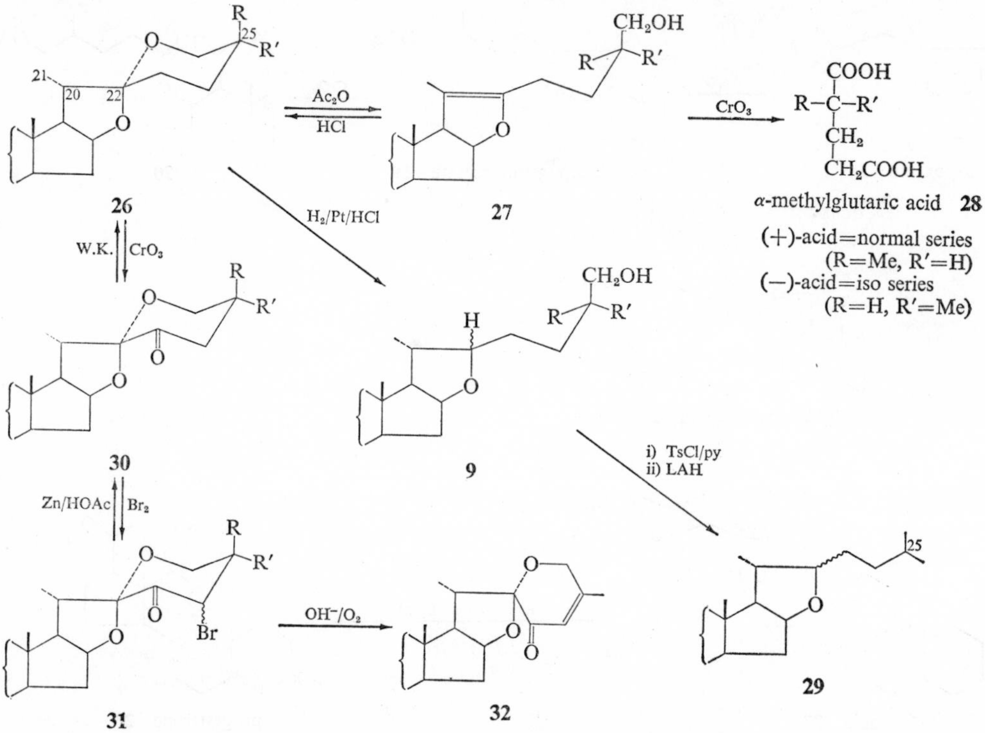

α-methylglutaric acid **28**

(+)-acid = normal series
(R = Me, R′ = H)
(−)-acid = iso series
(R = H, R′ = Me)

The following reversible oxidation-reduction mechanism has been proposed.[13] Asymmetry at C-25 is destroyed in the enolized aldehyde **37**, whereby a change in configuration at C-25 is accommodated. Support for the mechanism is reported by the Djerassi group.[14]

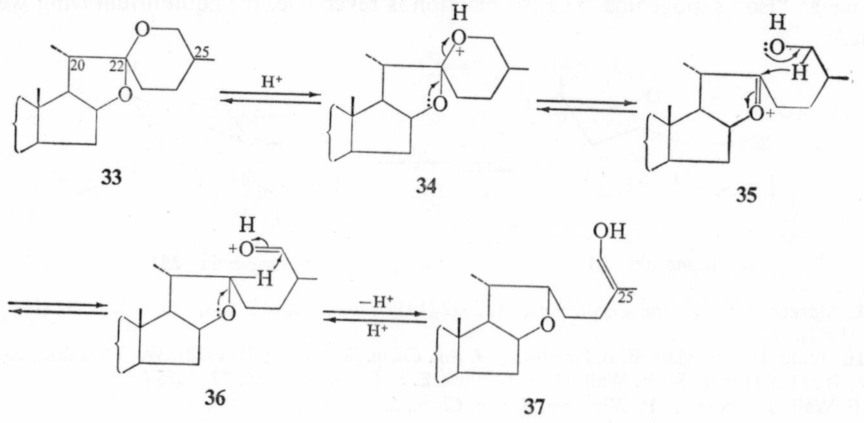

8) Configuration at C-20: Biogenetic considerations would suggest that the naturally occurring sapogenins have the same configuration at C-20 as most other steroids, i.e., that the 20-Me group is remote from the angular 13-Me group. An unambiguous support for this suggestion came from the observation[15] that lactone **A** derived from tigogenin acetate **2** is more stable than its 20-epimer **B**,[16] the latter being isomerized quantitatively to the former on treatment with base. The severe interaction between the 13-Me and 20-Me groups accounts for the instability of **39** with respect to **38**.

tigogenin acetate **2**   lactone **A 38**

lactone **B 39**   **40**

9) Configuration at C-22: A conformational analysis, based on the known greater stability of the iso series, has led to the 22-configuration shown in **24** and **25**. If the 22-configuration were reversed, the neo series would have been expected to be the more stable.[17] This was substantiated by x-ray studies.[17a]

neo sapogenin **24**   iso sapogenin **25**

10) Cyclopseudosapogenins: Although pseudosapogenins (**41** and **43**) can be recyclized by

9) I. Scheer, R. B. Kostic, E. Mosettig, *J. Am. Chem. Soc.*, **75**, 4891 (1953); *ibid.*, **77**, 641 (1955).
10) R. K. Callow, P. N. Massy-Beresford, *J. Chem. Soc.*, 4482 (1967).
11) V. H. T. James, *J. Chem. Soc.*, 637 (1955).
12) A. Fredga, *Arkiv. Kemi. Min. Geol.*, **19B**, 1 (1944).
13) R. B. Woodward, F. Sondheimer, Y. Mazur, *J. Am. Chem. Soc.*, **80**, 6693 (1968).
14) C. Djerassi, O. Halpern, G. R. Pettit, G. H. Thomas, *J. Org. Chem.*, **24**, 1 (1959).
15) J. W. Corcoran, H. Hirschmann, *J. Am. Chem. Soc.*, **78**, 2325 (1956).
16) N. Danieli, Y. Mazur, F. Sondheimer, *J. Am. Chem. Soc.*, **62**, 5889 (1960).
17) L. F. Fieser, M. Fieser, *Steroids*, p. 824, Reinhold, 1959.
17a) R. Callow, *J. Chem. Soc. C*, 288 (1966); see also discussions in A. H. Albert, G. R. Pettit, P. Brown, *J. Org. Chem.*, **38**, 2197 (1973).

refluxing in HCl/EtOH to the original sapogenins (**24** and **25**), gentler acid treatment produces the stereoisomeric cyclopseudosapogenins (**42** and **44**).[18,19] Each cyclopseudosapogenin under more vigorous acid catalysis is isomerized to the sapogenin (**24** and **25**) formed under the same conditions from the pseudo compound concerned.

neo **24**   pseudoneo **41**   cyclopseudoneo **42**

iso **25**   pseudoiso **43**   cyclopseudoiso **44**

The configurations of the spiroketal moiety of cyclopseudosapogenins were deduced as **43** and **44** by several approaches.[20] They all have a 20α-Me group and owe their instability to the 13-Me/20-Me interaction. The following results[21] support this suggestion. Namely, the cycle through **46**, **48** proves the presence of 20-OH (OsO₄ oxidation of **47** gave the five-membered 20-one), while the cycle **42→45→46→42** proves it to have a 20α-Me group (i.e., rear attack upon hydroxylation and hydrogenation).

cyclopseudotigogenin **42**   (strongly hydrogen bonded) **45**   pseudotigogenin **43**

46   **47**   **48**

18) M. E. Wall, C. R. Eddy, S. Serota, *J. Am. Chem. Soc.*, 76, 2849 (1954); M. E. Wall *et al., J. Am. Chem. Soc.* 76, 2850 (1954); *ibid.*, 77, 1230, 3086 (1955); R. K. Callow, V. H. T. James, *Chem. Ind.*, 691 (1954); R. K. Callow *et al., J. Chem. Soc.*, 1966 (1955); D. H. W. Dickson *et al., Chem. Ind.*, 692 (1954); J. B. Ziegler, W. E. Rosen, A. C. Shabica, *J. Am. Chem. Soc.*, 76, 3865 (1954); *ibid.*, 77, 1223 (1955).
19) D. A. H. Taylor, *Chem. Ind.*, 1066 (1954).
20) M. E. Wall, *Experientia*, 11, 340 (1955); W. E. Rosen, J. B. Ziegler, A. C. Shabica, J. N. Shoolery, *J. Am Chem. Soc.*, 81, 1687 (1959); H. Hirschmann, F. B. Hirschmann, J. W. Corcoran, *J. Org. Chem.*, 20, 572 (1955); H. Hirschmann, F. B. Hirschmann, *Tetr.*, 3, 2436 (1958).

Nmr studies[22] of the 13-, 20- and 25-Me groups of neo-, iso-, cyclopseudoiso- and cyclo-pseudoneo-sapogenins also support the respective configurations.

Remarks

The first source[23] of steroid saponins was *Digitalis purpurea*, more important as a source of the cardiac glycosides. Later, richer sources of saponins were found, many of them by Marker and his co-workers, in a number of other plants, notably among U.S. and Mexican members of the *Liliaceae, Dioscoreaceae* and *Scrophulariaceae* families.

The non-cardiac saponins from various plants have found some use as detergents, foaming agents in extinguishers and fish poisons. Fish are dazed or killed by saponins but are not rendered inedible, since saponins are not toxic to humans when taken orally, probably because they are not absorbed from the gut.

Contrary to the case of the sapogenins, the saponins are not easy to isolate in a pure state and they had been relatively little studied until recently. Their structures are established by conventional degradation methods. The completely elucidated structures of two saponins, sarsaporilloside[24] and dioscin[25,26] are shown here.

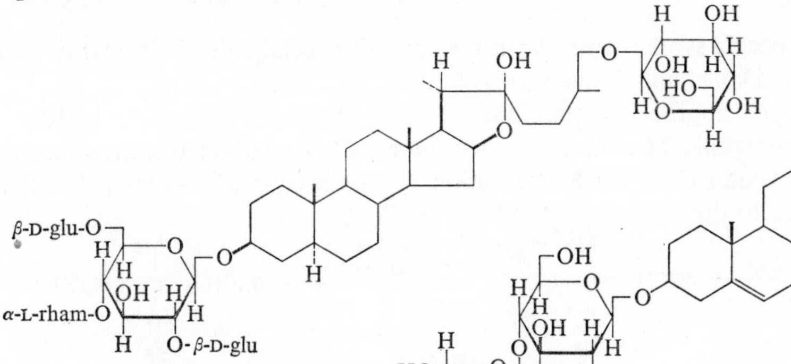

sarsaporilloside **49**

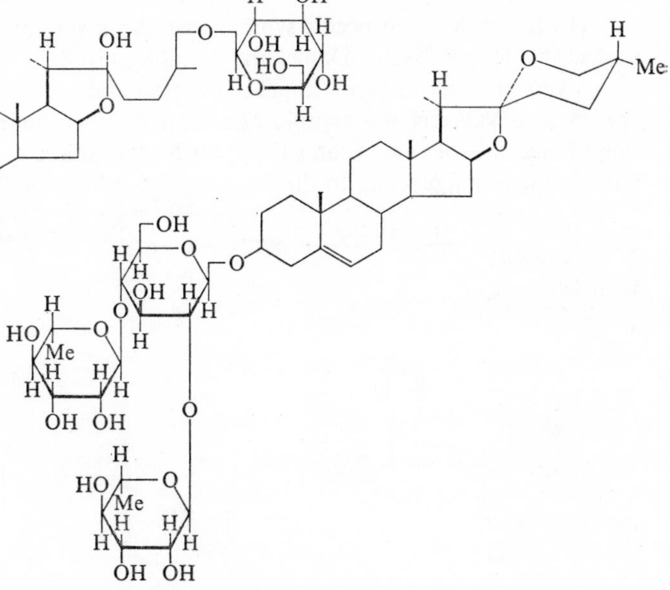

dioscin **50**

The sugars of saponins are the common ones, i.e., D-glucose, D-galactose, D-xylose, L-rhamnose and L-arabinose. These sugars, with few exceptions, are linked to the steroid sapogenins through the 3-OH group.[27]

Regarding nomenclature, the names "spirostan" and "furostan" are given to the skeletons 51 and 52, respectively.[28]

21) M. E. Wall, H. A. Welens, *J. Am. Chem. Soc.*, **77**, 5661 (1955); *ibid.*, **80**, 1984 (1958).
22) W. E. Rosen *et al., J. Am. Chem. Soc.*, **81**, 1687 (1959); G. F. H. Green, J. E. Page, V. S. Staniforth, *J. Chem. Soc.* (B), 807 (1966); D. H. Williams, N. S. Bhacca, *Tetr.*, **21**, 1641 (1965); J. P. Kutney, *Steroids*, **2**, 225 (1963).
23) O. Schmildeberg, *Arch. Exp. Parth. Pharmak*, **3**, 16 (1875).
24) R. Tschesche, G. Ludke, G. Wulff, *Tetr. Lett.*, 2785 (1967).
25) T. Tsukamoto *et al., J. Pharm. Soc.*, **56**, 802 (1936); *ibid.*, **77**, 1225 (1957); T. Tsukamoto *et al., Chem. Pharm. Bull.*, **4**, 350 (1956).
26) T. Kawasaki, T. Yamauchi, *Chem. Pharm. Bull.*, **10**, 703 (1962).

51   **52**

"Spirostan" specifies the configuration at all asymmetric centers except at C-5 and C-25. The configuration at C-5 and in the nucleus is defined by the usual α, β convention, while that at C-25 is determined by use of the sequence rule procedure[27] (R/S). The same sequence rule is applied to indicate configurations in the side chain.

The configuration at C-5 and in the nucleus of "furostan" is defined by the usual steroid convention, that at C-22 and side chain carbons is determined again by the sequence rule.

Synthesis

The first non-stereospecific synthesis of the sapogenin spiroketal side chain was accomplished in 1960 by Mazur, Daniel and Sondheimer.[29]

A versatile, stereospecific approach to steroidal sapogenins has been worked out by Kessar et al.[30] The synthesis of diosgenin **22** exemplifies their approach. The key step is the construction of elements of rings E and F by Michael addition of 1-acetoxy-5-nitro-2-methylpentane **56** of known configuration to the α, β-unsaturated ketone **57**.

53 (resolved via quinine salt)

$$\xrightarrow{\text{HBr/peroxide}}$$

$Br(CH_2)_3$—C—COOH   **54**

$$\xrightarrow[\text{ii) LAH/AlCl}_3]{\text{i) EtOH/HCl}}$$

$Br(CH_2)_3$—C—CH$_2$OH   **55**

i) AcCl
ii) NaNO$_2$/urea/DMF

58

$$\xrightarrow{\text{N}_2\text{H}_4/\text{KOH}}$$

59

$$\xrightarrow{\text{MnO}_2}$$

57   +   **56**

O_2N ... CH_2OAc

t-BuOK/t-BuOH

60

$$\xleftarrow[\text{ii) H}^+]{\text{i) NaBH}_4/\text{EtOH (}\Delta\text{, 3 hr)}}$$

diosgenin **22**

27) R. S. Cahn, C. Ingold, V. Prelog, *Angew. Chem. Intern. Ed.*, **5**, 385 (1966).
28) *Rodd's Chemistry of Carbon Compounds* (ed. S. Coffey), vol. II (E), 2nd ed., p. 18, Elsevier, 1971.
29) Y. Mazur, N. Danieli, F. Sondheimer, *J. Am. Chem. Soc.*, **82**, 5889 (1960).
30) S. V. Kessar, Y. P. Gupta, R. K. Mahajan, G. S. Joshi, A. L. Rampal, *Tetr.*, **24**, 899 (1968).

6.13 TOTAL SYNTHESIS OF EQUILENIN

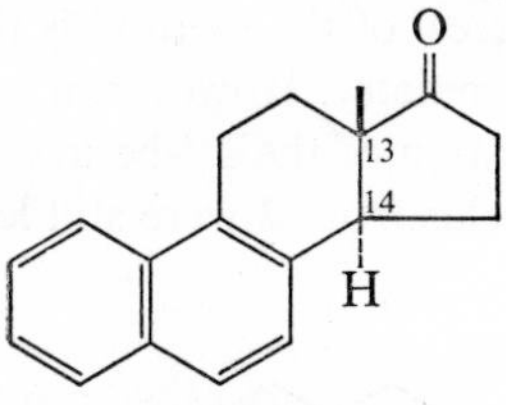

equilenin **1**

Equilenin **1**, having only two centers of asymmetry (C-13 and C-14), is the simplest representative of all natural steroids and thus was the first to be synthesized. Bachmann et al.[1] achieved the synthesis in 1939 of dl-**1** in twenty steps with a total yield of 2.7 % from **2** (Route I). The Reformatskii reaction was utilized to introduce C-15 and C-16, and the D ring was formed by a Dieckmann cyclization. The naturally occuring d isomer was obtained by resolution with 1-menthoxyacetic ester.

Route I

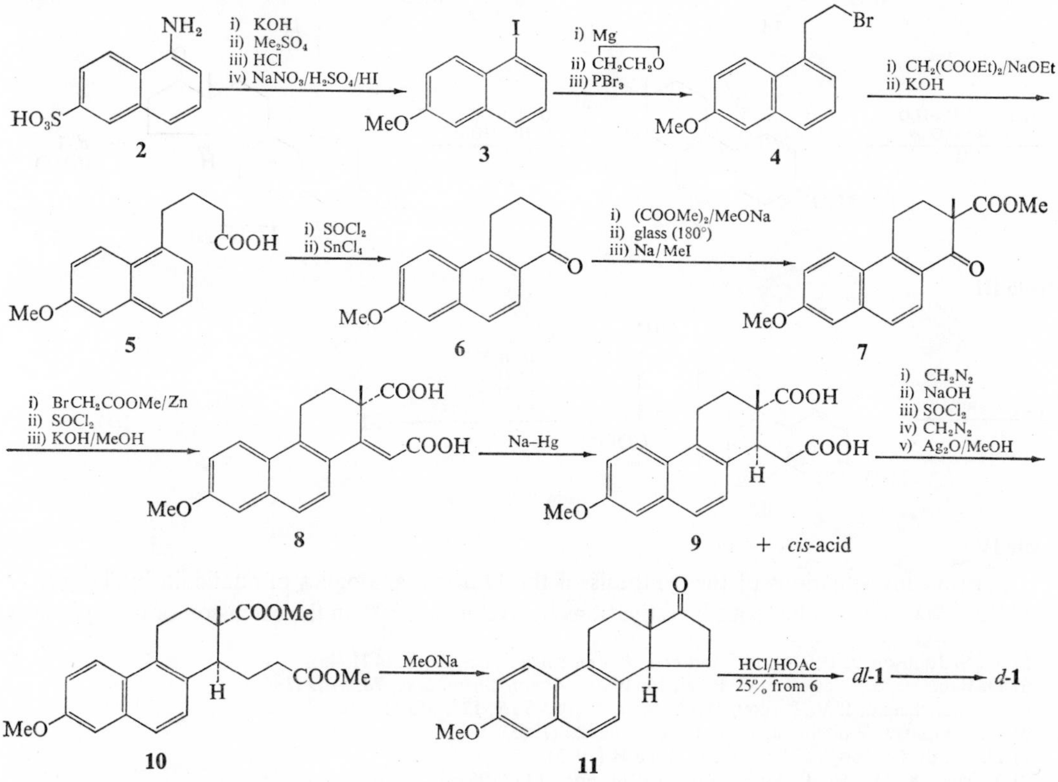

---

1) W. Bachmann, W. Cole, A. Wilds, *J. Am. Chem. Soc.*, **61**, 974 (1939).

Route II

Johnson et al.[2] in 1945 also utilized the tricyclic ketone **6** for the synthesis of dl-**1**, obtained in a yield of 30% from **6** (see below). This "isoxazole method" employs the Stobbe condensation of **14** with dimethylsuccinate to form the D ring directly. The main defect of the method is the absence of stereospecificity in the formation of the C-14 center of asymmetry. However, it has been found[3] that the reduction of the keto group in **15** prevents migration of the Δ^{14}-bond and thus a 34% yield of dl-**1** from **6** can be obtained (Route III). Total yields from **2** were still less than 4%.

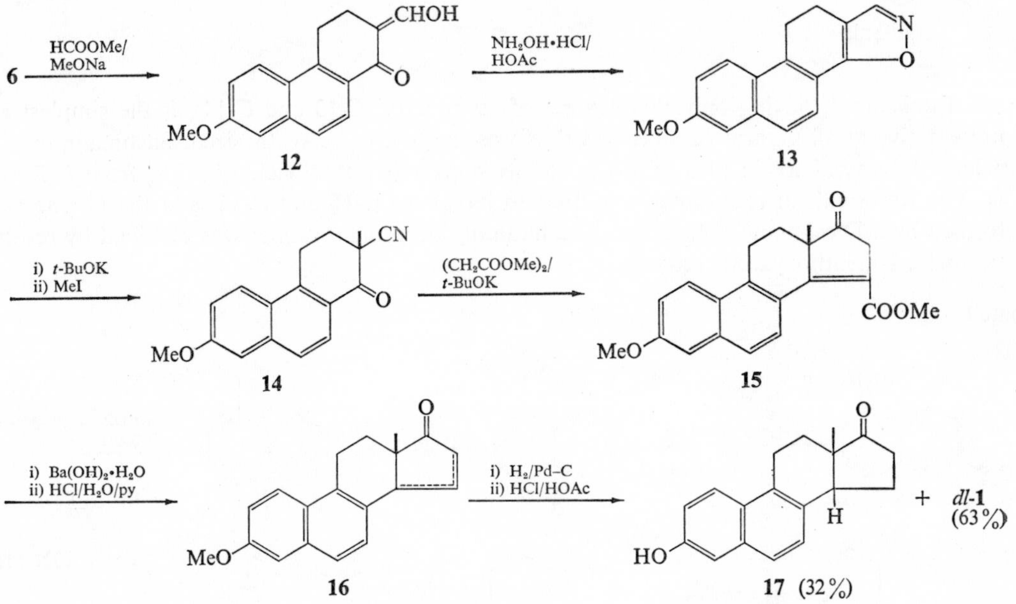

Route III

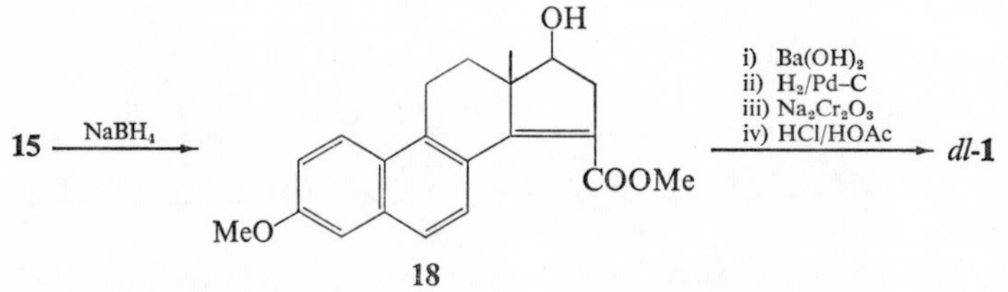

Route IV

From investigations of the synthesis of the D-homo analogues of equilenin by Torgov et al.[4] in 1960 and simultaneous investigations by five groups [5-9] on the extension of this synthesis

2) W. S. Johnson, J. Petersen, C. Lutsche, *J. Am. Chem. Soc.*, **67**, 2274 (1945).
3) D. Banerjee, S. Chatterjee, C. Pillai, M. Bhatt, *J. Am. Chem. Soc.*, **78**, 3769 (1956).
4) S. Ananchenko, I. V. Torgov, *Dokl. Akad. Nauk SSSR,* **127**, 553 (1959).
5) D. Criapin, J. Whitehurst, *Proc. Chem. Soc.*, 22 (1963).
6) H. Smith, C. Douglas, *Experientia,* **19**, 394 (1963).
7) T. Miki, K. Hiraya, T. Asako, *Proc. Chem. Soc.*, 139 (1963).
8) T. Wendholz, J. Fried, A. Patchett, *J. Org. Chem.*, **28**, 1092 (1963).
9) S. N. Anachenko, I. Torgov, *Tetr. Lett.*, 1553 (1963).

to the five-membered D ring, a synthesis of *dl*-1 in eight steps with an overall yield of 10.5% from **19** was evolved (below). The key step in this reaction scheme involved condensation of the vinyl carbinol **21** with 2-methyl-1,3-cyclopentanedione to form the ABD ring system, followed by closure of the C ring.

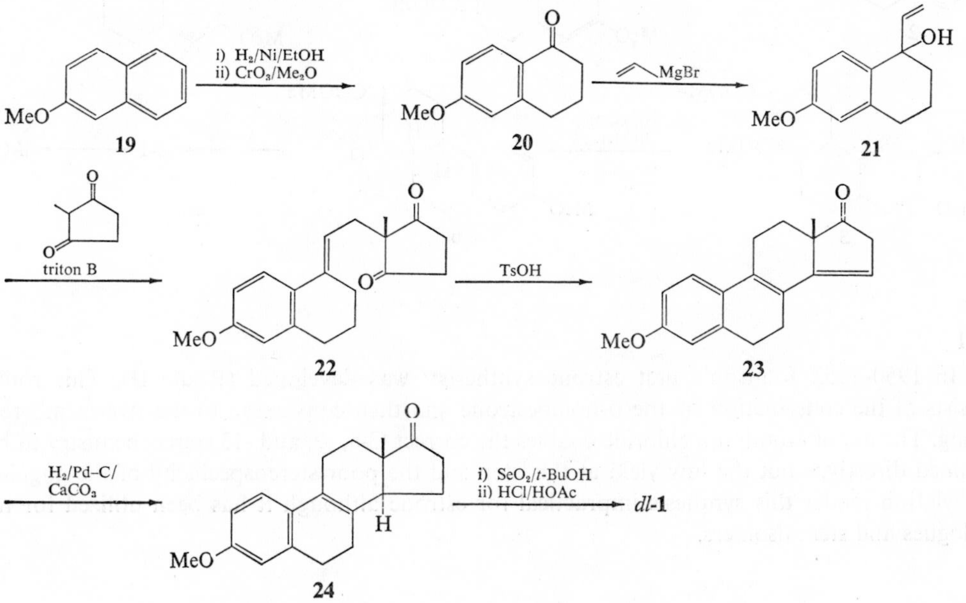

6.14   TOTAL SYNTHESIS OF ESTRONE

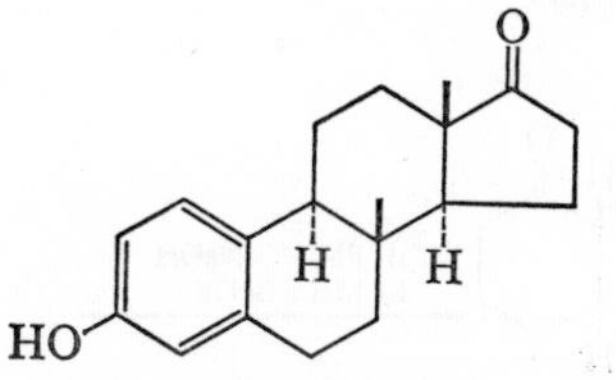

estrone **1**

Estrone **1**, was the second natural steroid to be synthesized, Anner and Mieschere[1] obtaining a 0.1% yield from **2** in eighteen steps in 1948.[2] As this molecule possesses four sites of asymmetry, the synthesis is accordingly more difficult than for equilenin. The original route (Route I) involved the tricyclic ketone **6** as a key intermediate which was separated with difficulty from its racemates. The formation of ring D essentially followed the procedure developed by Bachmann for equilenin[3] (see previous topic). *dl*-1 was resolved by means of the 1-menthoxyacetic ester to form the natural *d*-isomer.

1) G. Anner, H. Meischere, *Experientia*, **3**, 279 (1947).
2) A. L. Wilds, T. L. Johnson, *J. Am. Chem. Soc.*, **70**, 1166 (1948).
3) W. Bachmann, W. Cole, A. Wilde, *J. Am. Chem. Soc.*, **61**, 974 (1939).

Route I

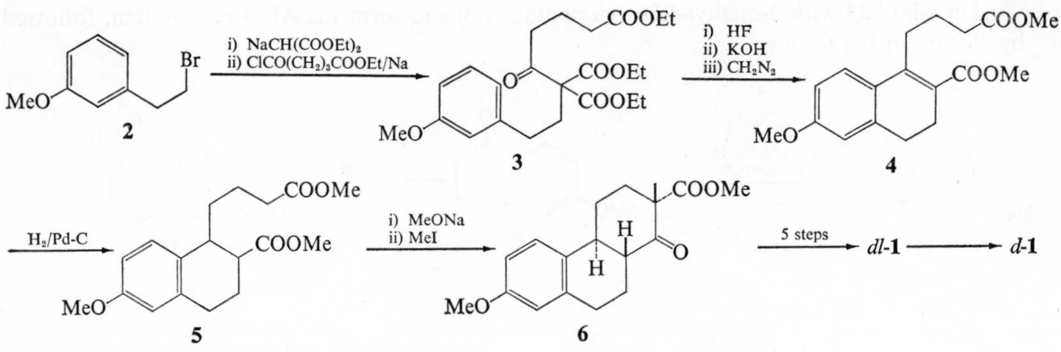

Route II

In 1950–1952 Johnson's first estrone synthesis[4] was developed (Route II). This route consists of the construction of the D-homoestrone and then conversion to the five-membered D ring. The use of aluminum chloride enables the correct C-8, -9, and -13 stereochemistry to be obtained directly,[5] but the low yield at this step and the poor stereospecificity of the angular methylation render this synthesis impractical for estrone although it has been utilized for its analogues and stereoisomers.

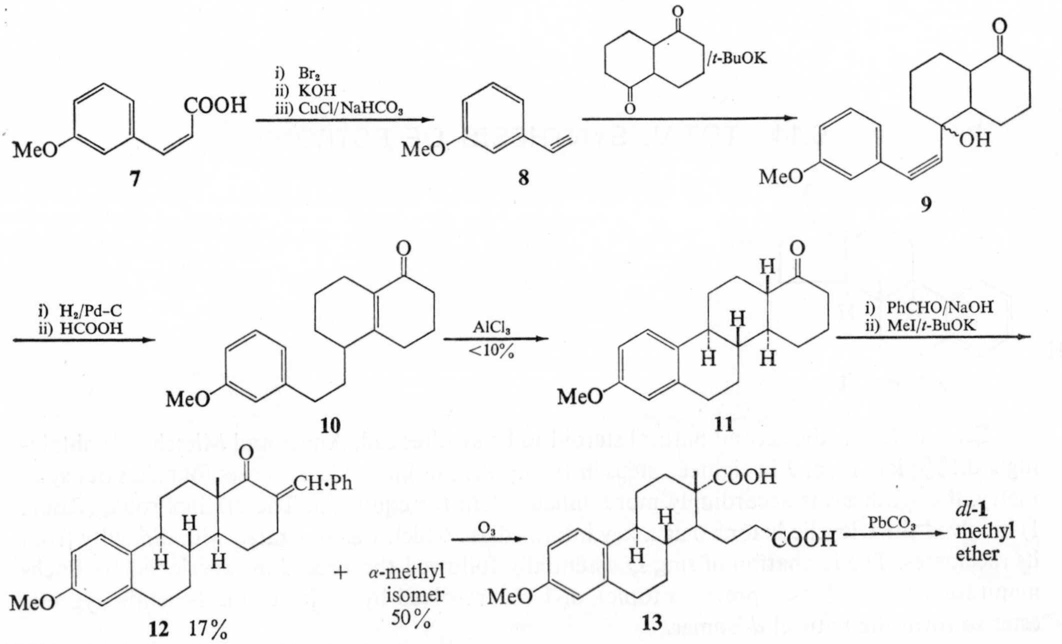

4) W. S. Johnson, B. Banerjee, W. Schneider, C. Gutsche, W. Shelberg, L. Chenn, *J. Am. Chem. Soc.*, **74**, 2832 (1952).

5) W. L. Meyer, D. D. Cameron, W. S. Johnson, *J. Org. Chem.*, **27**, 1130 (1962).

Route III

A more stereospecific route is Johnson's second synthesis[6] (1951) which was later perfected by Kipriyanov and Kutsenko[7] (1961) who obtained an overall yield of 4.8 % from **14**. The condensation of anisole **14** with glutaric anhydride and the subsequent Stobbe reaction followed by a Dieckmann cyclization gave the carbon skeleton for the ABC rings. The D ring was formed last in a manner analogous to that employed in the Bachmann synthesis.[3]

i) CO(CH₂)₃COO/AlCl₃
ii) MeOH/H₂SO₄

(CH₂COOMe)₂/t-BuOK

14   **15**   **16**

i) H₂/Ni/KOH
ii) MeOH/H₂SO₄
iii) Na
iv) CH₃I

i) BrZnCH₂COOMe
ii) HCOOH
iii) NaOH

17   **18**

i) SOCl₂
ii) AlCl₃

i) Na/NH₃
ii) NaBH₄
iii) py·HCl

dl-**1** methyl ether

19   **20**

Route IV

Johnson et al.[8] also investigated the Diels-Alder addition of **21** and **22** which yielded **23**, but the problem of the angular methyl group was still unresolved.

i) addition
ii) Zn/HOAc

21   **22**   **23**

Route V

A recent synthesis has been reported by Valenta and co-workers[9] by which dl-**1** is obtained in a 22% overall yield based on **24**. The key step involves the BF₃-catalyzed Diels-Alder reaction of **24** and **25**; this gave the cis-adduct **26** with the methyl at the desired C-13 rather than at C-14 as obtained by normal thermal reactions. Compound **26** was readily isomerized to the trans-epimer. The six-membered enedione D ring was then converted to the five-membered estrone D ring.

6) W. S. Johnson, R. G. Christiansen, *J. Am. Chem. Soc.*, **73**, 5511 (1951); W. S. Johnson, R. G. Christiansen, R. E. Ireland, *ibid.*, **79**, 1995 (1957).
7) G. I. Kipriyanov, L. M. Kutsenko, *Med. Prom. SSSR*, **2**, 43 (1951).
8) J. E. Cole, W. S. Johnson, P. A. Robins, J. Walker, *Proc. Chem. Soc.*, 114 (1958).
9) R. A. Dickinson, R. Kubela, G. A. MacAlpine, Z. Stojanac, Z. Valenta, *Can. J. Chem.*, **50**, 2377 (1972).

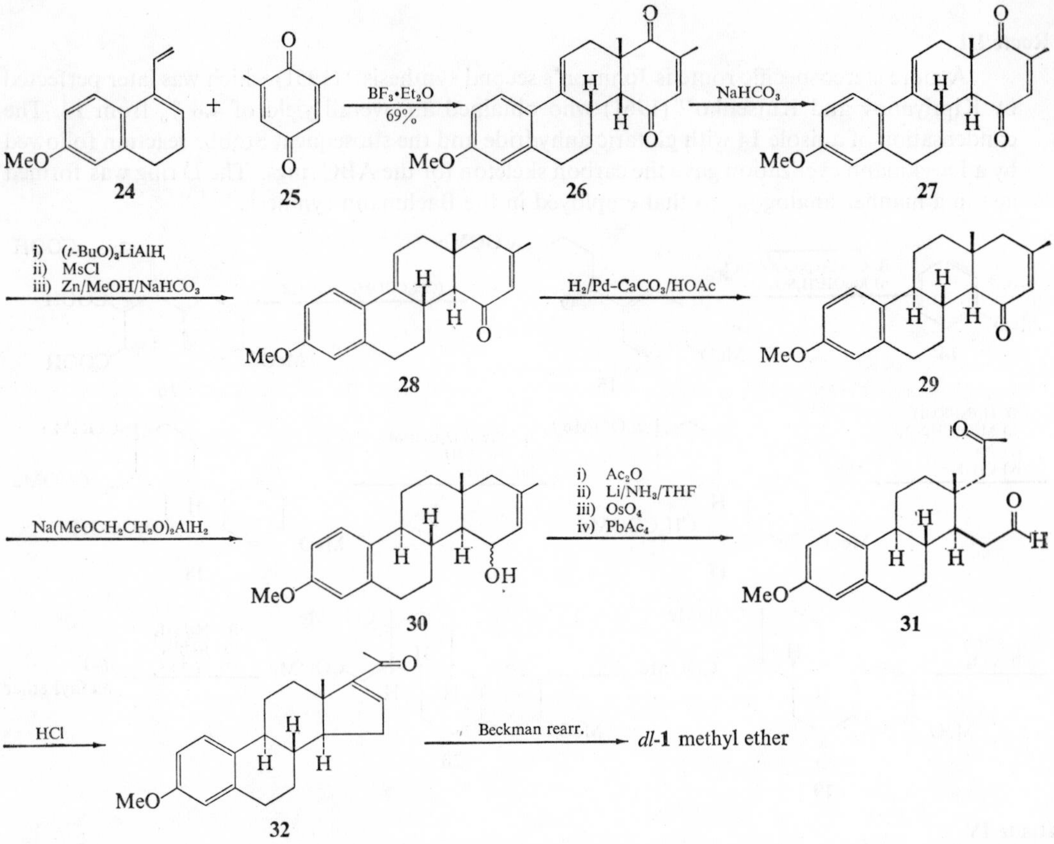

Route VI

The best synthesis to date for brevity and high yield is that developed by Torgov and other groups[10] in which a 27% yield of *dl*-1 is obtained in eight steps based on 2-methoxy napthalene 33. The compound 34, obtained as an intermediate in the synthesis of equilenin, is converted by reduction of the Δ^8-bond to *dl*-1.

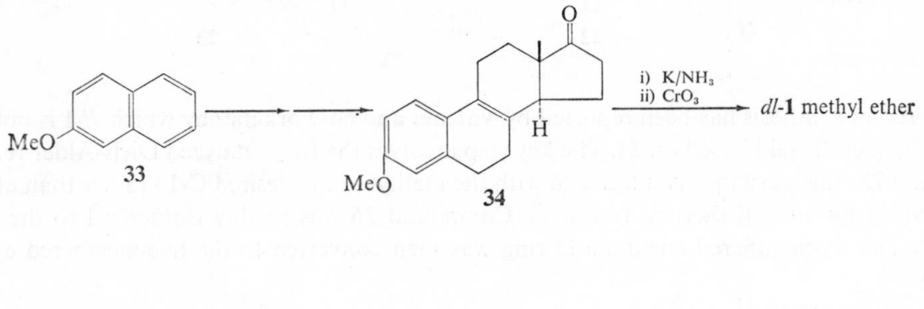

---

10) See "synthesis of equilenin", refs 4–9.

6.15 TOTAL SYNTHESIS OF NON-AROMATIC STEROIDS

The first total synthesis of a non-aromatic steroid was not reported until 1951. In that year both Robinson and Woodward reported syntheses of a non-aromatic steroid. Within the next decade the total syntheses of all the major non-aromatic steroids were achieved.[1]

Route I

For their syntheses of epiandrosterone **15**, Robinson, Cornforth and co-workers[2] utilized the tricyclic ketone **3** which was prepared by "Robinson's annelation." After reduction of the aromatic ring and introduction of the 13-Me group, ring D was built up by methods similar to those used by earlier workers for the equilenin syntheses. This route suffered from the disadvantages of most of the early syntheses: lack of stereospecificity in many steps, the use of protection groups which then had to be removed, and low yields. The overall yield for **15** is only about 0.00007% over forty steps based on **1**. However this synthesis has great historical value as it, with Woodward's work, first demonstrated the possibility of synthetic preparation of non-aromatic steroids.

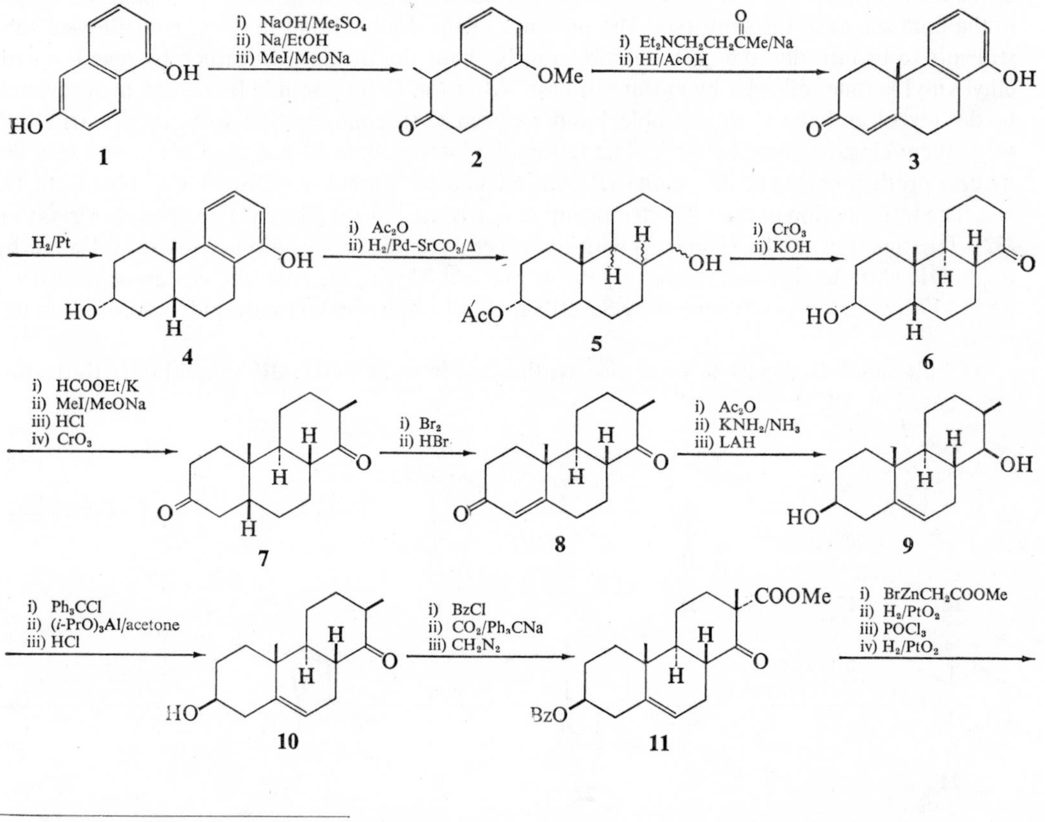

1) For a more extensive review of total syntheses, see A. A. Akhrem, Y. A. Titov, *Total Steroid Synthesis*, Plenum Press, 1970.
2) J. W. Cornforth, R. Robinson, *J. Chem. Soc.*, 676 (1946); H. M. E. Cardwell, J. W. Cornforth, S. R. Duff, H. Holtermann, R. Robinson, *Chem. Ind.*, 389 (1951).

i) K$_2$CO$_3$
ii) Ac$_2$O

12

i) (COCl)$_2$
ii) CH$_2$N$_2$
iii) Ag$_2$O
iv) KOH/MeOH

13

i) Ac$_2$O
ii) Δ

14

15

Route II

The total synthesis of cortisone **38** and cholestanol **41** by Woodward et al.[3] was also reported in 1951. The bicyclic compound **18** was converted into "Woodward's ketone" **21** which formed the C and D rings of the product. The decalin system was used to direct formation of the *trans* C/D junction and the D-homo ring was contracted to give the cyclopentane system in the final stages of the synthesis. The presence of the double bond in the C ring allowed substituents to be introduced at C-8 unambiguously. Thus the B ring was formed by reaction with ethyl vinyl ketone followed by closure in base. After the D-ring double bond was hydroxylated to the glycol and the C-ring double bond reduced (**24**), condensation with acrylonitrile and subsequent closure formed ring A. The tetracyclic intermediate **30** was then converted into the natural product cortisone **38** (Route III)[4] in a 0.000005% yield over forty-nine steps from **18**.

The introduction of the C-11 oxo group in cortisone **38** took place via an epoxide formation (**32**). Treatment of the hemiketal **33** with hydrogen bromide (12-bromo derivative) followed by Zn/AcOH gave the thermodynamically stable product **34** with the correct C-9 stereochemistry.[4] The C-17 hydroxy was introduced by Sarett's method using potassium cyanide and osmium tetroxide.[5]

Cholestanol **41** (Route IV) was also synthesized from **30** in 0.00019% yield over thirty-two steps based on **18**.

16   +   **17**   →   **18**

i) NaOH
ii) LAH

19

H$_2$SO$_4$

20

i) Ac$_2$O
ii) Zn/Ac$_2$O

21

i) HCOOEt/MeONa
ii) COEt/t-BuOK

22

KOH

23

i) OsO$_4$
ii) Me$_2$CO/CuSO$_4$
iii) H$_2$/Pd–SrCO$_3$

3) R. B. Woodward, F. Sondheimer, D. Taub, K. Heusler, W. M. MacLamore, *J. Am. Chem. Soc.*, **74**, 2403, 4057, 4223 (1952).
4) H. Heymann, L. F. Fieser, *J. Am. Chem. Soc.*, **73**, 4054, 5252 (1951); *ibid.*, **74**, 5938 (1952).
5) L. H. Sarett, *J. Am. Chem. Soc.*, **70**, 1454 (1948); *ibid.*, **71**, 2443 (1949).

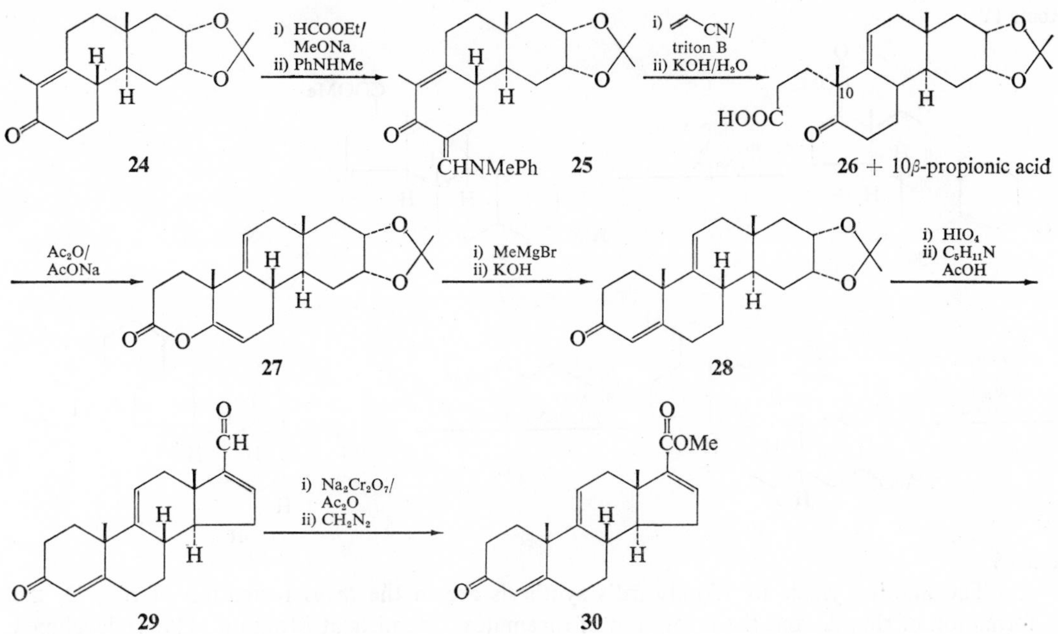

i) HCOOEt/ MeONa
ii) PhNHMe

24 → CHNMePh **25**

i) ⟍CN/ triton B
ii) KOH/H₂O

26 + 10β-propionic acid

Ac₂O/ AcONa

27

i) MeMgBr
ii) KOH

28

i) HIO₄
ii) C₅H₁₁N AcOH

29

i) Na₂Cr₂O₇/ Ac₂O
ii) CH₂N₂

30

Route III

30

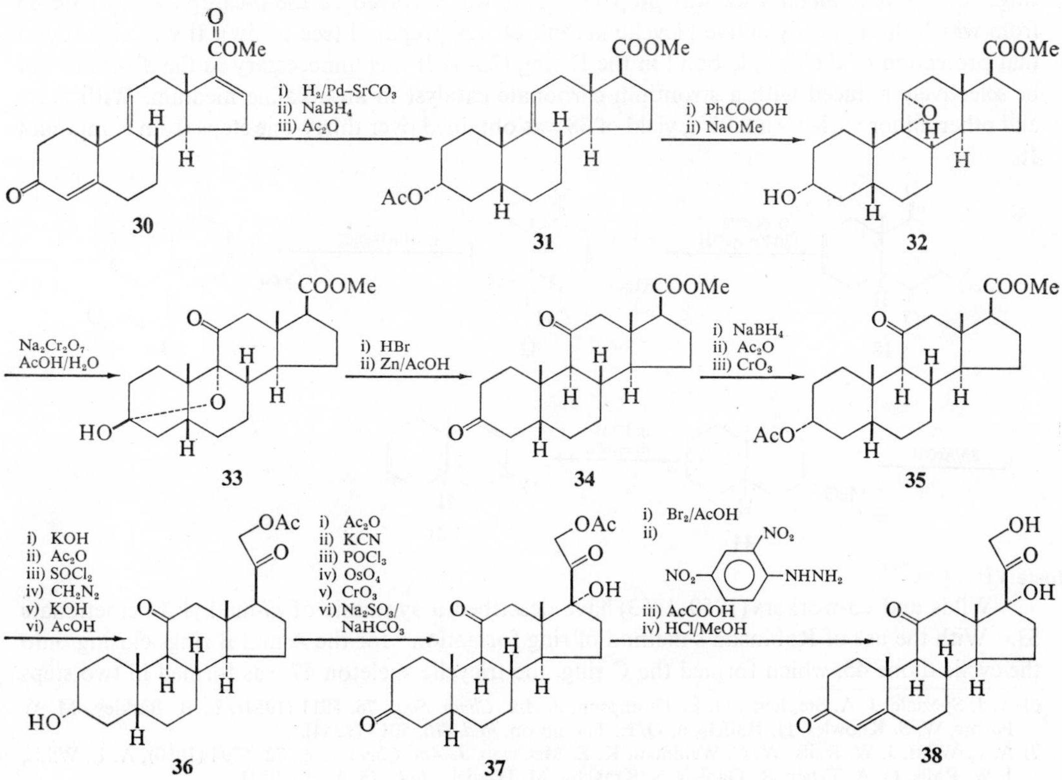

i) H₂/Pd–SrCO₃
ii) NaBH₄
iii) Ac₂O

31

i) PhCOOOH
ii) NaOMe

32

Na₂Cr₂O₇ AcOH/H₂O

33

i) HBr
ii) Zn/AcOH

34

i) NaBH₄
ii) Ac₂O
iii) CrO₃

35

i) KOH
ii) Ac₂O
iii) SOCl₂
iv) CH₂N₂
v) KOH
vi) AcOH

36

i) Ac₂O
ii) KCN
iii) POCl₃
iv) OsO₄
v) CrO₃
vi) Na₂SO₃/ NaHCO₃

37

i) Br₂/AcOH
ii) NO₂—⟨⟩—NHNH₂ with NO₂
iii) AcCOOH
iv) HCl/MeOH

38

Route IV

i) H₂/Pt/AcOH
ii) CrO₃
iii) NaBH₄

30 → **39**

i) KOH
ii) Ac₂O
iii) SOCl₂
iv) Me₂Cd

40

i) [image] MgBr
ii) Ac₂O
iii) H₂/Pt/AcOH

41

Route V

The limiting yields in Woodward's synthesis are in the *trans* formation of ring D, the formation of ring A, and the resolution of racemates. Chemists at Monsanto (1954) developed numerous variants of Woodward's cortisone synthesis which have enabled the total yield to be improved and the number of steps to be reduced.[6] In order to introduce resolution at an early stage, the bicyclic alcohol **42** was prepared. This was resolved as the *d*-camphorsulfonate **43** from which the optically active bicyclic ketone **44** was prepared (see below). It was also found that protection of the double bond in the D ring (**23**→**24**) was unnecessary as the Δ^{11}-bond can be selectively reduced with a strontium carbonate catalyst in an alkaline medium. With these and other minor variants a 0.07% yield of **38** was obtained over thirty-one steps from compound **18**.

18 → (i) NaOH, ii) Zn/AcOH) → **42** → (C₁₀H₁₅OSO₂Cl) → **43**

43 → (Zn/AcOH) → **44** → (i) LAH, ii) H₂SO₄) → **l-21**

Route VI

Wilds and co-workers (1950–1953) have described a synthesis of *dl*-methyl 3-ketoetionate **53**.[7] With the use of Robinson's method of ring formation[2] for the A and B rings closing onto the cyclic dione **45**, which formed the C ring, the tricyclic skeleton **47** was formed in two steps

6) A. J. Speziale, J. A. Stephens, Q. E. Thompson, *J. Am. Chem. Soc.*, **76**, 5011 (1954); L. B. Barkley, M. W Farrar, W. S. Knowles, H. Raffelson, Q. E. Thompson, *ibid.*, **76**, 5014 (1954).
7) A. L. Wilds, J. W. Ralls, W. C. Weldman, K. E. McCaleb, *J. Am. Chem. Soc.*, **72**, 5794 (1950); A. L. Wilds, J. W. Ralls, D. A. Tyner, R. Daniels, S. Kraychy, M. Harnek, *ibid.*, **75**, 4878 (1953).

(9.3% from **45**). Formation of the D ring mainly utilized methods developed for the aromatic steroids.

45 → NEt₂·MeI/MeONa → **46** → /MeONa → **47**

[reaction scheme: 45 (C) → 46 → 47]

H₂/Pd/KOH → **48** → i) (CH₂OH)₂/p-TsOH ii) HCOOEt/EtONa → **49** (CHOH)

i) MeI/MeONa ii) BrCH₂COOMe/Ph₃CNa → **50** (Me CH₂COOH) → i) (COCl)₂ ii) NaCH(COOt-Bu)₂ iii) HCl iv) MeONa → **51**

i) (CH₂OH)₂/p-TsOH ii) Me₂CO₃/Ph₃CNa iii) HCl/MeOH → **52** (COOMe) → i) H₂/Pt/AcOH ii) CrO₃ → **53** (COOMe)

Route VII

The total synthesis of cortisone **38** reported by Sarrett et al.[8] in 1952 is noteworthy due to the high degree of stereospecificity in each step and the obtaining of d-cortisone as well as the dl form. The Diels-Alder condensation of **54** with **55** gave the cis adduct **56**. Epimerization at C-8 to the anti-trans isomer, which has the correct stereochemistry for cortisone **38**, occurs in the selective Oppenauer oxidation of **58** to form **59**. An important feature of this route is the retention of the protective ketal of the C-3 oxo group. A novel introduction of the C-17, C-20 and C-21 fragment involved alkylation of **60** at C-13 with methallyl iodide which produced the correct stereochemistry. The C-17 hydroxyl group was introduced by an earlier method utilized in the Woodward synthesis.[3] An overall yield of dl-**38** of 1.3% in twenty-two steps based on **59** was achieved.

54 (EtO) + **55** → i) addition ii) H₂/Ni → **56** (EtO) → i) LAH ii) AcOH → **57** (HO, OH) → /triton B

8) L. H. Sarett, G. E. Arth, R. M. Lukes, R. E. Beyler, G. I. Poos, W. F. Johns, J. M. Constantin, J. Am. Chem Soc., **74**, 4974 (1952); L. H. Sarett, J. M. Vandergrift, G. E. Arth, ibid., **74**, 1393 (1952); L. H. Sarett, W. F. Johns, R. E. Beyler, R. M. Lukes, G. I. Poos, G. E. Arth, ibid., **75**, 2112 (1953).

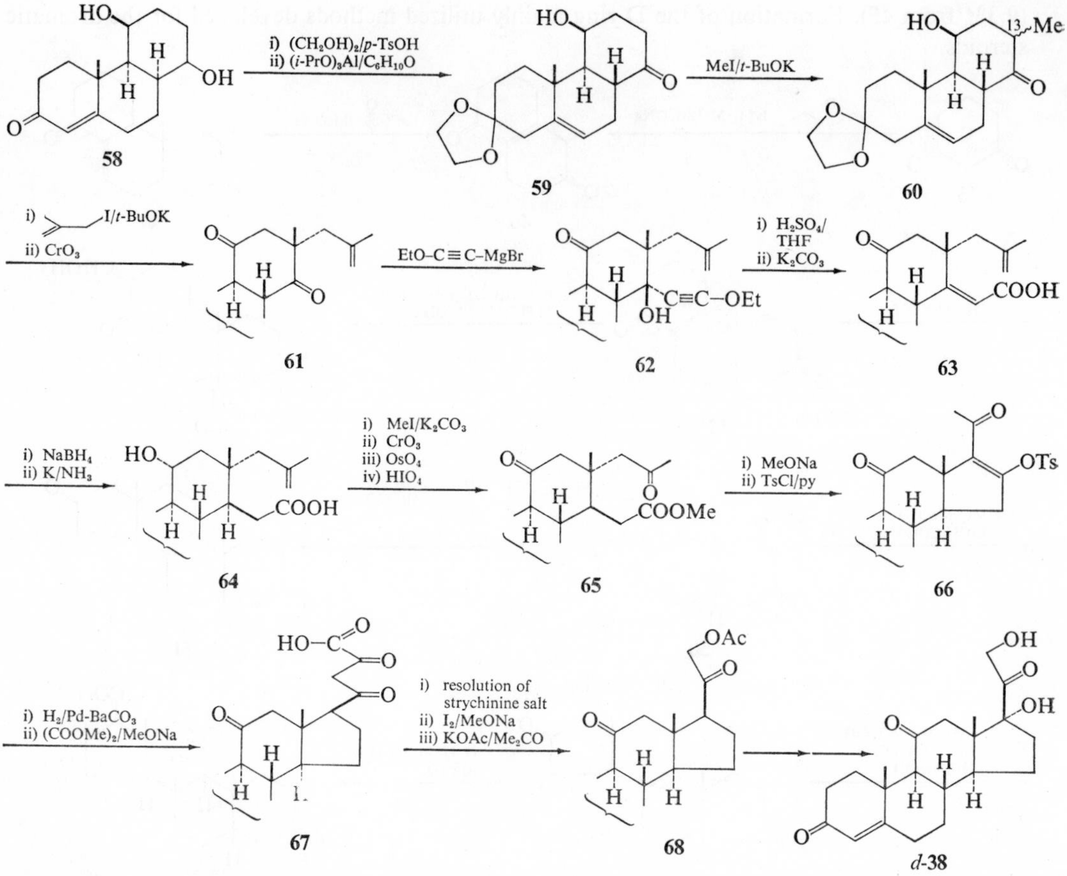

Route VIII

"Sarett's ketone" **59** was utilized for the synthesis of aldosterone **79** by three groups of workers.[9-11] A common feature of all these syntheses is the use of the C-11 center of asymmetry present in **59** to form the C-13 center of asymmetry. The most interesting of these syntheses is the route by Wettstein *et al.* (1957) employing the "geminal principle"[12] which consists of the introduction of two identical substituents at C-13 of **59**. The β group will be bound to the 11β-hydroxy group and the α group is utilized to form the D ring. In this manner complete stereospecificity in the formation of the C-13 center is assured. A yield of 3% in twenty steps from **59** was obtained.

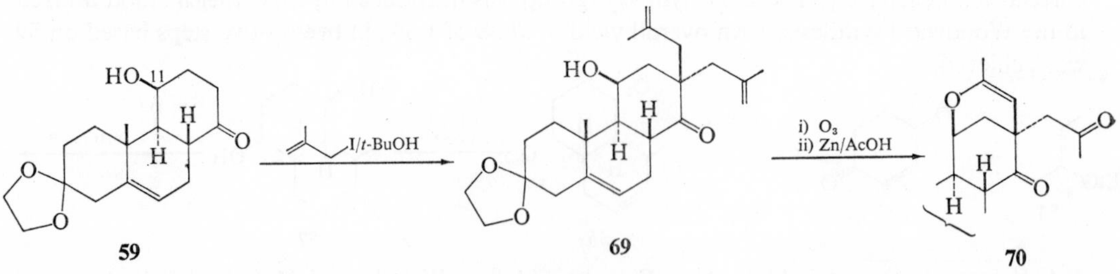

9) J. Schmidlin, G. Anner, J. R. Billeter, K. Heusler, H. Ueberwasser, P. Wieland, A. Wettstein, *Helv. Chim. Acta*, **40**, 1034 (1957).

Route IX

A number of steroids were synthesized by Johnson and colleagues (1956–1958) from 5-methoxy-2-tetralone **80** which formed the CD rings of the natural products. These syntheses were carried out in three main stages: Firstly, preparation of a highly unsaturated tetracyclic intermediate containing only one asymmetric center (compound **82**).[13] Secondly, stereodirected reduction of the unsaturated bonds with the formation of new centers of asymmetry (compound **85**), and finally appropriate structural modifications to the steroid skeleton to form the natural product (i.e., **91**).

Testosterone **90** (see below) was thus obtained as the *d*-enantiomer in an overall yield of 1.1% in sixteen steps based on **82**.[14]

Cholesterol,[15] conessine[16] and progesterone[17] have also been synthesized from **85** (1962–1964). Aldosterone **79** has been synthesized from compound **83** by a lengthy procedure[18] (Route X). A hydroxy group was introduced β to the aromatic ring at C-11 via a double bond (**93**). Alkylation with methacrylonitrile at C-13 followed by ozonolysis formed the lactone ring. The D ring was then closed and correctly functionalized to form aldosterone **79**.

---

10) W. J. van der Burg, D. A. van Dorp, O. Schindler, C. M. Segmann, S. A. Szpilfogel, *Rec. Trav. Chim.*, **77**, 157, 171 (1958).
11) A. Lardon, O. Schindler, T. Reichstein, *Helv. Chim. Acta*, **40**, 666 (1957).
12) K. Heusler, P. Wieland, H. Ueberwasser, A. Weltstein, *Chimica*, **12**, 121 (1958).
13) W. S. Johnson, J. Szmuszkowicz, E. R. Rogier, H. J. Hadler, H. Wynberg, *J. Am. Chem. Soc.*, **78**, 6235 (1956).
14) W. S. Johnson, A. D. Kemp, R. Pappo, J. Ackerman, W. F. Johns, *J. Am. Chem. Soc.*, **78**, 6312 (1956).
15) J. W. Keana, W. S. Johnson, *Steroids*, **4**, 457 (1964).
16) J. A. Marshall, W. S. Johnson, *J. Am. Chem. Soc.*, **84**, 1485 (1962).
17) W. S. Johnson, J. F. W. Keana, J. A. Marshall, *Tetr. Lett.*, 193 (1963).
18) W. S. Johnson, J. C. Collins, R. Pappo, M. B. Rubin, *J. Am. Chem. Soc.*, **80**, 2585 (1958)

Route X

i) O_3/H_2O_2
ii) NaOH

95 → **79**

Route XI

Nagata and co-workers[19] utilized 6-methoxy-2-tetralone **96** as the C/D fragment in a synthetic route very similar to that of Johnson (Route IX). This bicyclic compound has the advantage that the tetracyclic compound **97** thus formed will undergo Birch reduction more readily and unambiguously than the corresponding compound with the C-17 methoxy group in the Johnson synthesis. By hydrocyanation in the presence of triethylammonium salts, the dione **99** was selectively converted into the C/D-*trans*-13β-cyano derivative (**100**). Opening and reclosure of the D ring led to the *dl*-5β-$\Delta^{9(11),16}$-pregnadien-3α-ol-20-one **101** which can be used as an intermediate in the synthesis of cortico-steroids. However, the overall yield was low, ~0.03% from **96**.

96 → **97** → **98**

i) Br_2
ii) LiBr

i) $HC(OEt)_3$
ii) AcOH
iii) $HCN/AlEt_3/THF$

99 → **100** → **101**

Route XII

Velluz and co-workers (1960) carried out several syntheses of steroids from the tricyclic compound **105**.[20] This ketone is prepared from 6-methoxy-1-tetralone **96** and may be obtained in its optically active form by resolution of the acid **104** by both chemical[21] and microbiological methods.[22] The aromatic ring B is then reduced (**108**) and ring A constructed by use of 1,3-dichloro-2-butene; the compound **109** obtained has the desired 8β-configuration. The 9,10 position of the double bond, obtained after hydrolysis, was found to be extremely useful as it can be reduced, alkylated, oxidized or rearranged.

These workers' synthesis of cortisone **38**[23] is particularly outstanding for the improvement in yield and reduction of the number of steps from earlier routes (1% yield obtained over twenty-seven steps from **96**). The stereospecific methylation of **110** to give **111** was particularly

19) W. Nagata, T. Terasawa, *Chem. Pharm. Bull.*, **9**, 267 (1961); W. Nagata, *Tetr.*, **13**, 268, 278, 287 (1961).
20) L. Velluz, G. Mathieu, G. Monine, *Tetr. Suppl.*, **8**, 495 (1966).
21) Y. Kurosawa, H. Shimojima, Y. Osawa, *Steroids, Suppl. I*, 185 (1965).
22) L. Velluz, A. Amiard, R. Heymes, *Bull. Soc. Chim. Fr.*, 1015 (1954).
23) L. Velluz, G. Nomine, J. Mathieu, *Angew. Chem.*, **72**, 725 (1960); L. Velluz, G. Nomine, J. Mathieu, E. Toromanoff, D. Bertin, R. Bucourt, J. Tessier, *Compt. Rend.*, **250**, 1293 (1960).

advantageous. The 11-oxo group **113** was obtained from **112** via the $9\alpha,11\beta$-bromohydrin with subsequent oxidation of the 11-hydroxy group and debromination. The side chain is built up by specific addition of sodium acetylide to the C-17 oxo group.

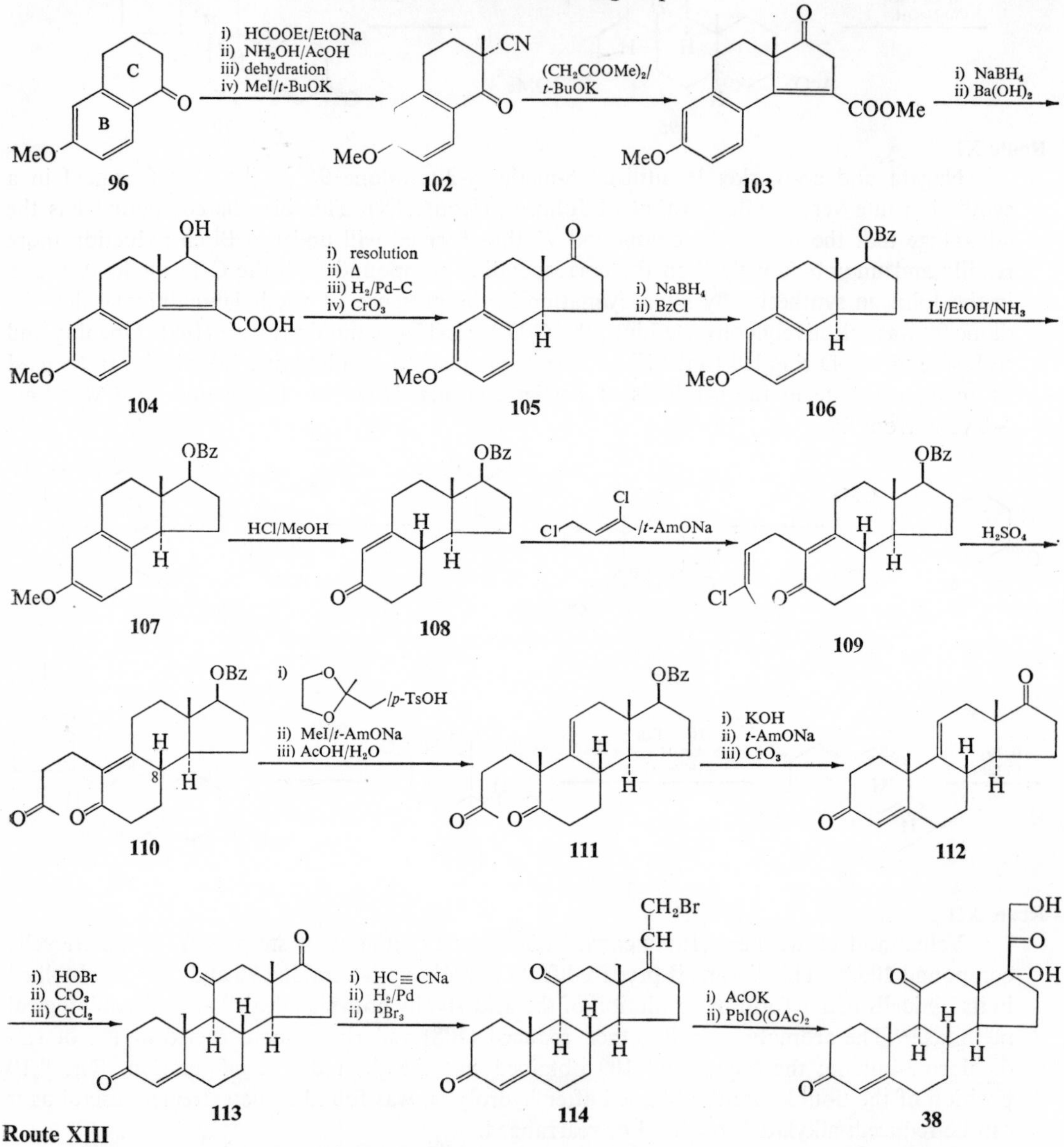

Route XIII

The synthesis of *dl*-progesterone **127** developed by Stork and co-workers[24] utilized both a new annelation reaction involving an isoxazole derivative for the construction of the A and B rings and a stereospecific introduction of the C-18 methyl. The enolate of 10-methyl $\Delta^{1,9}$-octalin-1,5-dione **119** was alkylated with the isoxazole **118** to form **120**. After reduction of the 17-oxo group and $\Delta^{8,14}$ bond, the isoxazole ring was hydrogenolyzed and then refluxed first with methanolic sodium methoxide and then with aqueous sodium hydroxide to produce the tri-

24) G. Stork, S. Danishefsky, M. Ohashi, *J. Am. Chem. Soc.*, **89**, 5459 (1967); G. Stork, J. E. McMurry, *ibid.*, **89**, 5463, 5464 (1967).

cyclic enone **122** in a 60% yield from **120**. The stereospecific methylation at C-10 was accomplished by an alkylation trapping method using lithium/ammonia, to form the β isomer **123** in almost quantitative yield. Deketalization and closure of ring A gave *dl*-D-homotestosterone **124** in 26% yield from **119**. The homo D ring was then converted into a five-membered ring with the correct C-17 side chain for *dl*-progesterone **127**.

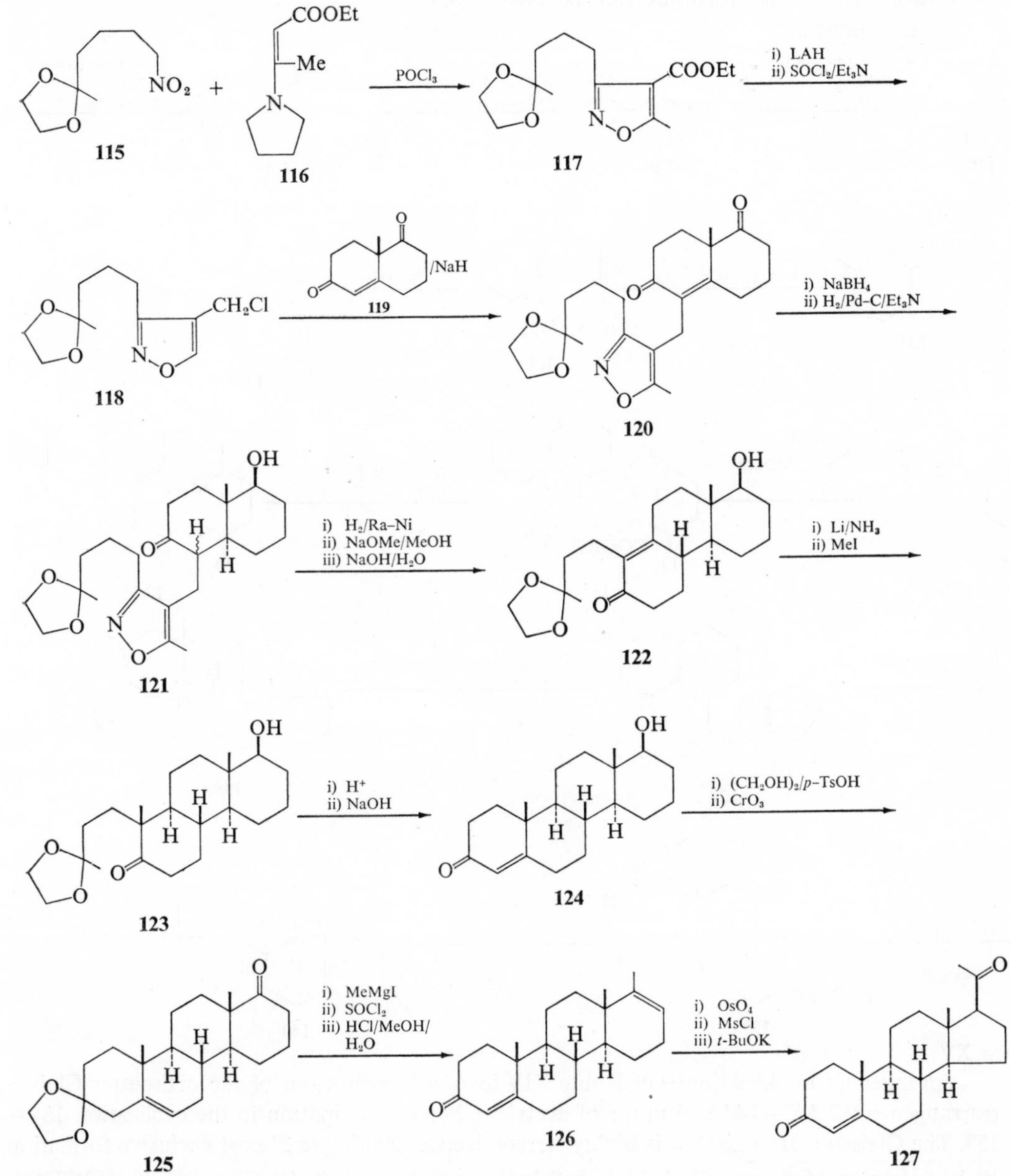

Route XIV

This scheme by Johnson *et al.* (1968) demonstrated a new approach to steroid synthesis employing a nonenzymatic biogenetic-like olefinic cyclization which generated five asymmetric centers stereospecifically. The key step is the treatment of the carbinol **137** with excess trifluoroacetic acid at $-78°$ followed by reduction with LAH to give the tetracyclic diene **138** in 30% yield from **137**. Since the carbinol **137** is very susceptible to dehydration, the authors suggest the stereoselective cyclization proceeds via a pentaene which has no asymmetric center. Both the A and D rings were then functionalized by treatment with osmium tetroxide and lead tetra-acetate. Reclosure of both rings is effected simultaneously by treatment with potassium hydroxide to form the *dl*-16-dehydropregesterone **140**.[25]

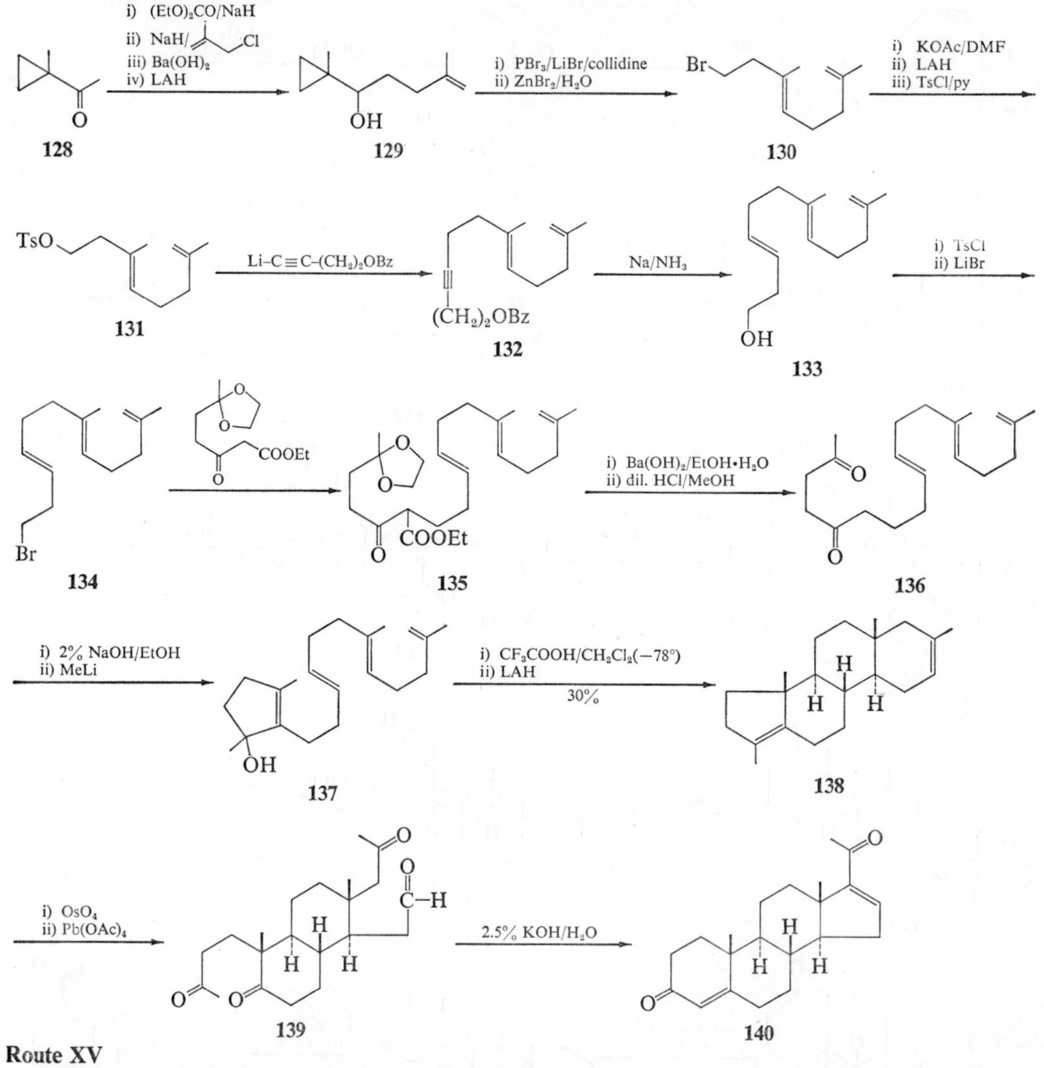

Route XV

An ingenous modification[26] of Route XIV involved application of the orthoester Claisen rearrangement[27] **143→144** and usage of acetylene bond participation in the cyclization **154→155**. The Claisen rearrangement is highly stereoselective, leading to almost exclusive formation

25) W. S. Johnson, M. F. Semmelhack, M. U. S. Sultanbawa, L. A. Dolak, *J. Am. Chem. Soc.*, 90, 2994 (1968); S. F. Brady, M. A. Ilton, W. S. Johnson, *ibid.*, 90, 2882 (1968).

26) W. S. Johnson, M. B. Gravestock, B. E. McCarry, *J. Am. Chem. Soc.*, 93, 4332 (1971); W. S. Johnson, M. B. Gravestock, R. J. Parry, R. F. Myers, T. A. Bryson, D. H. Miles, *ibid.*, 93, 4330 (1971).

27) W. S. Johnson, L. Werthemann, W. R. Bartlett, T. J. Brockson, T. -t Li, D. J. Faulkner, M. R. Peterson, *J. Am. Chem. Soc.*, 92, 741 (1970).

of *trans*-enyne aldehyde **145**; the stereoselectivity is attributed to nonbonding interaction between the OEt and R groups (**148**) that forces R to adopt an equatorial-like conformation, and hence ensures *trans* backbone arrangement, as in **144**.

A Wittig reaction between aldehyde **144** and ylide **151** gave the diketal dienyne **152**, which was cyclized first to **154** and then to **155** (5:1 mixture of 17β and 17α ketones). Crude **155** (85:15 mixture of 17β and 17α) was recrystallized to give pure *dl*-progesterone **155**.

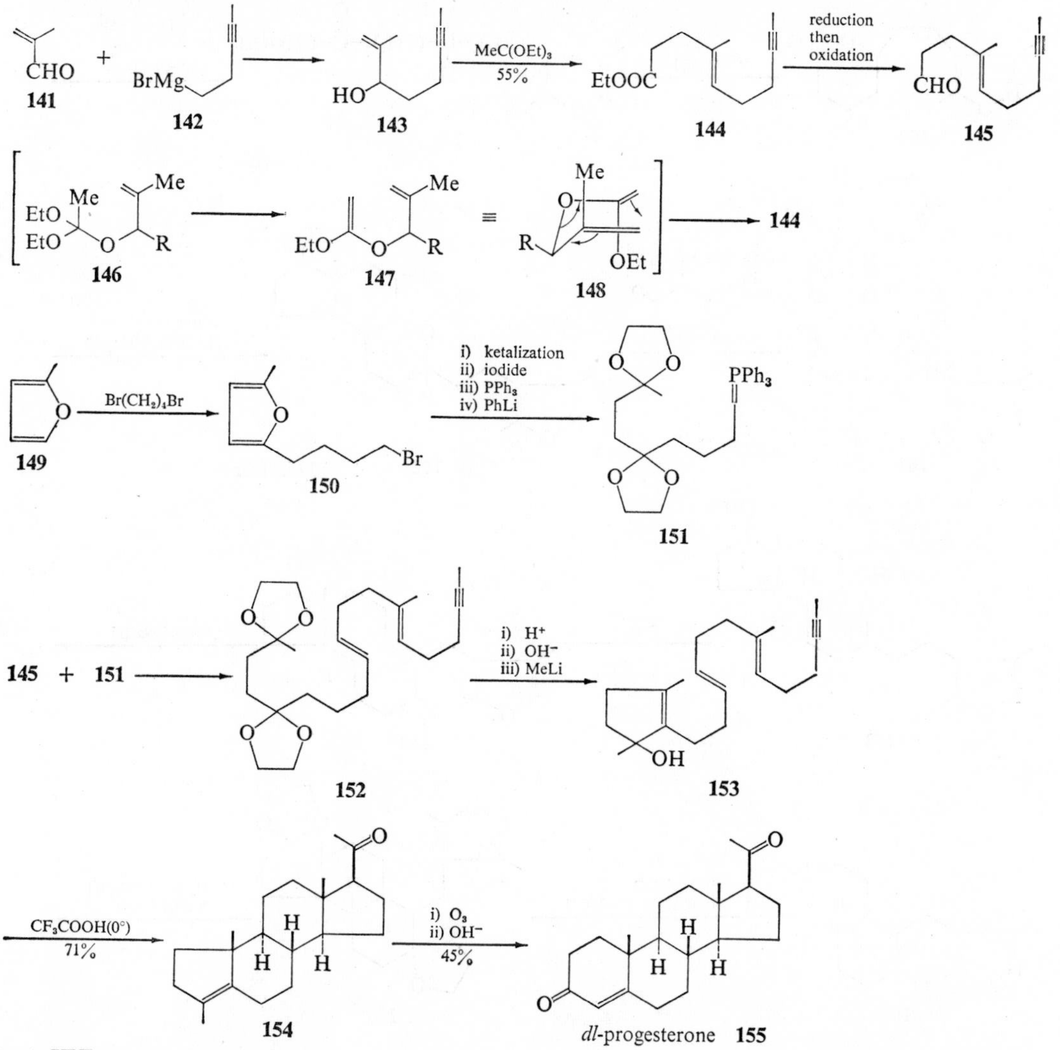

Route XVI

The 19-norsteroid **169** was synthesized by Saucy and co-workers (1972) in an elegant manner involving the use of the isoxazole moiety.[28] The complex isoxazole derivative **161** contained the carbon skeleton for the A, B and C rings, and the D ring was incorporated via the methylcyclopentanedione fragment. The overall yield of this route was high, **169** being obtained in a 10.2% yield from **162**. An additional advantage of this synthesis is the early resolution of the optical isomers at the Mannich base stage (**161**) before condensation with the cyclopentanedione.

28) J. W. Scott, G. Saucy, *J. Org. Chem.*, **37**, 1652 (1972); J. W. Scott, R. Borer, G. Saucy, *ibid.*, **37**, 1659 (1972).

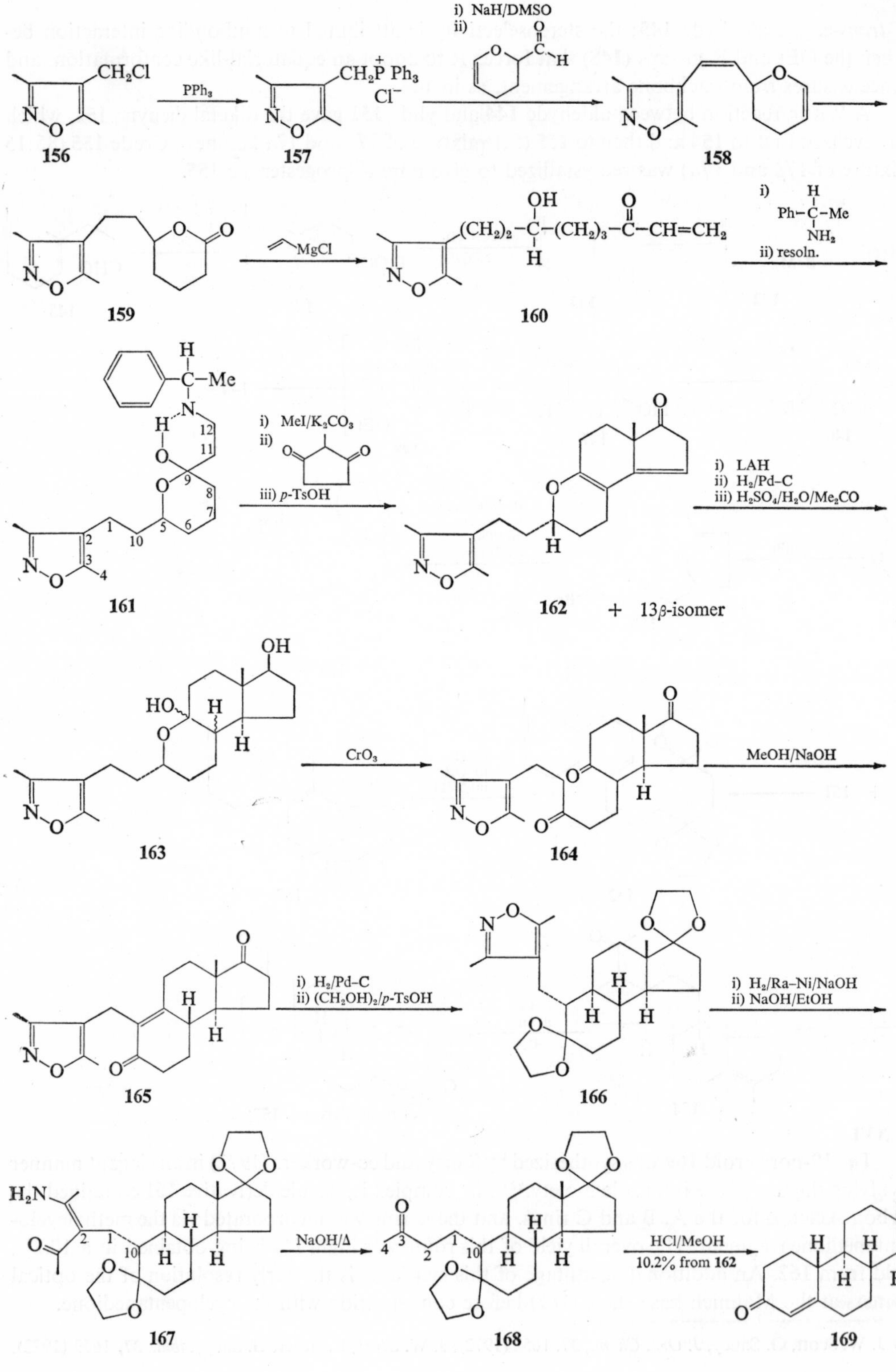

i) NaH/DMSO
ii)

156 → PPh₃ → **157** → **158**

159 → MgCl → **160** → i) Ph–C(H)(Me)NH₂ ii) resoln.

161 i) MeI/K₂CO₃ ii) iii) p-TsOH → **162** + 13β-isomer → i) LAH ii) H₂/Pd–C iii) H₂SO₄/H₂O/Me₂CO

163 → CrO₃ → **164** → MeOH/NaOH

165 → i) H₂/Pd–C ii) (CH₂OH)₂/p-TsOH → **166** → i) H₂/Ra–Ni/NaOH ii) NaOH/EtOH

167 → NaOH/Δ → **168** → HCl/MeOH 10.2% from **162** → **169**

6.16   STEROID BIOSYNTHESIS

The key compound in all terpenoid biosynthesis is mevalonic acid **1** (MVA), which in turn is derived from acetyl–CoA (Lynen[1]). It was discovered fortuitously by the Merck team in 1956.[2,3] during investigations of "vitamin B_{13}" from concentrates of dried distillers' solubles; isolation of a lactobacilli growth factor led to a β,δ-dihydroxy-β-methylvaleric acid (hence mevalonic acid) structure. At this time, the biosynthesis of squalene from precursors such as hydroxymethylglutaric acid (HMG, 5-COOH instead of 5-CH_2OH in **1**) and the cyclization of squalene to cholesterol were under intensive study (Bloch,[4] Cornforth and Popják[5]). Thus it did not take long for the Merck group[6,7] to discover that one enantiomeric form of *dl*-MVA, with concomittant decarboxylation, was very efficiently incorporated (86.8% as compared to 0.32% for HMG) into cholesterol by rat liver homogenate, and that MVA was the precursor of the C_5 units of the classical "isoprene rule."

It is also interesting to note that the trimethyl sterol lanosterol **30**, first isolated from wool fat in 1930[9] and structurally elucidated in 1952,[8] turned out to be another vital intermediate in steroid biosynthesis. The intermediacy of lanosterol, first hypothesized by Woodward and Bloch[10] in 1953, was proven by the *in vitro*[11] and *in vivo*[12] biosynthesis of lanosterol by rat tissue. Important cyclization mechanisms of squalene to lanosterol were also proposed.[13]

The more subtle stereospecific aspects of steroid biosynthesis going through the stages of (−)-mevalonic acid (MVA) **1**→geranyl pyrophosphate **20**→presqualene alcohol **29**→squalene **30**→lanosterol **34**→cholesterol **36**, are summarized below.[14-17]

Synthesis of stereospecifically labeled MVA

1) All six prochiral hydrogen atoms at C-2, C-4 and C-5 in MVA **1** have been replaced by deuterium or tritium (denoted by *H in the structures below) to give stereospecifically labeled MVA. All these are of vital importance in defining the stereochemical details of biosynthesis.

1) F. Lynen in *Biosynthesis of Terpenes and Sterols, A Ciba Foundation Symposium*, (ed. G. E. Wolstenholme, M. O'Conner), J. and A. Churchill, London, p. 95 (1959).
2) Ref. 1. p. 20; D. E. Wolff, C. H. Hoffman, P. E. Aldrich, H. R. Skegg, L. D. Wright, K. Folkers, *J. Am. Chem. Soc.*, **78**, 4499 (1956); *ibid.*, **79**, 1486 (1957).
3) Absolute configuration (by correlation with quinic acid); M. Eberle, D. Arigoni, *Helv. Chim. Acta*, **43**, 1508 (1960).
4) K. Bloch in *Vitamins and Hormones,* vol. 15, p. 119, Academic Press, 1957.
5) G. Popják, J. W. Cornforth in *Advances in Enzymology*, (ed. F. Nord), vol. 22, p. 281, Interscience, 1960.
6) P. A. Tavormina, M. H. Gibbs, J. W. Huff, *J. Am. Chem. Soc.*, **78**, 4498 (1956).
7) MVA was also isolated from *sake* as a growth factor of *hiochi* bacteria which spoil the rice wine. It was named hiochic acid and a 3,5-dihydroxyhexanoic acid structure was proposed, but its significance in cholesterol biosynthesis was not realized at the time of its characterization: G. Tamura, *J. Gen. Microbiol.*, **2**, 431 (1956).
8) H. Windaus, R. Tschesche, *Z. Physiol.*, **190**, 51 (1930).
9) R. G. Curtis, J. Fridrichsons, A. McL. Mathieson, *Nature*, **170**, 321 (1952); W. Voser, M. V. Mijovic, H. Heusser, O. Jeger, L. Ruzicka, *Helv. Chim. Acta*, **35**, 2414 (1952); C. S. Barnes, D. H. R. Barton, J. S. Fawcett, B. R. Thomas, *J. Chem. Soc.*, 576 (1953); J. F. McGhie, M. K. Pradham, J. F. Cavalla, *J. Chem. Soc.*, 3176 (1952).
10) R. B. Woodward, K. Bloch, *J. Am. Chem. Soc.*, **75**, 2023 (1953).
11) R. B. Clayton, K. Bloch, *J. Biol. Chem.*, **218**, 305, 319 (1956).
12) Schneider, R. B. Clayton, K. Bloch, *J. Biol. Chem.*, **224**, 175 (1957).
13) A. Eschenmoser, L. Ruzicka, O. Jeger, D. Arigoni, *Helv. Chim. Acta*, **38**, 1890 (1955); G. Stork, A. W. Burgstahler, *J. Am. Chem. Soc.*, **77**, 5068 (1955); T. T. Tchen, K. Bloch, *J. Biol. Chem.*, 226, 931 (1957).
14) Review: R. B. Clayton, *Quart. Rev.*, **19**, 168, 201 (1965).
15) Review: J. W. Cornforth, *Quart. Rev.*, **23**, 125 (1969).
16) L. J. Mulheirn, P. J. Ramm, *Chem. Soc. Rev.*, **1**, 259 (1972).

2) Elegant preparations of mevalonic acids carrying isotopic hydrogens at C-2 and C-4 are shown below.[17,18] The 5R-isomer was prepared by an interesting combination of enzymatic and chemical reactions,[19] and more recently the remaining 5S-isomer has also been made.[20]
3) Mevalonate kinase, which converts MVA into its 5-phosphate, the first intermediate in subsequent biosynthesis (see next subsection), only phosphorylates the natural 3R-isomers, **5**, **8**, **12** and **15**; hence in practice it usually suffices to use the racemic form.
4) The benzhydrylamide derivative (such as **2** and **9**) was in fact the first crystalline derivative of MVA.[2]
5) Epoxide cleavage **3**→**5** etc., by metal hydrides proceeds through inversion at C-4.
6) Conversion **5**→**7**, etc., involves formal interchange of C-2 and C-5.

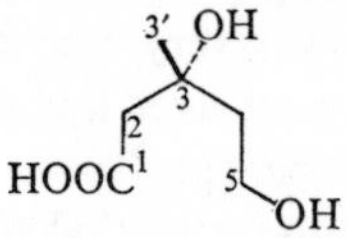
3'
OH
3
2
1
HOOC
5
OH

3R-(−)-mevalonic acid (MVA) **1**

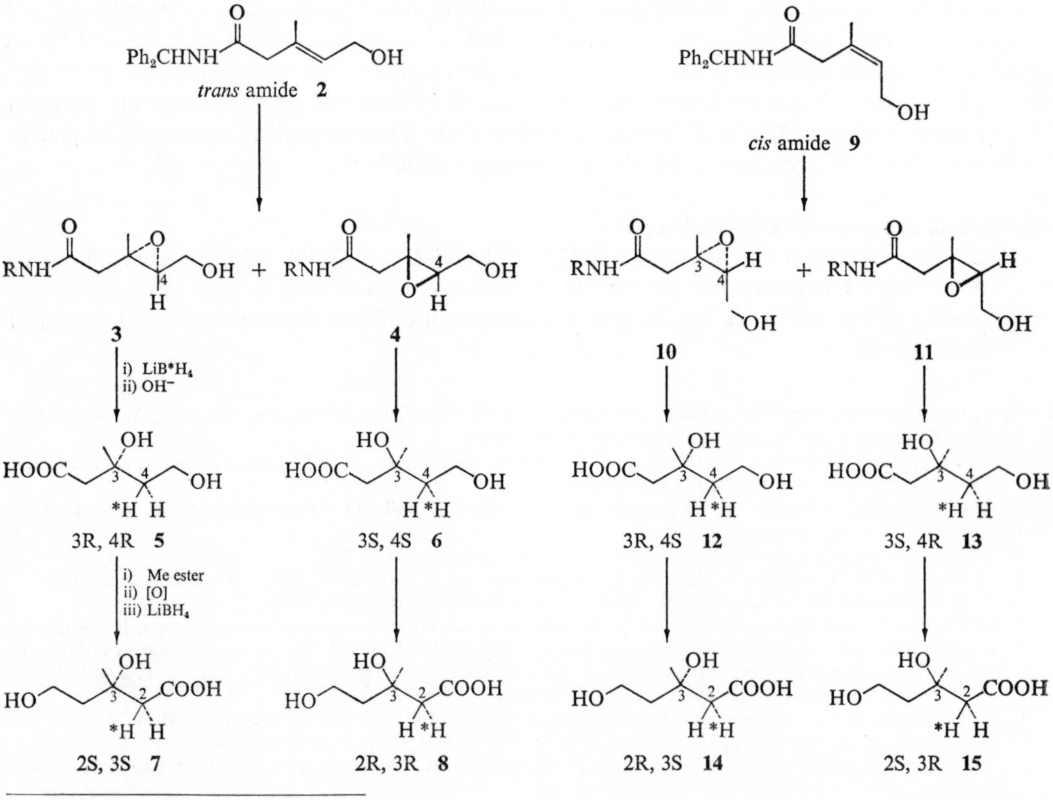
Ph2CHNH
OH
trans amide 2
cis amide 9
RNH
3
4
10
11
i) LiB*H4
ii) OH−
HOOC
*H H
H *H
3R, 4R 5
3S, 4S 6
3R, 4S 12
3S, 4R 13
i) Me ester
ii) [O]
iii) LiBH4
COOH
HO
2S, 3S 7
2R, 3R 8
2R, 3S 14
2S, 3R 15

17) Experimental techniques: R. B. Clayton (ed.) *Steroids and Terpenoids* in *Methods in Enzymology*, vol. 15, Academic Press, 1969.
18) J. W. Cornforth, R. H. Cornforth, C. Donninger, G. Popják, *Proc. Roy. Soc., B*, **163**, 492 (1966); J. W. Cornforth, R. H. Cornforth, G. Popják, L-Yengoyan, *J. Biol. Chem.*, **241**, 3970 (1966).
19) C. Donninger, G. Rybak, *Biochem. J.*, **91**, 11P (1964); C. Donninger, G. Popják, *Proc. Roy. Soc., B*, **163**, 465 (1966).
20) P. Blattmann, J. Retey, *Chem. Commun.*, 1394 (1970); J. W. Cornforth, F. P. Ross, *ibid.*, 1395; A. I. Scott, G. T. Phillips, P. B. Reinchardt, I. G. Sweeney, *ibid.*, 1396.

Conversion of MVA into farnesyl pyrophosphate

MVA 1 —— mevalonate kinase ——► MVA 5-phosphate —— phosphomevalonate kinase ——►

MVA 5-PP 16 —— decarboxylase ——► isopentenyl PP 17 ⇌ isomerase ⇌ γ,γ-dimethylally PP 18

18 + 17 X(nucleophile) —— geranyl transferase ——► 19

—— geranyl transferase ——► geranyl PP 20 farnesyl PP 21

PP=pyrophosphate
R,S=prochiral R and S hydrogens in MVA 16
Carbon numberings are those of MVA

17→18: 4-pro-S hydrogen is eliminated in the biosynthesis of *trans*-polyprenyls (polyisoprenes); however, in *cis*-polyprenyls such as rubber, it is the 4-pro-R hydrogen which is lost.

17+18→19: Nucleophile X attacks the sinistral side (or α-side when as depicted) of **17**, which in turn attacks C-5 of **18** with inversion.

19→20: Again the 4-pro-S hydrogen is lost to give the *trans*-polyprenyl, geranyl pyrophosphate **20** (monoterpenoid).

20→21: Linking of another MVA moiety to **20** with concomittant inversion at C-1 (geraniol numbering) gives the sesquiterpenoid farnesyl pyrophosphate **21** (sesquiterpenoid).

Conversion of MVA to squalene

1) The stereospecific fate of each hydrogen in MVA **22** is shown in structures **23** and **24**. Locations of C-2 carbons of MVA are also indicated (Cornforth, Popják and co-workers).[14,15]

2) Squalene biosynthesized from [5-^{2}H$_2$]-MVA by rat liver only contains 11 of the 12 expected deuterium atoms (mass spectrometry). The missing hydrogen was found to be supplied by the reduced coenzyme NADPH **25**.[21]

3) Squalene prepared from 5R-[5-^{2}H$_1$]-MVA was ozonized to yield acetone (2 moles), succinic acid **26** (1 mole) and levulinic acid **27** (4 moles). The dideuteriosuccinic acid **26** was optically inactive. This meant that the squalene central bond was formed by overall inversion at one C-5 center of MVA and retention at the other C-5 center (otherwise it would have been optically active).

4) Hypoiodite oxidation of levulinic acid **27** gave authentic R-deuteriosuccinic acid **28**; hence head-to-tail linkage of MVA units had occurred with C-5 inversions.

5) Squalene from 4S-[4-^{3}H$_1$, 2-^{14}C]-MVA had lost the tritium and only contained the ^{14}C label, whereas that from the 4R epimer retained the ^{3}H/^{14}C ratio of starting MVA. Accordingly the 4,3-double bonds are formed by loss of 4-pro-S hydrogens of MVA **22**.

21) G. Popják, DeW. S. Goodman, J. W. Cornforth, R. H. Cornforth, R. Ryhage, *J. Biol. Chem.*, **236**, 1934 (1961).

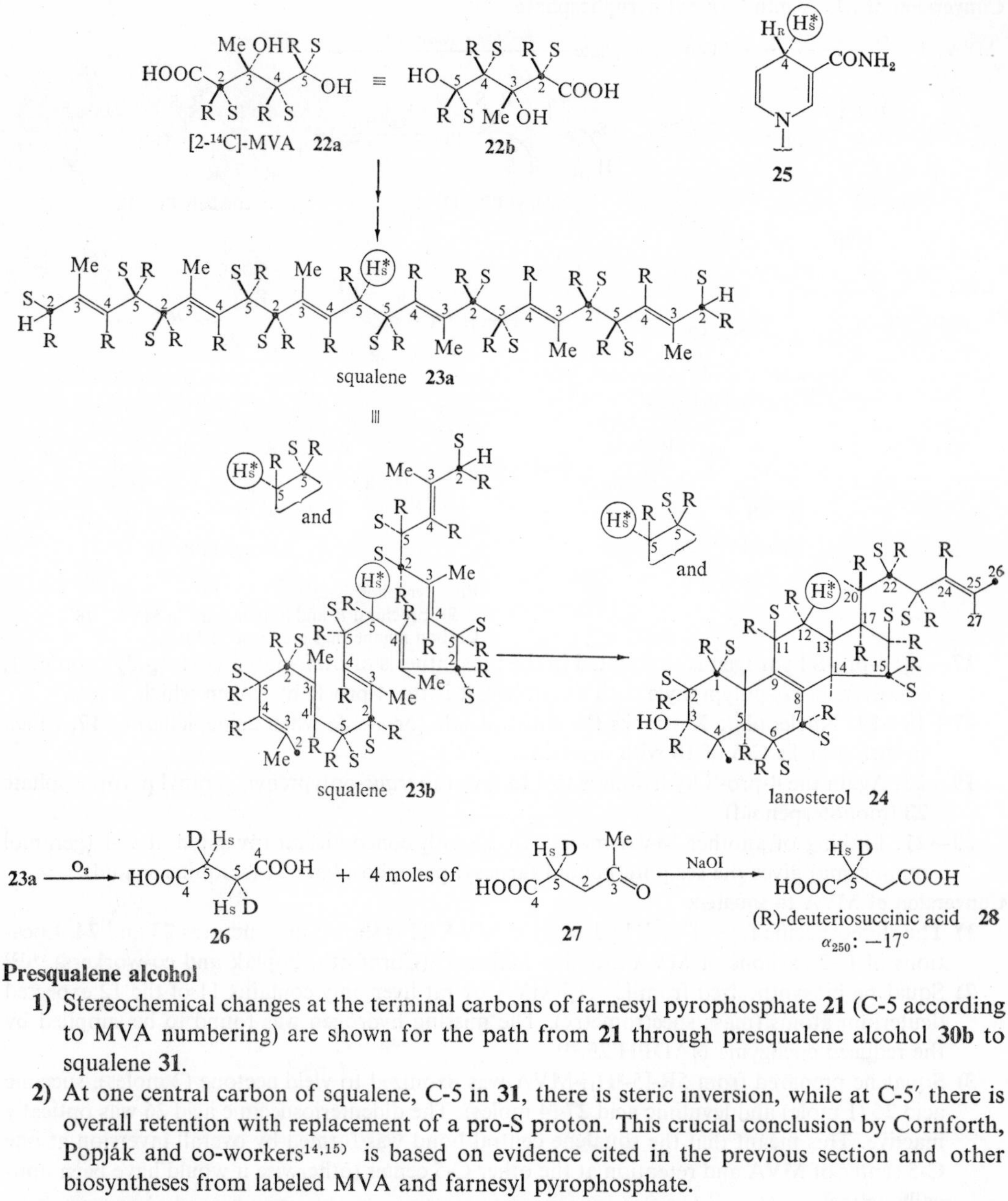

squalene **23a**

III

squalene **23b**   lanosterol **24**

23a → (O₃) → HOOC–CHD–CHD–COOH **26** + 4 moles of HOOC–CHD–CH₂–C(Me)=O **27** → (NaOI) → HOOC–CHD–CH₂–COOH

(R)-deuteriosuccinic acid **28**
$\alpha_{250}: -17°$

Presqualene alcohol

1) Stereochemical changes at the terminal carbons of farnesyl pyrophosphate **21** (C-5 according to MVA numbering) are shown for the path from **21** through presqualene alcohol **30b** to squalene **31**.

2) At one central carbon of squalene, C-5 in **31**, there is steric inversion, while at C-5' there is overall retention with replacement of a pro-S proton. This crucial conclusion by Cornforth, Popják and co-workers[14,15] is based on evidence cited in the previous section and other biosyntheses from labeled MVA and farnesyl pyrophosphate.

3) An important advance was the isolation[22] of the intermediate presqualene pyrophosphate **30b** from NADPH-starved yeast subcellular culture. It could be converted to squalene upon addition of the coenzyme. After several erroneous structures were proposed, **30** was formulated[23] and confirmed by three simultaneous syntheses.[24-26] The absolute configuration **30b**[27] is enantiomeric to that previously proposed.[25]

22) H. C. Rilling, *J. Biol. Chem.*, **241**, 3233 (1966).
23) H. C. Rilling, W. W. Epstein, *J. Am. Chem. Soc.*, **91**, 1041 (1969).

[5-³H]-farnesyl PP **21**

29

30a

presqualene alcohol pyrophosphate **30b**

30c

31

squalene **31** (from [5-³H]-farnesyl PP)

H$_S^*$=4-pro-S hydride from NADPH
R,S=prochiral R and S hydrogens in MVA
Carbon numberings are those of MVA

4) A hypothetical mechanism which takes into account the stereochemical details of farnesol, presqualene alcohol and squalene is shown.[25a] The C-5 of one farnesyl unit undergoes one inversion (**30b**→**30c**). The other C-5′ position is inverted once (**21**→**29**) but closure of the cyclopropyl ring to give presqualene pyrophosphate **30a** must somehow proceed with configurational retention at C-5 (**29**→**30a**); another inversion at C-5′ in **30c**, and hence overall retention at C-5′, leads to squalene **31**.

Cyclization of squalene

1) In 1966, Corey et al.[28] and van Tamelen et al.[29] discovered the key role of 2,3-oxidosqualene **32** by showing that rat liver enzyme preparations efficiently converted it to lanosterol.

2) Oxidosqualene **32** undergoes multicyclizations initiated by epoxide cleavage to give a hypothetical protosterol carbonium ion **33** (Eschenmoser et al.)[13] or its neutral equivalent (nucleophile at C-17),[31] which undergoes manifold hydrogen and methyl migrations to yield lanosterol **34**.

24) L. J. Altman, R. C. Kowerski, H. C. Rilling, J. Am. Chem. Soc., **93**, 1782 (1971).
25) H. C. Rilling, C. D. Poulter, W. W. Epstein, B. Larsen, J. Am. Chem. Soc., **93**, 1783 (1971).
25a) Modified from scheme shown in: J. Edmond, G. Popják, S. -M. Wong, V. P. Williams, J. Biol. Chem., **246**, 6254 (1971).
26) R. M. Coates, W. H. Robinson, J. Am. Chem. Soc., **93**, 1785 (1971).
27) G. Popják, J. Edmond, S. -M. Wong, J. Am. Chem. Soc., **95**, 2713 (1973).
28) E. J. Corey, W. E. Russey, P. R. Ortiz de Montellano, J. Am. Chem. Soc., **88**, 4750 (1966).
29) E. E. van Tamelen, J. D. Willett, R. B. Clayton, K. E. Lord, J. Am. Chem. Soc., **88**, 4752 (1966).
30) Review: E. E. van Tamelen, Acc. Chem. Res., **1**, 111 (1968); XXIIIrd International Congress of Pure and Applied Chemistry, vol. 5, p. 85, Butterworth, 1971.

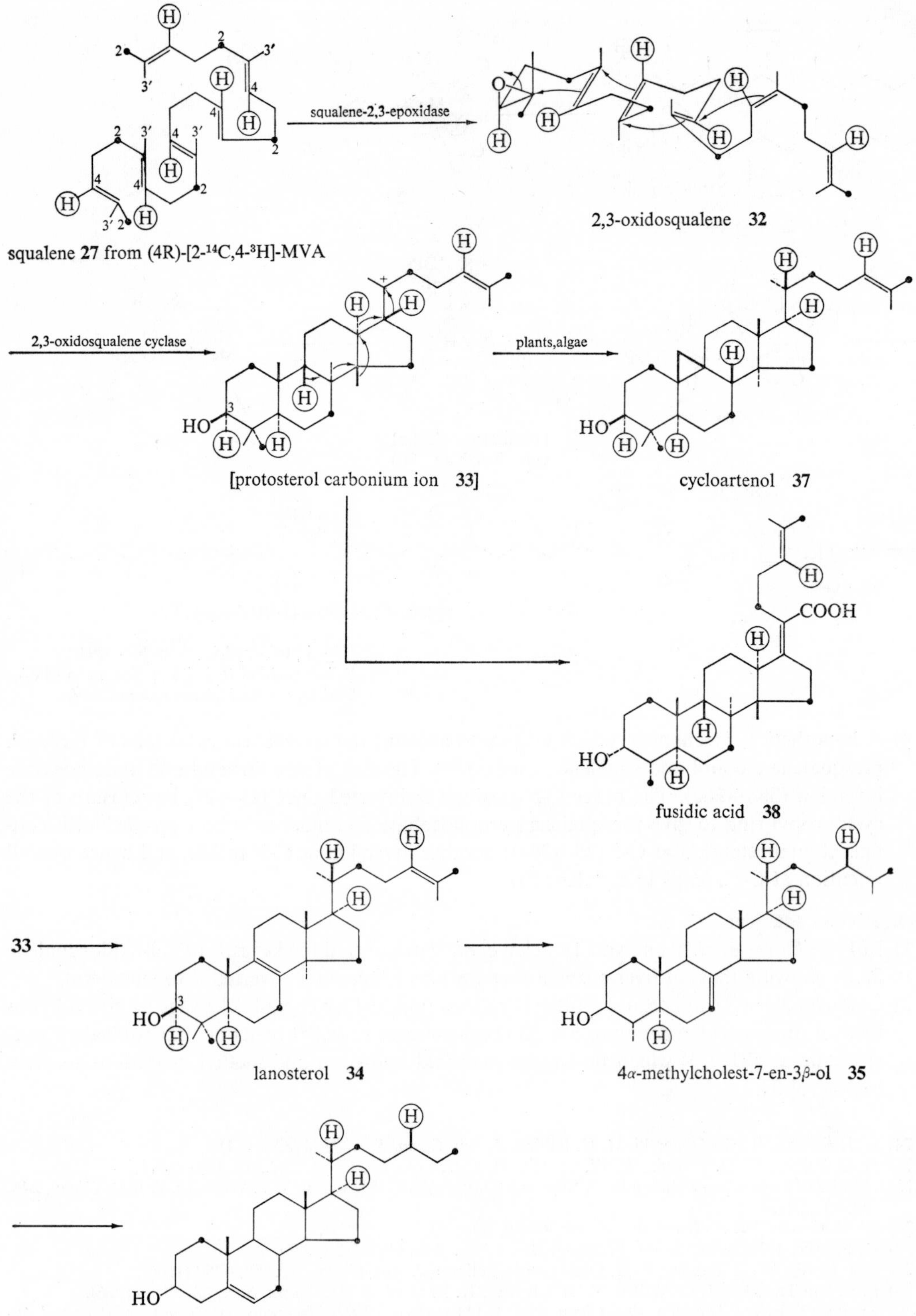

squalene **27** from (4R)-[2-¹⁴C,4-³H]-MVA

squalene-2,3-epoxidase

2,3-oxidosqualene **32**

2,3-oxidosqualene cyclase

[protosterol carbonium ion **33**]

plants,algae

cycloartenol **37**

fusidic acid **38**

33 ⟶

lanosterol **34**

4α-methylcholest-7-en-3β-ol **35**

cholesterol **36**

3) In some higher plants[16,32-34] and algae[16,35] cycloartenol **37** replaces lanosterol as the principal triterpene. A protosterol-type antibiotic fusidic acid **38** is produced by the fungus *Fussidium coccineum.*[36]

4) Removal of the lanosterol 4,4-gem-dimethyl group[16] proceeds through: oxidation of 4α-methyl to 4-hydroxymethyl and then to 4-carboxyl; oxidation of 3-hydroxyl to 3-ketone and 4α-decarboxylation; equilibration of 4β-methyl to 4α-methyl and reduction of 3-ketone to 3β-hydroxyl (formation of **35**); loss of 4α-methyl in **35** through a similar sequence.

5) The transformations described above have been established by extensive biosynthetic studies employing labeled precursors. The fate of 4-pro-R protons (labeled as ^{2}H or ^{3}H) and 2-^{14}C carbons in MVA is indicated by ⊕ in the scheme.

Hydrogenation of 24-ene

1) Sinistral hydrogenation at C-24: enzymatic cleavage of cholesterol **40** gave the acid **41** which was degraded to the alcohol **42** containing one tritium, as expected. Dehydrogenation of **42** with yeast alcohol dehydrogenase (YAD), which removes only 1-prochiral R protons of primary alcohols gave the aldehyde **43** with retention of tritium. This clarified the stereochemistry at C-24 in **40**.[37]

2) *Cis*-hydrogenation at C-24: *Mycobacterium smegmatis* converts cholesterol **40** to (25S)-26-hydroxycholest-3-en-4-one **44** (x-ray).[38] The fact that the hydroxylated carbon in **44** is derived from C-2 of MVA was shown by removal of C-26 by oxidation to aldehyde and decarbonylation. Since the 24-ene geometry of lanosterol is as in **39**, the *cis*-hydroxylation pathway by rat liver is established.[38]

lanosterol **39**
from (4R)-[2-^{14}C;4-^{3}H]-MVA

cholesterol **40**

(25S)-26-hydroxycholest-3-en-4-one **44**

41

42

43

31) J. W. Cornforth, *Angew. Chem. Intern. Ed.*, **7**, 903 (1968).
32) H. H. Rees, L. J. Goad, T. W. Goodwin, *Tetr. Lett.*, 723 (1968).
33) G. Ponsinet, G. Ourisson, *Phytochemistry*, **9**, 1499 (1970).
34) R. Heinz, P. Benveniste, *Phytochemistry*, **9**, 1499 (1970).
35) L. J. Goad, T. W. Goodwin, *European J. Biochem.*, **7**, 502 (1969); H. H. Rees, L. J. Goad, T. W. Goodwin, *Biochem. Biophys. Acta*, **176**, 892 (1969).

14-Demethylation and 5-ene formation

The scheme below summarizes the changes occurring in rings BCD in going from lanosterol **45** to cholesterol **52**. However, as it is unknown at which stage the 4-demethylation and saturation of the 24-ene take place, these structures are left unspecified.

1) 14-Demethylation: unlike 4-demethylation, which involves decarboxylation, it has been shown that this occurs at the oxidation state of an aldehyde **46** resulting in release of C-23 as formic acid.[39] An intermediate such as **47** has been postulated.

2) Participation of 8,14-diene **48**: this was established by incubations of 5α-cholesta-8,14-dien-3-ol and 24-dihydrolanosterol with rat liver, and trapping the diene by addition of unlabeled material.[40-43]

3) Loss of 15α-H (**45**→**49**): this was shown by loss of 15α-T originating from (2S)-[2-^{3}H$_1$]-MVA;[41-44] the 14-ene in **48** is reduced via a transhydrogenation[40,45] so that 15-H$_R$ which was β in **45** is now α.

4) Conversion of 8-ene **49** to 5-ene **52**: the 7β-H$_S$ of lanosterol **45** is lost in cholesterol **52**, but the 7α-H$_R$ is retained and now occupies the β-configuration.[46,47] Formation of the 5-ene (**50**→**51**) involves loss of 5α- and 6α-protons;[48] finally the 7-ene is reduced by NADPH from the α-side[49] to yield cholesterol **52**.

lanosterol **45**   **46**   **47**

48   **49**   **50**

36) W. O. Godtfredsen, H. Lorck, E. E. van Tamelen, J. D. Willett, R. B. Clayton, *J. Am. Chem. Soc.*, **90**, 208 (1968).

37) J. B. Greig, K. R. Varma, E. Caspi, *J. Am. Chem. Soc.*, **93**, 760 (1971).

38) D. J. Duchamp, C. G. Chidester, J. A. F. Wickramasinghe, E. Caspi, B. Yagen, *J. Am. Chem. Soc.*, **93**, 6283 (1971).

39) K. Alexander, M. Akhtar, R. B. Boar, J. F. McGhie, D. H. R. Barton, *Chem. Commun.*, 383 (1972).

40) M. Akhtar, A. D. Rahimtula, I. A. Watkinson, D. C. Wilton, K. A. Munday, *Chem. Commun.*, 1406 (1968); 149 (1969).

41) L. Canonica, A. Fiecchi, M. G. Kienle, A. Scala, G. Gali, E. G. Paoletti, R. Paoletti, *J. Am. Chem. Soc.*, **90** 3597, 6532 (1968); *Steroids*, **11**, 749 (1968); *ibid.*, **12**, 445 (1969).

42) B. N. Lutsky, G. J. Schroepfer, *Biochem. Biophys. Res. Commun.*, **33**, 492 (1968).

43) K. T. W. Alexander, M. Akhtar, R. B. Boar, J. F. McGhie, D. H. R. Barton, *Chem. Commun.*, 1479 (1971).

44) G. F. Gibbons, L. J. Goad, T. W. Goodwin, *Chem. Commun.*, 1458 (1968).

45) E. Caspi, P. J. Ramm, R. E. Gain, *J. Am. Chem. Soc.*, **91**, 4012 (1969).

46) E. Caspi, J. B. Greig, P. J. Ramm, K. R. Varma, *Tetr. Lett.*, 3829 (1968).

47) G. F. Gibbons, L. J. Goad, T. W. Goodwin, *Chem. Commun.*, 1212 (1968).

48) A. M. Paliokas, G. J. Schroepfer, Jr., *J. Biol. Chem.*, **243**, 453 (1968).

49) D. C. Wilton, K. A. Munday, S. J. M. Skinner, M. Akhtar, *Biochem. J.*, **106**, 803 (1968).

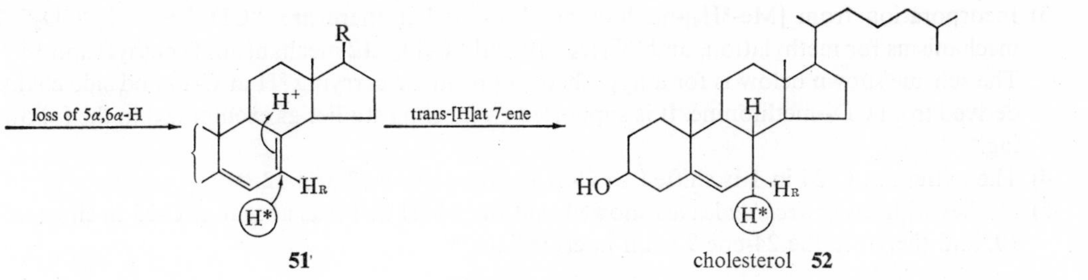

$H^* =$ pro-S hydride from NADPH
R,S = prochiralities in MVA
Structures of ring A and side chain are unspecified in intermediates

6.17   SIDE CHAIN ALKYLATION AND DEALKYLATION OF STEROLS

Biosynthesis of phytosterol side chains

1) A distinctive feature of many phytosterols is the presence of C_1 or C_2 groups at C-24. Whether the first C-alkylation in natural biosynthesis occurs at the C_{27} level or at the C_{30} level, however, is difficult to decide because enzymes modifying the side chain appear to be relatively non-specific regarding the tetracyclic structure, and *vice versa*.[1]

2) The C_2 groups are produced by double C-methylations from methionine. Namely, usage of [Me-^{14}C]-methionine showed both C-28 and C-29 to be radioactive.[2-4]

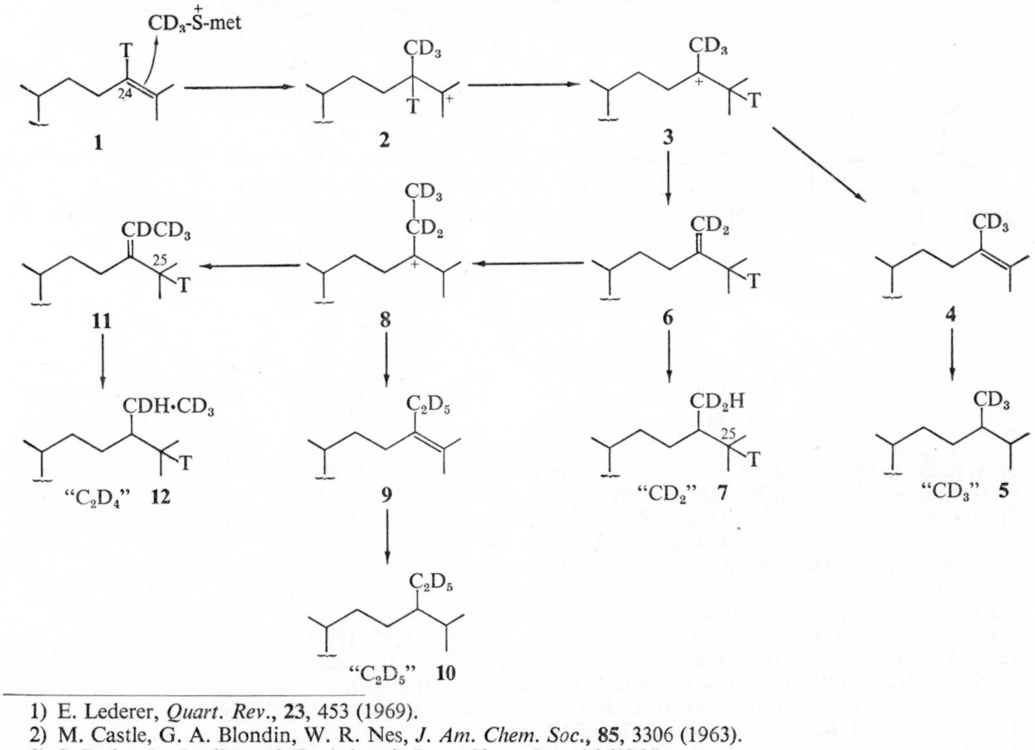

1) E. Lederer, *Quart. Rev.*, **23**, 453 (1969).
2) M. Castle, G. A. Blondin, W. R. Nes, *J. Am. Chem. Soc.*, **85**, 3306 (1963).
3) S. Bader, L. Gugliemetti, D. Arigoni, *Proc. Chem. Soc.*, 16 (1964).
4) V. R. Villanueva, M. Barbier, E. Lederer, *Bull. Soc. Chim. Fr.*, 1423 (1964).

3) Incorporation from [Me-^{2}H$_3$]-methionine showed that there are "CD$_3$" **5** and "CD$_2$" **7** mechanisms for methylation, and "C$_2$D$_5$" **10** and "C$_2$D$_4$" **12** mechanisms for ethylation.[1,5-8] The scheme shown below is for a hypothetical precursor carrying ^{3}H at C-24 and side chains derived from CD$_3$-methionine. It is supported by labeling studies as exemplified in the following.

4) The tritium at C-24 in **1** is shifted to C-25 in some sterols **7** and **12**.[9-11]

5) Studies with *Larix decidua* leaves showed that the 24-^{3}H in **1** was absent at C-25 in sitosterol **10**, and therefore the 24-ene **9** is an intermediate.[12]

Side chain demethylation of phytosterols in insects

1) Insects cannot biosynthesize cholesterol and hence they must rely on plants as a source for C$_{27}$ steroids, which include the vital ecdysones.

2) The 24-methylene sterol **14** has been identified as an intermediate in the conversion of campesterol **13** to **16** (tobacco hornworn).[13]

3) Fucosterol **18** and desmosterol **15** are intermediates in the conversion of β-sitosterol **17** to **16**.[14-16]

4) Epoxide **19** has been isolated by trapping experiments from silk-worm (*Bombyx mori*) fed with tritiated steroids.[17]

5) Feeding experiments with the yellow mealworm (*Tenebrio molitor*) showed that the 25-^{3}H label of isofucosterol **20** was retained in **16**. Hence the dealkylation must involve rearrangement of 25-^{3}H, presumably to C-24, as shown by **22**.[18]

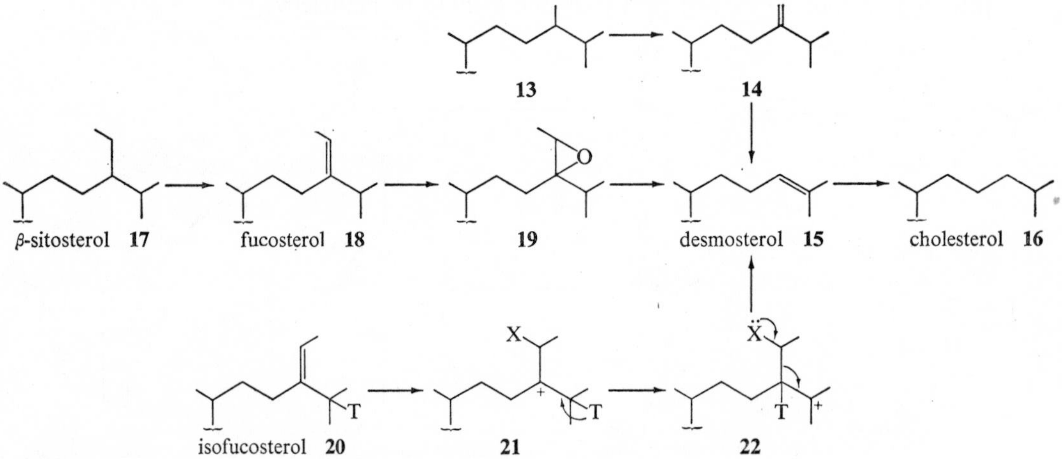

5) A. R. H. Smith, L. J. Goad, T. W. Goodwin, E. Lederer, *Biochem. J.*, **104**, 56C (1967).
6) J. Varenne, J. Polonsky, N. Cagnoli-Bellavita, P. Ceccherelli, *Biochimie*, **53**, 261 (1969).
7) M. Lenfant, R. Ellioz, B. C. Das, E. Zissman, E. Lederer, *European J. Biochem.*, **7**, 159 (1969).
8) Y. Tomita, A. Vomori, E. Sakurai, *Phytochemistry.*, **10**, 573 (1971).
9) M. Akhtar, P. F. Hunt, M. A. Parvez, *Chem. Commun.*, 565 (1966); *Biochem. J.*, **103**, 616 (1967).
10) K. H. Raab, N. J. de Sousa, W. R. Nes, *Biochem. Biophys. Acta*, **152**, 742 (1968).
11) L. J. Goad, T. W. Goodwin, *European J. Biochem.*, **7**, 502 (1969).
12) P. J. Randall, H. H. Rees, T. W. Goodwin, *Chem. Commun.*, 1295 (1972).
13) J. A. Svoboda, M. J. Thompson, W. E. Robbins, *Lipids*, **7**, 156 (1972).
14) F. J. Ritter and W. H. J. M. Wientjens, *TNO-Nieuws*, **22**, 381 (1967).
15) J. A. Svoboda, M. J. Thompson, W. E. Robbins, *Nature, New Biology*, **230**, 57 (1971).
16) J. P. Allais, M. Barbier, *Experientia*, **27**, 507 (1971).
17) M. Morisaki, H. Ohtake, M. Okubayashi, N. Ikekawa, *Chem. Commun.*, 1275 (1972).
18) P. J. Randall, J. G. Lloyd-Jones, I. F. Cook, H. H. Rees, T. W. Goodwin, *Chem. Commun.*, 1296 (1972).

6.18 YEAST STEROLS

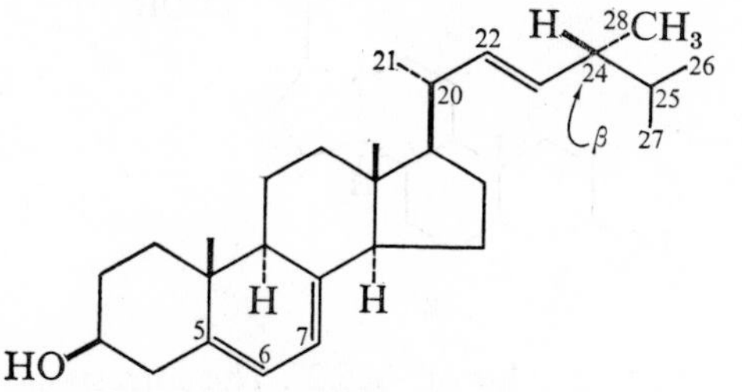

ergosterol **1**
mp: 174° α_D: −20°

Structure of ergosterol

1) Ergosterol **1**, the most abundant yeast sterol, occurs in yeast together with 5α,6-dihydro-ergosterol.

2) The presence of the 22-double bond and 24-methyl group was proven by ozonolysis which gave methylisopropylacetaldehyde.[1]

3) The nuclear skeleton was derived by saturation of the double bonds and subsequent oxidation to nor-5α-cholanic acid **2**.[2]

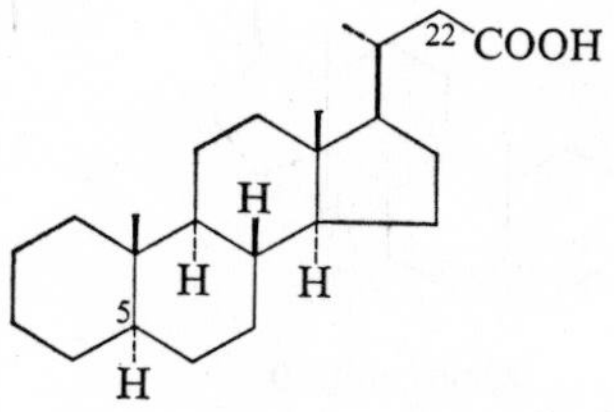

2

4) The presence of the 5,6-double bond was proven in a manner[3] analogous to cholesterol. That is, ergosterol was reacted with perbenzoic acid to give **3**, which was converted in 5 steps to give the dione **4**.

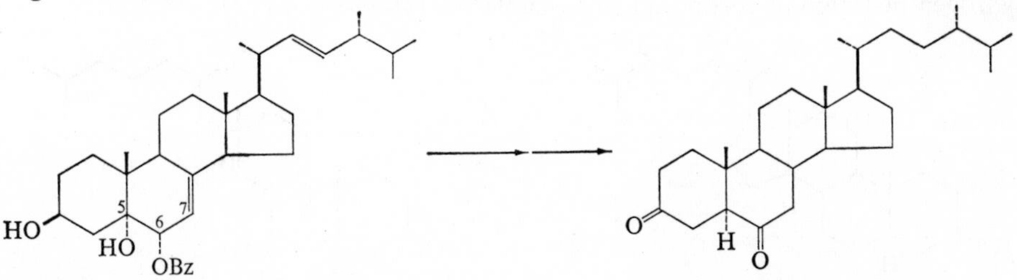

5) Evidence for the presence of conjugated nuclear double bonds comes from uv data[4,5] and the formation of a maleic anhydride adduct.[6]

1) A. Guiteras, *Ann. Chem.*, **494**, 116 (1932).
2) C. K. Chaung, *Ann. Chem.*, **500**, 270 (1933).
3) A. Windaus, H. H. Inhoffen, S. von Reichel, *Ann. Chem.*, **510**, 248 (1934).
4) I. M. Heilbron, R. A. Morton, W. A. Sexton, *J. Chem. Soc.*, 47, (1928).
5) K. Dimroth, G. Trautmann, *Chem. Ber.*, **69**, 669 (1936).
6) A. Windaus, A. Lüttringhaus, *Chem. Ber.*, **64**, 850 (1931).

6) The position and configuration of the 3β-hydroxy group was derived from oxidation of the acetate of 5α-ergostanol **5** to 3β-hydroxy-nor-5α-cholanic acid **6**.[7]

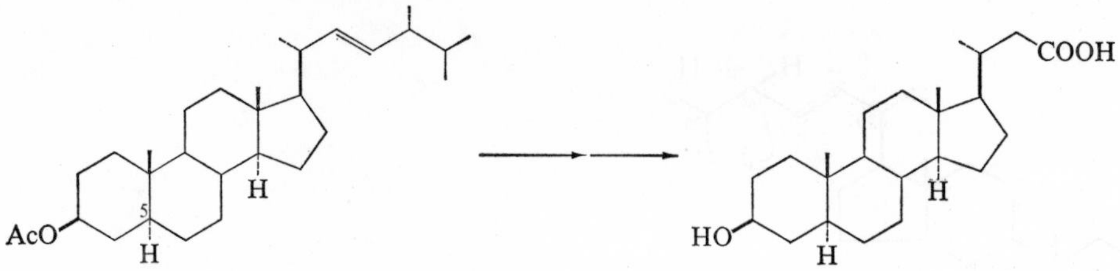

5   **6**

7) The aldehyde **7** resulting from ozonolysis of **1**[1] had an α_D value of $-65.2°$. The corresponding aldehyde derived from ozonolysis of stigmasterol **8**[8] was dextrorotatory and therefore of opposite absolute configuration. Since the 24-ethyl group of stigmasterol was proven[8] to be α, the 24-methyl group of ergosterol is β.

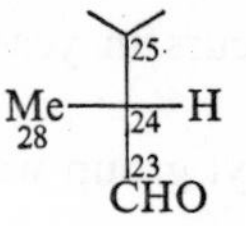

7 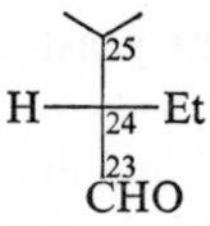   **8**

Structure of zymosterol

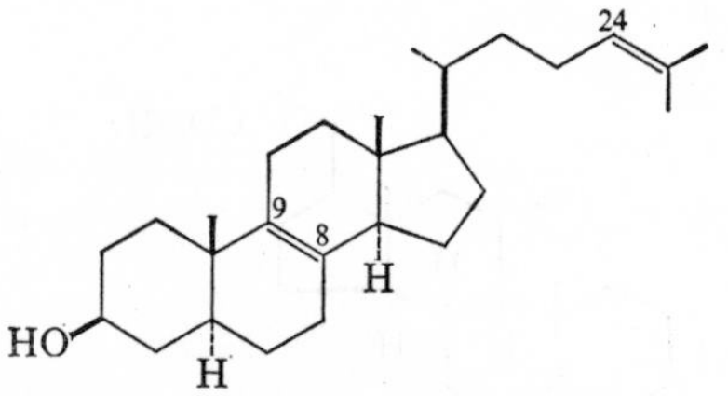

zymosterol **9**
mp : 110° α_D : 49°

1) Zymosterol **9**, is the second most abundant yeast sterol.[9,10]

2) The nuclear double bond is resistant to hydrogenation[11] or to bromination.[12]

3) Hydrogenation in neutral conditions affords **10** which can be isomerized with HCl or hydrogen-platinum in acetic acid to 5α-cholest-8(14)-en-3β-ol **11**.[13]

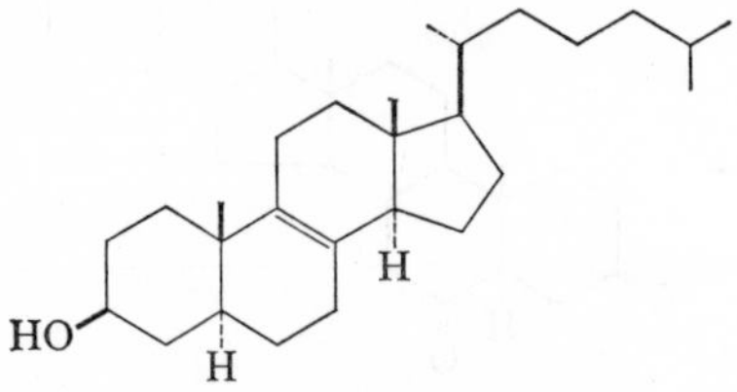

10 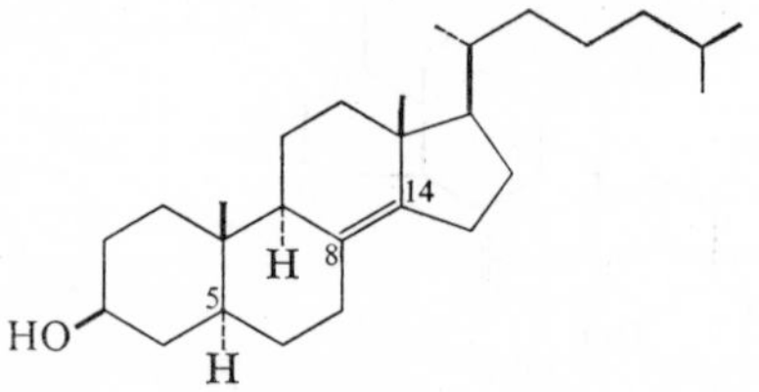   **11**

7) E. Fernholz, P. N. Chakrovorty, *Ann. Chem.*, **507**, 128 (1933); *Chem. Ber.*, **67**, 2021 (1934).
8) K. Tsuda, Y. Kishida, R. Hayatsu, *J. Am. Chem. Soc.*, **82**, 3396 (1960).
9) M. C. Hart, F. W. Heyl, *J. Biol. Chem.*, **95**, 311 (1933).
10) J. C. E. Simpson, *J. Chem. Soc.*, 730 (1937).
11) S. Kuwada, S. Yoshiki, *J. Pharm. Soc.*, (Japan), **60**, 407 (1940).
12) H. Dam, A. Geiger, J. Glavind, P. Karrer, W. Kaner, E. Rothschild, H. Salomon, *Helv. Chim. Acta*, **22**, 313 (1939).
13) H. Wieland, L. Görnhardt, *Ann. Chem.*, **557**, 248 (1947).

Ascosterol and fecosterol

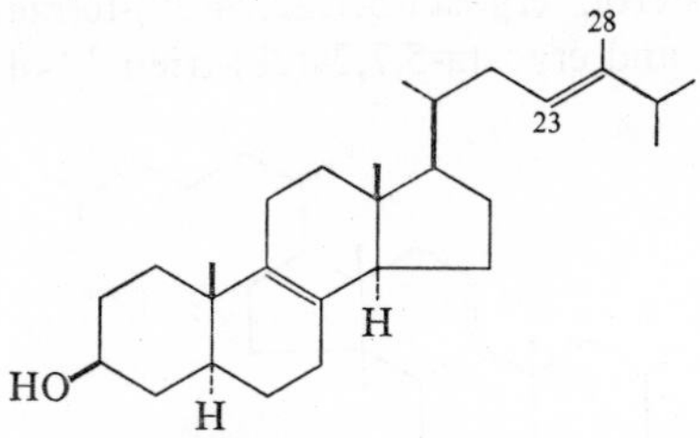

ascosterol **12**
mp : 142° α_D : 45°

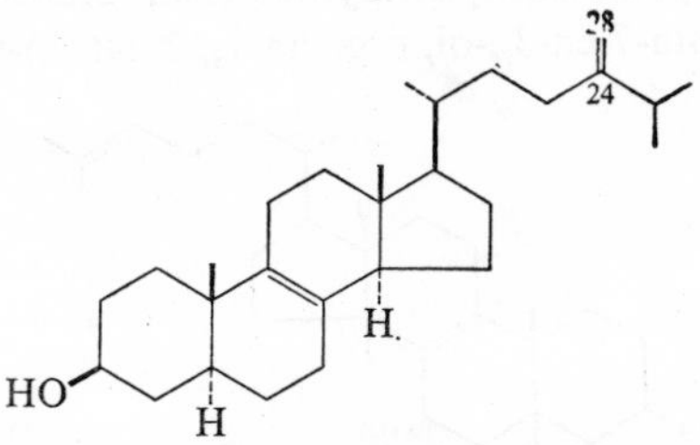

fecosterol **13**
mp : 162° α_D : 42°

1) Ascosterol **12** and fecosterol **13** are both minor yeast sterols and are believed to have structures as shown.[14]

2) Ascosterol can be converted to fecosterol by shaking with Pt in acetic acid under a nitrogen atmosphere.[15]

3) Both sterols can be converted to 5α-ergost-8(14)-en-3β-ol **14** upon hydrogenation in acetic acid.[15]

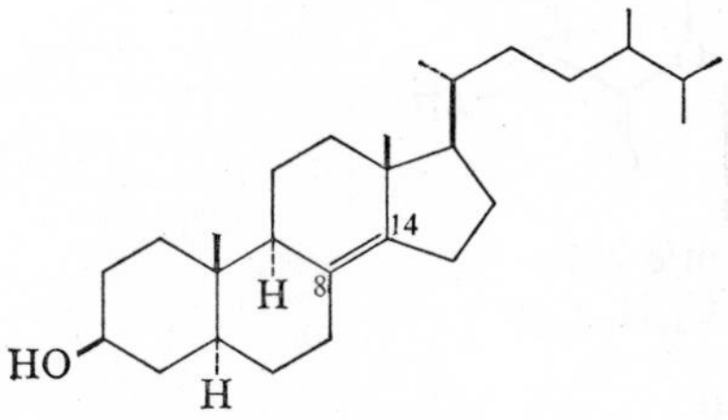

14

Episterol and other yeast sterols

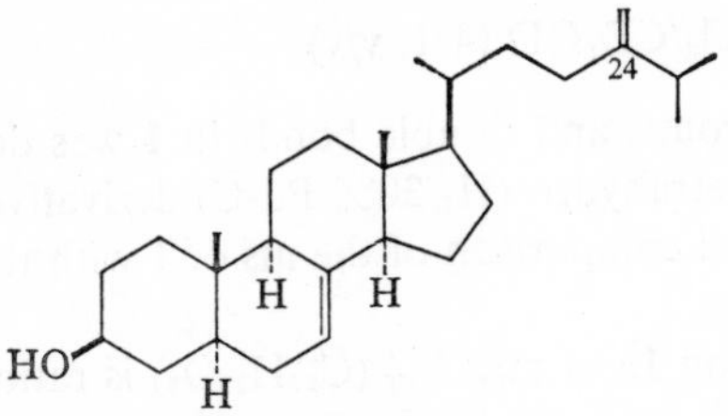

episterol **15**
mp : 151° α_D : -5°

1) Episterol **15**, yields formaldehyde upon ozonolysis[15] and is converted by hydrogenation to 5α-ergost-8(14)-en-3β-ol.[15]

2) Molecular rotation studies[16] (especially the low α_D) suggest that the double bond is at C-7.

3) In addition to the above sterols, the following have also been isolated from yeast:[17] lano-

14) L. F. Fieser, M. Fieser, *Natural Products Related to Phenanthrene*, p. 292, Reinhold, 1949.
15) H. Wieland, F. Rath, H. Hesse, *Ann. Chem.*, **548**, 34 (1941).
16) D. H. R. Barton, J. P. Cox, *J. Chem. Soc.*, 1354 (1948).
17) D. H. R. Barton, V. M. Kempe, D. A. Widdowson, *J. Chem Soc.*, 513 (1972).

sterol **16**, 14-demethyl-lanosterol, parkeol **17**, 4α-methylzymosterol, 24,25-dihydro-4α-methyl-24-methylenezymosterol, 5,6-dihydroergosterol, ergosta-5,7,22,24(28)-tetraen-3β-ol, ergosta-7-en-3β-ol, ergosta-7,22,24(28)-trien-3β-ol, and ergosta-5,7,24(28)-trien-3β-ol.

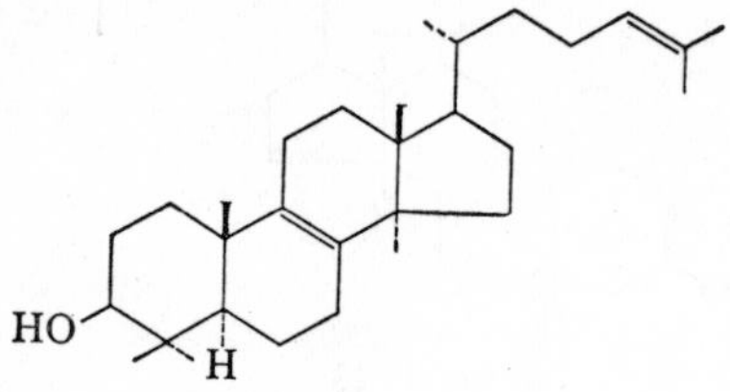

16

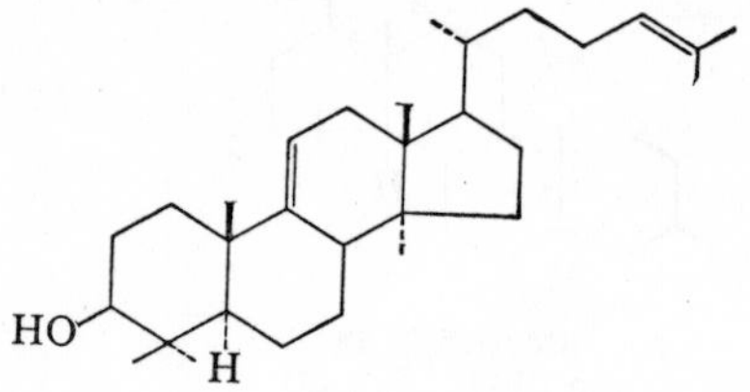

17

6.19 ANTHERIDIOL

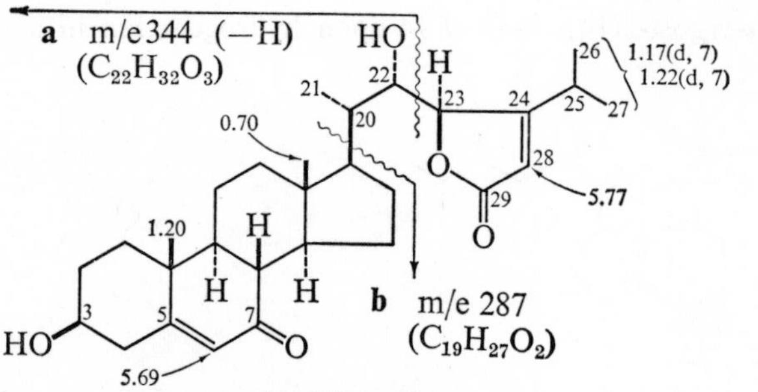

antheridiol **1**[1]
mp: 250–255° (dec.)
ms: 470 (M$^+$ $C_{29}H_{42}O_5$, high resoln.)
ir: 3390, 1742, 1672
uv: 220 (ϵ 17,000)
nmr: 4mg of **1** in $CDCl_3/CD_3OD$ (4:1, v/v)

1) The number of hydroxyl groups and double bonds in **1** was determined by formation of a diacetate (Ac$_2$O/py) and a tetrahydro (H$_2$/30% Pd-C) derivative, respectively.

2) Uv data of the dienone **2** and comparison of the ms of **1** with that of a model **3** led to the 3-hydroxy-5-en-7-one structure.

3) Fragmentation to yield an ion **1a** at m/e 344 ($C_{22}H_{32}O_3$) is rationalized by the presence of a double bond in the side chain.

$$\textbf{1} \xrightarrow[\text{(reflux)}]{\text{MeOH/HCl}} \textbf{2}, \text{ms}: 452 (M^+ C_{29}H_{40}O_4)$$
uv : 215 (ϵ 12,900), 278 (ϵ 23,700)

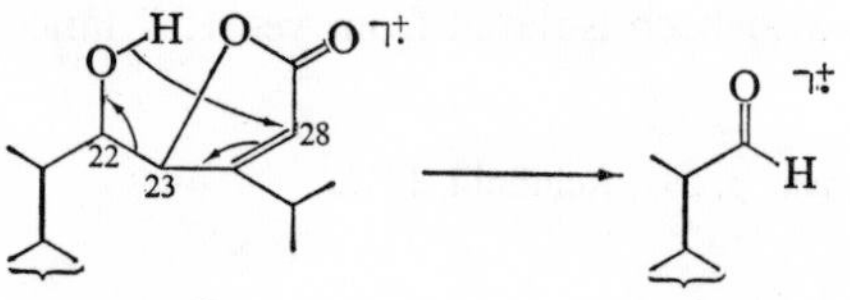

1   **1a** 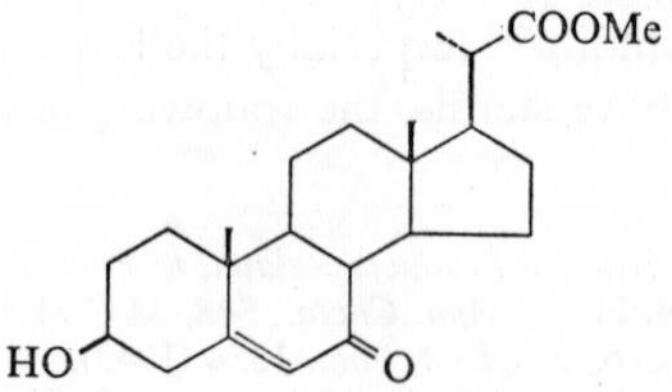   **3** 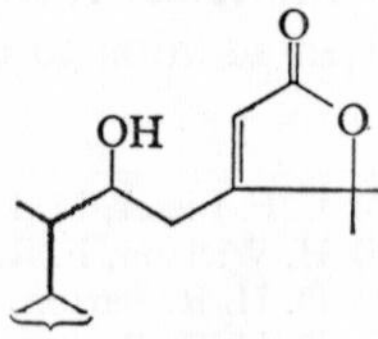   **4**

4) An alternative side-chain structure **4** was finally excluded by the synthesis of **1**.[2]

5) The 22S,23R-stereochemistry is based on cd data.[3]

Remarks[1]

Antheridiol **1** is one of the substances which govern the sexual reproduction in the aquatic fungus *Achlya bisexualis*. It is secreted by the female and induces formation of antheridial hyphae on the male.

Synthesis[2]

The first synthesis from **5** gave a small amount of antheridiol **1** together with its 22,23-isomer **12**.

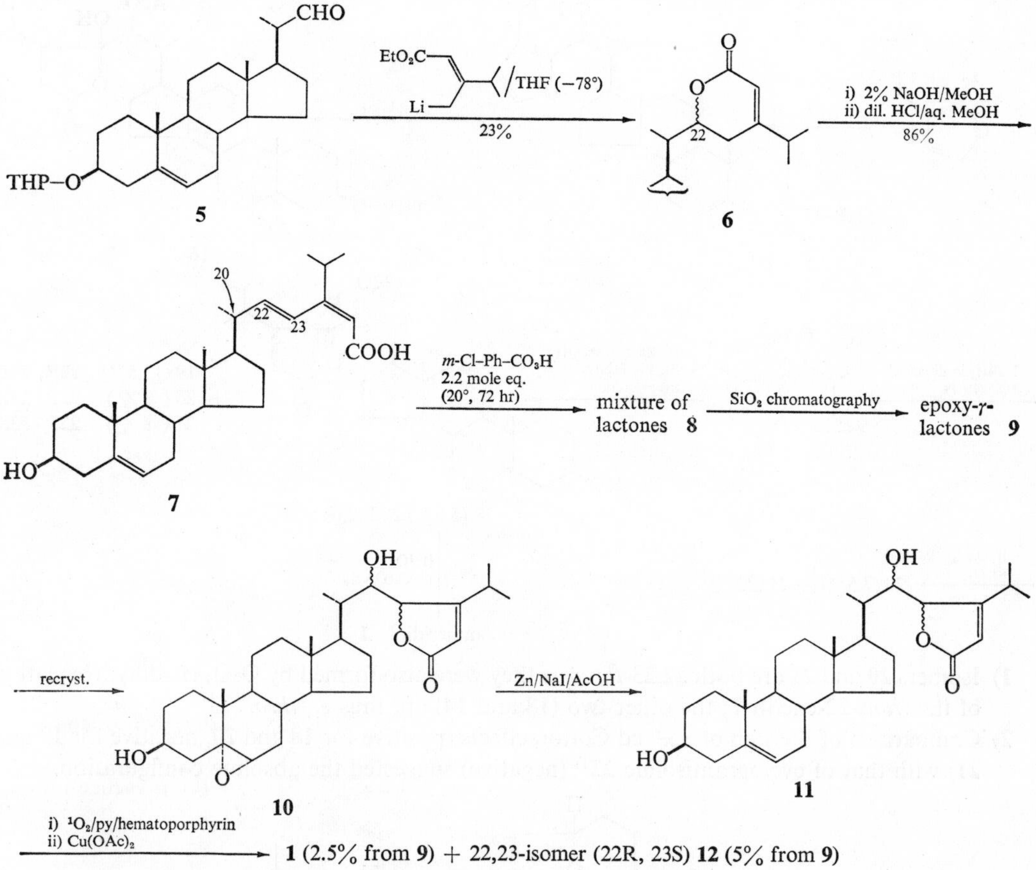

1) Configuration at C-20 in **7** (hence in **1**) was shown to be R by its conversion into **13**, which was also derived from stigmasterol.

2) Pure **10** (6% from **7**), obtained by repeated recrystallization of **9**, gave **11** (78%), which gave **12** (20%) as the sole product. However, the desired **1** was obtained in 2.5% yield together with **12** (5%) when "**10**" resulting from once-recrystallized **9** was employed.

1) G. P. Arsenault, K. Biemann, A. W. Barksdale, T. C. McMorris, *J. Am. Chem. Soc.*, **90**, 5635 (1968).
2) J. A. Edwards, J. S. Mills, J. Sundeen, J. H. Fried, *J. Am. Chem. Soc.*, **91**, 1248 (1969).
3) J. A. Edwards, J. Sundeen, W. Salmond, T. Iwadera, J. H. Fried, *Tetr. Lett.*, 791 (1972).

3) Both **1** and **12** are believed to possess a 22,23-*erythro* structure, since they were prepared by epoxidation of the *trans* 22-ene and subsequent ring opening (see **14**).

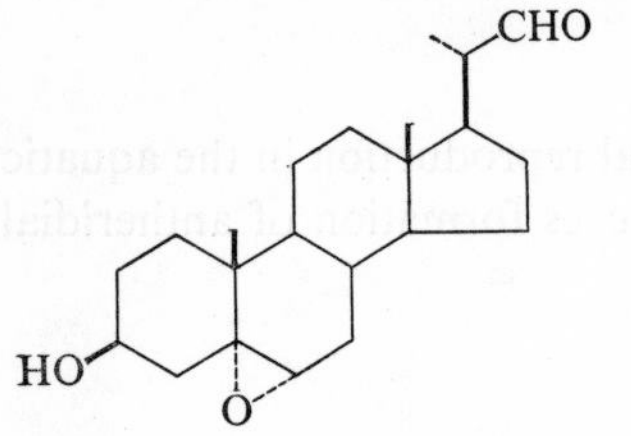

13 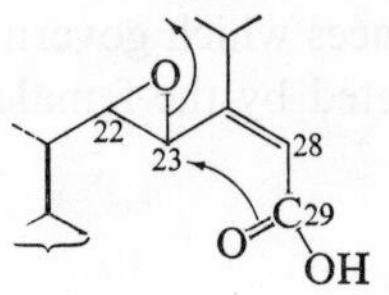   **14**

Modifications of the synthesis and stereochemistries at C-22 and C-23[3]

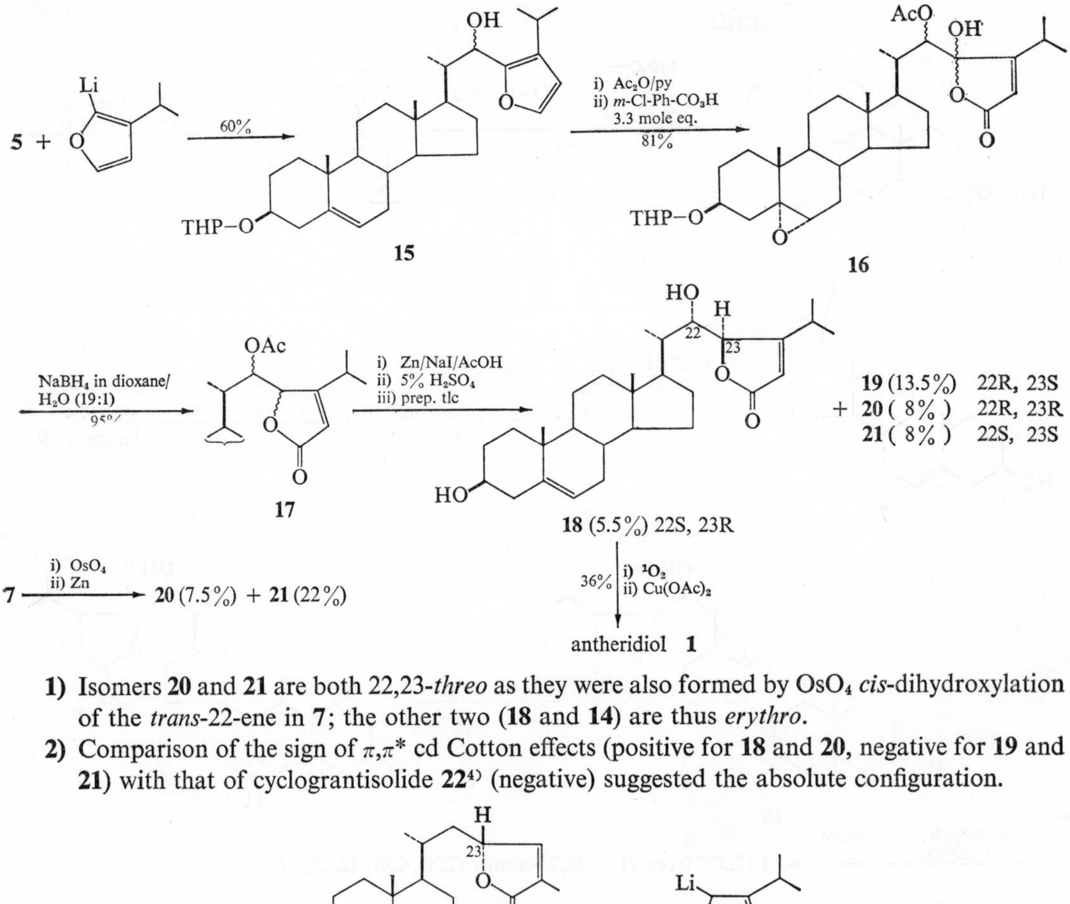

15   **16**

17   **18** (5.5%) 22S, 23R

19 (13.5%)   22R, 23S
+ **20** (8%)   22R, 23R
    21 (8%)   22S, 23S

$$7 \xrightarrow[\text{ii) Zn}]{\text{i) OsO}_4} \textbf{20}\,(7.5\%) + \textbf{21}\,(22\%)$$

36% │ i) ^{1}O$_2$
      │ ii) Cu(OAc)$_2$

antheridiol **1**

1) Isomers **20** and **21** are both 22,23-*threo* as they were also formed by OsO$_4$ *cis*-dihydroxylation of the *trans*-22-ene in **7**; the other two (**18** and **14**) are thus *erythro*.

2) Comparison of the sign of π,π^* cd Cotton effects (positive for **18** and **20**, negative for **19** and **21**) with that of cyclograntisolide **22**[4] (negative) suggested the absolute configuration.

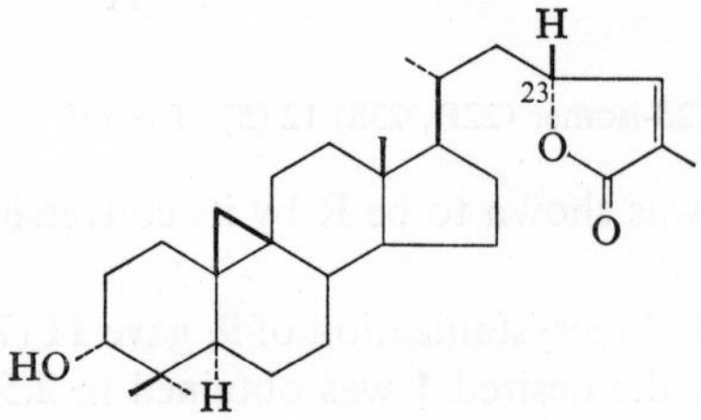

cyclograndisolide **22**
(23S : x-ray)   **23**

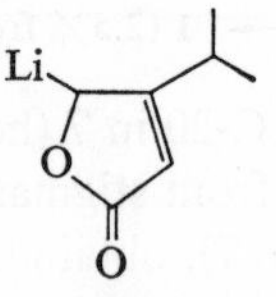

3) Addition of **23** to the aldehyde (Ac instead of THP in **5**) gave, in 70% yield, the 11-3-acetate

4) F. H. Allen, J. P. Kutney, J. Trotter, N. D. Westcott, *Tetr. Lett.*, 283 (1971).

(diastereomeric mixture), which was then converted into **18** (2/15 of total diastereomeric mixture).[5,6]

Related compound[7]

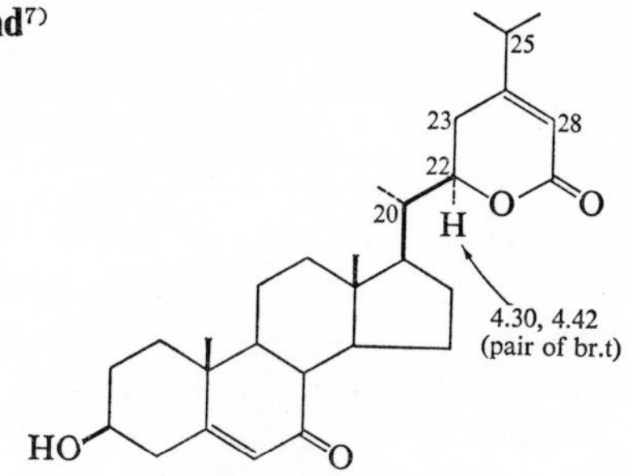

23-deoxyantheridiol **24**[7]
mp : 265–270°

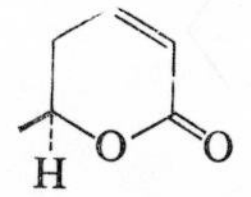

parasorbic acid **25**
cd $(\Delta\epsilon_{262})$: + 2.25

1) 23-Deoxyantheridiol **24** (3 mg) was isolated from *Achlya bisexualis* together with antheridiol **1** (2 mg). Its activity was 1/1000 that of **1**, "but this was probably due to contamination".

2) A 22R configuration **24** was suggested on the basis of a positive cd contribution (at ca. 235 nm) of the lactone and comparison of this with that of **25**.[8]

3) A synthesis was attempted from **6**. However, the nmr (see **24** and **27**) and cd (negative Cotton effect at ca. 255 nm) indicated it to be the 22-epimer (**27**).

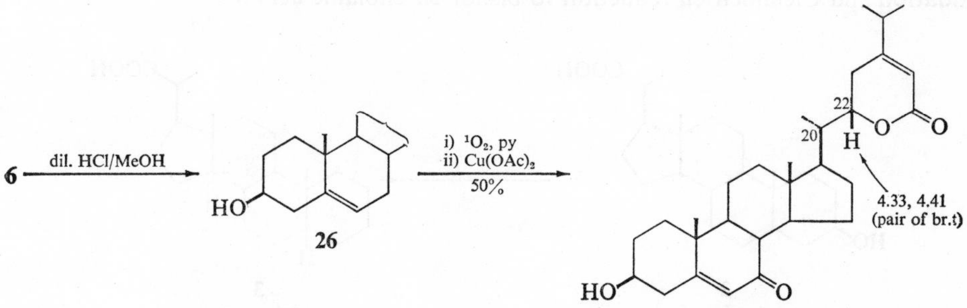

5) T. C. McMorris, R. Seshadri, *Chem. Commun.*, 1646 (1971).
6) T. C. McMorris, T. Arundachalam, R. Seshadri, *Tetr. Lett.*, 2677 (1972).
7) D. M. Green, J. A. Edwards, A. W. Barksdale, T. C. McMorris, *Tetr.*, **27**, 1199 (1971).
8) G. Snatzke, H. Schwang, P. Welzel, *Some Newer Physical Methods in Structural Chemistry* (ed. R. Bonnett, J. G. Davis), p. 159, United Trade Press, 1967.

6.20 STIGMASTEROL AND RELATED PHYTOSTEROLS

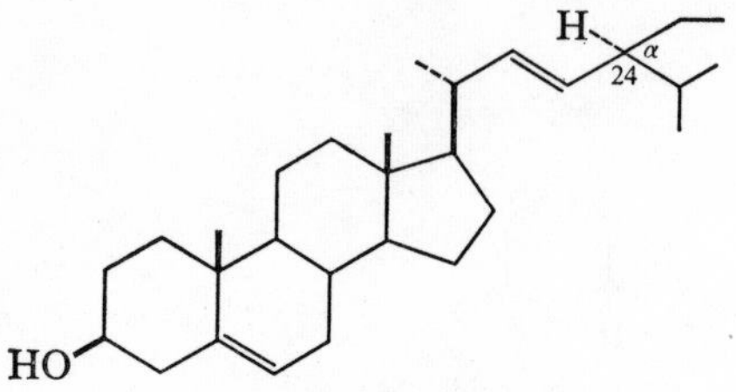

stigmasterol **1**
for tetrabromide:
mp: 211° α_D: −40°

1) Stigmasterol **1** is widely distributed in plants and is available especially from calabar[1] and soybeans. It is isolated as the sparingly soluble acetate tetrabromide.[1]

2) Ozonolysis[2] gave ethylisopropylacetaldehyde **9**, thus proving the presence of a C-24 ethyl group and Δ^{22} unsaturation.

3) Its acetate gave the 5,6-dibromide which, upon ozonolysis and debromination gave the acetate of 3β-hydroxy bisnorchol-5-enic acid **2**. This is converted by hydrogenation, hydrolysis, oxidation and Clemmensen reduction to bisnor-5α-cholanic acid **3**.[3]

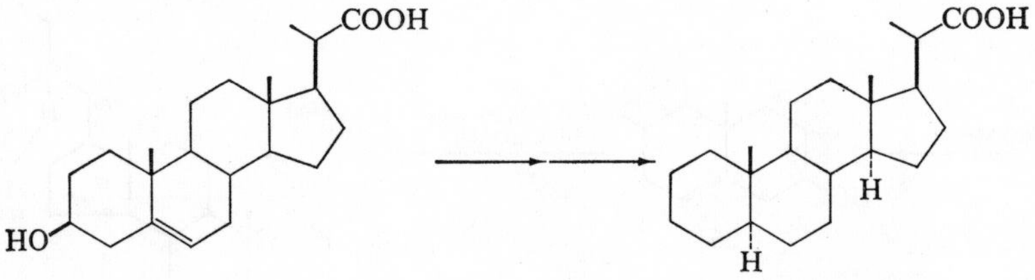

4) The position and configuration of the hydroxyl group was established[3] by hydrogenation to 5α-stigmastanol **4** and subsequent acetylation and chromium trioxide oxidation to 3β-hydroxynor-5α-cholanic acid **5**.

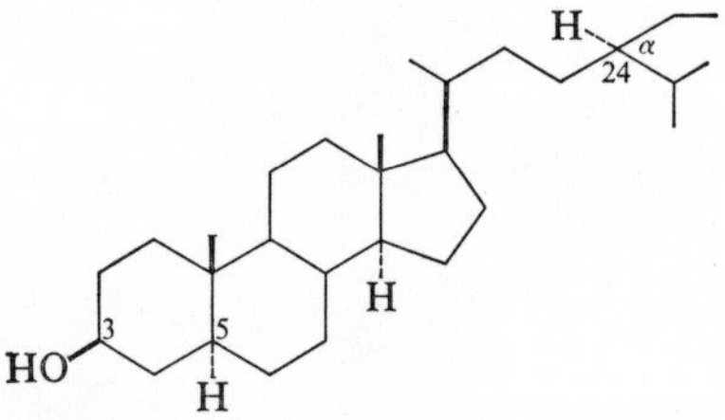

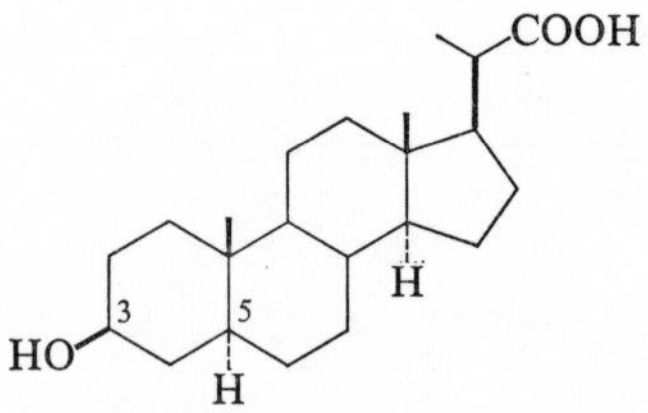

5) The position of the nuclear double bond was proven by use of the same sequence of reactions

1) A. Windaus, H. Hauth, *Chem. Ber.*, **39**, 4378 (1906).
2) A. Guiteras, *Ann. Chem.*, **494**, 116 (1932); *Z. Physiol. Chem.*, **214** 89 (1933).
3) E. Fernholz, P. N. Chakrovorty, *Ann. Chem.*, **507**, 188 (1933); *Chem. Ber.*, **67**, 2021 (1934)

as for cholesterol. Namely, the 5,6-oxide was formed, converted to the triol and then to the 3,6-dione **6**.[4] This was converted to the enone **7** and reduced to give Δ^{22}-stigmasten-3β-ol **8**.[5]

6   **7**   **8**

6) The side chain double bond was assigned as *trans* from the ir band at 970 cm^{-1}.
7) The 24-ethyl group was proven to be α as follows.[6] The ozonization product, aldehyde **9**, was converted by means of reduction, tosylation and reduction to the levorotatory hydrocarbon **10**. This had been earlier correlated[7] with (+)-isopropylsuccinic acid **11** of known configuration.

9   **10**   **11**

β-Sitosterol

1) β-Sitosterol **12** is the most common sterol among higher plants and is usually accompanied by its 24α-methyl isomer, campesterol.
2) Acetylation and subsequent chromium trioxide oxidation gives (+)-5-ethyl-6-methylheptan-2-one,[8] 3β-acetoxynorchol-5-enic acid **13**, and dehydroepiandrosterone **14**.[9]
3) It can be obtained from stigmasterol by partial side-chain hydrogenation.

β-sitosterol **12**   **13**   **14**
mp : 140° α_D : -36°

Brassicasterol

1) Brassicasterol **15** occurs in rape-seed oil and is isolated as the acetate tetrabromide.[10] It differs from stigmasterol only at C-24.

---

4) E. Fernholz, *Ann. Chem.*, **508**, 215 (1934).
5) R. E. Marker, E. L. Wittle, E. Rohrmann, *J. Am. Chem. Soc.*, **59**, 2704 (1937); *ibid.*, **60**, 1073 (1938).
6) K. Tsuda, Y. Kishida, R. Hayatsu, *J. Am. Chem. Soc.*, **82**, 3396 (1960).
7) K. Freudenberg, W. Lwowski, *Ann. Chem.*, **587**, 213 (1954).
8) O. Dalmer, F. von Werder, H. Honigmann, K. Hayns, *Chem. Ber.*, **68**, 1814, 1821 (1935).
9) W. Dirscherl, H. Nahm, *Ann. Chem.*, **555**, 57 (1943); *ibid.*, **558**, 231 (1947).
10) A. Windaus, A. Welsch, *Chem. Ber.*, **42**, 612 (1909).

2) Ozonolysis[11] gives methylisopropylacetaldehyde and 3β-hydroxybisnorchol-5-enic acid **16**.

3) Hydrogenation gave 5α-ergostanol **17**, a derivative of the common yeast sterol, ergosterol.

brassicasterol **15**
mp : 148° α_D : —64°

16

17

α-Spinasterol and cholesterol

1) α-Spinasterol **18** has been isolated from spinach,[12] senega root[13] and alfalfa.[14]

2) **18** can be obtained from 7-dehydrostigmasterol[15] under conditions in which the 7-ene is not isomerized.

3) **18** can be hydrogenated under non-isomerizing conditions to Δ^7-stigmasterol.[16]

α-spinasterol **18**
mp : 172° α_D : —4°

cholesterol **19**

4) Cholesterol **19**, of wide occurrence in animals, was first isolated from plants by Tsuda.[17] It has since been isolated from several algae,[17] fungi[18] and higher plants.[19]

11) E. Fernholz, H. E. Slavely, *J. Am. Chem. Soc.*, **62**, 428, 1875 (1940).
12) M. C. Hart, F. W. Heyl, *J. Biol. Chem.*, **95**, 311 (1932).
13) J. C. E. Simpson, *J. Chem. Soc.*, 730 (1937).
14) W. Karrer, E. Rothschild, H. Salomon, *Helv. Chim. Acta*, **22**, 313 (1939); E. Fernholz, M. L. Moore, *J. Am. Chem. Soc.*, **61**, 2467 (1939).
15) L. F. Fieser, M. Fieser, P. N. Chakravorty, *J. Am. Chem. Soc.*, **71**, 2226 (1949).
16) D. H. R. Barton, J. D. Cox, *J. Chem. Soc.*, 1354 (1948).
17) K. Tsuda, S. Akagi, Y. Kishida, *Chem. Pharm. Bull.*, **6**, 101 (1958); K. Tsuda, S. Akagi, Y. Kishida, R. Hayatsu, K. Sakai, *ibid.*, **6**, 724 (1958); G. F. Gibbons, L. J. Goad, T. W. Goodwin, *Phytochemistry*, **7**, 983 (1968); R. C. Reitz, J. G. Hamilton, *Comp. Biochem. Physiol.*, **25**, 401 (1968); N. Ikekawa, N. Morisaki, K. Tsuda, T. Yoshida, *Steroids*, **12**, 41 (1968).
18) Y. S. Chen, R. H. Haskins, *Can. J. Chem.*, **41**, 1647 (1963): K. Schubert, G. Rose, C. Horhold, *Biochim Biophys. Acta*, **37**, 168 (1967).
19) D. F. Johnson, R. D. Bennet, E. Heftmann, *Science*, **140**, 198 (1963).

6.21 ECDYSONES

α-ecdysone **1**

mp: 237–239°
ir (KBr): 3333, 1657
uv (MeOH): 242 (ε 12,400)
nmr in C₅D₅N

β-ecdysone **2**

mp: 241–242.5°
ir (KBr): 3370, 1653
uv (MeOH): 243 (ε 11,180)
nmr in C₅D₅N

I. ZOOECDYSONES

α-Ecdysone

1) Chemical studies arrived at the working structure **3**[2]; the full structure was established by x-ray analysis of **1** without usage of heavy atoms.[3]

2) Dehydration of **1** with HCl/MeOH gave two uv maxima (248, 295 nm) which are ascribable to 8,14-dien-6-one and 7,14-dien-6-one chromophores, respectively. This is a convenient diagnostic method for characterization of the allylic 14-OH in ecdysones.[4,5]

β-Ecdysone

1) Spectroscopic data (especially ms and nmr) led to the 20-hydroxy-α-ecdysone structure for β-ecdysone **2**.[5]

2) The side chain configurations (20R, 22R) are based[6] firstly on ord comparison of **7** derived from ponasterone A **5** (same 20,22-stereochemistry as **2**)[7] with the model compound **8** (derived from isoleucine); and secondly, the W-type long-range coupling between 21-H and 22-H in **5**-acetonide, which indicated a 20,22-*threo* structure, hence 20R.

1) For reviews, see P. Karlson, *Angew. Chem. Intern. Ed.*, **2**, 175 (1963); V. B. Wigglesworth, *Insect Hormones*, W. H. Freeman and Co., 1970; K. Nakanishi, *Pure Appl. Chem.*, **25**, 167 (1971); D. H. S. Horn, *Naturally Occurring Insecticides* (ed. M. Jacobson, D. G. Crisby, p. 330, Dekker, 1971; H. Hikino, Y. Hikino, *Fortschritte der Chemie Organischer Naturstoffe* (ed. L. Zechmeister), **28**, 256, Springer Verlag, 1970; C. E. Berkoff, *Quart. Rev.*, **23**, 372 (1969).
2) C. Rufer, H. Hoffmeister, H. Schairer, M. Trout, *Chem. Ber.*, **98**, 2385 (1965).
3) R. Huber, W. Hoppe, *Chem. Ber.*, **98**, 2403 (1965).
4) P. Karlson, H. Hoffmeister, W. Hoppe, R. Huber, *Ann. Chem.*, **662**, 1 (1963).
5) F. Hampshire, D. H. S. Horn, *Chem. Commun.*, 37 (1966).
6) M. Koreeda, D. A. Schooley, K. Nakanishi, H. Hagiwara, *J. Am. Chem. Soc.*, **93**, 4084 (1971).
7) G. Hüppi, J. B. Siddall, *Tetr. Lett.*, 1113 (1968).

4 (synthetic intermediate)   ponasterone A **5**

2,3,22-triacetate
of **5**

6

7 — Cotton effect (22R)   **8** + Cotton effect

3) An independent x-ray analysis on **2** also established its entire structure.[8]

Remarks

α- and β-ecdysones are insect (α- and β-) and crustacean (β-) molting hormones. The first isolation of α-ecdysone (25 mg) in crystalline form was accomplished in 1954 from 500 kg of silk-worm pupae, *Bombyx mori*.[9]

The following molting hormones from insects and crustaceans (zooecdysones) have since been identified (**9–12**). The ring structures are the same as those of α- and β-ecdysones, except for **10**. Only the sources for the first isolations are given.

9
26-hydroxy-β-ecydsone[10]
tobacco hornworm

10
2-deoxy-β-ecdysone[11]
crayfish

11
callinecdysone A[12]
(=inokosterone)

12
callinecdysone B[12]
(=makisterone A)

crab

Insects and crustaceans are incapable of biosynthesizing steroids from non-steroid precursors and thus require steroids in their diet.[1,13,14] Both α- and β-ecdysones are biosynthesized from cholesterol. Besides this, it has been shown that 7-dehydrocholesterol **13** is converted into ecdysones.[15]

blow fly,
Calliphora stygia

13 α- and β-ecdysones **1,2**

8) B. Dammeier, W. Hoppe, *Chem. Ber.*, **104**, 1660 (1971).
9) A. Butenandt, P. Karlson, *Z. Naturforsch.*, **9b**, 389 (1954).
10) M. J. Thompson, J. N. Kaplanis, W. E. Robbins, R. T. Yamamoto, *Chem. Commun.*, 650 (1967).
11) M. N. Galbraith, D. H. S. Horn, E. J. Middleton, R. J. Hackney, *Chem. Commun.*, 83 (1968).
12) A. Faux, D. H. S. Horn, E. J. Middleton, H. M. Fales, M. E. Lowe, *Chem. Commun.*, 175 (1969).
13) W. E. Robbins, J. N. Kaplanis, J. A. Svoboda, M. J. Thompson, *Ann. Rev. Entomol.*, **16**, 53 (1971).
14) F. J. Ritter, W. H. J. M. Wientjens, *TNO-Nieuws*, **22**, 381 (1967).

The results of catabolic studies on ecdysones in animals may be summarized as follows.[1]

Animals used:
a: tobacco hornworm
b: crayfish
c: crab
d: silk-worm
e: fly

Ecdysones have long been considered to be secreted from the prothoracic glands.[1] However, experiments such as the demonstration of ecdysone biosyntheses from cholesterol outside the glands (i.e., in isolated abdomen) have cast doubt upon such views.[17-20]

II. PHYTOECDYSONES

A group of steroids with molting hormone activity have now been found in the plant kingdom (phytoecdysones). The first independent isolations of ponasterones,[21] and inokosterone **11** and β-ecdysone **2**[22] were reported in 1966 and 1967, respectively, just after the structure determinations of α- and β-ecdysones. They are distributed widely and frequently in abundant quantities—e.g., dry leaves (1 kg) of *Podocarpus nakaii* gave 1–2 g of ponasterone A **5**.[23] A total of about forty phytoecdysones are known. All the zooecdysones excepting **9** have also been isolated from plants.

Accumulation of spectroscopic data of these ecdysones now makes it relatively easy to characterize the various ecdysone structures. For example, the chemical shifts of methyl groups

15) M. N. Galbraith, D. H. S. Horn, E. J. Middleton, J. A. Thomson, *Chem. Commun.*, 179 (1970).
16) M. Locke, *Tissue Cell.*, **1**, 103 (1969).
17) S. B. Wier, *Nature*, **228**, 580 (1970).
18) K. Nakanishi, H. Moriyama, T. Okauchi, S. Fujioka, M. Koreeda, *Science*, **176**, 51 (1972).
19) M. Gersch, J. Stürzebecher, *Experientia*, **27**, 1475 (1971).
20) P. E. Ellis, E. D. Morgan, A, P. Woodbridge, *Nature*, **238**, 274 (1972).
21) K. Nakanishi, M. Koreeda, S. Sasaki, M. L. Chang, H. Y. Hsu, *Chem. Commun.*, 915 (1966).
22) T. Takemoto, S. Ogawa, N. Nishimoto, *Yakugaku Zasshi*, **87**, 325 (1967).
23) For large-scale extraction and application of high-speed liquid chromatography, see D. A. Schooley, G Weiss, K. Nakanishi, *Steroids*, **19**, 377 (1972).

in pyridine provide important information. See **1** and **2**; all nmr data in pyridine-d_5.[1] The side chain branching and number of hydroxyls in nucleus and side chain are most conveniently diagnosed from ms.[1]

In spite of the variations in side chains, the 18-, 21- and 22-H nmr peaks show that the 20R, 22R-configurations are common to all, excepting shidasterone (C-20 and/or 22-isomer of β-ecdysone).[1]

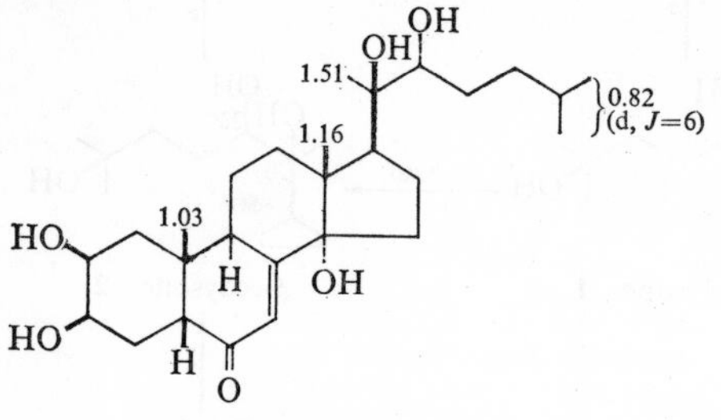

ponasterone A **5**

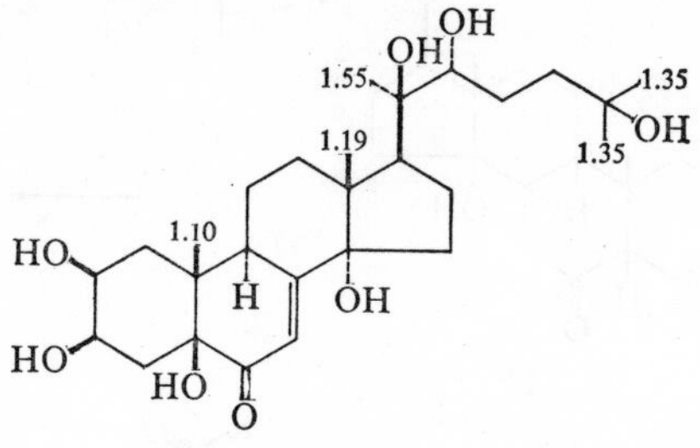

polypodine B **20**

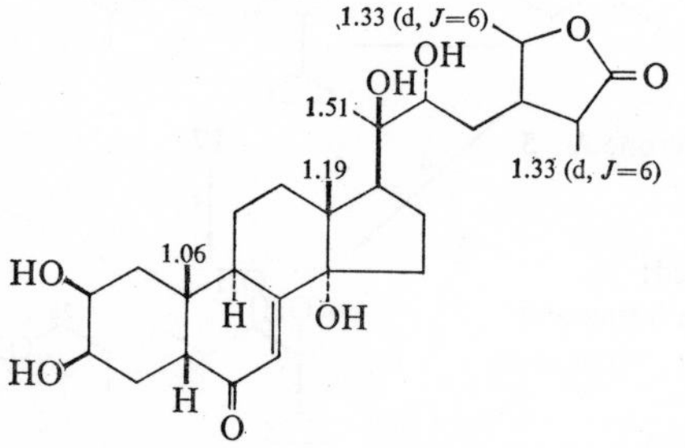

cyasterone **21**

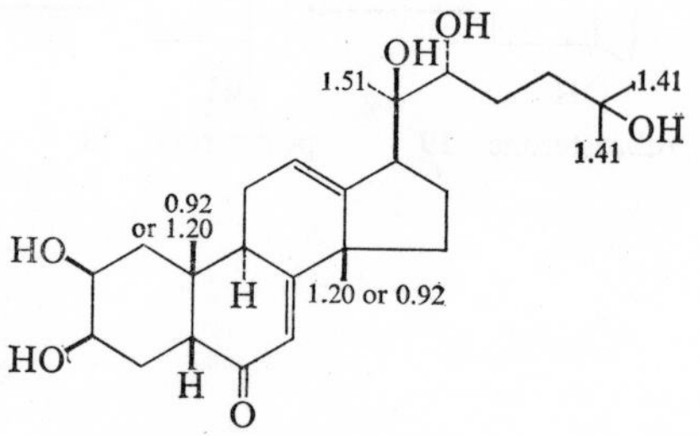

stachysterone A **22**

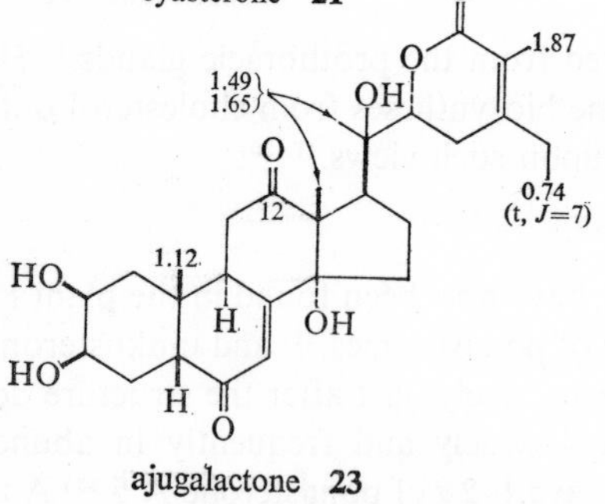

ajugalactone **23**

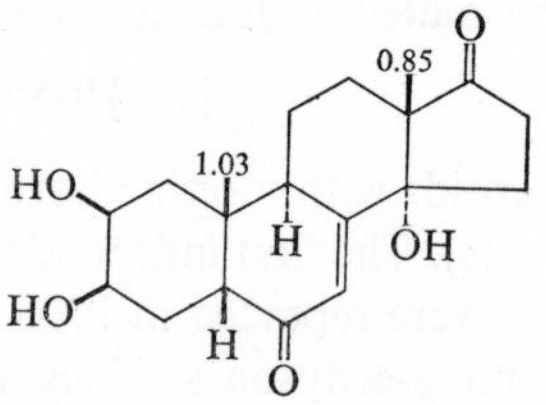

rubrosterone **19**

1) Polypodine B **20** is the first ecdysone with a 5β-OH.[24]
2) Cyasterone **21** has a uniquely strong activity in certain systems.[25]
3) Stachysterone A **22** with a rearranged methyl at C-14 still shows bioactivity.[26]
4) Ajugalactone **23** showed "anti-ecdysone" activity against ponasterone A, but not against β-ecdysone.[27] However, this antagonistic activity is so far limited to a dipping assay employing *Chilo suppressalis* (rice-stem borer).[27a]
5) The activity of the androstane type ecdysone, rubrosterone **19**, is very weak. It is assumed to be the metabolite of ecdysones both in animals and in plants (see **18→19**).[28]

24) J. Jizba, V. Herout, F. Sorm, *Tetr. Lett.,* 5139 (1967).
25) T. Takemoto, Y. Hikino, K. Nomoto, H. Hikino, *Tetr. Lett.,* 3191 (1967).
26) S. Imai, E. Murata, S. Fujioka, T. Matsuoka, M. Koreeda, K. Nakanishi, *J. Am. Chem. Soc.,* **92**, 7510 (1970).
27) M. Koreeda, K. Nakanishi, M. Goto, *J. Am. Chem. Soc.,* **92**, 7512 (1970).
27a) Y. Sato, M. Sakai, S. Imai, S. Fujioka, *Appl. Ent. Zool.,* **3**, 49 (1968).
28) T. Takamoto, Y. Hikino, H. Hikino, *Tetr. Lett.,* 3053 (1968).

III. Synthesis

The first two syntheses of α-ecdysone (Routes I and II) were followed by alternative routes.

Route I (Syntex group)[29]

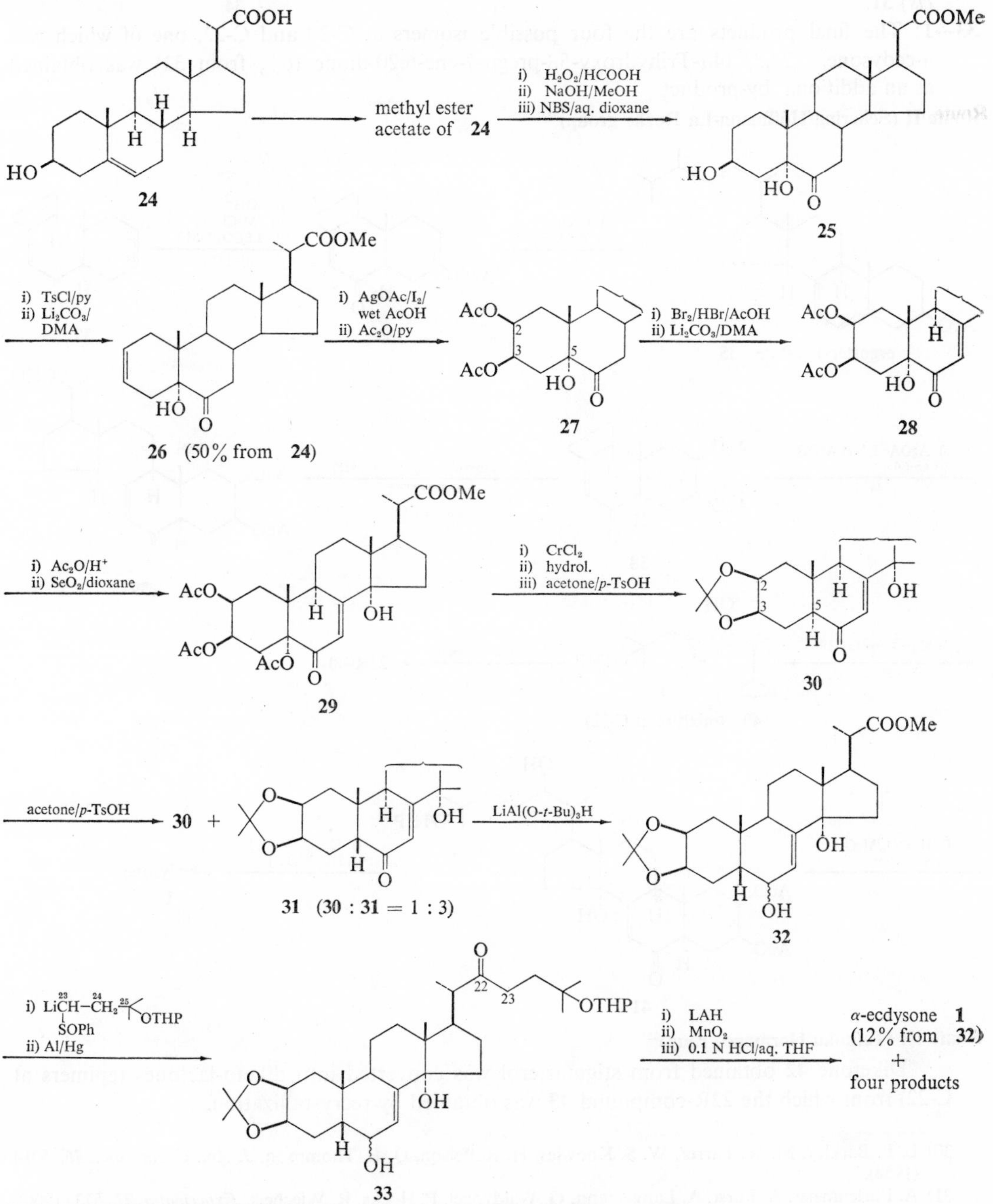

COOH
HO
24
methyl ester
acetate of 24
i) H_2O_2/HCOOH
ii) NaOH/MeOH
iii) NBS/aq. dioxane
COOMe
HO
HO
O
25
i) TsCl/py
ii) Li_2CO_3/DMA
COOMe
HO
O
26 (50% from 24)
i) AgOAc/I_2/wet AcOH
ii) Ac_2O/py
AcO
AcO
HO
O
27
i) Br_2/HBr/AcOH
ii) Li_2CO_3/DMA
AcO
AcO
HO
O
28
i) Ac_2O/H^+
ii) SeO_2/dioxane
COOMe
AcO
AcO
AcO
OH
O
29
i) $CrCl_2$
ii) hydrol.
iii) acetone/p-TsOH
OH
O
30
acetone/p-TsOH
30 +
OH
O
31 (30 : 31 = 1 : 3)
LiAl(O-t-Bu)$_3$H
COOMe
OH
OH
32
i) LiCH–CH$_2$–OTHP
SOPh
ii) Al/Hg
O
OTHP
OH
OH
33
i) LAH
ii) MnO_2
iii) 0.1 N HCl/aq. THF
α-ecdysone 1
(12% from 32)
+
four products

29) J. B. Siddall, A. D. Cross, J. H. Fried, *J. Am. Chem. Soc.*, 88, 862 (1966).

26→27: In reaction i) of this step, a cyclic oxonium intermediate **34** adds water to yield a 2β-acetoxy-3β-hydroxy structure.[30]

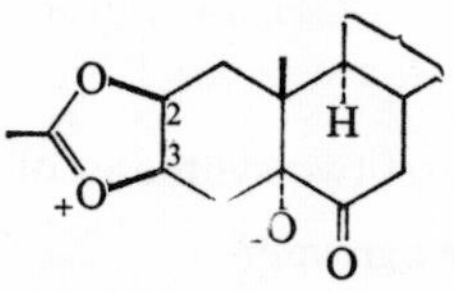

30→30+31: Steric repulsion between 19-H and the 2β-substituent favored 5β-H 7-en-6-one (A/B *cis*) **31**.

33→1: The final products are the four possible isomers at C-20 and C-22, one of which was α-ecdysone. 2β,3β,14α-Trihydroxy-5β-pregn-7-ene-6,20-dione (6% from **32**) was obtained as an addit'ona by-product.

34

Route II (Schering/Hoffmann-La Roche group)[31]

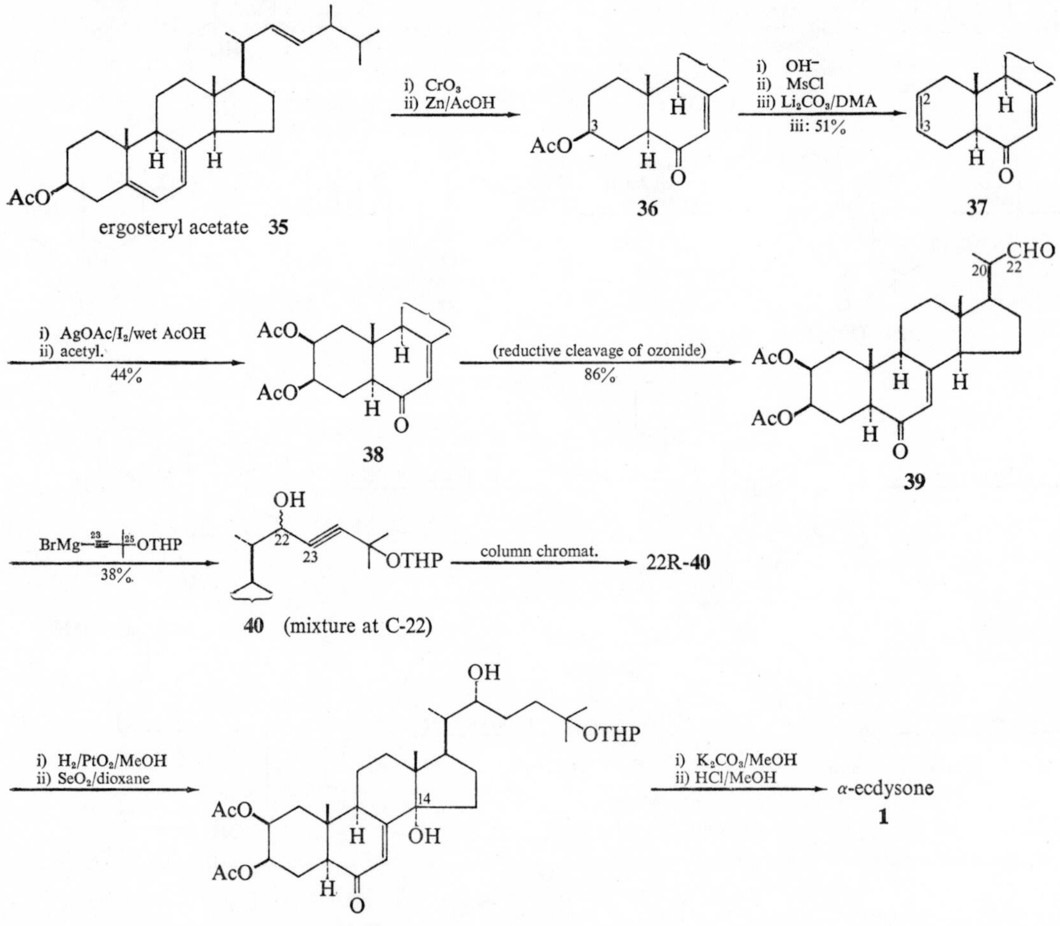

ergosteryl acetate **35**

i) CrO₃
ii) Zn/AcOH

36

i) OH⁻
ii) MsCl
iii) Li₂CO₃/DMA
iii: 51%

37

i) AgOAc/I₂/wet AcOH
ii) acetyl.
44%

38

(reductive cleavage of ozonide)
86%

39

BrMg—²³≡≡—|²⁵OTHP
38%.

40   (mixture at C-22)

column chromat.

22R-40

i) H₂/PtO₂/MeOH
ii) SeO₂/dioxane

41

i) K₂CO₃/MeOH
ii) HCl/MeOH

α-ecdysone

1

Route III (Teikoku Hormone group)[32]

Diketone **42** obtained from stigmasterol was converted into diketo-lactones (epimers at C-22) from which the 22R-compound **43** was obtained by recrystallization.

30) L. B. Barkley, M. W. Farrer, W. S. Knowles, H. Raffelson, Q. E. Thompson, *J. Am. Chem. Soc.*, **76**, 5014 (1954).

31) A. Furlenmeier, A. Furst, A. Langemann, G. Waldvogel, P. Hocks, R. Wiechert, *Experientia*, **22**, 573 (1966).

32) H. Mori, K. Shibata, K. Tsuneda, M. Sawai, *Chem. Pharm. Bull.*, **16**, 563, 2416 (1968).

The 14α-OH was introduced by perphthalic acid oxidation of the dienol acetate of the 7-en-6-one (6-acetoxy-6,8(14)-diene).

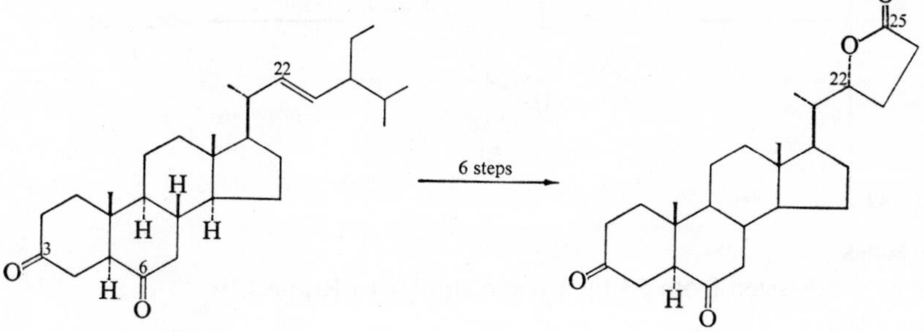
22
H
H
H
3
H
6
O
O
6 steps
O
25
O
22
O
H
O

42 43

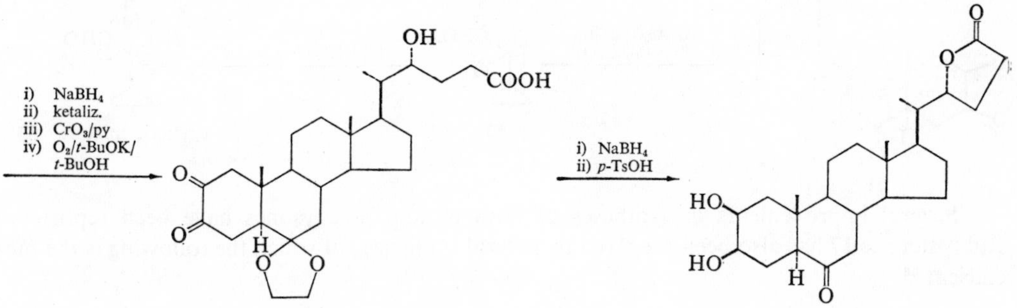
i) NaBH4
ii) ketaliz.
iii) CrO3/py
iv) O2/t-BuOK/t-BuOH
OH
COOH
O
O
H
O
O
i) NaBH4
ii) p-TsOH
O
O
HO
HO
H
O

44 45

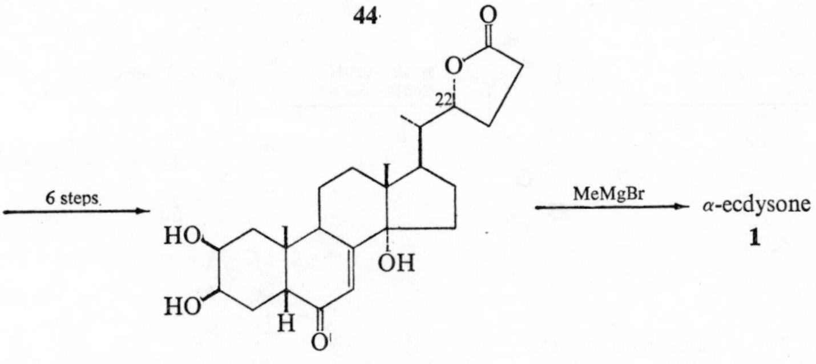
6 steps
O
O
22
HO
HO
H
O
OH
MeMgBr
α-ecdysone
1

46

Route IV (Barton, Sammes and co-workers)[33)]

The key step in this synthesis is direct formation of the 2,7-dien-6-one **48**, **49** from **47** via the 3β-tosylate. Oxidation with perphthalic acid was also employed in this synthesis for 14α-hydroxylation.

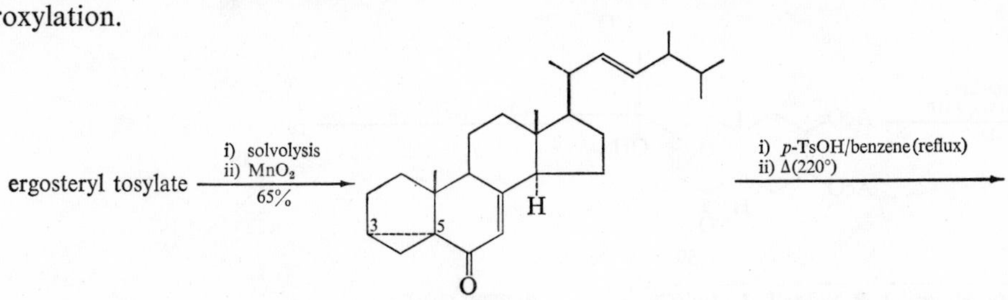
ergosteryl tosylate
i) solvolysis
ii) MnO2
65%
H
3
5
O
i) p-TsOH/benzene (reflux)
ii) Δ(220°)

47

33) D. H. R. Barton, P. G. Feakins, J. P. Poyser, P. G. Sammes, *J. Chem. Soc.* C, 1584 (1970).

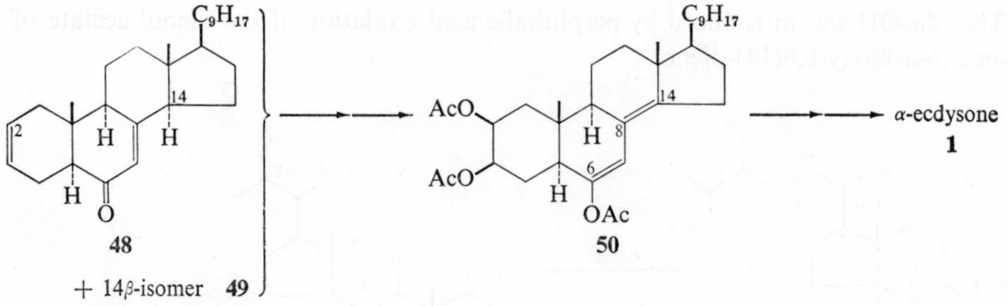

$+$ 14β-isomer **49**

Syntheses of other ecdysones

β-Ecdysone **2** was synthesized from **53** by a route similar to Route I.[34]

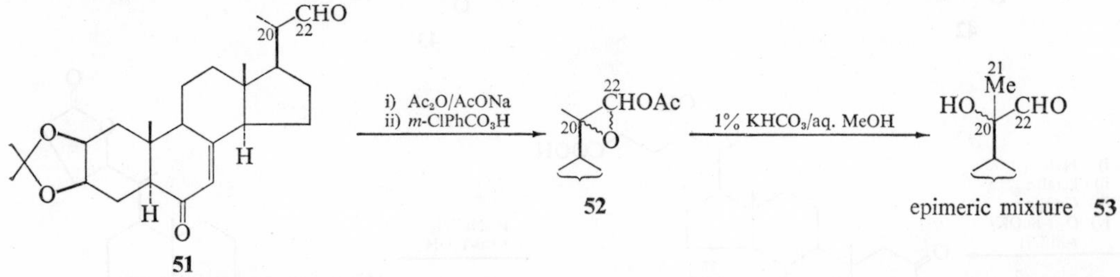

Several improvements in syntheses of both α- and β-ecdysones have been reported.[35] Rubrosterone **17** has also been prepared by several syntheses, of which the following is the most efficient.[36]

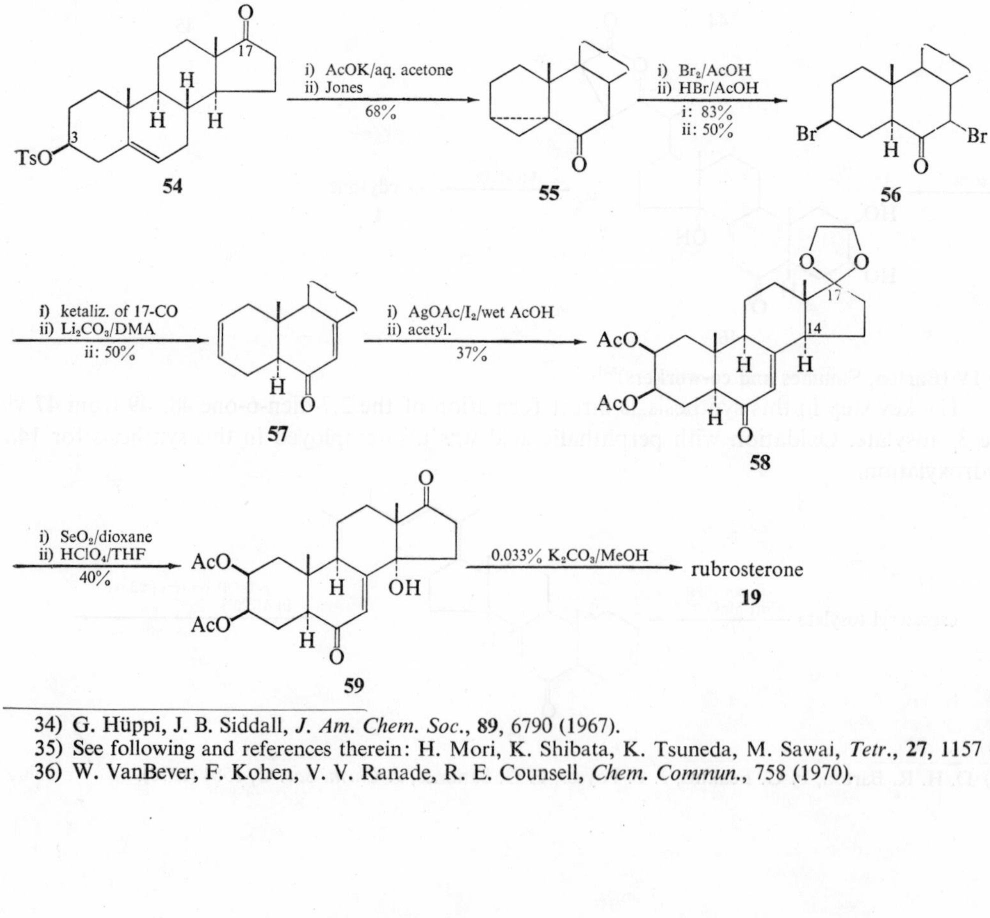

34) G. Hüppi, J. B. Siddall, *J. Am. Chem. Soc.*, **89**, 6790 (1967).
35) See following and references therein: H. Mori, K. Shibata, K. Tsuneda, M. Sawai, *Tetr.*, **27**, 1157 (1971).
36) W. VanBever, F. Kohen, V. V. Ranade, R. E. Counsell, *Chem. Commun.*, 758 (1970).

6.22 THE STRUCTURE OF CHIOGRALACTONE

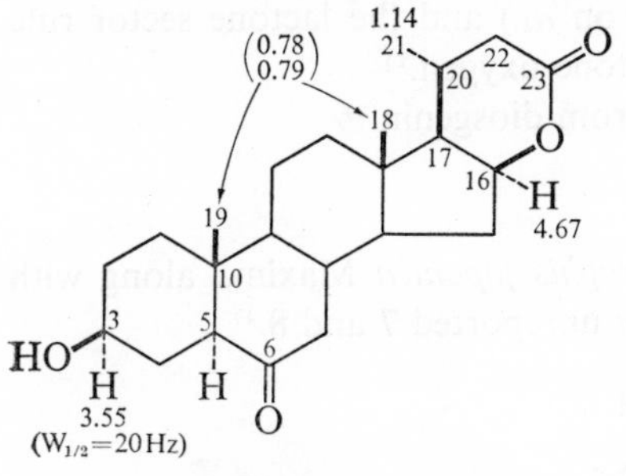

chiogralactone **1**
mp: 238–240° α_D: −113°
ir: 3730, 3540(OH), 1733
(δ-lactone), 1712(C=O)

1

1) The fact that **1** is isolated together with other steroidal sapogenins suggested a steroi da skeleton.

2) Formation of **1** monoacetate with no additional OH bands (ir) indicated the presence of only one hydroxyl group.

3) The ir of triketocarboxylic acid **2** exhibited bands corresponding to five- and six-membered ring ketones. The five-membered ring ketone should result from the hydroxyl group of the δ-lactone.

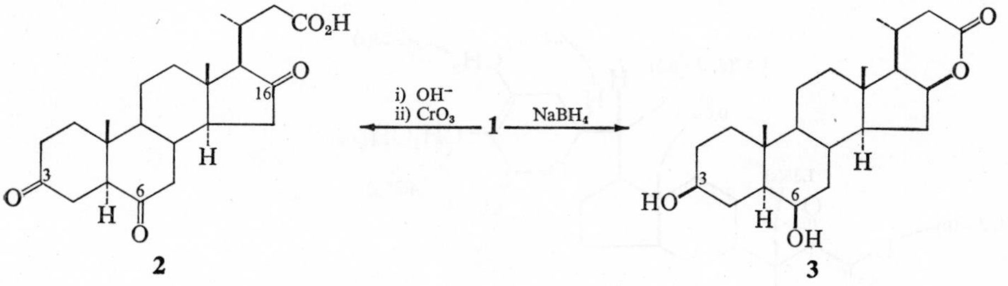

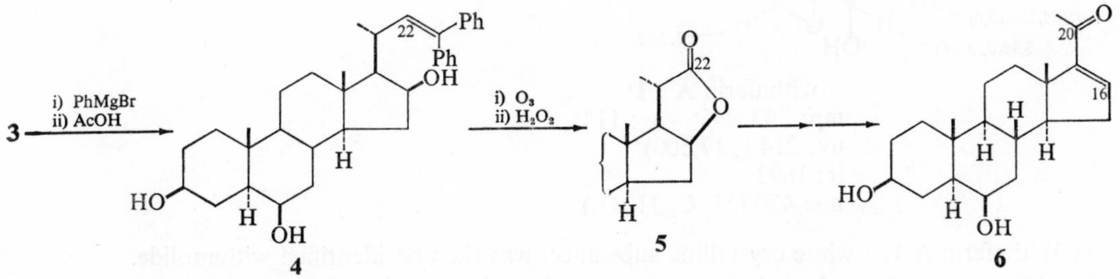

1) K. Takeda, M. Iwasaki, A. Shimaoka, H. Minato, *Tetr. Suppl.*, **8**, 123 (1966).
2) M. Iwasaki, *Tetr.*, **23**, 2145 (1967).
3) K. Takeda, A. Shimaoka, M. Iwasaki, H. Minato, *Chem. Pharm. Bull.*, **13**, 691 (1965).
4) K. Takeda, H. Minato, A. Shimaoka, *Chem. Commun.*, 105 (1968).

4) The chemical shifts of 19-H of **1** and **3** are in agreement with those of 3β-hydroxy-6-keto or 3β,6β-dihydroxy steroids.

5) Conversion of **1** to 3,6-dihydroxy-16-en-20-one **6** supported the previous assignment.

6) Application of the Hudson-Klyne lactone rule, (based on α_D) and the lactone sector rule (based on cd) suggested the β configuration of the δ-lactone oxygen.[1]

7) The structure of **1** has been confirmed by its synthesis from diosgenin.[2]

Remarks

Chiogralactone was isolated from the plant *Chionographis japonica* Maxim., along with nine other steroids[3] including diosgenin and the previously unreported **7** and **8**.[4]

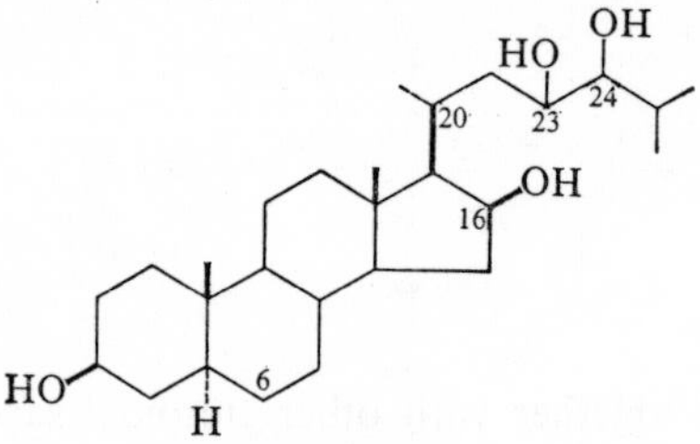

7 6-keto
8 6β-OH

6.23   WITHANOLIDES

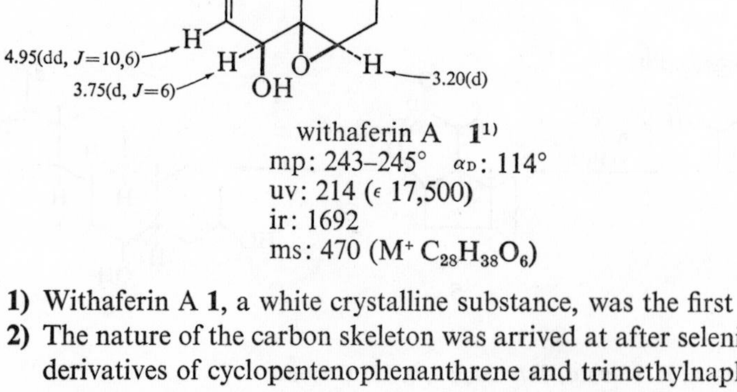

withaferin A   **1**[1]
mp: 243–245° α_D: 114°
uv: 214 (ϵ 17,500)
ir: 1692
ms: 470 (M$^+$ C$_{28}$H$_{38}$O$_6$)

1) Withaferin A **1**, a white crystalline substance, was the first identified withanolide.

2) The nature of the carbon skeleton was arrived at after selenium dehydrogenation which gave derivatives of cyclopentenophenanthrene and trimethylnaphthalene.[2]

1) D. Lavie, E. Glotter, Y. Shvo, *J. Org. Chem.*, **30**, 1774 (1965).
2) D. Lavie, E. Glotter, Y. Shvo, *J. Chem. Soc.* (C), 7517 (1965).

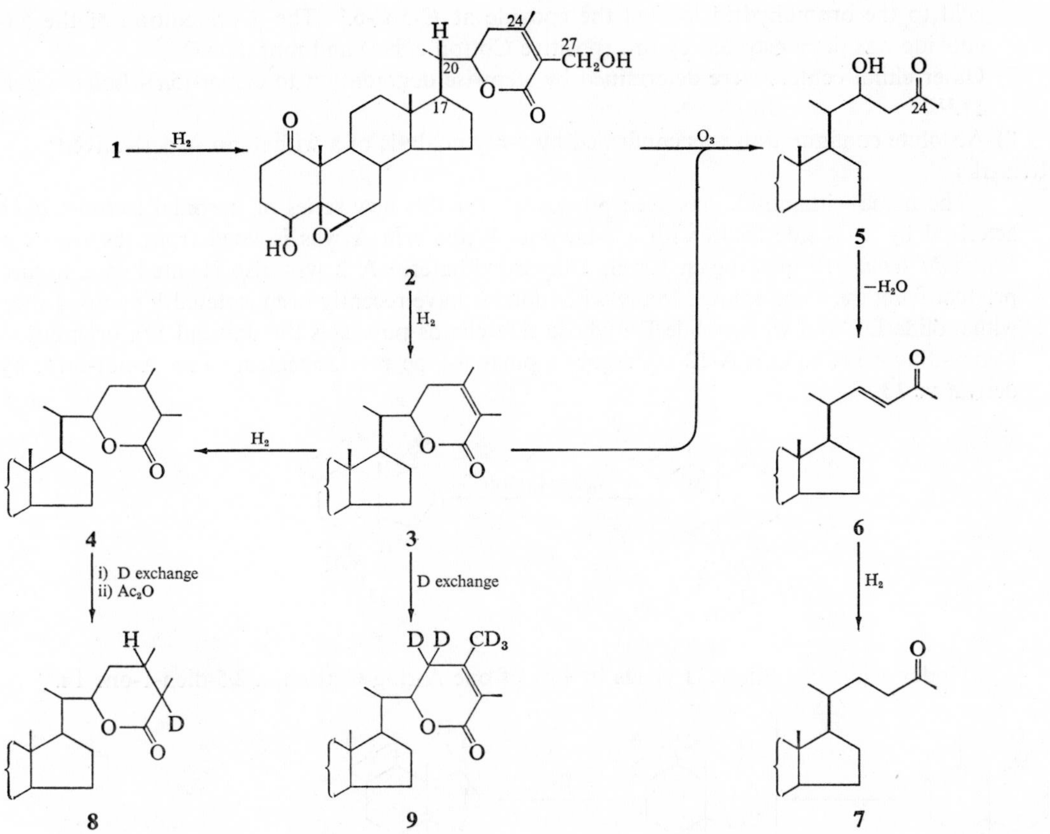

3) Stepwise hydrogenation of **1** led to **2** and through hydrogenolysis of the 27-OH to **3** and finally to **4**. This showed the presence of two double bonds, both conjugated to carbonyls.[1]

4) The side-chain structure was obtained by nmr[1] and by a series of reactions including D exchange (see **8** and **9**) and ozonolysis. The α,β-unsaturated lactone ring was confirmed by ozonolysis of **3** resulting in production of the β-hydroxyketone **5**, which established the position of the primary alcohol at C-27 of the lactone ring. Location of the secondary 20-Me group was deduced from ms and nmr data.[1]

5) The presence of a Δ^2-1,4-oxohydroxy system was inferred in **1** from nmr data.[2] The allylic position of the 4-hydroxyl group was ascertained by oxidation of **1** to the enedione **10**. In a steroid nucleus a hydroxy-enone system can be oxidized to a homo-cisoid enedione only if it is in ring A.[2]

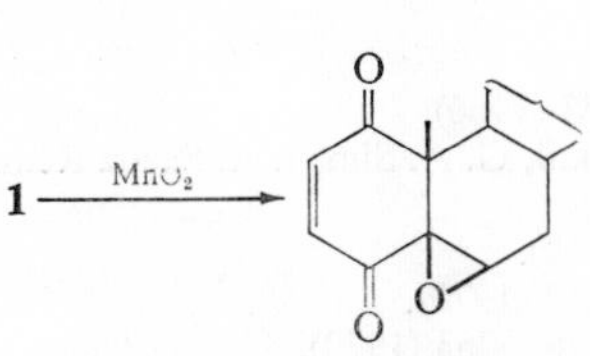

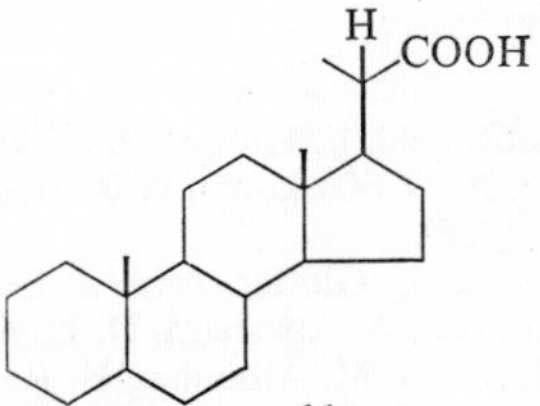

6) Epoxide functionality was indicated by nmr.[1] Opening of the epoxide with hydrobromic

acid to the bromohydrin located the epoxide at C-5/C-6.[2] The β orientation of the 5,6-epoxide was demonstrated by ord (positive Cotton effect) and nmr data.[3]

7) Other chiral centers were determined by stepwise degradation to bisnor-(5α)-cholanic acid **11**.[3]

8) Absolute configuration was confirmed by x-ray analysis of a withaferin A derivative.[4]

Remarks

The name withanolide has been proposed[5] for this new series of steroidal lactones characterized by a C_9 side chain with a δ-lactone. Withaferin A was isolated from the leaves of *Withania semnifera* growing in Israel. Dihydrowithaferin A **2** was also isolated as a natural product from the same source. Many withanolides have recently been isolated,[6,7] among them withanolide D[5] and withanolide E[8] whose side chains possesses the unusual 17α orientation. Deoxy-dihydrowithaferin A **12** undergoes a pinacol-type rearrangement to an A-nor-5-formyl derivative **13**.[9]

Under acidic conditions, **1** yields by loss of one A-ring C atom, a 2,5-dien-1-one **14**.[9]

The insect repellant plant *Nicandra physaloides* contains two closely related compounds **15** and **16** (structure by x-ray).[10]

15 R=H₂
16 R=O

3) D. Lavie, S. Greenfield, E. Glotter, *J. Chem. Soc.* (C), 1753 (1966).
4) S. M. Kupchan, R. W. Koskotch, P. Bollinger, A. J. McPhail, G. A. Sim, J. A. Saenz Renauld, *J. Am. Chem. Soc.*, **87**, 5805 (1965).
5) D. Lavie, I. Kirson, E. Glotter, *Israel J. Chem.*, **6**, 671 (1968).
6) I. Kirson, E. Glotter, A. Abraham, D. Lavie, *Tetr.*, **26**, 5062 (1970).
7) I. Kirson, D. Lavie, S. M. Albonico, H. R. Juliani, *Tetr.*, **26**, 5063 (1970).
8) D. Lavie, I. Kirson, E. Glotter, *J. Chem. Soc.* (C), 877 (1972).
9) D. Lavie, Y. Kashman, E. Glotter, N. Danieli, *J. Chem. Soc.* (C), 1757 (1966).
10) M. J. Begley, L. Crombie, P. J. Ham, D. A. Whiting, *Chem. Commun.*, 1108 (1972).

6.24 STRUCTURE OF PHYSALIN A

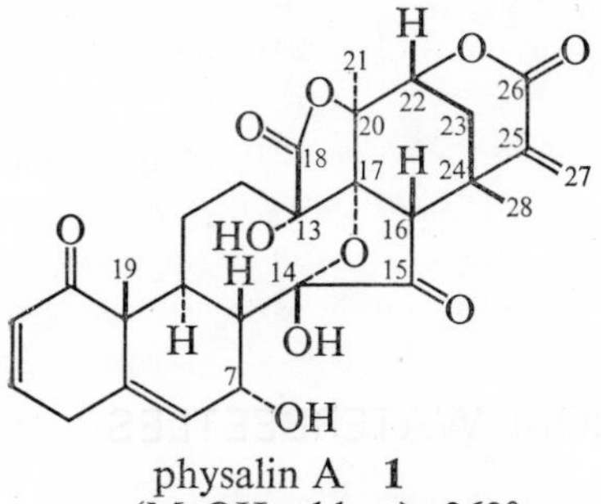

physalin A **1**
mp (MeOH adduct): 262°
uv: 218 (infl., ε 10,000)
ir: 1780 (γ-lactone)

1) Complicated ir absorption of **1** between 1600–1800 cm⁻¹ showed the existence of several carbonyl functions besides the γ-lactone. The nmr of **1** suggested the presence of five olefinic protons and three tertiary methyls.

2) Reduction of **1** to **2**, with the resulting appearance of a secondary methyl group (nmr) suggested the presence of an *exo* methylene group in **1**, which was confirmed by the formation of formaldehyde upon ozonolysis of **1**.

3) Selenium dehydrogenation of **2** gave only naphthalene derivatives, but no phenanthrene derivatives, indicating that **1** contains neither usual steroidal nor triterpenoid skeletons.[1]

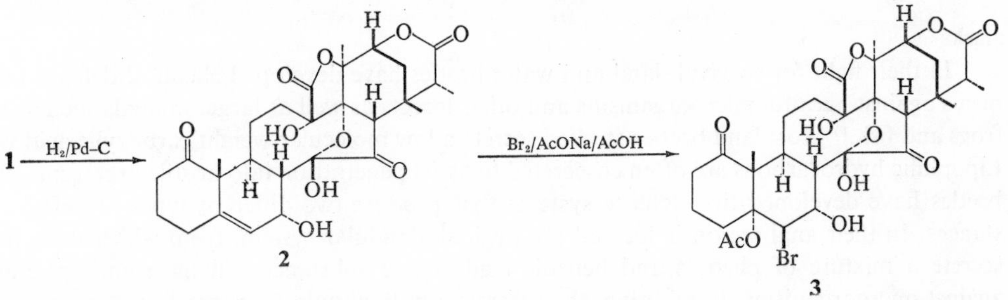

4) X-ray crystallographic analysis of the acetoxy bromide **3** established the planar structure and relative configuration of **2**.[2]

5) The structure of **1** can be completed by adding two double bonds to **2**. A terminal methylene group can only be located between C-25 and C-27. The uv absorption of **7** can be rationalized by an overlap of bands due to an enone and a conjugated lactone which fixes the location of the other double bond at C-2.

Remarks

Physalin A is the major bitter principle of *Physalis alkekengi* var. *francheti*. The congener physalin B **4**, a slightly bitter substance, has been isolated from this plant.[3]

1) T. Matsuura, M. Kawai, N. Nakashima, Y. Butsugan, *Tetr. Lett.*, 1083 (1969).
2) M. Kawai, T. Toga, K. Osaki, T. Matsuura, *Tetr. Lett.*, 1087 (1969).
3) T. Matsuura, M. Kawai, *Tetr. Lett.*, 1765 (1969).

physalin B **4**

6.25   STEROID DEFENSE SUBSTANCES FROM WATERBEETLES

cortexone **1**
mp: 135–140°
uv: 241(enone)
ir: 3475(OH), 1693(C=O),
    1613, 1668(enone)

1) Cd spectra showing extrema at 321 and 283 nm were indicative of a Δ^4-3-keto steroid and a saturated ketone, respectively.

2) The structural assignment was confirmed by comparison with authentic material.

Remarks

In their fight for survival, land and water beetles have developed chemical defence mechanisms against harmful microorganisms and other insects as well as larger animals such as birds, frogs and fish.[1] Most land beetles studied secrete a low molecular weight carboxylic acid vapor. Lipophilic hydrocarbons are often cosecreted to assist penetration into hostile organisms. Water beetles have developed two defense systems that produce two different types of defense substances. In their anal region is located the pygidal glandular system, from which these insects secrete a mixture of phenols and benzoic acid. These substances all have antiseptic action against microorganisms. In addition, they secrete a milky liquid from their prothoracic region, which as early as 1721 was recognized to poison fish and amphibians.

The first of these defense substances to be identified was cortexone **1** (11-deoxycorticosterone), isolated from the water beetle *Marginalis dytiscius*.[2] The amount of **1** contained in a single beetle is 0.4 mg, which is greater than that isolated from one thousand oxen. The insect steroids presumably disrupt the sodium-potassium balance of vertebrates preying upon the beetle, thereby stunning the host.

The prothoracic secretion from water beetles of various different "nationalities" have been shown to contain a number of related oxygenated pregnenones and 4,6-pregnadienones (**2–5**). For example, **4** has been isolated from both Mexican[3] and Indian[4] beetles.

1) H. Schildknecht, *Angew. Chem. Intern. Ed.*, **9**, 1 (1970).
2) H. Schildknecht, R. Siewerdt, U. Maschwitz, *Angew Chem. Intern. Ed.*, **5**, 421 (1966).
3) H. Schildknecht, W. Körnig, *Angew. Chem. Intern. Ed.*, **7**, 62 (1968).
4) M. S. Chadha, N. J. Joshi, V. A. Mamdapur, A. T. Sipahimalani, *Tetr.*, **26**, 2061 (1970).

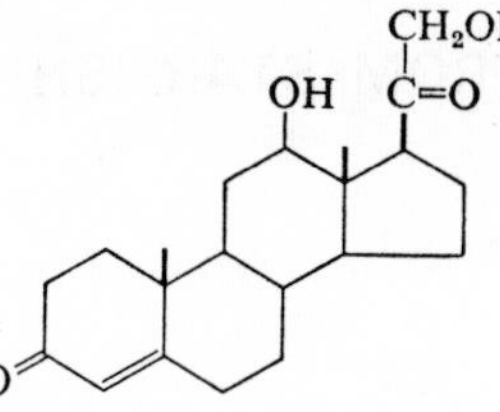

2 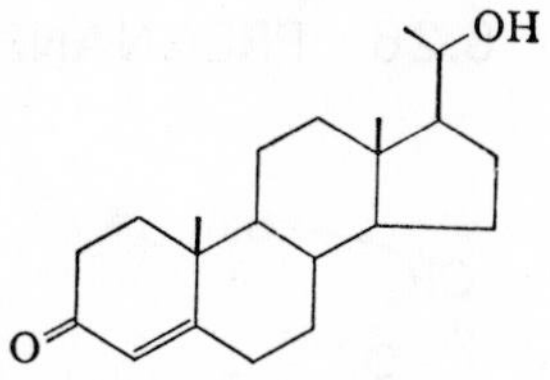   3

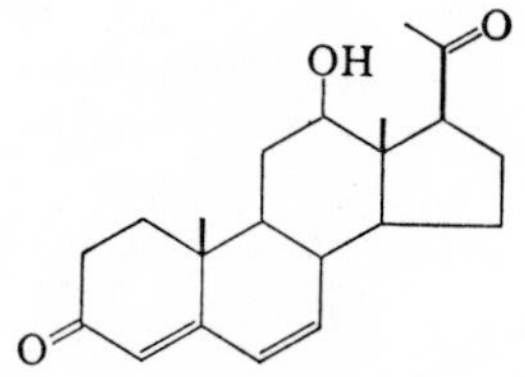

4 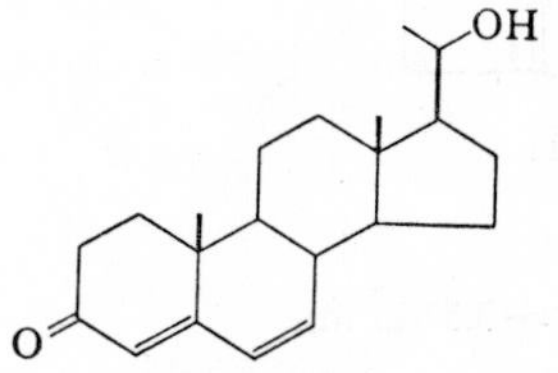   5

C-15 hydroxylated pregnadienones have also been isolated, some as their isobutyric esters (6).[5] More interesting was the discovery of the sex hormone testosterone 7[6] and estrone 8[7] and some additional C_{18}, C_{19} steroids from certain water beetles.

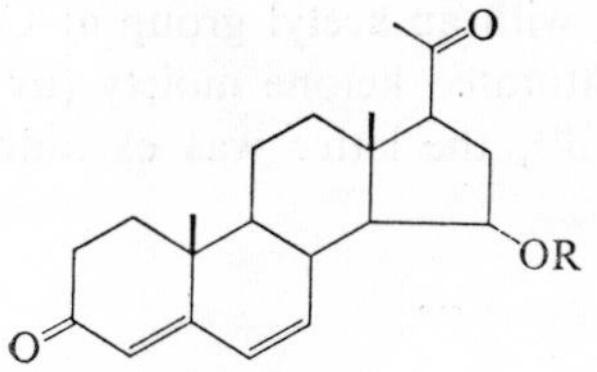

6a R = H
6b R = —C—⟨
        ‖
        O

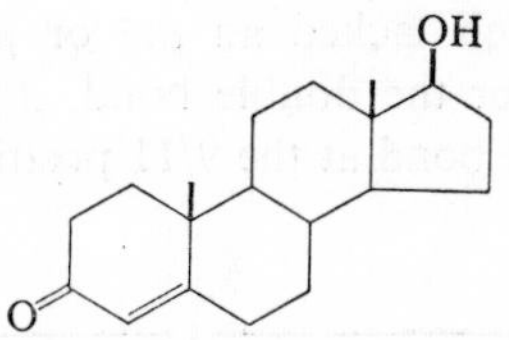

7 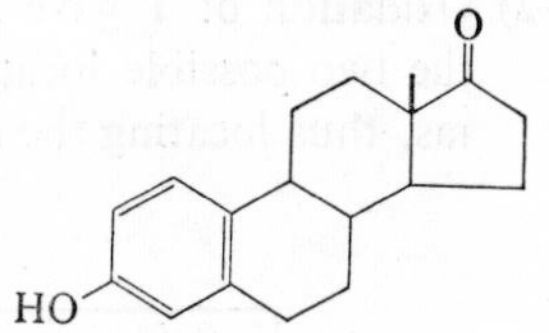   8

Biosynthetic studies of insect steroids have indicated that beetles are unable to convert mevalonate into steroids, whereas cholesterol and progesterone were incorporated.[1] Cholesterol is degraded by removal of either the entire side chain or part of it at C-20, forming pregnenolone and then progesterone.[8] The incorporation of cholesterol into the various enones and dienones is of the same order of magnitude, while progesterone was preferentially incorporated into dienones. This suggests different intermediates in the biosynthetic pathways from these precursors.

Labeling experiments with *Acilius sulcatus* showed the following results.

cholesterol →→ **1** + 6-dehydro-**1**
↓   **5** + 6,7-dihydro-**5**
progesterone →→ 6,7-dehydroprogesterone

5) H. Schildknecht, D. Holtz, *Proceedings of the Third International Congress on Hormonal Steroids* (ed. V. H. T. James, L. Martine), p. 158, Excerpta Medica, 1971.
6) H. Schildknecht, H. Birringer, U. Maschwitz, *Angew Chem. Intern. Ed.*, **6**, 558 (1967).
7) H. Schildknecht, H. Birringer, *Z. Naturforsch*, **24b**, 1929 (1969).
8) F. J. Ritter, W. H. J. M. Weintjens, *Steroid Metabolism of Insects*, p. 67, Centraal Laboratorium TNO, 1967.

6.26 PREGNANES FROM STARFISH

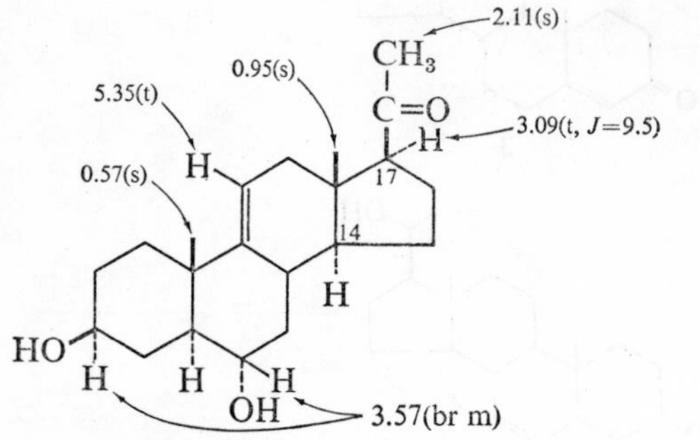

genin 1[1-3]

mp: 162–163° α_D: +65.2
ms: 332.23510 ($C_{21}H_{32}O_3$)
cd (θ_{287}): +7380
ir (nujol): 3240, 1700, 1040, 825

1) Ms data of genin diacetate[3] indicated a steroidal nucleus with an acetyl group at C-17.[1]
2) Oxidation of **1** gave **2**, which lacked an α,β or β,γ-unsaturated ketone moiety (uv). Of the two possible locations for the double bond, $\Delta^{9(11)}$ or Δ^{14}, the latter was excluded by ms, thus locating the double bond at the 9/11 position.[1]

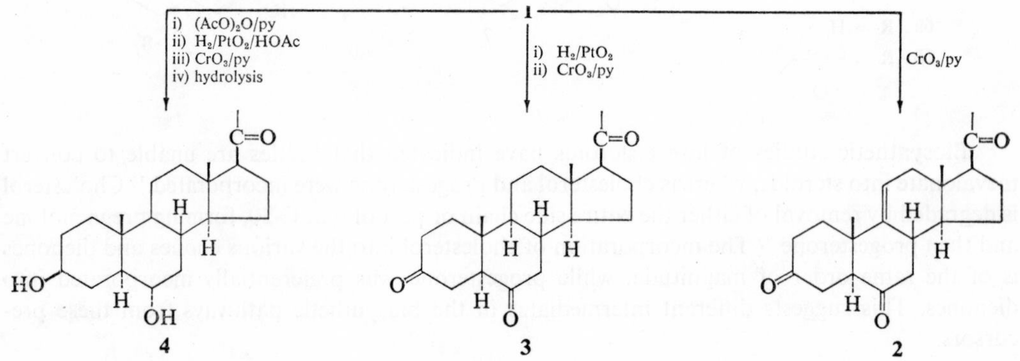

3) The 3- and 6-OH stereochemistry was determined by conversion to the known compound **4**.[1]
4) Structure **1** was confirmed by preparation of the 2,4-DNP hydrazone of **4** from the corresponding derivative of pregnenolone,[3] as well as conversion of **1** to the known triketone **3**.[1-4]

1) Y. M. Sheikh, B. M. Tursch, C. Djerassi, *J. Am. Chem. Soc.*, **94**, 3278 (1972).
2) Y. Shimizu, *J. Am. Chem. Soc.*, **94**, 4051 (1972).
3) S. Ikegami, Y. Kamiya, S. Tamura, *Tetr. Lett.*, **16**, 1601 (1972).
4) O. Mancera, G. Rosenkranz, C. Djerassi, *J. Org. Chem.*, **16**, 192 (1951).

Remarks

Genin **1** is the main aglycone of asterosaponins A and B isolated from the Japanese starfish, *Asterias amurensis*.[5] Both saponins were recognized to contain the same aglycones to which four or five sugars and a molecule of sulfuric acid (as sodium salt) were attached. The sugar components were identified as D-fucose, D-galactose, D-quinovose and D-xylose.[6,7] Progesterone, of which genin **1** may be a metabolite, has recently been discovered in the ovary of *A. amurensis*.[8] Genin **5**,[1] as well as genins **6** and **7**,[9] have recently been isolated from the starfish *Acanthaster planci*. Genin **1** has been shown to be a marine toxin.[2]

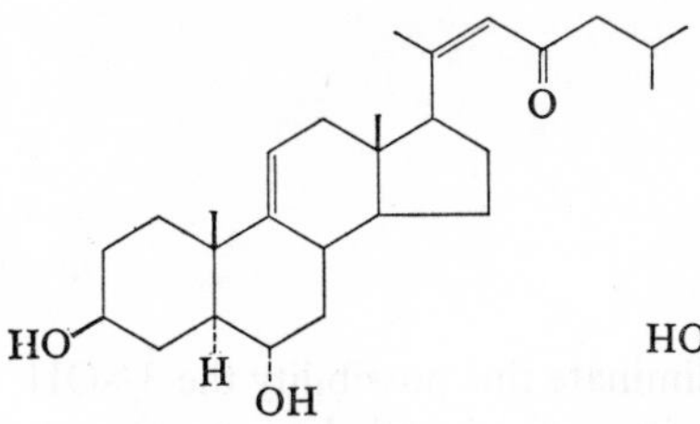

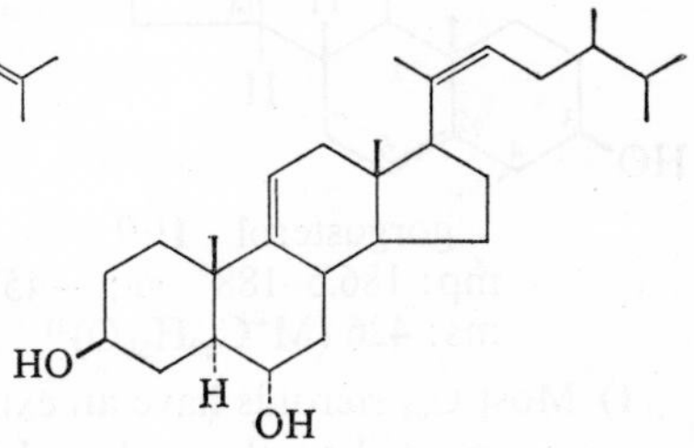

5) T. Yasumoto, T. Watanabe, Y. Hashimoto, *Bull. Jap. Soc. Sci. Fisheries*, **30**, 357 (1964).
6) T. Yasumoto, Y. Hashimoto, *Agr. Biol. Chem.* (Tokyo), **29**, 804 (1965).
7) T. Yasumoto, Y. Hashimoto, *Agr. Biol. Chem.* (Tokyo), **31**, 369 (1967).
8) S. Ikegami, H. Shirai, H. Kanatani, *Zool. Mag.* (Tokyo), **80**, 26 (1971).
9) Y. M. Sheikh, B. Tursch, C. Djerassi, *Tetr. Lett.*, **35**, 3721 (1972).

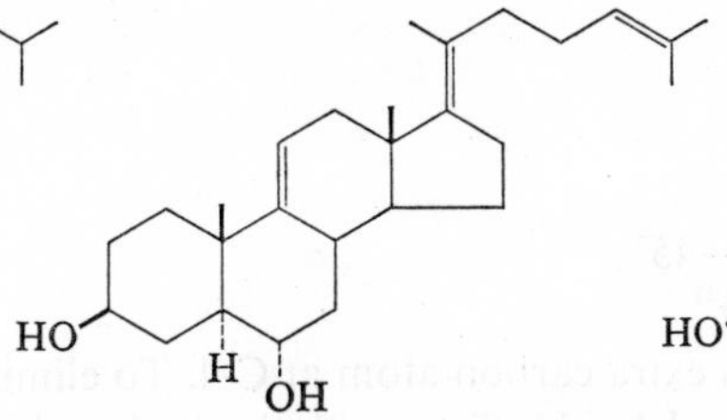

6.27 STRUCTURE OF GORGOSTEROL

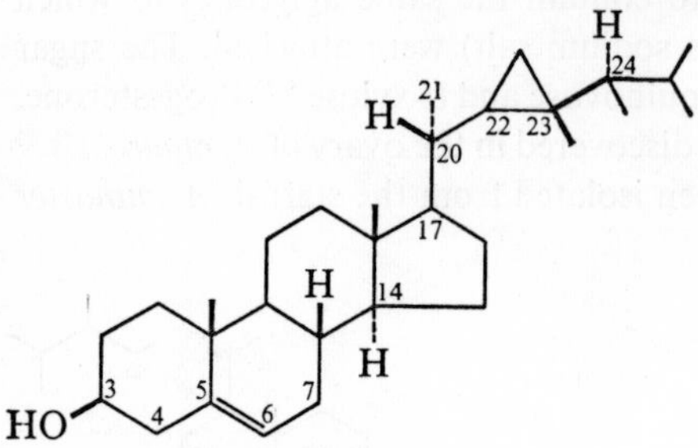

gorgosterol **1**[1,2]
mp: 186.5–188° α_D: −45°
ms: 426 (M⁺C₃₀H₅₀O)[3]

1) Most C_{30} steroids have an extra carbon atom at C-4. To eliminate this possibility the 3β-OH was converted to the carbonyl by Moffatt oxidation[4] then isomerized with base to the Δ^4-ketone **2**. Uv at 242 nm (ε 17,000) excluded methyl substitution at C-4.[5]

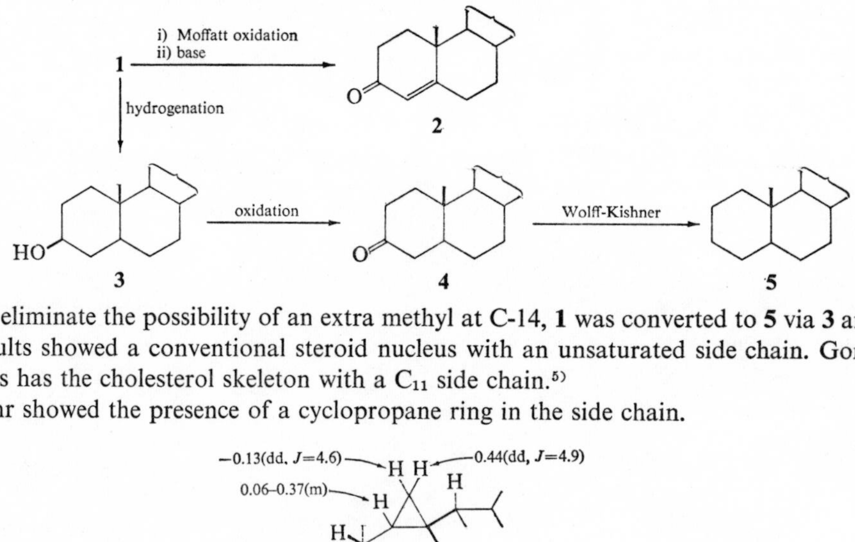

2) To eliminate the possibility of an extra methyl at C-14, **1** was converted to **5** via **3** and **4**; ms results showed a conventional steroid nucleus with an unsaturated side chain. Gorgosterol thus has the cholesterol skeleton with a C_{11} side chain.[5]

3) Nmr showed the presence of a cyclopropane ring in the side chain.

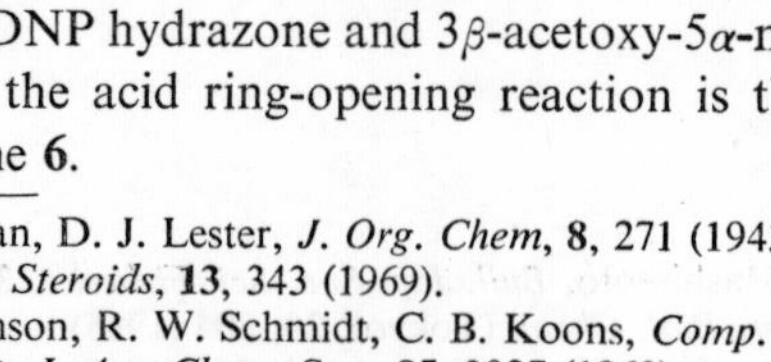

4) Cyclopropane ring opening by refluxing gorgosterol acetate in AcOH/HCl gave a mixture of tetrasubstituted olefins. Ozonolysis of this mixture and reaction with DNP yielded 3,4-dimethylpentan-2-one DNP hydrazone and 3β-acetoxy-5α-norcholan-22-one **7**. At least one of the components of the acid ring-opening reaction is therefore 3β-acetoxy-22,23,24-tri-methyl-Δ^{22}-5α-cholestene **6**.

1) W. Bergmann, M. J. McLean, D. J. Lester, *J. Org. Chem*, **8**, 271 (1943).
2) K. C. Gupta, P. J. Scheuer, *Steroids*, **13**, 343 (1969).
3) L. S. Ciereszko, M. A. Johnson, R. W. Schmidt, C. B. Koons, *Comp. Biochem. Physiol.*, **24**, 899 (1968).
4) K. E. Pfitzner, J. C. Moffatt, *J. Am. Chem. Soc.*, **85**, 3027 (1963).
5) R. H. Hale, J. Leclerq, B. Tursch, C. Djerassi, R. A. Gross, Jr., A. J. Weinheimer, K. Gupta, P. J. Scheuer, *J. Am. Chem. Soc.*, **92**, 2179 (1970).

6 → **7**

5) Ms data: intense peak at m/e 314 (in gorgosterol) is known to be associated with certain unsaturated side-chain steroids containing the steroid nucleus plus C-20, C-21 and C-22; this left only two possible positions for the cyclopropane ring, **8** or **9**.[5]

8   **9**

6) Final structure was confirmed by x-ray analysis of 3β-bromogorgostene.

Remarks

Gorgosterol was isolated from the gorgonia *Plexaura flexuosa*,[1] and the zoanthid *Palythoa* sp.,[2] and from other sources.[3] Gorgosterol was the first marine steroid isolated with a cyclopropane ring on its side chain as well as the first with C-22 and C-23 alkyl substituents. Characterization of this marine sterol provides the first evidence that cyclopropanes may be intermediates in the introduction of methyl substitutents in steroid side chains.[5] A Δ^{22}-23,24 dimethyl precursor has been postulated in the biosynthesis[6] although none has as yet been isolated. X-ray analysis has been performed on another marine steroid possessing a cyclopropane side-chain group, 23-demethyl gorgosterol, which may be a biosynthetic precursor to gorgosterol through a Δ^{22} intermediate.[7] Acanthasterol, a cyclopropane-containing marine steroid which is the Δ^7 analogue of gorgosterol has recently been characterized.[8] X-ray analysis has been undertaken of the marine sterol **10** containing the unusual 9,11-seco moiety and isolated from the gorgonian *Pseudopterogorgia americana*.[9]

10

6) N. C. Ling, R. L. Hale, C. Djerassi, *J. Am. Chem. Soc.*, **92**, 5281 (1970).
7) F. J. Schmitz, T. Pattabhiraman, *J. Am. Chem. Soc.*, **92**, 6073 (1970).
8) Y. M. Sheikh, C. Djerassi, B. Tursch, *Chem. Commun*, 217 (1971).
9) E. L. Enwall, D. van der Helm, T. Pattabhiraman, F. J. Schmitz, R. L. Spraggins, A. T. Weinheimer, *Chem. Commun.* 215 (1972).

6.28   SYNTHESIS OF PRECALCIFEROL

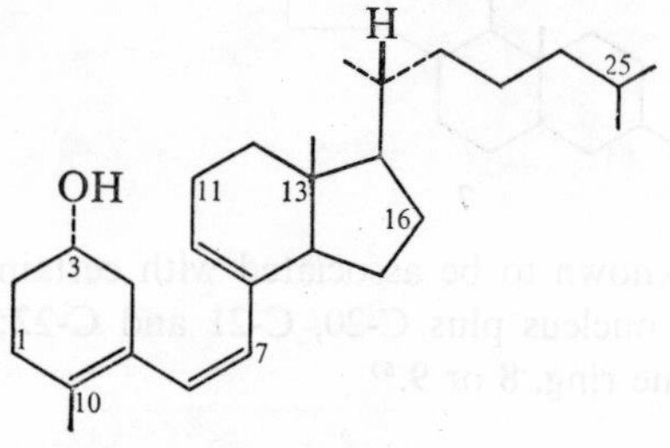

precalciferol **1**

Precalciferol **1** has been synthesized by Lythgoe[1] via a non-photochemical method using the scheme outlined below. Compound **2** was reacted with **3** in the orthacetate variant of the Claisen rearrangement[2] to give, after hydrolysis, **4**, in a stereospecific manner (i.e., 15α-OH leads to 13α substitution). Reaction of **4** with 1-dimethylamino-1-methoxyethylene, in a second Claisen rearrangement, gave **5** which, after hydrolysis, was cyclized to **6**. Precalciferol was then obtained from **6** by the scheme shown below.

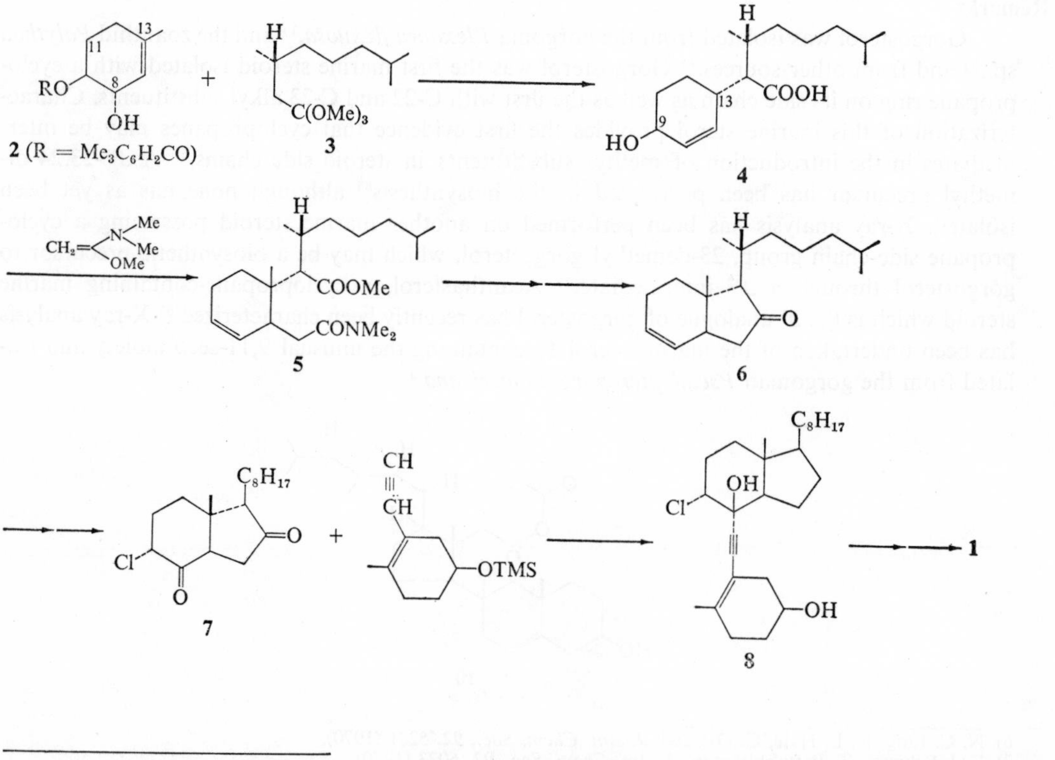

1) J. Dixon, P. S. Littlewood, B. Lythgoe, A. K. Saksena, *J.C.S. Chem. Comm.*, 993 (1970), and references therein.
2) W. S. Johnson, L. Werthemann, W. R. Bartlett, T. J. Brocksom, T. Li, D. J. Faulkner, M. R. Peterson, *J. Amer. Chem. Soc.*, **92**, 741 (1970).

6.29   1,25-DIHYDROXYCHOLECALCIFEROL

Recently, the presence of a new metabolite of vitamin D has been demonstrated. This has been identified as 1α,25-dihydorxycholecalciferol **1**.[1] This metabolite has high biological potency and is probably the form active in the intestine. Its formation *in vivo* from cholecalciferol has been demonstrated[2] by the use of $[1\text{-}^3H, 4\text{-}^{14}C]$- or $[1,2\text{-}^3H_2, 4\text{-}^{14}C]$-cholecalciferol. It has recently been synthesized by several workers, the key intermediate being 1α,25-dihydroxycholesterol **2**.

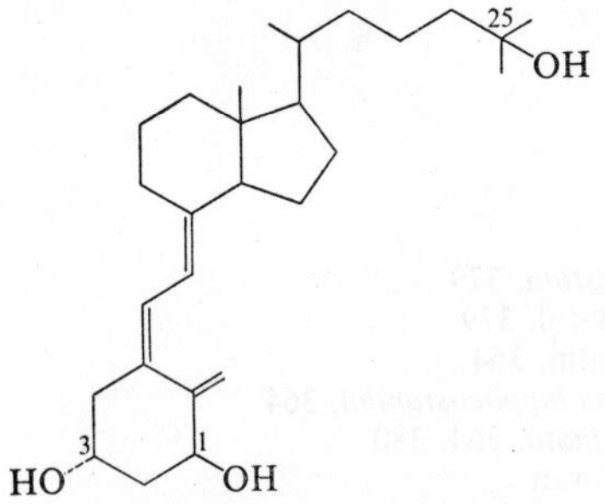

1,25-dihydroxycholecalciferol   **1** 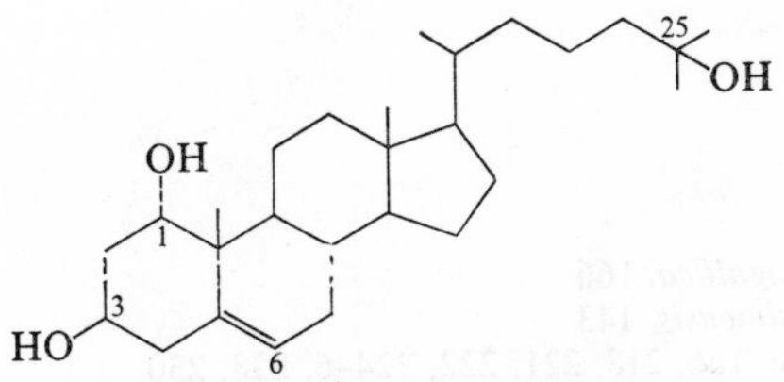   **2**

1) D. E. M. Lawson, P. A. Bell, P. W. Wilson, E. Kodicek, *Biochem. J.*, **121**, 673 (1971), and references therein.
2) E. B. Mawer, J. Backhause, G. A. Lumb, S. W. Stanbury, *Nature*, **232**, 188 (1971).

INDEX